# Process Plant Design

# Process Plant Design

## Robin Smith

Centre for Process Integration, Department of Chemical Engineering, The University of Manchester, UK

*Registered Offices*
John Wiley & Sons Inc., 111 River Street, Hoboken, NJ 07030, USA
John Wiley & Sons Ltd, The Atrium, Southern Gate, Chichester, West Sussex, PO19 8SQ, UK

For details of our global editorial offices, customer services, and more information about Wiley products visit us at www.wiley.com.

---

*Library of Congress Cataloging-in-Publication Data:*

Names: Smith, Robin (Chemical engineer), author. | John Wiley & Sons, publisher.
Title: Process plant design / Robin Smith.
Description: Hoboken, NJ : Wiley, 2024. | Includes index.
Identifiers: LCCN 2023017119 (print) | LCCN 2023017120 (ebook) | ISBN 9781119689911 (paperback) | ISBN 9781119690009 (adobe pdf) | ISBN 9781119689980 (epub)
Subjects: LCSH: Chemical plants–Design and construction. | Chemical process control.
Classification: LCC TH4524 .S65 2024 (print) | LCC TH4524 (ebook) | DDC 660/.2815–dc23/eng/20230520
LC record available at https://lccn.loc.gov/2023017119
LC ebook record available at https://lccn.loc.gov/2023017120

Cover Design: Wiley
Cover Image: © is1003/Shutterstock

Set in 9.5/12pt TimesLTStd-Roman by Straive, Pondicherry, India
Printed and bound by CPI Group (UK) Ltd, Croydon, CR0 4YY

C9781119689911_031123

*To George, Oliver, Ava, and Freya*

# Contents

# Preface

PREFACE

Chemical process projects involve the design, construction, commissioning, operation, and decommissioning of processes that involve physical, chemical, and biochemical change for the conversion of raw materials into useful chemical products on an industrial scale. A typical project goes through a series of stages, starting with the concept development in which process options are created, assessed, and screened for economic value, safety, environmental impact, sustainability, reliability, maintainability, availability, and risk. Once a process option is adopted, detail is added to the point where a decision can be made as to whether the project and investment should proceed. If the decision is made to proceed, then engineering, procurement, and construction are carried out. This leads to commissioning and then operation of the process plant to recoup the benefits of the investment. Finally, and all too often neglected until the end of the project life to the detriment of the project, the issues associated with decommissioning need to be addressed.

A number of process design texts are available to address the early stages of projects for the creation, assessment, and screening of process concepts. However, fewer texts are available to deal with the more detailed aspects of process plant design that follow after the development of process concepts.

This text starts by presenting general background from the early stages of chemical process projects and moves on to deal with the infrastructure required to support the operation of a process plant. The emphasis placed on the infrastructure is related to heating and cooling utilities and waste treatment.

Following this, reliability, maintainability, and availability issues related to process plant design are addressed. These issues are often left until late in the project, yet if considered early can have a major impact on the successful outcome of projects. There are a number of reasons that make it necessary to study reliability, maintainability, and availability, but the most important is safety. If the process needs to be shut down for maintenance, either unplanned or planned, the shut-down, maintenance, and subsequent start-up after maintenance create the potential for safety problems. Also, the design must be able to operate without frequent breakdowns interrupting production in order to meet production targets. System structures can be created that are resilient to system breakdown through the use of standby equipment, parallel production trains, and intermediate storage.

Process control is another major issue in the development of a process plant design. Whereas a great emphasis in process control has traditionally been placed on the analysis or simulation of control systems, the creation or synthesis of control systems has by comparison received little attention, especially when considering overall process control systems. This text provides a practical approach to the systematic synthesis of process control schemes for chemical processes.

Once the additional equipment for reliability, maintainability, and availability has been added and the control system synthesized, then piping and instrumentation diagrams can be developed. Piping and instrumentation diagrams (P&IDs) are probably the most important documents in process engineering. Much of the detailed information regarding the process is incorporated graphically in the P&ID to guide the engineering of the plant, analyze safety, operate, troubleshoot, maintain, and revamp the process. Whilst texts on process design list the features that need to be included and give completed examples, there is a lack of treatment of how P&IDs can be developed. This text goes into some detail on the development of preliminary P&IDs.

Choice of materials of construction for equipment and piping affects the resistance to corrosion, mechanical design, and the capital cost of equipment. Whilst the final mechanical design of vessels and equipment is normally carried out by specialist mechanical engineers, it is still necessary for process designers to have an understanding of mechanical design for a variety of reasons. The process design of equipment and vessels must be cognizant of the difficulties that the process design might impose on the consequent mechanical design. Even preliminary capital costing can require a preliminary mechanical design. Poor mechanical design can create major hazards, especially when installed equipment and piping is modified.

Finally, the text considers process plant layout. Process plant layout can be considered at the site, process, and equipment levels, which have important implications for process design, safety, environmental impact, and capital and operating costs.

This text aims to provide an introductory practical guide to process plant design suitable for undergraduate and postgraduate students of chemical engineering and for practicing process designers. For undergraduate studies the text assumes knowledge of basic thermodynamics, material and energy balances, fluid flow, heat transfer, and separation operations. Worked examples have been included throughout the text. These do not require the use of specialist software. A number of exercises have been included at the end of the chapters to allow the reader to practice the calculation procedures.

*Robin Smith*

# Acknowledgments
## ACKNOWLEDGMENTS

The author would like to express sincere gratitude to a number of people who have helped in the preparation of the text:

Jorge Alana (Johnson Matthey), Michael Binns (Department of Chemical and Biochemical Engineering, Dongguk University, South Korea), Mike Brown (independent consultant and formerly with BP International), Cheng-Liang Chen (Department of Chemical Engineering, National Taiwan University), Liuyi Cheng (ProAIM, PR China), Tony Kiss (University of Manchester and Delft University of Technology), Jiaxin Lu (Intern), Paul Oram (independent consultant and formerly with BP International), Qiying Yin (ProAIM, PR China).

Special thanks to Les Bolton (independent consultant and formerly with BP International).

# Nomenclature
## NOMENCLATURE

$a$    Cost law coefficient ($), or distance (m), or order of reaction (–)

$A$    Area ($m^2$), or availability (–)

$A_B$    Cross-sectional area of bolts at the bolt narrowest point ($m^2$)

$A_{CF}$    Annual cash flow ($ $y^{-1}$)

$A_{DCF}$    Annual discounted cash flow ($ $y^{-1}$)

$A_i$    Availability of Component $i$ (–)

$A_{IN}$    Inherent availability (–)

$A_{OP}$    Operational availability (–)

$A_P$    Projected area on a plane normal to the flow ($m^2$)

$A_{SYS}$    System availability (–)

$A_0$    Original cross-sectional area ($mm^2$, $m^2$)

$AF$    Annualization factor for capital cost (–)

$b$    Capital cost law coefficient (units depend on cost law), or breadth of a cross-section (m), or distance (m), or order of reaction (–)

$BOD$    Biological oxygen demand (kg $m^{-3}$, mg $l^{-1}$)

$c$    Capital cost law coefficient (–)

$c_D$    Drag coefficient (–)

$C$    Capacity ($m^3$ $h^{-1}$), or concentration (kg $m^{-3}$, kmol $m^{-3}$), or corrosion allowance (m)

$C_B$    Base capital cost of equipment ($)

$C_E$    Equipment capital cost ($)

$C_F$    Fixed capital cost of complete installation ($)

$C_P$    Specific heat capacity at constant pressure (kJ $kg^{-1}$ $K^{-1}$, kJ $kmol^{-1}$ $K^{-1}$)

$CO_{BIAS}$    Controller bias in process control (–)

$CO(t)$    Controller output at time $t$ (–)

$COD$    Chemical oxygen demand (kg $m^{-3}$, mg $l^{-1}$)

$COP$    Coefficient of performance (–)

$CP$    Heat capacity flowrate (kW $K^{-1}$, MW $K^{-1}$)

$CW$    Cooling water

$d$    Depth of a cross-section (m), or diameter (mm, m)

$D$    Demand rate on a safety system ($y^{-1}$), or diameter (mm, m)

$D_{EQ}$    Diameter of the equivalent cylinder to a conical frustum (m)

$D_I$    Inside diameter of vessel, pipe or tube (mm, m)

$DCFRR$    Discounted cash flow rate of return (%)

$e(t)$    Control error at time $t$ (–)

$E$    Activation energy of reaction (kJ $kmol^{-1}$), or ratio of stress to strain, or Young's Modulus, or Modulus of Elasticity (N $m^{-2}$)

$EP$    Economic potential ($ $y^{-1}$)

$f_i$    Capital cost installation factor for Equipment $i$ (–)

$f_P$    Capital cost factor to allow for design pressure (–)

$f_T$    Capital cost factor to allow for design temperature (–)

$f_M$    Capital cost correction factor for materials of construction (–)

$f(t)$    Probability density function in reliability (–)

$F$    Feed flowrate (kg $s^{-1}$, kg $h^{-1}$, kmol $s^{-1}$, kmol $h^{-1}$), or force (N), or future worth of a sum of money allowing for interest rates ($), or stream flowrate (kg $s^{-1}$, kg $h^{-1}$, kmol $s^{-1}$, kmol $h^{-1}$)

$F_i(t)$    Failure function for Component $i$ at time $t$ (–)

$F_T$    Correction factor for non-countercurrent flow in shell-and-tube heat exchangers (–)

$g$    Acceleration due to gravity (9.81 m $s^{-2}$)

$g(t)$    Repair time density function in reliability (–)

$G$    Gas flowrate (kg $s^{-1}$, kmol $s^{-1}$), or ratio of shear stress to shear strain, or modulus of rigidity (N $mm^{-2}$)

$GCV$    Gross calorific value of fuel (J $m^{-3}$, kJ $m^{-3}$, J $kg^{-1}$, kJ $kg^{-1}$)

$h$    Height or vertical distance (m)

$h_{CG}$    Distance from the tangent point of a cylindrical vessel to the center of gravity of the head along the center line of the vessel (m)

$H$    Hazard rate ($h^{-1}$, $y^{-1}$), or specific enthalpy (kJ $kg^{-1}$, kJ $kmol^{-1}$), or height (m)

$H_{BFW}$    Specific enthalpy of boiler feedwater (J $kg^{-1}$, kJ $kg^{-1}$)

$H_{COND}$    Specific enthalpy of saturated steam condensate (kJ $kg^{-1}$)

$H_{DESUP}$    Specific enthalpy of desuperheated steam (kJ $kg^{-1}$)

$H_f$    Specific enthalpy of saturated water (kJ $kg^{-1}$)

$H_{fg}$    Specific enthalpy of vaporization of saturated water (kJ $kg^{-1}$)

$H_g$    Specific enthalpy of saturated steam (kJ $kg^{-1}$)

$H_{in}$    Specific enthalpy of the inlet steam (kJ $kg^{-1}$)

$H_{IS}$    Enthalpy of steam at the outlet pressure having the same entropy as the inlet steam (kJ $kg^{-1}$)

$H_{out}$    Specific enthalpy of outlet steam (kJ $kg^{-1}$)

$H_{SH}$    Specific enthalpy of superheated steam (kJ $kg^{-1}$)

$H_{ST,in}$    Specific enthalpy of steam turbine inlet (kJ $kg^{-1}$)

$H_{ST,out}$    Specific enthalpy of steam turbine outlet (kJ $kg^{-1}$)

$H_{STEAM}$    Specific enthalpy of steam (J $kg^{-1}$, kJ $kg^{-1}$)

$H_{WET}$    Specific enthalpy of wet steam (J $kg^{-1}$, kJ $kg^{-1}$)

$HR$    Gas turbine heat rate (–)

$\Delta H_{COMB}$    Heat of combustion (J $kmol^{-1}$, kJ $kmol^{-1}$)

$\Delta H_{IS}$    Isentropic enthalpy change of an expansion (J $kmol^{-1}$, kJ $kg^{-1}$)

$\Delta H_{VAP}$    Latent heat of vaporization (kJ $kg^{-1}$, kJ $kmol^{-1}$)

$HP$    High pressure

| | |
|---|---|
| $i$ | Fractional rate of interest on money (–) |
| $I$ | Moment of inertia, or second moment of area (m$^4$) |
| $I_{XX}, I_{YY}$ | Moment of inertia around the $XX$ and $YY$ axis (m$^4$) |
| $J$ | Polar moment of inertia (m$^4$) |
| $k$ | Radius of gyration (m), or reaction rate constant (units depend on order of reaction), or step number in a numerical calculation (–) |
| $k_{XX}, k_{YY}$ | Radius of gyration around the $XX$ and $YY$ axis (m) |
| $k_0$ | Frequency factor for heat of reaction (units depend on order of reaction) |
| $K_D$ | Derivative gain in process control (s, min) |
| $K_i$ | Vapor-liquid equilibrium K-value for Component $i$ (–) |
| $K_I$ | Integral gain in process control (s$^{-1}$, min$^{-1}$) |
| $K_P$ | Proportional gain in process control (–) |
| $L$ | Length (m), or liquid flowrate (kg s$^{-1}$, kmol s$^{-1}$), or tangent-to-tangent length of a cylindrical vessel (m) |
| $L_C$ | Critical length of a cylindrical vessel subject to buckling (m) |
| $L_e$ | Effective length (m) |
| $L_0$ | Original length (m) |
| $LP$ | Low pressure |
| $m$ | Mass flowrate (kg s$^{-1}$), or molar flowrate (kmol s$^{-1}$) |
| $m_{BFW}$ | Mass flowrate of boiler feedwater (kg s$^{-1}$) |
| $m_C$ | Mass flowrate of cold stream (kg s$^{-1}$) |
| $m_{COND}$ | Mass flowrate of condensate (kg s$^{-1}$) |
| $m_{DESUP}$ | Mass flowrate of desuperheated steam (kg s$^{-1}$) |
| $m_H$ | Mass flowrate of hot stream (kg s$^{-1}$) |
| $m_{PROC}$ | Mass flowrate of process liquid (kg s$^{-1}$) |
| $m_{ST}$ | Mass flowrate through steam turbine (kg s$^{-1}$) |
| $m_{SUP}$ | Mass flowrate of superheated steam (kg s$^{-1}$) |
| $m_{STEAM}$ | Mass flowrate of steam (kg s$^{-1}$) |
| $M$ | Constant in capital cost correlations (–), or bending moment (N m), or molar mass (kg kmol$^{-1}$), or number of variables (–) |
| $M(t)$ | Probability of performing a successful repair action in time $t$ (–) |
| $M_{HYD}$ | Bending moment created by a hydraulic head (N m) |
| $MDT$ | Mean downtime for all maintenance actions including delays (h) |
| $MTBM$ | Mean uptime between maintenance actions (h) |
| $MTTF$ | Mean time to failure (h) |
| $MTTR$ | Mean time to repair (h) |
| $n$ | Number of items (–), or number of years (–), or stage number in an equilibrium cascade (–), or number of stages in an equilibrium cascade (–) |
| $N$ | Number of independent equations (–), or number of moles (kmol) |
| $Nc$ | Number of components in a multicomponent mixture (–) |
| $N_B$ | Number of bolts in a bolt circle (–) |
| $N_{CDOF}$ | Number of control degrees of freedom (–) |
| $N_{DOF}$ | Number of degrees of freedom (–) |
| $N_E$ | Number of inputs or outputs of energy (–) |
| $N_{Equations}$ | Number of independent equations describing the process system (–) |

| | |
|---|---|
| $N_{in}$ | Number of inlet streams (–) |
| $N_{INT}$ | Number of mass transfer equilibrium constrained phase interfaces (–) |
| $N_{out}$ | Number of outlet streams (–) |
| $N_P$ | Number of mass transfer equilibrium constrained phases (–) |
| $N_{Variables}$ | Number of variables or unknowns (–) |
| $NCV$ | Net calorific value of fuel (J m$^{-3}$, kJ m$^{-3}$, J kg$^{-1}$, kJ kg$^{-1}$) |
| $NPSH$ | Net positive suction head (m) |
| $NPV$ | Net present value ($) |
| $P$ | Present worth of a future sum of money ($), or pressure (N m$^{-2}$, bar) |
| $P_{CR}$ | Critical buckling pressure of a cylindrical vessel (N m$^{-2}$) |
| $P_H$ | Gauge pressure at height $H$ above the base of a tank (N m$^{-2}$) |
| $PFD$ | Probability of failure on demand (–) |
| $P_{VS}$ | Maximum gauge pressure in the vapor space above the liquid in a storage tank (N m$^{-2}$) |
| $Pr$ | Probability (–) |
| $PM$ | Preventive maintenance |
| $PV$ | Process variable (–) |
| $\Delta P$ | Pressure drop (N m$^{-2}$, bar) |
| $Q$ | Heat duty (kW, MW), or first moment of area about the neutral axis (m$^3$), or volumetric flowrate (m$^3$ s$^{-1}$, m$^3$ h$^{-1}$) |
| $Qc$ | Cooling duty (kW, MW) |
| $Q_{COND}$ | Condenser heat duty (kW, MW) |
| $Q_{EVAP}$ | Evaporator heat duty (kW, MW) |
| $Q_{FUEL}$ | Heat from fuel in a furnace, boiler, or gas turbine (kW, MW) |
| $Q_H$ | Heating duty (kW, MW) |
| $Q_{HP}$ | Heat duty on high-pressure steam (kW, MW) |
| $Q_{INPUT}$ | Heat input (W, kW) |
| $Q_{LP}$ | Heat duty on low-pressure steam (kW, MW) |
| $Q_{LOSS}$ | Stack loss from furnace, boiler, or gas turbine (kW, MW) |
| $Q_{OUTPUT}$ | Heat output to steam generation (kW, MW) |
| $Q_{PROC}$ | Process heat duty (kW, MW) |
| $Q_{STEAM}$ | Heat input for steam generation (kW, MW) |
| $r$ | Radius (m) |
| $r_i$ | Rate of reaction of Component $i$ (kmol$^{-1}$ s$^{-1}$) |
| $R$ | Reaction force from support (N), or universal gas constant (8314.5 N m kmol$^{-1}$ K$^{-1}$ = J kmol$^{-1}$ K$^{-1}$, 8.3145 kJ kmol$^{-1}$ K$^{-1}$) |
| $R_C$ | Crown radius (m) |
| $R_i$ | Molar flowrate of Component $i$ in the recycle (kmol s$^{-1}$), or reliability of Component $i$ (–) |
| $R_{EQ}$ | Radius of the equivalent cylinder to a conical frustum (m) |
| $R_i(t)$ | Reliability of Component $i$ at time $t$ (–) |
| $R_{SYS}(t)$ | System reliability at time $t$ (–) |
| $ROI$ | Return on investment (%) |
| $RR$ | Recycle ratio for a process recycle (–) |

| | |
|---|---|
| $S$ | Specific entropy (kJ kg$^{-1}$ K$^{-1}$) |
| $S_f$ | Specific entropy of saturated water (kJ kg$^{-1}$ K$^{-1}$) |
| $S_{fg}$ | Specific entropy of vaporization at saturated conditions (kJ kg$^{-1}$ K$^{-1}$) |
| $S_g$ | Specific entropy of saturated steam (kJ kg$^{-1}$ K$^{-1}$) |
| $S_{in}$ | Specific entropy of steam at an inlet (kJ kg$^{-1}$ K$^{-1}$) |
| $S_{out}$ | Specific entropy of steam at an outlet (kJ kg$^{-1}$ K$^{-1}$) |
| $S_{WET}$ | Specific entropy of wet steam (kJ kg$^{-1}$ K$^{-1}$) |
| $t$ | Thickness (mm, cm, m), or time (s, h, y) |
| $t_C$ | Thickness of a cylinder wall resisting a circumferential stress (m) |
| $t_L$ | Thickness of a cylinder wall resisting a longitudinal stress (m) |
| $t_{SPH}$ | Thickness of a sphere wall (m) |
| $T$ | Temperature (°C, K), or time to a dangerous undetected failure (h, y), or torque (N m) |
| $T_{BPT}$ | Normal boiling point (°C, K) |
| $T_C$ | Temperature of heat sink (°C, K) |
| $T_{COND}$ | Condenser temperature (°C, K) |
| $T_{EVAP}$ | Evaporator temperature (°C, K) |
| $T_{FEED}$ | Feed temperature (°C, K) |
| $T_H$ | Temperature of heat source (°C, K) |
| $T_{REB}$ | Reboiler temperature (°C, K) |
| $T_{SP}$ | Temperature setpoint (°C) |
| $T_T$ | Stream target temperature (°C) |
| $T_{TFT}$ | Theoretical flame temperature (°C, K) |
| $\Delta T$ | Temperature difference (°C, K) |
| $\Delta T_{LM}$ | Logarithmic mean temperature difference (°C, K) |
| $TAC$ | Total annual cost ($ y$^{-1}$) |
| $TOD$ | Total oxygen demand (kg m$^{-3}$, mg l$^{-1}$) |
| $U$ | Overall heat transfer coefficient (W m$^{-2}$ K$^{-1}$, kW m$^{-2}$ K$^{-1}$) |
| $v$ | Velocity (m s$^{-1}$) |
| $V$ | Molar volume (m$^3$ kmol$^{-1}$), or vapor flowrate (kg s$^{-1}$, kmol s$^{-1}$), or volume (m$^3$) |
| $V(t)$ | probability of not performing a successful repair action in time $t$ (−) |
| $VF$ | Vapor fraction (−) |
| $w$ | Force per unit length, or weight per unit length (N m$^{-1}$, kN m$^{-1}$) |
| $W$ | Shaft power (kW, MW), or shaft work (kJ, MJ), or weight (N, kN) |
| $x$ | Liquid-phase mole fraction (−) |
| $X$ | Reactor conversion (−) or dryness fraction of steam (−) |
| $y$ | Vapor-phase mole fraction (−) |
| $z$ | Elevation (m) |
| $z_i$ | Feed mole fraction of Component $i$ (−) |
| $Z$ | Section modulus (m$^3$) |

# Greek Letters

| | |
|---|---|
| $\alpha$ | Angle to the vertical (degrees) |
| $\beta$ | Fitting parameter in the Weibull Equation (−) |
| $\delta$ | Thickness (m) |
| $\varepsilon$ | Strain (−) |
| $\eta$ | Carnot factor (−), or efficiency (−) |

| | |
|---|---|
| $\eta_{BOILER}$ | Boiler efficiency (−) |
| $\eta_C$ | Carnot efficiency for refrigeration (−) |
| $\eta_{GT}$ | Gas turbine efficiency (−) |
| $\eta_{IS}$ | Steam turbine isentropic efficiency (−) |
| $\eta_J$ | Welded joint efficiency (−) |
| $\eta_{MECH}$ | Turbine mechanical efficiency (−) |
| $\eta_{ST}$ | Overall steam turbine efficiency (−) |
| $\theta$ | Angle (degrees, radians), or fitting parameter in the Weibull Equation (h) |
| $\lambda$ | Failure rate, assumed constant (h$^{-1}$, y$^{-1}$) |
| $\lambda(t)$ | Failure rate at time $t$ (h$^{-1}$, y$^{-1}$) |
| $\mu$ | Fluid viscosity (kg m$^{-1}$ s$^{-1}$, mN s m$^{-2}$ = cP), or repair rate, assumed constant (h$^{-1}$) |
| $\mu(t)$ | Repair rate at time $t$ (h$^{-1}$) |
| $\nu$ | Ratio of lateral strain to axial strain, or Poisson's Ratio (−) |
| $\rho$ | Density (kg m$^{-3}$, kmol m$^{-3}$) |
| $\sigma$ | Stress (N m$^{-2}$) |
| $\sigma_{ALL}$ | Allowable stress (N m$^{-2}$) |
| $\sigma_B$ | Longitudinal bending stress (N m$^{-2}$) |
| $\sigma_C$ | Circumferential (hoop) stress (N m$^{-2}$) |
| $\sigma_{BOLTS}$ | Stress in bolts (N m$^{-2}$) |
| $\sigma_L$ | Longitudinal stress (N m$^{-2}$) |
| $\sigma_{PROOF}$ | Proof stress of bolts (N m$^{-2}$) |
| $\sigma_{SKIRT}$ | Longitudinal stress in a vertical cylindrical vessel skirt (N m$^{-2}$) |
| $\sigma_x$ | Normal stress acting on the $x$-plane (N m$^{-2}$) |
| $\sigma_y$ | Normal stress acting on the $y$-plane (N m$^{-2}$) |
| $\sigma_{YS}$ | Tensile stress at the yield point (N m$^{-2}$) |
| $\sigma_1, \sigma_2, \sigma_3$ | Principal stresses (N m$^{-2}$) |
| $\tau$ | Shear stress (N m$^{-2}$) |
| $\tau_D$ | Dead time when a safety device is unavailable (h, y), or derivative time in process control (s, min) |
| $\tau_I$ | Integral time in process control (s, min) |
| $\tau_P$ | Interval between proof tests (h, y) |
| $\tau_{xy}$ | Shear stress acting on the $x$-plane in the direction of the $y$-axis (N m$^{-2}$) |
| $\tau_{yx}$ | Shear stress acting on the $y$-plane in the direction of the $x$-axis (N m$^{-2}$) |
| $\tau_{YS}$ | Shear stress at the yield point (N m$^{-2}$) |
| $\varphi$ | Angle (radians) |

# Subscripts

| | |
|---|---|
| $AIR$ | Air |
| $BFW$ | Boiler feedwater |
| $COND$ | Condenser conditions |
| $CW$ | Cooling water |
| $D$ | Downstream |
| $DISC$ | Bursting disc |
| $DS$ | Desuperheating |
| $EB$ | Electric boiler |
| $ELL$ | Ellipsoid |
| $EVAP$ | Evaporator conditions |
| $f$ | Property of saturated liquid |
| $fg$ | Refers to a change of phase at constant pressure |

| | | | |
|---|---|---|---|
| $F$ | Feed | $MOC$ | Material of construction |
| $FG$ | Flue gas | $out$ | Outlet |
| $FH$ | Flat head | $O$ | Outside |
| $g$ | Property of saturated vapor | $PROC$ | Process |
| $G$ | Gas phase | $PRV$ | Pressure relief valve |
| $H$ | Hot stream | $s$ | Saturated conditions |
| $HEAD$ | Vessel head | $SE$ | Semi-ellipsoid |
| $HP$ | High pressure | $STEAM$ | Steam |
| $HS$ | Hemispherical head | $SUP$ | Superheated conditions |
| $i$ | Component number | $TFT$ | Theoretical flame temperature |
| $in$ | Inlet | $TOR$ | Torispherical |
| $I$ | Inside | $U$ | Upstream |
| $IS$ | Isentropic | $V$ | Vapor phase |
| $L$ | Liquid phase | $WBT$ | Wet bulb conditions |
| $LP$ | Low pressure | $WG$ | Waste gas stream |
| $max$ | Maximum | 1 | Initial value |
| $min$ | Minimum | 2 | Final value |
| $M$ | Mean value | | |

# ABOUT THE COMPANION WEBSITE

This book is accompanied by a companion website:

**www.wiley.com/go/processplantdesign**

# Chapter 1

# Chemical Process Projects

## 1.1 The Process Plant Design Problem

Chemical process projects involve the design, construction, commissioning, operation, and decommissioning of processes that involve physical, chemical, and biochemical change for the conversion of raw materials into useful chemical products on an industrial scale. Process designs are likely to differ significantly, depending on the type of product being manufactured. There are three broad classes of product:

1) *Commodity* or *bulk* chemicals are produced in large volumes and purchased on the basis of their chemical composition, purity, and price. Examples are sulfuric acid, ethylene, benzene, propane, and nitrogen.

2) *Fine* chemicals are also purchased on the basis of their chemical composition purity and price, but produced in small volumes. Examples are *n*-butyric acid (used in beverages, flavorings, and fragrances) and barium titanate (used for the manufacture of electronic capacitors).

3) *Specialty* or *effect* or *functional* chemicals are purchased on the basis of their effect or function, rather than their chemical composition. Examples are pharmaceuticals, pesticides, and flavorings.

In a chemical process, the transformation of raw materials into desired chemical products usually cannot be achieved in a single step. Instead, the overall transformation is broken down into a number of steps that provide intermediate transformations. These are carried out through reaction, separation, mixing, heating, cooling, pressure change, or particle size reduction or enlargement for solids. Once individual steps have been selected, they must be interconnected to carry out the overall transformation (Figure 1.1a). Thus, the *synthesis* of a chemical process involves two broad activities. First, individual transformation steps are selected. Second, these individual transformations are interconnected to form a complete process that achieves the required overall transformation. A *flowsheet* or *process flow diagram* (*PFD*) is a diagrammatic representation of the process steps with their interconnections.

Once the flowsheet structure has been defined, a *simulation* of the process can be carried out. A simulation is a mathematical model of the process that attempts to predict how the process would behave if it was constructed (Figure 1.1b). Material and energy balances can be formulated to give a better definition to the inner workings of the process and a more detailed process design can be developed. Having created a model of the process, the flowrates, compositions, temperatures, and pressures of the feeds can be assumed. The simulation model then predicts the flowrates, compositions, temperatures, pressures, and properties of the products. It also allows the individual items of equipment in the process to be sized and predicts, for example, how much raw material is being used or how much energy is being consumed. The performance of the design can then be evaluated.

Figure 1.2 shows an example of a very simple process flow diagram. The process flow diagram shows only the main items of equipment and the normal process flows. From this a *piping and instrumentation diagram* (*P&ID*) can be developed to include all of the equipment (including multiple items of equipment represented as single items in a process flow diagram, standby equipment, utility, and effluent treatment equipment directly linked to the process), the design of the control system, all piping connections and fittings (including those used for start-up, shut-down, maintenance, and abnormal operation), safety, relief and blowdown systems. Figure 1.3 shows the process flow diagram for the chemical reactor from Figure 1.2. Figure 1.4 shows the corresponding piping and instrumentation diagram (P&ID). When complete, the P&ID gives a complete graphical documentation of the process. Further details will be discussed in Chapters 15 to 17.

*Process Plant Design*, First Edition. Robin Smith.
© 2024 John Wiley & Sons Ltd. Published 2024 by John Wiley & Sons Ltd.
Companion website: www.wiley.com/go/processplantdesign

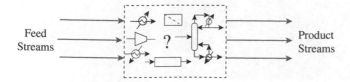

(a) Process design starts with the synthesis of a process to convert raw materials into desired products.

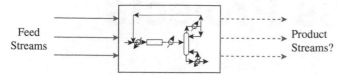

(b) Simulation predicts how a process would behave if it was constructed.

**Figure 1.1**

Synthesis is the creation of a process to transform feed streams into product streams. Simulation predicts how it would behave if it was constructed. Source: (Smith 2016), *Chemical Process Design and Integration*, 2nd Edition, John Wiley & Sons Ltd.

# 1.2   Continuous and Batch Processes

A fundamental decision that needs to be made early in a project is whether to adopt a continuous or batch design (Sharratt 1997). Small-scale processes, particularly those for specialty chemicals, pharmaceuticals, food, and beverages are often batch in nature. In principle, the process in Figure 1.2 could be continuous or batch in nature, depending on how it is operated. By contrast with the continuous process, in a batch process the main steps operate discontinuously. In contrast with a continuous process, a batch process does not deliver its product continuously but in discrete amounts.

This means that heat, mass, temperature, compositions, and other properties vary with time. In practice, most batch processes are made up of a series of batch and *semi-continuous* steps. A semi-continuous step runs continuously with periodic start-ups and shut-downs.

Batch processes are often multiproduct in which a number of different products are manufactured in the same equipment. Multi-product batch processes, with a number of different products manufactured in the same equipment, present even bigger challenges for design (Biegler, Grossman, and Westerberg 1997). Different products will demand different designs, different operating conditions, and, perhaps, different trajectories for the operating conditions through time. The design of equipment for multiproduct plants will thus require a compromise to be made across the requirements of a number of different products. The more flexible the equipment and the configuration of the equipment, the more it will be able to adapt to the optimum requirements of each product.

Batch processes:

- are economical for small volumes;
- are flexible in accommodating changes in product formulation;
- are flexible in changing the production rate by changing the number of batches made in any period of time;
- allow the use of standardized multipurpose equipment for the production of a variety of products from the same plant;
- are best if equipment needs regular cleaning because of fouling or needs regular sterilization;
- are amenable to direct scale-up from the laboratory, and
- allow product identification. Each batch of product can be clearly identified in terms of when it was manufactured, the feeds involved, and processing conditions. This is particularly important in industries such as pharmaceuticals and foodstuffs. If a problem arises with a particular batch, then all the products from that batch can be identified and withdrawn from the market. Otherwise, all the products available in the market would have to be withdrawn.

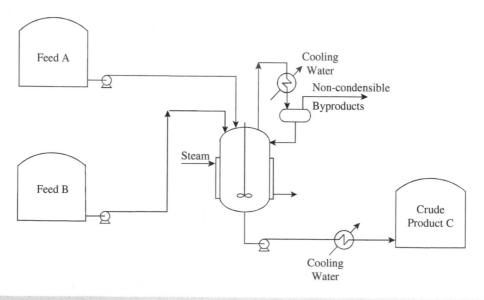

**Figure 1.2**

Process flow diagram for a simple process.

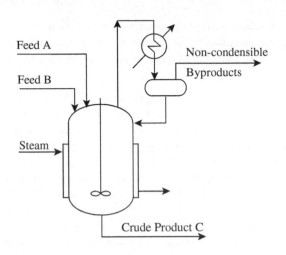

**Figure 1.3**

Process flow diagram representation for a reactor.

One of the major problems with batch processing is batch-to-batch conformity. Minor changes to the operation can mean slight changes in the product from batch to batch. Fine and specialty chemicals are usually manufactured in batch processes. However, these products often have very tight tolerances for impurities in the final product and demand batch-to-batch variation to be minimized.

# 1.3   New Design and Retrofit

There are two situations that can be encountered in process design:

**1)** *New build* or *greenfield* projects. New build projects require most of the deliverables to be generated from scratch.

**2)** *Retrofit* or *brownfield* projects. Retrofit projects might be an upgrade or production expansion of an existing facility or in extreme cases a complete rebuild of an existing facility.

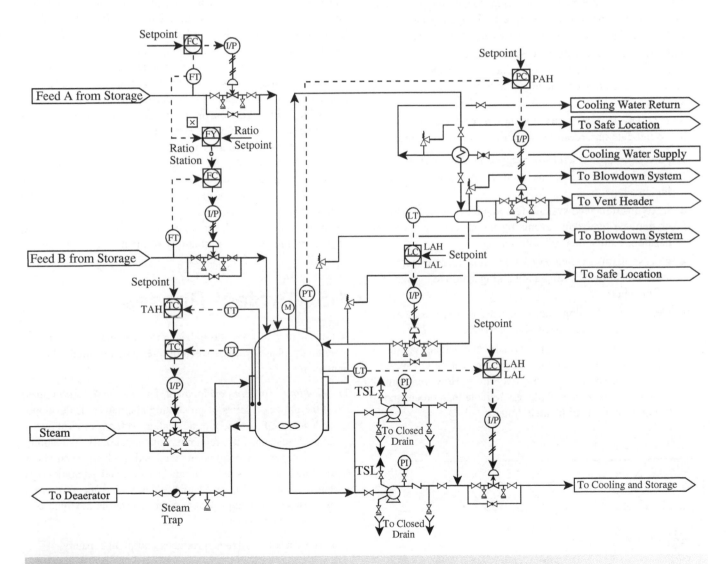

**Figure 1.4**

Piping and instrumentation diagram representation for the reactor from Figure 1.3.

New build (greenfield) projects and retrofit (brownfield) projects differ significantly. Retrofit (brownfield) projects require more effort at the outset to document what is currently installed and confirm the information used as the basis for the project. Retrofit projects are also highly constrained by the existing equipment. The objective is most often to maximize the use of existing equipment and minimize the introduction of new equipment. It is also necessary to minimize the downtime of the existing plant to implement the project.

# 1.4  Hazard Management in Process Plant Design

The major hazards of concern in process design are fire, explosion, and toxic release. To address these hazards, there is a hierarchy that should be followed for hazard management in process design, as illustrated in Figure 1.5:

a) *Inherent safety*. Make changes integral to the process to eliminate or reduce hazards at source, e.g. change from a process that uses a flammable solvent to a water-based process.

b) *Passive safety*. Incorporate design features that reduce the frequency or consequence of a hazard without the active functioning of a device, e.g. incorporate fire or blast protection walls.

c) *Active safety*. Requires active functioning of devices or personnel to mitigate hazardous conditions. This might require:

- process control systems to compensate for external disturbances and ensure safe operation within the normal operating envelope;

- safety instrumented systems (automated systems to prevent or mitigate hazardous events by taking the process from hazardous situations to a safe state);

- process alarms alerting process operators to abnormal operating situations, allowing a mitigating action to be taken by the operator;

- pressure relief using safety devices (e.g. pressure relief valve) to relieve any excessive pressure;

- active fire barriers requiring sprinkler and deluge systems to spray water onto a fire via a firewater distribution system. Human intervention is also active safety. However, human intervention is on the whole less reliable than process control and safety instrumented systems.

d) *Procedural safety*. Using administrative controls and emergency response to prevent safety incidents or minimize the effects of an incident e.g. use of safety permits to control maintenance work.

All four levels in the hierarchy in Figure 1.5 can contribute to the overall safety of the process. However, inherent safety differs in that it seeks to eliminate or reduce potential hazards at source. It starts at the very beginning of the project and runs through all phases of the project execution.

Early in the project, formal managerial processes can be used for the identification of hazards, known as *hazard identification* (*HAZID*) studies. This is a structured brainstorming technique. The HAZID study should not only identify potential hazards, but consider the consequences of hazards and identify safeguards for the elimination or mitigation of hazards.

Later, when P&IDs have been developed, a *hazard and operability study* (*HAZOP*) should be carried out. This is a structured, qualitative procedure that identifies potential safety and environmental hazards and major operability problems, assesses consequences and generates recommendations for action, but does not attempt to modify the design. In the HAZOP, P&IDs are analyzed by dividing them into *nodes*, consisting of one or more processing units with a common goal, e.g. distillation (including the reboiler and condenser). Within each node, process parameters are investigated for *deviations* using *guide words*, e.g. higher temperature. The HAZOP identifies hazardous scenarios, but should not deal with their prevention or mitigation within the study. This should be done outside of the HAZOP.

After hazardous scenarios have been identified by the HAZOP, and outside of the HAZOP, various actions need to be taken to mitigate the hazardous scenarios identified. A *layer of protection analysis* (*LOPA*) should be carried out after the HAZOP study has identified hazardous scenarios. LOPA is used to determine which independent protection layers (e.g. safety instrumented systems) are needed to reduce the probability of the hazard occurring.

# 1.5  Project Phases

A typical project goes through a series of phases. The names given to the phases vary between different organizations. Typical phases are:

1) *Selection – concept development*. During the *Selection* phase of the project, a series of processing alternatives is developed and assessed for economic value, safety, environmental impact, sustainability, reliability, maintainability, availability, and risk. Process options are screened, such as batch versus continuous processing, different reaction and separation systems, different utility provisions, etc. If viable, the most promising process concepts progress forward to the next phase.
   Selection includes:

- specification of feeds, products, and plant capacity;
- screening of existing process technologies;
- creation of process flow diagrams for the main process options;

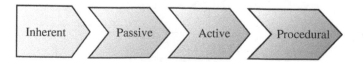

**Figure 1.5**

The hierarchy of hazard management. Source: (Smith 2016), *Chemical Process Design and Integration*, 2nd Edition, John Wiley & Sons Ltd.

- development of material and energy balances for the main process options;
- evaluation of plant location;
- hazard identification (HAZID);
- evaluation of environmental impact and sustainability;
- budgetary estimate of capital cost (see Chapter 2 for accuracy);
- estimation of operating costs and preliminary economic evaluation for main process options;
- identification of any new technology requirements;
- preparation of a statement of requirements.

2) *Definition – Front End Engineering Design*. The *Definition* phase is often referred to as *Front End Engineering Design* (*FEED*), *Front End Engineering* (*FEE*), or *Front End Loading* (*FEL*). The definition phase involves development of an engineering design and financial analysis from one or a small number of options created in the Selection Phase. The output is a design configuration, equipment details, and operating conditions needed to achieve the required economic criteria, safety standards, environmental performance, availability, and reliability. The design needs to be developed in enough detail and accuracy to provide the basis for the investment decision and, if approved, the commencement of the next phase. FEED studies can be carried out to different depths, depending on the requirements and time available. These can be defined as *light*, *normal*, or *extended*. The more work completed in the FEED phase, generally the more confidence there is in the next phase to be on budget and on schedule.

    FEED includes:

- finalize process selection;
- finalize process and utility flow diagrams;
- finalize integrated material and energy balances;
- finalize environmental impact and sustainability assessment;
- preparation of equipment sizing and specification sheets for the major items of equipment;
- design of the control system;
- design of the relief, blowdown, and flare systems
- design of electrical systems;
- development of P&IDs;
- finalize necessary geotechnical surveys (ground surveys for civil engineering);
- create site and process layouts;
- HAZOP and LOPA studies;
- acquire the required permits and planning permissions;
- capital cost estimate and economic evaluation to a high enough accuracy for the decision on whether to sanction the investment (see Chapter 2 for accuracy);
- create project execution plan,
- finalize the basis of design.

3) *Execution – Engineering Procurement and Construction*. The *Execution* of the *Engineering Procurement and Construction* (*EPC*) phase is normally conducted for the production company by an Engineering Contractor. Different types of contract can be used between the EPC Contractor and the production company. The EPC contractor might be required to execute and deliver the project within an agreed time and budget, commonly known as a *Lump Sum Turn Key* (*LSTK*) Contract. Alternatively, a reimbursable arrangement might be used, where costs are dictated by the actual work carried out.

    EPC includes:

- verify the basis of design;
- complete the detailed engineering design;
- finalize the control system design, including alarm and trip management;
- finalize the design of the electrical systems;
- create a 3-D model of the process;
- complete the piping design and create piping isometric drawings;
- finalize the details of the site and process layout;
- complete the civil and structural design;
- produce all construction documents and drawings;
- create a detailed bill of materials;
- produce an accurate capital cost estimate (see Chapter 2 for accuracy);
- create commissioning, operation, and maintenance manuals;
- project manage, and execute the design, procurement, and construction.

4) *Commissioning*. The *Commissioning* phase includes:

- test the integrity of all equipment and piping;
- test control systems and safety systems;
- start-up the process;
- operator training.

5) *Operation*. The *operation* phase includes:

- productive operation;
- operational optimization;
- asset management (maintenance and corrosion management).

6) *Decommissioning*. The *decommissioning* phase includes:

- safe disposal of residual process materials and chemicals;
- decontamination of equipment;
- dismantling and disposal of equipment and structures;
- site and ground remediation.

Decommissioning and end-of-life costs need to be considered early to ensure that excessive costs are not incurred at the end of the project.

# 1.6  Chemical Process Projects – Summary

A process design starts with the synthesis of a process flow diagram to transform raw materials into desired products. Once the process flow diagram has been fixed, piping and instrumentation diagrams (P&IDs) need to be developed. Small-scale processes, particularly those for specialty chemicals, pharmaceuticals, food, and beverage, are often batch in nature. Projects might be *new build* (*greenfield*) projects or *retrofit* (*brownfield*) projects. Early in the project, hazards should be identified by HAZID studies. Later, when the P&ID has been developed, a hazard and operability study (HAZOP) should be carried out followed by a layers of protection (LOPA) study.

A typical project goes through a series of phases:

- Selection – Concept Development;
- Definition – Front End Engineering Design (FEED);
- Execution – Engineering Procurement and Construction (EPC);
- Commissioning;
- Operation;
- Decommissioning.

# References

Biegler, L.T., Grossmann, I.E., and Westerberg, A.W. (1997). *Systematic Methods of Chemical Process Design*. Prentice Hall.

Sharratt, P.N. (1997). *Handbook of Batch Process Design*. Blackie Academic and Professional.

Smith, R. (2016). *Chemical Process Design and Integration*, 2e. Wiley.

# Chapter 2

# Process Economics

## 2.1 Capital Cost Estimates

The total investment required for a new design can be broken down into four main parts that add up to the total investment:

- battery limits investment;
- utility investment;
- off-site investment;
- working capital.

1) *Battery Limits Investment*. The battery limit is a geographic boundary that defines the manufacturing area of the process. This is that part of the manufacturing system that converts raw materials into products. It includes process equipment and buildings or structures to house it but excludes boilerhouse facilities, site storage, pollution control, site infrastructure, and so on. The battery limits investment required is the purchase of the individual plant items and their installation to form a working process.

2) *Utility Investment*. Capital investment in utility plant could include equipment for:

- electricity generation;
- electricity distribution;
- steam generation;
- steam distribution;
- process water;
- cooling water;
- firewater;
- effluent treatment;
- refrigeration;
- compressed air;
- inert gas (nitrogen).

The cost of utilities is considered from their sources within the site to the battery limits of the chemical process served.

3) *Off-site Investment*. Off-site investment includes:

- auxiliary buildings such as offices, medical, personnel changing and locker rooms, guardhouses, warehouses, and maintenance shops;
- roads and paths;
- railroads;
- fire protection systems;
- communication systems;
- waste disposal systems;
- storage facilities for raw materials and end products, water and fuel not directly connected with the process;
- plant service vehicles, unloading and loading facilities and weighing devices.

4) *Working Capital*. Working capital is what must be invested to get the plant into productive operation. This is money invested before there is a product to sell and includes:

- raw materials for plant start-up (including wasted raw materials);
- raw materials, intermediate and product inventories;
- cost of transportation of materials for start-up;
- money to carry accounts receivable (i.e. credit extended to customers) less accounts payable (i.e. credit extended by suppliers);
- money to meet payroll when starting up.

Theoretically, in contrast to fixed investment, much of this money is not lost but can be recovered when the plant is closed down. Stocks of raw materials and intermediate and product inventories often have a key influence on the working capital and are under the influence of the designer.

Different classes of capital cost estimates can be defined. These are summarized in Table 2.1 (ASTM Designation E2516-11 2011). The capital cost estimate class is normally

*Process Plant Design*, First Edition. Robin Smith.
© 2024 John Wiley & Sons Ltd. Published 2024 by John Wiley & Sons Ltd.
Companion website: www.wiley.com/go/processplantdesign

## Table 2.1

Different classes of capital cost estimates (ASTM Designation E2516-11 2011).

| Estimate Class | Level of project definition (Expressed as % of complete definition) | Typical purpose of estimate | Typical estimating method | Typical variation in low (L) and high (H) accuracy ranges |
|---|---|---|---|---|
| Class 5 | 0% to 2% | Concept screening | Based on previous cost data for similar projects, determined without the flowsheet, layout or equipment analysis by applying overall ratios to account for differences in scale of production, complexity and cost escalation factors. | L: −20% to −50% H: +30% to +100% |
| Class 4 | 1% to 15% | Study or feasibility | Based on a knowledge of the process flow diagram, major equipment items and geographical location of the facility. General correlations used for costing the major items of equipment at a given capacity and then factored to allow for materials of construction, operating conditions, installation, ancillaries, and utilities. | L: −15% to −30% H: +20% to +50% |
| Class 3 | 10% to 40% | Budget authorization or control | Based on process flow diagrams, utility flow diagrams, preliminary piping and instrumentation diagrams, preliminary layout, and budget vendor quotations for the major equipment items. Factoring methods used to estimate the less significant areas of the project. | L: −10% to −20% H: +10% to +30% |
| Class 2 | 30% to 70% | Control of costs and resources to monitor variations from budget | Based on process flow diagrams, utility flow diagrams, piping and instrumentation diagrams, final layout, equipment vendor quotations, and detailed project execution plan. For those areas of the project still undefined, use assumed design quantities rather than factoring methods. | L: −5% to −15% H: +5% to +20% |
| Class 1 | 70% to 100% | Final control baseline to monitor cost and resource variations from budget | Based on completed drawings and specifications, selected vendor's quotations, actual design quantities, site surveys, and project execution and commissioning plans. | L: −3% to −10% H: +3% to +15% |

defined by the *percentage of complete definition of the project*. However, it is the *confidence in the data* and its *maturity* that is the main determining factor in the accuracy. It is important to add *contingency* or *unallocated provision* to estimates to have a realistic final cost.

Starting with a Class 5 estimate, the accuracy is gradually increased as more design information becomes available.

## 2.2 Class 5 Capital Cost Estimates

Class 5 estimates are applicable before the complete process flow diagram has been developed and are mostly based on scaling of the capital cost from previous projects (either from within the company or published in the open literature) for a complete process that used the same (or similar) process technology:

$$\frac{Capital\ Cost_2}{Capital\ Cost_1} = \left(\frac{Capacity_2}{Capacity_1}\right)^M \tag{2.1}$$

where $M$ = constant (typically 0.6).

Investment per unit capacity decreases as the capacity increases (as long as the expansion in capacity can be achieved without complete duplication of the components). For example, for the case for process vessels, this can be understood by considering how they would scale. The cost of process vessels is related to the amount of material used in the fabrication and therefore the weight. As a first approximation, assume the thickness of the material is fixed for different sized vessels. Thus, the capital cost of a vessel would be related to its surface area. Assuming a vessel has a characteristic length $L$, surface area $A$, and volume $V$:

$$A \propto L^2$$
$$V \propto L^3 \tag{2.2}$$

Writing Eq. (2.2) in terms of vessels of two different sizes with Subscripts 1 and 2 and combining gives:

$$\frac{L_2}{L_1} = \left(\frac{A_2}{A_1}\right)^{1/2} = \left(\frac{V_2}{V_1}\right)^{1/3} \tag{2.3}$$

Rearranging Eq. (2.3):

$$\frac{A_2}{A_1} = \left(\frac{V_2}{V_1}\right)^{2/3} \qquad (2.4)$$

For example, if the volume of a vessel is doubled:

$$\frac{A_2}{A_1} = (2)^{2/3} = 1.6 \qquad (2.5)$$

Thus, if the volume is doubled, the surface area increases by 60%. If the capital cost is proportional to the weight of material used and the shell thickness is assumed constant, then the capital cost increases by 60% for a doubling of the volume. This corresponds with a value of $M$ in Eq. (2.1) of 0.67. This scaling law can easily be checked by, for example, considering the volume and area formulae for a spherical vessel. Of course, process plant will consist of equipment other than process vessels (e.g. pumps and compressors), which will not scale by the same law. Hence the value of $M$ can differ significantly from 0.67. Table 2.2 shows data for the battery limit capital costs of a variety of different chemical process technologies (Bauman 1964).

The cost might be adjusted empirically to allow for inflation, differences in process technology, complexity of the project, etc. Published data are often old, sometimes from a variety of sources, with different ages. Such data can be brought up-to-date and put on a common basis using cost indexes:

$$\frac{C_1}{C_2} = \frac{INDEX_1}{INDEX_2} \qquad (2.6)$$

where

$C_1$ = capital cost in Year 1
$C_2$ = capital cost in Year 2
$INDEX_1$ = cost index in Year 1
$INDEX_2$ = cost index in Year 2

Commonly used indices are the Chemical Engineering Indexes (1957–1959 index = 100) and Marshall and Swift (1926 index = 100), published in *Chemical Engineering* magazine, and the Nelson–Farrar Cost Indexes for refinery construction (1946 index = 100), published in the *Oil and Gas Journal*. The Chemical Engineering (CE) Indexes are particularly useful and can be applied to processes or individual types of equipment. Special care should be exercised when applying cost indexes to old cost data for complete processes. The process technology might have changed significantly through the introduction of new reactor technology, new environmental regulations might have forced significant changes to permissible discharges, and so on.

Care should also be taken to the geographic location. Costs to build the same plant can differ significantly between different locations, even within the same country. Such differences will result from variations in climate and its effect on the design requirements and construction conditions, transportation costs, local regulations, local taxes, availability and productivity of construction labor, and so on (Gary and Handwerk 2001). For example, in the United States of America, Gulf Coast costs tend to be the lowest, with costs in other areas typically 20–50% higher, and those in Alaska two or three times higher than the US Gulf Coast (Gary and Handwerk 2001). In Australia, costs tend to be the lowest in the region of Sydney and the other metropolitan cities, with costs in remote areas such as North Queensland typically 40–80% higher (Brennan 1998). Costs also differ from country to country. For example, relative to costs for a plant located in the US Gulf Coast, costs in India might be expected to be 20% cheaper, in Indonesia 30% cheaper, but in the United Kingdom 15% more expensive, because of labor costs, cost of land, and so on (Brennan 1998).

The cost of the utilities and off-sites (together sometimes referred to as *services*) ranges typically from 20–40% of the total installed cost of the battery limits plant (Bauman 1955). In general terms, the larger the plant, the larger will tend to be the fraction of the total project cost that goes to utilities and off-sites. In other words, a small project will require typically 20% of the total installed cost for utilities and off-sites. For a large project, the figure can be typically up to 40%. For an estimate of the working capital requirements, take either (Holland et al. 1983):

a) 30% of annual sales or

b) 15% of total capital investment.

Class 5 estimates are used in the earliest stages of the Select Phase of a project (see Chapter 1) for a very preliminary project evaluation.

## 2.3 Class 4 Capital Cost Estimates

Class 4 capital cost estimates are most often based on the capital cost of the main items of process equipment factored to allow for differences in materials of construction, process pressure and temperature, geographical location, installation of equipment, utilities, off-sites, contingency and working capital. This is applicable when the process flow diagram has been developed and the duty on the major items of equipment specified. The total capital cost of the process, services, and working capital can be obtained by applying multiplying factors or *installation factors* to the purchase cost of individual items of equipment (Lang 1947; Hand 1958):

$$C_F = \sum_i f_i C_{E,i} \qquad (2.7)$$

where

$C_F$ = fixed capital cost for the complete system
$C_{E,i}$ = cost of Equipment $i$
$f_i$ = installation factor for Equipment $i$

If an average installation factor for all types of equipment is to be used (Lang 1947),

$$C_F = f_I \sum_i C_{E,i} \qquad (2.8)$$

where $f_I$ = overall installation factor for the complete system.

## Table 2.2

Capital cost scaling for different process technologies.

| Plant | Size | Unit | 1962 Cost ($1000) | Range | Exponent |
|---|---|---|---|---|---|
| Acetylene (from natural gas) | 9.07 | t d$^{-1}$ | 2000 | 9.07–90.7 | 0.73 |
| Aluminum sulfate | 22.7 | t d$^{-1}$ | 375 | 22.7–90.7 | 0.71 |
| Ammonia (from natural gas) | 90.7 | t d$^{-1}$ | 3000 | 90.7–272 | 0.63 |
| Ammonium nitrate | 45.4 | t d$^{-1}$ | 630 | 45.4–272 | 0.65 |
| Ammonium sulfate | 90.7 | t d$^{-1}$ | 1100 | 90.7–272 | 0.65 |
| Butyl alcohol (n-butanol) | 3630 | t y$^{-1}$ | 3630 600 | $3.18 \times 10^5$ | 0.55 |
| Carbon dioxide | 45.4 | t d$^{-1}$ | 450 | 45.4–272 | 0.67 |
| Chlorine (electrolytic) | 4540 | t y$^{-1}$ | 4000 | $4540$–$2.72 \times 10^5$ | 0.38 |
| Ethanol (synthetic) | 1180 | t y$^{-1}$ | 1300 | $1180$–$9.07 \times 10^4$ | 0.60 |
| Ethylene | 9980 | t y$^{-1}$ | 2600 | $9980$–$3.36 \times 10^5$ | 0.71 |
| Ethylene oxide (including glycol mfg. facilities) | 1360 | t y$^{-1}$ | 830 | $1360$–$9.07 \times 10^4$ | 0.78 |
| Formaldehyde (100%) | 22.7 | t d$^{-1}$ | 900 | 22.7–90.7 | 0.56 |
| Hydrochloric acid (99.5%) | 9.07 | t d$^{-1}$ | 90 | 9.07–18.1 | 0.60 |
| Hydrocyanic acid | 9.07 | t d$^{-1}$ | 1800 | 9.07–45.4 | 0.52 |
| Hydrofluoric acid (anhydrous) | 9.07 | t d$^{-1}$ | 900 | 9.07–45.4 | 0.68 |
| Hydrogen (from natural gas) | $2.83 \times 10^4$ | Nm$^3$ d$^{-1}$ | 410 | $2.83 \times 10^4$–$2.83 \times 10^5$ | 0.57 |
| | $2.83 \times 10^5$ | Nm$^3$ d$^{-1}$ | 1400 | $2.83 \times 10^5$–$1.42 \times 10^6$ | 0.68 |
| Isopropyl alcohol | $1.44 \times 10^4$ | m$^3$ y$^{-1}$ | 2400 | $1.44 \times 10^4$–$4.54 \times 10^5$ | 0.60 |
| Maleic anhydride | 9.07 | t d$^{-1}$ | 4000 | 9.07–22.7 | 0.61 |
| Methanol | 45.4 | t d$^{-1}$ | 2000 | 45.4–272 | 0.75 |
| Nitric acid | 45.4 | t d$^{-1}$ | 820 | 45.4–272 | 0.55 |
| Oxygen | 18.1 | t d$^{-1}$ | 430 | 18.1–907 | 0.56 |
| Phosphoric acid (54%) | 45.4 | t d$^{-1}$ | 1100 | 45.4–272 | 0.63 |
| Sodium hydroxide | 3630 | t y$^{-1}$ | 3500 | $3630$–$2.72 \times 10^5$ | 0.38 |
| Styrene | 1810 | t y$^{-1}$ | 1150 | $1810$–$9.07 \times 10^4$ | 0.68 |
| Sulfuric acid | 45.4 | t d$^{-1}$ | 300 | 45.4–907 | 0.70 |
| Titanium dioxide | 9.07 | t d$^{-1}$ | 3400 | 9.07–90.7 | 0.61 |
| Urea | 136 | t d$^{-1}$ | 3700 | 136–272 | 0.70 |

Chemical Engineering Plant Cost Index for 1962 = 102.
Source: Adapted from Bauman HC, 1964, *Fundamentals of Cost Engineering in the Chemical Industry*, Reinhold, New York.

The overall installation factor for a new design is broken down in Table 2.3 into component parts according to the dominant phase being processed. The cost of the installation will depend on the balance of gas and liquid processing versus solids processing. If the plant handles only gases and liquids, it can be characterized as fluid processing. A plant can be characterized as solids processing if the bulk of the material handling is in the solid phase. For example, a solid processing plant could be an ore preparation plant. Between the two extremes of fluid processing and solids processing are processes that handle a significant amount of both solids and fluids. For example, a shale oil plant involves preparation of the shale oil followed by extraction of fluids from the shale

## Table 2.3

Typical factors for capital cost based on delivered equipment costs.

| Item | Type of process | |
|---|---|---|
| | Fluid processing | Solid processing |
| **Direct costs** | | |
| Equipment delivered cost | 1 | 1 |
| Equipment erection, $f_{ER}$ | 0.4 | 0.5 |
| Piping (installed), $f_{PIP}$ | 0.7 | 0.2 |
| Instrumentation and controls (installed), $f_{INST}$ | 0.2 | 0.1 |
| Electrical (installed), $f_{ELEC}$ | 0.1 | 0.1 |
| Utilities, $f_{UTIL}$ | 0.5 | 0.2 |
| Off-sites, $f_{OS}$ | 0.2 | 0.2 |
| Buildings (including services), $f_{BUILD}$ | 0.2 | 0.3 |
| Site preparation, $f_{SP}$ | 0.1 | 0.1 |
| *Total capital cost of installed equipment* | 3.4 | 2.7 |
| **Indirect costs** | | |
| Design, engineering and construction, $f_{DEC}$ | 1.0 | 0.8 |
| Contingency (about 10% of fixed capital costs), $f_{CONT}$ | 0.4 | 0.3 |
| *Total fixed capital cost* | 4.8 | 3.8 |
| **Working capital** | | |
| Working capital (15% of total capital cost), $f_{WC}$ | 0.7 | 0.6 |
| *Total capital cost, $f_I$* | 5.8 | 4.4 |

oil and then separation and processing of the fluids. For these types of plant, the contributions to the capital cost can be estimated from the two extreme values in Table 2.3 by interpolation in proportion of the ratio of major processing steps that can be characterized as fluid processing and solid processing. A number of points should be noted about the various contributions to the capital cost in Table 2.3. The values are:

- average values for all types of equipment, materials of construction, operating pressure and temperature whereas in practice the values will vary;
- only guidelines and the individual components will vary from project to project;
- applicable to new designs only.

The factored estimate approach requires capital cost of the main items of process equipment. Equipment costs may be obtained from equipment vendors or from published cost data. Care should be taken as to the basis of such cost data. What is required for cost

estimates is delivered cost, but cost is often quoted as FOB (Free On Board). Free on board means the manufacturer pays for loading charges onto a shipping truck, railcar, barge or ship, but not freight or unloading charges. To obtain a delivered cost requires typically 5–10% to be added to the FOB cost. The delivery cost depends on location of the equipment supplier, location of the site to be delivered to, size of the equipment, and so on.

The cost of a specific item of equipment will be a function of:

- size;
- materials of construction;
- design pressure;
- design temperature.

Cost data are often presented as cost versus capacity charts, or expressed as a power law of capacity:

$$C_E = C_B \left(\frac{Q}{Q_B}\right)^M \tag{2.9}$$

where

$C_E$ = equipment cost with capacity $Q$

$C_B$ = known base cost for equipment with capacity $Q_B$

$M$ = constant depending on equipment type

A number of sources of such data are available in the open literature (Guthrie 1969; Anson 1977; Hall et al. 1982; Ulrich 1984; Hall et al. 1988; Remer and Chai 1990; Gerrard 2000; Peters et al. 2003). Table 2.4 presents data for a number of equipment items on the basis of January 2000 costs (Gerrard 2000) (CE Composite Index = 391.1, CE Index of Equipment = 435.8).

Cost correlations for vessels are normally expressed in terms of the mass of the vessel. This means that not only a preliminary sizing of the vessel is required, but also a preliminary assessment of the mechanical design (Mulet et al. 1981a, 1981b). Mechanical design will be dealt with in detail in Chapter 19.

Materials of construction have a significant influence on the capital cost of equipment. Table 2.5 gives some approximate average factors to relate the different materials of construction for the equipment capital cost. Selection of materials of construction will be discussed in Chapter 18. It should be emphasized that the factors in Table 2.5 are average and only approximate and will vary, among other things, according to the type of equipment. As an example, Table 2.6 gives materials of construction cost factors for pressure vessels and distillation columns (Mulet et al. 1981a, 1981b). The cost factors for shell-and-tube heat exchangers are made more complex by the ability to construct the different components from different materials of construction. Table 2.7 gives typical materials of construction factors for shell-and-tube heat exchangers (Anson 1977).

The operating pressure also influences the equipment capital cost as a result of thicker vessel walls to withstand increased pressure. Table 2.8 presents typical factors to account for the pressure rating. As with materials of construction correction factors, the pressure correction factors in Table 2.8 are average and only approximate and will vary, among other things, according to the type of equipment.

## Table 2.4

Typical equipment capacity delivered capital cost correlations.

| Equipment | Material of construction | Capacity measure | Base size $Q_B$ | Base cost $C_B$ ($) | Size range | Cost exponent $M$ |
|---|---|---|---|---|---|---|
| Agitated reactor | CS | Volume (m$^3$) | 1 | $1.15 \times 10^4$ | 1–50 | 0.45 |
| Pressure vessel | SS | Mass (t) | 6 | $9.84 \times 10^4$ | 6–100 | 0.82 |
| Distillation column (empty shell) | CS | Mass (t) | 8 | $6.56 \times 10^4$ | 8–300 | 0.89 |
| Sieve trays (10 trays) | CS | Column diameter (m) | 0.5 | $6.56 \times 10^3$ | 0.5–4.0 | 0.91 |
| Valve trays (10 trays) | CS | Column diameter (m) | 0.5 | $1.80 \times 10^4$ | 0.5–4.0 | 0.97 |
| Structured packing (5 m height) | SS (low grade) | Column diameter (m) | 0.5 | $1.80 \times 10^4$ | 0.5–4.0 | 1.70 |
| Scrubber (including random packing) | SS (low grade) | Volume (m$^3$) | 0.1 | $4.92 \times 10^3$ | 0.1–20 | 0.53 |
| Cyclone | CS | Diameter (m) | 0.4 | $1.64 \times 10^3$ | 0.4–3.0 | 1.20 |
| Vacuum filter | CS | Filter area (m$^2$) | 10 | $8.36 \times 10^4$ | 10–25 | 0.49 |
| Dryer | SS (low grade) | Evaporation rate (kg H$_2$O h$^{-1}$) | 700 | $2.30 \times 10^5$ | 700–3000 | 0.65 |
| Shell-and-tube heat exchanger | CS | Heat transfer area (m$^2$) | 80 | $3.28 \times 10^4$ | 80–4000 | 0.68 |
| Air-cooled heat exchanger | CS | Plain tube heat transfer area (m$^2$) | 200 | $1.56 \times 10^5$ | 200–2000 | 0.89 |
| Centrifugal pump (small, including motor) | SS (high grade) | Power (kW) | 1 | $1.97 \times 10^3$ | 1–10 | 0.35 |
| Centrifugal pump (large, including motor) | CS | Power (kW) | 4 | $9.84 \times 10^3$ | 4–700 | 0.55 |
| Compressor (including motor) | | Power (kW) | 250 | $9.84 \times 10^4$ | 250–10 000 | 0.46 |
| Fan (including motor) | CS | Power (kW) | 50 | $1.23 \times 10^4$ | 50–200 | 0.76 |
| Vacuum pump (including motor) | CS | Power (kW) | 10 | $1.10 \times 10^4$ | 10–45 | 0.44 |
| Electric motor | | Power (kW) | 10 | $1.48 \times 10^3$ | 10–150 | 0.85 |
| Storage tank (small atmospheric) | SS (low grade) | Volume (m$^3$) | 0.1 | $3.28 \times 10^3$ | 0.1–20 | 0.57 |
| Storage tank (large atmospheric) | CS | Volume (m$^3$) | 5 | $1.15 \times 10^4$ | 5–200 | 0.53 |
| Silo | CS | Volume (m$^3$) | 60 | $1.72 \times 10^4$ | 60–150 | 0.70 |
| Package steam boiler (fire-tube boiler) | CS | Steam generation (kg h$^{-1}$) | 50,000 | $4.64 \times 10^5$ | 50 000–350 000 | 0.96 |
| Field erected steam boiler (water-tube boiler) | CS | Steam generation (kg h$^{-1}$) | 20,000 | $3.28 \times 10^5$ | 10 000–800 000 | 0.81 |
| Cooling tower (forced draft) | | Water flowrate (m$^3$ h$^{-1}$) | 10 | $4.43 \times 10^3$ | 10–40 | 0.63 |

CS = carbon steel; SS (low grade) = low-grade stainless steel, for example, type 304; SS (high grade) = high-grade stainless steel, for example, type 316.

Finally, its operating temperature also influences equipment capital cost. This is caused by, among other factors, a decrease in the allowable stress for materials of construction as the temperature increases (see Chapters 18 and 19). Table 2.9 presents typical factors to account for the operating temperature.

Thus, for a base cost for carbon steel equipment at moderate pressure and temperature, the actual cost can be estimated from:

$$C_E = C_B \left( \frac{Q}{Q_B} \right)^M f_M f_P f_T \qquad (2.10)$$

## Table 2.5

Typical average equipment materials of construction capital cost factors.

| Material | Correction factor $f_M$ |
|---|---|
| Carbon steel | 1.0 |
| Aluminum | 1.3 |
| Stainless steel (low grades) | 2.4 |
| Stainless steel (high grades) | 3.4 |
| Hastelloy C | 3.6 |
| Monel | 4.1 |
| Nickel and inconel | 4.4 |
| Titanium | 5.8 |

## Table 2.6

Typical materials of construction capital cost factors for pressure vessels and distillation columns (Mulet et al. 1981a, 1981b).

| Material | Correction factor $f_M$ |
|---|---|
| Carbon steel | 1.0 |
| Stainless steel (low grades) | 2.1 |
| Stainless steel (high grades) | 3.2 |
| Monel | 3.6 |
| Inconel | 3.9 |
| Nickel | 5.4 |
| Titanium | 7.7 |

## Table 2.7

Typical materials of construction capital cost factors for shell-and-tube heat exchangers (Adapted from Anson 1977).

| Material | Correction factor $f_M$ |
|---|---|
| CS shell and tubes | 1.0 |
| CS shell, aluminum tubes | 1.3 |
| CS shell, monel tubes | 2.1 |
| CS shell, SS (low grade) tubes | 1.7 |
| SS (low grade) shell and tubes | 2.9 |

where

$C_E$ = equipment cost for carbon steel at moderate pressure and temperature with capacity $Q$

$C_B$ = known base cost for equipment with capacity $Q_B$

$M$ = constant depending on equipment type

## Table 2.8

Typical equipment pressure capital cost factors.

| Design pressure (bar absolute) | Correction factor $f_P$ |
|---|---|
| 0.01 | 2.0 |
| 0.1 | 1.3 |
| 0.5–7 | 1.0 |
| 50 | 1.5 |
| 100 | 1.9 |

## Table 2.9

Typical equipment temperature capital cost factors.

| Design temperature (°C) | Correction factor $f_T$ |
|---|---|
| 0–100 | 1.0 |
| 300 | 1.6 |
| 500 | 2.1 |

$f_M$ = correction factor for materials of construction

$f_P$ = correction factor for design pressure

$f_T$ = correction factor for design temperature

Just like complete processes, historical data for equipment can be updated according to Eq. (2.6). The Chemical Engineering (CE) Indexes are particularly useful. CE Indexes are available for equipment covering:

- heat exchangers and tanks;
- pipes, valves, and fittings;
- process instruments;
- pumps and compressors;
- electrical equipment;
- structural supports and miscellaneous.

A combined CE Index of Equipment is also available.

In addition to the purchased cost of the equipment, investment is required to install the equipment. Installation costs include:

- cost of installation;
- piping and valves;
- control systems;
- foundations;
- structures;
- insulation;
- fire proofing;
- electrical;
- painting;
- engineering fees;
- contingency.

The total capital cost of the installed battery limits equipment in a new design will normally be two to four times the purchased cost of the equipment (Lang 1947; Hand 1958).

When equipment uses materials of construction other than carbon steel or operating temperatures are extreme, the capital cost needs to be adjusted accordingly. Whilst the equipment cost and its associated pipework will be changed, the other installation costs will be largely unchanged, whether the process equipment is manufactured from carbon steel or exotic materials of construction. Thus, the application of the factors from Tables 2.5 to 2.9 should only be applied to the equipment and pipework:

$$
\begin{aligned}
C_F = \sum_i [f_M f_P f_T (1 + f_{PIP})]_i C_{E,i} \\
+ (f_{ER} + f_{INST} + f_{ELEC} + f_{UTIL} + f_{OS} + f_{BUILD} \\
+ f_{SP} + f_{DEC} + f_{CONT} + f_{WS}) \sum_i C_{E,i}
\end{aligned}
\qquad (2.11)
$$

Thus, to estimate a Class 4 capital cost:

1) List the main plant items and estimate their size.
2) Estimate the equipment cost of the main plant items from Table 2.4.
3) Adjust the equipment costs to a common time basis using a cost index.
4) Convert the cost of the main plant items to carbon steel, moderate pressure, and moderate temperature.
5) Select the appropriate installation sub-factors from Table 2.3 and adjust for individual circumstances.
6) Select the appropriate materials of construction, operating pressure, and operating temperature correction factors for each of the main plant items.
7) Apply Eq. (2.11) to estimate the total fixed capital cost.

Equipment cost data used in the early stages of a design will by necessity normally be based on capacity, materials of construction, operating pressure, and operating temperature. However, in reality, the equipment cost will depend also on a number of factors that are difficult to quantify (Brennan 1998):

- multiple purchase discounts;
- buyer–seller relationships;
- capacity utilization in the fabrication shop (i.e., how busy the fabrication shop is);
- required delivery time;
- availability of materials and fabrication labor;
- special terms and conditions of purchase, and so on.

It should be emphasized that capital cost estimates using installation factors are at best uncertain. When preparing such an estimate, the designer spends most of the time on the equipment costs, which represents typically 20–40% of the total installed cost. The bulk costs (civil engineering, labor, etc.) are factored costs that lack definition. As given in Table 2.1, at best, this type of estimate can be expected to be accurate to $-15\%$ to $+20\%$ and could be as poor as $-30\%$ to $+50\%$. To obtain greater accuracy requires

detailed examination of all aspects of the investment. The shortcomings of capital cost estimates using installation factors are less serious in preliminary process design if used to compare options on a common basis. If used to compare options, the errors will tend to be less serious as the errors will tend to be consistent across the options.

**Example 2.1**    A new heat exchanger is to be installed as part of a large project. Preliminary sizing of the heat exchanger has estimated its heat transfer area to be 250 m². Its material of construction is low-grade stainless steel and its pressure rating is 5 bar. Estimate the contribution of the heat exchanger to the total cost of the project (CE Index of Equipment = 1037).

**Solution**

From Eq. (2.1) and Table 2.4, the capital cost of a carbon steel heat exchanger can be estimated from:

$$
\begin{aligned}
C_E &= C_B \left( \frac{Q}{Q_B} \right)^M \\
&= 3.28 \times 10^4 \left( \frac{250}{80} \right)^{0.68} \\
&= \$7.12 \times 10^4
\end{aligned}
$$

The cost can be adjusted to bring it up-to-date using the ratio of cost indexes:

$$
\begin{aligned}
C_E &= 7.12 \times 10^4 \left( \frac{1037}{435.8} \right) \\
&= \$1.69 \times 10^5
\end{aligned}
$$

Note that the cost was updated using the CE Index of Equipment. Alternatively, the cost index specifically for heat exchangers could have been used. The cost of a carbon steel heat exchanger needs to be adjusted for the material of construction. Because of the low-pressure rating, no correction for pressure is required (Table 2.8), but the cost needs to be adjusted for the materials of construction. From Table 2.7, $f_M = 2.9$, the total cost of the installed equipment can be estimated from Eq. (2.6) and Table 2.3. If the project is a completely new plant, the contribution of the heat exchanger to the total cost can be estimated to be:

$$
\begin{aligned}
C_F &= f_M (1 + f_{PIP}) C_E + (f_{ER} + f_{INST} + f_{ELEC} + f_{UTIL} + f_{OS} \\
&\quad + f_{BUILD} + f_{SP} + f_{DEC} + f_{CONT} + f_{WS}) C_E \\
&= 2.9(1 + 0.7)1.69 \times 10^5 + (0.4 + 0.2 + 0.1 + 0.5 \\
&\quad + 0.2 + 0.2 + 0.1 + 1.0 + 0.4 + 0.7) 1.69 \times 10^5 \\
&= 8.73 \times 1.69 \times 10^5 \\
&= \$1.48 \times 10^6
\end{aligned}
$$

Had the new heat exchanger been an addition to an existing plant that did not require investment in electrical services, utilities, off-sites, buildings, site preparation, or working capital, then the cost would be estimated from:

$$
\begin{aligned}
C_F &= f_M (1 + f_{PIP}) C_E + (f_{ER} + f_{INST} + f_{DEC} + f_{CONT}) C_E \\
&= 2.9(1 + 0.7) 1.69 \times 10^5 + (0.4 + 0.2 + 1.0 + 0.4) 1.69 \times 10^5 \\
&= 6.93 \times 1.69 \times 10^5 \\
&= \$1.17 \times 10^6
\end{aligned}
$$

Installing a new heat exchanger into an existing plant might require additional costs over and above those estimated here. Connecting new equipment to existing equipment, modifying or relocating existing equipment to accommodate the new equipment and downtime might all add to the costs.

Class 4 capital cost estimates are used in the Select Phase of a project (see Chapter 1) for preliminary project evaluation.

## 2.4 Class 3 to Class 1 Capital Cost Estimates

As the project progresses, there will be an increased availability of data. Historical and public data for the major items of equipment will be replaced by vendor cost data for the actual project. Also, as the project progresses there will be increased information of material quantities. To estimate the erection cost accurately requires knowledge of how much concrete will be used for foundations, how much structural steelwork is required, and so on.

Thus, as the project progresses, vendor equipment quotations gradually replace historical and public data for equipment and a detailed bill of materials gradually replaces factoring methods. Factoring methods might still be used in Class 3 capital cost estimates for the less significant areas of the process. However, it is preferable to replace factoring methods completely in Class 2 and Class 1 estimates, replaced by equipment quotes from vendors and knowledge of detailed or assumed material quantities.

Class 3 capital cost estimates are used in the later stages of the Select Phase of a project for preliminary project evaluation and the early stages of the Definition (FEED) Phase of a project (see Chapter 1). The decision to sanction projects in the Definition (FEED) Phase of a project will normally be based on a Class 2 capital cost estimate. Class 1 capital cost estimates are used to monitor cost and resource variations from the budget in the final stages of a project.

## 2.5 Capital Cost of Retrofit

Estimating the capital cost of a retrofit project is much more difficult than for a new design. In principle, the cost of individual items of new equipment will be the same, whether it is a grassroot design or a retrofit. By contrast, installation factors for equipment in retrofit can be completely different from grassroot design, and could be higher or lower. If the new equipment can take advantage of existing space, foundations, electrical cabling, and so on, the installation factor might in some cases be lower than in a new design. This will especially be the case for small items of equipment. However, most often, retrofit installation factors will tend to be higher than in grassroot design and can be very much higher. This is because existing equipment might need to be modified or moved to allow installation of new equipment. Also, access to the area where the installation is required is likely to be much more restricted in retrofit than in the phased installation of new plant.

**Table 2.10**

Modification costs for distillation column retrofit (Bravo 1997).

| Column modification | Cost of modification (multiply factor by cost of new hardware) |
| --- | --- |
| Removal of trays to install new trays | 0.1 for the same tray spacing<br>0.2 for different tray spacing |
| Removal of trays to install packing | 0.1 |
| Removal of packing to install new trays | 0.07 |
| Installation of new trays | 1.0–1.4 for the same tray spacing<br>1.2–1.5 for different tray spacing<br>1.3–1.6 when replacing packing |
| Installation of new structured packing | 0.5–0.8 |

Smaller projects (as the retrofit is likely to be) tend to bring a higher cost of installation per unit of installed equipment than larger projects.

As an example, one very common retrofit situation is the replacement of distillation column internals to improve the performance of the column. The improvement in performance sought is often an increase in the throughput. This calls for existing internals to be removed and then to be replaced with the new internals. Table 2.10 gives typical factors for the removal of old internals and the installation of new ones (Bravo 1997).

As far as utilities and off-sites are concerned, it is also difficult to generalize. Small retrofit projects are likely not to require any investment in utilities and off-sites. Larger-scale retrofit might demand a major revamp of the utilities and off-sites, which can be particularly expensive because existing equipment might need to be modified or removed to make way for new utilities and off-site equipment.

Working capital is also difficult to generalize. Most often, there will be no significant working capital associated with a retrofit project. For example, if a few items of equipment are replaced to increase the capacity of a plant, this will not significantly change the raw materials and product inventories, money to carry accounts receivable, money to meet the payroll, and so on. On the other hand, if the plant changes function completely, significant new storage capacity is added, and so there might be a significant element of working capital to be added to the retrofit capital cost.

One of the biggest sources of cost associated with retrofit can be the *downtime* (the period during which the plant will not be productive) required to carry out the modifications. The cost of lost production can be the dominant feature of retrofit projects. The cost of lost production should be added to the capital cost of a retrofit project. To minimize the downtime and cost of lost production requires that as much preparation as possible is carried out whilst the plant is operating. The modifications requiring the plant to be shut down should be minimized. For example, it might be possible for new foundations to be installed and new equipment put into place while the plant is still operating, leaving the final pipework and electrical modifications for the shut-down. Retrofit projects are often arranged such that the preparation is carried out prior to a regular maintenance shut-down, with the final modifications coinciding with the planned maintenance shut-down. Such considerations often dominate the decisions made as to how to modify the process for retrofit.

Because of all of these uncertainties, it is difficult to provide general guidelines for the capital cost of retrofit projects. The basis of the capital cost estimate should be to start with the required investment in new equipment. Installation factors for the installation of equipment for grassroot design from Table 2.3 need to be adjusted according to circumstances (often increased). If old equipment needs to be modified to take up a new role (e.g. move an existing heat exchanger to a new duty), then an installation cost must be applied without the equipment cost. In the absence of better information, the installation cost can be taken to be that required for the equivalent piece of new equipment. Some elements of the total cost breakdown in Table 2.3 will not be relevant and should not be included. In general, for the estimation of capital cost for retrofit, a detailed examination of the individual features of retrofit projects is necessary.

**Example 2.2** An existing heat exchanger is to be re-piped to a new duty in a retrofit project without moving its location. The only significant investment is piping modifications. The heat transfer area of the existing heat exchanger is 250 m². The material of construction is low-grade stainless steel and its design pressure is 5 bar. Estimate the cost of the project (CE Index of Equipment = 1037).

**Solution**

All retrofit projects have individual characteristics and it is impossible to generalize the costs. The only way to estimate costs with any certainty is to analyze the costs of all of the modifications in detail. However, in the absence of such details, a very preliminary estimate can be obtained by estimating the retrofit costs from the appropriate installation costs for a new design. In this case, piping costs can be estimated from those for a new heat exchanger of the same specification, but excluding the equipment cost. For Example 2.1, the cost of a new stainless steel heat exchanger with an area of 250 m² was estimated to be $1.69 \times 10^5$. The piping costs (stainless steel) can therefore be estimated to be:

$$Piping\ cost = f_M f_{PIP} C_E$$
$$= 2.9 \times 0.7 \times 1.69 \times 10^5$$
$$= 2.03 \times 1.69 \times 10^5$$
$$= \$3.43 \times 10^5$$

This estimate should not be treated with any confidence. It will give an idea of the costs and might be used to compare retrofit options on a like-for-like basis, but could be very misleading.

**Example 2.3** An existing distillation column is to be revamped to increase its capacity by replacing the existing sieve trays with stainless steel structured packing. The column shell is 46 m tall and 1.5 m diameter and is currently fitted with 70 sieve trays with a spacing of 0.61 m. The existing trays are to be replaced with stainless steel structured packing with a total height of 30 m. Estimate the cost of the project (CE Index of Equipment = 1037).

**Solution**

First, estimate the purchase cost of the new structured packing from Eq. (2.1) and Table 2.4, which gives the costs for a 5 m height of packing:

$$C_E = C_B \left(\frac{Q}{Q_B}\right)^M$$
$$= 1.8 \times 10^4 \times \frac{30}{5} \left(\frac{1.5}{0.5}\right)^{1.7}$$
$$= \$6.99 \times 10^5$$

Adjusting the cost to bring it up-to-date using the ratio of cost indexes:

$$C_E = 6.99 \times 10^5 \left(\frac{1037}{435 \times 8}\right)$$
$$= \$1.66 \times 10^6$$

From Table 2.10, the factor for removing the existing trays is 0.1 and that for installing the new packing is 0.5 to 0.8 (say 0.7). The estimated total cost of the project is:

$$= (1 + 0.1 + 0.7)\ 1.66 \times 10^6$$
$$= \$2.99 \times 10^6$$

# 2.6 Annualized Capital Cost

It is sometimes desirable to be able to express the cost of capital on an annual basis. To explore how this can be done meaningfully, start by considering where capital for new installations may be obtained from:

a) Loans from banks.

b) Issue by the company of common stock (ordinary shares), preferred stock (preference shares), or bonds.

c) Accumulated net cash flow arising from company profit over time.

Interest on loans from banks, preferred stock, and bonds is paid at a fixed rate of interest. A share of the profit of the company is paid as a dividend on common stock and preferred stock (in addition to the interest paid on preferred stock).

The cost of the capital for a project thus depends on its source. The source of the capital often will not be known during the early stages of a project and yet there is a need to select between process options and carry out optimization on the basis of both capital and operating costs. This is difficult to do unless both capital and operating costs can be expressed on a common basis. Capital costs can be expressed on an annual basis if it is assumed that the capital has been borrowed over a fixed period (for large projects this is typically 5–10 years, but might be much longer for strategic investments) at a fixed rate of interest, in which case the capital cost can be annualized according to (Smith 2016):

$$\text{Annualized capital cost} = \text{capital cost} \times \frac{i(1 + i)^n}{(1 + i)^n - 1} \quad (2.12)$$

where

$i$ = fractional interest rate per year

$n$ = number of years

As stated previously, the source of capital is often not known, and hence it is not known whether Eq. (2.12) is appropriate to represent the cost of capital. Equation (2.12) is, strictly speaking, only appropriate if the money for capital expenditure is to be borrowed over a fixed period at a fixed rate of interest. Moreover, if Eq. (2.12) is accepted, then the number of years over which the capital is to be annualized needs to be known, together with the rate of interest. However, the most important point is that, even if the source of capital is not known, and uncertain assumptions are necessary, Eq. (2.12) provides a common basis for the comparison of competing projects and design alternatives within a project.

---

**Example 2.4**   The purchased cost of a new distillation column installation is $1 million. Calculate the annual cost of installed capital if the capital is to be annualized over a five-year period at a fixed rate of interest of 5%.

**Solution**

First calculate the installed capital cost (Table 2.3):

$$
\begin{aligned}
C_F &= f_i\,C_E \\
&= 5.8 \times (1\,000\,000) \\
&= \$5\,800\,000
\end{aligned}
$$

$$
\begin{aligned}
\text{Annualization factor} &= \frac{i(1 + i)^n}{(1 + i)^n - 1} \\
&= \frac{0.05(1 + 0.05)^5}{(1 + 0.05)^5 - 1} \\
&= 0.2310
\end{aligned}
$$

$$\text{Annualized capital cost} = 5\,800\,000 \times 0.2310$$

$$= \$1\,340\,000\ \text{y}^{-1}$$

When using the annualized capital cost to carry out optimization, the designer should not lose sight of the uncertainties involved in the capital annualization. In particular, changing the annualization period can lead to very different results when, for example, carrying out a trade-off between energy and capital costs. When carrying out optimization, the sensitivity of the result to changes in the assumptions should be tested.

---

# 2.7  Operating Cost

## 2.7.1   Raw Materials Cost

In most processes, the largest individual operating cost is raw materials. The cost of raw materials and the product selling prices tend to have the largest influence on the economic performance of the process. The cost of raw materials and price of products depends on whether the materials in question are being bought and sold under a contractual arrangement (either within the same company or outside the company) or on the open market. Open market prices for some chemical products can fluctuate considerably with time. Raw materials might be purchased and products sold below or above the open market price when under a contractual arrangement, depending on the state of the market. Buying and selling on the open market may give the best purchase and selling prices but give rise to an uncertain economic environment. A long-term contractual agreement may reduce profit per unit of production but gives a degree of certainty over the project life.

The values of raw materials and products can be found in various on-line sources and trade journals. However, the values reported in such sources will be subject to short-term fluctuations and long-term forecasts will be required for investment analysis.

## 2.7.2   Catalysts and Chemicals Consumed in Manufacturing Other Than Raw Materials

Catalysts will need to be replaced or regenerated though the life of a process. The replacement of catalysts might be on a continuous basis if homogeneous catalysts are used. Heterogeneous catalysts might also be replaced continuously if they deteriorate rapidly, and regeneration cannot fully reinstate the catalyst activity. More often for heterogeneous catalysts, regeneration or replacement will be carried out on an intermittent basis, depending on the characteristics of the catalyst deactivation.

In addition to the cost of catalysts, there might be significant costs associated with chemicals consumed in manufacturing that do not form part of the final product. For example, acids and alkalis might be consumed to adjust the pH of streams. Such costs might be significant.

## 2.7.3   Utility Operating Cost

The utility operating cost is usually the most significant variable operating cost after the cost of raw materials. This is especially the case for the production of commodity chemicals. Utility operating cost includes:

- fuel;
- electricity;
- steam;
- cooling water;
- refrigeration;
- compressed air;
- inert gas.

Energy costs can vary enormously between different processing sites. Fuels required for heating by combustion in a furnace or for steam generation include:

- fossil fuels: coal, petroleum coke, oil, natural gas;
- biomass: wood from forestry, waste from wood processing, high yield woody crops, etc;
- biogas: gas from anaerobic digestion, syngas from biomass, etc.;
- waste to energy: agricultural waste, plastic waste, etc.

The price paid for fuel depends very much on how much is purchased, the pattern of usage, and contractual relationships. Fuels combusted in gas turbines or steam boilers will normally be part of a *cogeneration system*. Steam generation will often be at a high pressure and then expanded to lower pressures in steam turbines to generate power. Electricity is required to drive process machines (e.g., pumps and compressors) and might be used for heating. If electricity is generated from renewable sources (e.g., wind and solar), it can be used as a source of sustainable heating (e.g., steam generation in electric boilers). This will be discussed in more detail in Chapter 4.

When electricity is bought from external power-generation companies, the cost (and sales price if excess electricity is generated and exported) are normally subject to tariff variations. Electricity tariffs can depend on the season of the year (e.g., winter, summer, or transition), the time of the week (e.g., weekend versus weekday), and the time of day (e.g., night versus day). In hot countries, electricity is usually more expensive in the summer than in the winter because of the demand from air-conditioning systems. In cold countries, electricity is usually more expensive in the winter than in the summer because of the demand from space heating. The price structure for electricity can be complex. If electricity is purchased from a spot market in those countries that have such arrangements, then prices can vary wildly.

Steam costs in a cogeneration arrangement will depend on the price of fuel and electricity. If steam is only generated at low pressure and not used for power generation in steam turbines, then the cost of steam can be estimated from fuel costs (or electricity costs in electric boilers), assuming an efficiency of generation and distribution losses. The efficiency of generation depends on the boiler efficiency and the steam consumed in boiler feedwater production. Losses from the steam distribution system include heat losses from steam distribution and condensate return pipework to the environment, steam condensate lost to drain and not returned to the boilers, and steam leaks. The efficiency of steam generation (including auxiliary boiler-house requirements) is typically around 85–95% and distribution losses of perhaps another 5–10%, giving an overall efficiency for steam generation and distribution of typically around 75–90%. Care should be exercised when considering boiler efficiency and the efficiency of steam generation based on fuels. These figures might be quoted on the basis of *gross calorific value* of the fuel, which includes the latent heat of the water vapor from combustion. This latent heat is rarely recovered through condensation of the water vapor in the combustion gases. The *net calorific value* of the fuel assumes that the latent heat of the water vapor is not recovered and is therefore the most relevant value, yet figures are often quoted on the basis of gross calorific value.

If steam is generated from the combustion of fuel at high pressure, then the cost of steam should in general be related in some way to its capacity to generate power in a steam turbine rather than simply to its heating value. This will be discussed in more detail in Chapter 4. One simple way to cost steam is to calculate the cost of the fuel required to generate the high-pressure steam (including any losses), and this fuel cost is then the cost of the high-pressure steam. Although this neglects water costs, labor costs, and so on, it

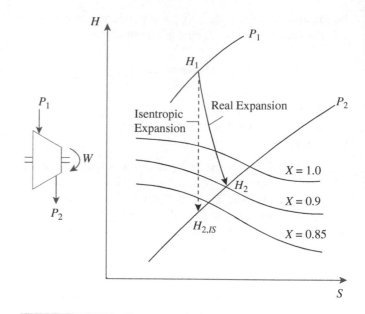

**Figure 2.1**

Steam turbine expansion. Source: Smith R, 2016, *Chemical Process Design and Integration*, 2nd Edition, John Wiley & Sons.

will be a reasonable estimate, as costs are normally dominated by fuel costs. Low-pressure mains have a value equal to that of the high-pressure mains minus the value of power generated by expanding the steam to low pressure in a steam turbine. To calculate the cost of steam that has been expanded though a steam turbine, the power generated in such an expansion can be calculated. The simplest way to do this is on the basis of a comparison between an ideal and a real expansion though a steam turbine. Figure 2.1 shows a steam turbine expansion on an enthalpy–entropy plot. In an ideal turbine, steam with an initial pressure $P_1$ and enthalpy $H_1$ expands isentropically to pressure $P_2$ and enthalpy $H_{2,IS}$. In such circumstances, the ideal work output is $(H_1 - H_{2,IS})$. Because of the frictional effects in the turbine nozzles and blade passages, the exit enthalpy is greater than it would be in an ideal turbine, and the work output is consequently less, given by $H_2$ in Figure 2.1. The actual work output is given by $(H_1 - H_2)$. The turbine isentropic efficiency $\eta_{IS}$ measures the ratio of the actual to ideal work obtained:

$$\eta_{IS} = \frac{H_1 - H_2}{H_1 - H_{2,IS}} \tag{2.13}$$

The output from the turbine might be superheated or partially condensed, as is the case in Figure 2.1. The following example illustrates the approach for high-pressure steam generated by combustion of fuel in a boiler.

**Example 2.5** The pressures of three steam mains have been set to the conditions given in Table 2.11. High-pressure (HP) steam is generated in boilers at 41 barg and superheated to 400 °C from the combustion of fuel. Medium-pressure (MP) and low-pressure (LP)

## Table 2.11

Steam mains pressure settings.

| Mains | Pressure (barg) |
|-------|-----------------|
| HP    | 41              |
| MP    | 10              |
| LP    | 3               |

steam are generated by expanding high-pressure steam through a steam turbine with an isentropic efficiency of 75%. The cost of fuel is $4.00\,GJ^{-1}$ and the cost of electricity is $0.07\,kWh^{-1}$. Boiler feedwater is available at 100 °C. Assuming an efficiency of steam generation of 90% and distribution losses of 10% of the fuel fired, estimate the cost of steam for the three levels. Steam properties can be taken from International Steam Tables (Kretzschmar and Wagner 2019).

### Solution

*Cost of 41 barg steam.* From steam tables, for 41 barg steam at 400 °C:

$$\text{Enthalpy} = 3211\,kJ\,kg^{-1}$$

For boiler feedwater from steam tables:

$$\text{Enthalpy} = 419\,kJ\,kg^{-1}$$

To generate 41 barg steam at 400 °C:

$$\text{Heat duty} = 3211 - 419 = 2792\,kJ\,kg^{-1}$$

For 41 barg steam:

$$\text{Cost} = 4.00 \times 10^{-6} \times 2792 \times \frac{1}{0.8}$$
$$= \$0.01396\,kg^{-1}$$
$$= \$13.96\,t^{-1}$$

*Cost of 10 barg steam.* Here 41 barg steam is now expanded to 10 barg in a steam turbine.

From steam tables, inlet conditions at 41 barg and 400 °C: are:

$$H_1 = 3211\,kJ\,kg^{-1}$$
$$S_1 = 6.744\,kJ\,kg^{-1}\,K^{-1}$$

Turbine outlet conditions for isentropic expansion to 10 barg are:

$$S_2 = 6.744\,kJ\,kg^{-1}\,K^{-1}$$
$$H_{2,IS} = 2872\,kJ\,kg^{-1}$$

For single-stage expansion with isentropic efficiency of 80%:

$$H_2 = H_1 - \eta_{IS}(H_1 - H_{2,IS})$$
$$= 3211 - 0.8\,(3211 - 2872)$$
$$= 2957\,kJ\,kg^{-1}$$

From steam tables, the outlet temperature is 258 °C, which corresponds to a saturation temperature of 184 °C and a superheat of 74 °C. Although steam for process heating is preferred close to saturated conditions, it is not desirable in this case to de-superheat by boiler feedwater injection to bring close to saturation conditions. If saturated steam is fed to the main, then the heat losses from the main will cause a large amount of condensation in the main, which is undesirable. Hence, it is standard practice to feed steam to the main with a superheat of at least 10–20 °C to avoid condensation in the main.

$$\text{Power generation} = 3211 - 2957$$
$$= 254\,kJ\,kg^{-1}$$
$$\text{Value of power generation} = 254 \times \frac{0.07}{3600}$$
$$= \$0.00494\,kg^{-1}$$
$$\text{Cost of 10 barg steam} = 0.01396 - 0.00494$$
$$= \$0.00902\,kg^{-1}$$
$$= \$9.02\,t^{-1}$$

*Cost of 3 barg steam.* Here, 10 barg steam from the exit of the first turbine is assumed to be expanded to 3 barg in another turbine.

From steam tables, inlet conditions of 10 barg and 258 °C are:

$$H_1 = 2957\,kJ\,kg^{-1}$$
$$S_1 = 6.910\,kJ\,kg^{-1}\,K^{-1}$$

Turbine outlet conditions for isentropic expansion to 3 barg are:

$$S_2 = 6.910\,kJ\,kg^{-1}\,K^{-1}$$
$$H_{2,IS} = 2746\,kJ\,kg^{-1}$$

For a single-stage expansion with isentropic efficiency of 75%:

$$H_2 = H_1 - \eta_{IS}(H_1 - H_{2,IS})$$
$$= 2957 - 0.75(2957 - 2746)$$
$$= 2799\,kJ\,kg^{-1}$$

From steam tables, the outlet temperature is 171 °C, corresponding with a saturation temperature of 144 °C which is superheated by 27 °C. Again, it is desirable to have some superheat for the steam fed to the low-pressure main.

$$\text{Power generation} = 2957 - 2799$$
$$= 158\,kJ\,kg^{-1}$$
$$\text{Value of power generation} = 158 \times \frac{0.07}{3600}$$
$$= \$0.00307\,kg^{-1}$$
$$\text{Cost of 3 barg steam} = 0.00902 - 0.00307$$
$$= \$0.00595\,kg^{-1}$$
$$= \$5.95\,t^{-1}$$

If the steam generated in the boilers is at a very high pressure and/or the ratio of power to fuel costs is high, then the value of low-pressure steam can be extremely low or even negative.

The operating cost for cooling water tends to be low relative to the value of both fuel and electricity. The principal operating cost associated with the provision of cooling water is the cost of power to drive the cooling tower fans and cooling water circulation pumps. Cooling water systems will be discussed in more detail in Chapter 5.

The cost of power required for a refrigeration system can be estimated as a multiple of the power required for an ideal system:

$$\frac{W_{IDEAL}}{Q_C} = \frac{T_H - T_C}{T_C} \tag{2.14}$$

where

$W_{IDEAL}$ = ideal power required for the refrigeration cycle

$Q_C$ = the cooling duty

$T_C$ = temperature at which heat is taken into the refrigeration cycle (K)

$T_H$ = temperature at which heat is rejected from the refrigeration cycle (K)

The ratio of ideal to actual power is often around 0.6. Thus:

$$W = \frac{Q_C}{0.6} \left( \frac{T_H - T_C}{T_C} \right) \tag{2.15}$$

where $W$ is the actual power required for the refrigeration cycle.

**Example 2.6** A process requires 0.5 MW of cooling at –20 °C. A refrigeration cycle is required to remove this heat and reject it to cooling water supplied at 25 °C and returned at 30 °C. Assuming a minimum temperature difference ($\Delta T_{min}$) of 5 °C and both vaporization and condensation of the refrigerant occur isothermally, estimate the annual operating cost of refrigeration for an electrically driven system operating 8000 hours per year. The cost of electricity is $0.07 kWh$^{-1}$.

**Solution**

From Eq. (2.15):

$$W = \frac{Q_C}{0.6} \left( \frac{T_H - T_C}{T_C} \right)$$
$$T_H = 30 + 5 = 35°C = 308 \text{ K}$$
$$T_C = -20 - 5 = -25°C = 248 \text{ K}$$
$$W = \frac{0.5}{0.6} \left( \frac{308 - 248}{248} \right)$$
$$= 0.202 \text{ MW}$$
$$\text{Cost of electricity} = 0.202 \times 10^3 \times 0.07 \times 8000$$
$$= \$113\,120 \text{ y}^{-1}$$

### 2.7.4 Labor Cost

The cost of labor is difficult to estimate. It depends on whether the process is batch or continuous, the level of automation, the number of processing steps, and the level of production. When synthesizing a process, it is usually only necessary to screen process options that have the same basic character (e.g. continuous), have the same level of automation, have a similar number of processing steps and have the same level of production. In this case, labor costs will be common to all options and hence will not affect the comparison.

If, however, options are to be compared that are very different in nature, such as a comparison between batch and continuous operation, some allowance for the difference in the cost of labor must be made. Also, if the location of the plant has not been fixed, the differences in labor costs between different geographical locations can be important.

### 2.7.5 Maintenance

The cost of maintenance depends on whether processing materials are solids on the one hand or gas and liquid on the other. Handling solids tends to increase maintenance costs. Highly corrosive process fluids increase maintenance costs. Average maintenance costs tend to be around 6% of the fixed capital investment (Peters et al. 2003).

## 2.8 Economic Evaluation

As the design progresses, more information is accumulated. The best methods of assessing the profitability of alternatives are based on projections of the cash flows during the project life (Allen 1980).

Figure 2.2 shows the cash flow pattern for a typical project (Allen 1980). The cash flow is a cumulative cash flow. Consider Curve 1 in Figure 2.2. From the start of the project at Point A, cash is spent without any immediate return. The early stages of the project consist of development, design and other preliminary work, which causes the cumulative curve to dip to Point B. This is followed by the main phase of capital investment in buildings, plant, and equipment, and the curve drops more steeply to Point C. Working capital is spent to commission the plant between Points C and D. Production starts at D, where revenue from sales begins. Initially, the rate of production is likely to be below design conditions until full production is achieved at E. At F, the cumulative cash flow is again zero. This is the project breakeven point. Toward the end of the project's life at G, the net rate of cash flow may decrease owing to, for example, increasing maintenance costs, increased competition in the market price for the same product, and so on.

Ultimately, the plant might be permanently shut down or given a major revamp. This marks the end of the project, H. If the plant is shut down, working capital is recovered and there may be salvage value, which would create a final cash inflow at the end of the project. On the other hand, decontamination and remediation might incur a final significant cash outflow at the end of the project.

The predicted cumulative cash flow curve for a project throughout its life forms the basis for a more detailed evaluation. Many quantitative measures or indices have been proposed. In each case, important features of the cumulative cash flow curve are identified and transformed into a single numerical measure as an index.

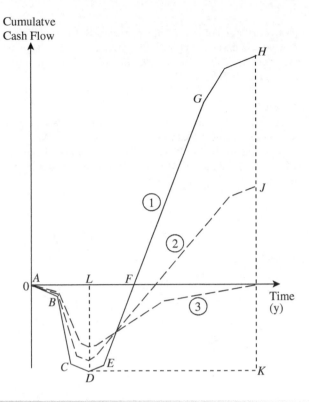

**Figure 2.2**

Cash flow pattern for a typical project. Source: Adapted from Allen DH, 1980, *A Guide to the Economic Evaluation of Projects*, IChemE.

**1)** *Payback Time*. Payback time is the time that elapses from the start of the project (*A* in Figure 2.2) to the breakeven point (*F* in Figure 2.2). The shorter the payback time, the more attractive is the project. Payback time is often calculated as the time to recoup the capital investment based on the mean annual cash flow. In retrofit, payback time is usually calculated as the time to recoup the retrofit capital investment from the mean annual improvement in operating costs.

**2)** *Return on Investment (ROI)*. Return on investment (ROI) is usually defined as the ratio of average yearly income over the productive life of the project to the total initial investment, expressed as a percentage. Thus, from Figure 2.2:

$$ROI = \frac{KH}{KD} \times \frac{100}{LD} \quad \% \text{per year} \qquad (2.16)$$

Payback and ROI select particular features of the project cumulative cash flow and ignore others. They take no account of the *pattern* of cash flow during a project. The other indices to be described, net present value and discounted cash flow return, are more comprehensive because they take account of the changing pattern of project net cash flow with time. They also take account of the *time value* of money.

**3)** *Net Present Value (NPV)*. Since money can be invested to earn interest, money received now has a greater value than money if received at some time in the future. The net present value of a project is the sum of the present values of each individual cash flow. In this case, the *present* is taken to be the start of a project.

Time is taken into account by discounting the annual cash flow $A_{CF}$ with the fractional rate of interest $i$ to obtain the annual discounted cash flow $A_{DCF}$. Thus, at the end of year 1:

$$A_{DCF1} = \frac{A_{CF1}}{(1 + i)}$$

at the end of year 2:

$$A_{DCF2} = \frac{A_{CF2}}{(1 + i)^2}$$

and at the end of year $n$:

$$A_{DCFn} = \frac{A_{CFn}}{(1 + i)^n} \qquad (2.17)$$

The sum of the annual discounted cash flows over $n$ years $\Sigma$ $A_{DCF}$ is known as the *net present value (NPV)* of the project:

$$NPV = \sum A_{DCF} \qquad (2.18)$$

The value of *NPV* is directly dependent on the choice of the fractional interest rate $i$ and project lifetime $n$.

Returning to the cumulative cash flow curve for a project, the effect of discounting is shown in Figure 2.2. Curve 1 is the original curve with no discounting, that is, $i = 0$, and the project *NPV* is equal to the final net cash position given by *H*. Curve 2 shows the effect of discounting at a fixed rate of interest and the corresponding project *NPV* is given by *J*. Curve 3 in Figure 2.2 shows a larger rate of interest, but it is chosen such that the *NPV* is zero at the end of the project.

The greater the positive *NPV* for a project, the more economically attractive it is. A project with a negative *NPV* is not a profitable proposition.

**4)** *Discounted Cash Flowrate of Return*. The discounted cash flow rate of return is defined as the discount rate $i$, which makes the *NPV* of a project equal to zero (Curve 3 in Figure 2.2):

$$NPV = \sum A_{DCF} = 0 \qquad (2.19)$$

The value of $i$ given by this equation is known as the *discounted cash flow rate of return (DCFRR)*. It may be found graphically or by trial and error.

**Example 2.7** A company has the alternative of investing in one of two projects, A or B. The capital cost of both projects is $10 million. The predicted annual cash flows for both projects are shown in Table 2.12. Capital is restricted, and a choice is to be made on the basis of the discounted cash flow rate of return, based on a five-year lifetime.

**Solution**

**Project A**

Start with an initial guess for *DCFRR* of 20% and increase as detailed in Table 2.13. Twenty percent is too low since $\Sigma A_{DCF}$ is positive at the end of year 5. Thirty percent is too high since $\Sigma A_{DCF}$ is negative at the end of year 5, as is the case with 25%. The answer must be between 20% and 25%. Interpolating on the basis of $\Sigma A_{DCF}$, the *DCFRR* ≈ 23%.

**Project B**

Again, start with an initial guess for *DCFRR* of 20% and increase as detailed in Table 2.14. From $\Sigma A_{DCF}$ at the end of year 5, 20% is too low, 40% too high, and 35% also too low. Interpolating on the basis of $\Sigma A_{DCF}$, the *DCFRR* ≈ 38%. Project B should therefore be chosen.

## Table 2.12

Predicted annual cash flows.

| Year | Cash flows ($10^6$) | |
| --- | --- | --- |
| | Project A | Project B |
| 0 | −10 | −10 |
| 1 | 1.6 | 6.5 |
| 2 | 2.8 | 5.2 |
| 3 | 4.0 | 4.0 |
| 4 | 5.2 | 2.8 |
| 5 | 6.4 | 1.6 |

## Table 2.13

Calculation of *DCFRR* for Project A.

| Year | $A_{CF}$ | DCF 20% | | DCF 30% | | DCF 25% | |
| --- | --- | --- | --- | --- | --- | --- | --- |
| | | $A_{DCF}$ | $\Sigma A_{DCF}$ | $A_{DCF}$ | $\Sigma A_{DCF}$ | $A_{DCF}$ | $\Sigma A_{DCF}$ |
| 0 | −10 | −10 | −10 | −10 | −10 | −10 | −10 |
| 1 | 1.6 | 1.33 | −8.67 | 1.23 | −8.77 | 1.28 | −8.72 |
| 2 | 2.8 | 1.94 | −6.73 | 1.66 | −7.11 | 1.79 | −6.93 |
| 3 | 4.0 | 2.31 | −4.42 | 1.82 | −5.29 | 2.05 | −4.88 |
| 4 | 5.2 | 2.51 | −1.91 | 1.82 | −3.47 | 2.13 | −2.75 |
| 5 | 6.4 | 2.57 | 0.66 | 1.72 | −1.75 | 2.10 | −0.65 |

## Table 2.14

Calculation of *DCFRR* for Project B.

| Year | $A_{CF}$ | DCF 20% | | DCF 40% | | DCF 35% | |
| --- | --- | --- | --- | --- | --- | --- | --- |
| | | $A_{DCF}$ | $\Sigma A_{DCF}$ | $A_{DCF}$ | $\Sigma A_{DCF}$ | $A_{DCF}$ | $\Sigma A_{DCF}$ |
| 0 | −10 | −10 | −10 | −10 | −10 | −10 | −10 |
| 1 | 6.5 | 5.42 | −4.58 | 4.64 | −5.36 | 4.81 | −5.19 |
| 2 | 5.2 | 3.61 | −0.97 | 2.65 | −2.71 | 2.85 | −2.34 |
| 3 | 4.0 | 2.31 | 1.34 | 1.46 | −1.25 | 1.63 | −0.71 |
| 4 | 2.8 | 1.35 | 2.69 | 0.729 | −0.521 | 0.843 | 0.133 |
| 5 | 1.6 | 0.643 | 3.33 | 0.297 | −0.224 | 0.357 | 0.490 |

# 2.9  Investment Criteria

Economic analysis should be performed at all stages of an emerging project as more information and detail become available. The decision as to whether to proceed with a project will depend on many factors. There is most often stiff competition within companies for any capital available for investment in projects. However, in an efficient capital market there should be plenty of capital available, providing the returns are high enough. The decision as to where to spend capital on a particular project will, in the first instance, but not exclusively, depend on the economic criteria discussed in the previous section. Criteria that account for the timing of the cash flows (the *NPV* and *DCFRR*) should be the basis of the decision making. The higher the value of the *NPV* and *DCFRR* for a project, the more attractive it is. The absolute minimum acceptable value of the *DCFRR* is the market interest rate. If the *DCFRR* is lower than the market interest rate, it would be better to put the money in the bank. However, since the bank account is less risky than most technical projects, it is prudent to set a return target of at least 5–10% higher than the return on a bank account. The essential distinction between *NPV* and *DCFRR* is:

- *Net present value* measures profit but does not indicate how efficiently capital is being used.
- *DCFRR* measures the rate of return that a project might achieve, but gives no indication of how competitive the project would be.

When choosing between competing projects, these will have different cash flow patterns and different capital investments. If the goal is to maximize profit, then projects with the highest *NPV* should be selected. This does not necessarily equate with selecting the project with the highest *DCFRR*. A lower *DCFRR* on a larger investment could yield a higher *NPV* than a higher *DCFRR* on a smaller investment. *NPV* gives a direct cash measure of the attractiveness of a project. For competing projects with different capital investments, a quick way to compare between the projects is the *investment efficiency*:

$$Investment\ Efficiency = \frac{NPV}{Capital\ Investment} \qquad (2.20)$$

Predicting future cash flows for a project is extremely difficult. There are many uncertainties, including the project life. Also, the appropriate interest rate will not be known with certainty. The acceptability of the rate of return will depend on the risks associated with the project and the company investment policy. For example, a *DCFRR* of 20% might be acceptable for a low-risk project. A higher return of, say, 30% might be demanded of a project with some risk, whereas a high-risk project with significant uncertainty might demand a 50% *DCFRR*.

The sensitivity of the economic analysis to the underlying assumptions should always be tested. A sensitivity analysis should be carried out to test the sensitivity of the economic analysis to:

- errors in the capital cost estimate;
- delays in the start-up of the project after the capital has been invested (particularly important for a high capital cost project);
- changes in the cost of raw materials;
- changes in the sales price of the product;
- reduction in the market demand for the product, and so on.

When carrying out an economic evaluation, the magnitude and timing of the cash flows, the project life, and interest rate are not known with any certainty. However, providing that consistent assumptions are made for projections of cash flows and the assumed rate of interest, the economic analysis can be used to choose between competing projects. It is important to compare different projects and options within projects on the basis of consistent assumptions. Thus, even though the evaluation will be uncertain in an absolute sense, it can still be meaningful in a relative sense for choosing between options. Because of this, it is important to have a reference against which to judge any project or option within a project.

However, the final decision to proceed with a project will be influenced as much by business strategy as by the economic measures described above. The business strategy might be to gradually withdraw from a particular market, perhaps because of adverse long-term projections of excessive competition, even though there might be short-term attractive investment opportunities. The long-term business strategy might be to move into different business areas, thereby creating investment priorities. Priority might be given to increasing market share in a particular product to establish business dominance in the area and achieve long-term global economies of scale in the business.

# 2.10  Process Economics – Summary

The total investment required for a new design can be broken down into four main parts that add up to the total investment: battery limits investment, utility investment, off-site investment, and working capital.

Different classes of capital cost estimates can be defined as summarized in Table 2.1. The capital cost estimate class is normally defined by the percentage of complete definition of the project. However, it is the confidence in the data and its maturity that is the main determining factor in the accuracy. It is important to add contingency or unallocated provision to estimates to have a realistic final cost.

Class 5 estimates are applicable before the complete process flow diagram has been developed and are mostly based on scaling of capital cost from previous projects. Class 4 capital cost estimates are most often based on the capital cost of the main items of process equipment factored to allow for differences in materials of construction, process pressure and temperature, geographical location, installation of equipment, utilities, off-sites, contingency, and working capital. As the project progresses, there will be an increased availability of data. Vendor equipment quotations gradually replace historical and public data for equipment and a detailed bill of materials gradually replaces factoring methods to obtain Class 3 to Class 5 capital cost estimates.

Utility investment, off-site investment, and working capital are also needed to complete the capital investment. The capital cost can be annualized by considering it as a loan over a fixed period at a fixed rate of interest.

The dominant operating cost is usually raw materials. However, other significant operating costs involve catalysts and chemicals consumed other than raw materials, utility costs, labor costs, and maintenance.

As a more complete picture of the project emerges, the cash flows through the project life can be projected. This allows more detailed evaluation of project profitability on the basis of cash flows. Net present value can be used to measure the profit, taking into account the time value of money. The discounted cash flow-rate of return measures how efficiently the capital is being used.

Overall, there are always considerable uncertainties associated with an economic evaluation. In addition to the errors associated with the estimation of capital and operating costs, the project life or interest rates are not known with any certainty.

# Exercises

1. Estimate the per cent increase in capital cost anticipated if the capacity of a process design is doubled. If the capital cost does not double explain why this is the case.

2. Using the data in Table 2.2, calculate a Class 5 capital cost estimate for a sulfuric acid process with a capacity of 150 tons per day. Assume the current Chemical Engineering Plant Cost Index to be 831.7. The Chemical Engineering Plant Cost Index corresponding with this data in Table 2.2 is 102. Starting from the battery limits cost, the other cost for the project can be assumed to be the following proportions of the battery limits cost:

| | |
|---|---|
| Outside battery limits capital cost | 40% |
| Engineering costs | 10% |
| Contingency | 10% |
| Working capital | 15% |

3. A new agitated reactor with a new external shell-and-tube heat exchanger and new centrifugal pump are to be installed in an existing facility. The agitated reactor is to be glass-lined, which can be assumed to have an equipment cost of three times the cost of a carbon steel vessel. The heat exchanger, pump, and associated piping are all high-grade stainless steel. The equipment is rated for moderate pressure. The reactor has a volume of 9 m³, the heat exchanger an area of 50 m², and the pump has a power of 5 kW. No significant investment is required in utilities, off-sites, buildings, site preparation, or working capital. Using Eq. (2.9) and Table 2.4 (extrapolating beyond the range of the correlation if necessary), calculate a Class 4 capital cost estimate for the project (CE Index of Equipment = 1037).

4. Steam is distributed on a site via high-pressure and low-pressure steam mains. The high-pressure main is at 40 bar and 350 °C. The low-pressure main is at 4 bar. The high-pressure steam is generated in boilers. The overall efficiency

## Table 2.15

Cash flows for two competing projects.

| | Cash flows $1000 | |
|---|---|---|
| Year | Project A | Project B |
| 0 | −1000 | −1000 |
| 1 | 150 | 500 |
| 2 | 250 | 450 |
| 3 | 350 | 300 |
| 4 | 400 | 200 |
| 5 | 400 | 100 |

of steam generation and distribution is 80%. The low-pressure steam is generated by expanding the high-pressure stream through steam turbines with an isentropic efficiency of 75%. The cost of fuel in the boilers is $3.5 GJ⁻¹ and the cost of electricity is $0.05 kWh⁻¹. The boiler feedwater is available at 100 °C. Estimate the cost of the high-pressure and low-pressure steam.

5. A refrigerated distillation condenser has a cooling duty of 0.75 MW. The condensing stream has a temperature of −10 °C. The heat from a refrigeration circuit can be rejected to cooling water at a temperature of 30 °C. Assuming a temperature difference in the distillation condenser of 5 °C and a temperature difference for heat rejection from refrigeration to cooling water of 10 °C, estimate the power requirements for the refrigeration.

6. A company has the option of investing in one of the two projects, A or B. The capital cost of both projects is $1 000 000. The predicted annual cash flows for both projects are shown in Table 2.15. For each project, calculate:

   a) The payback time for each project in terms of the average annual cash flow.
   b) The return on investment.
   c) The discounted cash flow rate of return.
   What do you conclude from the result?

7. A process has been developed for a new product for which the market is uncertain. A plant that plans to produce 50 000 t y⁻¹ requires an investment of $10 000 000 and the expected project life is five years. Fixed operating costs are expected to be $750 000 y⁻¹ and variable operating costs (excluding raw materials) are expected to be $40 t⁻¹ product. The stoichiometric raw material costs are $80 t⁻¹ product. The yield of product per ton of raw material is 80%. Tax is paid in the same year as the relevant profit is made at a rate of 20%. Calculate the selling price of the product to give a minimum acceptable discounted cash flowrate of return of 15% year.

8. How can the concept of simple payback be improved to give a more meaningful measure of project profitability?

9. It is proposed to build a plant to produce 170 000 t y⁻¹ of a commodity chemical. A study of the supply and demand projections for the product indicates that current installed capacity

in the industry is $6.8 \times 10^6 \, \text{t} \, \text{y}^{-1}$, whereas total production is running at $5.0 \times 10^6 \, \text{t} \, \text{y}^{-1}$. Maximum plant utilization is thought to be around 90%. If the demand for the product is expected to grow at 8% per year and it will take three years to commission a new plant from the start of a project, what do you conclude about the prospect for the proposed project?

# References

Allen, D.H. (1980). *A Guide to the Economic Evaluation of Projects*. Rugby, UK: IChemE.

Anson, H.A. (1977). *A New Guide to Capital Cost Estimating*. Rugby, UK: IChemE.

ASTM Designation E2516–11. (2011). *Standard Classification for Cost Estimate Classification System*. ASTM International.

Bauman, H.C. (1955). Estimating costs of process auxiliaries. *Chemical Engineering Progress* 51: 45.

Bauman, H.C. (1964). *Fundamentals of Cost Engineering in the Chemical Industry*. New York: Reinhold.

Bravo, J.L. (1997). Select structured packings or trays? *Chemical Engineering Progress* July: 36.

Brennan, D. (1998). *Process Industry Economics*. UK: IChemE.

Gary, J.H. and Handwerk, G.E. (2001). *Petroleum Refining Technology and Economics*, 4e. Marcel Dekker.

Gerrard, A.M. (2000). *Guide to Capital Cost Estimating*, 4e. UK: IChemE.

Guthrie, K.M. (1969). Data and techniques for preliminary capital cost estimating. *Chemical Engineering* 76: 114.

Hall, R.S., Matley, J., and McNaughton, K.J. (1982). Current costs of process equipment. *Chemical Engineering* 89: 80.

Hall, R.S., Vatavuk, W.M., and Matley, J. (1988). Estimating process equipment costs. *Chemical Engineering* 95: 66.

Hand, W.E. (1958). From flowsheet to cost estimate. *Petrol Refiner* 37: 331.

Holland, F.A., Watson, F.A., and Wilkinson, J.K. (1983). *Introduction to Process Economics*, 2e. New York: Wiley.

Kretzschmar, H.-J. and Wagner, W. (2019). *International Steam Tables – Properties of Water and Steam Based on the Industrial Formulation IAPWS-IF97*, 3e. Springer Veirweg.

Lang, H.J. (1947). Cost relationships in preliminary cost estimation. *Chemical Engineer* 54: 117.

Mulet, A., Corripio, A.B., and Evans, L.B. (1981a). Estimate costs of pressure vessels via correlations. *Chemical Engineering* Oct: 145.

Mulet, A., Corripio, A.B., and Evans, L.B. (1981b). Estimate costs of distillation and absorption towers via correlations. *Chemical Engineering* Dec: 77.

Peters, M.S., Timmerhaus, K.D., and West, R.E. (2003). *Plant Design and Economics for Chemical Engineers*, 5e. New York: McGraw-Hill.

Remer, D.S. and Chai, L.H. (1990). Design cost factors for scaling-up engineering equipment. *Chemical Engineering Progress* 86: 77.

Smith, R. (2016). *Chemical Process Design and Integration*, 2e. Wiley.

Ulrich, G.D. (1984). *A Guide to Chemical Engineering Process Design and Economics*. New York: Wiley.

**2**

# Chapter 3

# Development of Process Design Concepts

## 3.1 Formulation of Design Problems

The various stages for the execution of a project for a process plant design were discussed in Chapter 1. A typical project starts with the Selection or Concept Development Phase. However, before a process design can be started, the design problem must be formulated from the project specification. Formulation of the design problem requires a product specification. If a well-defined chemical product is to be manufactured, then the specification of the product might be straightforward (e.g. purity and impurity specifications). However, if a specialty product is to be manufactured, it is the functional properties that are important, rather than the chemical properties, and this might require a *product design* stage in order to specify the product (Seider et al. 2010; Cussler and Moggridge 2011).

The initial statement of the design problem is often ill-defined. Investigating design options should start by examining the problem at the highest level, in terms of its feasibility with the minimum of detail to ensure that the design option is worth progressing (Douglas 1985). Is there a large difference between the value of the product and the cost of the raw materials? If the overall feasibility looks attractive, then more detail can be added, the option re-evaluated, further detail added, and so on. Byproducts might play an important role in the economics. It might be that the process produces some byproducts that can be sold in small quantities to the market. However, as the process is expanded, there might be market constraints for an increased scale of production. If the byproducts cannot be sold, how does this affect the economics?

Starting from the original design specification, a series of plausible design options must be formulated to be screened by the methods of engineering and economic analysis. These design options are formulated into very specific design problems. In this way, the often initial ill-defined problem is turned into a series of well-defined design options for analysis.

Optional process flow diagrams for different process options can be developed to screen for different:

- raw materials;
- reaction paths to the same product;
- reaction, separation, and recycle systems for a given reaction path;
- utility systems;
- capacities.

The confidence with which process options can be screened depends very much on the quality and maturity of the process design data available.

## 3.2 Evaluation of Performance

Once design options have been created, they need to be evaluated. There are many facets to the evaluation of performance. Process options can be evaluated on the basis of:

1) *Capital cost*. Good economic performance is an obvious first criterion, but it is certainly not the only one. Capital cost in particular may be a constraint. There may be a limited amount of capital available in the company for the project or there may be a number of different projects in different business areas within the company competing for the same pool of capital.

2) *Operating cost*. Operating costs create a continuous burden on the annual cash flow of a company, as well as limiting the return on the capital investment. Some operating costs are variable and depend on the level of production (e.g. raw materials costs, utilities costs, maintenance costs incurred by operation), while others are fixed irrespective of the level of production (e.g. labor, safety services, laboratories, routine maintenance).

*Process Plant Design*, First Edition. Robin Smith.
© 2024 John Wiley & Sons Ltd. Published 2024 by John Wiley & Sons Ltd.
Companion website: www.wiley.com/go/processplantdesign

**3)** *Safety*. Process safety must secure the integrity of hazardous operations and processes by applying good design principles, engineering, and operating practices. Good safety practice should prevent and control incidents that have the potential to release hazardous materials or energy. Such incidents can cause fire, explosion, or toxic release and could ultimately result in serious injuries, property damage, lost production, and environmental impact. The aim should be to at least meet, and possibly exceed, regulatory standards.

**4)** *Flexibility*. Flexibility relates to the ability to operate under different conditions, such as differences in capacity, feedstock, and product specification.

**5)** *Reliability*. Reliability is the probability that the equipment or process has survived without any failures. Frequent failures of equipment or processes can result in excessive maintenance costs. However, the failures can also cause significant disruption in production, wasted process materials, safety problems, and possibly environmental problems.

**6)** *Maintainability*. Maintainability is the probability that equipment or processes can be maintained in a given time. Lengthy downtimes for maintenance can have a serious impact on process economics.

**7)** *Availability*. Availability is the probability that the process is performing the intended function and therefore achieves the required production capacity. Availability measures the portion of the total time that the process meets its production requirements. Unless a certain availability is achieved, production targets will not be met.

**8)** *Ease of control and operation*. Start-up, shut-down, emergency shut-down, and ease of control are important factors.

**9)** *Environmental discharges*. Gaseous, liquid, and solid waste should be minimized. Any waste discharges must be kept within regulatory limits, both in terms of concentrations and volume.

**10)** *Sustainability*. Chemical processes should be designed to maximize the *sustainability* of industrial activity. Maximizing sustainability requires that industrial systems should strive to satisfy human needs in an economically viable, environmentally benign, and socially beneficial way (Azapagic 2014). For a chemical process design, this means that processes should make use of materials of construction that deplete resources as little as practicable and minimize the environmental impact of their extraction and use. Process raw materials should be used as efficiently as is economic and practicable, both to prevent the production of waste that can be environmentally harmful and to preserve the reserves of manufacturing raw materials as much as possible. Processes should use as little energy as is economic and practicable. This not only reduces operating costs, but reduces emissions and preserves resources. Burning fossil fuels promotes the greenhouse effect and global warming. Thus, the use of fossil fuels should be minimized or eliminated by a combination of increased energy efficiency and substitution by renewable sources (such as wind and solar). Water must also be consumed in sustainable quantities that do not cause deterioration in the quality of the water source and the long-term quantity of the reserves. Solid waste to landfill must be avoided. Analysis of sustainability requires that the boundary of consideration should go beyond the immediate boundary of the manufacturing facility to maximize the benefit to society, to avoid adverse health effects, unnecessarily high burdens on transportation, odor, noise nuisances, and so on. Life cycle analysis (Azapagic and Perdan 2011) allows various categories of impact to be quantified in order to assess the sustainability of a project.

**11)** *Uncertainty and risk*. Uncertainty in the design, for example, resulting from poor design data, or uncertainty in the economic data, might guide the design away from certain attractive options. Even worse, the evaluation might suggest process options to be viable that would not be judged to be viable had more reliable data been available. The resulting uncertainty might not only create significant risk for the viability of the project, but in extreme cases risk for the viability of the business commissioning the project.

Some of these factors, such as economic performance, can be readily quantified; others, such as safety, often cannot. Evaluation of the factors that are not readily quantifiable, the intangibles, requires judgment.

# 3.3 Optimization of Performance

Once the basic performance of the design has been evaluated, changes can be made to improve the performance; the process is *optimized*. These changes might involve the synthesis of alternative structures in *structural optimization*. Thus, the process is simulated and evaluated again, and so on, optimizing the structure. Each structure can be subjected to *parameter optimization* by changing operating conditions within that structure. This is illustrated in Figure 3.1, which illustrates a sequential approach to process performance. From the project definition an initial design is synthesized. This can then be simulated and evaluated. Once evaluated, the design can be optimized in a parameter optimization by changing the continuous parameters of flowrate, composition, temperature, and pressure to improve the evaluation. However, this parameter optimization only optimizes the initial design configuration, which might not be an optimal configuration. The design process might therefore return to the synthesis stage to explore other configurations in a structural optimization. Also, if the parameter optimization adjusts the settings of the conditions to be significantly different from the original assumptions, then the design process might return to the synthesis stage to consider other configurations in the structural optimization. The different ways this design process can be followed will be considered next.

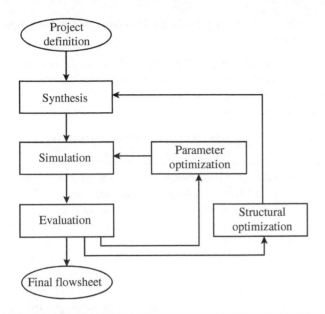

**Figure 3.1**

Optimization can be carried out as structural or parameter optimization to improve the evaluation of the design. Source: Adapted from (Smith 2016), *Chemical Process Design and Integration*, 2nd Edition, John Wiley & Sons Ltd.

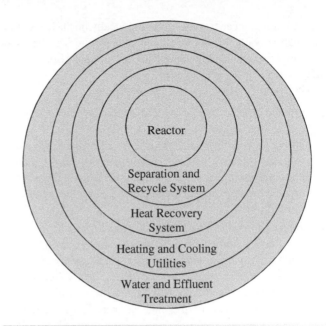

**Figure 3.2**

The onion model of process design. A reactor is needed before the separation and recycle system can be designed and so on. Source: (Smith 2016), *Chemical Process Design and Integration*, 2nd Edition, John Wiley & Sons Ltd.

## 3.4 Approaches to the Development of Design Concepts

This section will present a summary of the approaches to the development of a design concept. Further details are available elsewhere (Smith 2016). In broad terms, there are three approaches to the development of design concepts:

### 3.4.1 Creating an Irreducible Structure

If the process requires a reactor, this is where the design starts. This is likely to be the only place in the process where raw material components are converted into components for the products. The chosen reactor design produces a mixture of unreacted feed materials, products, and byproducts that need separating. Unreacted feed material should preferably be recycled. The reactor design dictates the separation and recycle problem. Thus, design of the separation and recycle system follows the reactor design. The reactor and separation and recycle system designs together define the process heating and cooling duties. Thus, the heat exchanger network design comes next. Those heating and cooling duties that cannot be satisfied by heat recovery dictate the need for external heating and cooling *utilities* (furnace heating, use of steam, steam generation, cooling water, air cooling, or refrigeration). Thus, utility selection and design follows the design of the heat recovery system. The selection and design of the utilities is made more complex by the fact that the process will most likely operate

within the context of a site comprising a number of different processes that are all connected to a common utility system. The process and the utility system will both need water, for example, for steam generation, and will also produce aqueous effluents that will need to be brought to a suitable quality for discharge. Thus, the design of the water and aqueous effluent treatment system comes last. Again, the water and effluent treatment system must be considered at the site level as well as the process level.

This hierarchy can be represented symbolically by the layers of the "onion diagram" shown in Figure 3.2 (Linnhoff et al. 1982). The diagram emphasizes the sequential, or hierarchical, nature of process design. Some processes do not require a reactor; for example, some processes just involve separation. Here, the design starts with the separation system and moves outward to the heat exchanger network, utilities, and so on. However, the same basic hierarchy prevails. Both continuous and batch process design follow this hierarchy, even though the time dimension in batch processes brings additional constraints in process design.

At each layer, decisions must be made on the basis of the information available at that stage. The ability to look ahead to the completed design might lead to different decisions. Unfortunately, this is not possible and, instead, decisions must be based on an incomplete picture.

This approach to creation of the design involves making a series of best local decisions. This might be based on the use of *heuristics* or *rules of thumb* developed from experience (Douglas 1985) or on a more systematic approach. Equipment is added only if it can be justified economically on the basis of the information available, albeit an incomplete picture. This keeps the structure *irreducible* and features that are technically or economically redundant are not included.

There are two drawbacks to this approach:

**a)** Different decisions are possible at each stage of the design. To be sure that the best decisions have been made, the other options must be evaluated. However, each option cannot be evaluated properly without completing the design for that option and optimizing the operating conditions. This means that many designs must be completed and optimized in order to find the best.

**b)** Completing and evaluating many options gives no guarantee of ultimately finding the best possible design, as the search is not exhaustive. Also, complex interactions can occur between different parts of a flowsheet. The effort to keep the system simple and not add features in the early stages of design may result in missing the benefit of interactions

between different parts of the flowsheet in a more complex system.

The main advantage of this approach is that the designer can keep control of the basic decisions and interact as the design develops. By staying in control of the basic decisions, the intangibles of the design can be included in the decision making.

### 3.4.2 Creating and Optimizing a Superstructure

In this approach, a *reducible* structure, known as a *superstructure*, is first created that has embedded within it all feasible process options and all feasible interconnections that are candidates for an optimal design structure. Initially, redundant features are built into the superstructure. As an example, consider Figure 3.3 (Kocis and Grossmann 1988). This shows one possible structure of a process for the manufacture of benzene from the reaction between toluene and hydrogen. In Figure 3.3, the hydrogen enters the process with a small amount of methane as an impurity. Thus, in Figure 3.3, the option of either purifying the hydrogen feed with

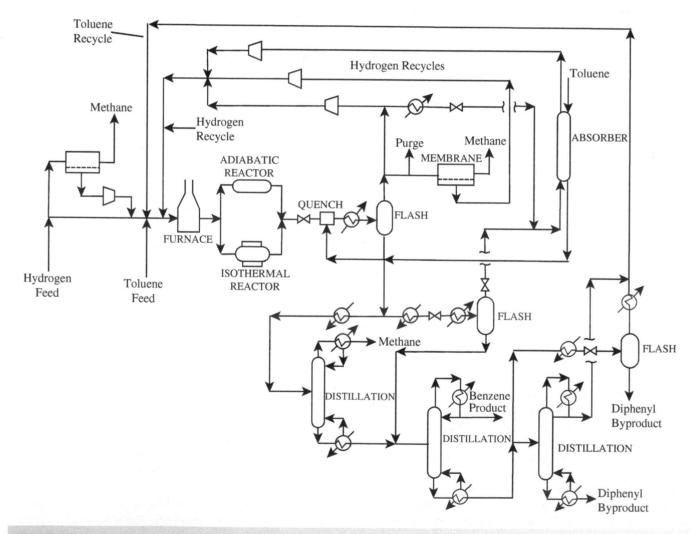

### Figure 3.3

A superstructure for the manufacture of benzene from toluene and hydrogen incorporating some redundant features. Source: Adapted from Kocis GR and Grossman IE, 1988, *Comp. Chem. Engng*, **13**: 797.

a membrane or of passing it directly to the process is embedded. The hydrogen and toluene are mixed and pre-heated to the reaction temperature. Only a furnace has been considered feasible in this case because of the high temperature required. Then the two alternative reactor options, isothermal and adiabatic reactors, are embedded, and so on. Redundant features have been included in an effort to ensure that all features that could be part of an optimal solution have been included.

The design problem is next formulated as a mathematical model. Some of the design features are continuous, describing the operation of each unit (e.g. flowrate, composition, temperature, and pressure) and its size (e.g. volume, heat transfer area, etc.). Other features are discrete (e.g. whether a connection in the flowsheet is included or not and whether a membrane separator is included or not). Once the problem is formulated mathematically, its solution is carried out through the implementation of an optimization algorithm. An *objective function* is maximized or minimized (e.g. profit is maximized or cost is minimized) in a *structural and parameter* optimization. The optimization justifies the existence of structural features and deletes those features from the structure that cannot be justified economically. In this way, the structure is reduced in complexity. At the same time, the operating conditions and equipment sizes are also optimized. In effect, the discrete decision-making aspects of process design are replaced by a discrete/continuous optimization. Thus, the initial structure in Figure 3.3 is optimized to reduce the structure to the final design shown in Figure 3.4

(Kocis and Grossmann 1988). In Figure 3.4, the membrane separator on the hydrogen feed has been removed by optimization, as has the isothermal reactor and many other features of the initial structure shown in Figure 3.3.

There are a number of difficulties associated with this approach:

a) The approach will fail to find the optimal structure if the initial structure does not have the optimal structure embedded somewhere within it. The more options included, the more likely it will be that the optimal structure has been included.

b) If the individual unit operations are represented accurately, the resulting mathematical model will be extremely large and the objective function that must be optimized will be extremely irregular. The profile of the objective function can be like the terrain in a range of mountains with many peaks and valleys. If the objective function is to be maximized (e.g. maximize profit), each peak in the mountain range represents a *local optimum* in the objective function. The highest peak represents the *global optimum*. Optimization requires searching around the mountains in a thick fog to find the highest peak, without the benefit of a map and only a compass to tell the direction and an altimeter to show the height. On reaching the top of any peak, there is no way of knowing whether it is the highest peak because of the fog. All peaks must be searched to find the highest. There are crevasses to fall into that might be impossible to climb out of.

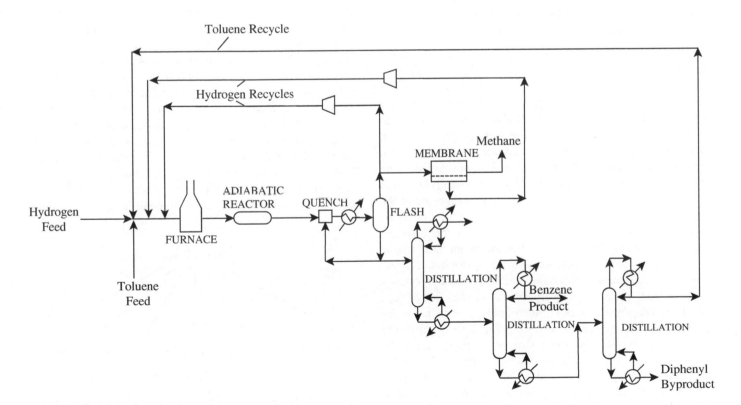

**Figure 3.4**
Optimization discards many structural features leaving an optimized structure. Source: Adapted from Kocis GR and Grossman IE, 1988, *Comp. Chem. Engng*, **13**: 797.

Such problems can be overcome in a number of ways. The first way is by changing the model such that the solution space becomes more regular, making the optimization simpler. This most often means simplifying the mathematical model. A second way is by repeating the search many times, but starting each new search from a different initial location. A third way exploits mathematical transformations and bounding techniques for some forms of mathematical expression to allow the global optimum to be found (Floudas 2000). A fourth way is by allowing the optimization to search the solution space so as to allow the possibility for an objective function being maximized of going downhill, away from an optimum point, as well as uphill. As the search proceeds, the ability of the algorithm to move downhill must be gradually taken away.

c) The most serious drawback of this approach is that the design engineer is removed from the decision making. Thus, the many intangibles in design, such as safety and layout, which are difficult to include in the mathematical formulation, cannot be taken into account satisfactorily.

On the other hand, this approach has a number of advantages. Many different design options can be considered at the same time. The complex multiple trade-offs usually encountered in chemical process design can be handled by this approach. Also, the entire design procedure can be automated and is capable of producing designs quickly and efficiently.

### 3.4.3 Creating an Initial Design and Evolving Through Structural and Parameter Optimization

The third approach is a variation on creating and optimizing a superstructure. In this approach, an initial design is first created. This is not necessarily intended to be an optimal design and does not necessarily have redundant features, but is simply a starting point. The initial feasible design is then subjected to *evolution* through structural and parameter optimization, using an optimization algorithm. As with the superstructure approach, the design problem is formulated as a mathematical model. The design is then evolved one step at a time. Each step is known as a *move*. Each move in the evolution might change one of the continuous variables in the flowsheet (e.g. flowrate, composition, temperature, or pressure) or might change the flowsheet structure. Changing the flowsheet structure might mean adding or deleting equipment (together with the appropriate connections), or adding new connections or deleting existing connections. At each move, the objective function is evaluated (e.g. profit or cost). New moves are then carried out with the aim of improving the objective function. Rules must be created to specify the moves. The same structural options might be allowed as included in the superstructure approach, but in this evolutionary approach the structural moves are carried out to add or delete structural features one at a time. In addition to the structural moves, continuous moves are also carried out to optimize the flowsheet conditions. The techniques used have been described in more detail elsewhere (Smith 2016).

The difficulties associated with this approach are similar to those for the creation and optimization of a superstructure:

a) The approach will fail to find the optimal structure if the structural moves have not been defined such that the sequence of moves can lead to the optimal structure from the initial design.

b) The objective function that must be optimized will be extremely irregular. Thus, again there are difficulties finding the global optimum. The optimization methods normally adopted for this approach allow the search to proceed even if the objective function deteriorates after a move. The ability of the algorithm to accept the deteriorating objective function is gradually taken away as the optimization proceeds.

c) Again, this approach has the disadvantage that the design engineer is removed from the decision making.

In summary, the three general approaches to chemical process design have advantages and disadvantages. Each approach can be used at different stages as the project progresses. However, whichever is used in practice, there is no substitute for understanding the problem.

## 3.5 Screening Design Options

A sequential approach to process design is most often used in practice. To develop a process design using a sequential approach requires design options to be first generated and then evaluated, as illustrated in Figure 3.5. There is a temptation to carry out a preliminary evaluation once a set of design options has been identified and eliminate options that initially appear to be unattractive, Figure 3.6. However, each decision point has consequences for decisions that need to be made later. This is illustrated in Figure 3.7. In Figure 3.7 a series of local decisions has been made at each stage in the development of the process design. The problem with this approach is that a series of local decisions have been made without fully understanding the consequences of those decisions as further details become available. Local decisions are being made on the basis of uncertain data. For example,

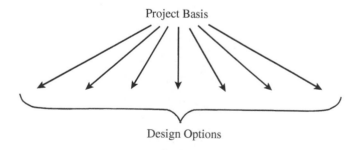

**Figure 3.5**

A project starts by generating design options.

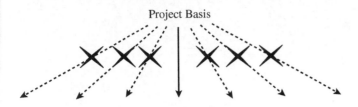

Project Basis

**Figure 3.6**

The options generated can be screened by evaluation.

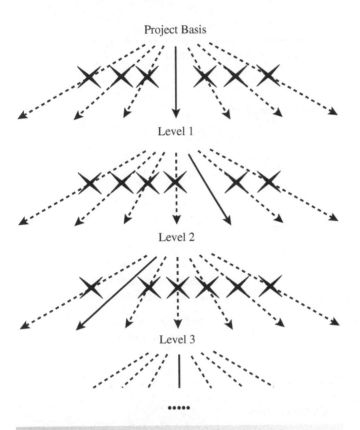

Project Basis

Level 1

Level 2

Level 3

•••••

**Figure 3.7**

There are different levels in the formation of any design concept allowing the options generated to be screened by evaluation.

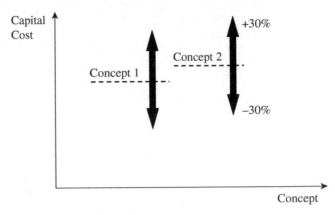

**Figure 3.8**

Errors in the evaluation are often too great to allow screening with confidence, particularly in the early stages of a project.

possible to foresee everything that lay ahead, decisions made early might well be different. There is a danger in focusing on one option without rechecking the assumptions later for validity when more information is available. The designer must not be boxed in early by preconceived ideas. All options should be considered, even if they appear unappealing at first. This means that design options should be left open as long as practicable until it is clear that options can be closed down.

## 3.6  Influencing the Design as the Project Progresses

Figure 3.9 presents an illustration of the project expenditure as the project progresses. In the early stages of a project in the Selection Phase and the early stages of the Definition Phase, the levels of expenditure are relatively modest. As the project progresses, the level of expenditure starts to increase significantly, especially during the Execution Phase, before leveling out in the Commissioning and Operation Phases. Also shown in Figure 3.9 is a profile of the ability to influence design changes. As the expenditure increases, the ability to influence design changes decreases significantly. The project becomes more and more committed to design options as the expenditure increases. Making design changes late in the project requires undoing features of the design that have been expensively put together, making late changes extremely expensive, which in turn makes it increasingly difficult to change the design late in the design project.

Because of this, it is good practice not to close down design options early, as discussed in the previous section. Figure 3.10 illustrates a project strategy that spends more time and expenditure in the Selection and Definition Phases of the project. This does not necessarily lead to a more expensive project overall and a longer project duration overall, but possibly a lower overall capital expenditure and a shorter project duration. This is made possible by less reworking of the design as a result of making poor decisions early on the basis of uncertain information.

Figure 3.8 shows two competing design options and their relative capital costs. In Figure 3.8, Concept 1 appears to be more attractive than Concept 2 on the basis of the capital cost estimate. However, the capital cost estimates might be subject to significant uncertainty, not allowing a choice to be made with any confidence. Such uncertainty not only applies to capital costs, but all aspects of the process design evaluation, such as safety, reliability, sustainability, and so on. The temptation to remove design options early must be avoided as much as possible. In the early stages of a design the uncertainties in the evaluation are often too serious for early elimination of options, unless it is absolutely clear that a design option is not viable. Initial cost estimates can be very misleading and the full safety and environmental implications of early decisions are only clear once detail has been added. If it were

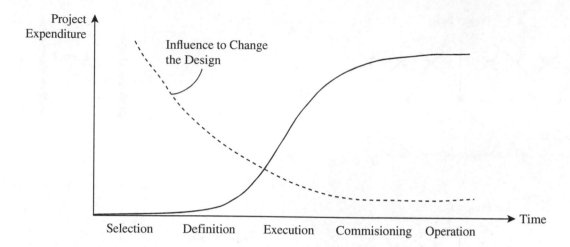

**Figure 3.9**

The opportunity to change the design in a project diminishes as the total project expenditure increases.

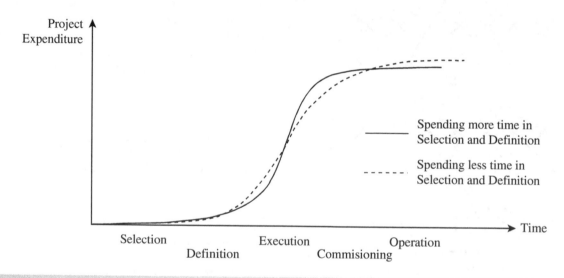

**Figure 3.10**

Spending more time in the Selection and Definition phases of a project can lead to both a lower overall expenditure and shorter overall timescale for the project due to less rework and recycling in the project.

# 3.7 Development of Process Design Concepts – Summary

The original design problem posed is often ill-defined, even if it appears on the surface to be well-defined. A series of well-defined design options must be formulated from the original ill-defined problem, and these must be compared on the basis of consistent criteria. Optional process flow diagrams for different process options can be developed and screened on the basis of:

- capital cost;
- operating cost;
- safety;
- flexibility;
- reliability;
- maintainability;
- availability;
- ease of control and operation;
- environmental discharges;

- sustainability;
- uncertainty and risk.

Both structural and parameter optimization can be carried out to improve the evaluation. There are three general approaches to process design:

- creating an irreducible structure;
- creating and optimizing a superstructure;
- creating an initial design and evolving through structural and parameter optimization.

Each of these approaches have advantages and disadvantages. Design options should be left open as far as practicable to avoid potentially attractive options being eliminated inappropriately on the basis of uncertain data. As the project progresses and the project expenditure increases, the ability to influence design changes decreases. Increasing the time spent and expenditure in the early stages of a project can decrease the overall project cost and duration through less reworking of the design as a result of making poor decisions early on the basis of uncertain information.

# References

Azapagic, A. and Perdam, S. (2011). *Sustainable Development in Practice: Case Studies for Scientists and Engineers*, 2e. Wiley-Blackwell.

Azapagic, A. (2014). Sustainability considerations for integrated biorefineries. *Trends in Biotechnology* 32: 1.

Cussler, E.L. and Moggridge, G.D. (2011). *Chemical Product Design*, 2e. Cambridge University Press.

Douglas, J.M. (1985). A hierarchical decision procedure for process synthesis. *AICHE Journal* 31: 353.

Floudas, C.A. (2000). *Deterministic Global Optimization: Theory, Methods and Applications*. Kluwer Academic Publishers.

Kocis, G.R. and Grossmann, I.E. (1988). A modelling and decomposition strategy for the MINLP optimization of process flowsheets. *Computers and Chemical Engineering* 13: 797.

Linnhoff, B., Townsend, D.W., Boland, D. et al. (1982). *A User Guide on Process Integration for the Efficient Use of Energy*. Rugby, UK: IChemE.

Seider, W.D., Seader, J.D., Lewin, D.R., and Widagdo, S. (2010). *Product and Process Design Principles*, 3e. Wiley.

Smith, R. (2016). *Chemical Process Design and Integration*, 2e. Wiley.

**3**

# Chapter 4

# Heating Utilities

In Chapter 2 it was pointed out that processes need to be serviced with a range of utilities (services). These include fuel, electricity, steam, process water, cooling water, firewater, refrigeration, compressed air, inert gas, and waste treatment. Prominent among the utilities are those that provide heating and cooling. Figure 4.1 shows a typical site utility system. Various processes operate on the site and are connected to a common steam system. In Figure 4.1, very high-pressure steam is generated in utility steam boilers. A gas turbine generates power and exhausts its flue gas to a *heat recovery steam generator* (HRSG) to generate high-pressure steam. Steam is expanded in steam turbines to high-, medium-, and low-pressure steam mains (or headers), producing power. The final exhaust steam from the steam turbines is expanded to vacuum conditions and condensed against cooling water. It may be that this power generation needs to be supplemented by the import of power from the electricity grid. It might also be the case that excess power is generated on the site and exported.

In Figure 4.1, three levels of steam are distributed around the site, and the various processes are connected to the steam mains. Some processes generate steam from waste heat, whilst other processes use the steam by drawing it from the steam headers. Thus, there is heat recovery between processes through the steam system. The processes also require the rejection of waste heat to the ambient, and this is achieved in Figure 4.1 using a cooling water circuit.

In Figure 4.1, there is *cogeneration* (combined heat and power generation) from steam turbines. Generation of power might be through a steam turbine or gas turbine coupled to an electric generator for the production of electricity in a *turbogenerator*, or a steam turbine or gas turbine coupled directly to a machine on a *direct drive* (e.g. a steam turbine driving a process compressor directly).

Steam is distributed around the site for various purposes:

1) *Steam heaters*. Steam is by far the most commonly used heating medium. Shell-and-tube heat exchangers are the most common, but for other heat exchanger designs, pipe coils and steam jackets are also used.

2) *Steam tracing*. Fluids in tanks, pipes, and other process equipment often have characteristics that cause them to freeze, become too viscous, or condense undesirably at ambient temperature, even if the tanks, pipes, and equipment are insulated. In order to prevent such problems, additional heat can be added, combined with insulation. A tube or small-diameter pipe carrying steam attached to the vessel or pipe for this purpose is referred to as *steam tracing*. Jackets can also be used instead of tubes.

3) *Space heating*. Production buildings, workshops, warehouses, and offices can require space heating. Pipes, radiators, and fan heaters can be used to create space heating from steam.

4) *Water heating through live steam injection*. Rather than use indirect steam heating in a heat exchange device, steam can be injected directly into water to provide heating. However, care must be exercised that any treatment chemicals in the steam do not cause unacceptable contamination of the water (e.g. in food processing operations).

5) *Boiler feedwater deaeration*. One of the operations required to prepare the water to be used for steam generation involves the use of steam to strip out gases dissolved in the water in a *deaeration* operation.

6) *Power generation in steam turbines*. A steam turbine converts the energy of steam into power by expanding the steam across rows of *blades* to create rotational power. Steam turbines will be discussed in more detail later in this chapter.

7) *Steam ejectors*. Steam can be used for the production of a vacuum in steam ejector systems.

8) *Flaring*. Combustible waste gases are often disposed of by combustion in flares. Flares will be considered in Chapter 6. Mixing of the waste gases with combustion air at the flare tip is often assisted by the use of steam jets. Large amounts of steam can be consumed, especially under upset process conditions.

*Process Plant Design*, First Edition. Robin Smith.
© 2024 John Wiley & Sons Ltd. Published 2024 by John Wiley & Sons Ltd.
Companion website: www.wiley.com/go/processplantdesign

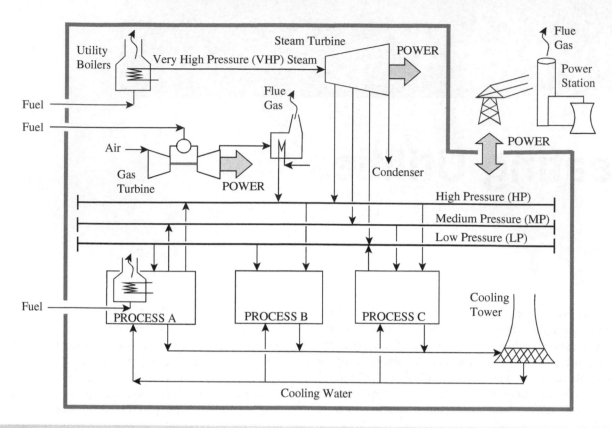

**Figure 4.1**

A typical site utility system. Source: Adapted from (Smith 2016), *Chemical Process Design and Integration*, 2nd Edition, John Wiley & Sons Ltd.

9) *Atomization of fuel oil in combustion operations.* The combustion of viscous fuel oil in burners often requires steam to atomize the fuel oil for combustion.

10) *Steam injection into combustion processes for NO$_x$ abatement.* Steam can be injected into combustion processes to lower emission of oxides of nitrogen (NO$_x$). The steam acts to decrease the flame temperature.

11) *Steam distillation.* Steam is added to some distillation operations to assist the separation. This is particularly important in crude oil distillation.

12) *Reactor dilution steam.* Steam is sometimes fed to reactors, along with the reactants in order to decrease the partial pressure of the reactants.

13) *Reactor decoking.* Reactors processing hydrocarbon feeds often suffer from the deposition of coke (carbon) on the reactor surfaces. Such deposits of coke can be removed by the application of steam. To achieve this, the reactor is taken off-line periodically and decoked by the application of steam jets.

14) *Soot blowing.* Furnace and boiler heat transfer surfaces exposed to the flue gas can become fouled through the deposition of soot and ash. This fouling reduces the efficiency of the heat transfer. *Soot blowing* removes the deposits by applying jets of steam periodically.

# 4.1 Process Heating and Cooling

Process streams need to be heated from their *supply* temperature to their *target* temperature. Priority normally should be given to heat recovery in order to minimize the energy consumption (Figure 4.2). Some heating and cooling utility is most often

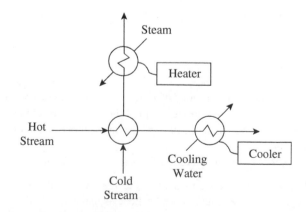

**Figure 4.2**

Heat recovery should take priority over the use of utilities.

**Table 4.1**

Commonly used heating and cooling utilities.

| Heating utilities | Cooling utilities |
|---|---|
| Steam | Steam generation |
| Furnace heat | Once through cooling water systems |
| Gas turbine exhaust | Recirculating cooling water systems |
| Hot oil | Air cooling |
| Hot water | Refrigeration |
| Heat pumping | Heat pumping |

required. A wide range of temperatures are required for heating and cooling utilities. The choice of utility mainly depends on the heat duty and the temperature. However, other factors contribute to the choice, including cost of fuel, cost of electricity, ambient temperature, and so on. Table 4.1 shows a list of the most commonly used heating and cooling utilities.

In Table 4.1, heat pumping appears both as a hot utility and a cold utility. How this is classified depends on the heat source and heat sink being matched.

The primary choice of heating and cooling utilities is based on cost. Generally, heating utilities become more expensive the higher the temperature of the heating utility. This means that steam should be used in preference to fired heating, low pressure steam should be used in preference to high pressure steam, and so on. For cooling utilities, it is generally the opposite way around. Generally, cooling utilities become more expensive the lower the temperature of the cooling utility. This means that steam generation should be used in preference to cooling water, cooling water used in preference to refrigeration, high temperature refrigeration used in preference to low temperature refrigeration, and so on.

## 4.2   Steam Heating

Steam heating can be supplied in a wide range of heat exchanger designs. Figure 4.3 shows examples of different applications of steam heating in shell-and-tube heat exchangers. Figure 4.3a shows a horizontal steam heater with steam condensing in the tubes in a *one pass* arrangement. A vertical arrangement is also possible. Figure 4.3b shows the corresponding horizontal design with the *two tube passes* arranged through *U-tubes*. Figure 4.3c shows a *kettle reboiler* for process fluid vaporization, which also features condensation on the tube side. Figure 4.3d shows an arrangement with steam condensing on the shell side of the heat exchanger and heating process fluid on the tube side. The shell-side features vertically mounted segmental baffles to direct the

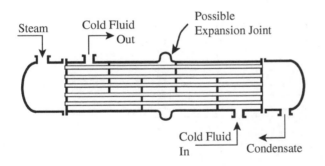

(a) One tube pass with fixed tube sheet.

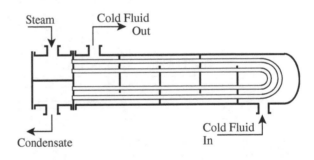

(b) Two tube passes with U-tube.

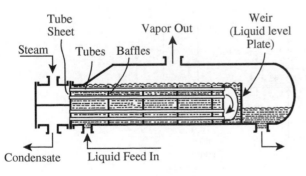

(c) Kettle reboiler.

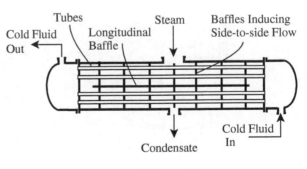

(d) Divided flow.

**Figure 4.3**

Examples of shell-and-tube steam heaters. Source: Adapted from (Smith 2016), *Chemical Process Design and Integration*, 2nd Edition, John Wiley & Sons Ltd.

condensing steam to move from side to side across the tubes. By contrast, Figure 4.4 shows a *gasketed plate and frame* heat exchanger used as a steam heater. This consists of a series of parallel plates with gaskets between the plates to provide the fluid seal. The plates are corrugated both to increase turbulence and to give mechanical rigidity. The plates are held together and compressed in a frame by use of lateral bolts. Each corrugated plate is provided with four ports. The gasket arrangement around the four ports directs the steam and cold fluids to flow down alternate channels.

Steam heating is commonly used for distillation reboilers. Figure 4.5 shows three different designs. In Figure 4.5a a *kettle reboiler* vaporizes the process fluid on the outside of the tubes immersed in a pool of boiling liquid. The bottom product is taken from an overflow from the liquid pool and there is no recirculation of liquid between the reboiler and the column. Figure 4.5b shows a *horizontal thermosyphon reboiler*, which also features vaporization on the outside of the tubes. However, in this case there is a

recirculation of liquid around the base of the column. A mixture of vapor and liquid leaves the reboiler and enters the base of the column, where it is separated. The column is arranged at such a height to provide a steady hydrostatic head for the circulation. The third type of reboiler, a *vertical thermosiphon reboiler*, is shown in Figure 4.5c. Again there is a recirculation of liquid around the base of the column, but this time the vaporization takes place inside the tubes.

Steam can be used for vaporization of a liquid when required for applications other than distillation. Two alternative designs are shown in Figure 4.6. Designs of heaters other than kettle designs will require a vapor–liquid separation device. Figure 4.6a shows a *natural circulation* arrangement in which hydrostatic head is used to create a flow through the heater. Figure 4.6b shows the corresponding design with a pump creating *forced circulation*.

When steam is used in a steam heater, some arrangement is required to prevent the steam from flowing through the heater

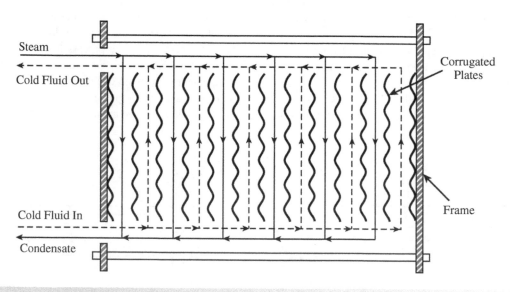

## Figure 4.4

Plate-and-frame steam heater. Source: (Smith 2016), *Chemical Process Design and Integration*, 2nd Edition, John Wiley & Sons Ltd.

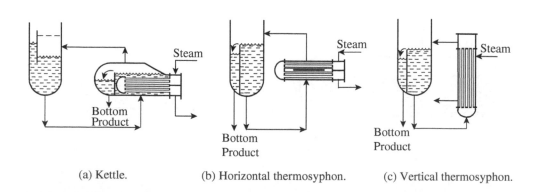

(a) Kettle.  (b) Horizontal thermosyphon.  (c) Vertical thermosyphon.

## Figure 4.5

Reboiler designs. Source: (Smith 2016), *Chemical Process Design and Integration*, 2nd Edition, John Wiley & Sons Ltd.

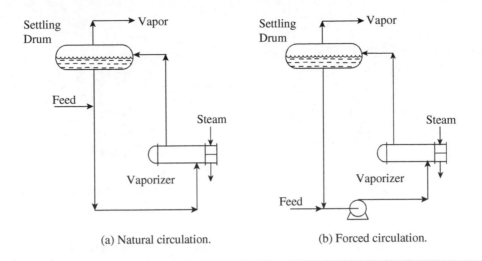

Settling Drum

Vapor

Feed

Steam

Vaporizer

(a) Natural circulation.

Settling Drum

Vapor

Feed

Steam

Vaporizer

(b) Forced circulation.

**4**

### Figure 4.6

Process vaporizer arrangements. Source: (Smith 2016), *Chemical Process Design and Integration*, 2nd Edition, John Wiley & Sons Ltd.

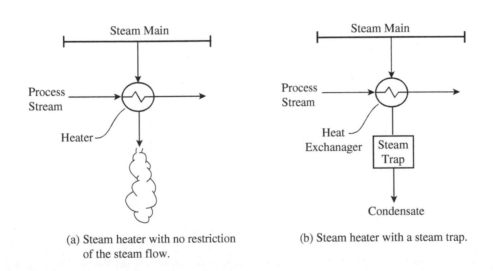

Steam Main

Process Stream

Heater

(a) Steam heater with no restriction of the steam flow.

Steam Main

Process Stream

Heat Exchanager

Steam Trap

Condensate

(b) Steam heater with a steam trap.

### Figure 4.7

A steam trap allows condensate to pass, but not steam.

and only partially condensing, as illustrated in Figure 4.7a. The most common way (but not the only way) to achieve this is to install a steam trap on the exit of the heater, as illustrated in Figure 4.7b. The steam trap is a valve that allows condensate to pass, but not steam. Figure 4.8 illustrates the design of two commonly used types of steam trap. Figure 4.8a shows a *float trap* commonly used on process heaters. This operates from the difference in density between steam and condensate. As condensate builds up within the steam trap, the float rises and lifts the valve off the seat, allowing condensate to pass. Once the condensate has passed, the float drops, closing the outlet. The design in Figure 4.8a incorporates a vent to allow any air present at start-up to be vented. Figure 4.8b shows an *inverted bucket trap*. The inverted bucket is attached by a lever to a valve. The inverted

bucket features a small vent hole in the top of the bucket. Condensate flows from the bottom of the bucket through to the outlet. Any steam entering the trap causes the bucket to become buoyant, which rises and closes the outlet. The trap remains closed until the steam in the bucket has condensed or bubbled through the vent hole. The bucket then sinks, opening the trap. Any air entering the trap at start-up will also give the bucket buoyancy and close the valve. The vent hole allows air to escape. Many other designs of steam trap are available.

For large heat duties, it is good practice to recover the steam that is flashed as the condensate is reduced in pressure through the steam trap. Such an arrangement is shown in Figure 4.9. Steam enters the steam heater and condensate (in practice, with some steam) passes through the trap. Flashing occurs before the mixture

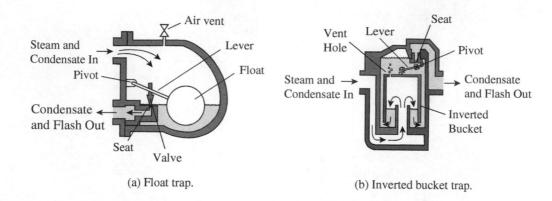

(a) Float trap.

(b) Inverted bucket trap.

**Figure 4.8**

Steam traps. Source: (Smith 2016), *Chemical Process Design and Integration*, 2nd Edition, John Wiley & Sons Ltd.

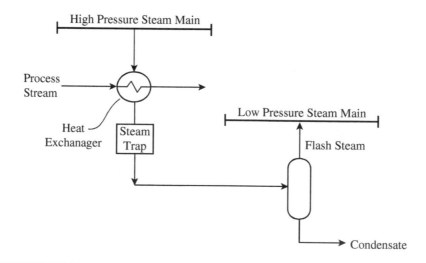

**Figure 4.9**

Flash steam recovery. Source: Adapted from (Smith 2016), *Chemical Process Design and Integration*, 2nd Edition, John Wiley & Sons Ltd.

enters a settling drum that allows the flash steam to be separated from the condensate. The flash steam can then be fed to a steam main at an appropriate pressure.

The condensate in the steam heater can in principle be subcooled, as shown in Figure 4.10. If subcooling is required, the heat exchanger can be divided into zones. Figure 4.10a shows subcooling on the shell side of a vertical heater. The subcooling arrangement in this example is achieved by using a loop seal to create a partially submerged tube bundle. Figure 4.10b illustrates subcooling on the shell side of a horizontal heater. The subcooling arrangement is again achieved in this example by using a loop seal to create a partially submerged tube bundle. Rather than use a loop seal, a dam baffle can be used to partially submerge the bundle (Kern 1950).

The steam heater might involve desuperheating of the steam as well as condensation. In this case, the steam heater can be divided into zones, as illustrated in Figure 4.11. The first section of the

heat exchange on the shell side desuperheats the incoming steam, followed by condensation of the steam. Desuperheating features a significantly lower heat transfer coefficient than condensation of the steam. Feeding steam with a high degree of superheat leads to an inefficient heat exchanger design, because of the lower heat transfer coefficient in desuperheating. The inefficiency in the heat exchanger design can be avoided by desuperheating the steam prior to use. As will be discussed later, it is bad practice to have saturated steam, or close to saturated steam, in the steam mains. However, desuperheating can be carried out locally for one or a number of heaters by the injection of treated water (boiler feedwater) into the steam, as illustrated in Figure 4.12. Treated water (boiler feedwater) is mixed with the superheated steam under temperature control, as shown in Figure 4.12. In this way, steam can be desuperheated local to where it will be used for heating to within 5 °C of saturation or less, prior to being used in a heat exchanger.

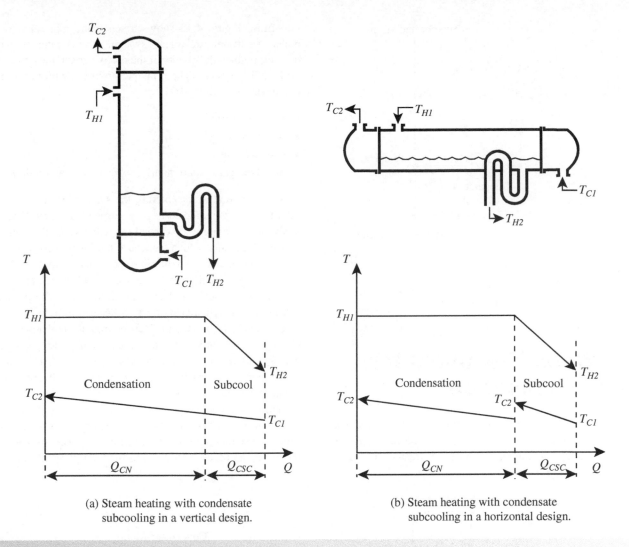

(a) Steam heating with condensate
subcooling in a vertical design.

(b) Steam heating with condensate
subcooling in a horizontal design.

**Figure 4.10**

Steam condensation with subcooling. Source: (Smith 2016), *Chemical Process Design and Integration*, 2nd Edition, John Wiley & Sons Ltd.

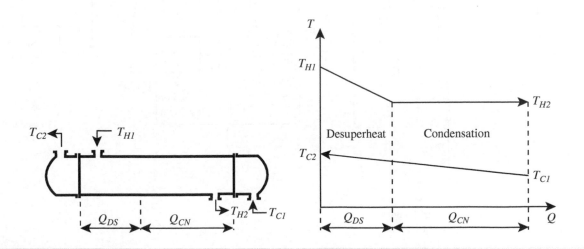

**Figure 4.11**

Steam condensation with desuperheating in a horizontal design. Source: (Smith 2016), *Chemical Process Design and Integration*, 2nd Edition, John Wiley & Sons Ltd.

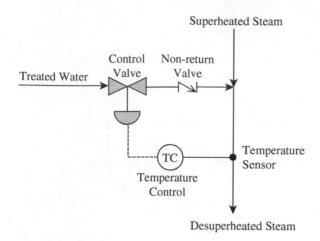

**Figure 4.12**

A desuperheater. Source: (Smith 2016), *Chemical Process Design and Integration*, 2nd Edition, John Wiley & Sons Ltd.

# 4.3 Water Treatment for Steam Generation

Before raw water entering the site from a reservoir, river, lake, borehole, or seawater desalination plant can be fed to a steam boiler for steam generation, it must be treated to remove impurities. Figure 4.13 shows a schematic representation of a boiler feedwater treatment system. The treatment required depends both on the quality of the raw water and the requirements of the utility system. The principal problems with raw water are (Kemmer 1988; Betz 1991):

- suspended solids;
- dissolved solids;
- dissolved salts;
- dissolved gases (particularly, oxygen and carbon dioxide).

Thus, the raw water entering the system in Figure 4.13 may need to be first filtered to remove suspended solids. If required, this would commonly be carried out using sand filtration.

Dissolved salts, which can cause fouling and corrosion in the steam boiler, then need to be removed. The principal problems are associated with calcium and magnesium ions, which would otherwise deposit as scale on heat transfer surfaces. The presence of silica is also a particular problem. Silica can form low conductivity deposits in the boiler and if carried from the boiler in water droplets or in a volatile form can cause damage to steam turbine blades, particularly the low-pressure section of steam turbines where some condensation can occur. The quality of feedwater required depends on the boiler operating pressure and boiler design. Sodium zeolite *softening* is the simplest process and is used to remove scale-forming ions, such as calcium and magnesium, sometimes referred to as *hardness*. The softened water produced can be used for low- to medium-pressure boilers. The

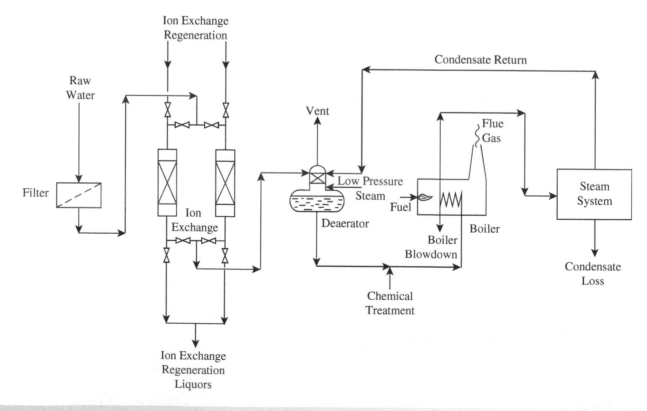

**Figure 4.13**

Boiler feedwater treatment. Source: (Smith 2016), *Chemical Process Design and Integration*, 2nd Edition, John Wiley & Sons Ltd.

softening process removes the sparingly soluble calcium and magnesium ions and replaces them with less objectionable sodium ions. Softening alone is not sufficient for most high-pressure boiler feedwaters. *Deionization*, in addition to removing hardness, removes other dissolved solids, such as sodium, silica, alkalinity, and mineral ions such as chloride, sulfate, and nitrate. Deionization is essentially the removal of all inorganic salts. As the ion-exchange beds approach exhaustion, they need to be taken offline and regenerated. Sulfuric acid and sodium hydroxide solution are used to regenerate exhausted ion-exchange beds. Rather than use ion exchange, water can be deionized using membrane processes. Both nano-filtration and reverse osmosis can be used, but only reverse osmosis is capable of producing high-quality boiler feedwater.

Having removed the suspended solids and dissolved salts, the water then needs to have the dissolved gases removed, principally, oxygen and carbon dioxide, which would otherwise cause corrosion in the steam boiler. The usual method to achieve this is deaeration, which removes dissolved gases by raising the water temperature and stripping the dissolved gases (Kemmer 1988; Betz 1991). Some form of packing of plates is normally used to assist the contact between the boiler feedwater and steam. Boiler feedwater is heated to within a few degrees of the saturation temperature of the steam. Most of the non-condensable gases (principally, oxygen and free carbon dioxide) are released into the steam. A small portion of the steam, which contains the non-condensable gases released from water, is vented to atmosphere.

Even after deaeration, there is some residual oxygen that needs to be removed by chemical treatment (Kemmer 1988; Betz 1991). After the deaerator, oxygen scavengers (e.g. hydroquinone) are added to react with the residual oxygen that would otherwise cause corrosion. Unfortunately, the boiler feedwater treatment does not remove all of the solids in the raw water, and the deposition of solids in the boiler is another problem. Phosphates can be added to precipitate any calcium and magnesium away from the heat transfer surfaces of the boiler. Polymer dispersant can also be added to help keep any precipitate dispersed.

The purity of the boiler feedwater and the treatment required depend on the pressure of the boiler. The higher the pressure of the boiler, the purer the feed used.

The deaerated treated boiler feedwater then enters the boiler. Evaporation takes place in the boiler and the steam generated is fed to the steam system. Solids (either suspended or dissolved) not removed by the boiler feedwater treatment build up in the boiler, along with products of corrosion. If the solids are allowed to build up they can lead to foaming and carryover of solids into the steam. The solids will also lead to scale formation in the boiler, resulting in the deterioration of performance and possibly localized overheating and boiler failure. The control of the concentration of solids is achieved by purging, known as *blowdown*.

The steam from the boiler goes to the steam system to perform various duties. The steam is ultimately condensed somewhere in the steam system. Some condensate is returned to the deaerator and some lost from the system to effluent. Condensate return as high as 90% or greater is possible, but return rates are often significantly lower than this. Higher levels of condensate return than 90% might not be able to be justified because of the capital cost of

the pipework required for condensate return or the possibility of some condensate being contaminated. This constant loss of condensate from the steam system means that there must be a constant make up with freshwater. To prevent corrosion in the condensate system, amine treatment chemicals are also added to the boiler feedwater. These are volatile and condense with steam, adjusting the pH on condensation to prevent corrosion in the condensate system.

## 4.4 Steam Generation from the Combustion of Fuels

For the generation of steam, first consider the generation of steam from the combustion of fuels. There are many types of combustion steam boilers, depending on the steam pressure, steam output, and fuel type (Dryden 1982). Pressures are normally of the following order.

- 100 barg, used for power generation;
- 40 barg is the normal maximum pressure for distribution (higher pressures require more expensive materials of construction);
- 10–40 barg are the conventional distribution pressure levels;
- 2–5 barg are typically the lowest pressures used for distribution.

Figure 4.14 illustrates a *fire-tube* (or *shell*) boiler. Water is preheated by the stack gas in an *economizer* and enters the boiler shell. The shell is a large metal cylinder housing a pool of boiling water that is vaporized by a hot flue gas flowing through the inside of tubes. The steam pressure is contained by the large cylindrical shell, which imposes mechanical limitations. A large-diameter shell requires a thicker wall to contain the same pressure than a small-diameter shell. The economic limit for such designs is around 20 barg (but they are available at higher pressures). Fire-tube boilers are normally used for small heating duties of low-pressure steam. Fuels are normally restricted to be natural gas, waste combustible gases from the process, biogas, gases from biomass gasification, renewable hydrogen, light fuel oil, or other low viscosity liquid fuel. One of the main disadvantages of the design is difficulty in providing superheat to the steam. Superheat can be added with a separate superheater.

Higher duty boilers use a *water-tube* arrangement, as shown in Figure 4.15. The water is first preheated in an economizer from the stack gas and introduced into the *steam drum* behind a baffle. The steam drum is connected to a lower *water* or *mud drum* by the boiler tubes. Because the cold water is dense, it descends in the *downcomer* toward the water drum. This causes hot water to flow upwards from the water drum in the front tubes closer to the flame. Continued heating of the water in the front tubes causes partial vaporization of the water. Steam is separated from the hot water in the steam drum. The steam separated in the steam drum enters a superheating section before entering the steam system. Water-tube boilers are suitable for high pressures and high steam output. Fuels can be natural gas, waste combustible gas from the process, biogas, gases from biomass gasification, renewable hydrogen, both low viscosity and high viscosity liquid fuel, or pulverized solid fuel. In the case of solid fuels, the boiler

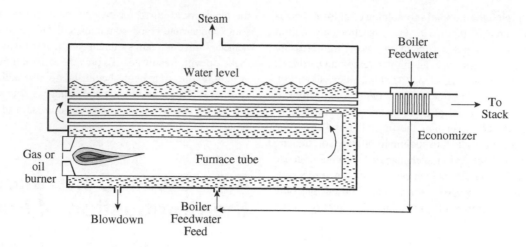

## Figure 4.14

Fire-tube (or shell) boiler. Source: Adapted from (Smith 2016), *Chemical Process Design and Integration*, 2nd Edition, John Wiley & Sons Ltd.

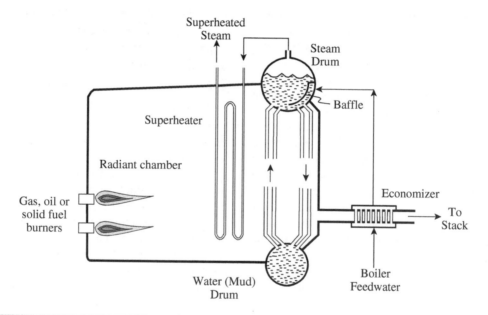

## Figure 4.15

Water-tube boiler. Source: Adapted from (Smith 2016), *Chemical Process Design and Integration*, 2nd Edition, John Wiley & Sons Ltd.

must have some mechanism for the removal of ash produced by combustion from the base of the radiant chamber.

Another design of boiler is illustrated in Figure 4.16, which uses a fluidized bed combustion system. This is mainly used for the combustion of solid fuels (coal, petroleum, coke, or biomass). The combustion air is typically preheated by the stack gas before entering the combustion zone. Fuel particles are suspended in a hot, bubbling fluidized bed of fuel and ash. The fuel burns throughout its volume with no flame. Light ash particles leave the top of the bed, together with waste gases. The fluidized bed creates good contact between the fuel and combustion air and combustion takes place at temperatures lower than a conventional

boiler, from 800 to 900 °C, with a resulting reduced formation of oxides of nitrogen. The solid fuel can have a high moisture and ash content and can contain significant amounts of sulfur as impurity. Limestone can be added to the combustion, which reacts with any sulfur to form calcium sulfate and leaves with the spent ash, rather than the sulfur being separated from the flue gas or discharged to atmosphere. Figure 4.16 illustrates a bubbling fluidized bed boiler. An alternative design uses a *circulating fluidized bed*. Circulating beds use a higher fluidizing velocity, causing the particles to pass through the main combustion chamber and into a cyclone, from which the larger particles are separated from the flue gas and returned to the combustion chamber.

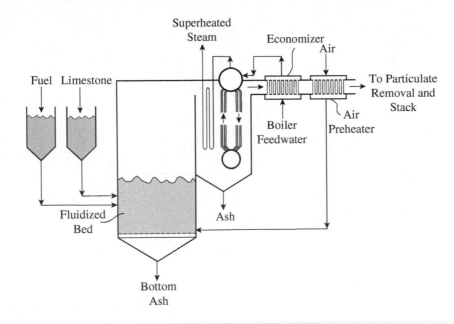

### Figure 4.16
Fluidized bed boiler. Source: Adapted from (Smith 2016), *Chemical Process Design and Integration*, 2nd Edition, John Wiley & Sons Ltd.

Other designs of boiler are available. The draft in boilers can be *forced draft*, in which a fan is used to blow air into the fired heater. The boiler is thus at a pressure slightly above atmospheric. Alternatively, in *induced draft* designs a fan is located in the duct for the waste combustion gases after the boiler before entering the stack. This creates a pressure in the boiler slightly below atmospheric. For large boilers requiring a significant amount of equipment in the flue to treat the exhaust gases to remove particulates, oxides of sulfur, or oxides of nitrogen, as well as equipment for heat recovery, a combination of forced and induced draft might need to be used in a *balanced draft* arrangement.

The performance of steam boilers is measured by boiler efficiency. Two different methods are used to calculate the boiler efficiency: the direct method and the indirect method. In the direct method, the efficiency is calculated directly from the heat output from the boiler and the fuel input:

$$\eta_{BOILER} = \frac{Q_{OUTPUT}}{Q_{INTPUT}} = \frac{m_{STEAM}(H_{STEAM} - H_{BFW})}{m_{FUEL}\, NCV} \qquad (4.1)$$

where

$\eta_{BOILER}$ = boiler efficiency (−)

$Q_{OUTPUT}$ = heat output to steam generation (W, kW)

$Q_{INPUT}$ = heat input from fuel (W, kW)

$m_{STEAM}$ = flowrate of steam (kg s$^{-1}$)

$m_{FUEL}$ = flowrate of fuel (m$^3$ s$^{-1}$, kg s$^{-1}$)

$H_{STEAM}$ = enthalpy of the steam (J kg$^{-1}$, kJ kg$^{-1}$)

$H_{BFW}$ = enthalpy of the boiler feedwater (J kg$^{-1}$, kJ kg$^{-1}$)

$NCV$ = net calorific value (J m$^{-3}$, kJ m$^{-3}$, J kg$^{-1}$, kJ kg$^{-1}$)

In the indirect method, the efficiency is calculated from the fuel input and the losses:

$$\eta_{BOILER} = 1 - \frac{Q_{LOSS}}{Q_{INTPUT}} = 1 - \frac{Q_{LOSS}}{m_{FUEL}\, NCV} \qquad (4.2)$$

where

$Q_{LOSS}$ = sum of the losses for the stack, unburnt combustible gas, unburnt combustible solid, or liquid fuel, ash, casing, blowdown, and atomizing steam (W, kW)

The definition used for boiler efficiency here is based on the *net calorific value* of the fuel.

Figure 4.17 shows the steam generation profile in the boiler as it is matched against the boiler flue gas. This starts at the *theoretical flame temperature*. The theoretical flame temperature or *adiabatic combustion temperature* is the temperature attained when fuel is burnt in air (or oxygen enriched air or oxygen) without loss or gain of heat. It can be calculated from an energy balance around the combustion process (Smith 2016). Although it does not represent the actual temperature in the combustion zone (principally because of heat loss from the flame), it does provide a convenient reference point for a simple model of the combustion process. The theoretical flame temperature with an appropriate amount of excess combustion air is typically around 1800 °C. This is cooled ultimately to ambient temperature. Figure 4.17 shows the steam being preheated from subcooled boiler feedwater to superheated steam. In turn, the flue gas is cooled to the stack temperature before release to the atmosphere. The flue gases are normally kept above their dew point before release to the atmosphere, particularly if there is any sulfur in the fuel. The heat loss in the stack $Q_{STACK}$ is the major inefficiency in the use of the fuel.

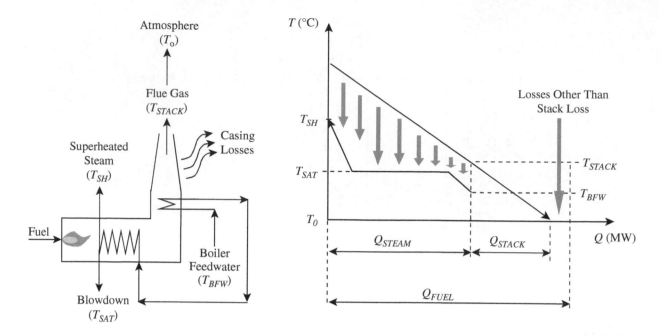

**Figure 4.17**

Model of steam boiler. Source: Adapted from (Smith 2016), *Chemical Process Design and Integration*, 2nd Edition, John Wiley & Sons Ltd.

Given that the largest loss from boilers using combustion of fuels is in the stack, reducing the stack loss has the biggest effect on the boiler efficiency. *Economizers* can be used to preheat the incoming boiler feedwater from the flue gas and decrease the stack temperature, as illustrated in Figures 4.14 to 4.17. These are heat exchangers between the incoming boiler feedwater and the hot flue gases before they are vented to the atmosphere (Dryden 1982).

Excess air also has a major influence on the boiler efficiency. Excess air is essential to maintain the efficiency of the combustion process. However, too much excess leads to an unnecessary stack loss. Boilers are typically controlled to have excess oxygen in the flue gas of around 3%, but can be lower. Every 1% above this will decrease the efficiency by 1%.

It should be noted that for a given combustion steam boiler there will be a loss of efficiency as the steam load on the boiler is decreased (Smith 2016).

# 4.5 Steam Generation from Electrical Energy

Electric steam boilers generate steam using electricity as an energy source, rather than fossil or biomass fuels. Use of electricity as a source of energy has the advantage that it can provide more sustainable process heating if the electricity is generated from a renewable source and is low-cost. Providing heating from steam on a sustainable basis using combustion is only possible if the fuel used in the boiler is biogas, gases from biomass gasification, renewable hydrogen, or biomass. There are two broad classes of electric steam boiler.

1) *Electric resistance steam boilers*. Electric resistance steam boilers produce steam using insulated electrical resistance elements (Figure 4.18). Heat is generated from the electrical resistance within the element. The boiler can be constructed in a horizontal or vertical cylindrical pressure vessel. Banks of resistance elements heat water in the pressure vessel. Individual resistance elements can be switched on and off to control the steam generation. Operating voltages are typically between 240 and 600 V, using single- or three-phase alternating current. Use of electric resistance boilers is typically limited to capacities less than 5 MW, but much larger capacities are available. Electric resistance steam boilers produce saturated steam, typically up to 200 °C. If superheated steam is required, then a separate electrically heated steam superheater can be added. The efficiency of such boilers is high, approaching 100%. There is no stack loss, which is the main loss from steam boilers using combustion, but there will still be heat losses from the surface of the boiler to the environment and losses from heat in the boiler blowdown, giving efficiencies of typically 95–98%. Unlike combustion steam boilers, electric resistance steam boilers do not suffer deterioration in efficiency with reduced load. If a larger capacity is required at higher pressures, then electrode steam boilers can be used.

2) *Electrode steam boilers*. Electrode steam boilers work on a different principle compared with electric resistance steam boilers. Electrode steam boilers produce steam by passing an electric current directly through the water between high-voltage electrodes. The resistance to the electrical current from the water generates heat directly in the water, producing steam. The heat generated is proportional to the flow of electrical current. The conductivity of the water allows a high current

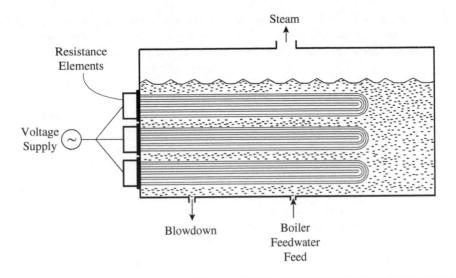

**Figure 4.18**

Electric resistance steam boiler.

density. A high-voltage three-phase alternating current in the range 6–24 kV is used. Different boiler configurations are available.

Figure 4.19 shows a design that consists of a vertical cylindrical pressure vessel. Three electrodes are immersed in the boiler feedwater, each using a different phase from the three-phase electrical supply. The internal circulation system feeds water to the electrodes, which are immersed in water in a header. Only a small portion of the circulation water flow is converted to steam. The header is electrically insulated from the outer shell. The water

and the header form an insulated zero point for the electric current. The steam generation is controlled by a throttle valve that regulates the level in the header, which controls the area of contact between the water and electrodes and the rate of steam generation.

Figure 4.20 shows an alternative design of electrode steam boiler. The boiler design again consists of a vertical cylindrical pressure vessel. A circulating pump produces a flow of water into a nozzle header. Boiler feed water is sprayed through the nozzles, striking the electrodes and falling downwards onto a zero point electrode. Only a small portion of the water flow through the nozzle header is converted to steam. Control of the steam generation

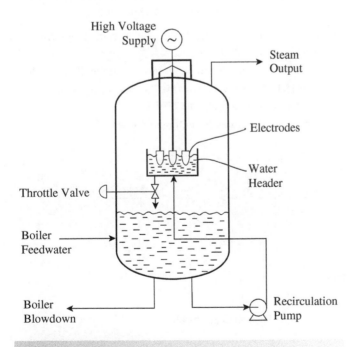

**Figure 4.19**

Electrode steam boiler using immersed electrodes.

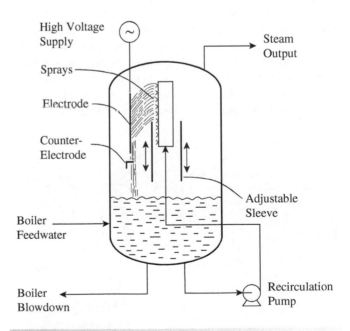

**Figure 4.20**

Electrode steam boiler using spray contacting.

can be through a movable sleeve that controls the number of water jets allowed to contact the electrodes, as shown in Figure 4.20. It is also possible to control the steam generation through the water recirculation rate.

Electrode steam boilers with a pressure of up to 85 barg and capacity up to 70 MW are available. Saturated steam is produced. If superheated steam is required, then a separate low voltage electric steam superheater needs to be added. The efficiency of such boilers is high, approaching 100%. As with electric resistance heaters, there is no stack loss compared with steam boilers using combustion, but there will still be heat losses from the surface of the boiler to the environment and losses from the boiler blowdown, giving efficiencies of typically 95–98%. Again, in contrast with combustion steam boilers, electrode steam boilers do not suffer from a deterioration of efficiency with load. Because electrode steam boilers depend on the electrical resistance of the water, the conductivity of the water in the boiler needs to be carefully controlled through the appropriate boiler feedwater treatment and the boiler blowdown. Appropriate boiler feedwater treatment is also necessary to prevent the build-up of scale on the electrodes. The optimum conductivity varies with the boiler voltage and temperature. Conductivity must be high enough to allow development of the required output of the boiler at its designed operating pressure.

# 4.6 Gas Turbines

In its simplest form, the gas turbine consists of a *compressor* and a *turbine* (or *expander*). In a *single-shaft* gas turbine, these are mechanically connected and are rotating at the same speed, as shown in Figure 4.21a. Air from ambient is compressed and raised in temperature as a result of the compression. A portion of the compressed air enters a combustion chamber where fuel is fired to raise the temperature of the gases. Most of the air from the compressor provides cooling for the combustor walls. The hot, compressed mixture of air and combustion gases then flows to the inlet of the turbine. Combustor outlet temperatures range between 800 °C to over 1600 °C. In the turbine, the gas is expanded to develop power to drive the compressor. Around two-thirds of the power produced by the turbine is needed to drive the

compressor. However, by virtue of the energy imported in the combustion process, excess power is produced. The higher the expander inlet temperature, the better is the performance of the machine. Temperatures in the turbine are limited by turbine blade materials. Different machine configurations are possible with gas turbines. Figure 4.21b shows a *split-shaft* or *twin-shaft* arrangement. This is mechanically more complex. The first turbine provides the power necessary to drive the compressor. The second turbine provides the power for the external load.

The efficiency of the gas turbine can be defined as:

$$\eta_{GT} = \frac{W}{Q_{FUEL}} \qquad (4.3)$$

where

$\eta_{GT}$ = gas turbine efficiency (–)

$W$ = turbine shaft power (kW)

$Q_{FUEL}$ = heat released from fuel (kW)

The efficiency of gas turbines can alternatively be expressed as a *heat rate*, which is simply the inverse of efficiency:

$$HR = \frac{Q_{FUEL}}{W} \qquad (4.4)$$

where

$HR$ = gas turbine heat rate (–)

Depending on the definition of $W$ and $Q_{FUEL}$ in Eq. (4.4), the heat rate often quoted is units dependent. The performance of the machine is normally specified at maximum load at International Standards Organization (ISO) conditions of 15 °C, 1.013 bar, and 60% relative humidity.

The basic characteristics of gas turbines are:

- sizes are restricted to standard frame sizes ranging from under 500 kW to over 300 MW;
- *micro-turbines* are available in the size range 25–500 kW;
- electrical efficiency is 30–45% (increasing to over 50% for designs with cooled turbine blades);
- exhaust temperatures in the range 450–600 °C;

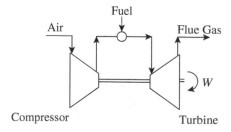

(a) Single-shaft gas turbine.

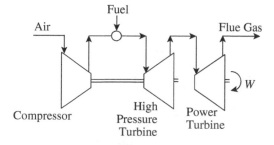

(b) Twin-shaft gas turbine.

**Figure 4.21**

Gas turbine configurations. Source: (Smith 2016), *Chemical Process Design and Integration*, 2nd Edition, John Wiley & Sons Ltd.

- fuels need to be at a high pressure;
- fuels must be free of particulates and sulfur.

The most common fuels used are gas (natural gas, biogas, methane, propane, synthesis gas, hydrogen) and light fuel oils. Contaminants such as ash, alkalis (sodium and potassium), and sulfur result in deposits, which degrade performance and cause corrosion in the hot section of the turbine. Total alkalis and total sulfur in the fuel should both be less than 10 ppm. Gas turbines can be equipped with *dual firing* to allow the machine to be switched between fuels (e.g. natural gas and light fuel oil).

Gas turbines can be classified as *industrial* or *frame* machines and *aero-derivative* machines, which are mechanically more complex. The aero-derivative machines are lighter weight units with designs derived from aircraft engines.

Gas turbines require a start-up device, which is usually an electric motor. The power of the start-up device can be up to 15% of the gas turbine power, depending on the size and design of the machine. Once the gas turbine has been started, the motor can be switched off, run as a helper motor to boost the power from the gas turbine, or switched to be a generator to generate electricity from the gas turbine.

Figure 4.22 shows an integrated gas turbine arrangement in which the exhaust from the gas turbine is used to generate steam in a *heat recovery steam generator* (HRSG) before being vented to atmosphere. It is possible to fire fuel after the gas turbine to increase the temperature of the gas turbine exhaust before entering the HRSG, as gas turbine exhaust is still rich in oxygen (typically, 15% $O_2$). This is *supplementary* or *auxiliary* firing. The steam from the HRSG might be used directly for process heating or be expanded in a steam turbine system to generate additional power. The steam turbine exhaust can either be back-pressure or condensing. There are three firing modes for gas turbines:

1) *Unfired HRSG*. An unfired HRSG uses the sensible heat in the gas turbine exhaust to raise steam.
2) *Supplementary fired HRSG*. Supplementary firing raises the temperature by firing fuel to use a portion of the oxygen in the exhaust. Supplementary firing uses convective heat transfer, and temperatures are limited to a maximum of 925 °C by ducting materials. However, the supplementary firing

temperature can be increased to 1200–1300 °C if the walls of the HRSG are water cooled.

3) *Fully fired HRSG*. Fully fired HRSGs make full utilization of the excess oxygen to raise the maximum amount of steam in the HRSG. Full firing means reducing the excess oxygen to a minimum of around the 3% normally demanded by all combustion processes to ensure efficient combustion. However, this means that radiant heat transfer will result from full firing. Essentially, fully firing means that the gas turbine exhaust is used as preheated combustion air to a steam boiler. Alternatively, gas turbines can exhaust into a process furnace, rather than a steam boiler.

## 4.7 Steam Turbines

A steam turbine converts the energy of steam into power by expanding the steam across rows of *blades* mounted on *wheels*, causing the wheels to rotate by extracting energy from the steam (Elliot 1989). The blades are mounted on a rotating shaft. The rotating shaft is supported within a *casing*. The casing supports the bearings for the rotating shaft, the stationary blades, and the steam inlet nozzles.

Steam turbine sizes vary from under 100 kW to over 250 MW. Figure 4.23 shows the basic ways in which a steam turbine can be configured. Steam turbines can be divided into two basic classes:

- *back-pressure* turbines, which exhaust steam at pressures higher than atmospheric pressure;
- *condensing* turbines, which exhaust steam at pressures lower than atmospheric pressure.

The exhaust pressure of a steam turbine is fixed by the operating pressure of the downstream equipment. Figure 4.23a shows a back-pressure turbine operating between high-pressure and low-pressure steam mains. The pressure of the low-pressure steam mains will be controlled elsewhere (see later). Figure 4.23b shows a condensing turbine. Three types of condensers are used in practice as follows:

1) *Direct contact*, in which cooling water is sprayed directly into the exhaust steam.

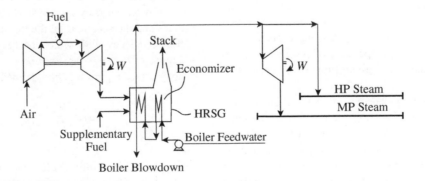

**Figure 4.22**

Gas turbine with single pressure heat recovery steam generator (HRSG). Source: Adapted from (Smith 2016), *Chemical Process Design and Integration*, 2nd Edition, John Wiley & Sons Ltd.

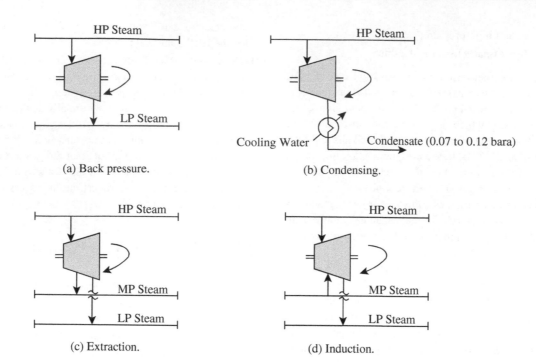

(a) Back pressure.

(b) Condensing.

(c) Extraction.

(d) Induction.

## Figure 4.23

Configuration of steam turbines. Source: (Smith 2016), *Chemical Process Design and Integration*, 2nd Edition, John Wiley & Sons Ltd.

2) *Surface condensers*, in which the cooling water and exhaust steam remain separate. Rather than using cooling water, boiler feedwater can be preheated to recover the waste heat.

3) *Air coolers* again use surface condensation but reject the heat directly to ambient air (see Chapter 5).

Of the three types, surface condensers using cooling water are by far the most common. When a volume filled with steam is condensed, the resulting condensate occupies a smaller volume and a vacuum is created. Any non-condensable gases remaining after the condensation are removed by steam ejectors or liquid ring pumps. The pressure at the exhaust of the condensing turbine is maintained at a reasonable temperature above cooling water temperature (e.g. 40 °C at 0.07 bara or 50 °C at 0.12 bara, depending on the temperature of the cooling water). The higher the pressure difference across the steam turbine, the more the energy that can be converted from the turbine inlet steam into power. For a given temperature difference between the condensing steam and the cooling medium, the lower the temperature of cooling, the more power that can be generated for a given flowrate of steam with fixed inlet conditions.

Both back-pressure and condensing turbines can be categorized further by the steam that flows through the machine. Figure 4.23c shows an *extraction* machine. Extraction machines bleed off part of the main steam flow at one or more points. The extraction might be uncontrolled and the flows dictated by the pressures at the inlets and outlets and the pressure drops in the sections of the turbine. Alternatively, the extraction might be controlled by internal control valves. Figure 4.23c shows a single extraction with both the extraction and the exhaust being fed to

steam mains. The exhaust could have been taken to vacuum conditions and condensed, as in Figure 4.23b.

Figure 4.23d shows an *induction* turbine. Induction turbines work like extraction machines, except in reverse. Steam at a higher pressure than the exhaust is injected into the turbine to increase the flow part-way through the machine to increase the power production. In a situation like the one shown in Figure 4.23d, an excess of medium-pressure (MP) steam generation over and above that for process heating is used to produce power and exhaust into a low-pressure steam, where there is a demand for the low-pressure steam for process heating.

Any given machine will have minimum and maximum allowable steam flowrates. In the case of extraction and induction machines, there will be minimum and maximum flowrates allowable in each turbine section. These minimum and maximum flowrates are determined by the physical characteristics of individual turbines and are specified by the turbine manufacturer.

The efficiency with which energy is extracted from steam is characterized by the *isentropic efficiency* introduced in Chapter 2 (see Figure 2.1) and defined as:

$$\eta_{IS} = \frac{H_{in} - H_{out}}{H_{in} - H_{IS}} = \frac{\Delta H}{\Delta H_{IS}} \tag{4.5}$$

where

$\eta_{IS}$ = turbine isentropic efficiency (–)

$H_{in}$ = specific enthalpy of the inlet steam (kJ kg$^{-1}$)

$H_{out}$ = specific enthalpy of the outlet steam (kJ kg$^{-1}$)

$H_{IS}$ = enthalpy of steam at the outlet pressure having the same entropy as the inlet steam (kJ kg$^{-1}$)

The isentropic efficiency reflects the inefficiency associated with the flow of the steam through the turbine and characterizes the efficiency at which energy is extracted from the steam. The *mechanical efficiency* reflects the efficiency with which the energy that is extracted from steam is transformed into useful power and accounts for machine frictional losses and heat losses. The mechanical efficiency is high (typically 0.97–0.99) (Siddhartha and Rajkumar 1999). The overall steam turbine efficiency can be defined as:

$$\eta_{ST} = \eta_{IS}\ \eta_{MECH} \tag{4.6}$$

where

$\eta_{ST}$ = overall steam turbine efficiency

$\eta_{MECH}$ = mechanical efficiency

**Example 4.1**   A steam turbine is required to generate 750 kW power. It has an inlet pressure of 20 bar and an exhaust pressure of 5 bar. If the turbine has an isentropic efficiency of 0.75 and a mechanical efficiency of 100%, calculate the flowrate of steam required for:

a) an inlet temperature of 350 °C.
b) an inlet temperature of 275 °C (it should be noted in this case that an isentropic expansion will result in an outlet enthalpy below steam saturation temperature and wet steam).

Steam properties can be taken from International Steam Tables (Kretzschmar and Wagner 2019).

**Notation**

$H_f$ = specific enthalpy of saturated water (kJ kg$^{-1}$)

$H_g$ = specific enthalpy of saturated steam (kJ kg$^{-1}$)

$H_{fg}$ = specific enthalpy of vaporization (kJ kg$^{-1}$)

$H_{in}$ = specific enthalpy of the inlet steam to the steam turbine (kJ kg$^{-1}$)

$H_{IS}$ = specific enthalpy of steam from the steam turbine at the outlet pressure having the same entropy as the inlet steam (kJ kg$^{-1}$)

$H_{out}$ = specific enthalpy of the outlet steam from the steam turbine (kJ kg$^{-1}$)

$S_f$ = specific entropy of saturated water (kJ kg$^{-1}$ K$^{-1}$)

$S_g$ = specific entropy of saturated steam (kJ kg$^{-1}$ K$^{-1}$)

$S_{fg}$ = specific entropy of vaporization (kJ kg$^{-1}$ K$^{-1}$)

$S_{in}$ = specific entropy of steam at the inlet of the steam turbine (kJ kg$^{-1}$ K$^{-1}$)

$S_{out}$ = specific entropy of steam at the outlet of the steam turbine (kJ kg$^{-1}$ K$^{-1}$)

$X$ = dryness fraction (–)

$$= \frac{\text{Mass of dry steam}}{\text{Mass of dry steam} + \text{Mass of saturated water}}$$

$\eta_{IS}$ = steam turbine isentropic efficiency (–)

**Solution**

a) For an inlet pressure of 20 bar and temperature 350 ° C, turbine inlet conditions from steam tables:

$$H_{in} = 3138\,\text{kJ kg}^{-1}$$
$$S_{in} = 6.958\,\text{kJ kg}^{-1}\,\text{K}^{-1}$$

Turbine outlet conditions for isentropic expansion:

$$S_{out} = 6.958\,\text{kJ kg}^{-1}\,\text{K}^{-1}$$

Outlet enthalpy at 5 bar for this entropy from steam tables:

$$H_{IS} = 2808\,\text{kJ kg}^{-1}$$

The real enthalpy outlet is given by Eq. (4.5):

$$H_{out} = H_{in} - \eta_{IS}(H_{in} - H_{IS})$$
$$= 3138 - 0.75(3138 - 2808) = 2891\,\text{kJ kg}^{-1}$$

Shaft power generation by enthalpy difference across the turbine

$$= 3138 - 2891 = 247\,\text{kJ kg}^{-1}$$

Steam flowrate for 750 kW shaft power generation

$$= \frac{750}{247}$$
$$= 3.04\,\text{kg s}^{-1}$$

b) For an inlet pressure of 20 bar and temperature 275 ° C, turbine inlet conditions from steam tables:

$$H_{in} = 2965\,\text{kJ kg}^{-1}$$
$$S_{in} = 6.663\,\text{kJ kg}^{-1}$$

Turbine outlet conditions for isentropic expansion:

$$S_{out} = 6.663\,\text{kJ kg}^{-1}\,\text{K}^{-1}$$

Checking steam tables, this entropy is below steam saturation, which means that an isentropic expansion would lead to wet steam at the outlet. To determine the enthalpy corresponding with this entropy at 5 bar, the dryness fraction must first be determined. The entropy of wet steam is given by:

$$S_{WET} = XS_g + (1 - X)S_f$$
$$= S_f + XS_{fg}$$

Taking the values of $S_f$ and $S_{fg}$ from steam tables at saturation conditions at 5 bar:

$$S_{WET} = S_f + XS_{fg}$$
$$6.663 = 1.861 + X \times 4.960$$

Solving gives $X = 0.9681$. Then the enthalpy for wet steam is given by:

$$H_{WET} = XH_g + (1 - X)H_f$$
$$= H_f + XH_{fg}$$

Substituting $X = 0.9681$ and the values of $H_f$ and $H_{fg}$ from steam tables at saturation conditions at 5 bar:

$$H_{WET} = H_f + XH_{fg}$$

$$= 640 + 0.9681 \times 2108$$

$$= 2681 \text{ kJ kg}^{-1}$$

This is the isentropic outlet enthalpy. The real outlet enthalpy is given by Eq. (4.5):

$$H_{out} = H_{in} - \eta_{IS}(H_{in} - H_{IS})$$

$$= 2965 - 0.75(2965 - 2681) = 2752 \text{ kJ kg}^{-1}$$

Check for any wetness in the real outlet steam. The outlet enthalpy is slightly higher than the saturation enthalpy and thus the steam should be dry. If there is any wetness, it should not be too high, as excessive wetness will damage the turbine. In this case, the enthalpy is above saturation. Any wetness should be acceptable for the steam turbine. The acceptable wetness depends on the design of the machine. However, it should also be noted that if the steam turbine exhausts into a steam main, then excessive condensate might create problems in the steam main. Shaft power generation by enthalpy difference across the turbine

$$= 2965 - 2752 = 213 \text{ kJ kg}^{-1}$$

Steam flowrate for 750 kW shaft power generation

$$= \frac{750}{213}$$

$$= 3.52 \text{ kg s}^{-1}$$

Thus, the lower temperature inlet steam requires a larger flowrate to produce the same power. Also, importantly, the steam conditions at the turbine outlet are significantly different.

**Example 4.2** A compressor drive requires a shaft power input of 1.5 MW, which is to be supplied by a steam turbine with an isentropic efficiency of 75%. Assume the mechanical efficiency is 100%. The inlet steam is at 40 bar and 400 °C. The outlet from the steam turbine is at a pressure of 15 bar.

a) Calculate the flowrate of steam required for the steam turbine.
b) Calculate how much heat is available in the exhaust steam for process heating, assuming that the heating value of the steam is the sum of the superheat and latent heat.
c) Part of the steam turbine exhaust is to be fed to a steam heater with a heat duty of 1 MW. Prior to being fed to the heater, the steam is to be desuperheated to within 5 °C of the saturation temperature by the injection of boiler feedwater at 15 °C. Calculate the flowrate of boiler feedwater for desuperheating and the flowrate of the turbine exhaust steam.

Steam properties can be taken from International Steam Tables (Kretzschmar and Wagner 2019).

**Notation**

$H_{BFW}$ = specific enthalpy of the boiler feedwater (kJ kg$^{-1}$)

$H_{DESUP}$ = specific enthalpy of the desuperheated steam (kJ kg$^{-1}$)

$H_{in}$ = specific enthalpy of the inlet steam to the steam turbine (kJ kg$^{-1}$)

$H_{IS}$ = specific enthalpy of steam from the steam turbine at the outlet pressure having the same entropy as the inlet steam (kJ kg$^{-1}$)

$H_{out}$ = specific enthalpy of the outlet steam from the steam turbine (kJ kg$^{-1}$)

$H_{SH}$ = specific enthalpy of superheated steam (kJ kg$^{-1}$)

$m_{BFW}$ = flowrate of boiler feedwater (kg s$^{-1}$)

$m_{DESUP}$ = flowrate of desuperheated steam (kg s$^{-1}$)

$m_{SUP}$ = flowrate of superheated steam (kg s$^{-1}$)

$S_{in}$ = specific entropy of steam at the inlet of the steam turbine (kJ kg$^{-1}$ K$^{-1}$)

$S_{out}$ = specific entropy of steam at the outlet of the steam turbine (kJ kg$^{-1}$ K$^{-1}$)

$\eta_{IS}$ = steam turbine isentropic efficiency (−)

**Solution**

a) Turbine inlet conditions at 40 bar and 400 °C from steam tables:

$$H_{in} = 3214 \text{ kJ kg}^{-1}$$
$$S_{in} = 6.771 \text{ kJ kg}^{-1} \text{ K}^{-1}$$

Turbine outlet conditions for isentropic expansion:

$$S_{out} = 6.771 \text{ kJ kg}^{-1} \text{ K}^{-1}$$

Outlet enthalpy at 15 bar for this entropy from steam tables:

$$H_{IS} = 2956 \text{ kJ kg}^{-1}$$

From Eq. (4.5):

$$H_{out} = H_{in} - \eta_{IS}(H_{in} - H_{IS})$$

$$= 3214 - 0.75(3214 - 2956) = 3021 \text{ kJ kg}^{-1}$$

Shaft power generation by enthalpy difference across the turbine

$$= 3214 - 3021 = 193 \text{ kJ kg}^{-1}$$

Steam flowrate for 1.5 MW shaft power generation

$$= \frac{1500}{193}$$

$$= 7.77 \text{ kg s}^{-1}$$

**b)** Enthalpy of superheated steam at 15 bar from Part (a)

$$= 3021 \, \text{kJ kg}^{-1}$$

Temperature of saturated condensate at 15 bar from steam tables

$$= 198.3°C$$

Enthalpy of saturated condensate at 15 bar from steam tables

$$= 845 \, \text{kJ kg}^{-1}$$

Total heating value of the exhaust steam from actual temperature to saturated condensate at 15 bar

$$= 7.77 \times (3021 - 845)$$
$$= 16\,908 \, \text{kW}$$

**c)** Temperature of the steam to the heater (saturation from steam tables plus 5 °C)

$$= 198.3 + 5 = 203.3°C$$

Enthalpy of steam of desuperheated steam at this temperature and 15 bar from steam tables:

$$H_{DESUP} = 2805 \, \text{kJ kg}^{-1}$$

Heat available in desuperheated steam from actual temperature to saturated condensate at 15 bar

$$= 2805 - 845 = 1960 \, \text{kJ kg}^{-1}$$

Flowrate of steam for 1 MW heat duty

$$= \frac{1000}{1960} = 0.510 \, \text{kg s}^{-1}$$

Enthalpy of boiler feedwater at 15 °C and 15 bar from steam tables

$$= 63 \, \text{kJ kg}^{-1}$$

Calculate the flowrate of boiler feedwater injected from an energy balance around boiler feedwater injection:

$$m_{SH}H_{SH} + m_{BFW}H_{BFW} = m_{DESUP}H_{DESUP}$$
$$(0.510 - m_{BFW})3021 + m_{BFW} \times 63 = 0.510 \times 2805$$
$$m_{BFW} = \frac{1540.7 - 1430.6}{3021 - 63}$$
$$= 0.0372 \, \text{kg s}^{-1}$$

Flowrate of superheated steam required from the exit of the turbine to the steam heater:

$$m_{SUP} = 0.510 - 0.0372$$
$$= 0.473 \, \text{kg s}^{-1}$$

## 4.8 Steam Distribution

Steam is distributed around the site for various purposes, as discussed earlier. When using steam in steam heaters, steam tracing, and space heating devices, it is preferred to feed the steam close to saturation conditions. Superheated steam is normally not preferred for process heating, as desuperheating to saturation before condensation as part of the process duty involves a poor heat transfer coefficient. Also, the higher temperature from the superheat might cause fouling or product degradation on the process side. Desuperheating in a heat exchanger prior to condensation is illustrated in Figure 4.11. Steam is often desuperheated locally to within 5 °C of saturation before entering the steam heater by mixing the steam with boiler feedwater under temperature control, as illustrated in Figure 4.12.

Steam condensate should be recovered and returned to the deaerator for boiler feedwater wherever possible. Return to the boiler might be via feedwater treatment for *polishing* if there is some danger of contamination and then deaeration. Steam condensate from steam heaters, steam tracing, and space heating devices should normally be returned to the deaerator, possibly via a boiler feedwater polishing step. If steam is injected directly into distillation operations to assist the distillation, then this is ultimately condensed, but the condensate is highly contaminated and either treated and disposed of, or treated and reused (not necessarily as boiler feedwater). Similarly, condensate from dilution steam used in reactors and recovered within the process will be too contaminated for direct reuse as boiler feedwater. In some cases, reactor dilution steam is generated within the process in a heat exchanger, rather than a steam boiler (e.g. from heat recovery), fed to the reactor and the condensate then recycled directly in a closed loop. This is possible, because the dilution steam can be contaminated by process components without causing problems in the reactor. Any steam used for flaring, fuel oil atomization, $NO_x$ abatement in combustion processes, reactor decoking, and soot blowing operations is lost.

Given the various users of steam on the site, a distribution system is required. *Steam mains* or *headers* distribute steam around the site at various pressures. The number of steam headers and their pressures depend on the various process requirements and the requirements to generate power from steam turbines centrally or locally. Figure 4.24 shows a schematic of a typical steam distribution system based on steam generation from the combustion of fuels. Such systems normally feature cogeneration using steam turbines. Raw water is treated before entering the boilers that in this case fire fuel to generate high-pressure (HP) steam. It is common to have at least three levels of steam. On larger sites, steam may also be generated at a very high pressure (typically 100 bar), which will only be used for power generation in steam turbines in the boiler house. Steam would then be distributed around the site, which for larger sites would typically be at three pressures. On small sites there might be only a single pressure of steam distributed around the site. Back-pressure turbines let steam down from the high-pressure mains to the lower-pressure mains to generate power. Supplementary power may be generated using condensing turbines. Operating a condensing turbine

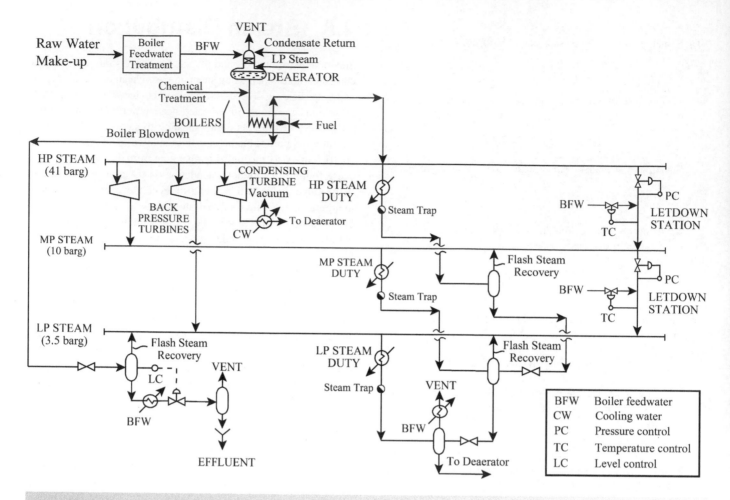

## Figure 4.24

Features of a typical steam distribution system. Source: (Smith 2016), *Chemical Process Design and Integration*, 2nd Edition, John Wiley & Sons Ltd.

from the highest-pressure inlet maximizes the power production. The system in Figure 4.24 shows flash steam recovery into the medium-pressure and low-pressure mains. Also, as shown in Figure 4.24, the boiler blowdown is flashed and flash steam is recovered before being used to preheat incoming boiler feedwater and the boiler blowdown being sent to the effluent. Whether flash steam recovery is economic is a matter of economy of scale.

Also shown in Figure 4.24 are *letdown stations* between the steam mains to control the mains pressures via a pressure control system. In some cases, the letdown stations also have desuperheaters, as illustrated in Figure 4.24. When steam is let down from a high to a low pressure under adiabatic conditions in a valve, the amount of superheat increases at the lower pressure. Desuperheating, if carried out, is achieved by the injection of boiler feedwater under temperature control, which evaporates and reduces the superheat. It should be noted that there will be significant heat losses in the distribution system that might be 10% of the energy supplied in the boilers.

There are two important factors determining the desirable amount of superheat in the steam mains.

a) Steam heating is most efficiently carried out using the latent heat from condensation, rather than having to desuperheat before condensing the steam. Thus, the design of steam heaters benefits from having no superheat in the steam. However, having no superheat in the steam mains is undesirable, as this would lead to excessive condensation in the steam mains. It is desirable to have at least 10–20 °C superheat in steam mains to avoid excessive condensation in the mains. The amount of superheat necessary will depend on the scale and mechanical condition of the steam system.

b) In addition to using steam for steam heating, it is also used for power generation by expansion through steam turbines. Steam turbines might generate electricity centrally that is distributed to motors around the site. Alternatively, steam turbines might be used to drive machines directly. Expansion in steam turbines reduces the superheat in the steam as it is reduced in pressure. If there is not enough superheat in the inlet steam, then condensation can take place in the steam turbine. While a small amount of condensation in the machine might be acceptable in some designs, excessive condensation can be damaging to the machine. Also, if the steam turbine is exhausting to a steam main, then it is desirable to have some degree of superheat in the outlet to maintain some superheat in the outlet low-pressure steam main.

Thus, there is a conflict between the requirements of steam heaters and steam turbines. If a high degree of superheat is maintained in the steam headers, then local desuperheating before entering steam heaters might be justified.

The general policy on steam usage for heating is that low-pressure steam should be used in preference to high-pressure steam. Using low-pressure steam for steam heating:

- allows power generation in steam turbines from the high-pressure steam;
- provides a higher latent heat in the steam for the steam heater;

- leads to lower capital cost heat transfer equipment due to the lower pressures.

It should be noted that the steam system in Figure 4.24 is based on steam generation from the combustion of fuels. In such an arrangement, the most efficient system adopts cogeneration using steam turbines. However, if steam generation is from electric boilers, the most appropriate configuration will be quite different. There is no point using electricity to generate heat in the form of steam to then turn the heat back into power using steam turbines. A steam system in which all of the steam is generated from electrical energy would not be expected to feature steam turbines. This is illustrated in the following examples.

**4**

**Example 4.3**   Figure 4.25 shows an existing steam system servicing a process site. Two high-pressure steam boilers generate steam at 40 bar and 350 °C from boiler feedwater at 105 °C at a pressure above saturation pressure, which is fed into a high-pressure steam main. This high-pressure steam is distributed around the site to service a total site heating duty at that pressure of 5 MW. Steam from the high-pressure main is let down through a passout steam turbine with an isentropic efficiency of 0.8 to a medium-pressure steam main operating at 20 bar. Steam is also expanded from the high-pressure to the medium-pressure steam main through a letdown station. Medium-pressure steam is distributed around the site to service a total heating duty at that pressure of 8 MW. The passout steam turbine exhausts to a low-pressure steam main operating at 10 bar. There is also a letdown station from the medium-pressure steam main to the low-pressure steam main. Low-pressure steam is distributed around the site to service a total heating duty at that pressure of 17 MW. The flowrate through the letdown stations is operated to be a minimum of 5 t h$^{-1}$ to maintain pressure control of the steam mains at the outlet and to keep the equipment hot. In addition to the high-pressure steam being expanded in a passout steam turbine, high-pressure steam is also expanded in a condensing steam turbine with an isentropic efficiency of 0.75. The steam turbine condenser operates at 0.1233 bar, corresponding with a steam condensate temperature of 50 °C.

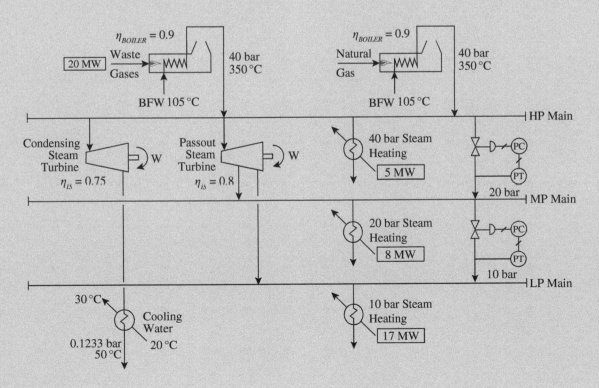

**Figure 4.25**

A steam system with two boilers.

One of the steam boilers uses waste gases as a fuel with an efficiency of 0.9. These waste gases are produced from various processes on the site as byproducts of reaction and cannot be released directly to the atmosphere. The waste gases are therefore combusted in a steam boiler, generating useful steam before safe release to the environment. The total heat release from combustion of the gases before steam generation is 20 MW. The other boiler shown in Figure 4.25 also has an efficiency of 0.9 and uses natural gas as fuel.

To make the existing utility system more sustainable, it is necessary to decarbonize the system as much as possible and decrease the carbon dioxide emissions. As much of the combustion fuel as possible is to be replaced by renewable electricity. Figure 4.26 shows a first suggested arrangement in which the natural gas boiler has been replaced by an electric boiler. The waste gas boiler has been retained in order to safely dispose of the waste gases. The electric boiler generates steam at 40 bar with an efficiency of 0.98, but is saturated. However, it is desirable to maintain at least 20 °C of superheat in the steam mains. Whilst it is possible to add a superheater powered by electricity to the electric steam boiler outlet, Figure 4.26 shows an arrangement in which part of the superheated steam from the waste gas boiler is used to provide the superheat to the high-pressure steam. To maintain the high-pressure main at 40 bar, the waste gas boiler has been upgraded to increase its output steam pressure slightly to 42 bar at 350 °C. Figure 4.26 shows part of the high-pressure steam from the waste

gas boiler mixed with the electric boiler outlet steam under temperature control to increase the temperature of the high-pressure steam to 270 °C. Figure 4.26 shows that the extraction turbine has been decommissioned, but the condensing turbine is retained. Medium- and low-pressure steam is created by letting down high-pressure steam through the letdown stations.

a) Calculate the flowrate of steam produced by the waste gas steam boiler in Figure 4.26.
b) From the enthalpy of the steam in the high-pressure steam, calculate the flowrates of high, medium, and low-pressure steam required to satisfy the respective steam heating duties. Assume that the steam heating is supplied from the enthalpy of the superheated steam to the enthalpy of saturated condensate at the steam pressure of the heating.
c) From an energy balance around the mixing junction of the waste gas steam and the outlet steam from the electric boiler, calculate the flowrates of steam that are required from the waste gas and electric boilers to provide the steam heating on the site.
d) Assuming that the balance of steam from the waste gas boiler not being fed to the mixer at the outlet of the electric boiler is fed to the condensing turbine, calculate the power generation from the condensing steam turbine.
e) Calculate the power required from the electric boiler and the net power required accounting for the power generation in the condensing steam turbine.

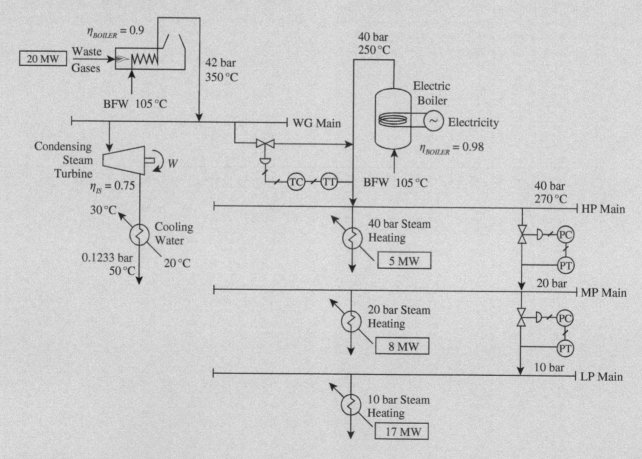

**Figure 4.26**

A modified steam system with an electric boiler to reduce the greenhouse gas emissions.

**f)** Instead of using the waste gas steam to superheat the high-pressure steam in Figure 4.26, an electric superheater could have been used with an efficiency of 0.98, allowing greater power generation from the condensing steam turbine. Calculate the net power requirement if an electric superheater is used, assuming the waste gas boiler retains its operating pressure of 42 bar and 350 °C.

**g)** Returning to the case in Figure 4.26 when high-pressure steam is superheated by the waste gas steam, if the low-pressure steam is desuperheated to within 5 °C of saturation, prior to being used for low-pressure steam heating using boiler feed-water at 105 °C, calculate the steam generation in the electric boiler and the resulting net power use. Give a brief explanation of the consequences of the calculation result.

Steam properties can be taken from International Steam Tables (Kretzschmar and Wagner 2019).

## Notation

$H_{BFW}$ = specific enthalpy of the boiler feedwater (kJ kg$^{-1}$)

$H_{DESUP}$ = specific enthalpy of the desuperheated steam (kJ kg$^{-1}$)

$H_f$ = specific enthalpy of saturated liquid (kJ kg$^{-1}$)

$H_{fg}$ = specific enthalpy of vaporisation (kJ kg$^{-1}$)

$H_g$ = specific enthalpy of saturated steam (kJ kg$^{-1}$)

$H_{IS}$ = specific enthalpy of steam from the steam turbine at the outlet pressure having the same entropy as the inlet steam (kJ kg$^{-1}$)

$H_{SH}$ = specific enthalpy of superheated steam (kJ kg$^{-1}$)

$H_{ST,in}$ = specific enthalpy of steam turbine inlet (kJ kg$^{-1}$)

$H_{ST,out}$ = specific enthalpy of steam turbine outlet (kJ kg$^{-1}$)

$H_{WET}$ = specific enthalpy of wet steam (kJ kg$^{-1}$)

$m_{BFW}$ = mass flowrate of boiler feedwater (kg s$^{-1}$)

$m_{DESUP}$ = mass flowrate of desuperheated steam (kg s$^{-1}$)

$m_{EB}$ = mass flowrate of steam from the electric boiler (kg s$^{-1}$)

$m_{STEAM}$ = mass flowrate of steam (kg s$^{-1}$)

$m_{ST}$ = mass flowrate through steam turbine (kg s$^{-1}$)

$m_{SUP}$ = mass flowrate of superheated steam (kg s$^{-1}$)

$m_{WG}$ = mass flowrate of steam from the waste gas boiler (kg s$^{-1}$)

$S_f$ = specific entropy of saturated water (kJ kg$^{-1}$ K$^{-1}$)

$S_{fg}$ = specific entropy of vaporization (kJ kg$^{-1}$ K$^{-1}$)

$S_g$ = specific entropy of saturated steam (kJ kg$^{-1}$ K$^{-1}$)

$S_{in}$ = specific entropy of steam at the inlet of the steam turbine (kJ kg$^{-1}$ K$^{-1}$)

$S_{out}$ = specific entropy of steam at the outlet of the steam turbine (kJ kg$^{-1}$ K$^{-1}$)

$S_{WET}$ = specific entropy of wet steam (kJ kg$^{-1}$ K$^{-1}$)

$T$ = temperature ( C)

$X$ = dryness fraction (−)

$$= \frac{\text{Mass of dry steam}}{\text{Mass of dry steam + Mass of saturated water}}$$

$\eta_{IS}$ = steam turbine isentropic efficiency (−)

## Subscripts

$EB$ = electric boiler

$f$ = property of saturated liquid

$g$ = property of saturated vapor

$fg$ = refers to a change of phase at constant pressure

$HP$ = high pressure

$MP$ = medium pressure

$LP$ = low pressure

$s$ = saturation state

$WG$ = waste gas generated steam

## Solution

**a)** Enthalpy of steam from waste gas boiler from steam tables:

$$= 3089 \text{ kJ kg}^{-1}$$

Enthalpy of boiler feedwater at 105 °C from steam tables:

$$= 440 \text{ kJ kg}^{-1}$$

Steam generation from waste gas boiler

$$m_{WG} = \frac{20 \times 10^3 \times 0.9}{(3089 - 440)}$$

$$= 6.795 \text{ kg s}^{-1}$$

**b)** Enthalpy of high-pressure steam main at 40 bar and 270 °C from steam tables:

$$= 2871 \text{ kJ kg}^{-1}$$

Enthalpy of saturated condensate at 40 bar from steam tables:

$$= 1087 \text{ kJ kg}^{-1}$$

Flowrate of high-pressure steam heating:

$$= \frac{5.0 \times 10^3}{(2871 - 1087)}$$

$$= 2.803 \text{ kg s}^{-1}$$

**4**

Enthalpy of medium pressure steam at 20 bar, given an isenthalpic expansion from high pressure:

$$= 2871 \text{ kJ kg}^{-1}$$

Enthalpy of saturated condensate at 20 bar from steam tables:

$$= 909 \text{ kJ kg}^{-1}$$

Flowrate of medium pressure steam heating:

$$= \frac{8.0 \times 10^3}{(2871 - 909)}$$

$$= 4.077 \text{ kg s}^{-1}$$

Enthalpy of low-pressure steam at 10 bar, given an isenthalpic expansion from medium pressure:

$$= 2871 \text{ kJ kg}^{-1}$$

Enthalpy of saturated condensate at 10 bar from steam tables:

$$= 763 \text{ kJ kg}^{-1}$$

Flowrate of low-pressure steam heating:

$$= \frac{17.0 \times 10^3}{(2871 - 763)}$$

$$= 8.065 \text{ kg s}^{-1}$$

c) From steam tables, the enthalpy of saturated steam for the electric boiler at 40 bar is 2801 kJ kg$^{-1}$. An enthalpy balance around the superheating mixing junction from mixing $m_{WG,HP}$ steam from the waste gas boiler is given by:

$$m_{WG,HP} H_{WG} + m_{EB} H_{EB} = (m_{WG,HP} + m_{EB}) H_{HP}$$

$$\frac{m_{WG,HP}}{m_{EB}} = \frac{H_{HP} - H_{EB}}{H_{WG} - H_{HP}}$$

$$= \frac{2871 - 2801}{3089 - 2871}$$

$$= 0.3211$$

Total flowrate of heating steam:

$$= m_{EB} + m_{WG,HP}$$

$$= m_{EB} + 0.3211 m_{EB}$$

$$= 1.3211 m_{EB}$$

From Part (b) above, total flowrate of heating steam: Thus, combining these expressions:

$$m_{EB} = \frac{14.945}{1.3211}$$

$$= 11.313 \text{ kg s}^{-1}$$

$$m_{WG,HP} = 14.945 - 11.313$$

$$= 3.632 \text{ kg s}^{-1}$$

Thus, the electric boiler generates 11.313 kg s$^{-1}$ 40 bar steam at 250 °C and this is mixed with 3.632 kg s$^{-1}$ of 40 bar steam at 350 °C from the waste gas boiler to generate 14.945 kg s$^{-1}$ of 40 bar steam 270 °C.

d) Flowrate of waste gas steam to the condensing steam turbine:

$$m_{ST} = 6.795 - 3.632$$

$$= 3.163 \text{ kg s}^{-1}$$

Inlet entropy to steam turbine at 42 bar and 350 °C from steam tables:

$$S_{in} = 6.556 \text{ kJ kg}^{-1} \text{ K}^{-1}$$

$$S_{out} = 6.556 \text{ kJ kg}^{-1} \text{ K}^{-1}$$

Checking steam tables, this entropy is below steam saturation, which means that an isentropic expansion would lead to wet steam at the outlet. To determine the enthalpy corresponding with this entropy at 0.1233 bar, the dryness fraction must first be determined. The entropy of wet steam is given by:

$$S_{WET} = X S_g + (1 - X) S_f$$

$$= S_f + X S_{fg}$$

Taking the values of $S_f$ and $S_{fg}$ from steam tables at saturation conditions at 0.1233 bar:

$$S_{WET} = S_f + X S_{fg}$$

$$6.556 = 0.703 + X \times 7.373$$

Solving gives $X = 0.7938$. Then the enthalpy for wet steam is given by:

$$H_{WET} = X H_g + (1 - X) H_f$$

$$= H_f + X H_{fg}$$

Substituting $X = 0.7938$ and the values of $H_f$ and $H_{fg}$ from steam tables at saturation conditions at 0.1233 bar:

$$H_{WET} = H_f + X H_{fg}$$

$$= 209 + 0.7938 \times 2382$$

$$= 2100 \text{ kJ kg}^{-1}$$

This is the isentropic outlet enthalpy. From Eq. (4.5):

$$H_{ST,out} = H_{ST,in} - \eta_{IS}(H_{ST,in} - H_{IS})$$

$$= 3089 - 0.75(3089 - 2100)$$

$$= 2347 \text{ kJ kg}^{-1}$$

Check the outlet condition of the steam for wetness:

$$H_{WET} = H_f + X H_{fg}$$

$$2347 = 209 + X \times 2382$$

$$X = 0.8976$$

This level of wetness might be too high for the steam turbine, as wetness can be damaging to the machine. For fixed inlet conditions, the dryness at the turbine outlet can be increased by increasing the steam condensing temperature and pressure. However, this decreases the power generation correspondingly. In this case, the wetness is considered acceptable.

Shaft power generation by enthalpy difference across the turbine:

$$= m_{STEAM}(H_{ST,in} - H_{ST,out})$$
$$= (6.795 - 3.632)(3089 - 2347)$$
$$= 2347 \text{ kW}$$

e) Power to electric boiler:

$$= \frac{m_{EB}(H_{EB} - H_{BFW})}{0.98}$$
$$= \frac{11.313(2801 - 440)}{0.98}$$
$$= 27\,255 \text{ kW}$$

Net power consumption:

$$= 27\,255 - 2347$$
$$= 24\,908 \text{ kW}$$

f) Instead of using waste heat boiler steam to superheat the electric boiler steam, use an electric superheater to increase the outlet temperature of the steam from the electric boiler from 250 °C to 270 °C with 98% efficiency.

Power for electric boiler plus superheater:

$$= \frac{14.945(2871 - 440)}{0.98}$$
$$= 37\,073 \text{ kW}$$

Power from condensing steam turbine:

$$= 6.795(3089 - 2347)$$
$$= 5042 \text{ kW}$$

Net power demand:

$$= 37\,074 - 5042$$
$$= 32\,032 \text{ kW}$$

Thus, the net electric power demand increases significantly if an electric superheater is used, rather than mixing with the superheated steam from the waste gas steam boiler. The use of steam from the waste gas boiler provides heat that would otherwise be supplied by the electric superheater.

g) Now the low-pressure steam is desuperheated to within 5 °C of saturation, prior to being used for low-pressure steam heating, using boiler feedwater at 105 °C. Calculate the effect of desuperheating the LP steam before use for heating.

Temperature of the steam to the low pressure heaters (saturation plus 5 °C):

$$= 179.9 + 5 = 184.9 \, ^\circ\text{C}$$

Enthalpy of steam of desuperheated steam at this temperature and 10 bar from steam tables:

$$H_{DESUP} = 2790 \text{ kJ kg}^{-1}$$

Heat available in desuperheated steam from inlet temperature to saturated condensate at 10 bar from steam tables:

$$= 2790 - 763 = 2027 \text{ kJ kg}^{-1}$$

Flowrate of desuperheated steam for 17 MW heat duty:

$$= \frac{17\,000}{2027} = 8.387 \text{ kg s}^{-1}$$

Enthalpy of boiler feedwater at 105 °C

$$= 440 \text{ kJ kg}^{-1}$$

Calculate the flowrate of boiler feedwater injected from an energy balance around boiler feedwater injection:

$$m_{SH}H_{SH} + m_{BFW}H_{BFW} = m_{DESUP}H_{DESUP}$$
$$(8.387 - m_{BFW})2871 + m_{BFW} \times 440 = 8.387 \times 2790$$
$$m_{BFW} = \frac{24\,079 - 23\,400}{2871 - 440}$$
$$= 0.279 \text{ kg s}^{-1}$$

Flowrate of superheated steam required from the LP steam main:

$$m_{SUP} = 8.387 - 0.279$$
$$= 8.108 \text{ kg s}^{-1}$$

This compares with a flowrate 8.065 kg s$^{-1}$ if the steam is not desuperheated:

$$m_{EB} = \frac{2.803 + 4.077 + 8.108}{1.3211}$$
$$= 11.345 \text{ kg s}^{-1}$$

Power for electric boiler:

$$= \frac{m_{EB}(H_{EB} - H_{BFW})}{0.98}$$
$$= \frac{11.345(2801 - 440)}{0.98}$$
$$= 27\,332 \text{ kW}$$

**4**

Power from condensing steam turbine:

$$= (6.795 - 0.3211 \times 11.345)(3089 - 2347)$$

$$= 2339 \text{ kW}$$

Net power demand:

$$= 27\,332 - 2339$$

$$= 24\,993 \text{ kW}$$

Desuperheating the low-pressure steam decreases its enthalpy for heating, increasing the flowrate required for a fixed heat duty. The production of low-pressure steam from the electric boiler increases slightly compared with the non-desuperheating case. Fundamentally, the energy in the low-pressure steam is degraded by mixing with boiler feeedwater for the sake of improved heat transfer in the heaters. The result is that the net power demand increases slightly.

**Example 4.4** Instead of the configuration in Figure 4.26, the back-pressure steam turbine could have been retained in the design and the condensing turbine decommissioned instead, as shown in Figure 4.27. Now there will be mixing of steam with two different enthalpies in the LP main. Thus, the flowrate of the LP heating steam changes relative to the case in Figure 4.26, as the enthalpy of the LP steam changes. The mixing is between the letdown from MP to LP mains and the exhaust steam from the back-pressure turbine. Now the mass flowrate of steam from the electric boiler cannot be calculated directly, as the enthalpy of the LP steam is unknown.

a) Calculate the enthalpy of the exhaust steam from the steam turbine.
b) From an enthalpy balance around the LP steam main, calculate the steam flowrate from the electric boiler.

c) Calculate the flowrate of steam to the steam turbine, the letdown flowrate from the MP main to the LP main, and the resulting enthalpy of the LP main assuming perfect mixing in the steam main. In practice, especially for large complex steam piping systems, the mixing will be far from perfect and vary significantly with location in the same header. However, there is insufficient information to account for this and perfect mixing can be assumed.
d) Calculate the power to the electric boiler, the shaft power generated by the steam turbine and the net power demand.

Steam properties can be taken from International Steam Tables (Kretzschmar and Wagner 2019).

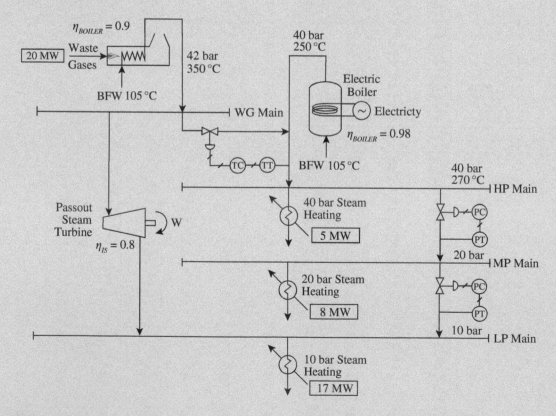

**Figure 4.27**

An alternative modified steam system with an electric boiler to reduce the greenhouse gas emissions.

## Solution

Retaining the backpressure turbine instead of the condensing turbine, the flowrates of the HP and MP steam are the same. However, the flowrate of the LP steam changes as the enthalpy of the LP steam changes. This results from the mixing of the letdown from MP to LP with the exhaust steam from the back-pressure turbine. Now the mass flowrate of steam from the electric boiler cannot be calculated directly, as the enthalpy of the LP steam is unknown.

a) First calculate the enthalpy of the exhaust steam from the steam turbine.

Inlet entropy to steam turbine at 42 bar and 350 °C from steam tables:

$$S_{in} = 6.556 \text{ kJ kg}^{-1} \text{ K}^{-1}$$

For an isentropic expansion:

$$S_{out} = 6.556 \text{ kJ kg}^{-1} \text{ K}^{-1}$$

Checking steam tables, this entropy is below steam saturation, which means that an isentropic expansion would lead to wet steam at the outlet. To determine the enthalpy corresponding with this entropy at 10 bar, the dryness fraction must first be determined. Taking the values of $S_f$ and $S_{fg}$ from steam tables at saturation conditions at 10 bar:

$$S_{WET} = S_f + X S_{fg}$$

$$6.556 = 2.138 + X \times 4.447$$

Solving gives $X = 0.9935$. Together with the values of $H_f$ and $H_{fg}$ from steam tables at saturation conditions at 10 bar, the enthalpy is given by:

$$H_{WET} = H_f + X H_{fg}$$

$$= 763 + 0.9935 \times 2014$$

$$= 2764 \text{ kJ kg}^{-1}$$

This is the isentropic outlet enthalpy. From Eq. (4.5):

$$H_{ST,out} = H_{ST,in} - \eta_{IS}(H_{ST,in} - H_{IS})$$

$$= 3089 - 0.8(3089 - 2764)$$

$$= 2829 \text{ kJ kg}^{-1}$$

This is above the saturation enthalpy of steam at 10 bar.

b) Given that all of the steam generated by both boilers in Figure 4.27 must ultimately flow to the LP main, minus the HP and MP heating steam use, then:

$$m_{LP} = m_{WG} + m_{EB} - m_{HP} - m_{MP}$$

An energy balance around the LP main gives:

$$m_{ST} H_{ST,out} + m_{HP,LD} H_{MP} = Q_{LP} + m_{LP} H_{COND,LP}$$

From Example 4.3:

$$m_{ST} = m_{WG} - 0.3211 m_{EB}$$

$$m_{MP,LD} = 1.3211 m_{EB} - m_{HP} - m_{MP}$$

Substituting in the LP energy balance gives:

$$(m_{WG} - 0.3211 m_{EB}) H_{ST,out} + (1.3211 m_{EB} - m_{HP} - m_{MP}) H_{MP}$$

$$= Q_{LP} + (m_{WG} + m_{EB} - m_{HP} - m_{MP}) H_{COND,LP}$$

Rearranging:

$$m_{EB} = \frac{Q_{LP} + m_{WG}(H_{COND,LP} - H_{ST,out}) + (m_{HP} + m_{MP})(H_{MP} - H_{COND,LP})}{1.3211 H_{MP} - 0.3211 H_{ST,out} - H_{COND,LP}}$$

$$= \frac{17 \times 10^3 + 6.795(763 - 2829) + (2.803 + 4.077)(2871 - 763)}{1.3211 \times 2871 - 0.3211 \times 2829 - 763}$$

$$= 8.232 \text{ kg s}^{-1}$$

c) Next calculate the steam flowrate to the steam turbine:

$$m_{ST} = m_{WG} - 0.3211 m_{EB}$$

$$= 6.795 - 0.3211 \times 8.232$$

$$= 4.152 \text{ kg s}^{-1}$$

$$m_{LP} = m_{WG} + m_{EB} - m_{HP} - m_{MP}$$

$$= 6.795 + 8.232 - 2.803 - 4.077$$

$$= 8.147 \text{ kg s}^{-1}$$

$$m_{MP,LD} = 8.147 - 4.152$$

$$= 3.995 \text{ kg s}^{-1}$$

Assuming perfect mixing of the steam turbine outlet and the MP letdown in the LP main:

$$H_{LP} = \frac{m_{MP,LD} \times H_{MP,LD} + m_{ST} \times H_{ST,out}}{m_{MP,LD} + m_{ST}}$$

$$= \frac{3.995 \times 2865 + 4.152 \times 2829}{3.995 + 4.152}$$

$$= 2847 \text{ kJ kg}^{-1}$$

Check the heating duty of the LP steam:

$$Q_{LP} = 8.147(2847 - 763)$$

$$= 17 \times 10^3 \text{ kW within rounding error}$$

d) Power to the electric boiler:

$$= \frac{m_{EB}(H_{EB} - H_{BFW})}{0.98}$$

$$= \frac{8.232(2801 - 440)}{0.98}$$

$$= 19\,436 \text{ kW}$$

Shaft power from steam turbine:

$$= m_{ST}(H_{ST,in} - H_{ST,out})$$

$$= 4.152(3089 - 2829)$$

$$= 1080 \text{ kW}$$

Net power demand:

$$= 19\,436 - 1080$$

$$= 18\,356 \text{ kW}$$

Thus, the net power demand for the scheme that uses combined heat and power from an extraction turbine is significantly lower than the one that uses a condensing turbine. When using a condensing turbine, the exhaust heat from the turbine is lost to cooling water and the environment. The use of a back-pressure turbine, rather than a condensing turbine, eliminates the loss of heat to cooling water.

## 4.9 Steam Heating Limits

When choosing a utility for heating, the first choice is generally steam. As pointed out earlier, low-pressure steam should be used in preference to high-pressure steam. High-pressure steam can be used for power generation and is more valuable. The choice of which level to use is dictated by the pressures of the steam mains available on the site. Generally, the choice of which level to use is dictated by the saturation temperature of the steam, rather than its actual temperature in the steam main. This is because it is preferable to use the latent heat for heating, rather than the superheat. In steam heaters it is usual practice to utilize only the superheat in the steam and the latent heat. Thus, a steam main with a saturation temperature of typically 10 °C above the process heating duty would be appropriate to allow a reasonable temperature difference for heat transfer in process heaters.

The maximum temperature that can be satisfied by steam will be dictated by the highest pressure steam main on the site. The highest pressure steam distributed around sites is normally around 40 bar. Given that the saturation temperature of 40 bar steam is 250 °C, this presents an approximate limit on the maximum temperature that can normally be achieved by steam heating. The precise maximum temperature depends on the pressure of the highest pressure steam main, which might be slightly higher than 40 bar. However, the limit for steam heating should normally be set at around 250 °C, depending on the actual conditions of the steam mains on the site.

## 4.10 Fired Heaters

In some situations, process heat needs to be supplied:

a) at a high temperature (e.g. at a temperature above which heat can be supplied by steam);

b) with a high heat flux (e.g. to supply heat of reaction in situations where a short residence time in the reactor is required);

c) with a very high heat duty that would otherwise be difficult to design as a steam heater (e.g. a very large reboiler).

In such cases, radiant heat transfer can be used from the combustion of fuel in a *fired heater* or *furnace*. The fuel might be gaseous (e.g. natural gas, biogas, renewable hydrogen, process waste gas), or liquid (e.g. fuel oil). Sometimes the function is to purely provide heat; sometimes the fired heater is also a reactor and provides heat of reaction.

Fired heater designs vary according to the function, heating duty, type of fuel, and the method of introducing combustion air. Figure 4.28 shows some typical fired heater designs. Combustion of fuel takes place in the *radiant section* or *firebox*, which is refractory lined. Most often, the walls are lined with one or two rows of tubes mounted horizontally or vertically through which the fluid that needs to be heated passes. Heat transfer in the radiant section is mainly by radiation, with a small contribution by convection. After the flue gas leaves the radiant section, most furnace designs extract further heat from the flue gas in horizontal banks of tubes in a *convection section*, before the flue gas is vented to the atmosphere through the *stack*. Heat transfer in the convection section is by both radiation and convection, but convection effects dominate.

The simplest design of fired heater is shown in Figure 4.28a featuring a cylindrical design with the tubes mounted vertically in a circle around the inner walls of the combustion chamber. The design in Figure 4.28a does not feature a convection section and would be an inefficient design, appropriate only for small duties. Figure 4.28b shows the corresponding cylindrical design but with a convection section. The convection section consists of a chamber with a square or rectangular cross-section located on top of the cylindrical radiant section. Cylindrical fired heater designs are used for duties less than 60 MW. Larger duties are suited to *box* designs. Such fired heaters have a rectangular cross-section. The design in Figure 4.28c shows a box furnace with horizontal tubes in the combustion chamber. The convection section consists of a chamber with a rectangular cross-section located on top of the cylindrical radiant section. Figure 4.28d shows a box furnace with dual firing. In this case, burners are located on either side of the radiation tubes. Dual firing allows a higher and more uniform heat flux and is therefore suited to the requirements of endothermic reactions. In this case, burners could be in principle mounted on the floor or the walls of the radiant section. Figure 4.28e shows a box furnace with dual firing, but with a refractory *bridgewall* dividing the combustion chamber into two parts. Such a design allows some independence of the heat transfer arrangements on either side of the bridgewall. A more extreme version is shown in Figure 4.28f, which features a box furnace with two *cells*. Rather than the horizontal tube arrangement in Figure 4.28f, the tubes can be mounted vertically in the center of the cells with burners mounted on the

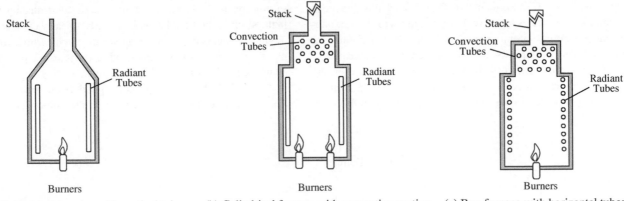

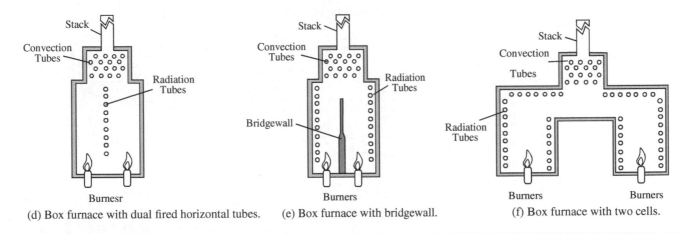

(a) Cylindrical furnace with vertical tubes.   (b) Cylindrical furnace with convection section.   (c) Box furnace with horizontal tubes.

(d) Box furnace with dual fired horizontal tubes.   (e) Box furnace with bridgewall.   (f) Box furnace with two cells.

**Figure 4.28**

Different furnace configurations. Source: Adapted from (Smith 2016), *Chemical Process Design and Integration*, 2nd Edition, John Wiley & Sons Ltd.

walls in a dual fired arrangement. Such arrangements are often used to carry out endothermic reactions within the tubes.

Flow of combustion air to the fired heater is created in one of three ways:

1) *Natural draft.* In natural draft designs the air flows into the fired heater as a result of the density difference between the hot gases in the stack and that of the surrounding air. Thus, the height of the stack provides both the driving force for air flow and the mechanism to disperse the waste combustion gases

2) *Forced draft.* In forced draft designs a fan is used to blow air into the fired heater. The fired heater is thus at a pressure slightly above atmospheric. The height of the stack is dictated purely by the dispersion requirements of the waste combustion gases.

3) *Induced draft.* In induced draft designs a fan is located in the duct for the waste combustion gases after the fired heater, before entering the stack. This creates a pressure in the fired heater slightly below atmospheric. As with forced draft arrangements, the height of the stack is dictated purely by the dispersion requirements of the waste combustion gases.

For large fired heaters requiring a significant amount of equipment in the flue to treat the exhaust gases in order to remove oxides of sulfur and/or oxides of nitrogen, as well as equipment for

heat recovery, a combination of forced and induced draft might need to be used in a *balanced draft* design.

Figure 4.29 shows a typical fired heater configuration for a process heating duty. The feed is split into parallel flows, known as *passes*. The example in Figure 4.29 features four passes. Many different pass arrangements are possible (Nelson 1958; Martin 1998). Figure 4.29 simply shows one possible example. The feed is first preheated in the *convection section* where the tubes usually have extended surfaces to increase the rate of heat transfer from the flue gas. The feed then passes to the *shield section*, which comprises plain tubes, known as *shock tubes* or *shield tubes*. These tubes need to be robust enough to be able to withstand high temperatures and receive significant radiant heat from the radiant section. The tubes in the convection section and shield section are horizontal and there are typically between 4 and 12 tubes in each row. After the shield section, the fluid passes to the tubes of the *radiant section* where the heat transfer is completed. The radiant tubes in Figure 4.29 are horizontal and feature a single row. For process heating duties the tubes are formed into loops, as shown in Figure 4.29. A *header box* encloses the ends of the loops. This is an insulated compartment separated from the flue gas. Thermal expansion of the furnace tubes is facilitated using pipe sleeves in the furnace wall to guide expansion. In other designs the tubes are oriented vertically. Vertically mounted tubes are supported from the roof of the furnace by heat-resistant metal alloy brackets. The bottoms of the tubes are provided with guides to allow expansion, but preventing lateral movement. For either horizontal or vertical orientation, two tube rows on a triangular pitch can also be used. After combustion of the fuel in the radiant section, the hot flue gases pass through the shield and convective sections before being collected in *breeching* for release through the stack

to atmosphere. The flowrate of the gases is controlled by a *damper* in the stack. If the fired heater is required to carry out a reaction, it might be necessary to have an arrangement with many more passes than used for process heating only, using parallel flows connected to inlet and outlet manifolds.

Figure 4.30 shows a simple conceptual model for a fired heater. The flue gas (combustion gases) from the combustion zone are

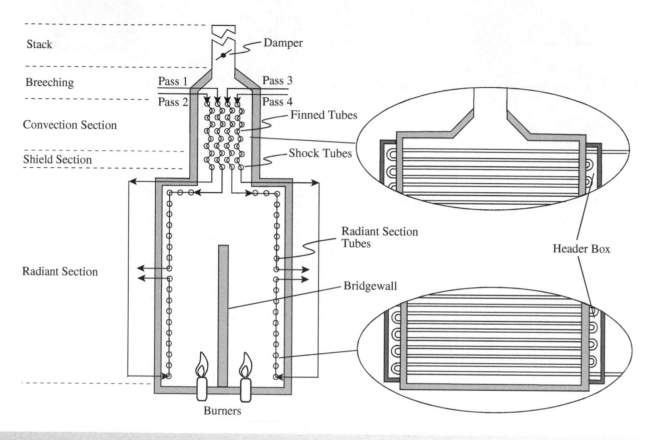

**Figure 4.29**

A typical furnace arrangement. Source: Adapted from (Smith 2016), *Chemical Process Design and Integration*, 2nd Edition, John Wiley & Sons Ltd.

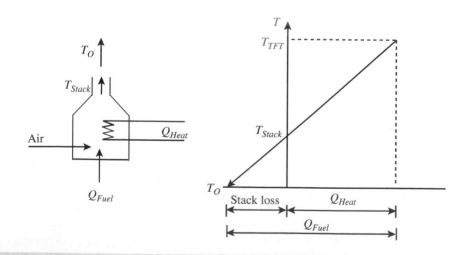

**Figure 4.30**

A simple model of furnace efficiency. Source: Adapted from (Smith 2016), *Chemical Process Design and Integration*, 2nd Edition, John Wiley & Sons Ltd.

modeled as a straight line from the theoretical flame temperature or adiabatic combustion temperature (Smith 2016) through the radiant zone and possibly convection zone to the stack temperature, where the gases are discharged to the atmosphere. The flue gas ultimately ends up at the ambient temperature. The heat released from the flue gas between the stack temperature and ambient temperature is the stack loss. The heat released from fuel is the sum of the process heat $Q_{Heat}$ and the stack loss (Smith 2016). The furnace efficiency can be defined as:

$$\text{Furnace efficiency} = \frac{\text{Heat to the process}}{\text{Heat released from combustion}} \quad (4.7)$$

In addition to using the convection bank to extract heat from the flue gas, another technique that can be used to extract additional heat is air preheating. This is illustrated in Figure 4.31. Figure 4.32 shows the conceptual model of the furnace with the air preheat. Preheating the air adds heat to the combustion process, increasing the theoretical flame temperature (Smith 2016). If the

**4**

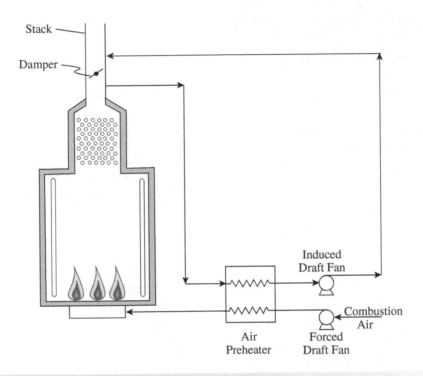

### Figure 4.31

Fired heater with air preheat. Source: Adapted from (Smith 2016), *Chemical Process Design and Integration*, 2nd Edition, John Wiley & Sons Ltd.

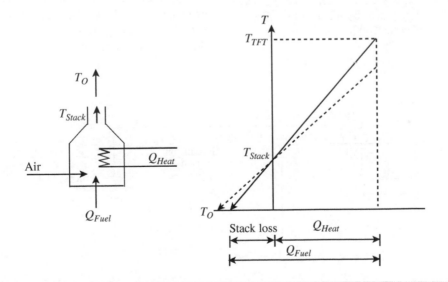

### Figure 4.32

Air preheat increases the theoretical flame temperature and furnace efficiency for the same stack temperature. Source: Adapted from (Smith 2016), *Chemical Process Design and Integration*, 2nd Edition, John Wiley & Sons Ltd.

stack temperature is maintained to be the same as the case without air preheat, then the increase in the theoretical flame temperature allows a steeper line to be drawn for the flue gas, reducing the stack loss and increasing the furnace efficiency. Higher flame temperatures increase the formation of oxides of nitrogen, which are environmentally harmful.

**Example 4.5**   A process heat duty of 20 MW requires heating from 280 °C to 360 °C. This is to be carried out in a furnace. Heat is to be transferred with a temperature difference of 20 °C. Ambient temperature is 10 °C.

a) For an assumed theoretical flame temperature of 1800 °C, calculate the furnace efficiency.
b) It is required to increase the furnace efficiency of the furnace by preheating the combustion air. It can be assumed that the combustion air and the furnace flue gas have the same heat capacity flowrate. If the combustion is preheated from 10 °C to 210 °C before the combustion, and the stack temperature is maintained to be the same as in Part (a), calculate the furnace efficiency.

**Solution**

a) Calculate the heat capacity flowrate of the flue gas:

$$CP_{FG} = \frac{15\,000}{1800 - 300}$$

$$= 10.0 \text{ kW K}^{-1}$$

$$Q_{FUEL} = 10 \times (1800 - 10)$$

$$= 17\,900 \text{ kW}$$

$$\eta = \frac{15\,000}{17\,900}$$

$$= 0.838$$

b) If the combustion air and the furnace flue gas have the same heat capacity flowrate, then an increase in temperature of the combustion air of 200 °C will increase the theoretical flame temperature by 200 °C to reach 2000 °C:

$$CP_{FG} = \frac{15\,000}{2000 - 300}$$

$$= 8.824 \text{ kW K}^{-1}$$

$$Q_{FUEL} = 8.824 \times (2000 - 10)$$

$$= 17\,559.8 \text{ kW}$$

$$\eta = \frac{15\,000}{17\,559.8}$$

$$= 0.854$$

## 4.11   Other Heat Carriers

Whilst steam is the first choice for the supply of heating from a hot utility, the upper temperature limit for steam use requires either a furnace or an alternative heat carrier to supply heat. Sometimes multiple process duties require heat at a temperature above that which can be supplied by steam. Sometimes layout issues prevent a fired heater being used directly for a process duty. In such cases an alternative heat carrier can be used.

Figure 4.33 shows a hot oil circuit used to supply heat above the steam temperature. In this arrangement a closed loop of hot oil supplies heat to a process duty (or duties) indirectly from a fired heater. A number of different heat transfer oils are available commercially (e.g. a mixture of diphenyl and diphenyl oxide). Mineral oils derived from crude oil have a maximum recommended temperature of between 270 °C and 315 °C. Synthetic thermal fluids have maximum fluid temperatures of between 315 °C and 400 °C. Rather than use a furnace to supply heat to the hot oil circuit, electric resistance heating can be used from renewable electricity.

If a higher temperature is required than can be supplied by hot oil, then a molten salt can be used. The major disadvantage related to the use of molten salt is the melting point being at a relatively high temperature. Lowering of the molten salt melting point can be brought about by using a eutectic mixture of different salts. Molten salts can operate up to a maximum temperature when

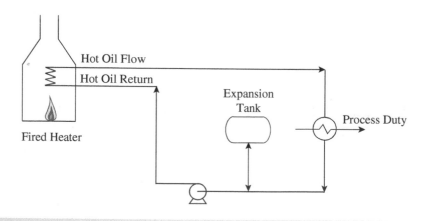

**Figure 4.33**
Hot oil circuit.

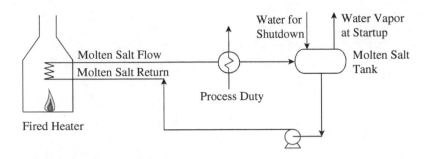

## Figure 4.34

Molten salt circuit.

the salt starts to decompose. A commonly used molten salt is a eutectic mixture of potassium nitrate and sodium nitrate. This has a melting point of 222 °C and can be used up to a maximum temperature of around 550 °C. The individual salts have melting points above 222 °C. Adding sodium nitrite to form a eutectic mixture of potassium nitrate, sodium nitrate, and sodium nitrite lowers the melting point to 142 °C and can be used up to a maximum temperature of around 450 °C. A mixture of potassium nitrate, sodium nitrate, and lithium nitrate lowers the melting point to 120 °C and can be used up to a maximum temperature of around 500 °C. Mixtures of chloride, fluoride, and carbonate salts can be used at maximum temperatures in excess of 700 °C. A fundamental problem with the use of molten salt as a heat carrier is start-up and shut-down. Lowering the temperature of the molten salt for shut-down (e.g. for maintenance) will cause the molten salt to solidify and potentially block equipment and pipes. One approach for simple configurations is to use a holding tank for the molten salt and design the system in such a way that the molten salt self-drains into the tank for shut-down. The holding tank can be supplied with heating (e.g. electric heating) to maintain the salt molten in the tank, or for complete shut-down it can be allowed to solidify and then to be re-melted in the tank for start-up. An alternative for shut-down of more complex configurations is to mix the molten salt with water as it cools to keep the system fluid. For start-up, the water can then be evaporated by the addition of heat. Figure 4.34 shows a molten salt circuit for process heating using a fired heater that uses water injection. A holding tank in the circuit provides a buffer for the molten salt. If the system needs to be shut down, then water can be added as the temperature is decreased below the melting point to prevent the salt solidifying. When the system is started up again the water is first boiled off using heat from the fired heater and the salt re-melted. Rather than use a fired heater to supply heat to the molten salt circuit, electric resistance heating can be used from renewable electricity.

**Example 4.6** A conceptual design is required for the local utility system of a process for the manufacture of phthalic anhydride. Figure 4.35 shows the separation section of the process. Crude phthalic anhydride is produced in the reaction section of the process and fed to the separation section shown in Figure 4.35. The crude phthalic anhydride feed to the separation section is heated and held in a holding tank to allow the residual reactions to go to completion. It is then separated into two distillation columns, both of which operate at high temperatures. The heating duties in the separation section are either at the temperature limit of what can be serviced by steam or are too high a temperature for steam heating. In Figure 4.35 the solution adopted to provide the heat has been to use a hot oil circuit with heat from a fired heater. The supply temperature of the hot oil from the furnace is 360 °C and the return temperature is 310 °C. The phthalic anhydride process is also serviced by a local steam system, shown in Figure 4.36. There are two local steam headers. The high-pressure steam header operates at a pressure of 20 bar and the low-pressure steam header at a pressure of 10 bar. The process has a recirculating cooling water system supplying cooling water at 20 °C and returns to the cooling tower at 30 °C. The condensers of the two distillation columns are at a high enough temperature to generate steam. Column 1 condenser can generate saturated steam at 20 bar from boiler feedwater fed at 90 °C. Column 2 condenser can generate saturated

steam at 10 bar, also from boiler feedwater fed at 90 °C. Heat transfer can be carried out with a minimum temperature difference of 10 °C, except in the case of furnace heat transfer, which requires a minimum of 20 °C.

The high-pressure steam header at 20 bar is fed by saturated steam generated from the condenser of Column 1. It is also fed with a supply of steam generated from the exothermic heat rejected by the phthalic anhydride reactor elsewhere in the process. The heat of reaction in the phthalic anhydride reactor is removed by a molten salt circuit, which in turn generates superheated steam at 20 bar and 270 °C. The boiler feedwater supplied to the steam generation from molten salt has been preheated by heat recovery elsewhere to 160 °C. In Figure 4.36 steam is expanded from the high-pressure header to the low-pressure header through a steam turbine to generate power. Steam is also expanded through a letdown station to control the pressure of the low-pressure header. In order to maintain control action and to keep the letdown station equipment hot, a minimum flowrate of $0.5 \text{ kg s}^{-1}$ is required through the letdown station.

a) Calculate the heat duty from fuel in kW and the furnace efficiency of the fired heater in Figure 4.35, assuming the theoretical flame temperature (adiabatic combustion temperature) of fuel combusted in the appropriate amount of excess air is

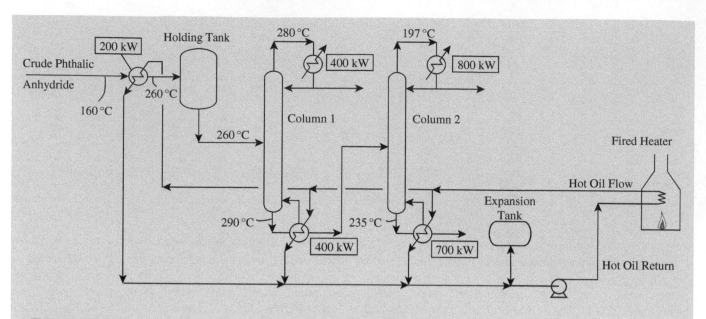

**Figure 4.35**

Phthalic anhydride separation system.

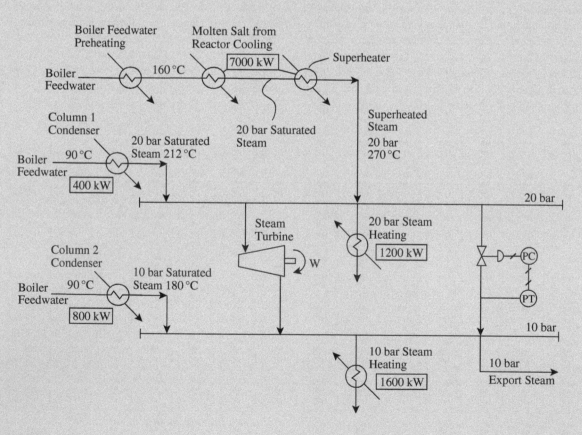

**Figure 4.36**

Phthalic anhydride steam system.

1800 °C. The resulting flue gas can be cooled to heat the hot oil with a minimum temperature difference of 20 °C. The ambient temperature is 10 °C. It can be assumed that the specific heat capacity of the flue gas $C_{P,FG} = 1.21$ kJ kg$^{-1}$ K$^{-1}$. Neglect heat losses from the casing of the furnace and the hot oil distribution system.

b) Concern over the furnace efficiency of the fired heater has led to the suggestion that the combustion air should be preheated from the stack gases. The preheated air can be heated to within 20 °C of the flue gas temperature as it exits the heating coil for the hot oil. Calculate the increase in the theoretical flame temperature, heat duty from fuel in kW, and furnace efficiency with the air preheat. Assume that the temperature of the flue gases exiting the heating coil for the hot oil is the same as in Part (a). As a first estimate assume that the mass flowrate and specific heat capacity of the combustion air and the flue gases are the same. Neglect heat losses from the casing of the furnace and the hot oil distribution system.

c) Repeat the calculation from Part (b) more accurately by no longer assuming that the flowrate and heat capacity of the flue gas and air are the same. The specific heat capacity of the flue gas can be assumed to be $C_{P,FG} = 1.21$ kJ kg$^{-1}$ K$^{-1}$ and the specific heat capacity for combustion air to be $C_{P,AIR} = 1.10$ kJ kg$^{-1}$ K$^{-1}$. For the combustion of fuel in the appropriate combustion air, the ratio of mass flowrate of flue gas $m_{FG}$ to mass flowrate of combustion air $m_{AIR}$ can be assumed to be $m_{FG}/m_{AIR} = 1.06$. Calculate the theoretical flame temperature, heat released from fuel, and furnace efficiency with preheat and compare with the approximate calculation in Part (b) above. Again, neglect heat losses from the casing of the furnace and the hot oil distribution system.

d) Calculate the mass flowrate of saturated steam generated from the condenser of Column 1 and condenser of Column 2 assuming that the boiler feedwater is fed at 90 °C.

e) Calculate the mass flowrate of superheated steam generated from the molten salt cooling for the reactor assuming that the preheated boiler feedwater is fed at 160 °C.

f) Calculate the enthalpy of the steam in the high-pressure header assuming perfect mixing in the header. In practice, especially for large complex steam piping systems, the mixing will be far from perfect and may vary significantly with location in the same header. However, there is insufficient information to account for this and perfect mixing can be assumed.

g) Calculate the flowrate of steam to the high-pressure steam heaters and the power generation in the steam turbine. It can be assumed that only the superheat and latent heat in the steam is extracted for process heating, that the steam turbine has an isentropic efficiency of 0.75, and that the flowrate through the letdown station is 0.5 kg s$^{-1}$.

h) Calculate the enthalpy of the steam in the low-pressure header, again assuming perfect mixing in the header.

i) Calculate the flowrate of steam required for the low-pressure steam heating, assuming that only the superheat and latent heat in the steam is extracted for process heating, and the resulting low-pressure steam export from the phthalic anhydride steam system shown in Figure 4.36.

Steam properties can be taken from International Steam Tables (Kretzschmar and Wagner 2019).

**Notation**

$C_P$ = specific heat capacity (kJ kg$^{-1}$ K$^{-1}$)

$H_{in}$ = specific enthalpy of the inlet steam to the steam turbine (kJ kg$^{-1}$)

$H_{IS}$ = specific enthalpy of steam from the steam turbine at the outlet pressure having the same entropy as the inlet steam (kJ kg$^{-1}$)

$H_{out}$ = specific enthalpy of the outlet steam from the steam turbine (kJ kg$^{-1}$)

$m$ = flowrate (kg s$^{-1}$)

$Q$ = heat duty (kJ s$^{-1}$)

$S$ = specific entropy (kJ kg$^{-1}$ K$^{-1}$)

$S_{in}$ = specific entropy of steam at the inlet of the steam turbine (kJ kg$^{-1}$ K$^{-1}$)

$S_{out}$ = specific entropy of steam at the outlet of the steam turbine (kJ kg$^{-1}$ K$^{-1}$)

$T$ = temperature (°C)

$\Delta T$ = Temperature difference (°C)

$W$ = power (kW)

$\eta$ = furnace efficiency (–)

$\eta_{IS}$ = steam turbine isentropic efficiency (–)

**Subscripts**

$AIR$ = air

$f$ = property of saturated liquid

$FG$ = flue gas

$g$ = property of saturated vapor

$fg$ = refers to a change of phase at constant pressure

$PROC$ = process

$STEAM$ = steam

$s$ = saturation state

$TFT$ = theoretical flame temperature

**Solution**

a) Figure 4.37 shows the simple model for the furnace. Process heating duty for the furnace from the three process duties on the hot oil circuit in Figure 4.35:

$$Q_{PROC} = 200 + 400 + 700$$

$$= 1300 \text{ kW}$$

4

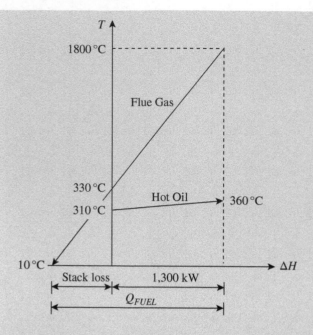

**Figure 4.37**

Furnace efficiency for hot oil circuit.

$$Q_{PROC} = m_{FG} C_{P,FG} \Delta T$$

$$1300 = m_{FG} \times 1.21 \times (1800 - 330)$$

$$m_{FG} = \frac{1300}{1.21 \times 1470}$$

$$= 0.7309 \ \text{kg s}^{-1}$$

Fuel heating duty:

$$Q_{PROC} = 0.7309 \times 1.21 \times (1800 - 10)$$

$$= 1583.1 \ \text{kW}$$

Furnace efficiency:

$$\eta = \frac{1300}{1583.1}$$

$$= 0.821$$

b)  The preheated air can be heated to within 20 °C of the flue gas temperature as it exits the heating coil for the hot oil, which is a pre-heated air temperature of 310 °C. Noting that the heat released by combustion is captured by the inlet gases reaching the theoretical flame temperature of 1800 °C without preheat, an energy balance around the combustion gives:

$$m_{FG} C_{P,FG}(T_{TFT} - 10) = m_{FG} C_{P,FG}(1800 - 10)$$
$$+ m_{AIR} C_{P,AIR}(310 - 10)$$

Assuming the flue gas and combustion air have the same specific heat capacity and flowrate:

$$m_{FG} C_{P,FG}(T_{TFT} - 10) = m_{FG} C_{P,FG}(1800 - 10)$$

$$+ m_{FG} C_{P,FG}(310 - 10)$$

$$T_{TFT} = 2100 \ °\text{C}$$

$$m_{FG} = \frac{1300}{1.21 \times (2100 - 330)}$$

$$= 0.6070 \ \text{kg s}^{-1}$$

Fuel heating duty:

$$Q_{PROC} = 0.6070 \times 1.21 \times (2100 - 10)$$

$$= 1535.0 \ \text{kW}$$

Furnace efficiency:

$$\eta = \frac{1300}{1535.0}$$

$$= 0.847$$

c)  Energy balance around combustion:

$$m_{FG} C_{P,FG}(T_{TFT} - 10) = m_{FG} C_{P,FG}(1800 - 10)$$
$$+ m_{AIR} C_{P,AIR}(310 - 10)$$

Given $C_{P,AIR} = 1.10$ kJ kg$^{-1}$ K$^{-1}$ and $m_{AIR} = \dfrac{m_{FG}}{1.06}$:

$$m_{FG} \times 1.21 \times (T_{TFT} - 10) = m_{FG} \times 1.21 \times (1800 - 10)$$

$$+ \frac{m_{FG}}{1.06} \times 1.10 \times (310 - 10)$$

$$T_{TFT} = 2057 \ °\text{C}$$

$$m_{FG} = \frac{1300}{1.21 \times (2057 - 330)}$$

$$= 0.6221 \ \text{kg s}^{-1}$$

Fuel heating duty:

$$Q_{PROC} = 0.6221 \times 1.21 \times (2057 - 10)$$

$$= 1540.9 \ \text{kW}$$

Furnace efficiency:

$$\eta = \frac{1300}{1540.9}$$

$$= 0.844$$

Thus, introducing air preheat increases the furnace efficiency from 82.1% to 84.4%. However, the approximation that the specific heat capacity and the flowrate of the combustion air are equal to that of the flue gas is a close approximation to the more accurate calculation.

d)  Column 1 Condenser:

$$Q = 400 \ \text{kW}$$

Enthalpy of saturated steam at 20 bar from steam tables:

$$H_{out} = 2798 \text{ kJ kg}^{-1}$$

Enthalpy of boiler feedwater:

$$H_{in} = 377 \text{ kJ kg}^{-1}$$

$$m_{STEAM} = \frac{400}{(2798 - 377)}$$

$$= 0.165 \text{ kg s}^{-1}$$

Column 2 Condenser:

$$Q = 800 \text{ kW}$$

Enthalpy of saturated steam at 10 bar from steam tables:

$$H_{out} = 2777 \text{ kJ kg}^{-1}$$

Enthalpy of boiler feedwater from above:

$$H_{in} = 377 \text{ kJ kg}^{-1}$$

$$m_{STEAM} = \frac{800}{(2777 - 377)}$$

$$= 0.333 \text{ kg s}^{-1}$$

**e)** Reactor cooling:

$$Q = 7000 \text{ kW}$$

Enthalpy of steam at 20 bar and 270 °C from steam tables:

$$H_{out} = 2953 \text{ kJ kg}^{-1}$$

Enthalpy of preheated boiler feedwater at 160 °C:

$$H_{in} = 676 \text{ kJ kg}^{-1}$$

$$m_{STEAM} = \frac{7000}{(2953 - 676)}$$

$$= 3.074 \text{ kg s}^{-1}$$

**f)** Enthalpy of high-pressure header assuming perfect mixing:

$$H_{HP} = \frac{0.165 \times 2798 + 3.074 \times 2953}{0.165 + 3.074}$$

$$= 2945 \text{ kJ kg}^{-1}$$

**g)** High-pressure steam heating:

$$Q = 1200 \text{ kW}$$

Enthalpy of saturated condensate at 20 bar from steam tables:

$$H_{out} = 909 \text{ kJ kg}^{-1}$$

Flowrate of steam for high-pressure steam heating:

$$m_{STEAM} = \frac{1200}{(2945 - 909)}$$

$$= 0.589 \text{ kg s}^{-1}$$

Inlet entropy of steam turbine at 20 bar and enthalpy of 2945 kJ kg$^{-1}$ from steam tables:

$$S_{in} = 6.626 \text{ kJ kg}^{-1} \text{ K}^{-1}$$

$$S_{out} = 6.626 \text{ kJ kg}^{-1} \text{ K}^{-1}$$

Isentropic outlet enthalpy of the steam at 10 bar with entropy of from 6.626 kJ kg$^{-1}$ K$^{-1}$ from steam tables:

$$H_{IS} = 2796 \text{ kJ kg}^{-1}$$

Outlet enthalpy from the turbine isentropic efficiency model is given by Eq. (4.5):

$$H_{out} = H_{in} - \eta_{IS}(H_{in} - H_{IS})$$

$$= 2945 - 0.75(2945 - 2796)$$

$$= 2833 \text{ kJ kg}^{-1}$$

Flowrate through the turbine, assuming a flowrate of steam of 0.5 kg s$^{-1}$ through the letdown:

$$= 0.165 + 3.074 - 0.589 - 0.5$$

$$= 2.150 \text{ kg s}^{-1}$$

Power from the turbine:

$$W = 2.150(2945 - 2833)$$

$$= 240.8 \text{ kW}$$

**h)** Enthalpy of low-pressure header assuming perfect mixing:

$$H_{LP} = \frac{0.333 \times 2777 + 2.150 \times 2833 + 0.5 \times 2945}{0.333 + 2.150 + 0.5}$$

$$= 2846 \text{ kJ kg}^{-1}$$

**i)** Low-pressure steam heating:

$$Q = 1600 \text{ kW}$$

Enthalpy of saturated condensate at 10 bar from steam tables:

$$H_{out} = 763 \text{ kJ kg}^{-1}$$

Flowrate of steam:

$$m_{STEAM} = \frac{1600}{(2846 - 763)}$$

$$= 0.768 \text{ kg s}^{-1}$$

Low-pressure surplus steam export:

$$= (0.333 + 2.150 + 0.5) - 0.768$$

$$= 2.215 \text{ kg s}^{-1}$$

If additional power generation is required, then the export steam could be expanded in a condensing steam turbine. The condensing steam turbine outlet would need to be checked for wetness. The wetness can be adjusted by adjusting the outlet pressure from such a condensing turbine.

**4**

# 4.12 Heating Utilities – Summary

When heating and cooling process streams, priority should be given to heat recovery. However, some heating and cooling utility is most often required. The choice of heating or cooling utility depends mainly on heat duty and temperature. Heating utilities include steam, furnace heat, gas turbine exhaust, molten salt, hot oil, hot water, and heat pumping. Cooling utilities include steam generation, once through cooling water systems, recirculating cooling water systems, air cooling, refrigeration, and heat pumping. Generally, the higher the temperature of heating utilities the more expensive they are. Generally, the lower the temperature of cooling utilities the more expensive they are.

Steam heaters require some mechanism to prevent the steam blowing through the heat exchanger without completely condensing. This is most often achieved using a steam trap that allows condensate to pass, but not steam. Flash steam recovery after steam heaters is good practice when servicing large heating duties.

Raw water requires treatment to become boiler feedwater prior to steam generation. Treatment is required for the raw water to remove suspended solids, dissolved solids, dissolved salts, and dissolved gases (particularly oxygen and carbon dioxide). The raw water entering the site is commonly filtered and dissolved salts removed using softening or ion exchange. The principal problems are calcium and magnesium ions. The water is then stripped to remove the dissolved gases and treated chemically before being used for steam generation.

There are many types of steam boilers, depending on the steam pressure, steam output, and source of energy. Blowdown is required to remove the dissolved solids not removed in the boiler feedwater treatment.

Gas turbines consist of a compressor and a turbine mechanically connected to each other. Release of energy to the compressed gases after the compressor, from the combustion of fuel, creates a net production of power. Gas turbines are available in a wide range of sizes but are restricted to standard frame sizes. The hot exhaust gas is useful for raising steam and the temperature of the exhaust gas can be increased by using supplementary firing.

Steam turbines are used to convert part of the energy of the steam into power and can be configured in different ways. Steam turbines can be divided into two basic classes: back-pressure turbines and condensing turbines. The performance can be modeled using an isentropic efficiency.

Steam system configurations using fuel combustion to generate the steam normally generate steam at high pressure. The highest-pressure steam main is often used for power generation rather than process heating. Steam is expanded from high to lower levels by either steam turbines or expansion valves. Steam expanded across letdown valves can be used to provide additional superheat to the low-pressure main or desuperheated by the injection of boiler feedwater. It is also good practice to recover the flash steam from large flowrates of high-pressure steam condensate. Whilst such cogeneration systems are common for steam systems generating steam from the combustion of fuels, systems using electric boilers lead to quite different configurations.

The supply of heat at high temperature, very high heat duty, or high heat flux (e.g. in a reactor) requires radiant heat transfer from the combustion of fuel in a fired heater or furnace. Various designs of fired heater are used, depending on the application, duty, and fuel. If multiple process duties require a temperature above steam, or layout issues prevent the fired heater being used directly for a process duty, then a hot oil circuit can be used.

# Exercises

1. A steam turbine operates with inlet steam conditions of 40 barg and 420 °C and can be assumed to operate with an isentropic efficiency of 80% and a mechanical efficiency of 95%. Using steam properties from Table 4.2, calculate the power production for a steam flowrate of 10 kg s$^{-1}$ and the heat available per kg in the exhaust steam (i.e. superheat plus latent heat) for outlet conditions of:

   a) 20 barg.
   b) 10 barg.
   c) 5 barg.
   d) What do you conclude about the heat available in the exhaust steam as the outlet pressure varies?

## Table 4.2

Steam properties for Exercise 1.

| | $H_{SUP}$ (kJ kg$^{-1}$) | $H_L$ (kJ kg$^{-1}$) | $S_{SUP}$ (kJ kg$^{-1}$ K$^{-1}$) | $\Delta H_{IS}$ (kJ kg$^{-1}$) |
|---|---|---|---|---|
| 420 °C, 40 barg | 3260 | 1095 | 6.828 | 0 |
| 20 barg | — | 920 | — | 190 |
| 10 barg | — | 781 | — | 346 |
| 5 barg | — | 671 | — | 474 |

2. A steam turbine is to exhaust directly to provide 1.5 MW of heat in a steam heater. The inlet steam to the steam turbine is at 20 bar and 350 °C. The outlet from the steam turbine and the corresponding inlet to the steam heater is 5 bar. Calculate the amount of power generated in the steam turbine in the following steps:

   a) Determine the specific enthalpy of the steam turbine outlet for an isentropic expansion from 20 to 5 bar using the data from steam tables.
   b) Calculate the real outlet enthalpy of the steam turbine assuming an isentropic efficiency of 0.75.
   c) Calculate the flowrate of steam required to provide the steam heater duty of 1.5 MW.
   d) Calculate the power generation from the steam turbine.

3. Figure 4.38 shows a steam system in which steam is generated at 15 bar and 250 °C and expanded to 5 bar in a steam turbine to generate 500 kW power. In parallel with the steam turbine, a letdown station expands steam in a valve in order to control the conditions of the 5 bar exhaust steam.

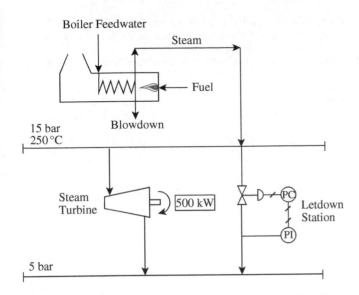

**Figure 4.38**

A steam turbine and bypass system.

a) Using steam tables, calculate the specific outlet enthalpy in kJ kg$^{-1}$ for an isentropic expansion in the steam turbine. It should be noted that an isentropic expansion will lead an outlet enthalpy below steam saturation temperature and wet steam.

b) The steam turbine is required to produce 500 kW of power. Calculate the flowrate of steam in kilograms per second, assuming it has an isentropic efficiency of 0.75.

c) The outlet steam main needs to be controlled with a super-heat of 20 °C (that is, the temperature of the 5 bar main should be 20 °C above the saturation steam temperature at 5 bar). Determine the specific enthalpy in kJ kg$^{-1}$ of the 5 bar steam main for this superheat of 20 °C from steam tables.

d) Control of the 5 bar main to have a superheat of 20 °C can be achieved by allowing steam from the 15 bar main to expand in the letdown station in Figure 4.38. Assuming the expansion in the letdown station is isenthalpic, calculate the flowrate of steam through the letdown station in kilograms per second to maintain a superheat of 20 °C in the 5 bar main.

**4.** The following data are given for a gas turbine:

| Power generation | $W$ | 15 MW |
|---|---|---|
| Power efficiency | $\eta_{GT}$ | 32.5% |
| Exhaust flow | $m_{EX}$ | 58.32 kg s$^{-1}$ |
| Exhaust temperature | $T_{EX}$ | 488 °C |
| Exhaust heat capacity | $C_{P,EX}$ | 1.1 kJ kg$^{-1}$ K$^{-1}$ |
| Stack temperature | $T_{STACK}$ | 100 °C |

a) Calculate the fuel consumption $Q_{FUEL}$.

b) Assuming 72% of the heat available in the exhaust can be recovered for steam generation, how much steam ($Q_{STEAM}$) can be generated from an unfired HRSG in MW?

c) With supplementary firing ($T_{SF} = 800$ °C), calculate the fuel consumption $Q_{SF}$ for the supplementary firing, available heat from the exhaust $Q_{EX}$, and steam generation for the same HRSG with an 83% efficiency.

**5.** A gas turbine has the following performance data:

| Electricity output | 13.5 MW |
|---|---|
| Heat rate | 10 810 kJ kWh$^{-1}$ |
| Exhaust flow | 179 800 kg h$^{-1}$ |
| Exhaust temperature | 480 °C |

The exhaust is to be used to generate steam in an unfired HRSG with a minimum stack temperature of 150 °C. The specific heat capacity of the exhaust is 1.1 kJ kg$^{-1}$ K$^{-1}$. Enthalpy data for steam are given in Table 4.3. Calculate the following.

**Table 4.3**

Enthalpy data for Exercise 5.

| | Enthalpy at 10 bar (kJ kg$^{-1}$) |
|---|---|
| Steam at 200 °C | 2827 |
| Saturated steam (179.9 °C) | 2776 |
| Saturated condensate (179.9 °C) | 763 |
| Water at 100 °C | 420 |

a) Fuel requirement in MW.

b) The efficiency of the gas turbine.

c) Amount of steam at 10 bara and 200 °C that can be produced from boiler feedwater at 100 °C in an unfired HRSG. Assume $\Delta T_{min} = 20$ °C for the heat recovery steam generator.

**6.** A process stream with a flowrate of 40 kg s$^{-1}$ and heat capacity of 3300 J kg$^{-1}$ K$^{-1}$ needs to be heated from 240 °C to 300 °C using hot oil. The temperature rise of the hot oil in the furnace can be assumed to be 90 °C. The minimum temperature difference between the process fluid and the hot oil can be assumed to be 10 °C. It can be assumed that the theoretical flame temperature in the furnace is 1800 °C and ambient air temperature is 15 °C.

a) Calculate the process heating duty.

b) Assuming a minimum temperature difference between the furnace flue gas and the hot oil is 20 °C, calculate the heat duty from fuel.

c) Calculate the furnace efficiency.

# References

Betz (1991). *Handbook of Industrial Water Conditioning*, 9e.

Dryden, I.G.C. (1982). *The Efficient Use of Energy*. Butterworth Scientific.

Elliot, T.C. (1989). *Standard Handbook of Powerplant Engineering*. McGraw-Hill.

Kemmer, F.N. (1988). *The Nalco Water Handbook*, 2e. McGraw-Hill.

Kern, D.Q. (1950). *Process Heat Transfer*. McGraw-Hill.

Kretzschmar, H.-J. and Wagner, W. (2019). *International Steam Tables – Properties of Water and Steam Based on the Industrial Formulation IAPWS-IF97*, 3e. Springer Vieweg.

Martin, G.R. (1998). Heat-flux Imbalances in Fired Heaters Cause Operating Problems, *Hydrocarbon Processing*, May: 103.

Nelson, W.L. (1958). *Petroleum Refinery Engineering*, 4e. McGraw-Hill.

Siddhartha, M. and Rajkumar, N. (1999). Performance enhancement in coal fired thermal power plants. Part II: Steam turbines. *International Journal of Energy Research* **23**: 489.

Smith, R. (2016). *Chemical Process Design and Integration*, 2e. Wiley.

# Chapter 5

# Cooling Utilities

In the previous chapter it was noted that when heating and cooling process streams from their supply to their target temperature, priority should be given to heat recovery. However, once the opportunity for heat recovery has been exhausted, there is inevitably some need for a cooling utility for some of the process streams. It was also noted that generally the lower the temperature of a cold utility, the more expensive it is. If the cooling utility is required, then an above ambient temperature utility should be used in the first instance. This could be:

- steam generation for heating purposes or power generation elsewhere;
- generation of hot water for heating purposes elsewhere;
- once-through water cooling;
- recirculating cooling water systems;
- air coolers.

If heat rejection is required below that which can normally be achieved using cooling water or air cooling, then refrigeration is required. A refrigeration system is a heat pump with the purpose of providing cooling below ambient temperature. This means that heat must be rejected at a higher temperature, to ambient via an external cooling utility (e.g. cooling water), a heat sink within the process, or to another refrigeration system.

Start by considering above-ambient temperature cooling.

## 5.1 Waste Heat Steam Generation

If high-temperature waste heat cannot be used for heat recovery within a process, it can be used to generate steam. This steam could be fed into a steam main and used by another process on the site. In this way, heat is recovered between processes on the site through the steam system. If the steam is to be fed into a steam

main, then the process stream providing the waste heat should preferably be at least 10–20 °C above the saturation temperature of one of the steam mains. Sometimes, other uses can be found for the steam other than feeding into a steam main, such as creation of vacuum in steam ejectors, used as dilution steam in chemical reactors or providing heat for an absorption refrigeration system if the process requires a moderate level of refrigeration (see later). Figure 5.1 shows two heat exchangers that could be used for steam generation from waste heat. The first in Figure 5.1a shows a waste heat steam generator that is very similar to a kettle reboiler. The high-temperature process stream is fed through the tubes, which are immersed in a pool of boiling boiler feedwater. Overflow from the pool of boiling water can be used for blowdown to prevent the build-up of solids in the boiling water. Figure 5.1b shows a similar design, but with an alternative arrangement for taking the blowdown. The steam generated in heat exchangers such as those illustrated in Figure 5.1 will be saturated at the pressure of the heat exchanger. If some degree of superheat is required, this can be obtained by providing extra heat in a separate superheater heat exchanger, or by mixing saturated steam with superheated steam, or by generating the saturated steam at a slightly higher pressure than required and then decreasing the pressure of the steam leaving the heat exchanger in a pressure letdown valve.

In some situations, particularly for very high temperature cooling, it is necessary to provide process cooling with a heat carrier in a closed loop. The closed loop then rejects the heat to steam generation. The heat carrier might be a hot oil or molten salt, as described in Chapter 4, but this time the function of the hot oil or molten salt is to remove process heat and transfer the heat to steam generation.

## 5.2 Once-Through Cooling Water Systems

Once-through cooling systems take water from a river, canal, lake, or the sea, use it for cooling, and then return it to its source. Heat rejection to the environment in this way does have environmental consequences and can have an impact on ecosystems. For heat

*Process Plant Design*, First Edition. Robin Smith.
© 2024 John Wiley & Sons Ltd. Published 2024 by John Wiley & Sons Ltd.
Companion website: www.wiley.com/go/processplantdesign

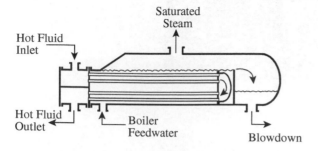

(a) Heat recovery steam generator with weir overflow.

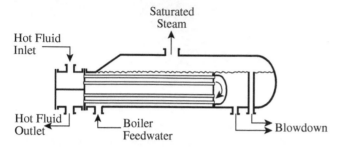

(b) Heat recovery steam generator with pipe overflow.

**Figure 5.1**

Steam generation from waste heat recovery.

rejection to rivers, canals, and lakes, any change of temperature in surface and groundwater resulting from waste heat rejection affects the chemical, biochemical, and hydrological properties of the water and potentially has an impact on the overall ecosystem. Once-through cooling systems require large amounts of water to be used on a single-use basis. This method is simple and cheap but has environmental consequences and problems can be created in the process from fouling unless the water is treated before use.

When the use of fresh cold water is limited, or environmental regulations limit the heat rejection to rivers, canals, lakes, and the sea, then recirculating cooling water systems or air cooling must be used. Recirculating cooling water systems are the most common method used for heat rejection to the environment.

## 5.3 Recirculating Cooling Water Systems

Figure 5.2 illustrates the basic features of a recirculating cooling water system (Kröger 1998). Cooling water from the cooling tower is pumped to heat exchangers where waste heat needs to be rejected from the process to the environment. The cooling water is in turn heated in the process cooling duty and returned to the cooling tower. The hot water returned to the cooling tower flows down over packing and is contacted counter-currently or in cross-flow with air. The packing should provide a large interfacial area for heat and mass transfer between the air and the water. The air is humidified and heated, and rises through the packing. The water is cooled mainly by evaporation as it flows down through the packing. The evaporated water leaving the top of the cooling tower reflects the cooling duty that is being performed. Water is also lost through *drift* or *windage*. Drift is droplets of water entrained in the air leaving the top of the tower. Drift water has the same composition as the recirculating water and is different from evaporation. Drift should be minimized because it wastes water and can also cause staining of nearby buildings, and so on, that are some distance from the cooling tower. *Drift loss* is around 0.1–0.3% of the water circulation rate. *Blowdown*, as shown in Figure 5.2, is necessary to prevent the build-up of contamination in the recirculation. *Makeup water* is required to compensate for the loss of water from evaporation, drift, and blowdown. The makeup water contains solids that build up in the recirculation as a result of the evaporation. The blowdown purges these, along with products of corrosion and microbiological growth. Both corrosion and microbiological growth need to be inhibited by chemical dosing of the recirculation system. *Dispersants* are added to prevent deposit of solids, *corrosion inhibitors* to prevent corrosion, and *biocides* to inhibit biological growth. In Figure 5.2, the blowdown is shown to be taken from the cooling tower *basin*. It can be taken as cold blowdown, as shown in Figure 5.2. Alternatively, it can be taken from the hot recirculation water before it is returned to the cooling tower as *hot blowdown*. Taking hot blowdown might be helpful in increasing the heat rejection from the cooling system, but might not be acceptable environmentally from the point of view of the resulting increase in effluent temperature.

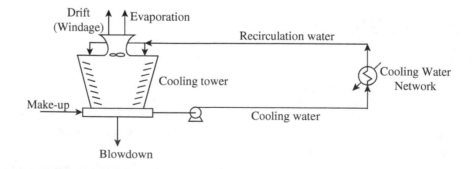

**Figure 5.2**

Recirculating cooling water system. Source: (Smith 2016), *Chemical Process Design and Integration*, 2nd Edition, John Wiley & Sons Ltd.

Many different designs of cooling tower are available. These can be broken down into two broad classes as follows.

1) *Natural draft.* Natural draft cooling towers consist of an empty shell, usually hyperbolic in shape and constructed in concrete. The upper, empty portion of the shell merely serves to increase the draft. The lower portion is fitted with the packing. The draft is created by the difference in density between the warm humid air within the tower and the denser outside ambient air.

2) *Mechanical draft cooling towers.* Mechanical draft cooling towers use fans to move the air through the cooling tower. In a *forced draft* design, fans push the air into the bottom of the tower. *Induced draft* cooling towers have a fan at the top of the cooling tower to draw air through the tower. The tower height for mechanical draft towers does not need to extend much beyond the depth of the packing. Mechanical draft cooling towers for large duties often comprise a series of rectangular *cells* constructed together, but operating in parallel, each with its own fan.

The type of packing used can be as simple as *splash bars* but is more likely to be packing, similar in form to that used in absorption and distillation towers. The temperature limitation of the packing needs careful attention. Plastic packing has severe temperature limitations, as far as the cooling water return temperature is concerned. If the temperature is too high, the plastic packing will deform and this will result in a deterioration of cooling tower performance. Polyvinylchloride is limited to a maximum temperature of around 50 °C. Other types of plastic packing can withstand temperatures up to around 70 °C.

There are some general trends that can be observed for the design of cooling towers in terms of the temperature and flowrate of the inlet cooling water to the tower. Increasing the temperature of the inlet to a given cooling tower design for a fixed flowrate increases the performance of the cooling tower and allows more heat to be removed (Kim and Smith 2001). On the other hand, if the flowrate to the inlet of a given cooling tower is decreased for a fixed inlet temperature, then the performance of the cooling tower improves, allowing more heat to be removed (Kim and Smith 2001). Thus, the performance of a cooling tower is maximized by maximizing the inlet temperature to the cooling tower and minimizing the inlet flowrate (Kim and Smith 2001). There will be constraints on the maximum return temperature determined by the maximum temperature allowable for the cooling tower packing and fouling and corrosion issues, as discussed above.

Various parameters need to be defined in order to specify the performance of the cooling tower.

1) *Dry-bulb temperature.* Dry-bulb temperature is the temperature of the ambient air indicated by a dry-bulb thermometer.

2) *Wet-bulb temperature.* The wet-bulb temperature is the temperature to which air can be cooled adiabatically to saturation by the addition of water vapor. It is measured with a thermometer whose mercury reservoir is covered by porous cloth that is soaked in water with the thermometer placed in an air stream. The water evaporates, extracting heat from the environment, including the mercury reservoir, and the measured temperature

decreases compared with a dry-bulb thermometer. The wet-bulb temperature is always lower than the dry-bulb temperature. The measured wet-bulb temperature is a function of relative humidity and ambient air temperature (dry-bulb temperature). The wet-bulb temperature of the inlet air to the cooling tower represents the lowest temperature that can be achieved due to evaporation.

3) *Temperature approach.* The temperature approach is the temperature difference between the cooled water leaving the cooling tower and the wet bulb temperature of the air entering the tower. It is a function of the design of the cooling tower. The temperature of the cooling water leaving the cooling tower is typically 3 °C above the wet bulb temperature.

4) *Temperature range.* The temperature range is the temperature difference between the hot water at the inlet of the cooling tower and the cold water at the outlet. This is typically 10 °C.

These parameters are illustrated graphically in Figure 5.3.

As evaporated water is pure, solids are left behind in the recirculating water, making it more concentrated than the makeup water. Blowdown purges the solids from the system. Note that the blowdown has the same chemical composition as the recirculated water. *Cycles of concentration* is a comparison of the dissolved solids in the blowdown compared with that in the makeup water. For example, at three cycles of concentration, the blowdown has three times the solids concentration as the makeup water. For calculation purposes, blowdown is defined to be all non-evaporative water losses (drift, leaks, and intentional blowdown). In principle, any soluble component in the makeup and blowdown can be used to define the concentration for the cycles; for example, chloride and sulfate being soluble at high

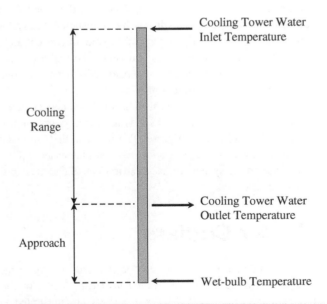

**Figure 5.3**

Cooling water performance.

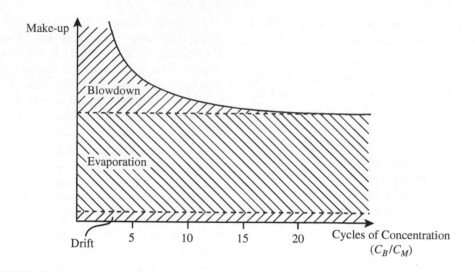

## Figure 5.4

Relationship between makeup water and blowdown for cooling towers. Source: (Smith 2016), *Chemical Process Design and Integration*, 2nd Edition, John Wiley & Sons Ltd.

concentrations can be used. The cycles of concentration are thus defined to be:

$$CC = \frac{C_B}{C_M} \tag{5.1}$$

where

$CC$ = cycles of concentration

$C_B$ = concentration in blowdown

$C_M$ = concentration in makeup

Figure 5.4 shows the relationship between the makeup, blowdown, evaporation, and drift versus the cycles of concentration. For a given design of cooling tower, fixed heat duty, and fixed conditions, the evaporation and drift will be constant as the cycles of concentration increase. However, as the cycles of concentration increase, the blowdown decreases and hence the makeup water decreases. This can be an important consideration from the point of view of site water consumption.

The consequences of an increase in the cycles of concentration are that, as the level of dissolved solids increases, corrosion and deposition tendencies also increase. The result is that, although increasing the cycles of concentration decreases the water requirements of the cooling system, the required amount of chemical dosing also increases.

## 5.4   Air Coolers

Heat above ambient temperature can be rejected directly to the environment by the use of *air-cooled heat exchangers* or *fin-fan exchangers* (Kröger 1998). Two designs are illustrated in Figure 5.5. In these designs, the fluid to be cooled passes through the inside of tubes and ambient air is blown across the outside by the use of fans. The outside surface most often features extended

surfaces to enhance the heat transfer. Figure 5.6 shows three typical air cooler layouts from the many possibilities. The tubes are grouped into rectangular *bundles*, typically 1–2 m wide. The bundles are factory assembled. Bundles are grouped into *bays* containing one or more bundles in parallel, with each bundle connected to inlet and outlet headers. Each bay can be serviced by one to three fans, depending on the geometry of the tube layout.

The headers can be fitted with pass partition plates in order to create different tube pass arrangements, in a similar way to shell-and-tube heat exchangers. Figure 5.5a and b show two pass arrangements, which are most common, but anywhere between one and four pass arrangements are used. In Figure 5.5a the tubes are mounted horizontally and the air is drawn across the tubes in an *induced draft* arrangement. In Figure 5.5b the tubes are also mounted horizontally but the air is driven across the tubes in a *forced draft* arrangement.

The horizontal arrangements in Figure 5.5a and b are suitable if no condensation is taking place, as would be the case for cooling of liquids or cooling of gas without condensation. If the cooling duty is condensing, then the *A-frame* arrangement in Figure 5.7 is used, in which the tubes are sloped to allow separation of the condensed liquid and vapor.

For a given mass flowrate of air, an induced draft fan in principle requires slightly greater power than a forced draft fan, as the induced draft fan is compressing hot air with a lower density and higher volume than a forced draft equivalent. However, induced draft arrangements provide a more uniform air distribution over the tube bundle. Forced draft arrangements have the disadvantage that hot air can be recirculated back to the air inlet. The performance of the units is affected significantly both by ambient conditions and by fouling of the outside surfaces. The air flow across the tubes can be controlled by the use of variable pitch fan blades and variable speed electric motors to drive the fans. The outside surfaces should be cleaned regularly to maintain performance,

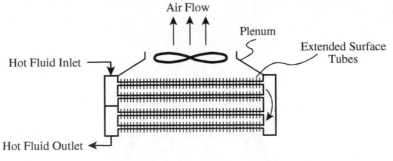

(a) Induced draft air-cooled heat exchanger.

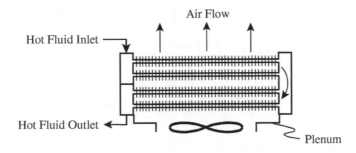

(b) Forced draft air-cooled heat exchanger.

**Figure 5.5**

Air-cooled heat exchangers. Source: (Smith 2016), *Chemical Process Design and Integration*, 2nd Edition, John Wiley & Sons Ltd.

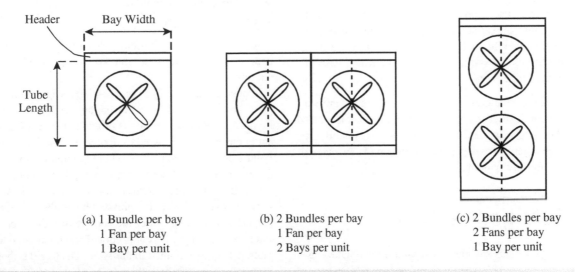

| (a) 1 Bundle per bay | (b) 2 Bundles per bay | (c) 2 Bundles per bay |
| 1 Fan per bay | 1 Fan per bay | 2 Fans per bay |
| 1 Bay per unit | 2 Bays per unit | 1 Bay per unit |

**Figure 5.6**

Some typical layouts of the air-cooled heat exchangers. Source: Adapted from (Smith 2016), *Chemical Process Design and Integration*, 2nd Edition, John Wiley & Sons Ltd.

typically by water sprays. Spraying water on the outside surface can also be used to temporarily enhance the heat transfer when the ambient temperature is high. If the exchangers have to work in very low temperatures under winter conditions, then they can be erected within enclosures equipped with louvers to allow some recirculation of hot air and prevent freezing.

General guidelines for the design of air-cooled heat exchangers are (Kröger 1998; Serth 2007; Hudson Products 2013; Smith 2016):

- Air temperature is normally taken to be the temperature only exceeded for 5% of the year. This means that the cooler will

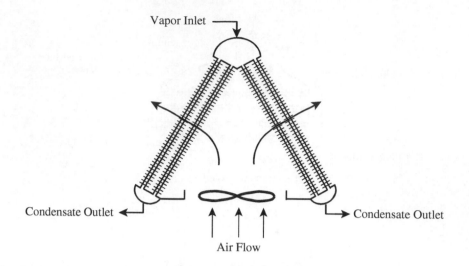

Vapor Inlet

Condensate Outlet

Condensate Outlet

Air Flow

**Figure 5.7**

A-frame air-cooled heat exchanger. Source: Adapted from (Smith 2016), *Chemical Process Design and Integration*, 2nd Edition, John Wiley & Sons Ltd.

be overdesigned for 95% of the year and will require appropriate control systems to avoid overcooling. An air temperature near, but below, the maximum to be encountered needs to be chosen.

- Recirculation in congested areas can increase the effective inlet air temperature 1–2 °C compared with ambient air temperatures in uncongested areas.
- Minimum temperature approach (the difference between the process fluid outlet temperature and the design air temperature) is not normally lower than 10 °C and is often taken to be 20 °C. More than one tube pass might be used, in which case this applies for each tube pass.

Air-cooled heat exchangers of the type in Figure 5.5 are very common in some industries, particularly when the plant is located in a region where water is scarce for cooling tower systems. Air-cooled heat exchangers also have the environmental advantage of eliminating cooling tower plumes.

## 5.5   Refrigeration

There are many processes such as gas separation and natural gas liquefaction that operate below ambient temperature. If heat rejection is required below ambient temperature, then refrigeration is required. A refrigeration system is a heat pump that provides cooling at temperatures below that which can normally be achieved using cooling water or air cooling (Gosney 1982; Isalski 1989; Dossat 1991). Thus, the removal of heat using refrigeration leads to its rejection at a higher temperature. This higher temperature might be to an external cooling utility (e.g. cooling water), to a heat sink within the process, or to another refrigeration system. Refrigeration is important in chemical engineering operations where low temperatures are required to condense gases (Isalski 1989), crystallize solids, control reactions, and so on, and for the preservation of foods and air conditioning. Generally, the

lower the temperature of the cooling required to be serviced by the refrigeration system and the larger its duty, the more complex the refrigeration system. Processes involving the liquefaction and separation of gases (e.g. ethylene, liquefied natural gas, etc.) are serviced by complex refrigeration systems.

There are two broad classes of refrigeration system:

- compression refrigeration;
- absorption refrigeration.

Compression refrigeration cycles are by far the most common and absorption refrigeration is only applied in special circumstances.

Consider first a simple compression refrigeration cycle using a pure component as the refrigerant fluid, as illustrated in Figure 5.8a. Process cooling is provided by a cold liquid refrigerant in the *evaporator*. A mixture of vapor and liquid refrigerant at Point 1 enters the evaporator where the liquid vaporizes and produces the cooling effect before leaving the evaporator at Point 2. The refrigerant vapor is then compressed to Point 3. At Point 3, the vapor is not only at a higher pressure but is also superheated by the compression process. After the compressor, the vapor refrigerant enters a *condenser* where it is cooled and condensed to leave as a saturated liquid at Point 4 in the cycle. The liquid is then expanded in an expansion valve (or *throttle*) to a lower pressure to Point 1 in the cycle. The expansion process partially vaporizes the liquid refrigerant across the expander, producing a cooling effect to provide refrigeration, and the cycle continues. Figure 5.8b shows the cycle on a temperature-entropy diagram. The diagram shows a two-phase envelope, inside of which the refrigerant is present as both vapor and liquid. To the left of the two-phase envelope, the refrigerant is liquid and to the right of the two-phase envelope, the refrigerant is vapor. Starting at Point 1 with a two-phase mixture, the refrigerant enters the evaporator where there is an isothermal vaporization to Point 2. From Point 2 to Point 3, there is a compression process finishing at Point 3. On a temperature-entropy diagram, as shown in Figure 5.8b, the compression

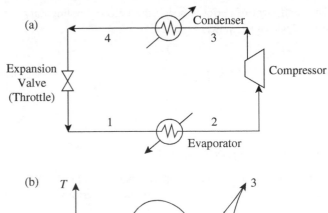

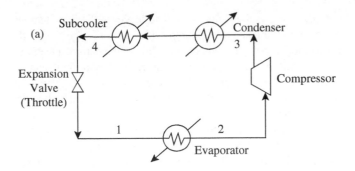

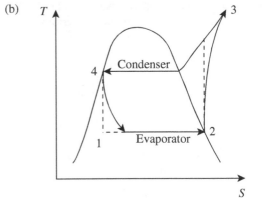

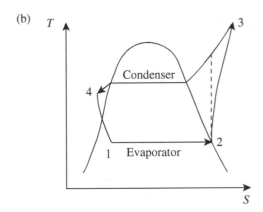

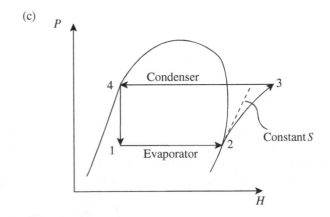

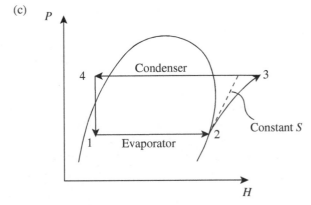

## Figure 5.8

Simple compression cycle with expansion valve and non-ideal compression.

## Figure 5.9

Simple compression cycle with subcooled condensate.

process is non-ideal and features an increase in entropy. From Point 3, the refrigerant is first cooled to remove the superheat and is then condensed to Point 4, where there is a saturated liquid. The refrigerant is expanded from Point 4 to Point 1. The expansion features an increase in entropy. In Figure 5.8c, the same cycle is represented on a pressure–enthalpy diagram. Again the two-phase envelope is evident. From Point 1 to Point 2, the enthalpy of the refrigerant increases in the evaporator. From Point 2 to Point 3, the pressure is increased in the compressor with a corresponding increase in the enthalpy. From Point 3 to Point 4, there is assumed to be no change in pressure but a decrease in the enthalpy in the desuperheating and the condensation. Between Point 4 and Point 1, there is a decrease in pressure in the throttle valve.

The expansion between Points 4 and 1 is assumed to be isenthalpic, leading to a vertical line on the pressure–enthalpy diagram.

Figure 5.9a shows the corresponding simple compression cycle, but with subcooled condensate. The cycle is basically the same as before, except that the liquid leaving the condenser is now subcooled rather than being saturated. The subcooling effect can be seen on the temperature–entropy diagram in Figure 5.9b and the pressure–enthalpy diagram in Figure 5.9c. Subcooling the liquid in the way shown in Figure 5.9 brings a benefit to the cycle. The fact that the liquid is subcooled when entering the expansion process means that there is less vaporization in the expansion process to produce refrigerant at the required temperature. This leads to an increase in the proportion of the liquid refrigerant at Point 1 entering the evaporator. The liquid refrigerant provides the cooling and the vapor present does not provide a cooling effect as the vapor enters and leaves at the same temperature without a change in enthalpy. If a greater proportion of the refrigerant entering the evaporator is liquid, it means that the flowrate around the cycle for a given refrigeration duty can be decreased. This, in turn, reduces the compression duty. Subcooling the liquid in this way therefore increases the overall efficiency of the cycle.

In Figures 5.8a and 5.9a, the expansion of the liquid refrigerant produces a two-phase vapor–liquid mixture that is fed directly to the evaporator to provide process cooling. For a vapor–liquid mixture of a pure refrigerant at saturated conditions, the vapor provides no cooling, as it enters and leaves the evaporator at the same temperature without a change in enthalpy (neglecting any pressure drop). Only the vaporization of the liquid produces cooling. Some designs of refrigeration use a vapor–liquid separation drum after expansion of the refrigerant. The vapor from the separation drum is then fed directly to the compressor. The liquid from the separation drum is fed to the evaporator and then to the compressor after vaporization. This arrangement with a vapor–liquid separator is thermodynamically equivalent to the design without a separator (neglecting pressure drops). The separation drum adds capital cost and complexity to the design without bringing any thermodynamic benefit. Whether such a separation drum is used largely depends on the type of heat exchanger used for the evaporator. For example, if the evaporation takes place on the shell side of a kettle reboiler, then feeding a two-phase vapor–liquid mixture does not present any problems and a vapor–liquid separator would not normally be justified. However, if the evaporation takes place in the channels of a plate-fin heat exchanger (Smith 2016), then feeding a two-phase vapor–liquid mixture would present problems, as it is difficult to distribute a two-phase mixture evenly across a manifold of channels. Under such circumstances, a vapor–liquid separation drum would normally be used.

The performance of refrigeration cycles is measured as a *coefficient of performance* (*COP*<sub>*REF*</sub>), as illustrated in Figure 5.10.

The coefficient of performance is the ratio of cooling duty performed per unit of power required:

$$COP_{REF} = \frac{Q_C}{W} \tag{5.2}$$

where

$Q_C$ = cooling duty

$W$ = refrigeration power required

The higher the coefficient of performance, the more efficient is the refrigeration cycle. An ideal coefficient of performance can be defined by:

$$\text{Ideal } COP_{REF} = \frac{Q_C}{W} = \frac{T_{EVAP}}{T_{COND} - T_{EVAP}} \tag{5.3}$$

where

$T_{EVAP}$ = evaporation temperature (K)
$T_{COND}$ = condensing temperature (K)

Actual performance can be predicted by introducing a Carnot efficiency into Eq. (5.3):

$$COP_{REF} = \frac{Q_C}{W} = \eta_C \frac{T_{EVAP}}{T_{COND} - T_{EVAP}} \tag{5.4}$$

where

$\eta_C$ = Carnot Efficiency (−)

The Carnot Efficiency is a function of the physical properties of the working fluid, the cycle configuration, system pressures in the cycle, and the compressor performance. The Carnot Efficiency of the refrigeration cycle can be determined by a simulation of the

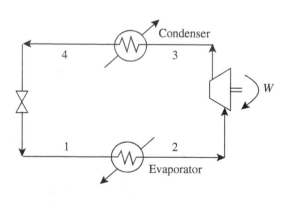

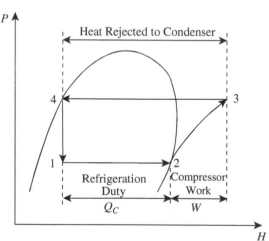

$$\text{Coefficient of Performance } (COP) = \frac{Q_C}{W}$$

**Figure 5.10**

Performance of practical refrigeration cycles. Source: (Smith 2016), *Chemical Process Design and Integration*, 2nd Edition, John Wiley & Sons Ltd.

cycle. An approximate value of 0.6 can be used for a first estimate of cycle performance.

It is obvious from Eq. (5.4) that the larger the temperature difference across the refrigeration cycle ($T_{COND} - T_{EVAP}$), the lower will be the coefficient of performance and the higher will be the power requirements for a given cooling duty.

---

**Example 5.1**  Estimate the power requirement and heat rejected to cooling water with a cooling tower return temperature of 25 °C for refrigeration cycles cooling process duties of 3 MW at process temperatures of –20, –30, and –40 °C. Assume heat can be transferred with $\Delta T = 10$ °C for heat exchange with cooling water, $\Delta T = 5$ °C for heat exchange with refrigeration, and a Carnot efficiency of 60%.

**Solution**

Temperatures for the condenser and evaporator must allow for the appropriate $\Delta T$. For a process temperature of –20 °C:

$$T_{COND} = 25 + 10 = 35°C = 308 \text{ K}$$
$$T_{EVAP} = -20 - 5 = -25°C = 248 \text{ K}$$

From Eq. (5.4):

$$W = \frac{Q_C(T_{COND} - T_{EVAP})}{0.6 T_{EVAP}}$$

$$= \frac{3(308 - 248)}{0.6 \times 248} = 1.21 \text{ MW}$$

Calculate the heat rejection by energy balance:

$$Q_{CW} = 3 + 1.21 = 4.21 \text{ MW}$$

Now repeat for the other temperatures:

| $Q_{PROC}$ (°C) | $W$ (MW) | $Q_{CW}$ (MW) |
| --- | --- | --- |
| –20 | 1.21 | 4.21 |
| –30 | 1.47 | 4.47 |
| –40 | 1.75 | 4.75 |

The increase in power required and heat rejected increase with a decreasing process cooling temperature and is a non-linear change.

5

Simple cycles like the ones discussed so far can be used to provide cooling as low as typically −40 °C. For lower temperatures, complex cycles are normally used.

One way to reduce the overall power requirement of a refrigeration cycle is to introduce multistage compression and expansion, as shown in Figure 5.11a. The expansion is carried out in two stages with a vapor–liquid separator between the two stages, often called an *economizer*. Vapor from the economizer passes directly to the high-pressure compression stage. Liquid from the economizer passes to the second expansion stage. The introduction of an economizer reduces the vapor flow in the low-pressure compression stage through reduction in the vapor volume. This reduces the overall power requirement. Figure 5.11a also shows an intercooler for the vapor between the low-pressure and the high-pressure compression stages. This intercooler reduces further the compressor power in the high-pressure compression stage, reducing the overall power requirement. The cycle is shown as a pressure-enthalpy diagram in Figure 5.11b.

Figure 5.12a shows another way to reduce the overall power requirement of a refrigeration cycle by introducing multistage compression and expansion. Again the expansion is carried out

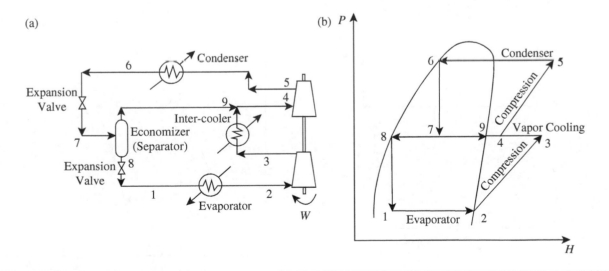

**Figure 5.11**

Multistage compression and expansion with an economizer. Source: (Smith 2016), *Chemical Process Design and Integration*, 2nd Edition, John Wiley & Sons Ltd.

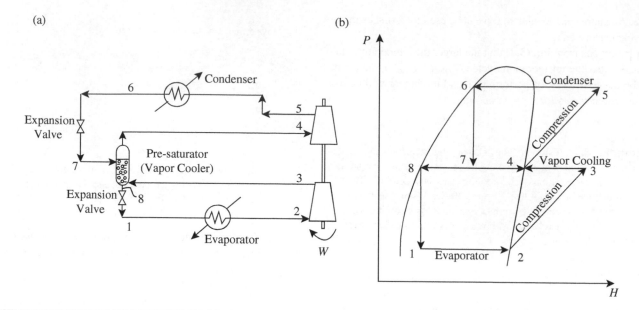

## Figure 5.12

Multistage compression and expansion with a pre-saturator. Source: (Smith 2016), *Chemical Process Design and Integration*, 2nd Edition, John Wiley & Sons Ltd.

in two stages with a vapor–liquid separator between the two. However, this time the cooled liquid and vapor from the first expansion stage is contacted directly with the compressed vapor from the low-pressure compression stage in a *presaturator*. The presaturator cools the vapor from the low-pressure compression stage by direct contact and acts as a vapor–liquid separator between the stages. The vapor leaving the presaturator being passed to the high-pressure compression stage is saturated. Liquid

from the presaturator passes to the second expansion stage. Again, the introduction of a presaturator reduces the vapor flow in the low-pressure compression stage, reducing the overall power requirement. The cycle is shown in Figure 5.12b as a pressure-enthalpy diagram.

Figure 5.13a shows a *cascade cycle*. This involves two refrigerant cycles linked together. Each cycle will use a different refrigerant fluid. The low-temperature cycle provides the process

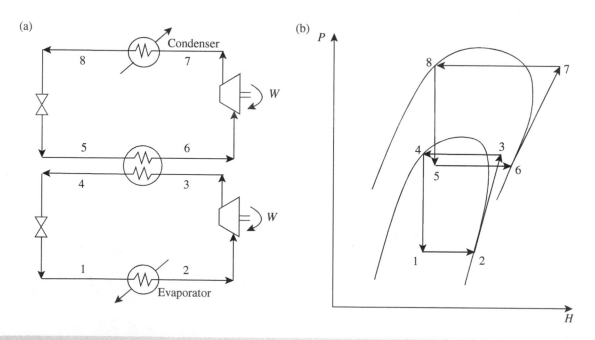

## Figure 5.13

A cascade cycle. Source: (Smith 2016), *Chemical Process Design and Integration*, 2nd Edition, John Wiley & Sons Ltd.

cooling in the evaporator and rejects its heat to the other cycle, which rejects its heat in the condenser to cooling water. Cascade cycles are used to provide very low temperature refrigeration where a single refrigerant fluid would not be suitable to operate across such a wide temperature range. In the cascade in Figure 5.13a, each cycle comprises a simple cycle. The cascade cycle is represented in Figure 5.13b in a pressure–enthalpy diagram. The temperature of the interface between the two cycles is normally specified in terms of the temperature of the evaporator of the upper (high-temperature) cycle and is known as the

*partition temperature*. The partition temperature is an important degree of freedom in the design.

Complex refrigeration cycles can be built up in different ways. Figure 5.14 illustrates an example of a cascade cycle with multistage compression and expansion of the high-temperature cycle.

Figure 5.15a shows a *two-level* refrigeration cycle. Level 1 and Level 2 provide process cooling at different temperatures. This reduces the refrigerant flow through the low-pressure compression stage and reduces the overall power requirements. It is represented in Figure 5.15b in a pressure–enthalpy diagram.

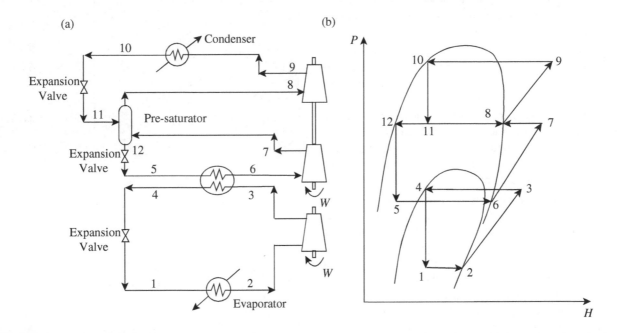

**Figure 5.14**

Cascade cycle with multistage compression and expansion. Source: (Smith 2016), *Chemical Process Design and Integration*, 2nd Edition, John Wiley & Sons Ltd.

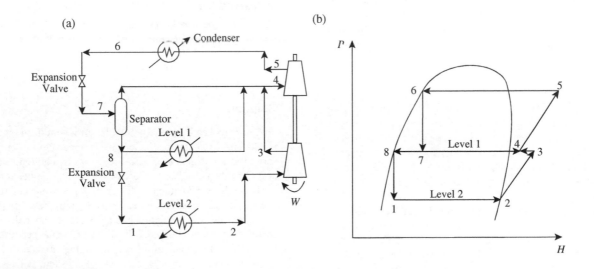

**Figure 5.15**

A two-level refrigeration cycle. *Source*: (Smith 2016), *Chemical Process Design and Integration*, 2nd Edition, John Wiley & Sons Ltd.

# 5.6    Choice of a Single Component Refrigerant for Compression Refrigeration

Now consider the main factors affecting the choice of refrigerant for compression refrigeration, but for now restricting consideration to single components.

1) *Freezing point*. Firstly, the evaporator temperature should be well above the freezing temperature at the operating pressure. The freezing points of some common refrigerants at atmospheric pressure are given in Table 5.1.

2) *Vacuum operation*. The second consideration, illustrated in Figure 5.16, is that at the evaporator temperature, evaporator pressure below atmospheric pressure should be avoided. An evaporator pressure above atmospheric pressure avoids potential problems with the ingress of air into the cycle, which can cause performance and safety problems. However, special designs can use evaporator pressures below atmospheric. The boiling points of some common refrigerants are given in Table 5.1 at atmospheric pressure.

3) *Latent heat of vaporization*. It is desirable to have a refrigerant with a high latent heat. A high latent heat will lead to a lower flowrate of refrigerant around the loop and reduce the power requirements correspondingly. The shape of the two-phase envelope is such that as the temperature of the refrigerant increases, the latent heat decreases as the critical temperature is approached. It is therefore desirable to have evaporator and condenser temperatures significantly below the critical temperature. Also, as the condenser temperature approaches the critical point, a greater portion of the heat is extracted in the desuperheating, relative to condensation. This reduces the coefficient of performance as it increases the average heat rejection temperature. It also increases the heat transfer area in the condenser. Given that it is desirable to operate away from the critical point where the latent heat of vaporization is high, for a given refrigerant fluid, how close would it be desirable to go to the critical temperature? A value of 50% of the latent heat at the normal boiling point is probably as low as would be desirable for many applications.

## Table 5.1

Freezing and normal boiling points for some common refrigerants.

| Refrigerant | Freezing point at atmospheric pressure (°C) | Boiling point at atmospheric pressure (°C) |
| --- | --- | --- |
| Ammonia | −78 | −33 |
| Chlorine | −101 | −34 |
| *n*-butane | −138 | 0 |
| *i*-butane | −160 | −12 |
| Ethylene | −169 | −104 |
| Ethane | −183 | −89 |
| Methane | −182 | −161 |
| Propane | −182 | −42 |
| Propylene | −185 | −48 |
| Nitrogen | −210 | −196 |

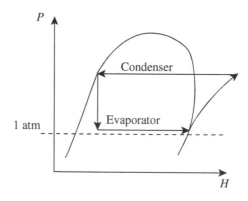

## Figure 5.16

Choice of refrigerant for pressure. Source: (Smith 2016), *Chemical Process Design and Integration*, 2nd Edition, John Wiley & Sons Ltd.

**Example 5.2**    Figure 5.17 gives the operating ranges of a number of refrigerants. The upper limit of the operating range in Figure 5.17 has been set to ambient temperature if the critical temperature is well above ambient temperature, or to a temperature corresponding with a heat of vaporization of 50% of that at atmospheric pressure. The lower temperature in Figure 5.17 has been limited by the normal boiling point to avoid vacuum conditions in the refrigeration loop. It should be noted that the operating range of the refrigerants can be extended at the lower limit by operation under vacuum conditions and at the upper limit by relaxing the restriction that the latent heat should not be lower than 50% of that at atmospheric pressure. The refrigerants in Figure 5.17 have been placed in approximate order of power requirement for a given refrigeration duty. Four process streams given in Table 5.2 require refrigeration cooling. Given the information in Figure 5.17, make an initial choice of refrigerants for each of the streams in Table 5.2.

**Solution**

*Stream 1*. At 175 K, the only suitable refrigerant from Figure 5.17 is ethylene. Methane is too close to its upper limit. If ethylene is chosen as a refrigerant and heat rejection is required to ambient at 313 K, then it needs to be cascaded with another refrigerant for heat rejection to ambient. In principle, ethylene could be cascaded with chlorine, ammonia, *i*-butane, propylene, or propane. However, the overlap with *i*-butane is small and will require a small temperature difference in the heat exchanger linking the two cycles. The choice of refrigerant for the cascade will depend on a number of factors. Choosing chlorine will minimize the power for the system, but introduces significant safety problems. Choosing a component already in the process would be desirable in order to avoid introducing new safety problems (e.g. propylene).

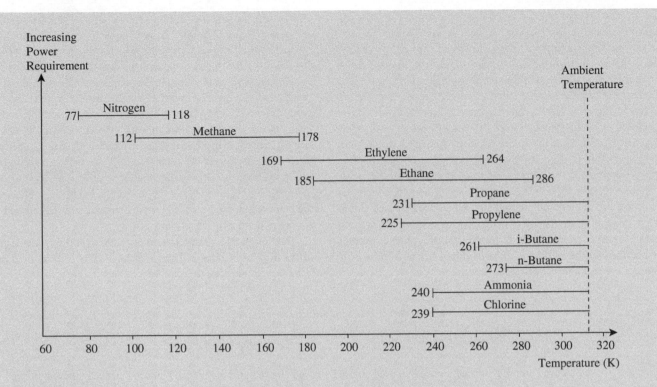

**Figure 5.17**

Operating ranges of refrigerants. Source: (Smith 2016), *Chemical Process Design and Integration*, 2nd Edition, John Wiley & Sons Ltd.

**Table 5.2**

Process streams to be cooled by refrigeration.

| Stream no. | Temperature (K) |
|---|---|
| 1 | 175 |
| 2 | 200 |
| 3 | 230 |
| 4 | 245 |

*Stream 2.* At 200 K, either ethane or ethylene would be suitable refrigerants, with ethane being slightly better from the point of view of the power requirements. As for Stream 1, a cascade system would be required for heat rejection to ambient temperature.

*Stream 3.* At 230 K, propylene, ethane, and ethylene would all be suitable refrigerants. Given that propylene requires the lowest power and does not require a cascade for heat rejection to ambient, this would be a suitable choice.

*Stream 4.* At 245 K, chlorine, ammonia, propylene, and propane could all be chosen. In principle, ethane and ethylene could also have been included but at 245 K they are too close to their critical temperature and would require a cascade design. The safety problems associated with chlorine are likely to be greater than ammonia. Thus, ammonia might be a suitable choice of refrigerant. Choosing a component already in the process would be desirable.

It should be noted that the lower and upper bounds for the temperatures in Figure 5.17 are not rigid, but only guidelines. The lower bound could be at a lower temperature if vacuum operation in the evaporator is acceptable. The upper bound could be at a higher temperature if a lower latent heat of vaporization is acceptable, which would lead to higher refrigerant flowrates for the same cooling duty.

# 5.7   Mixed Refrigerants for Compression Refrigeration

It was discussed above how pure refrigerants have a restricted temperature range over which they can operate. The working range of refrigerant fluids can be extended and modified by using a mixture for the refrigerant rather than a pure component. Mixed refrigerants can then be applied in simple, multistage, or cascade refrigeration systems. However, unlike pure refrigerants and azeotropic mixtures, the temperature and vapor and liquid compositions of non-azeotropic mixtures do not remain constant at constant pressure as the refrigerants evaporate or condense. Use of mixed refrigerants can be particularly effective if the cooling duty involves a significant change in temperature. The composition of the mixture is selected such that the liquid refrigerant evaporates over a temperature range similar to that of the process cooling demand. Figure 5.18 compares a pure refrigerant (or an azeotropic mixture) and a mixed refrigerant, both cooling a low-temperature heat source that changes temperature significantly. In Figure 5.18a, it can be seen that a pure refrigerant (or azeotropic mixture) evaporates at a constant temperature, leading to a small difference in temperature at one end of the heat exchanger, but a large difference in temperature at the other end. By contrast, a mixed refrigerant, as shown in Figure 5.18b, evaporates over a range of temperature and follows more closely the low-temperature cooling profile. Small temperature differences throughout the heat exchange in low-temperature systems lead to lower refrigeration power requirements. A higher average temperature for the refrigeration evaporation means a lower temperature lift and a lower power requirement.

Figure 5.19 illustrates a simple refrigeration cycle using a mixed refrigerant. It also shows temperature enthalpy profiles for both the process cooling demand and the refrigerant evaporation profile. It can be seen in the refrigerant evaporation profile for the process that cooling demand tends to follow the process cooling demand with a much smaller temperature difference than would be the case for a pure refrigerant. The match between the process cooling and the refrigerant evaporation would normally be carried out in a plate-fin heat exchanger in a countercurrent match with small temperature differences. It should be noted that it is normal practice in such cycles to feature some superheating of the refrigerant, as illustrated in Figure 5.19. Superheating the refrigerant before it is returned to the compressor protects the compressor against potential damage from liquid entering the compressor.

The design of a mixed refrigerant system requires the degrees of freedom to be manipulated simultaneously. These are as follows (Lee et al. 2003).

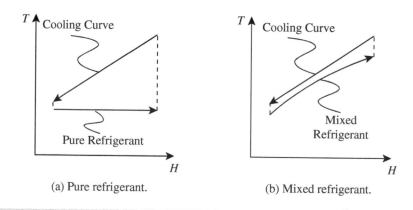

(a) Pure refrigerant.    (b) Mixed refrigerant.

**Figure 5.18**

Pure and mixed refrigerants. Source: (Smith 2016), *Chemical Process Design and Integration*, 2nd Edition, John Wiley & Sons Ltd.

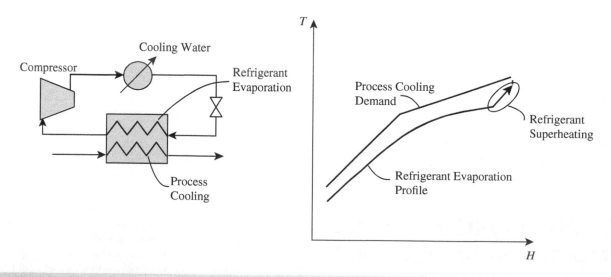

**Figure 5.19**

A simple refrigeration cycle using a mixed refrigerant. Source: Adapted from (Smith 2016), *Chemical Process Design and Integration*, 2nd Edition, John Wiley & Sons Ltd.

1) *Pressure level*. The choice of pressure level for the mixed refrigerant evaporation affects the temperature difference between the process cooling curve and refrigerant evaporation curve. This is illustrated in Figure 5.20, where a natural gas stream is cooled and liquefied using a mixed refrigerant. The mixed refrigerant evaporation profile shown in Figure 5.20a is infeasible as a result of a temperature across. Increasing the compressor discharge pressure, as in Figure 5.20b creates feasible temperature differences throughout the natural gas liquefaction. However, this is at the expense of increased compressor power. Increasing the overall temperature difference will increase the refrigeration power requirements.

2) *Refrigerant flowrate*. Increasing the refrigerant flowrate can widen the temperature difference between the process cooling curve and refrigerant evaporation curve and vice versa. This is illustrated in Figure 5.21, again for a natural gas liquefaction process. The mixed refrigerant evaporation profile shown in Figure 5.21a is infeasible as a result of a temperature cross. Increasing the refrigerant flowrate, as illustrated in Figure 5.21b, creates feasible temperature differences throughout the natural gas liquefaction. However, increasing the refrigerant flowrate increases the compressor power. If the refrigerant flowrate is too high, there might be some wetness in the inlet stream to the compressor, which should be avoided. Therefore, the refrigerant flowrate can be changed only within a feasible range.

3) *Refrigerant composition*. The composition of the mixture can be varied to achieve the required characteristics. Introduction of new components or replacing existing components by new

ones provides additional freedom to achieve better performance. Unlike adjusting the refrigerant pressure level and flowrate, optimization of the composition does not inevitably mean that increasing the temperature difference in some part of the heat exchange will result in higher refrigeration power requirements. This is illustrated in Figure 5.22 for the same natural gas liquefaction. Figure 5.22a shows a mixed refrigerant evaporation profile that features a temperature cross. In this case, the refrigerant composition can be changed to alter the mixed refrigerant evaporation profile. Figure 5.22b shows an alternative refrigerant evaporation profile through changing the composition that features feasible temperature differences throughout. In this case, change actually lowers the compressor power. Since pressure levels and refrigerant flowrate can only be changed within certain ranges, the refrigerant composition is the most flexible and significant variable when designing mixed refrigerant systems.

The major difficulty in selection of refrigerant composition for mixed refrigerants comes from the interactions between the variables and the small temperature difference between the process cooling curve and refrigerant evaporation curve. Any change in the refrigerant composition, evaporating pressure or the refrigerant flowrate will change the shape and position of the refrigerant evaporation curve.

Optimization can be used to vary the refrigerant composition, flowrate, and pressure level to obtain the desired refrigerant properties and conditions. Ideal matching between the process cooling curve and refrigerant evaporation curve would lead to the evaporation curve tending to follow the general shape of the cooling curve, but with the two curves not necessarily being parallel with

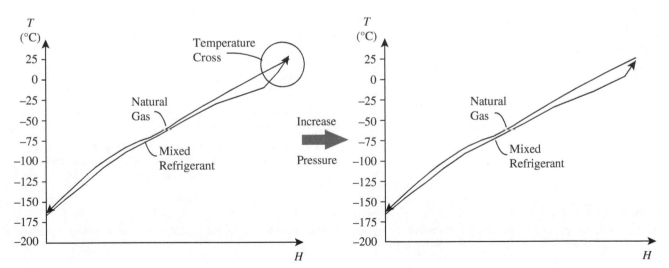

(a) Liquefaction of a natural gas stream exhibiting a temperature cross.

(b) Increasing the compressor discharge pressure with the same suction pressure creates feasible temperature differences throughout at the expense of increased compressor power.

## Figure 5.20

Effect of pressure on temperature feasibility for liquefaction of a natural gas stream using a mixed refrigerant. Source: Adapted from (Smith 2016), *Chemical Process Design and Integration*, 2nd Edition, John Wiley & Sons Ltd.

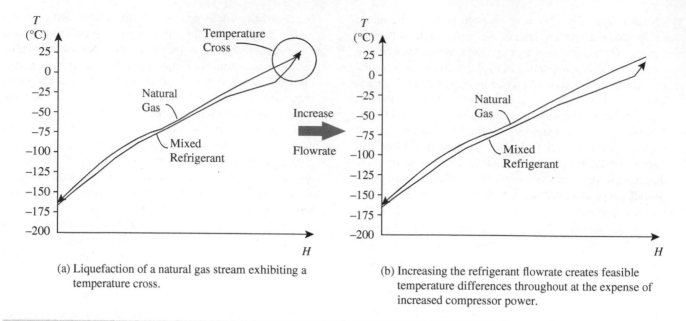

(a) Liquefaction of a natural gas stream exhibiting a temperature cross.

(b) Increasing the refrigerant flowrate creates feasible temperature differences throughout at the expense of increased compressor power.

## Figure 5.21

Effect of refrigerant flowrate on temperature feasibility for liquefaction of a natural gas stream using a mixed refrigerant. Source: Adapted from (Smith 2016), *Chemical Process Design and Integration*, 2nd Edition, John Wiley & Sons Ltd.

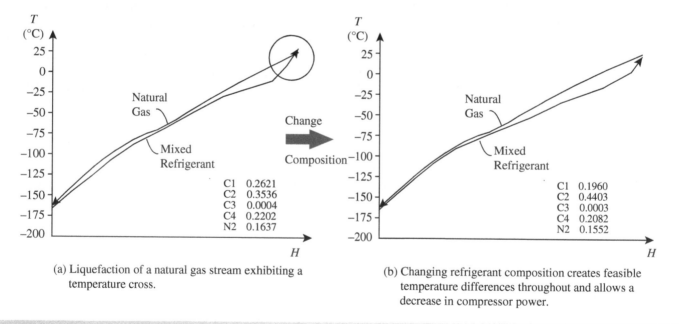

(a) Liquefaction of a natural gas stream exhibiting a temperature cross.

(b) Changing refrigerant composition creates feasible temperature differences throughout and allows a decrease in compressor power.

## Figure 5.22

Effect of refrigerant composition on temperature feasibility for liquefaction of a natural gas stream using a mixed refrigerant. Source: Adapted from (Smith 2016), *Chemical Process Design and Integration*, 2nd Edition, John Wiley & Sons Ltd.

each other, and the refrigerant profile will normally feature some superheating.

To determine the optimum settings for the compressor suction and discharge pressures, the refrigerant flowrate and composition requires a simulation model that is subjected to optimization. The degrees of freedom can be manipulated using nonlinear programming (e.g. sequential quadratic programming (SQP), see Smith 2016) to perform the optimization by manipulating the composition, flowrate, and pressures (Lee et al. 2003). The objective is normally to minimize the compressor power. However, this approach is vulnerable to the highly nonlinear characteristics of the optimization. Alternatively, a stochastic search method can be used at the expense of the additional computational requirements. Genetic algorithms (Smith 2016) have proven to be successful in developing good solutions (Wang 2004; Del Nogal et al. 2008). Great care must be taken to ensure temperature feasibility throughout the cooling process. The temperatures need to be checked at discrete points along the cooling profile to ensure that the temperature difference is greater than some practical minimum.

# 5.8 Absorption Refrigeration

Consider now absorption refrigeration. Compression refrigeration is powered by a compressor compressing refrigerant vapor (Figure 5.23a). The basic problem with this is the high compression costs. Figure 5.23b illustrates an alternative way to bring about the compression. A refrigerant vapor is first absorbed in a solvent. The resulting liquid solution can have its pressure increased using a pump. The compressed refrigerant is then separated from the solvent in a stripper (regenerator). The pump requires significantly less power to bring about the increase in pressure compared with the corresponding gas compression. The overall effect is to increase the pressure of the refrigerant with far less power. The drawback is that a heat supply is needed for the stripper (regenerator).

Figure 5.24 shows a typical absorption refrigeration arrangement. To the left of the cycle in Figure 5.24 is the absorber and stripper (regenerator) arrangement. Low-pressure refrigerant vapor from the evaporator is first absorbed in a solvent, which is then increased in pressure and then increased in temperature in a heat exchanger. The refrigerant then enters the vapor generator where the refrigerant is stripped from the solvent. Heat is input to the vapor generator and the solvent is cooled in the heat exchanger, decreased in pressure, and returned to the absorber. The high-pressure vapor from the vapor generator is condensed in the condenser, expanded in the expansion valve to produce the cooling effect and then enters the evaporator to provide process cooling.

The features of absorption refrigeration are that there is a low power requirement relative to compression refrigeration, but heat input to the vapor generator (stripper) is required. Heat output from the absorber is required, usually to cooling water. Absorption refrigeration is only useful for moderate temperature refrigeration.

The most common working fluids for absorption refrigeration are given in Table 5.3, together with the working range.

When should absorption refrigeration be used rather than compression refrigeration? There are two important criteria. The first is that absorption refrigeration can only be used when moderate levels of refrigeration are required. Second, it should only be used when there is a large source of waste heat available for the vapor generator. This must be at a temperature greater than 90 °C, but the higher the better.

# 5.9 Indirect Refrigeration

Indirect refrigeration is often used. Figure 5.25 shows a liquid intermediate being recycled to provide the cooling to a process load. Heat is removed from the liquid intermediate by a refrigeration cycle. This arrangement is used for distribution of

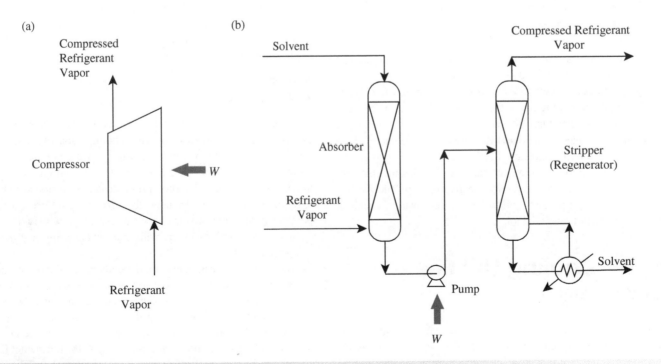

**Figure 5.23**

Compression versus absorption refrigeration. Source: (Smith 2016), *Chemical Process Design and Integration*, 2nd Edition, John Wiley & Sons Ltd.

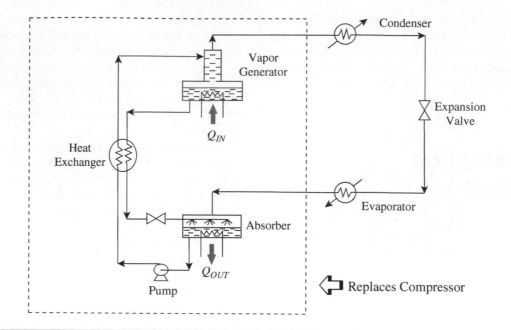

### Figure 5.24

A typical absorption refrigeration arrangement. Source: (Smith 2016), *Chemical Process Design and Integration*, 2nd Edition, John Wiley & Sons Ltd.

### Table 5.3

Common working fluids for absorption refrigeration.

| Refrigerant | Solvent | Lower temperature limit (°C) |
|---|---|---|
| Water | Lithium bromide | 5 |
| Ammonia | Water | −40 |

refrigeration across a number of process loads, or when contact between the refrigerant and process fluids is unacceptable for safety or product contamination reasons. The arrangement leads to higher power requirements than direct refrigeration because of the extra temperature lift required in the heat exchanger between the refrigerant fluid and the intermediate liquid.

The liquid intermediates used are typically water solutions of various concentrations, such as salts (e.g. calcium chloride, sodium chloride), glycols (e.g. ethylene glycol, propylene glycol), and alcohols (e.g. methanol, ethanol), or pure substances such as acetone, methanol, and ethanol.

# 5.10 Cooling Utilities – Summary

Rejection of waste heat is a feature of most chemical processes. High-temperature cooling utilities should be used in preference to low-temperature utilities. High-temperature waste heat should be used to generate steam. Cooling water using once-through cooling systems based on river water, and so on, can be used. However, such once-through systems require large amounts of

water, and the waste heat rejected to aquatic systems can create environmental problems.

The most common way to reject waste heat above ambient temperature is through recirculating cooling water systems. Both natural draft and mechanical draft cooling towers are used. Heat is rejected through the evaporation of water, but other losses of water result from the need for blowdown to prevent build-up of undesired contaminants in the recirculation and drift losses.

Refrigeration is required to produce cooling below ambient temperature. There are two broad classes of refrigeration:

- compression refrigeration;
- absorption refrigeration.

Compression refrigeration is by far the most common. Simple cycles can be employed to provide cooling typically as low as −40 °C. For lower temperatures, complex cycles are used. Economizers, presaturators and the introduction of multiple refrigeration levels allow multistage compression and expansion to be used to reduce the power requirements for refrigeration. Very low temperature systems require the use of cascade cycles with two refrigerant cycles linked together, with each cycle using a different refrigerant fluid.

The choice of refrigerant fluid depends on a number of issues. Use of an evaporator pressure below atmospheric pressure is normally avoided. It is desirable to have a high latent heat at the evaporator conditions in order to reduce the flowrate around the refrigeration loop in order to reduce the power requirements correspondingly. Various other factors relating to the refrigerant fluid also affect the choice of refrigerant, such as the toxicity, flammability, corrosivity, and environmental impact.

Mixed refrigerant systems use a mixture as refrigerant instead of pure refrigerants. Evaporation of the refrigerant mixture over a

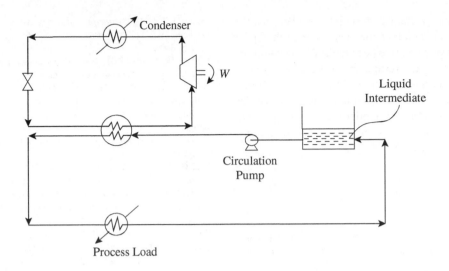

## Figure 5.25

Indirect refrigeration. Source: (Smith 2016), *Chemical Process Design and Integration*, 2nd Edition, John Wiley & Sons Ltd.

**5**

range of temperature allows a better match between the refrigerant duty and refrigerant evaporation if the refrigeration duty varies significantly in temperature. When designing mixed refrigerant systems, important degrees of freedom include:

- composition of the refrigerant mixture;
- pressure levels;
- refrigeration flowrate.

Absorption refrigeration is much less common than compression refrigeration. Absorption refrigeration works by compressing the refrigerant fluid dissolved in a solvent using a pump, rather than gas compression.

# Exercises

1. A refrigeration process is to reject heat to cooling water from the refrigeration condenser with a temperature difference of 5 °C between the cooling water and the refrigerant fluid. The summer wet bulb temperature can be assumed to be 22 °C. Assuming an approach temperature for the cooling tower of 3 °C and a cooling range for the cooling water of 10 °C, determine the condensing temperature of the refrigerant fluid.

2. Cooling water is being circulated at a rate of 20 m$^3$ min$^{-1}$ to the cooling network. The cooling water from the cooling tower is at a temperature of 25 °C and is returned at 40 °C. Measurements on the concentrations of the feed and circulating water indicate that there are five cycles of concentration. Assuming that the heat capacity of the water is 4.2 kJ kg$^{-1}$ K$^{-1}$ and the makeup water is at a temperature of 10 °C, calculate:

   a) the rate of evaporation assuming the latent heat of vaporization to be 2423 kJ kg$^{-1}$ at the conditions in the cooling tower;

   b) the makeup water requirement, assuming drift losses are negligible.

3. What options might be considered to accommodate the new cooling duty without having to invest in a new cooling tower if the cooling tower is near maximum capacity?

4. For cooling duties at:

   a) –40 °C,

   b) –60 °C,

   estimate the power required to reject 1 MW of cooling duty to cooling water with a return temperature of 40 °C. Assume the Carnot Efficiency to be 0.6, $\Delta T_{min}$ = 5 °C for the refrigeration evaporator and $\Delta T_{min}$ = 10 °C for the condenser. Suggest suitable refrigeration fluids for these two duties using Figure 15.17.

5. A process requires a low-temperature cooling duty of 1.5 MW at a process temperature at a constant temperature of –60 °C. This is to be achieved by a cascade refrigeration cycle consisting of two simple cycles, each using a pure refrigerant. The lower cycle removes heat from the process and rejects it to the upper cycle. The upper cycle rejects heat to cooling water with a return temperature of 30 °C. Heat is to be transferred in each evaporator and condenser with a temperature difference $\Delta T$ = 5 °C. The partition temperature between the cycles (evaporation temperature of the upper cycle that rejects heat to cooling water) is –20 °C. The power required for the refrigeration process for each cycle can be calculated by a Carnot efficiency model with an efficiency of 0.6.

   a) Calculate the power required for the lower cycle in MW.
   b) Calculate the heat rejection from the lower cycle to the upper cycle in MW.
   c) Calculate the power required for the upper cycle rejecting heat to cooling water in MW.

6. A cascade refrigeration system is to be used to service a cooling duty with an evaporation temperature of –90 °C and rejecting heat ultimately to cooling water with a rejection temperature of

30 °C. The partition temperature of the interface between the two cycles is the temperature of the evaporator of the upper (high-temperature) cycle and is to be optimized. The minimum temperature allowed for the refrigerant in the upper cycle is 225 K and the maximum temperature allowed for the refrigerant in the lower cycle is 264 K. Assuming the cycles can be modeled by a Carnot efficiency model with an efficiency of 0.6, determine the partition temperature to minimize the total power input. Assume $\Delta T_{min} = 5$ °C.

# References

Del Nogal, F., Kim, J.K., Perry, S., and Smith, R. (2008). Optimal design of mixed refrigerants. *Industrial and Engineering Chemistry Research* 47: 8724.

Dossat, R.J. (1991). *Principles of Refrigeration*, 3e. Prentice Hall.

Gosney, W.B. (1982). *Principles of Refrigeration*. Cambridge University Press.

Hudson Products (2013). *Basics of Air-cooled Heat Exchangers*. http://www.hudsonproducts.com/products/finfan.

Isalski, W.H. (1989). *Separation of Gases*. Oxford Science Publications.

Kim, J.K. and Smith, R. (2001). Cooling water system design. *Chemical Engineering Science* 56: 3641.

Kröger, D.G. (1998). *Air-Cooled Heat Exchangers and Cooling Towers*. Department of Mechanical Engineering, University of Stellenbosch, South Africa.

Lee, G.C., Smith, R., and Zhu, X.X. (2003). Optimal synthesis of mixed-refrigerant systems for low-temperature processes. *Industrial and Engineering Chemistry Research* 41: 5016.

Serth, R.W. (2007). *Process Heat Transfer: Principles and Applications*. Academic Press.

Smith, R. (2016). *Chemical Process Design and Integration*, 2e. Wiley.

Wang, J. (2004). *Synthesis and Optimization of Low Temperature Gas Separation Processes*, PhD Thesis, UMIST, UK.

# Chapter 6

# Waste Treatment

In addition to heating and cooling utilities, another important utility is waste treatment. When considering heating and cooling utilities, it was noted that emphasis should be given to efficient heat recovery in order to minimize the amount of heating and cooling utilities required by a process plant. When considering waste treatment, a high emphasis should be given to the efficient use of raw materials, including water. Minimizing waste, and therefore minimizing emissions, will almost always lead to a better outcome than providing complex waste treatment systems (Smith 2016). Once waste has been created, it cannot be destroyed. It can be concentrated or diluted, its physical or chemical form can be changed, but it cannot be destroyed. Waste treatment systems do not so much solve the problem, but move it from one place to another. However, once waste minimization has been taken to an economic level, the treatment of the resulting waste must be considered.

## 6.1 Aqueous Emissions

Figure 6.1 shows a schematic of a greatly simplified water system for a processing site. Raw water enters the processing system and might need some raw water treatment. This might be something as simple as sand filtration. In many instances, the raw water supply is good enough to enter the processes directly. Water is used in various operations as a reaction medium, solvent in extraction processes, for cleaning, and so on. Water becomes contaminated and is discharged to effluent. Also, as shown in Figure 6.1, some of the raw water is required for the steam system. It first requires upgrading in boiler feed water treatment to remove suspended solids, dissolved salts, and the dissolved gases before being fed to the steam boiler, as discussed in Chapter 4. Deionized water might also be required for process use. The steam from steam boilers is distributed and some of the steam condensate is not returned to boilers, but is lost to effluent. The boiler requires blowdown to remove the build-up of solids, as discussed in Chapter 4. Also, as discussed in Chapter 4, the ion-exchange beds used to remove dissolved salts and soluble ions need to be regenerated by saline solutions or acids and alkalis and this goes to effluent. Finally, as shown in Figure 6.1, water is used in evaporative cooling systems to make up for the evaporative losses and blowdown from the cooling water circuit, as discussed in Chapter 5. All of the effluents tend to be mixed together, along with contaminated storm water, treated centrally in a wastewater treatment system, and discharged to the environment. It is usual for most of the water entering the processing system to leave as wastewater. If the use of water can be reduced, then this will reduce the cost of water supplied. However, it will also reduce the cost of effluent treatment, as a result of most of the water entering the processing environment leaving as effluent. There is thus a considerable incentive to reduce both fresh water consumption and wastewater generation.

Figure 6.2a shows three operations, each requiring fresh water and producing wastewater. By contrast, Figure 6.2b shows an arrangement where there is a *reuse* of water from Operation 2 to Operation 1. Reusing water in this way reduces both the volume of the fresh water and the volume of wastewater, as the same water is used twice. However, for such an arrangement to be feasible, any contamination level at the outlet of Operation 2 must be acceptable at the inlet of Operation 1. Not all operations require the highest quality of water (Smith 2016).

Figure 6.2c and d both show arrangements involving *regeneration*. Regeneration is a term used to describe any treatment process that regenerates the quality of water such that it is acceptable for further use. Figure 6.2c shows *regeneration reuse* where the outlet water from Operation 2 is too contaminated to be used directly in Operation 3. A regeneration process between the two allows reuse to take place. Regeneration reuse reduces both the volume of the fresh water and the volume of the wastewater, as with reuse, but also removes part of the *effluent load* (i.e. kilograms of contaminant). The regeneration, in addition to allowing a reduction in the water volume, also removes part of the contaminant load that would have to be otherwise removed in the final effluent treatment before discharge.

A third option is shown in Figure 6.2d where a regeneration process is used on the outlet water from the operations and the

*Process Plant Design*, First Edition. Robin Smith.
© 2024 John Wiley & Sons Ltd. Published 2024 by John Wiley & Sons Ltd.
Companion website: www.wiley.com/go/processplantdesign

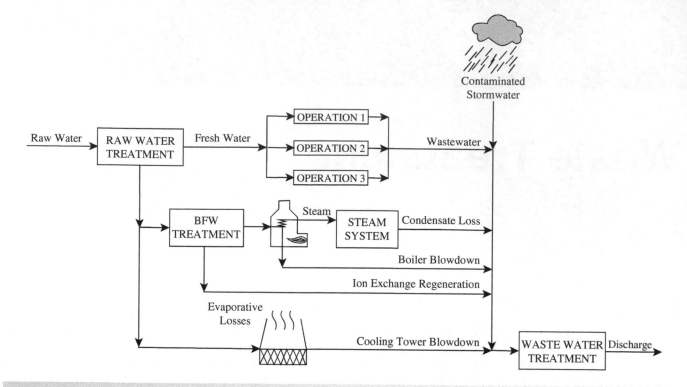

## Figure 6.1

A typical water and effluent treatment system. Source: (Smith 2016), *Chemical Process Design and Integration*, 2nd Edition, John Wiley & Sons Ltd.

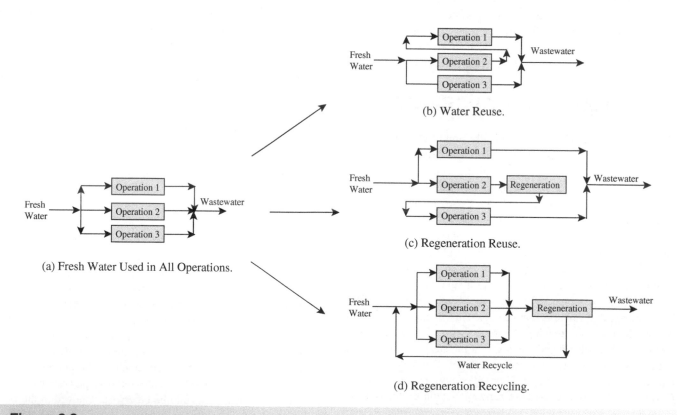

(a) Fresh Water Used in All Operations.

(b) Water Reuse.

(c) Regeneration Reuse.

(d) Regeneration Recycling.

## Figure 6.2

Water re-use and regeneration. Source: (Smith 2016), *Chemical Process Design and Integration*, 2nd Edition, John Wiley & Sons Ltd.

water is *recycled*. The distinction between the regeneration reuse shown in Figure 6.2c and the *regeneration recycling* shown in Figure 6.2d is that in regeneration reuse the water only goes through any given operation once. Figure 6.2c shows that the water goes from Operation 2 to regeneration, then to Operation 3, and then discharge. By contrast, in Figure 6.2d, the water can go through the same operation many times. Regeneration recycling reduces the volume of fresh water and wastewater and also reduces the effluent load by virtue of the regeneration process taking up part of the required effluent treatment load.

As shown in Figure 6.1, all of the wastewater was collected together and treated centrally before discharge. Another way to deal with the effluent treatment is by *distributed effluent treatment* or *segregated effluent treatment*. The basic concept is illustrated in Figure 6.3. In addition to some reuse of water between Operation 2 and Operation 1, some local treatment (distributed treatment) is taking place on the outlet of Operation 1 and on the outlet of Operation 3 before going to final wastewater treatment and discharge. Another feature of the arrangement in Figure 6.3 is that a number of the effluents are no longer being treated before discharge. The arrangement in Figure 6.1 is such that the more contaminated effluents from the outlet of Operations 1, 2, and 3 need to be diluted by the less contaminated streams before being treated and discharged. By contrast, as shown in Figure 6.3, the pretreatment on the outlet of Operation 1 and Operation 3 is such that the dilution is no longer required before final effluent treatment and discharge.

This means that the streams with only light contamination now no longer need to be treated.

The capital cost of aqueous waste treatment operations is generally proportional to the total flow of wastewater and the operating cost generally increases with decreasing concentration for a given mass of contaminant to be removed. Thus, if two streams require different treatment operations, it makes no sense to mix them and treat the two streams in both treatment operations. This will increase both capital and operating costs. The concept of distributed effluent treatment is one that tends to treat effluents before they are mixed together. Treatment is made specific to individual (or small numbers of ) contaminants while still concentrated. The benefit is that, by avoiding mixing, this increases the potential to recover material, leading to less waste and a lower cost of raw materials. However, the overriding benefit is usually that the effluent volume to be treated is reduced significantly, leading to lower effluent treatment costs overall.

Figure 6.4 shows an arrangement for the segregation of different types of effluent. Aqueous and organic effluents are kept separate. Aqueous effluents can be most often collected and run in open sewers to treatment, or directly to discharge if uncontaminated. Organic effluents are not only kept separate, but also in closed sewers. Closed sewers prevent atmospheric pollution from the evaporation of organic components. The organic waste would usually need a different treatment from that for aqueous waste.

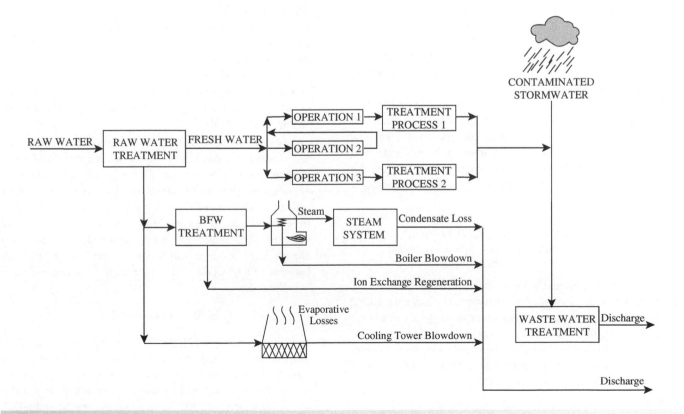

**Figure 6.3**

Distributed effluent treatment. Source: (Smith 2016), *Chemical Process Design and Integration*, 2nd Edition, John Wiley & Sons Ltd.

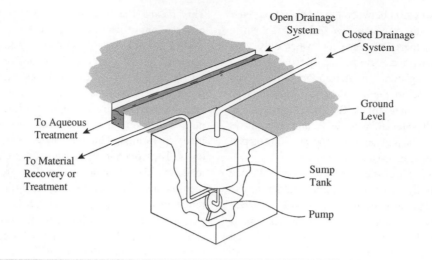

## Figure 6.4

Segregation of liquid effluents using separate drains. Source: (Smith 2016), *Chemical Process Design and Integration*, 2nd Edition, John Wiley & Sons Ltd.

Effluents must comply with environmental regulations before discharge. The environmental effects of pollution can be direct, such as toxic emissions providing a fatal dose of toxicant to fish, animal life, and even human beings. The effects can also be indirect. Toxic materials that are non-biodegradable, such as insecticides and pesticides, if released to the environment, are absorbed by bacteria and enter the food chain. These compounds can remain in the environment for long periods of time, slowly being concentrated at each stage in the food chain until ultimately they prove fatal, generally to predators at the top of the food chain, such as fish or birds. The concentration, and perhaps load, of contamination of various specified contaminants must be less than the regulatory requirements.

Consider first aqueous emissions of organic waste material. When this is discharged to the receiving water, bacteria feed on the organic material. This organic material will eventually be oxidized to stable end products. Carbon in the molecules will be converted to $CO_2$, hydrogen to $H_2O$, nitrogen to $NO_3^-$, sulfur to $SO_4^{2-}$, and so on. As an example, consider the degradation of urea:

$$CH_4N_2O + 9/2\,O_2 \longrightarrow CO_2 + 2H_2O + 2NO_3^-$$
$$\text{urea} \quad \text{oxygen} \quad \text{carbon-} \quad \text{water} \quad \text{nitrate}$$
$$\text{dioxide}$$

This equation indicates that every molecule of urea requires 9/2 molecules of oxygen for complete oxidation. The oxygen required for the reactions depletes receiving water of oxygen, causing the death of aquatic life. Standard tests have been developed to measure the amount of oxygen required to degrade a sample of wastewater (Tchobanoglous and Burton 1991; Berne and Cordonnier 1995; Eckenfelder and Musterman 1995).

1) *Biochemical oxygen demand (BOD)*. A standard test has been devised to measure biological oxygen demand (BOD) in which the oxygen utilized by microorganisms in contact with the wastewater over a five-day period at 20 °C is measured (usually termed $BOD_5$). The period of the test can be extended to a much longer period (in excess of 20 days) to measure the *ultimate demand*. While the $BOD_5$ test gives a good indication of the effect the effluent will have on the environment, it requires five days to carry out (or longer for the ultimate BOD). Other tests have been devised to accelerate the oxidation process.

2) *Chemical oxygen demand (COD)*. In the chemical oxidation demand (COD) test, oxidation with acidic potassium dichromate is used. A catalyst (silver sulfate) is required to assist oxidation of certain classes of organic compounds. Chemical oxygen demand results are generally higher than $BOD_5$, since the COD test oxidizes materials that are only slowly biodegradable. Although the COD test provides a strong oxidizing environment, certain organic compounds are oxidized only slowly, or not at all.

3) *Total oxygen demand (TOD)*. The total oxygen demand (TOD) test measures the oxygen consumed when a sample of wastewater is oxidized in a stream of air at high temperature (up to 1200 °C) in a furnace. Under these harsh conditions, all the carbon is oxidized to $CO_2$. The oxygen demand is determined from the change in oxygen in the carrier gas. The resulting value of TOD embraces oxygen required to oxidize both organic and inorganic substances present. The relationship between $BOD_5$, COD, and TOD for the same organic waste is in the order:

$$BOD_5 < COD < TOD$$

4) *Total organic carbon (TOC)*. The total organic carbon (TOC) test measures the carbon dioxide produced when a sample of wastewater is subjected to a strongly oxidizing environment. One option is to oxidize the sample in a stream of air at a high temperature (around 1000 °C) in a furnace, similar to the TOD test, but measuring the change in $CO_2$ rather than the change in $O_2$. To allow complete oxidation of all carbon compounds also requires a catalyst such as copper oxide or platinum. Rather than using high temperature in a furnace, other strongly

oxidizing environments can also be used. For example, UV light with sodium persulfate as a digesting reagent can be used. Other strong chemical oxidizing environments can be used. To obtain the TOC requires that inorganic carbon compounds be removed prior to test, or results corrected for their presence. The inorganic carbon can be removed prior to test by adding acid to convert the inorganic carbon to carbon dioxide that must be stripped from the sample by a sparge carrier gas.

The relationships between $BOD_5$, COD, and TOD are not easy to establish, since different materials will oxidize at different rates. To compound the problem, many wastes contain complex mixtures of oxidizable materials, perhaps together with chemicals that *inhibit* the oxidation reactions.

Effluent treatment regulations might specify a level of $BOD_5$, COD, or both. Other contaminants that might be specified are:

- specifically nominated contaminants (e.g. phenol, benzene, etc.);
- heavy metals (e.g. chromium, cobalt, vanadium, etc.);
- halogenated organic compounds;
- organic nitrogen;
- organic sulfur;
- nitrates;
- phosphates;
- suspended solids;
- pH, and so on.

Nitrates and phosphates in particular can cause *eutrophication* of the receiving water, either terrestrial or marine water. Nitrates and phosphates act as nutrients that lead to enhanced growth of aquatic vegetation or *phytoplankton* and *algal blooms*. This disrupts normal functioning of the ecosystem, causing a variety of problems such as a lack of oxygen needed for aquatic life to survive, decreased biodiversity, changes in species composition and dominance, and toxicity effects.

In addition to the levels of contamination, typically specified as parts per million (ppm) or milligrams per liter (mg $l^{-1}$), there might be regulations for the total discharge of a contaminant in

kilograms per day (kg $d^{-1}$) or tons per year (t $y^{-1}$). There might also be a specification for the maximum effluent temperature.

Water treatment processes can be classified in order as (Tchobanoglous and Burton 1991):

- primary (or pretreatment);
- secondary (or biological);
- tertiary (or polishing).

## 6.2  Primary Wastewater Treatment Processes

*Primary* or *pretreatment* of wastewater can involve either physical or chemical treatment, depending on the nature of the contamination, and serves three purposes:

- allows reuse or recycling of water;
- recovers useful material from the wastewater where possible;
- prepares the aqueous waste for biological treatment by removing excessive load or contaminants that will inhibit the biological processes in biological treatment.

The pretreatment processes may be most effective when applied to individual waste streams from particular processes or process steps before effluent streams are combined for biological treatment.

A brief review of the primary treatment methods will now be given (Tchobanoglous and Burton 1991; Berne and Cordonnier 1995; Eckenfelder and Musterman 1995; Smith 2016). Pretreatment usually starts with phase separation if the effluent is a heterogeneous mixture.

1) *Solids separation*. The most commonly used technique for separation of solids is *sedimentation*. Figure 6.5 shows a *clarifier*. The effluent containing solids is fed at the center of the vessel below the surface of the liquid. Clear liquid overflows from the top edge of the vessel. A slowly removing rake serves

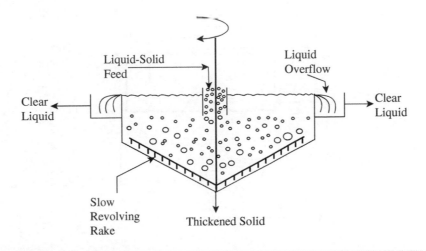

**Figure 6.5**

A clarifier for liquid–solid separation. Source: (Smith 2016), *Chemical Process Design and Integration*, 2nd Edition, John Wiley & Sons Ltd.

to scrape the thickened solids toward the center of the base for removal. As an example of the effectiveness of clarifiers, in petroleum refinery applications, they are capable of removing typically 50–80% of suspended solids, together with 60–95% of dispersed hydrocarbon (which rises to the surface), 30–60% of $BOD_5$, and 20–50% of COD (Betz Laboratories 1991). The performance depends on the design and the effluent being treated.

2) *Coalescence.* Coalescence by gravity in simple settling devices can often be used to separate immiscible liquid–liquid mixtures. In petroleum and petrochemical plants, a common device used for the separation of dispersed hydrocarbon liquids from aqueous effluent is the American Petroleum Institute (API) separator, as illustrated in Figure 6.6. This is a simple settling device in which the effluent enters a large volume. The resulting low velocity allows light particles of hydrocarbon to rise to the surface and any heavy solid particles to settle to the base of the device. Rakes might be employed to move the light material along the surface and the heavy material along the base for collection. The light material can be removed using a *scum skimmer*. This, for example, could be

as simple as a horizontal pipe with a slot in its upper surface. Rotation of the device, such that the slot is just below the surface of the liquid, allows the light material (hydrocarbon) to overflow into the pipe for collection. In petroleum refinery applications, API separators typically remove 60–99% of dispersed hydrocarbon liquids, together with 10–50% of suspended solids, 5–40% of $BOD_5$, and 5–30% of COD (Betz Laboratories 1991). It is a very simple device used for the first stage of treatment. The performance depends on the design and the effluent being treated, and the performance can be enhanced by the use of chemical additives.

3) *Flotation.* Flotation can be used to separate solid or immiscible liquid particles from aqueous effluent. These particles should have a lower density than water and be naturally hydrophobic. Gas bubbles are generated in the water and become attached to solid particles or immiscible liquid droplets, causing the particles or droplets to rise to the surface for collection. Direct air injection of air is usually not the most effective method of flotation. Dissolved-air flotation is usually superior. A typical arrangement is shown in Figure 6.7. This shows some of the effluent water being recycled and air being dissolved in the

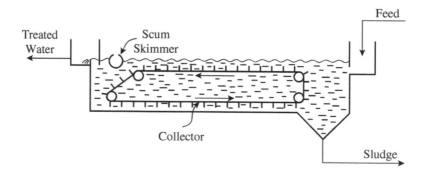

**Figure 6.6**

A typical API (American Peroleum Institute) separator. Source: (Smith 2016), *Chemical Process Design and Integration*, 2nd Edition, John Wiley & Sons Ltd.

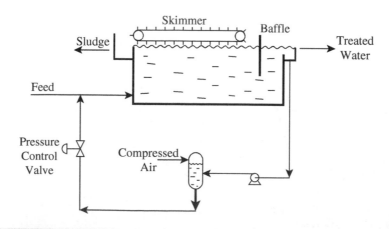

**Figure 6.7**

Dissolved air flotation (DAF). Source: (Smith 2016), *Chemical Process Design and Integration*, 2nd Edition, John Wiley & Sons Ltd.

recycle under pressure. The pressure of the recycle is then reduced, releasing the air from solution as a mist of fine bubbles. This is then mixed with the incoming feed that enters the cell. Low-density material floats to the surface with the assistance of the air bubbles and is collected by skimming. Flotation is particularly effective for the separation of very small and light particles. Typical applications include removal of dispersed hydrocarbon liquids from petroleum and petrochemical effluents and removal of fibers from pulp and paper effluents. As an example of the performance of such units, when applied to effluent from a petroleum refinery, dissolved-air flotation can typically remove 70–85% of dispersed oil and 50–85% of suspended solids, together with 20–70% of $BOD_5$ and 10–60% of COD (Betz Laboratories 1991). The performance depends on the equipment design and the effluent being treated. In such applications, it is normally used after an API separator has been used for preliminary treatment.

4) *Stripping*. Volatile organic compounds (VOCs) and dissolved gases can be stripped from wastewater. The usual arrangement would involve wastewater being fed down through a column with packing or trays and the stripping agent (usually steam or air) fed to the bottom of the column. If steam is used as a stripping agent, either live steam or a reboiler can be used. The use of live steam increases the effluent volume. The volatile organic materials are taken overhead, condensed, and, if possible, recycled to the process. If recycling is not possible, then further treatment or disposal is necessary. A common application of steam stripping is the stripping of hydrogen sulfide and ammonia from water in petroleum refinery operations to regenerate contaminated water for reuse. Steam stripping is capable of removing typically 90–99% $H_2S$, 90–97% $NH_3$, and 75–99% organic materials. It should be noted though, that some organic materials are resistant to steam stripping and it is thus not a universal solution to contamination with organic materials. If air is used as a stripping agent, further treatment of the stripped material will be necessary. The gas might be fed to a thermal oxidizer or some attempt made to recover material by use of adsorption.

5) *Liquid–liquid extraction*. With liquid–liquid extraction, wastewater containing organic waste is contacted with a solvent in which the organic waste is more soluble. The waste is then separated from the solvent by evaporation or distillation and the solvent recycled. One common application of liquid–liquid extraction is the removal and recovery of phenol and compounds of phenol from wastewaters. Although phenol can be removed by biological treatment, only limited levels can be treated biologically. Variations in phenol concentration are also a problem with biological treatment, since the biological processes take time to adjust to the variations.

6) *Adsorption*: Adsorption can be used for the removal of organic compounds (including many toxic materials) and heavy metals (especially when complexed with organic compounds). Activated carbon is primarily used as the adsorbent, although synthetic resins are also used. Both fixed and moving beds can be used, but fixed beds are by far the most commonly used arrangement. For activated carbon, the removal of organic compounds depends, amongst other things, on the molar mass and the polarity of the molecule. A general trend is that nonpolar molecules (e.g. benzene) tend to adsorb more readily than polar molecules (e.g. methanol) on activated carbon. As an example, adsorption with activated carbon when applied to petroleum refinery effluent can remove typically 70–95% $BOD_5$ and 70–90% COD, depending on the effluent being treated (Betz Laboratories 1991).

As the adsorbent becomes saturated, regeneration is required. When removing organic materials, activated carbon can be regenerated by steam stripping or heating in a furnace. Stripping allows recovery of material, whereas thermal regeneration destroys the organic material. Thermal regeneration requires a furnace with temperatures above 800 °C to oxidize the adsorbates. This causes a loss of carbon of around 5–10% per regeneration cycle.

7) *Ion exchange*. Ion exchange is used for selective ion removal and finds some application in the recovery of specific materials from wastewater, such as heavy metals. As with adsorption processes, regeneration of the medium is necessary. Ion exchange resins are regenerated chemically, which produces a concentrated waste stream requiring further treatment or disposal.

8) *Wet oxidation*. In wet oxidation, an aqueous mixture is heated in the liquid phase under pressure in the presence of air or pure oxygen, which oxidizes the organic material. The efficiency of the oxidation process depends on the reaction time and pressure.

Temperatures of 150–300 °C are used, together with pressures of 3–200 bar, depending on the process and the nature of the waste being treated and whether a catalyst is used. Oxidation at low temperatures and pressures is only possible if a catalyst is used. Carbon is oxidized to $CO_2$, hydrogen to $H_2O$, chlorine to $Cl^-$, nitrogen to $NH_3$ or $N_2$, sulfur to $SO_4^{2-}$, phosphorous to $PO_4^{2-}$, and so on.

Wet oxidation is particularly effective in treating aqueous wastes containing organic contaminants with a COD up to 2% prior to biological treatment. Chemical oxygen demand can be reduced by up to 95% and organic halogen compounds by up to 95%. Organic halogen compounds are particularly resistant to biological degradation. If designed for a specific organic contaminant (e.g. phenol), removal can be 99% or greater. Wet oxidation is often used prior to biological treatment to pretreat wastes that would otherwise be resistant to biological oxidation.

A basic flowsheet for a wet oxidation process is shown in Figure 6.8. Although the oxidation reactions release heat, the process might still require a net input of heat from an external source (e.g. steam or hot oil). In some cases, the heat release can be high enough to avoid the need for an external source of heat. The largest cost associated with the process is the capital cost of the high-pressure reactor, which normally has an internal titanium cladding.

9) *Chemical oxidation*. Chemical oxidation can be used for the oxidation of organic contaminants that are difficult to treat biologically. It can be used to kill microorganisms by oxidation.

**6**

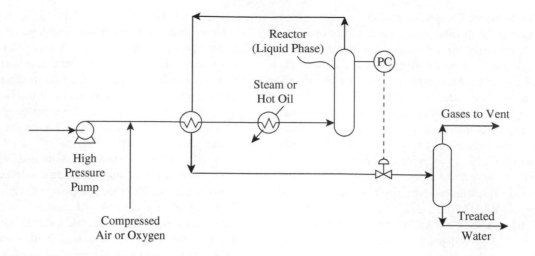

### Figure 6.8

A typical wet oxidation process. Source: (Smith 2016), *Chemical Process Design and Integration*, 2nd Edition, John Wiley & Sons Ltd.

When used before biological treatment, organic pollutants that are difficult to treat biologically can be oxidized to simpler, less refractory organic compounds. Chlorine (as gaseous chlorine or hypochlorite ions) can be used. In solution, chlorine reacts with water to form hypochlorous acid (HOCl) according to:

$$Cl_2 + H_2O \rightarrow HOCl + HCl$$

$$HOCl \rightleftharpoons H^+ + OCl^-$$

Then, for example, the reaction for nitrogen removal is:

$$2NH_3 + 3HOCl \rightarrow N_2 + 3H_2O + 3HCl$$

Heterogeneous solid catalysts can enhance the performance. Chlorination has the disadvantage that small quantities of undesirable halogenated organic compounds can be formed.

Hydrogen peroxide is another common oxidizing agent. It is used as a 30–70% solution for the treatment of:

- cyanides;
- formaldehyde;
- hydrogen sulfide;
- hydroquinone;
- mercaptans;
- phenol;
- sulfites, and so on.

Ozone can also be used as an oxidizing agent, but because of its instability, it must be generated on-site. It is a powerful oxidant for some organic materials, but others are oxidized only slowly or not at all. Ozone is only suitable for low concentrations of oxidizable materials. A common use is for the sterilization of water.

The effectiveness of these chemical oxidizing agents is enhanced by the presence of ultraviolet (UV) light and solid catalysts. Chemical oxidation is also sometimes applied after biological treatment.

10) *pH adjustment*. The pH of the wastewater often needs adjustment prior to reuse, discharge, or biological treatment. For biological treatment, the pH is normally adjusted to between 8 and 9. Bases used include sodium hydroxide, calcium oxide, and calcium carbonate. Acids used include sulfuric acid and hydrochloric acid.

11) *Chemical precipitation*: Chemical precipitation followed by solids separation is particularly useful for separating heavy metals. The heavy metals of particular concern in the treatment of wastewaters include cadmium, chromium, copper, lead, mercury, nickel and zinc. This is a particular problem in the manufacture of dyes and textiles and in metal processes such as pickling, galvanizing, and plating.

Heavy metals can often be removed effectively by chemical precipitation in the form of carbonates, hydroxides, or sulfides. Sodium carbonate, sodium bisulfite, sodium hydroxide and calcium oxide can be used as precipitation agents. The solids precipitate as a *floc* containing a large amount of water in the structure. The precipitated solids need to be separated by thickening or filtration and recycled if possible. If recycling is not possible, then solids are usually disposed of to landfill.

The precipitation process tends to be complicated when a number of metals are present in solution. If this is the case, then the pH must be adjusted to precipitate out the individual metals, since the pH at which precipitation occurs depends on the metal concerned.

## 6.3 Biological Wastewater Treatment Processes

In *secondary* or *biological treatment*, a concentrated mass of microorganisms is used to break down organic matter into stabilized wastes. The degradable organic matter in the wastewater is used as food by the microorganisms. Biological growth requires supplies of oxygen, carbon, nitrogen, phosphorus, and inorganic ions such as calcium, magnesium, and potassium. Domestic sewage satisfies the requirements, but industrial wastewaters may lack nutrients and this can inhibit biological growth. In such circumstances, nutrients may need to be added. As the waste treatment progresses, the microorganisms multiply, producing an excess of this *sludge*, which cannot be recycled.

There are three main types of biological process (Tchobanoglous and Burton 1991; Berne and Cordonnier 1995; Eckenfelder and Musterman 1995):

1) *Aerobic.* Aerobic reactions take place only in the presence of free oxygen and produce stable, relatively inert end products such as carbon dioxide and water. Aerobic reactions are by far the most widely used. The oxidation reactions are of the type:

$$\begin{bmatrix} C \\ O \\ H \\ N \\ S \end{bmatrix} + O_2 + [\text{Nutrients}] \rightarrow CO_2 + H_2O + NH_3 + [\text{Cells}] + [\text{Other end products}]$$

Organic matter

*Endogenous respiration* reactions also occur, which reduce the sludge formation:

$$[\text{Cells}] + O_2 \rightarrow CO_2 + H_2O + NH_3 + [\text{Energy}]$$

*Nitrification* reactions can occur, in which organic nitrogen and ammonia are converted to nitrate:

$$\text{Organic Nitrogen} \rightarrow NH_4{}^+$$

$$NH_4{}^+ + CO_2 + O_2 \rightarrow NO_2{}^- + H_2O + [\text{Cells}] + [\text{Other end products}]$$

$$NO_2{}^- + CO_2 + O_2 \rightarrow NO_3{}^- + [\text{Cells}] + [\text{Other end products}]$$

The nitrification reactions are inhibited by high concentrations of ammonia.

2) *Anaerobic.* Anaerobic reactions function without the presence of free oxygen and derive their energy from organic compounds in the waste. Anaerobic reactions proceed relatively slowly and lead to end products that are unstable and contain considerable amounts of energy, such as methane and hydrogen sulfide:

$$\begin{bmatrix} C \\ O \\ H \\ N \\ S \end{bmatrix} + [\text{Nutrients}] \rightarrow CO_2 + CH_4 + [\text{Cells}] + [\text{Other end products}]$$

Organic matter

3) *Anoxic.* Anoxic reactions also function without the presence of free oxygen. However, the principal biochemical pathways are not the same as in anaerobic reactions, but are a modification of aerobic pathways and hence are termed anoxic. Anoxic reactions are used for *denitrification* to convert nitrate to nitrogen:

$$NO_3{}^- + BOD \rightarrow N_2 + CO_2 + H_2O + OH^- + [\text{Cells}]$$

Various methods are used to contact the microorganisms with the wastewater. In a completely mixed system, the hydraulic residence time of the wastewater and the solids residence time of the microorganisms would be the same. Thus, the minimum hydraulic residence time would be defined by the growth rate of the microorganisms. Since the crucial microorganisms can take a considerable time to grow, this would lead to hydraulic residence times that would be prohibitively long. To overcome this, a number of methods have been developed to decouple the hydraulic and solids residence times.

1) *Aerobic digestion*

   a) *Suspended growth methods*

   The *suspended growth* or *activated sludge* method is illustrated in Figure 6.9. Biological treatment takes place in a tank where the waste is mixed with a flocculated biological sludge. To maintain aerobic conditions, the tank must be aerated. Sludge separation from effluent is normally achieved by gravity sedimentation. Part of the sludge is recycled and excess sludge is removed. The hydraulic flow pattern in the aeration tank can vary between extremes of mixed flow and plug flow. For mixed-flow reactors, the wastewater is rapidly dispersed throughout the reactor and its concentration is reduced. This feature is advantageous at sites where periodic discharges of more concentrated waste are received. The rapid dilution of the waste means that the

**6**

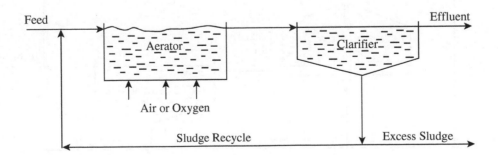

## Figure 6.9

Suspended growth aerobic digestion. Source: (Smith 2016), *Chemical Process Design and Integration*, 2nd Edition, John Wiley & Sons Ltd.

concentration of any toxic compounds present will be reduced, and thus the microorganisms within the reactor may not be affected by the toxicant. Thus, mixed-flow reactors produce an effluent of uniform quality in response to fluctuations in the feed.

Plug-flow reactors have a decreasing concentration gradient from inlet to outlet, which means that toxic compounds in the feed remain undiluted during their passage along the reactor, and this may inhibit or kill many of the microorganisms within the reactor. The oxygen demand along the reactor will also vary. On the other hand, the increased concentration means that rates of reaction are increased, and for two reactors of identical volume and hydraulic retention time, a plug-flow reactor will show a greater degree of $BOD_5$ removal than a mixed-flow reactor.

The biochemical population can be specifically adapted to particular pollutants. However, in the majority of cases, a wide range of organic materials must be dealt with and mixed cultures are used.

Nitrification–denitrification reactions can be carried out in suspended growth by separating the reactor into different cells. A typical arrangement would control the first cell under anoxic conditions to carry out the denitrification reactions. These reactions require organic carbon and this will already be present in mixed effluents. If there is insufficient organic carbon in the feed, then this must be added (e.g. adding methanol). The water from the anoxic cell would then overflow into a cell maintained under aerobic conditions in which the nitrification reactions are carried out, along with the oxidation of organic carbon. The nitrification reactions produce nitrate and this requires a recycle back to the anoxic cell.

Suspended growth aerobic processes are capable of removing up to 95% of BOD. The inlet concentration is restricted to a maximum BOD of around $1\ kg\ m^{-3}$ (1000 $mg\ l^{-1} \approx 1000$ ppm) or COD of around $3.5\ kg\ m^{-3}$ (3500 $mg\ l^{-1} \approx 3500$ ppm). Chloride (as $Cl^-$) should be less than $8$–$15\ kg\ m^{-3}$. The performance of the process can be enhanced by the use of pure oxygen, rather than air.

This reduces the plant size and increases the allowable inlet concentration of COD to around $10\ kg\ m^{-3}$, but does not have a significant effect on the overall removal.

**b)** *Attached growth (film) methods*

Here the wastewater is trickled over a packed bed through which air is allowed to percolate. A *biological film* or *slime* builds up on the packing under aerobic conditions. Oxygen from the air and biological matter from the wastewater diffuses into the slime. As the biological film grows, it eventually breaks its contact with the packing and is carried away with the water. Packing material varies from pieces of stone to preformed plastic packing. Figure 6.10 shows a typical attached growth arrangement. Alternatively, the wastewater can be used to fluidize a bed of carbon or sand onto which the film is attached.

Attached growth processes are capable of removing up to 90% of $BOD_5$ and are thus less effective than suspended growth methods. Nitrification–denitrification reactions can also be carried out in attached growth processes.

**2)** *Anaerobic digestion*

With wastewaters containing a high organic content, the oxygen demand may be so high that it becomes extremely difficult and expensive to maintain aerobic conditions. In such circumstances, anaerobic processes can provide an efficient means of removing large quantities of organic material. Anaerobic processes tend to be used when $BOD_5$ levels exceed $1\ kg\ m^{-3}$ (1000 $mg\ l^{-1}$). However, they are not capable of producing very high-quality effluents and further treatment is usually necessary.

The inability to produce high-quality effluents is one significant disadvantage. Another disadvantage is that anaerobic processes must be maintained at temperatures between 35 and 40 °C to get the best performance. If low-temperature waste heat is available from the production process, then this is not a problem. One advantage of anaerobic reactions is that the methane produced can be a useful source of energy. This can be fed to steam boilers or burnt in a heat engine to produce power.

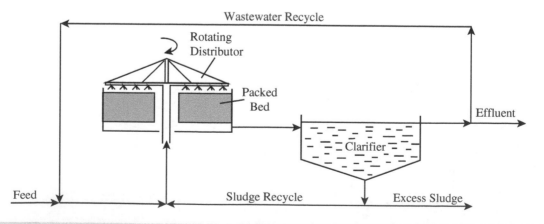

## Figure 6.10

Attached growth aerobic digestion. Source: (Smith 2016), *Chemical Process Design and Integration*, 2nd Edition, John Wiley & Sons Ltd.

**a)** *Suspended growth methods*

The contact type of anaerobic digester is similar to the activated sludge method of aerobic treatment. The feed and microorganisms are mixed in a tank (this time closed). Mechanical agitation is usually required since there is no air injection. The sludge is separated from the effluent by sedimentation or filtration, with part of the sludge recycled and excess sludge removed.

Another method is the *upflow anaerobic sludge blanket* illustrated in Figure 6.11. Here the sludge is contacted by upward flow of the feed at a velocity such that the sludge is not carried out of the top of the digester.

A third method of contact known as an *anaerobic filter* also uses upward flow but keeps the sludge in the digester by a physical barrier such as a grid.

**b)** *Attached growth (film) methods*

As with aerobic digestion, the microorganisms can be encouraged to grow attached to a support medium such as plastic packing or sand. In anaerobic attached growth digestion, the bed is usually fluidized rather than a fixed-bed arrangement, as shown in Figure 6.12.

Anaerobic processes typically remove 75–85% of COD (Tchobanoglous and Burton 1991; Eckenfelder and Musterman 1995).

**3)** *Reed beds*

Reed bed processes require reeds (phragmites) to be planted in soil, sand, or gravel in the wastewater. Figure 6.13 shows a typical reed bed arrangement. Oxygen from air is transported through the leaves, stems, and rhizomes to high concentrations of microorganisms in the root zone. Aerobic treatment takes place in the region of the rhizomes, with anoxic and anaerobic treatment in the surrounding soil. Different flow arrangements other than that shown in Figure 6.13 can also be used. Reed beds are capable of removing 60–80% of $BOD_5$ (Schierup and Brix 1990). They are also capable of removing 25–50% of total nitrogen and 20–40% of total phosphorus (Schierup and Brix 1990). They have the advantage that there is no sludge disposal. The maximum inlet concentration can be as high as $3\,kg\,m^{-3}$ $BOD_5$. A significant disadvantage of reed beds is the time taken for them to become fully established and they need between six months and two years for this. The difference between winter and summer performance needs to be considered for such arrangements. Another significant disadvantage is that the reed beds need to be refurbished every 7–10 years.

Excess sludge is produced in most biological treatment processes, which must be disposed of. The treatment and disposal of sludge is a major problem that can be costly to deal with. Anaerobic processes have the advantage here, since they produce considerably less sludge than aerobic processes (of the order of 5% of aerobic processes for the same throughput). Sludge disposal can typically be responsible for 25–40% of the operating costs of an aerobic biological treatment system. Treatment of sludge is primarily aimed at reducing its volume. This is because the sludge is usually 95–99% water and the cost of disposal is closely linked to its volume. The water is partly free, partly

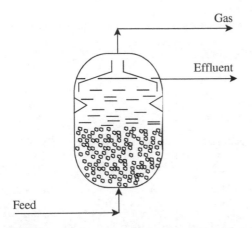

**Figure 6.11**

Suspended growth anaerobic digestion using an upward flow anaerobic sludge blanket. Source: (Smith 2016), *Chemical Process Design*, 2nd Edition, John Wiley & Sons Ltd.

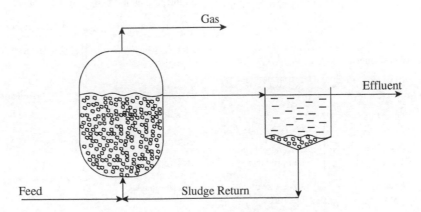

**Figure 6.12**

Attached growth anaerobic digestion using a fluidized anaerobic bed. Source: (Smith 2016), *Chemical Process Design*, 2nd Edition, John Wiley & Sons Ltd.

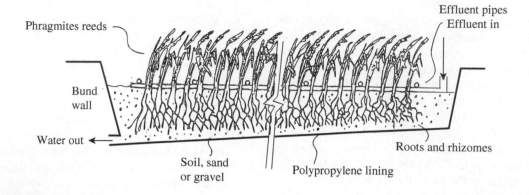

**Figure 6.13**

A reed bed. Source: (Smith 2016), *Chemical Process Design*, 2nd Edition, John Wiley & Sons Ltd.

trapped in the flocs and partly bound in the microorganisms. Anaerobic digestion of the sludge can be used, followed by dewatering. The dewatering can be by filtration or centrifugation. Alternatively, filtration or centrifugation can be used directly to carry out the dewatering. For centrifugation, the dewatering process can be enhanced by the addition of clay. Adding powdered activated carbon to aerobic suspended growth processes can also help with sludge dewatering. The resulting water content after these processes is reduced to typically 60–85%. The water content can be reduced to perhaps 10% by drying. The sludge may finally be used for agricultural purposes (albeit a poor fertilizer) or thermally oxidized.

Large sites might require their own biological treatment processes for final treatment before discharge. Smaller sites might rely on local municipal treatment processes, which treat a mixture of industrial and domestic effluent, for final effluent discharge.

Table 6.1 provides a summary of the main features of biological wastewater treatment. Table 6.2 summarizes the treatment processes that can be used for various types of contamination.

**Table 6.1**

Comparison of biological wastewater treatments.

| Aerobic | Anaerobic | Reed beds |
|---|---|---|
| $BOD_5 < 1$ kg m$^{-3}$ (higher if $O_2$ used) | $BOD_5 > 1$ kg m$^{-3}$ | $BOD_5 < 3.5$ kg m$^{-3}$ |
| Stable end products ($CO_2$, $H_2O$, etc.) | Unstable end products ($CH_4$, $H_2S$, etc.) | Stable and unstable end products |
| $BOD_5$ removal up to 95% | $BOD_5$ removal 75–85% | $BOD_5$ removal 60–80% |
| High sludge formation | Low sludge formation | No sludge disposal |

**Table 6.2**

Summary of treatment processes for some common contaminants.

| Suspended solids | Dispersed oil | Dissolved organic |
|---|---|---|
| Gravity separation | Coalescence | Biological oxidation aerobic, anaerobic, reed beds |
| Centrifugal separation | Centrifugal separation | Chemical oxidation |
| Filtration | Flotation | Activated carbon |
| Membrane filtration | Wet oxidation | Wet oxidation |
| | Thermal oxidation | Thermal oxidation |
| **Ammonia** | **Phenol** | **Heavy metals** |
| Steam stripping | Solvent extraction | Chemical precipitation |
| Air stripping | Biological oxidation (aerobic) | Ion exchange |
| Biological nitrification | Wet oxidation | Adsorption |
| Chemical oxidation | Activated carbon | Nano-filtration |
| Ion exchange | Chemical oxidation | Reverse osmosis |
| | | Electrodialysis |
| **Dissolved solids** | **Neutralization** | **Sterilization** |
| Ion exchange | Acid | Heat treatment |
| Reverse osmosis | Base | UV light |
| Nano-filtration | | Chemical oxidation |
| Electrodialysis | | |
| Crystallization | | |
| Evaporation | | |

# 6.4 Tertiary Wastewater Treatment Processes

*Tertiary treatment* or *polishing treatment* prepares the aqueous waste for final discharge. The final quality of the effluent depends on the nature and flow of the receiving water. Table 6.3 gives an indication of the final quality required (Tebbutt 1990).

Aerobic digestion is normally capable of removing up to 95% of the BOD. Anaerobic digestion is capable of removing less, in the range 75–85%. With municipal treatment processes, which treat a mixture of domestic and industrial effluent, some disinfection of the effluent might be required to destroy any disease-causing organisms before discharge to the environment. Tertiary treatment processes vary, but constitute the final stage of effluent treatment to ensure that the effluent meets specifications for disposal. Tertiary processes used include:

1) *Filtration.* Examples of such processes are microstrainers (a fine screen with openings) and sand filters. They are designed to improve effluents from biological treatment processes by removing suspended material and, with it, some of the remaining $BOD_5$. Sand filtration can remove effectively all of the remaining $BOD_5$ in many circumstances.

2) *Ultrafiltration.* Ultrafiltration was described under pretreatment methods. It is used to remove finely divided suspended solids and when used as a tertiary treatment can in many circumstances remove virtually all the $BOD_5$ remaining after biological treatment.

3) *Adsorption.* Some organic materials are not removed in biological systems operating under normal conditions. Removal of residual organic material can be achieved by adsorption. Both activated carbon and synthetic resins are used. As described earlier under pretreatment methods, regeneration of activated carbon in a furnace can cause carbon losses of perhaps 5–10%.

4) *Nitrogen and phosphorus removal.* Since nitrogen and phosphorus are essential for growth of the microorganisms, the effluent from secondary treatment will contain some nitrogen and phosphorus. The amount that is discharged to receiving waters can cause eutrophication and have a considerable effect on the growth of algae. If discharge is to a high-quality receiving water and/or dilution rates are low, then removal

## Table 6.3

Typical effluent quality for various receiving waters (Adapted from Tebbutt 1990).

| Receiving water | Typical effluent | |
|---|---|---|
| | $BOD_5$ (mg $l^{-1}$) | Suspended solids (mg $l^{-1}$) |
| Tidal estuary | 150 | 150 |
| Lowland river | 20 | 30 |
| Upland river | 10 | 10 |
| High-quality river with low dilution | 5 | 5 |

may be necessary. Nitrogen principally occurs as ammonium ($NH_4^+$), nitrate ($NO_3^-$), and nitrite ($NO_2^-$) ions. Phosphorus principally occurs as orthophosphate ($PO_4^{3-}$) ions. A variety of biological and chemical processes are available for the removal of nitrogen and phosphorus (Tchobanoglous and Burton 1991; Eckenfelder and Musterman 1995). These processes produce extra biological and inorganic sludge that requires disposal.

5) *Disinfection.* Chlorine, as gaseous chlorine or as the hypochlorite ion, is widely used as a disinfectant. However, its use in some cases can lead to the formation of toxic organic chlorides and the discharge of excess chlorine can be harmful. Hydrogen peroxide and ozone are alternative disinfectants that lead to products that have a lower toxic potential. Treatment is enhanced by ultraviolet light. Indeed, disinfection can be achieved by ultraviolet light on its own.

# 6.5 Atmospheric Emissions

There are many types of emissions to atmosphere, and these can be characterized as particulate (solid or liquid), vapor, and gaseous. Overall, the control of atmospheric emissions is difficult because the majority of emissions come from small sources that are difficult to regulate and control. Legislators therefore control emissions from sources that are large enough to justify monitoring and inspection. Industrial emissions of major concern are as follows.

1) $PM_{10}$. Particulate material less than 10 μm in diameter is formed as a byproduct of combustion processes through incomplete combustion and through reactions between sulfur and nitrogen compounds in the atmosphere. Such emissions can also be formed by solids processing operations. $PM_{10}$ is a particular problem as it causes damage to the human respiratory system.

2) $PM_{2.5}$. Particulate material less than 2.5 μm diameter forms in the same way as $PM_{10}$, but it can penetrate deeper into the respiratory system than $PM_{10}$ and can therefore be more dangerous.

3) $O_3$. Ozone is a very reactive compound present in the upper atmosphere (stratosphere) and the lower atmosphere (troposphere). Whilst ozone is vital in the stratosphere, its presence at ground levels is a danger to human health and contributes to the formation of other pollutants.

4) *VOCs.* A volatile organic compound (VOC) is any compound of carbon, excluding carbon monoxide, carbon dioxide, carbonic acid, metal carbides or carbonates, and ammonium carbonate, which participate in atmospheric photochemical reactions (US Environmental Protection Agency 1992). VOCs are precursors to ground-level ozone production and various photochemical pollutants. They are major components in the formation of smog through photochemical reactions (Harrison 1992; De Nevers 1995). There are many potential sources of VOCs, as will be discussed later.

5) $SO_X$. Oxides of sulfur ($SO_2$ and $SO_3$) are formed by the combustion of fuels containing sulfur and as a byproduct of the production of chemicals.

**6**

**6)** $NO_X$. Oxides of nitrogen (principally NO and $NO_2$) are formed in combustion processes and as a byproduct of the production of chemicals.

**7)** *CO*. Carbon monoxide is formed by the incomplete combustion of fuel and as a byproduct of the production of chemicals.

**8)** $CO_2$. Carbon dioxide is formed principally by the combustion of fuel but also as a byproduct of the production of chemicals.

**9)** *Dioxins and furans*. Dioxins and furans are the abbreviated names for a family of over 200 different toxic substances that share similar chemical structures. They are not commercial chemical products but are trace-level unintentional byproducts from thermal oxidation, the processing of metals, and paper manufacture. Dioxins and furans are also released naturally from forest fires and volcanoes. They are widely distributed throughout the environment at extremely low concentrations, but resist degradation and accumulate. They are considered to be human carcinogens and can promote adverse non-cancer health effects.

Atmospheric emissions are controlled by legislation because of their damaging effect on the environment and human health. These effects can be categorized into their local effects and global effects. There are six main problems associated with atmospheric emissions.

**1)** *Urban smog*. Urban smog is commonly found in modern cities, especially where air is trapped in a basin. It is observable as a brownish colored air. The formation of urban smog is through complex photochemical reactions involving sunlight (*hf*) that can be characterized by:

$$VOCs + NO_x + O_2 \xrightarrow{hf} O_3 + \text{other photochemical pollutants}$$

Photochemical pollutants such as ozone, aldehydes, and peroxynitrates such as peroxyethanoyl (or peroxyacetyl) nitrate (PAN) are formed. Ozone and other photochemical pollutants have harmful effects on living organisms. High levels of these pollutants can cause breathing difficulties and bring on asthma attacks in humans. Warm weather and still air exacerbate the problem. In addition to VOCs and $NO_x$, the problem of urban smog is made worse by particulate emissions and carbon monoxide from incomplete combustion of fuel.

**2)** *Acid rain*. Natural (unpolluted) rain and other forms of precipitation are naturally acidic with a pH often in the range of 5–6 caused by carbonic acid from dissolved carbon dioxide and sulfurous and sulfuric acids from natural emissions of $SO_x$ and $H_2S$. Human activity can reduce the pH very significantly down to the range 2–4 in extreme cases, mainly caused by emissions of oxides of sulfur, but also oxides of nitrogen. Because atmospheric pollution and clouds travel over long distances, acid rain is not a local problem. The problem may manifest itself a long way from the source. Acid rain leads to the *acidification* of oceans, freshwater, and soil. Problems associated with acid rain include:

- damage to plant life, particularly in forests;
- acidification of soil leading to a decline in crop and pasture production, as some nutrients become less available, while other components in the soil may reach toxic levels;
- acidification of water, leading to dead lakes and streams, loss of aquatic life, and possible damage to the human water supply;
- corrosion of buildings, particularly those made of marble and sandstone.

**3)** *Ocean acidification*. Ocean acidification is the decrease in pH of the oceans, caused by the uptake of carbon dioxide from the atmosphere. A significant proportion of the anthropogenic carbon dioxide emissions dissolve in oceans. When carbon dioxide enters the ocean, it combines with seawater to produce carbonic acid, which increases the acidity of the water, lowering its pH. A consequence of the oceans becoming more acidic is the binding up of carbonate ions, which are used by marine creatures to make their calcium carbonate shells and skeletons. As the availability of carbonate ions decreases, it becomes more difficult for these animals to build their calcium carbonate structures. When shelled organisms are at risk, the entire food chain is also at risk.

**4)** *Eutrophication*. Eutrophication occurs when a body of terrestrial or marine water suffers an increase in nutrients that leads to enhanced growth of aquatic vegetation or *phytoplankton* and *algal blooms*. This disrupts normal functioning of the ecosystem, causing a variety of problems, such as a lack of oxygen needed for aquatic life to survive, decreased biodiversity, changes in species composition and dominance, and toxicity effects. $NO_x$ emissions contribute to eutrophication.

**5)** *Ozone layer destruction*. The upper atmosphere contains a layer rich in ozone. Whilst ozone in the lower levels of the atmosphere is harmful, ozone in the upper levels of the atmosphere is essential as it absorbs considerable amounts of ultraviolet light that would otherwise reach the Earth's surface. The destruction of ozone is catalyzed by oxides of nitrogen in the upper atmosphere:

$$NO^{\bullet} + O_3 \longrightarrow NO_2^{\bullet} + O_2$$
$$NO_2^{\bullet} + O \longrightarrow NO^{\bullet} + O_2$$
$$NO_2^{\bullet} \xrightarrow{hf} NO^{\bullet} + O$$

Destruction is also initiated by certain halocarbon compounds, such as:

$$CCl_2F_2 \xrightarrow{hf} {}^{\bullet}CClF_2 + Cl^{\bullet}$$
$$Cl^{\bullet} + O_3 \longrightarrow ClO^{\bullet} + O_2$$
$$ClO^{\bullet} + O \longrightarrow Cl^{\bullet} + O_2$$

The Cl• can then react with further ozone. Destruction of ozone has led to thinning of the ozone layer, especially over the North and South Poles, where the ozone layer is thinnest. The result of ozone layer destruction is increased ultraviolet light reaching the Earth, potentially increasing skin cancer and endangering polar species. The thinning that leads to lower absorption of ultraviolet light also causes a cooling of the stratosphere and disruption of weather patterns.

**6)** *The Greenhouse Effect*. Gases such as $CO_2$, $CH_4$, $N_2O$, and $H_2O$ are present in low concentrations in the Earth's atmosphere.

These gases reduce the Earth's emissivity and reflect some of the heat radiated by the Earth. Thus, the effect is to create a "blanket" to keep the Earth warmer than it would otherwise be. The problem arises mainly from burning fossil fuels and clearing forests by burning. The result is that global temperatures increase, leading to melting of the polar ice caps and glaciers, rising sea levels, desertification of areas, thawing of permafrost, increased weather disruptions, and changes to ocean currents.

When applying legislation to atmospheric emissions, the regulatory authorities can either control emissions from individual points of release or combine all of the releases together as a combined release from a "bubble" around the manufacturing facility.

One of the major problems with atmospheric emissions is the number of potential sources. Solid emissions arise from:

- solids drying operations;
- kilns used for high-temperature treatment of solids;
- metal manufacture;
- crushing, grinding, and screening operations for solids;
- any solids handling operation open to the atmosphere;
- incomplete combustion or fuel ash from furnaces, boilers, and thermal oxidizers;
- incomplete combustion in flares, and so on.

Vapor emissions are even more difficult to deal with as they have more sources. Some are point sources, some are small, slow leakages from, for example, pipe flanges and valve and pump seals, known as *fugitive* emissions. Vapor emission sources include:

- condenser vents;
- venting of pipes and vessels;
- inert gas purging of pipes and vessels;
- process purges to atmosphere;
- drying operations;
- application of solvent-based surface coatings;
- open operations such as filters, mixing vessels, and so on, leading to evaporation of VOCs;
- drum emptying and filling, leading to evaporation of VOCs;
- spillages of VOCs;
- process plant ventilation of buildings processing VOCs;
- storage tank loading and cleaning;
- road, rail, and barge tank loading and cleaning;
- fugitive emissions through gaskets and shaft seals
- fugitive emissions from sewers and effluent treatment;
- fugitive emissions from process sampling points;
- incomplete combustion of fuel in furnaces, boilers, and thermal oxidizers;
- incomplete combustion in flares, and so on.

In larger plants, significant reductions in VOC emissions can usually be made by controlling major sources, such as tank venting, condensers, and purges, and by inspection and maintenance of gaskets and rotating shaft seals.

The largest volume of atmospheric emissions from process plants occurs from combustion emissions. Such emissions are created from:

- furnaces, boilers, and thermal oxidizers;
- gas turbine exhausts;
- flares;
- process operations where coke needs to be removed from catalysts (e.g. fluid catalytic cracking regeneration in refineries), and so on.

In addition to the gaseous emissions from the combustion of fuel, gaseous emissions are also produced by chemical production, for example, $SO_x$ from sulfuric acid production, $NO_x$ from nitric acid production, HCl from chlorination reactions, and so on.

# 6.6 Treatment of Solid Particulate Emissions to Atmosphere

The selection of equipment for the treatment of solid particle emissions to atmosphere depends on a number of factors (Stenhouse 1981; Svarovsky 1981; Rousseau 1987; Dullien 1989; Schweitzer 1997):

- size distribution of the particles to be separated;
- particle loading;
- gas throughput;
- permissible pressure drop;
- temperature.

There is a wide range of equipment available for the control of emissions of solid particles to atmosphere. These methods are classified in broad terms in Table 6.4 (Stenhouse 1981).

1) *Gravity settlers.* A simple gravity settler is illustrated in Figure 6.14. This functions on the basis of density difference and is used to collect coarse particles. The gas carrying the solid particles enters a chamber in which the velocity of the gas must be decreased to be below the settling velocity of

**Table 6.4**

Methods of control of emissions of solid particles.

| Equipment | Approximate particle size range (mm) |
|---|---|
| Gravity settlers | > 100 |
| Inertial separators | > 50 |
| Cyclones | > 5 |
| Scrubbers | > 3 |
| Venturi scrubbers | > 0.3 |
| Bag filters | > 0.1 |
| Electrostatic precipitators | > 0.001 |

the particles. Simple gravity settlers may be used as prefilters. Only particles in excess of 100 μm can reasonably be removed (Stenhouse 1981; Rousseau 1987; Dullien 1989; Schweitzer 1997).

2) *Inertial separators*. Inertial separators improve the efficiency of gas solid settling devices by giving the particles downward momentum, in addition to the gravitational force. Figure 6.15 illustrates three possible types of inertial separator. The particles are given a downward momentum to assist the settling. Only particles in excess of 50 μm can be reasonably removed (Stenhouse 1981; Rousseau 1987; Dullien 1989; Schweitzer 1997). Like gravity settlers, inertial collectors are widely used as prefilters.

3) *Cyclones*. Cyclones are also primarily used as prefilters (Figure 6.16). The particle-laden gas enters tangentially and

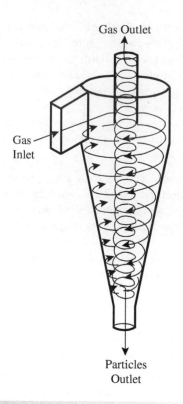

**Figure 6.16**

A cyclone generates centrifugal force by the fluid motion. Source: (Smith 2016), *Chemical Process Design*, 2nd Edition, John Wiley & Sons Ltd.

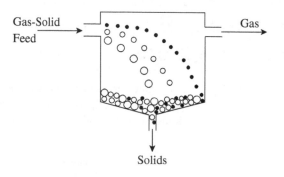

**Figure 6.14**

Gravity settler for the separation of gas–solid mixtures. Source: Adapted from (Smith 2016), *Chemical Process Design*, 2nd Edition, John Wiley & Sons Ltd.

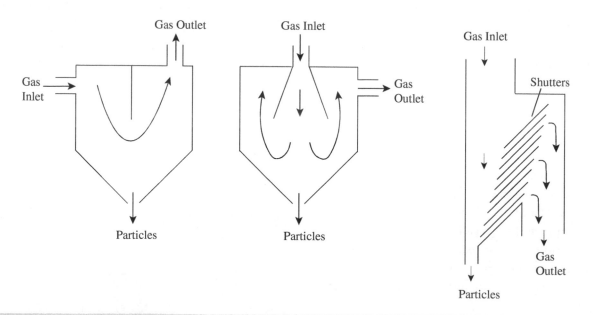

**Figure 6.15**

Inertial separators increase the efficiency of separation by giving the particles downward momentum. Source: (Smith 2016), *Chemical Process Design*, 2nd Edition, John Wiley & Sons Ltd.

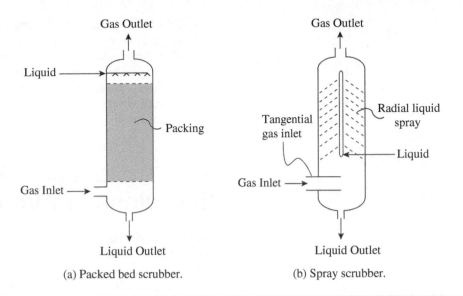

(a) Packed bed scrubber.

(b) Spray scrubber.

**Figure 6.17**

Various scrubber designs can be used to separate solid from gas or vapor. Source: Redrawn from (Teja et al., 1981) / Society of Chemical Industry.

**6**

spins downwards and inwards, ultimately leaving the top of the unit. Particles are thrown radially outwards to the wall by the centrifugal force and leave at the bottom. Cyclones can be used under conditions of high particle loading. They are cheap, simple devices with low maintenance requirements. Problems occur when separating materials that have a tendency to stick to the cyclone walls.

4) *Scrubbers*. Scrubbers are designed to contact a liquid with the particle-laden gas and entrain the particles with the liquid. They offer the obvious advantage that they can be used to remove gaseous as well as particulate pollutants. The gas stream may need to be cooled before entering the scrubber. Two of the more common types of scrubbers are shown in Figure 6.17. Packed columns are widely used in gas absorption, but particulates are also removed in the process (Figure 6.17a). The main disadvantage of packed columns is that the solid particles will often accumulate in the packing and require frequent cleaning. Some designs of packed column allow the packing to self-clean through inducing movement of the bed from the upflow of the gas. Spray scrubbers are less prone to fouling and use a tangential inlet to create a swirl to enhance the separation, as illustrated in Figure 6.17b.

5) *Venturi scrubbers*. An increase in the efficiency of particulate removal can be achieved at the expense of increased pressure drop using a venturi scrubber (Figure 6.18).

6) *Bag filters*. Bag filters are illustrated in Figure 6.19 and are probably the most common method of separating particulate materials from gases. A cloth or felt filter material is used, which is impervious to the particles. Bag filters are suitable for use in very high dust load conditions. They have a high efficiency but suffer from the disadvantage that the pressure drop across them may be high. As particles build up on the inside of the thimble, the unit is periodically taken off-line and the flow

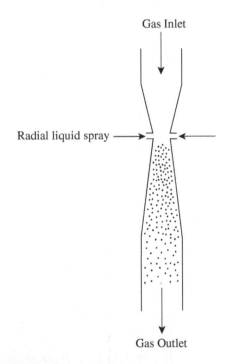

**Figure 6.18**

Venturi scrubber designs can be used to separate solid from gas or vapor. Source: Stenhouse JIT, 1981, Pollution control, in Teja AS, *Chemical Engineering and the Environment*, Blackwell Scientific Publications.

reversed to recover the filtered particles. The separation of solid particles from gases, conventional filter media can be used up to a temperature of around 250 °C. Higher temperatures require ceramic or metallic filter media.

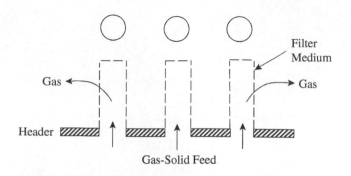

**Figure 6.19**

A bag filter arrangement. Source: (Smith 2016), *Chemical Process Design*, 2nd Edition, John Wiley & Sons Ltd.

7) *Electrostatic precipitators*. Electrostatic precipitators are used where collection of fine particles at a high efficiency coupled with a low-pressure drop is necessary. The arrangement is illustrated in Figure 6.20. Particle-laden gas enters a number of tubes or passes between parallel plates. An electrostatic field is produced between wires or grids and the collection surface by applying a high voltage between the two. As the particle laden gas passes through the space, the electrostatic field ionizes molecules of gas such as $O_2$ and $CO_2$ present in the gas stream. These charged molecules attach themselves to

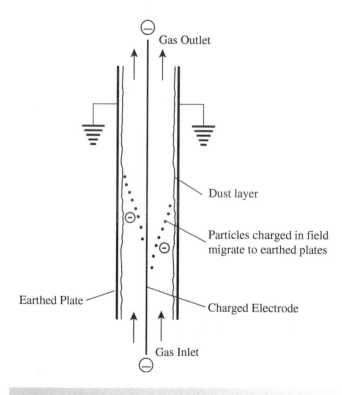

**Figure 6.20**

Electrostatic precipitation. Source: Stenhouse JIT, 1981, Pollution control, in Teja AS, *Chemical Engineering and the Environment*, Blackwell Scientific Publications.

particulate matter, thereby charging the particles. The oppositely charged collection plates attract the particles. The charged particles are deposited on the earthed plates or tube walls. The walls are mechanically *rapped* periodically to remove the accumulated dust layer. Electrostatic precipitators can be used to treat large gas flows, e.g. boilers, cement plants, and waste thermal oxidizers.

# 6.7 Treatment of VOC Emissions to Atmosphere

The first consideration for dealing with VOC emissions should be to eliminate or reduce those emissions at source through changes in design or operation (Smith 2016). Release limits for VOCs are set for either specific components (e.g. benzene, carbon tetrachloride) or as VOCs for organic compounds with a lower environmental impact and classed together and reported, for example, as toluene equivalent.

The first class of the VOC treatment processes allow the VOC to be recovered:

1) *Condensation*. Condensation and recovery of the VOCs can be accomplished by an increase in pressure or a decrease in temperature. Most often, a decrease in temperature is preferred. Figure 6.21 illustrates two ways in which VOC recovery can be accomplished from an airstream used in a process. Figure 6.21a shows a process in which air picks up VOCs and enters a condenser for recovery of the VOCs before being vented directly or passed to secondary treatment before venting. Unfortunately, condensation of VOCs usually requires refrigeration (Figure 6.21b). If the gas stream contains water vapor or organic compounds with high freezing points, then these will freeze in the condenser, causing it to block up. It is therefore often necessary to precool the gas stream to remove water before condensing the VOCs. Figure 6.21b shows an arrangement in which a refrigeration loop is first used to pre-cool the gas stream to the region of 2–4 °C to remove water vapor and high-boiling organics. A second condenser using a second refrigeration loop then condenses the VOCs. It is likely that the VOC condensers will have to be taken off-line periodically to remove frozen water and organic material that will accumulate over time.

The problem of freezing on the condenser surface can be avoided by using direct contact condensation, as shown in Figure 6.22. A secondary refrigerant is contacted directly with the vent stream containing VOC. A refrigeration loop is used to cool the secondary refrigerant. The mixture of secondary refrigerant and condensed VOC then needs to be separated and the secondary refrigerant recycled.

2) *Membranes*. VOCs can be recovered using an organic-selective membrane that is more permeable to organic vapors than permanent gases. Figure 6.23 shows one possible arrangement for recovery of VOCs using a membrane (Hydrocarbon Processing's Environmental Processes 1998). A vent gas is

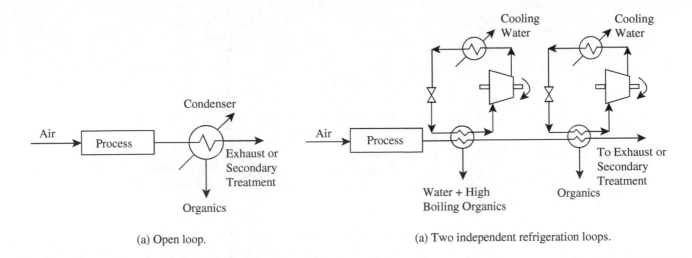

(a) Open loop.                         (a) Two independent refrigeration loops.

**6**

### Figure 6.21

Refrigeration is usually required for the condensation of vapor emissions. Source: Adapted from (Smith 2016), *Chemical Process Design*, 2nd Edition, John Wiley & Sons Ltd.

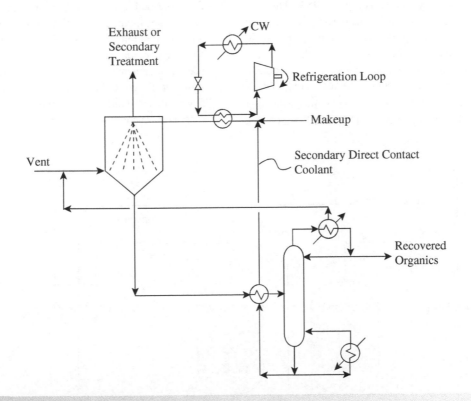

### Figure 6.22

Direct contact condensation. Source: (Smith 2016), *Chemical Process Design*, 2nd Edition, John Wiley & Sons Ltd.

compressed and enters a condenser in which VOCs are recovered. The gases from the condensers then enter a membrane unit, in which the VOC permeates through the membrane. A VOC-enriched permeate is then recycled back to the compressor inlet. Recovery rates for such an arrangement can be as high as 90–99%.

3) *Absorption*. Physical absorption can be used for the recovery of VOCs. The solvent used should be regenerated and recycled, with the VOC being recovered if at all possible. If the VOC is water-soluble (e.g. formaldehyde), then water can be used as the solvent. However, high-boiling organic solvents are most often used for VOC absorption. Efficiency of

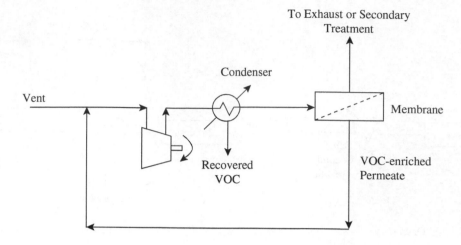

**Figure 6.23**

Recovery of VOCs using membranes. Source: (Smith 2016), *Chemical Process Design*, 2nd Edition, John Wiley & Sons Ltd.

recovery depends on the VOC, solvent, and absorber design, but efficiencies of recovery can be as high as 95%. The efficiency of recovery increases with decreasing temperature and increases with increasing pressure. When using an organic solvent, care must be exercised in the selection of the solvent to avoid vaporization of the solvent into the gas stream, creating a new environmental problem to replace the recovered VOC.

4) *Adsorption*. Adsorption of VOCs is most often carried out using activated carbon in fixed beds. The VOC's molecules become attached to the surface of the activated carbon bed. Once the bed is at its adsorption capacity, it must be taken off-line and regenerated. The most commonly used method for regeneration is to inject steam in situ to recover the VOC and regenerate the bed. A number of different fixed bed arrangements are used. One possible arrangement is shown in Figure 6.24. This involves a three-bed system. The first bed encountered by the VOC-laden gas stream is the primary bed. The gas from the primary bed enters the secondary

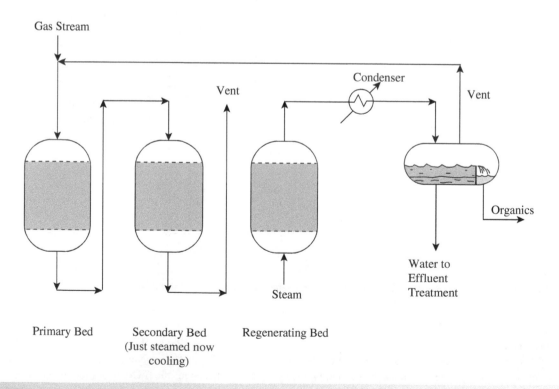

**Figure 6.24**

Adsorption with a three-bed system. Source: (Smith 2016), *Chemical Process Design*, 2nd Edition, John Wiley & Sons Ltd.

bed, which will have just been regenerated using steam and is now cooling. The third bed is off-line, being regenerated using steam. The steam from the regenerating bed is condensed and the condensate separated to recover the organics. The vent from the condenser receiver will normally be too concentrated to vent directly to atmosphere and therefore will normally be connected back to the inlet of the primary bed. Once the primary bed has become saturated, the beds are switched. The current secondary bed becomes the new primary bed. The regenerated bed becomes the secondary bed and the previous primary bed is now regenerated. The switching cycle is normally based on a timing arrangement.

Once the potential for the recovery of VOCs has been exhausted, then any remaining VOC emissions must be destroyed. There are two methods of VOC destruction (US Environmental Protection Agency 1992):

- thermal oxidation (including catalytic thermal oxidation);
- biological treatment.

1) *Flare stacks.* Flares use an open combustion process with oxygen for the combustion provided by air around the flame. Good combustion depends on the flame temperature, residence time in the combustion zone, and good mixing to complete the combustion. The flare stack can be dedicated to a specific vent stream or for a combination of streams via a header. Flares can be categorized by the:

- height of the flare tip (elevated versus ground flares);
- method of enhancing mixing at the flare tip (steam assisted, air assisted, unassisted).

The most common type used in the process industries is the steam-assisted elevated flare. This is illustrated in Figure 6.25. Combustion takes place at the flare tip. The flare tip requires pilot burners, an ignition device, and fuel gas to maintain the pilot flame. In steam-assisted flares, steam nozzles around the perimeter of the flare tip are used to assist the mixing. As shown in Figure 6.25, before the vent stream enters the flare stack, a knockout drum is required to remove any entrained liquid. Liquids must be removed as they can extinguish the fame or lead to irregular combustion in the flame. Also, there is a danger that the liquid might not be completely combusted, which can result in liquid reaching the ground and creating hazards. The vent header needs to be protected against the flame propagating back into the header. A *gas barrier* or a *flame arrester* prevents the flame passing down into the vent header when the vent stream flowrate is low. A *flashback seal drum* containing water provides additional protection. In some designs the flashback seal drum is incorporated into the base of the flare stack.

Flare performance can achieve a destruction efficiency of up to 98% with steam-assisted flares. Halogenated and sulfur-containing compounds should not be flared. Flares are applicable to almost any VOC-laden stream and can be used when there are large variations in the composition, heat content, and flowrate of the vent stream. However, flares should only be used for

**6**

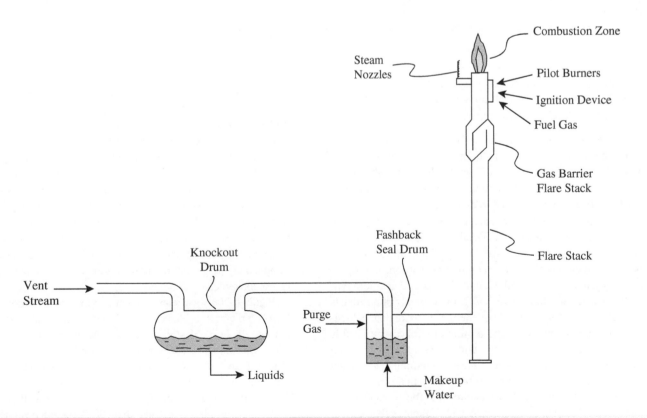

**Figure 6.25**

An elevated steam-assisted flare stack. Source: Adapted from (Smith 2016), *Chemical Process Design*, 2nd Edition, John Wiley & Sons Ltd.

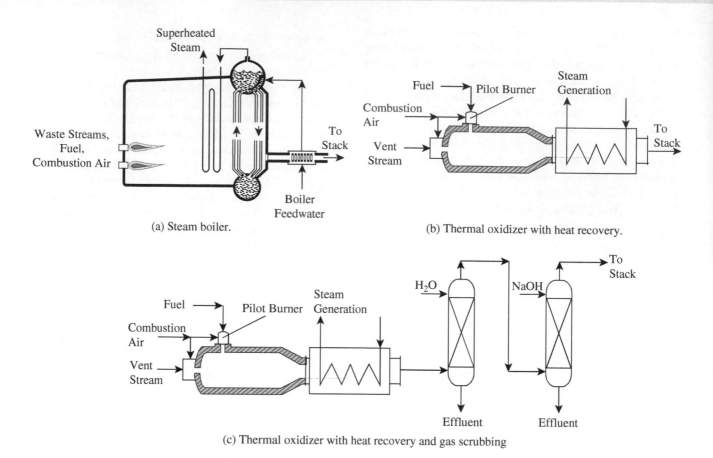

(a) Steam boiler.

(b) Thermal oxidizer with heat recovery.

(c) Thermal oxidizer with heat recovery and gas scrubbing

## Figure 6.26

Different arrangements for thermal oxidation. Source: (Smith 2016), *Chemical Process Design*, 2nd Edition, John Wiley & Sons Ltd.

abnormal operation or emergency upsets when there is a short-term requirement to deal with an abnormal flowrate of VOC-laden gases. For normal operation, other methods should be used.

2) *Thermal oxidation.* In some cases, vent streams can be directed to be the inlet combustion air to steam boilers in which the VOCs are oxidized to $CO_2$ and $H_2O$, as shown in Figure 6.26a. However, it is often necessary to use a *thermal oxidizer* specifically designed to deal with waste streams containing VOCs. Conditions in the thermal oxidizer are typically 25% oxygen over and above stoichiometric requirements to ensure complete oxidation to $CO_2$ and $H_2O$. Minimum temperatures required for VOCs comprising carbon, hydrogen, and oxygen are around 750 °C, but 850–900 °C are more typical. The residence time of the VOCs in the combustion zone is typically 0.55–1 second. When thermally oxidizing halogenated organic compounds, the minimum temperature used is in the range of 1100–1300 °C, with a residence time of up to two seconds. When oxidizing waste containing halogenated materials, it is important to avoid conditions that form dioxins and furans. Dioxins and furans can form as the gases from the thermal oxidation are cooled, especially if there is flyash present. The greatest formation of dioxins and furans is as the gases are cooled between 500 and 200 °C. Gases can be cooled quickly by quenching (water injection) to reduce formation.

Typical conditions for thermal oxidation of halogenated materials are greater than 1200 °C, cooling to 400 °C in a waste heat boiler and then quench to 70 °C. Permitted emission levels of dioxins and furans are typically less than 0.1 ng m$^{-3}$. Adsorption on powdered activated carbon or a carbon filter can be used for treatment of the exhaust gases for dioxins and furans. Alternatively, catalytic oxidation using a honeycomb catalyst can be used. Supplementary fuel is required for start-up and is required if the feed concentration of organic material is low or the feed concentration varies. Supplementary fuel is usually required for vent streams, as processed vents are normally designed to operate below concentrations in which they would be combustible.

Figure 6.26b shows an arrangement in which the hot gases from the thermal oxidizer are used to generate steam for process use. If the vent stream contains halogenated material or sulfur, the exhaust gases must be scrubbed before release. Figure 6.26c shows the corresponding arrangement with scrubbing before release.

The performance of thermal oxidizers usually gives destruction efficiency greater than 98%. When designed to do so, destruction efficiency can be virtually 100% (depending on temperature, residence time, and mixing in the thermal oxidizer). Thermal oxidizers can be designed to handle minor

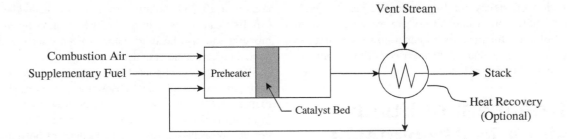

**Figure 6.27**

Catalytic thermal oxidation. Source: (Smith 2016), *Chemical Process Design*, 2nd Edition, John Wiley & Sons Ltd.

fluctuations but are not suitable for large fluctuations in flowrate.

**3)** *Catalytic thermal oxidation.* A catalyst can be used to lower the combustion temperature for thermal oxidation and to save fuel. Figure 6.27 shows a schematic of a catalytic thermal oxidation arrangement. Combustion air and supplementary fuel, along with preheated VOC-laden stream, enter the preheater before the catalyst bed. Heat recovery, as illustrated in Figure 6.27, is not always used. The catalysts are typically platinum or palladium on an alumina support or metal oxides such as chromium, manganese, or cobalt. The vent stream should be well outside the range where it would be combustible. The catalyst can be deactivated by compounds containing sulfur, bismuth, phosphorus, arsenic, antimony, mercury, lead, zinc, tin or halogens. Catalysts are gradually deactivated over time, and their life is generally two to four years. Operating temperature for the catalyst should not normally exceed 500–600 °C; otherwise the catalyst can suffer from sintering. Typical operating conditions are between 200 and 480 °C with a residence time of 0.1 seconds. Catalytic thermal oxidation is not suitable for halogenated compounds because of the lower combustion temperature used. The destruction efficiency is up to 99.9%, but the arrangement is not suitable for large fluctuations in the flowrate.

**4)** *Biological treatment.* Microorganisms can be used to oxidize VOCs (Hydrocarbon Processing's Environmental Processes 1998). The VOC-laden stream is contacted with microorganisms. The organic VOC is food for the microorganisms. Carbon in the VOC is oxidized to $CO_2$, hydrogen to $H_2O$, nitrogen to nitrate, and sulfur to sulfate. Biological growth requires an ample supply of oxygen and nutrients. Figure 6.28 shows a *bio-scrubber* arrangement. The VOC-laden gas stream enters a chamber and rises up through a packing material, over which water flows. Nutrients (phosphates, nitrates, potassium, and trace elements) need to be added to promote biological growth. The microorganisms grow on the surface of the packing material. The water from the exit of the tower needs to be adjusted for pH before recycle to prevent excessively low pH. As the film of microorganism grows on the surface of the packing through time, eventually it becomes too thick and detaches from the packing. This excess *sludge* must be removed from the recycle and disposed of.

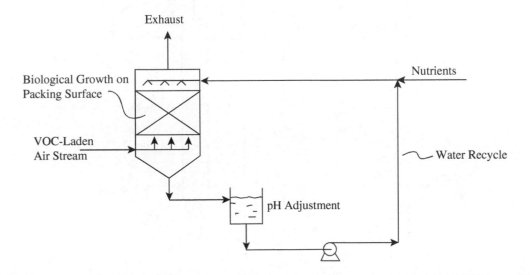

**Figure 6.28**

A bioscrubber for treatment of VOC. Source: (Smith 2016), *Chemical Process Design*, 2nd Edition, John Wiley & Sons Ltd.

Destruction efficiency for biological treatment is typically up to 95%, but some VOCs are very difficult to degrade. Biological treatment is limited to low concentration streams (typically less than 1000 ppm) with flowrates typically less than $100\,000\,\mathrm{m^3\,h^{-1}}$.

# 6.8 Treatment of Sulfur Emissions to Atmosphere

Emissions of sulfur to atmosphere are mostly in the form of hydrogen sulfide $H_2S$ or sulfur dioxide $SO_2$ and sulfur trioxide $SO_3$, together referred to as $SO_x$. Generally, sulfur in the form of $H_2S$ is more straightforward to deal with than $SO_x$. This is because it is generally easier to separate $H_2S$ from gas streams than $SO_x$. Chemical absorption using amines is a commonly used method. For example, monoethanolamine in a 20–30% solution by mass can be used according to the reaction:

$$\underset{\text{monoethanolamine}}{HOCH_2CH_2NH_2} + H_2S \underset{\text{Regeneration}}{\overset{\text{Absorption}}{\rightleftharpoons}} \underset{\substack{\text{monoethanolamine} \\ \text{hydrogen sulfide}}}{HOCH_2CH_2NH_3HS}$$

Figure 6.29 shows the process arrangement. The gas containing $H_2S$ is contacted with a monoethanolamine solution in the absorber and reacts to form monoethanolamine hydrogen sulfide. This is carried out at low temperature. The spent solvent is then heated and sent to the stripper (or regenerator) where the reaction reverses under the influence of heat from the reboiler and the $H_2S$ is sent for further treatment. Different amines can be used other than monoethanolamine (MEA). Diethanolamine (DEA) can be used in a 20–35% solution by mass or methyldiethanolamine (MDEA) can be used in a 35–50% solution by mass. Solvent mixtures can also be used. Other solvents such as hot aqueous potassium carbonate can also be used.

Rather than use chemical absorption for the separation of $H_2S$ from gas streams, physical absorption can be used as shown in Figure 6.30. Various solvents can be used such as refrigerated methanol typically in the range of −40 to −70 °C (the Rectisol Process) or a mixture of dimethyl ethers of polyethylene glycol (the Selexol Process). The absorption is carried out at high pressure and low temperature. Pressure is then let down, releasing some of the dissolved $H_2S$, before passing to the stripper (or regenerator), where the remaining $H_2S$ is stripped by reboiling the solvent. The pressure letdown is typically using two stages with flash drums.

Having separated sulfur in the form of $H_2S$, it cannot be vented to atmosphere. The sulfur from $H_2S$ is normally recovered as elemental sulfur by partial oxidation in a Claus Process. Figure 6.31 shows the outline of a Claus Process. Hydrogen sulfide in air enters a reaction furnace where a partial combustion takes place. After cooling the gases by steam generation, the gases enter a converter. The chemistry is:

Burner $\quad 2H_2S + 3O_2 \rightarrow 2H_2O + 2SO_2$

Converter $\ 2H_2S + SO_2 \rightarrow 2H_2O + 3S$

Overall $\quad H_2S + \frac{1}{2}O_2 \rightarrow S + H_2O$

After the converter in Figure 6.31, the sulfur is recovered by condensation. Conversion is limited by reaction equilibrium to about 60%, so several (often three, but possibly four) stages of

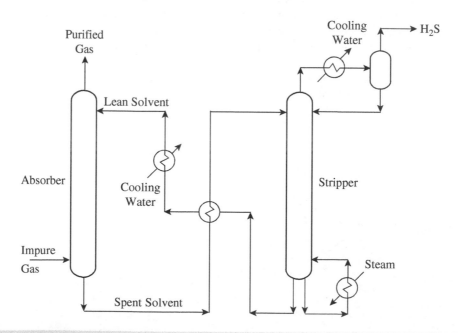

## Figure 6.29

Chemical absorption for acid gases. Source: Adapted from (Smith 2016), *Chemical Process Design*, 2nd Edition, John Wiley & Sons Ltd.

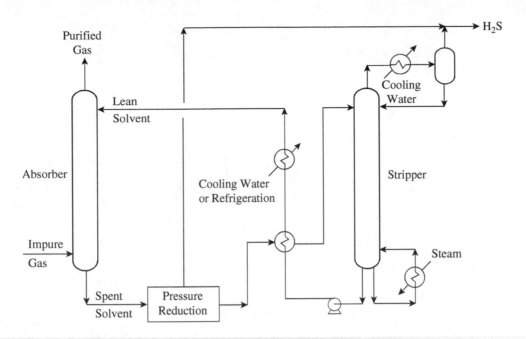

## Figure 6.30

Physical absoption for acid gases.

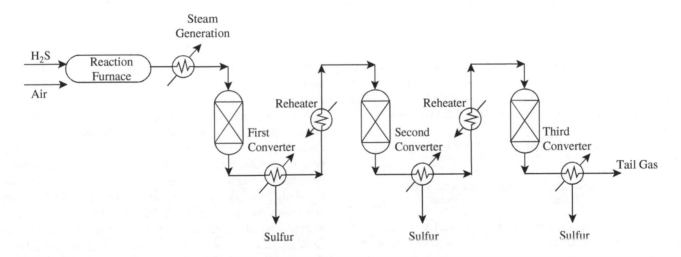

## Figure 6.31

Removal of H$_2$S using the Claus process. Source: (Smith 2016), *Chemical Process Design*, 2nd Edition, John Wiley & Sons Ltd.

conversion are used, as shown in Figure 6.31. After each conversion stage, the sulfur is condensed and the gas is reheated before passing to the next conversion stage. The tail gas contains CO$_2$, H$_2$S, SO$_2$, CS$_2$, COS, and H$_2$O and cannot be released directly to atmosphere. Various processes are available for the cleanup of Claus Process tail gas. Figure 6.32 shows a process in which the sulfur compounds are converted to H$_2$S by reaction with hydrogen. Hydrogen is generated by partial oxidation of fuel gas. Sulfur and sulfur compounds are hydrogenated to hydrogen sulfide in the hydrogenation reactor. Gases from the reactor are cooled by generating steam and then in a quench column by recirculating water (Figure 6.32). Hydrogen sulfide is then separated from the other gases by absorption (e.g. in an amine solution). The solvent from the absorber is regenerated in the stripper and the hydrogen sulfide is recycled to the Claus Process.

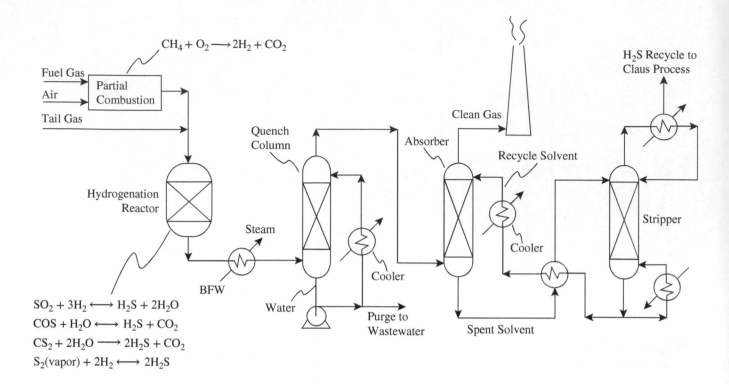

$$CH_4 + O_2 \longrightarrow 2H_2 + CO_2$$

$$SO_2 + 3H_2 \longleftrightarrow H_2S + 2H_2O$$
$$COS + H_2O \longleftrightarrow H_2S + CO_2$$
$$CS_2 + 2H_2O \longrightarrow 2H_2S + CO_2$$
$$S_2(vapor) + 2H_2 \longleftrightarrow 2H_2S$$

**Figure 6.32**

Claus process tail gas clean-up. Source: (Smith 2016), *Chemical Process Design*, 2nd Edition, John Wiley & Sons Ltd.

So far, the techniques to recover sulfur have all related to sulfur in the form of $H_2S$. Consider now how sulfur in the form of $SO_2$ can be dealt with. $SO_2$ can be removed from gas streams using sodium hydroxide solution. In an oxidizing environment:

$$\underset{\text{sodium hydroxide}}{2NaOH} + SO_2 \rightarrow \underset{\text{sodium sulfite}}{Na_2SO_3} + H_2O$$

$$\underset{\text{sodium sulfite}}{Na_2SO_3} + \tfrac{1}{2}O_2 \rightarrow \underset{\text{sodium sulfate}}{Na_2SO_4}$$

However, this is only economic for small-scale applications, as the sodium hydroxide is expensive relative to other reagents that can be used. A less expensive reagent to remove sulfur dioxide is limestone, which can be reacted to produce solid calcium sulfate (gypsum), according to (Crynes 1977):

$$\underset{\text{limestone}}{CaCO_3} + SO_2 \rightarrow \underset{\text{calcium sulfite}}{CaSO_3} + CO_2$$

$$CaSO_3 + \tfrac{1}{2}O_2 + 2H_2O \rightarrow \underset{\text{Gypsum}}{CaSO_4 \cdot 2H_2O}$$

The flowsheet for the removal of $SO_2$ using wet limestone scrubbing is shown in Figure 6.33 (Crynes 1977). The gas stream containing $SO_2$ enters a scrubbing chamber into which a limestone slurry is sprayed. The slurry is then collected, separated by thickening and filtration. This creates a solid material (gypsum) that

does have some potential for use as a building material. However, the solid product is usually of low value and often needs to be disposed of.

Whilst wet limestone scrubbing can be effective for large-scale utility systems such as centralized power stations, it is not readily applied to smaller-scale processes, such as utility boilers on processing sites. Rather than use wet limestone scrubbing, as illustrated in Figure 6.33, "dry" scrubbing can be used. Dry scrubbing refers to the state of the waste byproduct being dry, rather than the wet byproduct from wet limestone scrubbing. The process is illustrated in Figure 6.34. An aqueous slurry of calcium hydroxide (or sometimes sodium carbonate) is sprayed into a flue gas. The reaction is:

$$\underset{\text{calcium hydroxide}}{Ca(OH)_2} + SO_2 \rightarrow \underset{\text{calcium sulfite}}{CaSO_3} + H_2O$$

$$CaSO_3 + \tfrac{1}{2}O_2 \rightarrow \underset{\text{calcium sulfate}}{CaSO_4}$$

Water evaporates and reaction products are removed as dry solids. Advantages over wet scrubbing include simplicity, cheaper construction, a dry waste product, and no wastewater. Disadvantages relative to wet scrubbing include less efficient use of reagent and lower $SO_2$ removal efficiency. $SO_2$ removal efficiency is typically 70–90%. Around 2.2 tons of dry waste are produced per ton of $SO_2$ recovered.

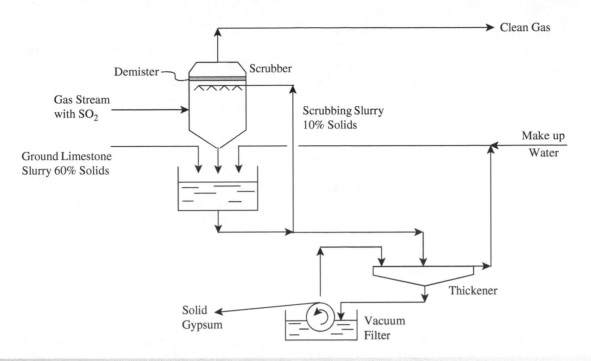

**Figure 6.33**

Removal of $SO_2$ using wet limestone scrubbing. Source: (Smith 2016), *Chemical Process Design*, 2nd Edition, John Wiley & Sons Ltd.

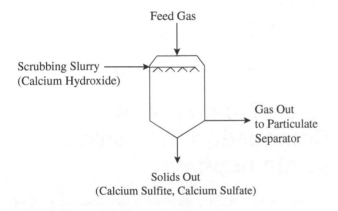

**Figure 6.34**

Removal of $SO_2$ using dry limestone scrubbing. Source: Adapted from (Smith 2016), *Chemical Process Design*, 2nd Edition, John Wiley & Sons Ltd.

# 6.9  Treatment of Oxides of Nitrogen Emissions to Atmosphere

Nitrogen forms eight oxides, but the principal concern is with the two most common ones:

- nitric oxide (NO);
- nitrogen dioxide ($NO_2$).

These are collectively referred to as $NO_x$. $NO_x$ emissions are produced from:

- chemicals production (e.g. nitric acid production, nitration reactions, etc.);
- use of nitric acid in metal and mineral processing;
- combustion of fuels.

$NO_x$ from the combustion of fuels is formed initially as NO, which subsequently oxidizes to $NO_2$ according to:

$$NO + \tfrac{1}{2}O_2 \rightleftharpoons NO_2$$

$NO_x$ formation in combustion depends on the fuel and the type of combustion device. $NO_x$ formation in combustion processes can be reduced by various measures, which all reduce the flame temperature (see Smith 2016). Once $NO_x$ formation has been minimized at source, if the $NO_x$ levels still do not achieve the required environmental standards, then treatment must be considered.

The first option that might be considered is absorption into water. The problem is that $NO_2$ is soluble in water, but NO is only sparingly soluble. To absorb NO, a complexing agent or oxidizing agent must be added to the solvent. One method that can be used is to absorb into a solution of hydrogen peroxide according to:

$$2NO + 3H_2O_2 \rightarrow 2HNO_3 + 2H_2O$$

$$2NO_2 + H_2O_2 \rightarrow 2HNO_3$$

This allows nitric acid to be recovered from the oxides of nitrogen. It can be a useful technique, for example, in metal finishing

6

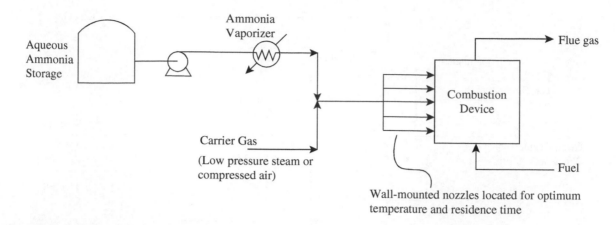

**Figure 6.35**

Removal of NO*x* using selective non-catalytic reduction. Source: (Smith 2016), *Chemical Process Design*, 2nd Edition, John Wiley & Sons Ltd.

operations where nitric acid is used. More commonly, $NO_x$ is removed using reduction. Ammonia is usually used as the reducing agent, according to the reactions:

$$6NO + 4NH_3 \rightarrow 5N_2 + 6H_2O$$

$$4NO + 4NH_3 + O_2 \rightarrow 4N_2 + 6H_2O$$

$$2NO_2 + 4NH_3 + O_2 \rightarrow 3N_2 + 6H_2O$$

This can be carried out without a catalyst in a narrow temperature range (850–1100 °C) and is known as *selective non-catalytic reduction* (Hydrocarbon Processing's Environmental Processes 1998). Below 850 °C, the reaction rate is too slow. Above 1100 °C, the dominant reaction becomes:

$$NH_3 + O_2 \rightarrow NO + \tfrac{3}{2}H_2O$$

Thus, the technique can become counterproductive. A typical arrangement for selective non-catalytic reduction is shown in Figure 6.35. Aqueous ammonia is vaporized and mixed with a carrier gas (low-pressure steam or compressed air) and injected into nozzles located in the combustion device for optimum temperature and residence time (Hydrocarbon Processing's Environmental Processes 1998). $NO_x$ reduction of up to 75%

can be achieved. However, *slippage* of excess ammonia must be controlled carefully.

Rather than selective non-catalytic reduction, the reduction can be carried out over a catalyst (typically vanadium pentoxide and tungsten trioxide dispersed on titanium-dioxide) at 250–450 °C. This is known as a selective catalytic reduction. Figure 6.36 shows a typical selective catalytic reduction arrangement. Either anhydrous or aqueous ammonia can be used. This is mixed with air and injected into the flue gas stream upstream of the catalyst. Removal efficiency of up to 95% is possible. Again, slippage of excess ammonia needs to be controlled.

# 6.10 Treatment of Combustion Emissions to Atmosphere

The major emissions from the combustion of fuel are $CO_2$, $SO_x$, $NO_x$, and particulates (Glassman 1987). The products of combustion are best minimized by making the process efficient in its use of energy through efficient heat recovery and avoiding unnecessary thermal oxidation of waste through minimization of process waste.

1) *Treatment of $SO_x$.* The various techniques that can be considered for the treatment of $SO_x$ are:

   - absorption into NaOH, $CaCO_3$, and so on;
   - oxidation and conversion to $H_2SO_4$;
   - reduction by conversion to $H_2S$ and then sulfur.

2) *Treatment of $NO_x$.* The techniques used for treatment of $NO_x$ are:

   - absorption into an NO-complexing agent or oxidizing agent;
   - reduction using selective non-catalytic or catalytic processes.

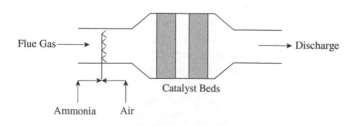

**Figure 6.36**

Removal of NO*x* using selective catalytic reduction. Source: (Smith 2016), *Chemical Process Design*, 2nd Edition, John Wiley & Sons Ltd.

3) *Treatment of particulates*. The various methods for treatment of particulates reviewed earlier include:

- scrubbers;
- inertial collectors;
- cyclones;
- bag filters;
- electrostatic precipitators.

4) *Treatment of CO$_2$*. If CO$_2$ needs to be separated from the other combustion products for the purpose of reducing greenhouse gas emissions, there are different ways this can be done, depending on the combustion arrangement. There are two general combustion arrangements; post-combustion and pre-combustion separation (Steeneveldt et al. 2006).

a) Post-combustion CO$_2$ Separation

Figure 6.37 shows two arrangements for post-combustion CO$_2$ separation. The first arrangement in Figure 6.37a shows a standard combustion arrangement with air used for combustion of the fuel. The separation following the combustion is suited to chemical absorption because of the low partial pressure of the CO$_2$ in the combustion gases. The same processes discussed earlier for the separation of H$_2$S can also be used for the separation of CO$_2$. An amine or mixture of amines in an aqueous solution can be used. The amines used are monoethanolamine (MEA), diethanolamine (DEA), or methyldiethanolamine (MDEA). Rather than use an amine or mixture of amines, a hot aqueous solution of potassium carbonate (K$_2$CO$_3$) can be used.

The other post-combustion separation arrangement shown in Figure 6.37b is oxy-combustion. The process starts with an air separation plant to provide pure oxygen for the combustion. This removes the nitrogen from the combustion process, giving products of combustion that are essentially CO$_2$ and water vapor. The CO$_2$ can then be separated by condensation of the water vapor. However, use of pure oxygen alone can create temperatures from combustion of hydrocarbon in excess of 3000 °C, which is problematic for the design of the combustion chamber. Thus, an inert dilution stream needs to be added to the combustion chamber to control the flame temperature. This can be a recycle of CO$_2$ from after the separation, or in principle steam can be added.

b) Pre-combustion CO$_2$ Separation

Rather than wait until after the combustion process to separate CO$_2$, various processes can be used to first produce synthesis gas (a mixture of H$_2$ and CO), the CO converted to CO$_2$ and the CO$_2$ separated, leaving H$_2$ for combustion and/or chemical use. Figure 6.38 shows four arrangements for such pre-combustion. The four arrangements differ in the way the synthesis gas is produced. Figure 6.38 shows two different *catalytic reformer* processes. In a steam reformer the dominant reaction for a general hydrocarbon feed is:

$$C_nH_m + nH_2O \rightleftharpoons nCO + (n + m/2)H_2$$

The *steam reforming* reaction is strongly endothermic and takes place in a fired heater. Steam reforming is the preferred method for feeds with light hydrocarbons, but is difficult to scale down for small scales.

An alternative reforming process for light hydrocarbon feeds shown in Figure 6.38 is *autothermal reforming*. This utilizes the heat of a partial oxidation reaction to drive the steam reforming reaction and is better suited to smaller scale applications than steam reforming. An autothermal reformer uses two reaction zones, a combustion zone and a catalytic zone. The overall reaction is exothermic. Fuel, steam, and oxygen are mixed and combusted in a substoichiometric flame in the combustion zone according to:

$$C_nH_m + (n/2 + m/4)O_2 \rightleftharpoons nCO + m/2H_2O$$

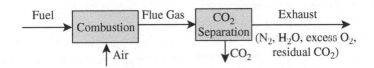

(a) Post-combustion separation of CO$_2$ with air

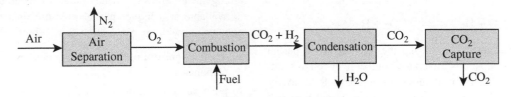

(b) Oxy-combustion separation of CO$_2$.

## Figure 6.37

Post-combustion arrangements for the separation of CO$_2$ with air and oxygen combustion. Source: Adapted from (Smith 2016), *Chemical Process Design*, 2nd Edition, John Wiley & Sons Ltd.

**6**

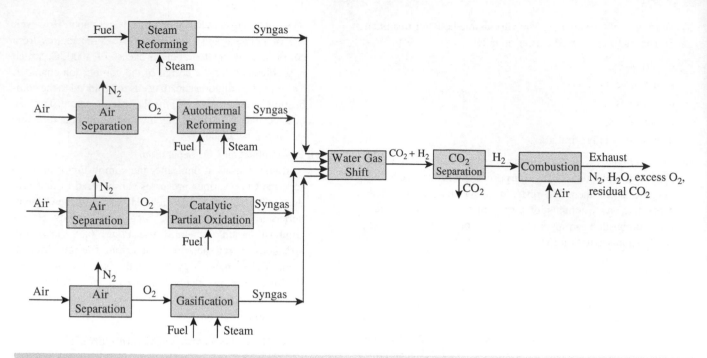

**Figure 6.38**

Pre-combustion arrangements for the separation of $CO_2$. Source: Adapted from (Smith 2016), *Chemical Process Design*, 2nd Edition, John Wiley & Sons Ltd.

In the catalytic zone, the reaction proceeds as the steam reforming reaction above.

The other two options illustrated in Figure 6.38 are partial oxidation reactions. Catalytic partial oxidation, which can also be used for light hydrocarbon feeds, uses oxygen (but can use air) according to:

$$C_nH_m + n/2O_2 \rightleftharpoons nCO + m/2H_2$$

The final option in Figure 6.38 is gasification. Although the reactions in gasification are complex, the dominant reaction is the partial oxidation reaction above. Gasification can be used for a wide range of solid, liquid, or gas feeds. It is normally used for the more difficult feeds of biomass, coal, petroleum residue, and other heavy hydrocarbons. Because of the nature of the feeds normally used, particulates need to be removed from the gas stream after the gasification.

All of the pre-combustion arrangements in Figure 6.38 also feature the *water gas shift reaction*:

$$CO + H_2O \rightleftharpoons CO_2 + H_2$$

Whilst this reaction occurs during the reforming and partial oxidation reaction steps, it is forced to a high conversion in a catalytic water gas shift reactor, as shown in Figure 6.38. In principle, the CO can be forced to a high conversion. Whether a high conversion is justified depends on the ultimate use of the synthesis gas and the method of separation to be used after the water gas shift. If the synthesis gas is to be used for chemical conversion (as well as combustion) then a high conversion might be justified. Alternatively, if only a bulk separation of $CO_2$ is required before combustion to reduce environmental discharge, then a high conversion might not be justified. The conversion of CO in the water gas shift reaction is favored by low temperature. Thus, to obtain a high conversion requires two water gas shift reaction steps, one at a high temperature and the second at a low temperature with intermediate cooling. Different catalysts are used in the high and low temperature shifts. The high temperature shift operates in the range of 300–450 °C and takes advantage of high reaction rates at high temperature and can give conversion to below 3% carbon monoxide on a dry basis at the reactor exit. The low temperature shift operates in the temperature range 180–250 °C and can reduce the CO content in the product to as low as 0.2% on a dry basis at the reactor exit. Whether one or two shifts are used, the resulting gas is essentially a mixture of hydrogen, carbon dioxide, and water vapor with small amounts of CO (and $H_2S$ if there is sulfur in the feed).

The $CO_2$ then needs to be separated (and $H_2S$ if there is sulfur in the feed), leaving essentially pure hydrogen for combustion and/or chemical use. Separation of the $CO_2$ in pre-combustion arrangements allows a wider range of techniques, as the $CO_2$ is available at a higher partial pressure than the case of post-combustion with air. Separation can be by:

- reactive absorption, e.g. using amines or hot potassium carbonate;
- physical absorption, e.g. using refrigerated methanol (the Rectisol Process) or a mixture of the dimethyl ethers of polyethylene glycol (the Selexol Process);
- pressure swing adsorption, e.g. using alumina or zeolite;
- membranes;
- cryogenic separation by condensing the $CO_2$ at low temperature.

Once the $CO_2$ is separated, the remaining gas is essentially pure hydrogen that can be used for chemical purposes or combusted to provide heat, or power, or both in cogeneration systems. Subsequent combustion produces mainly water vapor. The separated $CO_2$ can then be used or disposed of. The options are:

- Small-scale use in food and beverage products, fire extinguishers, enhanced growth in greenhouses, etc. However, such uses are too small to make a significant difference to the global effects of $CO_2$ emissions.

- Conversion of the $CO_2$ to chemical products (e.g. carbon monoxide, methane, methanol, formic acid, and so on). However, conversion to chemical products requires energy input, which might create additional $CO_2$ emissions depending on the source of the energy.

- Separated $CO_2$ can be dehydrated, compressed to liquid, and deposited in geological storage or used for enhanced gas/oil recovery. Such *sequestration* requires a stable location for long-term disposal and is an expensive way to deal with $CO_2$ emissions.

If it is important to significantly reduce $CO_2$ emissions to the atmosphere. Overwhelmingly the best solution is demand reduction through increased energy efficiency.

# 6.11 Atmospheric Dispersion

The objective must be to reduce atmospheric emissions to a minimum or at least below legislative requirements. However, there is inevitably some residual emission and this must be safely dispersed in the environment. The factors that affect the dispersion of gases to atmosphere are (De Nevers 1995):

1) *Temperature*. Temperature is a critical factor for atmospheric dispersion. Generally, the temperature of the atmosphere decreases with height, and the actual change of temperature with height is known as the *environmental lapse rate*. In the lower levels of the atmosphere, the environmental lapse rate changes with the time of day. An inversion can occur in which air temperature at low levels increases with height, rather than decreasing. This creates a strong resistance to upward motion of the gases released within the inversion, which inhibits atmospheric dispersion and can cause high concentrations of dispersed gases at low levels.

2) *Wind*. Not only does wind direction change but also the wind speed increases with height above the ground to a maximum at a height at which speed is equal to the *free air* or *geostrophic* wind speed. The rate of change of wind speed with height is affected by the change in air temperature with height. It is also affected by the terrain. The buildings in urban areas, for example, slow the air close to the ground, meaning that the maximum speed occurs at a greater height than, for example, over level country.

3) *Turbulence*. The third factor affecting dispersion is turbulence. Mechanical turbulence is caused by the roughness of the Earth's surface. Away from the surface, convective turbulence (heated air rising and cooler air falling) becomes increasingly important. The amount of turbulence and the height to which it operates depends on the surface roughness, wind speed, and change in air temperature with height.

The problem for the designer is to determine the appropriate stack height. This is illustrated in Figure 6.39. This shows that the effective stack height is a combination of the *actual stack height* and the *plume rise*. The plume rise is a function of discharge velocity, temperature of emission, and change in air temperature with height (De Nevers 1995).

The emissions from the stack itself must comply with environmental regulations relating to concentration and flowrate of pollutants. However, the stack must also be high enough such that any pollutant reaching the ground must be lower than ground level concentrations specified by the regulatory authorities. Pollution concentration at ground level depends on many factors, the most important being:

- height of the emitting stack;
- velocity and temperature of the emission from the stack;
- wind speed;

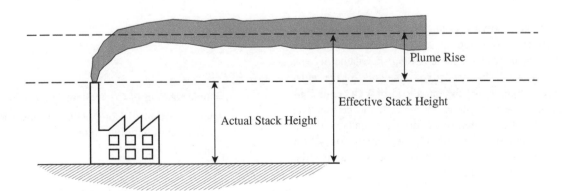

**Figure 6.39**

Stack height. Source: (Smith 2016), *Chemical Process Design*, 2nd Edition, John Wiley & Sons Ltd.

- change in air temperature with height;
- precipitation;
- topology of the surrounding terrain.

Calculation of pollution concentration at ground level requires specialized modeling techniques. These need to consider meteorological conditions over an extended period, often several years, to ensure that ground level concentrations remain below safe levels under all potential weather scenarios. These considerations are outside the scope of this text.

# 6.12   Waste Treatment – Summary

Water consumption and wastewater generation can be reduced through reuse, regeneration reuse, and regeneration recycling. Distributed effluent treatment requires that a philosophy of design be adopted that segregates effluent for treatment wherever appropriate, and combines it for treatment where appropriate. Various primary, secondary and tertiary treatment processes are available to achieve the required discharge concentrations.

Industrial emissions to atmosphere of major concern are VOCs, $NO_x$, $SO_x$, CO, $CO_2$, dioxins and furans, and particulates, $PM_{10}$ and $PM_{2.5}$. Urban smog, acid rain, ocean acidification, eutrophication, ozone layer destruction, and the greenhouse effect are environmental problems caused by atmospheric emissions. Treatment processes to deal with atmospheric emissions include:

- removal of particulates;
- condensation;
- membranes;
- adsorption;
- physical absorption;
- chemical absorption;
- thermal oxidation;
- gas dispersion.

# Exercises

1. Toluene is used as a solvent for the application of surface coatings. The solvents evaporate as a result of the application, creating a problem with for the emission of volatile organic compounds (VOCs). The legislative framework for the emission of VOCs requires that the mass load of VOC emissions allowed to be released to the atmosphere should be less than 60% of the mass of solids deposited during the coating process. The coating operations are currently depositing $50 \, t \, y^{-1}$ of solids. The concentration of the solids in the coating material is 20%, the remainder being toluene.

   a) Calculate the mass of toluene released during the coating operations and the mass that needs to be recovered or destroyed as a result of the legislation.

   b) It is proposed to solve the emissions problem by installing a ventilation system using air to collect the vapors and to destroy these using thermal oxidation. For safety reasons, the concentration of the flammable material in the ventilation system must be less than 30% of the lower flammability limit. The lower flammability limit for toluene in air is 1.2% by volume. The release of VOCs can be assumed to be evenly distributed over 8000 hours per year of operation. Calculate the air flowrate to the thermal oxidizer in $m^3 \, h^{-1}$ required to collect the VOCs that need to be destroyed.

2. A flue gas with a flow of $10 \, Nm^3 \, s^{-1}$ ($Nm^3$ = normal $m^3$) contains 0.1% vol. $NO_x$ (expressed as $NO_2$ at 0 °C and 1 atm) and 3% vol. oxygen. It is proposed to remove $NO_x$ by absorption in water to 100 ppmv for discharge. Whilst $NO_2$ is highly soluble, NO is only sparingly soluble and there is a reversible reaction in the gas phase between the two according to:

$$NO + \tfrac{1}{2} O_2 \rightleftharpoons NO_2$$

The equilibrium relationship for the reaction is given by:

$$K_a = \frac{p_{NO_2}}{p_{NO} p_{O_2}^{0.5}}$$

where $K_a$ is the equilibrium constant for the reaction and $p$ the partial pressure. At 25 °C, the equilibrium constant is $1.4 \times 10^6$ and at 725 °C is 0.14. The molar masses of NO and $NO_2$ are 30 and 46 kg $kmol^{-1}$, respectively.

   a) Calculate the mole ratio of $NO_2$ to NO assuming chemical equilibrium at 25 °C and at 725 °C.

   b) Calculate the flowrate of $NO_2$ and NO assuming chemical equilibrium at 25 °C and at 725 °C. Assume the kilogram molar mass occupies $22.4 \, m^3$ at standard conditions.

   c) It is proposed to remove the $NO_x$ by absorption in water at 25 °C. Do you consider this to be a good option if chemical equilibrium is assumed?

   d) In the worst case, if equilibrium is not attained, none of the NO would react to $NO_2$. Calculate the flowrate of water to remove the NO to a concentration of 100 ppmv by absorption in water at 25 °C and 1 atm. The solubility of NO in water can be assumed to follow Henry's Law:

$$x_i = \frac{y_i^* P}{H}$$

   where

   $x_i$ = mole fraction of Component $i$ in the liquid phase

   $y_i^*$ = mole fraction of Component $i$ in the vapor phase in equilibrium with the liquid

   $P$ = total pressure

   $H$ = Henry's Law constant

   At 25 °C, $H = 28$ atm for NO. If the concentration in the gas phase at the exit of the absorber is assumed to be 80% of equilibrium and gas phase is assumed to be ideal, calculate the flowrate of water required.

   e) Suggest other methods to treat the $NO_x$.

# References

Berne, F. and Cordonnier, J. (1995). *Industrial Water Treatment*. Gulf Publishing.

Betz Laboratories (1991). *Betz Handbook of Industrial Water Conditioning*, 9e.

Crynes, B.L. (1977). *Chemical Reactions as a Means of Separation – Sulfur Removal*. Marcel Dekker Inc.

De Nevers, N. (1995). *Air Pollution Control Engineering*. New York: McGraw Hill.

Dullien, F.A.L. (1989). *Introduction to Industrial Gas Cleaning*. Academic Press.

Eckenfelder, W.W. and Musterman, J.L. (1995). *Activated Sludge Treatment of Industrial Wastewater*. Technomic Publishing.

Glassman, J. (1987). *Combustion*, 2e. Academic Press.

Harrison, R.M. (1992). *Understanding Our Environment: An Introduction to Environmental Chemistry and Pollution*. Cambridge, UK: Royal Society of Chemistry.

Hydrocarbon Processing's Environmental Processes '98, *Hydrocarbon Process*, August: 71 (1998).

Rousseau, R.W. (1987). *Handbook of Separation Process Technology*. New York: Wiley.

Schierup, H.H. and Brix, H. (1990). Danish experience with Emergent Hydrophyte Treatment Systems (EHTS) and prospects in the light of future requirements on outlet water quality. *Water Science and Technology* 22: 65.

Schweitzer, P.A. (1997). *Handbook of Separation Process Techniques for Chemical Engineers*, 3e. New York: McGraw-Hill.

Smith, R. (2016). *Chemical Process Design and Integration*, 2e. Wiley.

Steeneveldt, R., Berger, B., and Torp, T.A. (2006). $CO_2$ capture and storage closing the knowing – doing gap. *Chemical Engineering Research and Design* 84 (A9): 739.

Stenhouse, J.I.T. (1981). Pollution control. In: *Chemical Engineering and the Environment* (ed. A.S. Teja). Blackwell Scientific Publications.

Svarovsky, L. (1981). *Solid-Gas Separation*. New York: Elsevier Scientific.

Tchobanoglous, G. and Burton, F.L. (1991). *Metcalf & Eddy Wastewater Engineering—Treatment, Disposal and Reuse*. McGraw-Hill.

Tebbutt, T.H.Y. (1990). *BASIC Water and Wastewater Treatment*. Butterworths.

US Environmental Protection Agency (1992). *Hazardous Air Pollution Emissions from Units in the Synthetic Organic Chemical Industry—Background Information for Proposed Standards*, vol. 1B. Springfield, VA: Control Technologies, US Department of Commerce.

**6**

# Chapter 7

# Reliability, Maintainability, and Availability Concepts

## 7.1 Reliability, Maintainability, and Availability

Figure 7.1 shows the operating profile of a chemical process. During a given time horizon when it is "up" and operating, it is able to make a product and potentially make a profit. However, there are going to be periods when the process is "down" for maintenance and not operating. These periods when maintenance is required not only incur maintenance costs, but also incur the cost of lost production and the waste of materials during the process shut-down and start-up after maintenance. The shut-down and start-up also potentially expose the process to significant safety and environmental problems.

The period over which the process is in operation is characterized by its *reliability*. The reliability defines how long a system can *survive*. *Maintainability* defines how long the system is down for maintenance. Taking the reliability and maintainability together, the *availability* defines the percentage of *uptime* over a time horizon, as shown in Figure 7.1.

There are a number of reasons that make it necessary to study reliability, maintainability, and availability:

1) *Safety*. If the process needs to be shut down for maintenance, either unplanned or planned, the shut-down and subsequent start-up after maintenance create potential for safety problems. Over 50% of process safety incidents occur during start-ups, shut-downs, and other events that infrequently occur (Duguid 1998a, 1998b, 1998c). This is because the start-ups and shut-downs involve many non-routine procedures, and these periods can result in unexpected and unusual situations. The maintenance work might be unplanned, due to an unexpected failure, or planned as part of a routine maintenance schedule. The plant shut-down might thus be an emergency shut-down, a planned shut-down, a partial shut-down of a section of the process, or the process paused temporarily keeping the process operating, but without making products. Each of these different cases will have different safety implications. The process plant design should therefore consider reliability and maintainability to include features in the design that prevent unnecessary shut-downs, both planned and unplanned.

2) *Environmental damage*. The start-ups and shut-downs associated with maintenance requirements not only bring safety problems, but also bring the potential for significant environmental damage. Disrupted conditions within the process during start-up and shut-down result in off-specification products and part processed materials that might need to be disposed of. The maintenance itself requires process materials to be cleared from at least that part of the process requiring maintenance, followed by purging to make that part of the process safe for maintenance procedures. The waste created by the need for shut-down for maintenance can be environmentally damaging through waste of materials.

3) *Cost of lost production*. The process needs to be kept in operation with the necessary capacity for long enough to meet production targets. The process must therefore achieve a certain *availability* at the required *capacity* in order to meet production targets. That means avoiding unnecessary shut-downs and ensuring that planned shut-downs are short enough to allow production targets to be met. Avoiding unnecessary shut-downs also reduces the cost associated with waste materials generated in the shut-down and start-up periods.

4) *Reliability of safety systems*. Safety systems such as relief valves and bursting discs for the relief of overpressure (see Chapter 15) stay in the background until an abnormal situation requires some action to keep the system safe. A dangerous undetected failure of the safety system might not be detected until there is a demand on the system. It is therefore necessary to ensure that the *probability of failure on demand* of the safety system is low enough to be acceptable for process safety.

5) *Standby equipment*. One way to avoid a process shut-down in the event of equipment failure, or the need for equipment maintenance, is to install standby equipment. This involves the use

*Process Plant Design*, First Edition. Robin Smith.
© 2024 John Wiley & Sons Ltd. Published 2024 by John Wiley & Sons Ltd.
Companion website: www.wiley.com/go/processplantdesign

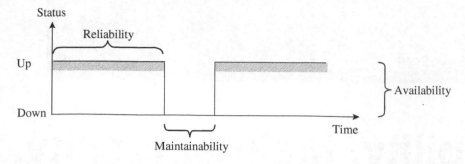

## Figure 7.1

Reliability, maintainability, and availability.

of additional components or units beyond the minimum number required for operation with the purpose of avoiding unnecessary process shut-downs. Figure 7.2 shows the example of a pump that has a standby, which is not operated until the online pump fails or requires maintenance. The location and number of items of standby equipment need to be quantified in a reliability, maintainability, and availability analysis.

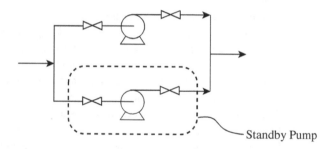

## Figure 7.2

Standby equipment to maintain production in the event of equipment failure.

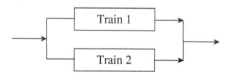

## Figure 7.3

Parallel trains can be used to maintain production in the event of failure.

6) *Parallel production trains.* Rather than the process be designed with a single production train, it can be designed with smaller parallel trains that together maintain the overall required capacity, as illustrated in Figure 7.3. Operation of the parallel trains independently can be used to maintain production in the event of failure of one of the trains, or the need for maintenance of one of the trains. In the event of failure of one of the trains, production will be maintained, but most likely at a lower capacity.

7) *Intermediate storage.* Process plant can often be divided into sections that can be operated independently of the other sections for short periods. If this is the case, then intermediate storage, as illustrated in Figure 7.4, can be used to gain a number of advantages. It can make the process easier to operate, particularly for start-up and shut-down. It can also make the overall process control more straightforward. However, within the present context, intermediate storage can be used to maintain production in the event of failure, or the need for maintenance, increasing availability and capacity. However, the location and capacity of the intermediate storage needs to be quantified in a reliability, maintainability, and availability analysis.

In process plant design it is therefore necessary to consider reliability, maintainability, and availability as an integral part of the process design. The individual concepts of reliability, maintainability, and availability first need to be established before applying them in the process plant design. The concepts are not deterministic but are based on probability. According to Benjamin Franklin:

*"In this world nothing can be said to be certain except death and taxes."*

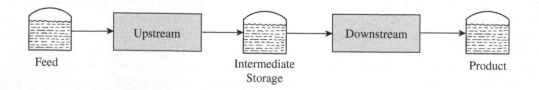

## Figure 7.4

Intermediate storage can be used to maintain production in the event of failure.

Everything else is a balance of probabilities. The analysis of reliability, maintainability, and availability will be a balance of probabilities.

# 7.2 Reliability

If a number of components with the *same design* are put under test at the *same specified conditions* for an *indefinite period of time*, the individual components will fail at different times. This is because the components are not absolutely identical and the conditions under which each operates will also not be absolutely identical. The materials used to fabricate the components will have minor defects that will differ from component to component. The manufacturing process for each component will have minor differences. The conditions under which each component operates will have minor differences. The installation of the components will have been subject to slight differences. If the components are such that they are repairable and have been repaired, the maintenance history of the components will be different. These various effects will come together with the result that components with seemingly the same design and operated under seemingly the same conditions will ultimately fail at different times. The fraction that failed or survived up to time *t* can be plotted. Figure 7.5a shows a typical plot of the fraction of components that failed up to time *t* when operated under stated conditions, known as the *cumulative distribution function* or *failure function*. The curve starts at zero and approaches unity as time progresses and represents the probability that failure occurs before time *t*. Rather than plot the fraction of components that failed, the fraction that *survived* when operated under stated conditions can be plotted versus time *t*, as shown in Figure 7.5b. This plot of the fraction of components that survived versus time is known as the *reliability function*. This time the curve starts at unity and approaches zero as time progresses.

The two plots of the failure function and reliability function are mirror images of each other, as illustrated in Figure 7.6. For the reliability function $R(t)$:

$$R(0) = 1, R(\infty) = 0 \qquad (7.1)$$

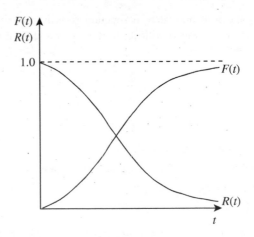

**Figure 7.6**

The failure and reliability functions are mirror images of each other.

Reliability can be defined as *the probability that a component or system will perform a required function for a given period of time when used under stated conditions*. It is the probability of non-failure over time (Beasley 1991; Billinton and Allan 1983; Davidson 1994; Ebeling 1997). For the *failure function F(t)*:

$$F(t) = 1 - R(t)$$
$$F(0) = 0, F(\infty) = 1 \qquad (7.2)$$

Another useful function is the *probability density function*, which is defined by taking the derivative of $F(t)$:

$$\frac{dF(t)}{dt} = f(t) \qquad (7.3)$$

where $f(t)$ = probability density function.
From Eq. (7.3):

$$f(t)dt = dF(t) = F(t + dt) - F(t) \qquad (7.4)$$

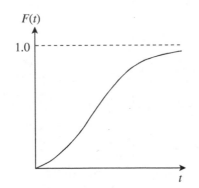

(a) Failure function measures the fraction of identical components that have failed at time *t*.

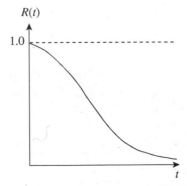

(b) Reliabilty function measures the fraction of identical components that have survived at time *t*.

**Figure 7.5**

Failure and reliability functions.

Thus, the probability density function $f(t)$ is the probability that a component or system will fail in the interval $(t, t + dt)$. Given Eqs. (7.2) and (7.3):

$$f(t) = \frac{dF(t)}{dt} = -\frac{dR(t)}{dt} \qquad (7.5)$$

The properties of the probability density function are given by:

1) $f(t)$ is always positive or zero.
2) $\int_{t_1}^{t_2} f(t)\, dt = R(t_1) - R(t_2)$ = fraction of failures ocurring in the interval $t_1$ to $t_2$.
3) $\int_0^\infty f(t)\, dt = R(0) - R(\infty) = 1$.

The probability density function $f(t)$ is illustrated in Figure 7.7.
The probability that a component or system will fail per unit time can be quantified by the failure rate $\lambda(t)$:

$\lambda(t)$ = probability that a component or system experiences a failure per unit time at time $t$, *given that it has survived to time t*

$$= \frac{\text{probability of failure in the interval } (t, t+dt)}{\text{reliability at time } t}$$

$$= \frac{f(t)}{R(t)}$$

$$= -\frac{dR(t)}{dt}\frac{1}{R(t)} \qquad (7.6)$$

Integrating with $R(0) = 1$ as the lower bound of the integral:

$$\int_0^{t'} \lambda(t)\, dt = \int_1^{R(t')} -\frac{dR(t)}{R(t)} \qquad (7.7)$$

$$= -\ln R(t')$$

Thus, after rearranging:

$$R(t') = \exp\left[-\int_0^{t'} \lambda(t)\, dt\right] \qquad (7.8)$$

Figure 7.8 shows a typical variation of the failure rate through time. The curve can be separated into three different periods. Together, the three periods characterize the 'bathtub curve'. In the first phase, *early* or *infant* failure is characterized by a decrease in the failure rate. This decrease results from problems that are identified and corrected:

- design flaws;
- incorrect selection of a component;
- fabrication errors;
- installation errors;

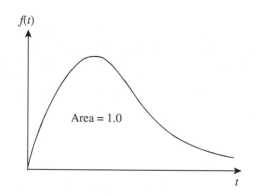

**Figure 7.7**

The probability density function.

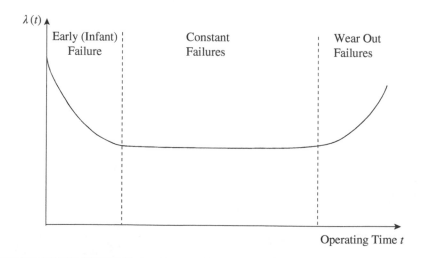

**Figure 7.8**

The variation of failure rate through time tends to follow a "bathtub" profile.

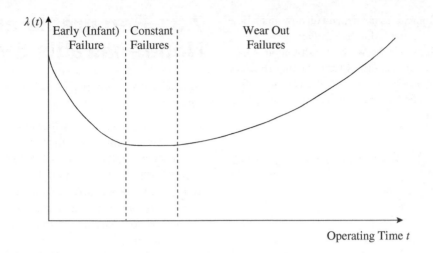

## Figure 7.9
Mechanical systems might show early onset of wear-out.

- instrumentation errors of surrounding equipment;
- operational errors.

After the early failure phase, the failure rate becomes constant. This is referred to as the *normal* period with a low failure rate, which is usually the basis for the selection of components. After the constant failure period, the failure rate starts to increase in the *wear-out* period beyond the normal operating period. However, for mechanical systems, there might be an early onset of wear-out, as illustrated in Figure 7.9. It should be emphasized that entering the wear-out period does not necessarily mean a component is at the end of its useful life, but might need more careful management.

A model is required for the representation of the failure rate through time in order to quantify the reliability of components. The simplest assumption is to assume that the failure rate is constant and equal to $\lambda$. Substitute the constant failure rate in Eq. (7.8):

$$
\begin{aligned}
\ln R(t') &= -\int_0^{t'} \lambda \, dt \\
&= -[\lambda t]_0^{t'} \\
&= -\lambda t'
\end{aligned}
\tag{7.9}
$$

Rearranging:

$$
R(t') = \exp[-\lambda t']
\tag{7.10}
$$

Figure 7.10 shows a typical reliability function for a constant failure rate. Note that the exponential function in Eq. (7.10) predicts $R(0) = 1$ and $R(\infty) = 1$. Equation (7.10) is referred to the

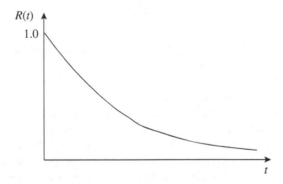

## Figure 7.10
Reliability function with a constant failure rate.

*exponential distribution* for reliability. An important feature of the exponential function with a constant failure rate is that it does not have a *memory*; it is *memoryless*. Consider the failure of a component that has survived to time $t_0$. Having survived to time $t_0$, the probability of survival at time $t_0 + t$ is given by the conditional reliability $R(t$ given $t_0)$, written as $R(t|t_0)$:

$$
\begin{aligned}
R(t|t_0) &= \frac{R(t_0 + t)}{R(t_0)} = \frac{\exp[-\lambda(t_0 + t)]}{\exp[-\lambda(t_0)]} \\
&= \frac{\exp[-\lambda(t_0)]\exp[-\lambda(t)]}{\exp[-\lambda(t_0)]} \\
&= \exp[-\lambda(t)] = R(t)
\end{aligned}
\tag{7.11}
$$

If the failure rate is constant, the time to failure depends only on the operating time and not the current age. The probability of failure happening in the future does not depend on how much time has elapsed. This means that the failure is *random*.

A more general distribution is the Weibull distribution, which is the most useful general purpose model for reliability (Beasley 1991; Billinton and Allan 1983; Davidson 1994; Ebeling 1997). In the Weibull distribution, the failure rate is assumed to be a function of time $t$ according to:

$$\lambda(t) = \frac{\beta}{\theta}\left(\frac{t}{\theta}\right)^{\beta-1} \quad \theta > 0, \beta > 0, \ t \geq 0 \qquad (7.12)$$

where

$\beta, \theta$ = fitting parameters

$\beta$ = shape parameter

$\beta$ = 1 reduces to the exponential distribution

Substituting Eq. (7.12) into Eq. (7.8) gives (see Appendix A):

$$R(t) = \exp\left[-\left(\frac{t}{\theta}\right)^{\beta}\right] \qquad (7.13)$$

By changing the fitting parameters $\beta$ and $\theta$, different variations of failure rate through time can be created. This is illustrated in Figure 7.11. In Figure 7.11, $\theta$ has been assumed to have a value of 100 hours. Different values of $\beta$ can then create a failure rate that increases or decreases through time. Assuming $\beta = 1$ gives a constant failure rate and the exponential distribution. Many other forms of distribution have been proposed (Beasley 1991; Billinton and Allan 1983; Davidson 1994; Ebeling 1997). However, the exponential and Weibull distributions are the most commonly used in the process industries.

# 7.3 Repairable and Non-repairable Systems

Generally, two categories of failure can be identified requiring repair:

1) *Non-repairable*. Non-repairable systems exhibit only one failure and are removed permanently and the unit replaced by a similar unit. Examples would be light bulbs, electronic components, and bearings in a machine.

2) *Repairable*. Repairable systems can be restored to satisfactory operation through maintenance to replace parts or to make changes to adjustable settings. Examples are pumps and compressors.

For non-repairable systems a *mean time to failure* (*MTTF*) can be defined. Figure 7.12 illustrates the time to failure of a number of components. The mean time to failure is defined by:

$$MTTF = \frac{t_1 + t_2 + t_3 + \cdots + t_n}{n} \qquad (7.14)$$

where

$MTTF$ = mean time to failure (h)

$t_i$ = time to failure of Component $i$ (h)

$n$ = number of components under test (–)

For the failure of repairable systems the picture is more complex. This is illustrated in Figure 7.13. In this case, the component is operated until it fails. It is then repaired and brought back online, until it fails again. A time between consecutive failures can be defined, as illustrated in Figure 7.13. However, this can be a source of confusion, as it includes both failure and repair characteristics. It is more meaningful to measure the time to failure and

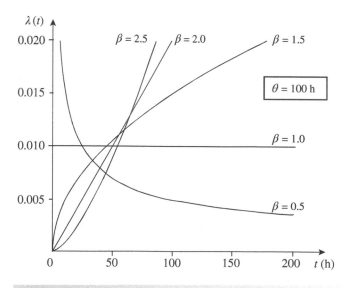

**Figure 7.11**

The Weibull distribution allows decreasing, increasing, or constant failure rates to be represented by adjusting the parameters.

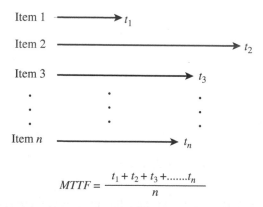

**Figure 7.12**

A mean time to failure (MTTF) can be defined for non-repairable systems.

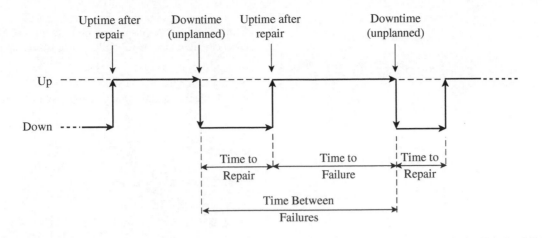

**Figure 7.13**

Failure of repairable systems.

separately the time to repair. The mean time to failure for repairable systems can then be defined by:

$$MTTF = \text{mean uptime}$$

$$= \frac{\text{uptime}}{\text{number of system failures}} \tag{7.15}$$

In order to apply Eq. (7.15) it is necessary to quantify this in terms of reliability distributions. This can be done by first referring back to the probability density function given in Eq. (7.5):

$$f(t)\mathrm{d}t = F(t + \mathrm{d}t) - F(t)$$

$$= \text{probability of failure at time } t \tag{7.16}$$

Then the *MTTF* is given by the integration across time with a probability of failure at time $t$:

$$MTTF = \int_0^\infty t\, f(t)\, \mathrm{d}t$$

$$= \int_0^\infty -t\, \frac{\mathrm{d}R(t)}{\mathrm{d}t}\, \mathrm{d}t$$

$$= [-tR(t)]_0^\infty + \int_0^\infty R(t)\, \mathrm{d}t \quad \text{integrating by parts}$$

$$= \int_0^\infty R(t)\, \mathrm{d}t \quad \text{given that } t{\cdot}R(t) \to 0 \text{ as } t \to \infty \tag{7.17}$$

Whilst the limit in Eq. (7.17) applies to the reliability functions used in practice, it does not necessarily apply to all mathematical functions.

Thus:

$$MTTF = \int_0^\infty R(t)\, \mathrm{d}t \tag{7.18}$$

A number of things need to be noted about *MTTF*. For repairable systems, it assumes that repairs are to the original state, which is not always true. Many texts restrict *MTTF* to only be applicable to non-repairable systems and define a mean time between failure

(*MTBF*) for repairable systems. However, *MTBF* includes both failure and repair characteristics. It is more meaningful to measure the mean time to failure (*MTTF*) and separately the mean time to repair (*MTTR*). There are many different interpretations of *MTBF* and it is best not used.

For a system with a constant failure rate:

$$MTTF = \int_0^\infty \exp(-\lambda t)\, \mathrm{d}t$$

$$= \left[ -\frac{1}{\lambda} \exp(-\lambda t) \right]_0^\infty \tag{7.19}$$

$$= \frac{1}{\lambda}$$

Thus, for systems with a constant failure rate, assuming repair to the original state:

$$MTTF = \frac{1}{\lambda} \tag{7.20}$$

Note that there is no guarantee that a unit with a constant failure rate will survive to *MTTF*. Substituting Eq. (7.20) into Eq. (7.10):

$$R(t) = \exp[-\lambda t] = \exp\left[ -\frac{1}{MTTF} t \right] \tag{7.21}$$

Various multiples of *MTTF* can be substituted for the time $t$ into Eq. (7.21) to obtain the reliability. This is shown in Table 7.1. It can be seen from Table 7.1 that there is no certainty of survival at any multiple of *MTTF*. This is because the assumed constant failure rate implies completely random failure. Even at $1 \times MTTF$ there is only a 37% chance of survival. It must be emphasized that the time to the next failure is not always the same.

If the Weibull distribution is assumed from Eq. (7.13) and substituted into Eq. (7.18), the corresponding *MTTF* is given by (see Appendix B):

$$MTTF = \theta\, \Gamma\left(1 + \frac{1}{\beta}\right) \tag{7.22}$$

where $\Gamma$ = gamma function

## Table 7.1

Probability that a unit will be working for multiples of *MTTF*.

| After a time $t$ | Probability that the unit will still be working $R(t)$ |
|---|---|
| $0.01 \times MTTF$ | 0.99 |
| $0.1 \times MTTF$ | 0.90 |
| $0.5 \times MTTF$ | 0.61 |
| $1 \times MTTF$ | 0.37 |
| $2 \times MTTF$ | 0.14 |

The gamma function is defined by:

$$\Gamma(x) = \int_0^\infty y^{x-1} \exp(-y)dy \qquad (7.23)$$

where

$$x = \left(1 + \frac{1}{\beta}\right)$$

$$y = \left(\frac{t}{\theta}\right)^\beta$$

$\beta, \theta$ = Weibull fitting parameters

The gamma function is a standard mathematical function, values of which are tabulated, or can be evaluated readily in a spreadsheet or general purpose mathematical software.

**Example 7.1**   A component experiences a pattern of random failure with a mean time to failure (*MTTF*) of 1500 hours.

a) Calculate the reliability for a mission time of 500 hours.
b) Calculate the expected life for a reliability of 0.9.

**Solution**

a)  $\lambda = \dfrac{1}{MTTF} = \dfrac{1}{1500} = 6.67 \times 10^{-4}\,\mathrm{h}^{-1}$

$R = \exp(-\lambda t)$

$= \exp(-6.67 \times 10^{-4} \times 500)$

$= 0.72$

b)  $0.9 = \exp(-\lambda t)$

$= \exp(-6.67 \times 10^{-4}t)$

$t = -\dfrac{\ln 0.9}{6.67 \times 10^{-4}}$

$= 158\,\mathrm{h}$

**Example 7.2**   An instrument on a compressor has exhibited a constant failure rate of 150 failures per $10^6$ operating hours.

a) Calculate the *MTTF*.
b) Represent this as a reliability function.
c) Calculate the reliability after a 5000 hour continuous operating period.

**Solution**

a)  $MTTF = \dfrac{1}{\lambda} = \dfrac{10^6}{150} = 6667\,\mathrm{h}$

b)  $R(t) = \exp\left[-1.5 \times 10^{-4}\,t\right]$

c)  $R(5000) = \exp\left[-1.5 \times 10^{-4} \times 5000\right] = 0.47$

Thus, there is a 47% probability that the instrument will survive a 5000 hour operating period, or a 53% probability of failure.

**Example 7.3**   A compressor is experiencing an increasing failure rate that can be characterized by a Weibull distribution with $\beta = 2.5$ and $\theta = 7000$ hours.

a) Calculate the reliability for a time horizon of 5000 hours.
b) Calculate the mean time to failure *MTTF*.

**Solution**

a) From Eq. (7.13):

$$R(t) = \exp\left[-\left(\frac{t}{\theta}\right)^\beta\right]$$

$$R(5000) = \exp\left[-\left(\frac{5000}{7000}\right)^{2.5}\right]$$

$$= 0.650$$

b) The mean time to failure *MTTF* is given by Eq. (7.22):

$$MTTF = \theta\,\Gamma\left(1 + \frac{1}{\beta}\right)$$

From tables of the Gamma Function:

$$\Gamma\left(1 + \frac{1}{\beta}\right) = \Gamma\left(1 + \frac{1}{2.5}\right) = 0.8873$$

$$MTTF = \theta \times 0.8873$$

$$= 7000 \times 0.8873$$

$$= 6211\,\mathrm{h}$$

## 7.4   Reliability Data

The data required to use the reliability equations developed so far cannot be determined fundamentally, but requires historical records of equipment performance, either from plant operations or equipment manufacturers. There are many sources of reliability data, but these depend on the type of equipment (process plant, electronic equipment, safety equipment, military applications, nuclear plant, etc). For process plant useful sources are:

1) *Offshore & Onshore Reliability Data (OREDA) Handbook* (SINTEF and NTNU 2015). These data were collected from offshore and subsea equipment for North Sea oil and gas exploration operations.
2) *Guidelines for Process Equipment Reliability Data* (CCPS 1989).
3) *The Reliability Data Handbook* (Moss 2005).
4) *Lees' Loss Prevention in the Process Industries* (Mannan 2012).
5) Historical data/maintenance records from operating plant. Historical data are fitted to a reliability distribution.

Consider a simple example of fitting data to a reliability distribution. Historical failure data can be fitted to a reliability model. If only a few data points are available, fitting is not so straightforward. Table 7.2 gives five data points for failure of a component listed in rank order (Beasley 1991). Reliability $R(t)$ is defined as the fraction of units surviving at $t_i$:

$$R(t_i) = \frac{n-i}{n} \tag{7.24}$$

where

$i$ = rank order of the data point (–)
$n$ = number of data points (–)

The fraction failed $F(t)$ at $t_i$ is given by:

$$F(t_i) = 1 - R(t_i) = \frac{i}{n} \tag{7.25}$$

The cumulative fraction that failed at a given time $t$ is plotted in Figure 7.14a. The plot implies that 20% of the total population will

### Table 7.2

Component failure data.

| Rank order of data $i$ | Failure time $t_i$ (h) |
|---|---|
| 1 | 350 |
| 2 | 1050 |
| 3 | 2300 |
| 4 | 4200 |
| 5 | 7300 |

fail before the first data point and no failures of the total population will occur after the last data point. This seems unreasonable. Instead of plotting the data in this way, the data could be adjusted on an arbitrary basis to plot (Beasley 1991):

$$F(t_i) = \frac{i-1}{n} \tag{7.26}$$

This is shown in Figure 7.14b. Now it appears that there are no failures in the total population before the first data point and 20% of the failures from the total population will occur after the last data point. This seems equally unreasonable. The problem with this dataset is that it is too small in number to give a reasonable picture of the total population. It is necessary to represent the whole problem, rather than a small sample.

Rather than take either of these extremes an average could be taken and a plot made of the average of $1/n$ and $(i-1)/n$ (Beasley 1991):

$$F(t_i) = \frac{i-0.5}{n} \tag{7.27}$$

This is shown in Figure 7.14c and seems to give a more reasonable representation of the approximate probability that a value is less than $F_i$. More accurate methods for the plotting position are available (Ebeling 1997). A better approximation to the $i$th median rank for $n$ data points is given by (Ebeling 1997):

$$F(t_i) = \frac{(i-0.3)}{(n+0.4)} \tag{7.28}$$

Now try fitting an exponential model to the data in Figure 7.14c. Taking natural logarithms of both sides of Eq. (7.10) gives:

$$-\ln R(t) = \lambda t \tag{7.29}$$

Figure 7.15 shows a plot of $-\ln R(t)$ versus $t$, assuming from Eq. (7.27) that $R(t) = 1 - (i - 0.5)/n$. This shows a good fit to a straight line and therefore an exponential distribution with constant $\lambda$. The slope defines $\lambda = 0.000311$ failures h$^{-1}$.

Similarly, to fit a Weibull model:

$$R(t) = \exp\left[-\left(\frac{t}{\theta}\right)^{\beta}\right] \tag{7.30}$$

Take natural logarithms of both sides:

$$\ln R(t) = -\left(\frac{t}{\theta}\right)^{\beta} \tag{7.31}$$

Take natural logarithms of both sides again:

$$\ln[-\ln R(t)] = \ln\left(\frac{t}{\theta}\right)^{\beta} \tag{7.32}$$

Rearranging:

$$\ln[-\ln R(t)] = \beta \ln(t) - \beta \ln(\theta) \tag{7.33}$$

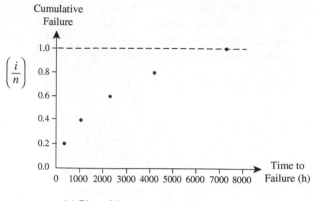

(a) Plot of the cumulative failure data.

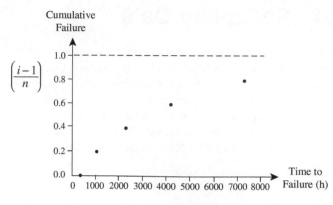

(b) Adjusting the cumulative failure data down.

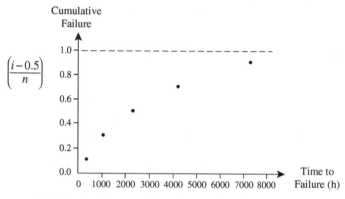

(c) Adjusting the cumulative failure data to a mean position

## Figure 7.14

Failure data for a system.

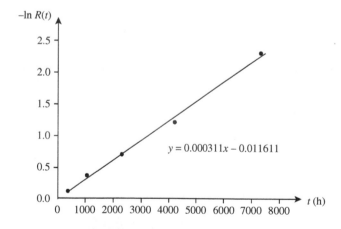

$$y = 0.000311x - 0.011611$$

## Figure 7.15

A plot of the data assuming a constant failure rate shows an acceptable fit to a straight line.

Thus, to fit a Weibull model requires a plot of $\ln[-\ln R(t)]$ versus $\ln(t)$. For a good fit to a Weibull distribution, the data will follow a straight line with slope $\beta$ and intercept $-\beta \ln(\theta)$.

The data analysis so far assumes the dataset is *complete*, that is all information is available. This is illustrated in Figure 7.16 in which all of the components have failed, and at a time that has been recorded. However, the dataset might be *incomplete* or *censored* or *suspended*. If the dataset is censored, this means that some of the information is missing. Figure 7.17 shows a dataset in which some of the components have failed, but some are still running and the failure times for these is unknown. The data shown in Figure 7.17 is referred to as *right censored*. A common mistake in the analysis of such data is to leave out the data for the components that are still functioning and analyze only the ones that have failed. Another example of censored data is shown in Figure 7.18. This is referred to as *interval censored*. In this case some of the components have failed but the exact time of failure is unknown, perhaps because of non-constant monitoring. A third example of censored data is shown in Figure 7.19 and referred to as *left censored* data. Here, components have failed before a certain time, but it is not known exactly when. Again, this can be a result of non-constant monitoring. If censored data are not included in the data analysis the results can be completely erroneous. Special techniques are required to analyze sensor data, which are outside the scope of this text (Ebeling 1997).

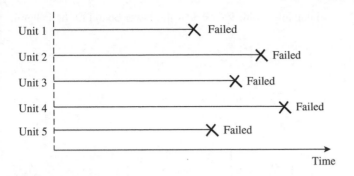

**Figure 7.16**

The data analysis so far assumes the data set is complete with all data available.

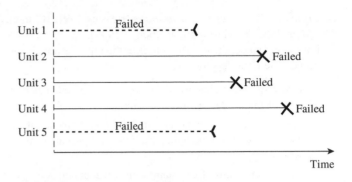

**Figure 7.19**

Left censored data feature units that have failed before a certain time, but it is not known exactly when.

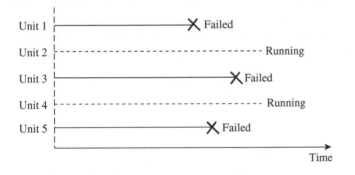

**Figure 7.17**

Right censored data feature units still running and the failure time is unknown.

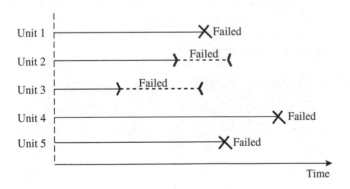

**Figure 7.18**

Interval censored data feature units that have failed but the exact time is unknown because of non-constant monitoring.

## 7.5  Maintainability

**7**

For repairable systems, three categories of maintenance are carried out:

1) *Corrective maintenance (CM)*. Corrective maintenance is usually carried out as a result of equipment failure. It is carried out only after fault recognition and is intended to return the system to a state that will allow the required function to be performed. Corrective maintenance can lead to significant disruption and safety problems through allowing equipment to fail. It can also lead to long downtimes for correction.

2) *Preventive maintenance (PM)*. Preventive maintenance is a well-defined set of tasks that is carried out at predetermined intervals, or according to prescribed criteria. It is intended to reduce the probability of failure or degradation of the performance of the equipment. However, preventive maintenance brings the possibility of unnecessary maintenance and unnecessary downtime for that maintenance.

3) *Predictive maintenance*. Predictive maintenance predicts through diagnostic tools or measurements when equipment is near failure and should be repaired or replaced. Measurements could be, for example, measuring vibration on the shaft of a rotating machine such as a compressor, measuring metallic debris in lubricating oils, measuring noise, or process measurements such as temperature and pressure. The objective would be to relate the measurement to the condition of the equipment, which requires a good understanding of the equipment. In this way, the maintenance is performed before failure, but not regularly and only when needed. However, it is necessary to know which are the critical items of equipment and how to monitor the condition of that equipment.

An effective maintenance policy should both avoid unnecessary failures and unnecessary maintenance.

*Maintainability* is defined to be the probability that a component or system will be restored or repaired to a *specified condition within a period of time when maintenance is performed in accordance with prescribed procedures* (Beasley 1991; Billinton and Allan 1983; Davidson 1994; Ebeling 1997).

For maintenance, the time to complete the repair is not always the same. The time depends on:

- the time required to shut down the process for maintenance, or the part of the process requiring maintenance;
- the time required to isolate and make the equipment safe to perform maintenance (including draining and purging of process materials);
- availability and skill of the maintenance personnel;
- deterioration of the component or system through time requiring longer repair times;
- improved maintenance procedures shortening repair time through time;
- ease of access to the equipment requiring maintenance;
- construction of temporary scaffolding to allow access where necessary;
- heavy lifting requirements;
- availability of spare parts.

Maintainability data must be obtained from historical maintenance records. Defining a failure to repair function $V(t)$ and a maintainability function $M(t)$:

$V(t) =$ probability of *not* performing a successful repair action in time

$M(t) =$ probability of performing a successful repair action in time $t$

$\quad = 1 - V(t)$

Taking the derivative of the maintainability function:

$$\frac{\mathrm{d}M(t)}{\mathrm{d}t} = g(t) \qquad (7.34)$$

where

$g(t) =$ repair time density function

$\quad =$ probability that a component or system will be repaired in the interval $(t, t + \mathrm{d}t)$

The probability that a component item of equipment or system will be repaired can be quantified by the repair rate $\mu(t)$:

$\mu(t) =$ probability that a component or system will be repaired at time $t$, given that it has not been repaired to time $t$

$\quad = \dfrac{\text{probability of repair in the interval } (t, t + \mathrm{d}t)}{\text{probability of failure to repair at time } t}$

$\quad = \dfrac{g(t)}{V(t)}$

$\quad = \dfrac{\mathrm{d}M(t)}{\mathrm{d}t} \cdot \dfrac{1}{V(t)}$

$\quad = -\dfrac{\mathrm{d}V(t)}{\mathrm{d}t} \cdot \dfrac{1}{V(t)}$

$$(7.35)$$

Integrating with $V(0) = 1$ as the lower bound of the integral:

$$\int_0^{t'} \mu(t)\, \mathrm{d}t = \int_1^{V(t')} -\frac{\mathrm{d}V(t)}{V(t)} \qquad (7.36)$$

$$= -\ln V(t')$$

Thus, after rearranging:

$$V(t') = \exp\left[-\int_0^{t'} \mu(t)\, \mathrm{d}t\right] \qquad (7.37)$$

In terms of the maintainability function:

$$M(t') = 1 - \exp\left[-\int_0^{t'} \mu(t)\, \mathrm{d}t\right] \qquad (7.38)$$

The repair rate $\mu(t)$ might, for example, decrease through time as a result of increasing repair times as an asset ages. Alternatively, the repair rate might increase through time, shortening the repair time, as maintenance procedures are improved. If the repair rate is assumed constant and equal to $\mu$, then Eq. (7.38) becomes:

$$1 - M(t') = \exp\left[-\int_0^{t'} \mu(t)\, \mathrm{d}t\right]$$

$$\ln[1 - M(t')] = -\int_0^{t'} \mu(t)\, \mathrm{d}t \qquad (7.39)$$

$$= -[\mu t]_0^{t'}$$

$$= -\mu t'$$

Rearranging gives:

$$M(t') = 1 - \exp[-\mu t'] \qquad (7.40)$$

A mean time to repair (*MTTR*) can be determined from historical records:

$$MTTR = \frac{\text{corrective maintenance downtime}}{\text{number of system failures}} \qquad (7.41)$$

In the absence of other information, it is often assumed that $MTTR = 8$ hours for straightforward repairs (the normal duration of a maintenance shift). $MTTR$ can be defined in terms of probability distributions given:

$$MTTR = \int_0^\infty t\, g(t)\, \mathrm{d}t$$

$$= \int_0^\infty t\, \frac{\mathrm{d}M(t)}{\mathrm{d}t}\, \mathrm{d}t$$

$$= \int_0^\infty -t\, \frac{\mathrm{d}V(t)}{\mathrm{d}t}\, \mathrm{d}t$$

$$= [-t\, V(t)]_0^\infty + \int_0^\infty V(t)\, \mathrm{d}t \qquad \text{integrating by parts}$$

$$= \int_0^\infty V(t)\, \mathrm{d}t \qquad \text{given that } t\, V(t) \to 0 \text{ as } t \to \infty$$

$$(7.42)$$

Whilst the limit in Eq. (7.42) applies to the repair functions used in practice, it does not necessarily apply to all mathematical functions.

Thus, in terms of the maintainability function:

$$MTTR = \int_0^\infty \left(1 - M(t)\right) dt \qquad (7.43)$$

If an exponential distribution is assumed for the maintainability, with a constant repair rate, defined in Eq. (7.40):

$$MTTR = \int_0^\infty \exp(-\mu t)\, dt$$

$$= \left[ -\frac{1}{\mu} \exp(-\mu t) \right]_0^\infty \qquad (7.44)$$

$$= \frac{1}{\mu}$$

Thus, for systems with a constant repair rate:

$$MTTR = \frac{1}{\mu} \qquad (7.45)$$

Table 7.3 presents a comparison of maintainability and reliability for a constant repair rate and constant failure rate. Table 7.4 shows data for MTTF and MTTR for process equipment (Koolen 2001).

---

**Example 7.4**   The repair times for a pump have a mean of eight hours. Assuming an exponential distribution for the maintainability, calculate the probability of completing a repair in six hours.

**Solution**

$$M(t) = 1 - \exp[-\mu \cdot t]$$

$$= 1 - \exp\left[ -\frac{t}{MTTR} \right]$$

$$= 1 - \exp\left[ -\frac{6}{8} \right]$$

$$= 0.528$$

Thus, there is a 52.8% probability the repair will be completed in six hours.

---

## Table 7.3

Comparison of maintainability and reliability.

| Maintainability | Reliability |
|---|---|
| Downtime | Uptime |
| Repair rate | Failure rate |
| $M(t) = 1 - \exp[-\mu \cdot t]$ | $F(t) = 1 - \exp[-\lambda \cdot t]$ |
| Repair rate $\mu = \dfrac{1}{MTTR}$ | Failure rate $\lambda = \dfrac{1}{MTTF}$ |

## Table 7.4

Reliability and maintainability data for process equipment (Adapted from Koolen 2001).

| Equipment | MTTF (h) | MTTR (h) |
|---|---|---|
| Drum/vessel | 920 000 | 60 |
| Distillation/absorption column | 162 000 | 120 |
| Compressor | 13 000 | 32 |
| Pump (liquid seal)[a] | 9900 | 12 |
| Pump (gas seal)[a] | 19 800 | 12 |
| Pump (seal-less)[a] | 12 600 | 12 |
| Vacuum pump | 13 500 | 12 |
| Heat exchanger | 108 500 | 20 |
| Condenser | 107 700 | 20 |
| Reboiler | 150 000 | 20 |
| Air cooler | 29 300 | 12 |

[a]A mechanical seal is required where a rotating shaft passes through a stationary housing. Mechanical seals consist of two seal rings mated together to create a seal at their interface. One seal ring rotates with the shaft and the other is stationary. The mating faces of these seal rings require lubrication to separate the faces. This is achieved using the process liquid, a barrier liquid, or nitrogen in the case of gas seals. Seal-less pumps offer complete containment of process liquid, utilizing bearings lubricated by the product and a magnetic coupling to transmit torque across a containment shell.

## 7.6  Availability

Given the capability to repair components that will restore the system to an operating state after maintenance, another measure of system performance is required. This will depend on both reliability and maintainability.

*Availability* is the probability that a component or system is performing its required function over a stated period of time when operated and maintained in a prescribed manner (Beasley 1991; Billinton and Allan 1983; Davidson 1994; Ebeling 1997). Figure 7.20 illustrates the operating profile of a process, showing periods when it is operating and when it is down for repair. The availability can be defined as:

$$A = \frac{\text{uptime}}{\text{uptime} + \text{downtime}} \qquad (7.46)$$

where $A$ = availability (–)

Different definitions of uptime and downtime are possible.

1) *Inherent availability*. Inherent availability includes the downtime for corrective maintenance (CM), but excludes preventive maintenance (PM), and all other operational delays.

$$A_{IN} = \frac{MTTF}{MTTF + MTTR} \qquad (7.47)$$

**7**

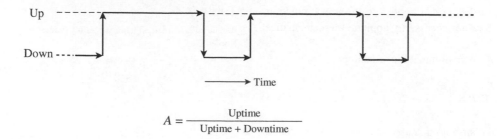

$$A = \frac{\text{Uptime}}{\text{Uptime} + \text{Downtime}}$$

**Figure 7.20**

To define the availability requires the mean uptime and downtime to be known.

where

$$A_{IN} = \text{inherent availability } (-)$$

$$MTTF = \frac{\text{uptime}}{\text{number of system failures}}$$

$$MTTR = \frac{\text{CM downtime}}{\text{number of system failures}}$$

Inherent availability for exponential reliability and maintainability distributions is given by:

$$A_{IN} = \frac{MTTF}{MTTF + MTTR}$$

$$= \frac{1/\lambda}{1/\lambda + 1/\mu} \qquad (7.48)$$

$$A_{IN} = \frac{\mu}{\lambda + \mu}$$

where

$\lambda$ = failure rate (h$^{-1}$)

$\mu$ = repair rate (h$^{-1}$)

**2)** *Operational availability*. Operational availability includes the time required for corrective maintenance (CM) procedures, preventive maintenance (PM) procedures, delays of maintenance procedures (e.g. unavailability of a maintenance crew), delays from unavailability of spare parts, isolation and decontamination of equipment for maintenance, time required for safety procedures associated with maintenance (e.g. issue of safety permits), process shut-down and process start-up after maintenance if the process requires shut-down for the maintenance actions.

$$A_{OP} = \frac{MTBM}{MTBM + MDT} \qquad (7.49)$$

where

$$A_{OP} = \text{operational availability } (-)$$

$MTBM$ = mean uptime between maintenance actions
(CM and PM)

$$= \frac{\text{uptime}}{\text{number of system failures} + \text{number of PM actions}}$$

$MDT$ = mean downtime for all CM and PM procedures and all delays associated with CM and PM actions

$$= \frac{\text{CM downtime} + \text{PM downtime} + \text{all operational delays}}{\text{number of system failures} + \text{number of PM actions}}$$

In the early stages of process design there is likely to be a shortage of information on PM and operational delays associated with CM and PM. However, the operational delays associated with maintenance are likely to be significant, and in many cases dominant.

# 7.7 Process Shut-down for Maintenance

If maintenance is required on an item of equipment, it needs to be isolated, decontaminated, and made safe for the maintenance procedures. To isolate the equipment might require part of the process, or the whole process, to be shut down. Processes cannot be shut down and started up instantly. The period during which the process is shut down and started up after maintenance represents an additional loss of production to the time taken for the maintenance itself. Also, even after the plant has been shut down, the equipment requiring maintenance might require safe isolation from the rest of the process, draining or venting of process fluids and decontamination from process fluids before the maintenance can be initiated. That means the $MTTR$ used to assess availability must include an allowance for delays in production caused by shut-down, isolation, and decontamination for maintenance, issue of safety permits to allow maintenance to proceed, reversing the maintenance isolation after maintenance and plant start-up, as well as the maintenance procedure itself.

If an item of equipment is allowed to fail requiring corrective maintenance, then the process will need to be shut down immediately, if shut-down is necessary to repair the equipment. If the maintenance is anticipated through preventive (planned) maintenance or predictive maintenance (condition monitoring), then the shut-down can be planned and will be less disruptive. There are two general classes of shut-down:

1) *Planned.* A *planned* or *normal* shut-down follows a planned sequence of events. The operating limits of the equipment must not be exceeded, and dangerous mixtures must not be allowed to form as a result of abnormal conditions in the transient process. Transient conditions might have the requirement for additional utilities and produce additional effluent. Emergency relief systems must also be able to cope with transient conditions. The start-up and shut-down might be flexible enough to be conducted in a number of different ways. However, a system of interlocks might be incorporated in the control system in order to prevent a hazardous sequence of operations being initiated.

2) *Emergency.* Emergency shut-down is a forced shut-down necessitated by the detection of unsafe conditions within the process. Such unsafe conditions might be caused by failure of equipment (e.g. a compressor failure or a loss of containment caused by failure of a pipe or vessel). Emergency shut-down will normally be carried out automatically and can be initiated automatically or shut down can be by the operator through the control panel.

The extent to which a process needs to be shut down for maintenance cannot be generalized and needs to be taken on a case-by-case basis.

1) *Standby equipment.* If the item of equipment to be maintained has a spare, which is installed in parallel with the online equipment (redundancy, see Chapter 8), it might not be necessary to shut the plant down. Operation can be switched to the spare while the failed equipment is repaired (see Chapter 8).

2) *Partitioning of the process.* In a large complex process it is often possible for part of the process to be run independently for short periods. This is made possible by the use of intermediate storage (buffer storage, see Chapter 9). Thus, it might be possible to shut down only part of the process for a short period for maintenance. Intermediate storage can in some cases be used to maintain production, even in the event of partial shut-down of the process (see Chapter 9).

3) *Hot standby.* Complete shut-down of the process might not be necessary. In some cases, it might be possible to pause operation of the process and keep it on *hot standby* by maintaining process conditions, but without throughput (see Chapter 8). For example, in a separation section of the process, it might be possible to keep distillation in operation under total reflux. This can allow the part of the process requiring maintenance to be isolated and maintained, allowing a very rapid start-up after the maintenance is complete.

Any interruption in the production greater than the time required for maintenance must be allowed for in the assessment of the availability. The time that needs to be added to the downtime required for maintenance actions for shut-down and start-up and preparation for maintenance will depend very much on the individual case. Shut-down and start-up can vary between a few hours and a few days, depending on the size and complexity of the process and how much the disruption to the process operation can be minimized.

# 7.8  Reliability, Maintainability, and Availability Concepts – Summary

Reliability, maintainability, and availability have an important influence on:

- safety;
- environmental damage;
- cost of lost production;
- reliability of safety systems;
- need for standby equipment;
- appropriate use of parallel production trains;
- storage requirements;
- maintenance strategy.

Reliability measures the probability of a component or system being able to perform its required function. Failure rate measures the probability of failure per unit time, which can vary through time according to the bathtub curve.

For repairable systems, mean time to failure ($MTTF$) assumes repair to the original state. For a constant failure rate, mean time to failure $MTTF = 1/\lambda$.

Maintainability measures the probability that a repair can be achieved within a certain period. The repair rate measures the probability of repair being successfully completed per unit time. The mean time to repair for a constant repair rate is given by $MTTR = 1/\mu$.

Availability of process systems measures the fraction of time a component or system can perform its required function. Measures of availability include:

- inherent availability;
- operational availability.

# Exercises

1. Two competing replacement components have constant failure rates. Component $A$ is specified to have a mean time to failure $MTTF$ of 5000 hours. Component $B$ is specified to have a reliability of 0.95 at a mission time of 1000 hours. Which component should be chosen?

**2.** An agitator has a constant failure rate of 30 failures per $10^6$ hours. It has been in operation for 5000 hours. Calculate the probability of failure:

**a)** in the next 1000 hours;
**b)** in the next 5000 hours.

**3.** A compressor is experiencing a failure rate that can be characterized as linearly increasing in terms of time $t$. The failure rate can be expressed as $\lambda(t) = t \times 10^{-7}$ failures per hour. Calculate the reliability for a time horizon of 3000 hours.

**4.** A sensor has been characterized to have a Weibull failure distribution with $\beta = 2.0$ and $\theta = 8000$ h.

**a)** Calculate the reliability for a time horizon of 5000 hours.
**b)** Calculate the mean time to failure (*MTTF*).

**5.** A pump has a constant failure rate. The mean time to failure is 8000 hours and the mean time to repair is 8 hours. In addition to the time required for repair, failure of the pump necessitates shut-down of part of the process, isolation and decontamination of the pump and pipework, the issue of safety permits, and subsequent start-up. These additional delays require 16 hours. Calculate the pump:

**a)** Inherent availability.
**b)** Operational availability.

## References

Beasley, M. (1991). *Reliability for Engineers*. Macmillan.
Billinton, R. and Allan, R.N. (1983). *Reliability Evaluation of Engineering Systems*, 2e. Plenum Press.
CCPS (1989). *Guidelines for Process Equipment Reliability Data*. AIChE.
Davidson, J. (1994). *The Reliability of Mechanical Systems*. London: The Institution of Mechanical Engineers.
Duguid, I.M. (1998a). Analysis of Past Incidents in the Oil, Chemical and Petrochemical Industries, *Loss Prevention Bulletin*, No. 142, Institution of Chemical Engineers, Rugby, UK.
Duguid, I.M. (1998b). Analysis of Past Incidents in the Oil, Chemical and Petrochemical Industries, *Loss Prevention Bulletin*, No. 143, Institution of Chemical Engineers, Rugby, UK.
Duguid, I.M. (1998c). Analysis of Past Incidents in the Oil, Chemical and Petrochemical Industries, *Loss Prevention Bulletin*, No. 144, Institution of Chemical Engineers, Rugby, UK.
Ebeling, C.E. (1997). *An Introduction to Reliability and Maintainability Engineering*. McGraw-Hill.
Koolen, J.L.A. (2001). *Design of Simple and Robust Process Plants*. Wiley VCH.
Mannan, S. (2012). *Lees' Loss Prevention in the Process Industries*, 4e, vol. 3. Butterworth-Heinemann Appendix 14.
Moss, T.R. (2005). *The Reliability Data Handbook*. Professional Engineering Publishing Limited.
SINTEF and NTNU (2015). *Offshore & Onshore Reliability Data (OREDA) Handbook*, 6e, OREDA Participants.

# Chapter 8

# Reliability, Maintainability, and Availability of Systems

## 8.1 System Representation

Before applying the concepts developed in the previous chapter to systems, an approach to the representation of systems needs to be developed. There are two general ways to represent systems for the analysis of reliability, maintainability, and availability.

1) *Fault Tree*. A fault tree is a graphical top-down deductive analysis of events that can consider failures or other events acting alone or in combination. Pathways connect contributory events and conditions using standard logic symbols. Figure 8.1 shows four of the many logic symbols that can be used. Figure 8.1a shows an AND gate, for which an output event exists only when all input events exist. Figure 8.1b shows an OR gate, for which an output event exists if one or more input events exist. Figure 8.1c shows a RESULTANT EVENT, which gives an event resulting from the logical combination of other fault events. Finally, Figure 8.1d shows a BASIC FAULT EVENT, which is a fault event that ends the analysis, requiring no further development.

2) *Reliability Block Diagram*. A reliability block diagram is a graphical representation of a system to show the dependency of the system reliability to the reliability of the individual components. It is drawn as a series of blocks connected in parallel or series. Each block represents a component of the system. Parallel paths are redundant, which means that all parallel parts must fail for that parallel network to fail. Any failure along a series path causes the entire series path to fail. The diagram should have a single inlet node and a single outlet node.

Figure 8.2 shows some examples comparing a fault tree with a reliability block diagram. Figure 8.1a shows a *series structure* as both a fault tree and a reliability block diagram. In this example, if Component 1, 2, *or* 3 fails, the system fails. Figure 8.2b shows a *parallel structure* in both representations. This time, Components 1, 2, *and* 3 must all fail for the system to fail. The third example shown in Figure 8.2c is a series–parallel structure. In this case,

Component 1 *or* Component 2 *together with* Component 3 must fail for the system to fail. Both fault trees and reliability block diagrams are used in practice. However, the reliability block diagram (RBD) is simpler to construct and more easily interpreted for many process engineering applications and will be adopted for the rest of the text.

## 8.2 Reliability of Series Systems

Consider first the series system, as illustrated in Figure 8.3. From probability theory, the *product rule of independent events* states (Chatfield 1999):

$$\text{Probability (Event } A \text{ and Event } B)$$
$$= \text{Probability (Event } A) \times \text{Probability (Event } B) \qquad (8.1)$$

Thus, the probability of survival of a series system at time $t$ is given by (Beasley 1991; Billinton and Allan 1983; Davidson 1994; Ebeling 1997):

$$R_{SYS}(t) = R_1(t) \times R_2(t) \times \cdots \times R_n(t) \qquad (8.2)$$

where

$R_{SYS}(t)$ = system reliability at time $t$

$R_i(t)$ = reliability of Component $i$ at time $t$

In terms of probability, the product rule can be interpreted as:

$$\begin{bmatrix} \text{Probability} \\ \text{of System} \\ \text{Survival} \end{bmatrix} = \begin{bmatrix} \text{Probability} \\ \text{of Survival of} \\ \text{Component 1} \end{bmatrix} and$$

$$\begin{bmatrix} \text{Probability} \\ \text{of Survival of} \\ \text{Component 2} \end{bmatrix} and \ldots and \begin{bmatrix} \text{Probability} \\ \text{of Survival of} \\ \text{Component } n \end{bmatrix}$$

Equation (8.2) can be generalized to (Beasley 1991; Billinton and Allan 1983; Davidson 1994; Ebeling 1997):

$$R_{SYS}(t) = \prod_i^n R_i(t) \qquad (8.3)$$

*Process Plant Design*, First Edition. Robin Smith.
© 2024 John Wiley & Sons Ltd. Published 2024 by John Wiley & Sons Ltd.
Companion website: www.wiley.com/go/processplantdesign

(a) AND gate output event exists only when all input events exist.

(b) OR gate output event exists if one or more of the input events exist.

(c) RESULTANT EVENT gives an event resulting from the logical combination of other fault events.

(d) BASIC FAULT EVENT is a fault event that ends the analysis, requiring no further development.

## Figure 8.1

Four of the logic symbols used in fault tree analysis.

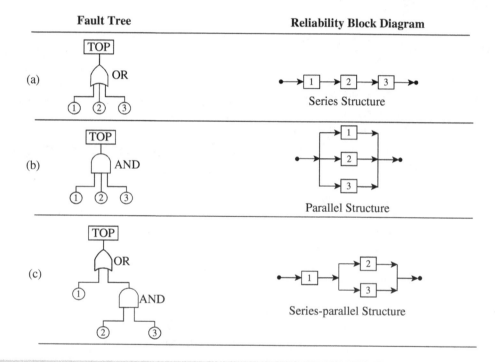

## Figure 8.2

Comparison between fault trees and reliability block diagrams.

## Figure 8.3

Series systems.

 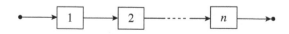

This is often referred to as the *product rule*. For a series system with constant failure rates for the components, following Eq. (7.10):

$$R_{SYS}(t) = R_1 \times R_2 \times R_3 \times \cdots \times R_n$$
$$= \exp[-\lambda_1 t] \times \exp[-\lambda_2 t] \times \exp[-\lambda_3 t] \times \cdots \times \exp[-\lambda_n t]$$
$$= \exp\left[\sum_{i=1}^{n} -\lambda_i t\right] \tag{8.4}$$

Thus, for a series system with constant failure rates for the components:

$$R_{SYS}(t) = \exp\left[\sum_{i=1}^{n} -\lambda_i t\right] \tag{8.5}$$

The mean time to failure (MTTF) for a series system with constant failure rates for the components is given by substituting Eq. (8.5) into Eq. (7.18):

$$MTTF = \int_0^\infty R(t)\, dt$$
$$= \int_0^\infty \exp\left[\sum_{i=1}^{n} -\lambda_i t\right] dt$$
$$= \left[\frac{1}{\sum_{i=1}^{n} -\lambda_i} \exp\left[\sum_{i=1}^{n} -\lambda_i t\right]\right]_0^\infty \tag{8.6}$$
$$= \frac{1}{\sum_{i=1}^{n} \lambda_i}$$

Thus, for a series system of components with constant failure rates:

$$MTTF = \frac{1}{\sum_{i=1}^{n} \lambda_i} \tag{8.7}$$

## 8.3 Reliability of Parallel Systems

Figure 8.4 shows a system involving two components in parallel. Success of 1 *or* 2 *or* both constitutes success of the system. The system fails when both 1 *and* 2 fail. If the probabilities of failure of the two components in Figure 8.4 are independent (Beasley 1991; Billinton and Allan 1983; Davidson 1994; Ebeling 1997):

$$R_{SYS}(t) = 1 - F_1(t) \times F_2(t)$$
$$= 1 - [1 - R_1(t)] \times [1 - R_2(t)] \tag{8.8}$$

where

$R_{SYS}(t)$ = system reliability at $t$ (−)
$F_i(t)$ = failure function for Component $i$ at time $t$ (−)
$R_i(t)$ = reliability of Component $i$ at time $t$ (−)

Generalizing Eq. (8.8) for $n$ components in parallel:

$$R_{SYS}(t) = 1 - \prod_{i}^{n} [1 - R_i(t)] \tag{8.9}$$

It must be emphasized that the parallel system has not failed as long as at least one of the components has not failed. For the two-component parallel system in Figure 8.4, the probability of both surviving is $R_1(t) \times R_2(t)$. For $n$ components in parallel the probability of all surviving is given by Eq. (8.3).

If the failure rate of the components is constant for the parallel system in Figure 8.4:

$$R_{SYS} = 1 - [1 - R_1(t)] \times [1 - R_2(t)]$$
$$= 1 - [1 - R_1(t) - R_2(t) + R_1(t)R_2(t)]$$
$$= R_1(t) + R_2(t) - R_1(t)R_2(t)$$
$$= \exp[-\lambda_1 t] + \exp[-\lambda_2 t] - \exp[-(\lambda_1 + \lambda_2)t] \tag{8.10}$$

Thus, for a parallel system of two components with constant failure rates:

$$R_{SYS} = \exp[-\lambda_1 t] + \exp[-\lambda_2 t] - \exp[-(\lambda_1 + \lambda_2)t] \tag{8.11}$$

The corresponding mean time to failure for two parallel components with constant failure rates is obtained by substituting Eq. (8.11) into Eq. (7.18):

$$MTTF = \int_0^\infty R(t)\, dt$$
$$= \int_0^\infty \exp[-\lambda_1 t] + \exp[-\lambda_2 t] - \exp[-(\lambda_1 + \lambda_2)t]\, dt$$
$$= \left[-\frac{1}{\lambda_1} \exp(-\lambda_1 t) - \frac{1}{\lambda_2} \exp(-\lambda_2 t) + \frac{1}{\lambda_1 + \lambda_2} \exp(-[\lambda_1 + \lambda_2]t)\right]_0^\infty$$
$$= \frac{1}{\lambda_1} + \frac{1}{\lambda_2} - \frac{1}{\lambda_1 + \lambda_2} \tag{8.12}$$

Thus, for a parallel system of two components with constant failure rates:

$$MTTF = \frac{1}{\lambda_1} + \frac{1}{\lambda_2} - \frac{1}{\lambda_1 + \lambda_2} \tag{8.13}$$

A system with three parallel components is illustrated in Figure 8.5. The reliability for this parallel system is given by:

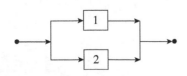

**Figure 8.4**
Parallel system of two components.

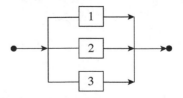

**Figure 8.5**
Parallel system of three components.

$$R_{SYS} = 1 - [1 - R_1(t)][1 - R_2(t)] \times [1 - R_3(t)]$$
$$= 1 - [1 - R_1(t) - R_2(t) + R_1(t)R_2(t)] \times [1 - R_3(t)]$$
$$= R_1(t) + R_2(t) + R_3(t) - R_1(t)R_2(t) - R_1(t)R_3(t) - R_2(t)R_3(t)$$
$$+ R_1(t)R_2(t)R_3(t)$$

Assuming constant failure rates gives:

$$R_{SYS} = \exp[-\lambda_1 t] + \exp[-\lambda_2 t] + \exp[-\lambda_3 t] - \exp[-(\lambda_1 + \lambda_2)t]$$
$$- \exp[-(\lambda_1 + \lambda_3)t] - \exp[-(\lambda_2 + \lambda_3)t] + \exp[-(\lambda_1 + \lambda_2 + \lambda_3)t]$$
$$\tag{8.14}$$

The corresponding mean time to failure for this parallel system with three components is given by substituting Eq. (8.14) into Eq. (7.18):

$$MTTF = \int_0^\infty R(t)\, dt$$
$$= \int_0^\infty \exp[-\lambda_1 t] + \exp[-\lambda_2 t] + \exp[-\lambda_3 t]$$
$$- \exp[-(\lambda_1 + \lambda_2)t]$$
$$- \exp[-(\lambda_1 + \lambda_3)t] - \exp[-(\lambda_2 + \lambda_3)t] \tag{8.15}$$
$$+ \exp[-(\lambda_1 + \lambda_2 + \lambda_3)t]\, dt$$
$$= \left[\frac{1}{\lambda_1} + \frac{1}{\lambda_2} + \frac{1}{\lambda_3}\right] - \left[\frac{1}{\lambda_1 + \lambda_2} + \frac{1}{\lambda_1 + \lambda_3} + \frac{1}{\lambda_2 + \lambda_3}\right]$$
$$+ \left[\frac{1}{\lambda_1 + \lambda_2 + \lambda_3}\right]$$

**Example 8.1**   A system consists of four components, each with a reliability of 0.95. Figure 8.6 shows three different configurations.

a) Figure 8.6a shows a series configuration.
b) Figure 8.6b shows two parallel trains, each with four components.
c) Figure 8.6c shows a parallel pair for each of the four components in a series configuration.

Calculate the reliability of each of the configurations.

**Solution**

a) For the series system in Figure 8.6a the system reliability is given by Eq. (8.3):

$$R_{SYS} = 0.95 \times 0.95 \times 0.95 \times 0.95 = 0.815$$

b) For the two parallel trains in Figure 8.6b, the reliability of each parallel path is that calculated in Part (a). Then for the two paths in parallel from Eq. (8.9):

$$R_{SYS} = 1 - [1 - 0.815] \times [1 - 0.815] = 0.966$$

c) For each parallel pair in Figure 8.6c, from Eq. (8.9):

$$R_{PARALLEL\ PAIR} = 1 - [1 - 0.95] \times [1 - 0.95] = 0.998$$

Then for the four parallel pairs in series from Eq. (8.3):

$$R_{SYS} = 0.998 \times 0.998 \times 0.998 \times 0.998 = 0.992$$

Note that the parallel arrangements have a very significant effect on the reliability of the system. A parallel arrangement for each component individually brings the biggest benefit.

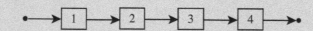

(a) A series system of 4 components each with a reliability of 0.95.

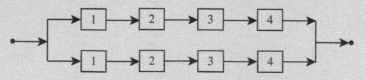

(b) A system of two parallel trains, each with 4 components having a reliability of 0.95.

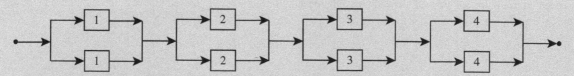

(c) A series system of 4 parallel pairs of 2 components, each having a reliability of 0.95.

**Figure 8.6**

Reliability of different series–parallel arrangements.

**Example 8.2**    Figure 8.7a shows the configuration of a pumping system. Two pumps, each in operation and with a reliability of 0.9, are arranged in parallel, followed by a filter with a reliability of 0.95. Calculate the system reliability.

**Solution**

Calculate the reliability of the two pumps in parallel, from Eq. (8.9):

$$R_{SYS} = 1 - (1 - R_A) \cdot (1 - R_B)$$
$$= 1 - (1 - 0.9) \cdot (1 - 0.9)$$
$$= 0.99$$

Now calculate the reliability of the series arrangement between the pumps and the filter from Eq. (8.3):

$$R_{SYS} = 0.99 \times 0.95 = 0.941$$

Thus, the system reliability is 0.941.

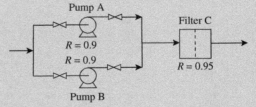

(a) Configuration of the pumping system.

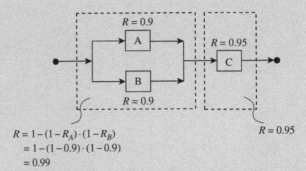

$$R = 1 - (1 - R_A) \cdot (1 - R_B)$$
$$= 1 - (1 - 0.9) \cdot (1 - 0.9)$$
$$= 0.99$$

(b) Reliability block diagram for the pumping system.

**Figure 8.7**

A simple example of a process flow diagram with two pumps and a filter.

**8**

For more complex reliability block diagrams, the diagram can in many cases be decomposed into simple series and parallel arrangements. The reliability of these simple arrangements can then be calculated and combined to give the equivalent single block. This process is repeated until a simple series or parallel arrangement is obtained that is equivalent to the original diagram, and the reliability of the initial complex diagram can be calculated. The procedure is illustrated in Example 8.3.

**Example 8.3**    Figure 8.8a shows a more complex reliability block diagram. The individual component reliabilities are also shown in Figure 8.8. Calculate the system reliability by decomposing the reliability block diagram into pairs of components either in series or parallel.

**Solution**

Figure 8.8b shows how the reliability block diagram can be decomposed into pairs of components in series and parallel. Block A comprises only Component 1:

$$R_A = 0.95$$

Block B is a parallel pair comprising Components 2 and 3 and from Eq. (8.9):

$$R_B = 1 - (1 - 0.95)(1 - 0.95) = 0.9975$$

Block C is a series pair of Block B and Component 4. From Eq. (8.3):

$$R_C = 0.9975 \times 0.95 = 0.9476$$

Block D is a series pair comprising Components 5 and 6. From Eq. (8.3):

$$R_D = 0.9 \times 0.9 = 0.81$$

Block E is a parallel pair comprising Blocks C and D. From Eq. (8.9):

$$R_E = 1 - (1 - 0.9476)(1 - 0.81) = 0.99$$

Block F is a parallel pair comprising Components 7 and 8. From Eq. (8.9):

$$R_F = 1 - (1 - 0.92)(1 - 0.92) = 0.9936$$

Finally, the system reliability comprises a series of Blocks A, E and F. From Eq. (8.3):

$$R_{SYS} = 0.95 \times 0.99 \times 0.9936 = 0.9345$$

Thus, the system reliability is 0.9345.

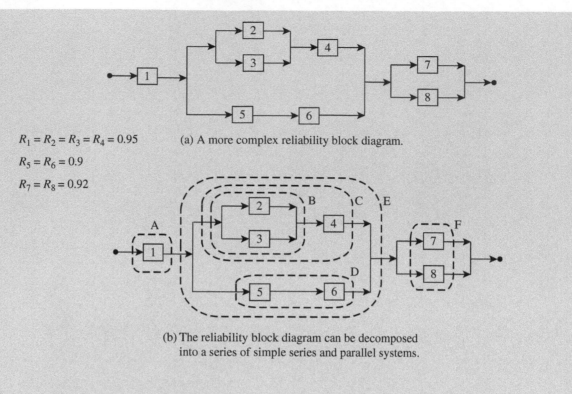

$R_1 = R_2 = R_3 = R_4 = 0.95$

$R_5 = R_6 = 0.9$

$R_7 = R_8 = 0.92$

(a) A more complex reliability block diagram.

(b) The reliability block diagram can be decomposed
into a series of simple series and parallel systems.

**Figure 8.8**

Reliability of a more complex system can be determined by decomposition.

Whilst the approach for the evaluation of complex networks illustrated in Example 8.3 can be applied to many reliability block diagrams, there are also cases when the reliability block diagram cannot be decomposed into simple series and parallel arrangements. If this is the case, different ways to decompose the diagram can sometimes be exploited (Ebeling 1997). Example 8.4 illustrates an alternative way to decompose such diagrams.

**Example 8.4** Figure 8.9a shows a reliability block diagram that cannot be decomposed into simple series and parallel arrangements. Decompose the diagram into two subnetworks corresponding with Component $D$ assumed to be functioning and another assuming Component $D$ to have failed and calculate the system reliability.

**Solution**

The original diagram in Figure 8.9a can be decomposed into two subnetworks as shown in Figure 8.9b and c. In Figure 8.9b, Component $D$ is assumed to be functioning, with a probability of Component $D$ functioning of $R_D$. In Figure 8.9c, Component $D$ is assumed to have failed, with a probability of Component $D$ having failed of $(1 - R_D)$. In Figure 8.9b, Components $B$ and $C$ are in parallel with a branch that does not fail and has a reliability of 1. This means that the reliability of the $BC$ parallel arrangement from Eq. (8.8) is given by $1 - (1 - R_B R_C)(1 - 1) = 1$. Thus, if Component $D$ is assumed not to fail, then the reliability of Subnetwork $F$ in Figure 8.9b is a series arrangement with the reliability given by Eq. (8.3)

$$R_F = R_A \times 1 \times R_E$$
$$= 0.95 \times 1 \times 0.93$$
$$= 0.8835$$

If Component $D$ is assumed to have failed, then the reliability of Subnetwork $G$ in Figure 8.9c is a series arrangement and the reliability is given by Eq. (8.3)

$$R_G = R_A R_B R_C R_E$$
$$= 0.95 \times 0.9 \times 0.9 \times 0.93$$
$$= 0.7156$$

The reliability of the system is obtained by accounting for the reliability of Component $D$:

$$R_{SYS} = R_D R_F + (1 - R_D) R_G$$
$$= 0.93 \times 0.8835 + (1 - 0.93)0.7156$$
$$= 0.8717$$

Thus, the system reliability is 0.8717.

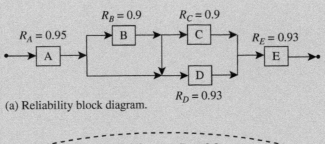

(a) Reliability block diagram.

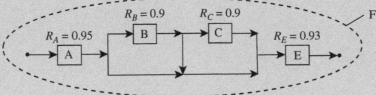

(b) Reliability block diagram when Component $D$ does not fail.

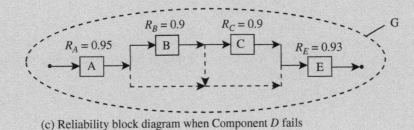

(c) Reliability block diagram when Component $D$ fails

**Figure 8.9**

Reliability block diagram for a complex parallel arrangement.

Other methods are available to calculate the reliability of reliability block diagrams (Henley and Kumamoto 1981; Ebeling 1997), but all have limitations. A widely used method to solve even the most complex reliability block diagrams is Monte Carlo Simulation, which will be discussed later.

## 8.4  Availability of Parallel Systems

Since availability is the probability that a component or system is up at any instant in time, then the same approach can be used for calculating the availability of systems as used for system reliability. The availability of parallel systems is then given by (Beasley 1991; Billinton and Allan 1983; Davidson 1994; Ebeling 1997):

$$A_{SYS} = 1 - \prod_{i}^{n} [1 - A_i] \qquad (8.16)$$

where

$A_{SYS}$ = system availability (–)

$A_i$ = availability of Component $i$ (–)

$n$ = number of components (–)

If availability is defined to be the inherent availability and both reliability and maintainability follow exponential functions:

$$A_{SYS} = 1 - \prod_{i}^{n} \left[ 1 - \frac{MTTF_i}{MTTF_i + MTTR_i} \right]$$

$$= 1 - \prod_{i}^{n} \left[ 1 - \frac{1/\lambda_i}{1/\lambda_i + 1/\mu_i} \right] \qquad (8.17)$$

Thus, from Eq. (8.17):

$$A_{SYS} = 1 - \prod_{i}^{n} \left[ \frac{\lambda_i}{\lambda_i + \mu_i} \right] \qquad (8.18)$$

## 8.5  Availability of Series Systems

For series systems the product rule for availability would imply for $n$ independent components (Beasley 1991; Billinton and Allan 1983; Davidson 1994; Ebeling 1997):

$$A_{SYS} = \prod_{i}^{n} A_i \qquad (8.19)$$

However, for the product rule, each component being up or down is assumed to be independent of the others. Strictly, the component availabilities are not independent of each other in this case.

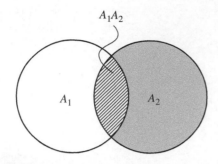

**Figure 8.10**

A Venn diagram representing the availability of two components in series.

If a component is in the failed state, the other components are not at risk of failure until the failed component has been repaired and the system is up again. Figure 8.10 shows a Venn diagram representing the availability of two components in series (Sherwin and Bossche 2012). The total area in the Venn diagram represents the probability that one *or* both of the units are operating. The intersection represents the probability that both $A_1$ *and* $A_2$ are operating:

$$A_1 \text{ and } A_2 = A_1 A_2$$

For $A_1$ *or* $A_2$ to be operating:

$$A_1 \text{ or } A_2 = A_1 + A_2 - A_1 A_2$$

The union reduces the probability space from 1 to $(A_1 + A_2 - A_1 A_2)$. The product probability needs to be conditioned by the reduced probability space when $A_1$ and $A_2$ are not at risk:

$$A_{SYS} = A_{12} = \frac{A_1 A_2}{A_1 + A_2 - A_1 A_2} \tag{8.20}$$

The product rule underpredicts the true availability.

For more than two components the formula can be applied recursively, as illustrated in Figure 8.11:

$$A_{SYS} = A_{123} = \frac{A_{12} A_3}{A_{12} + A_3 - A_{12} A_3} \tag{8.21}$$

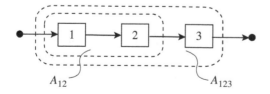

**Figure 8.11**

Availability for series systems can be determined by decomposing the system.

Combining Eqs. (8.20) and (8.21):

$$
\begin{aligned}
A_{SYS} = A_{123} &= \frac{A_{12} A_3}{A_{12} + A_3 - A_{12} A_3} \\
&= \frac{\left(\dfrac{A_1 A_2}{A_1 + A_2 - A_1 A_2}\right) A_3}{\left(\dfrac{A_1 A_2}{A_1 + A_2 - A_1 A_2}\right) + A_3 - \left(\dfrac{A_1 A_2}{A_1 + A_2 - A_1 A_2}\right) A_3} \\
&= \frac{1}{1 + \left(\dfrac{1 - A_1}{A_1}\right) + \left(\dfrac{1 - A_2}{A_2}\right) + \left(\dfrac{1 - A_3}{A_3}\right)}
\end{aligned}
\tag{8.22}
$$

In general, for $n$ units in series (Sherwin and Bossche 2012):

$$A_{SYS} = \frac{1}{\left[1 + \sum_i^n \dfrac{1 - A_i}{A_i}\right]} \tag{8.23}$$

If availability is the inherent availability and both reliability and maintainability follow exponential functions:

$$
A_{SYS} = \frac{1}{\left[1 + \sum_i^n \dfrac{1 - \dfrac{MTTF_i}{MTTF_i + MTTR_i}}{\dfrac{MTTF_i}{MTTF_i + MTTR_i}}\right]}
= \frac{1}{\left[1 + \sum_i^n \dfrac{1 - \left[\dfrac{1/\lambda_i}{1/\lambda_i + 1/\mu_i}\right]}{\left[\dfrac{1/\lambda_i}{1/\lambda_i + 1/\mu_i}\right]}\right]}
\tag{8.24}
$$

Thus, after simplifying:

$$A_{SYS} = \frac{1}{\left[1 + \sum_i^n \dfrac{\lambda_i}{\mu_i}\right]} \tag{8.25}$$

Comparing the product rule from Eq. (8.19) with the more accurate Eq. (8.23):

$$\prod_i^n A_i < \frac{1}{\left[1 + \sum_i^n \dfrac{1 - A_i}{A_i}\right]} \tag{8.26}$$

Thus, the product rule will give an underestimate of the actual availability. However, if $n$ is small (less than 10) and all $A_i$ are large (greater than 0.99), then the product rule gives errors significantly less than 1% (Sherwin and Bossche 2012). If even one of the components has $A_i$ less than 0.9, then the errors can be more serious (Sherwin and Bossche 2012).

**Example 8.5** Determine the error from using the product rule to predict the system availability for 10 components in series, each with an inherent availability of:

a) 0.99
b) 0.90

## Solution

a) The product rule is given by Eq. (8.19):

$$A_{SYS} = \prod_i^n A_i = 0.99^{10} = 0.904$$

The accurate figure is given by Eq. (8.23):

$$A_{SYS} = \cfrac{1}{\left[1 + \sum_i^n \frac{1-A_i}{A_i}\right]} = \cfrac{1}{\left[1 + 10 \times \left(\frac{1-0.99}{0.99}\right)\right]} = 0.908$$

b) Repeating the calculation for $A_i = 0.9$:

$$A_{SYS} = \prod_i^n A_i = 0.90^{10} = 0.349$$

$$A_{SYS} = \cfrac{1}{\left[1 + \sum_i^n \frac{1-A_i}{A_i}\right]}$$

$$= \cfrac{1}{\left[1 + 10 \times \left(\frac{1-0.90}{0.90}\right)\right]} = 0.474$$

Thus, the product rule predicts a lower availability compared with the accurate figure. The error increases as the availability decreases and depends on the number of components in series. The greater the number, the greater the error.

**Example 8.6** A pump and filter arrangement is shown in Figure 8.12. Failure rates can be assumed to be constant. Failure and repair data for the components are given in Table 8.1:

a) Represent the system as a reliability block diagram.
b) Calculate the reliability for the system over a one-year time horizon, assuming the process operates for 8400 h y$^{-1}$.
c) Calculate the inherent availability of the system.

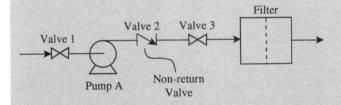

## Figure 8.12

Availability of a pumping system.

## Table 8.1

Failure and repair data for Example 8.6.

| Component | Type | Failure rate (failures per $10^6$ h) | MTTR (h) |
|---|---|---|---|
| Pump | Centrifugal | 125 | 12 |
| Filter | Liquid | 65 | 24 |
| Valve 1 | Plug | 3 | 4 |
| Valve 2 | Non-return | 4 | 4 |
| Valve 3 | Plug | 3 | 4 |

## Solution

a) Figure 8.13 shows the reliability block diagram. It is a series system. If any of the components fail, the system fails.
b) From Eq. (7.10):

$$R(t) = \exp[-\lambda t]$$

Calculated component reliabilities are given in Table 8.2. From Eq. (8.3):

$$R_{SYS} = 0.3499 \times 0.5793 \times 0.9751 \times 0.9670 \times 0.9751$$

$$= 0.1864$$

c) Component repair rates can be calculated from Equation 7.45 are given in Table 8.3. Then, from Eq. (8.25):

$$A_{SYS} = \cfrac{1}{\left[1 + \sum_i^n \frac{\lambda_i}{\mu_i}\right]}$$

$$= \cfrac{1}{\left[1 + \frac{125 \times 10^{-6}}{0.08333} + \frac{65 \times 10^{-6}}{0.04167} + \frac{3 \times 10^{-6}}{0.2500} + \frac{4 \times 10^{-6}}{0.2500} + \frac{3 \times 10^{-6}}{0.2500}\right]}$$

$$= 0.9969$$

Although the system has a high availability, it will suffer frequent breakdowns (mainly from pump failures). This has the potential to introduce significant safety problems, waste of process materials, and environmental damage. No allowance has been made for the delay caused by shut-down and start-up, which would result from a breakdown, along with isolation of equipment, decontamination for maintenance, and issue of safety permits for maintenance work, which will extend the downtime and decrease the system availability.

8

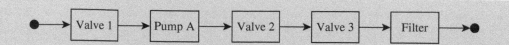

**Figure 8.13**

Availability of a pumping system.

**Table 8.2**

Component reliabilities for Example 8.6.

| Component | Failure rate (failures per $10^6$ h) | $R_i$ |
|---|---|---|
| Pump | 125 | 0.3499 |
| Filter | 65 | 0.5793 |
| Valve 1 | 3 | 0.9751 |
| Valve 2 | 4 | 0.9670 |
| Valve 3 | 3 | 0.9751 |

**Table 8.3**

*MTTR* and repair rates for Example 8.6.

| Component | Failure rate (failures per $10^6$ h) | *MTTR* (h) | $\mu$ ($h^{-1}$) |
|---|---|---|---|
| Pump | 125 | 12 | 0.08333 |
| Filter | 65 | 24 | 0.04167 |
| Valve 1 | 3 | 4 | 0.25000 |
| Valve 2 | 4 | 4 | 0.25000 |
| Valve 3 | 3 | 4 | 0.25000 |

# 8.6  Redundancy

Redundancy involves the use of additional components or units beyond the minimum number required for operation with the purpose of improving reliability or availability of the system. A series system does not have redundant components. Parallel, series–parallel and parallel–series have redundant components. Redundancy can be divided into three classes:

1) *Active Redundancy*: For active redundancy all components are operating simultaneously in parallel and sharing the load. For active redundancy there are two different redundancy levels, as illustrated in Figure 8.14. Low-level redundancy

pairs each of the components, as shown in Figure 8.14a. High-level redundancy involves the uses of two parallel trains, as shown in Figure 8.14b. As explored previously in Example 8.1, low-level redundancy will result in a greater system reliability:

$$R_{LOW} = \left[1 - (1 - R_A)^2\right] \times \left[1 - (1 - R_B)^2\right] \quad (8.27)$$

$$R_{HIGH} = 1 - (1 - R_A R_B)^2 \quad (8.28)$$

2) *Non-active Redundancy*: Non-active redundancy involves the use of components that do not share any of the load. Redundant components only take load when one or more

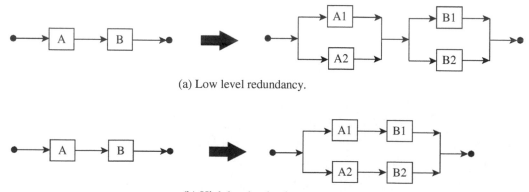

(a) Low level redundancy.

(b) High level redundancy.

**Figure 8.14**

Different redundancy levels.

operating components fail. It is usually applied when the start-up time for the component is unacceptably long. Non-active redundant components can be kept on *hot standby* or *warm standby*:

a) *Hot standby* components are kept in operational mode, but with no load. Upon failure of the primary system, the hot standby system immediately takes over, replacing the primary system. The standby component is assumed to have the same failure rate as when it is in operation. In Figure 8.15a, even though the hot standby component takes no load, it still has a system reliability given by:

$$R_{HOT\ STANDBY} = 1 - (1 - R_A)^2 \qquad (8.29)$$

b) *Warm standby* components are kept close to operational mode, but with no load. The standby component has a failure rate lower than that for hot standby, but is still

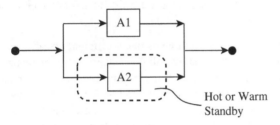

(a) Hot or warm standby redundancy.

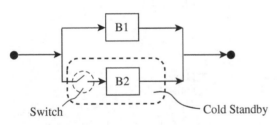

(b) Cold standby redundancy.

**Figure 8.15**

Hot and cold standby.

susceptible to failure. Thus the system reliability for the system in Figure 8.15a on warm standby will be higher than the hot standby case.

3) *Passive Redundancy*: For passive redundancy or *cold standby* the redundant components are switched off completely (Figure 8.15b). When the operating component fails, operation is switched to a standby component. Assume there are $k$ identical units in parallel, of which one is on-line. When the on-line unit fails the next unit is switched to be on-line, and so on. If the failure rate is constant and switching is assumed to be perfect (see Appendix C for the derivation):

$$R(t) = \exp(-\lambda t) \sum_{k=0}^{n-1} \frac{(\lambda t)^k}{k!} \qquad (8.30)$$

$$MTTF = \frac{n}{\lambda} \qquad (8.31)$$

where $n$ = number of units in parallel (1 on-line and $n - 1$ on standby). It is common to operate with two identical units, one on cold standby. For $n = 2$ (noting $0! = 1$):

$$R(t) = (1 + \lambda t) \exp(-\lambda t) \qquad (8.32)$$

$$MTTF = \frac{2}{\lambda} \qquad (8.33)$$

Cold standby will give the highest reliability, followed by warm standby, with hot standby having the lowest. However, hot or warm standby might be necessary if the start-up time for cold standby components is unacceptably long.

4) *Availability*: Redundancy not only increases the reliability of systems, but also increases the availability. If switching load to the standby component does not involve any delay, then there is no interruption in the availability on switching. However, as soon as the standby unit is switched to be the online unit, then it is subject to possible failure. For many components in the process industries, MTTR is much shorter than MTTF. This is illustrated in Figure 8.16. If the failed unit is repaired without delay, this means that the availability is *effectively* 100%. However, this is not strictly true given the random nature of failures. The online unit might in

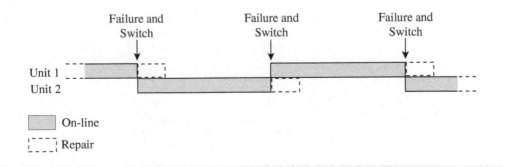

**Figure 8.16**

Availability of standby systems.

principle fail before the failed unit has been repaired, even though this might be unlikely. It is also assumed that the repair of the failed unit is started immediately on switch-over to the standby unit. In addition, the standby unit might fail to start (imperfect switching). However, as long as $MTTR \ll MTTF$, there is no delay in repairing the failed unit and there is perfect switching, it is then reasonable to assume the availability is 100% as a first approximation.

As an example of standby options consider steam generation in a complex steam system, of the type discussed in Chapter 4. Steam systems are subject to major changes in demand on a day-to-day and hour-to-hour basis. The steam system must satisfy the site requirements, even if the steam system equipment fails; otherwise, there will be major disruption in production and possible safety problems due to unstable operation of processes. As a consequence, steam systems always have significant redundancy. There will often be multiple high-pressure steam boilers on large steam systems. Boilers can fail for a wide variety of reasons. It is rarely a catastrophic failure, but is more often associated with failure of instrumentation or peripheral equipment, such as fans. Figure 8.17 shows a typical arrangement in which four boilers are on line, but operating at a reduced load. If one of the boilers fails for some reason, then the remaining boilers can quickly increase firing to make up the shortfall in steam generation. However, the disadvantage of operating with the boilers operating at low load for extended periods is that the efficiency of the boilers decreases with decreasing load, as shown in Figure 8.17 (Smith 2016). This creates inefficiency in the use of energy in the boilers. If not operated actively on partial load, boilers can be kept on standby under three different conditions:

1) *Cold Standby.* On cold standby the boiler is shut down and is not subject to failure. However, it might take three to six hours to bring it on-line. This is because boilers can only be heated up or cooled down at a rate of no greater than 50–100 °C per hour; otherwise, there will be mechanical damage from thermal stresses.

2) *Warm Standby.* In warm standby the boiler is typically maintained within 55 °C of the operating temperature by use of a steam coil in the lower drum (see Figure 4.15), with steam from another boiler. As a result, in warm standby it takes of the order of 30 minutes to bring it on-line

3) *Hot Standby.* In hot standby the boiler is kept at the operating temperature and pressure with no steam output. Conditions are maintained by a pilot burner or intermittent firing of the main burners. As a result, in hot standby it takes of the order of five minutes to bring it on-line.

Figure 8.18 shows three of many different possible operating modes. Figure 8.18a shows all four boilers operating on part load. Figure 8.18b shows three of the boilers operating on a higher load, with one of the boilers on cold standby. Figure 8.18c shows again three boilers operating, but now with one of the boilers on hot standby. The actual mode of operation adopted will require detailed examination, balancing energy efficiency against the consequences of a boiler failure.

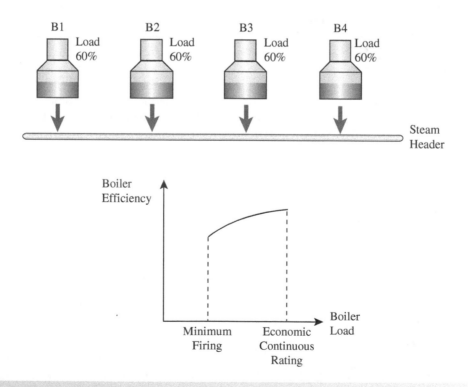

**Figure 8.17**

Redundancy for a steam boiler system with all boilers on part load.

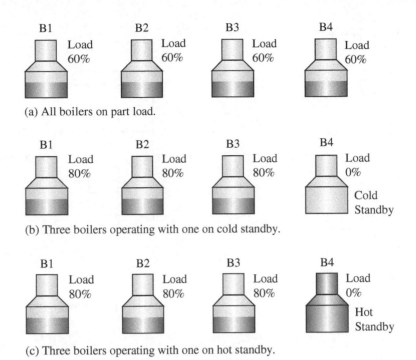

(a) All boilers on part load.

(b) Three boilers operating with one on cold standby.

(c) Three boilers operating with one on hot standby.

### Figure 8.18

Different operating strategies for boiler redundancy.

**8**

**Example 8.7**   A pump has a failure rate of 100 failures per $10^6$ hours. Assume the operation is for 8400 h $y^{-1}$:

a) Calculate the reliability for a single pump over the time horizon.
b) Calculate the reliability of the system over the time horizon if an identical pump is placed on hot standby.
c) Calculate the reliability of the system over the time horizon if an identical pump is placed on cold standby.

**Solution**

a) For a single pump, from Eq. (7.10):

$$R = \exp(-\lambda t)$$
$$= \exp\left(-\frac{100}{10^6} \times 8400\right)$$
$$= 0.432$$

b) For a pump with a hot standby, from Eq. (8.29):

$$R = 1 - (1-R)^2$$
$$= 1 - (1-0.432)^2$$
$$= 0.677$$

c) For a pump with a cold standby from Eq. (8.32):

$$R = (1 + \lambda t)\exp(-\lambda t)$$
$$= \left(1 + \frac{100}{10^6} \times 8400\right)\exp\left(-\frac{100}{10^6} \times 8400\right)$$
$$= 0.794$$

The reliability on cold standby is greater than hot standby, as the cold standby pump is not subject to failure whilst on standby, unlike the hot standby case. However, this assumes perfect switching for the cold standby case. Unlike equipment such as steam boilers, cold standby pumps can be brought into operation effectively immediately.

## 8.7   *k*-out-of-*n* Systems

When considering the reliability of parallel systems, it was assumed that the system functions if only one of the parallel components has not failed. However, more generally a parallel system might require more than one of the parallel components to function for the system to function. For example, the supply of cooling water from a cooling tower might use four pumps, all operating and in parallel. It might be that the system can still function if one of the pumps fails, but if more than one fails, then the pumping might be inadequate. This situation can be generalized as *n* components operating in parallel with *k* out of *n* components required to function for the system to function. Then if *k* = 1 and *n* > 1 this becomes a simple parallel system. If *k* = *n* it becomes a series system. If *k* = 3 and *n* = 4, it is a three-out-of-four system requiring three of the four components to function for the system to function. In this analysis it will be assumed that all of the components are identical and independent. Figure 8.19 shows the representation of *k*-out-of-*n* redundancy in a reliability block diagram. A node is used to signify the k-out-of-n redundancy.

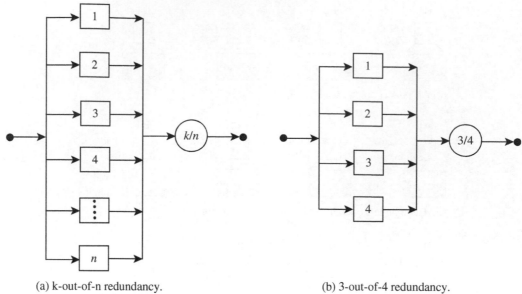

(a) k-out-of-n redundancy.                    (b) 3-out-of-4 redundancy.

## Figure 8.19

Reliability block diagram for *k*-out-of-*n* redundancy.

Figure 8.19a shows a general *k*-out-of-*n* redundancy and Figure 8.19b shows an example of three-out-of-four redundancy.

In general, if each component has the same probability of survival, then the probability of *x* components operating $Pr(x)$ is given by (Ebeling 1997):

$$Pr(x) = \frac{n!}{x!(n-x)!} R^x (1-R)^{n-x} \qquad (8.34)$$

where

$x$ = number of components operating (–)

$Pr(x)$ = probability that *x* components are operating (–)

$n$ = number of components (–)

$R$ = component reliability (–)

$R^x$ = probability of *x* components surviving (–)

$\dfrac{n!}{x!(n-x)!}$ = number of combinations in which *x* components survive from *n* components (–) (Chatfield 1999)

$(1-R)^{n-x}$ = probability of $(n-x)$ failures (–)

The first term on the right-hand side of Eq. (8.34) determines the number of ways in which *x* functional units can be obtained from *n* components. The second term on the right-hand side is the probability of *x* elements functioning successfully. The third term on the right-hand side is the probability of failure for the remaining $(n-x)$ components. The probability of *k* or more successes from *n* components is given by (Ebeling 1997):

$$\begin{aligned} R_{SYS} &= \sum_{x=k}^{n} Pr(x) \\ &= \sum_{x=k}^{n} \frac{n!}{x!(n-x)!} R^x (1-R)^{n-x} \end{aligned} \qquad (8.35)$$

where

$R_{SYS}$ = reliability of the system (–)

$k$ = minimum number of components required for successful operation (–)

If the failure rate is constant, giving the reliability an exponential distribution, Eq. (8.35) becomes:

$$R_{SYS}(t) = \sum_{x=k}^{n} \frac{n!}{x!(n-x)!} \exp(-x\lambda t)[1 - \exp(-x\lambda t)]^{n-x} \qquad (8.36)$$

where

$\lambda$ = failure rate (h$^{-1}$)

The mean time to failure (MTTF) for a *k*-out-of-*n* system for a constant failure rate is given by (Jumonville and Lesso 1969):

$$MTTF = \int_{0}^{\infty} R_{SYS}(t)\, dt = \frac{1}{\lambda} \sum_{x=k}^{n} \frac{1}{x} \qquad (8.37)$$

An important application for *k*-out-of-*n* systems is in process safety systems, where *voting* is applied to determine a safety action. The voting arrangement employs the use of redundant equipment for the purpose of being able to tolerate failure of a component and still have a safety system perform its action. In the process industries there are several common voting arrangements. The most common voting arrangements are:

- one-out-of-one (1oo1);
- one-out-of-two (1oo2);
- two-out-of-two (2oo2);
- two-out-of-three (2oo3).

The first number in the arrangement is the number of devices that must vote to cause a safety action to occur. The second number is the total number of devices. For example, a safety system might have three sensors, and for a two-out-of-three voting system, a signal from any two of the sensors might action a shut-down.

For a two-out-of-three system with the devices having the same reliability $R$:

$$
\begin{aligned}
R_{SYS} &= \sum_{x=k}^{n} \frac{n!}{x!(n-x)!} R^x (1-R)^{n-x} \\
&= \frac{3!}{2!(3-2)!} R^2 (1-R)^{3-2} + \frac{3!}{3!(3-3)!} R^3 (1-R)^{3-3} \\
&= \frac{6}{2(1)} R^2 (1-R)^1 + \frac{6}{6(1)} R^3 (1-R)^0 \\
&= 3R^2 (1-R) + R^3 \\
&= 3R^2 - 2R^3
\end{aligned}
\tag{8.38}
$$

Since availability is the probability that a component or system is up at any instant in time, then Eq. (8.35) can also be applied to the calculation of system availability, providing the availability is the same for the system components. This will be applied later when considering the capacity of systems.

---

**Example 8.8**   Cooling water is pumped from a cooling tower around a processing site using four identical pumps operating continuously in parallel. Successful operation requires at least three of the four pumps to operate. The pumps have a constant failure rate of 50 per $10^6$ hours. Calculate the reliability of at least three pumps operating after 8400 hours.

**Solution**

From Eq. (8.36):

$$
\begin{aligned}
R_{SYS}(t) &= \sum_{x=k}^{n} \frac{n!}{x!(n-x)!} \exp(-x\lambda t)[1 - \exp(-x\lambda t)]^{n-x} \\
&= \frac{4!}{3!(4-3)!} \exp\left(-3 \times \frac{50}{10^6} \times 8400\right) \left[1 - \exp\left(-\frac{50}{10^6} \times 8400\right)\right]^{4-3} \\
&+ \frac{4!}{4!(4-4)!} \exp\left(-4 \times \frac{50}{10^6} \times 8400\right) \left[1 - \exp\left(-\frac{50}{10^6} \times 8400\right)\right]^{4-4} \\
&= 4 \times 0.28365 \times 0.34295 + 1 \times 0.18637 \times 1 \\
&= 0.575
\end{aligned}
$$

---

# 8.8  Common Mode Failure

*Common mode failures* (CMF) cause simultaneous failure of two or more components or systems from a common cause. The results of a common mode failure might be the complete loss of a protective function at a critical time. Common mode failure mechanisms may be the result of failure of:

- a common utility (e.g. power);
- common specification errors or omissions;
- common component design flaws;
- common component manufacturing problems;
- common installation faults;
- repeated maintenance errors;
- quality control failures, etc.

An example of a reliability block diagram in which a common mode failure causes total system failure is shown in Figure 8.20. Two examples of common mode failure are:

1) *Fukushima Daiichi Nuclear Disaster*. In March 2011 an earthquake caused shutdown of the six boiling water nuclear reactors at the Fukushima Daiichi Nuclear Power Plant in Japan. After shutdown a nuclear reactor continues to generate heat at a rate of around 6% of the heat from operational generation. This results from the heat generated by the decay of unstable isotopes and continues for several days, requiring cooling for several days even after shutdown. Power is needed to maintain the flow of cooling water. To protect against mains electricity failure there were two backup systems in the Fukushima Daiichi Plant (Figure 8.21a). However, all three systems were vulnerable to common mode failure from flooding caused by a large tsunami (Figure 8.21b). The tsunami that followed the earthquake caused the loss of the mains electricity and the two backup systems. The resulting loss of coolant caused the meltdown of three of the nuclear reactors.

The failure of the backup systems was not inevitable. The backup diesel generators and the batteries could have been tsunami-proofed by moving to higher ground and encasing in high concrete walls. This would have removed the common mode failure (Figure 8.21c).

**8**

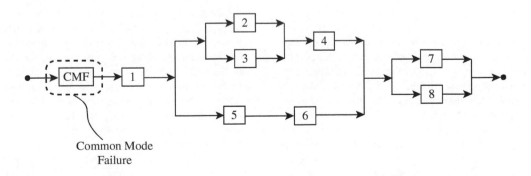

## Figure 8.20
Example of a common mode failure that causes the system to fail.

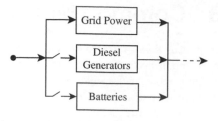

(a) Power supply system from grid power with two standby options.

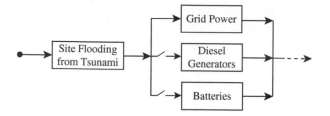

(b) A common mode failure for the power supply system.

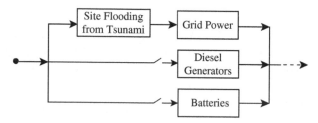

(c) Overcoming the common mode failure for the power supply system.

## Figure 8.21

Fukushima Daiichi Nuclear power plant electricity supply.

2) *Three Mile Island Nuclear Disaster.* In March 1979 one of the pressurized water reactors (PWRs) at the Three Mile Island Nuclear Power Plant in Pennsylvania, USA was shut down automatically as a result of an improper maintenance procedure and faulty valves. Heat continued to be generated after the shut-down, caused by the decay of unstable isotopes,

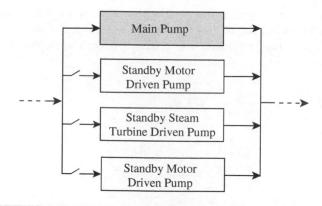

## Figure 8.22

Three Mile Island power plant standby pumping system.

requiring continued cooling for several days. Heat was removed in a closed primary pressurized water loop that was used to generate steam in a steam generator. The main boiler feedwater pump had three backup pumps (Figure 8.22). After the system was shut down the control system automatically started the standby pumps for boiler feedwater to the steam generator to maintain cooling. However, no boiler feedwater entered the steam generator to remove the residual heat from the closed reactor cooling circuit, the pressure rose in the primary reactor cooling circuit, and its relief valve opened. Unfortunately, the relief valve failed to close after the pressure was released. There was a large loss of the reactor coolant and the reactor core melted down.

Why did the standby pumps fail to supply boiler feedwater when started? Prior to the incident the standby pumps had been isolated for routine maintenance and testing. Following the testing, the isolation valves had not been reopened by the maintenance crew. This is a common mode failure caused by the same mistake being made by the maintenance crew.

**Example 8.9**   Figure 8.23 shows an environmental protection system for a process to safely vent a tail gas (Henley and Kumamoto 1981). The temperature of the tail gas is first decreased by quenching with water. The cooled tail gas is then passed to a pre-scrubber that removes solid particles using a spray of feedwater and a spray of recirculated water. A purge is taken from the recirculated water to remove solid particles separated from the tail gas. A mesh pad at the top of the pre-scrubber prevents entrainment of droplets and particles from the pre-scrubber. The tail gas is then passed to a scrubber where the gaseous contaminants are removed by absorption in a combination of feedwater and recirculated water. The recirculated water is cooled using cooling water. A purge is taken from the recirculated water to remove the gases separated from the tail gas. The tail gas from the absorber is vented to the atmosphere. The fan and pumps shown in Figure 8.23 are driven by electric motors from a mains power supply.

Failure rates and repair times for the equipment in the environmental protection system are given in Table 8.4. If equipment failure causes the system to fail, then the process must be shut down temporarily. The shut-down of the process, isolation and decontamination of the failed equipment for maintenance, issue of the required safety permit, and subsequent start-up after the repair of the equipment require a total of 24 hours in addition to the repair time for the equipment. It is anticipated that there will be no planned maintenance other than during an annual planned shut-down. The process is shut down every year for two weeks to carry out routine maintenance and safety checks. This creates a time horizon for reliability and availability of 8400 hours.

a) Draw a reliability block diagram to represent the system.
b) Calculate the reliability of the system for a time horizon of 8400 hours.

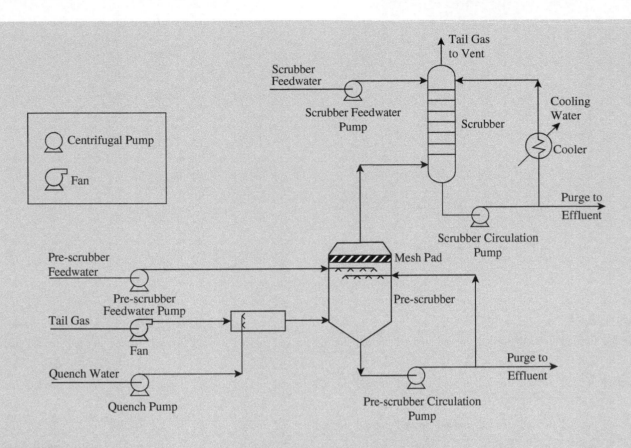

## Figure 8.23

Exhaust gas pre-scrubbing system. Source: Adapted from (Henley and Kumamoto 1981), *Reliability Engineering and Risk Assessment*, Prentice-Hall.

## Table 8.4

Failure rates and repair times for the equipment in the environmental protection system.

| Equipment | Failure rate (failures per $10^6$ h) | Repair time (h) |
|---|---|---|
| A) Fan | 25 | 24 |
| B) Quench pump | 50 | 12 |
| C) Pre-scrubber feedwater pump | 50 | 12 |
| D) Pre-scrubber circulation pump | 100 | 12 |
| E) Pre-scrubber | 20 | 20 |
| F) Scrubber feedwater pump | 50 | 12 |
| G) Scrubber circulation pump | 100 | 12 |
| H) Cooler | 30 | 36 |
| I) Scrubber | 20 | 120 |

c) Calculate the operational availability of the system components and the system operational availability assuming that the product rule is accurate enough for series systems.

d) Calculate the reliability and operational availability of the system if identical pumps are installed on cold standby for the two most vulnerable pumps. It can be assumed that the product rule is accurate enough for the availability of series systems.

e) Calculate the reliability at 8400 h and the operational availability of the system if identical pumps on cold standby are installed for all pumps. It can again be assumed that the product rule is accurate enough for the availability of series systems.

f) Calculate how many additional hours of operation are anticipated per year if cold standby pumps are installed for all of the pumps, instead of just the two most vulnerable ones.

g) Describe the advantages of having the additional standby pumps for all pumps and how the extra investment in standby pumps might be justified economically.

## Solution

a) Figure 8.24a shows the reliability block diagram for the process in Figure 8.23. The system is series in nature because failure of any of the components causes the system to fail.

b) The reliability for each Component $i$ is given by Eq. (7.10):

$$R_i(t) = \exp[-\lambda_i t] = \exp[-\lambda_i \times 8400]$$

The calculated reliabilities are given in Table 8.5.

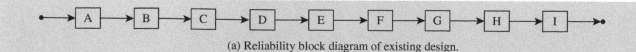

(a) Reliability block diagram of existing design.

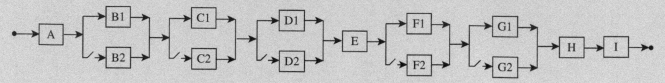

(b) Reliability block diagram with standby on the most vulnerable pumps.

(c) Reliability block diagram with standby on all pumps.

## Figure 8.24

Reliability block diagram of vent gas environmental system.

## Table 8.5

Failure rates and component reliabilities.

| Equipment | Failure rate (failures per $10^6$ h) | $R_i$ (–) |
|---|---|---|
| A) Fan | 25 | 0.81058 |
| B) Quench pump | 50 | 0.65705 |
| C) Pre-scrubber feedwater pump | 50 | 0.65705 |
| D) Pre-scrubber circulation pump | 100 | 0.43171 |
| E) Pre-scrubber | 20 | 0.84535 |
| F) Scrubber feedwater pump | 50 | 0.65705 |
| G) Scrubber circulation pump | 100 | 0.43171 |
| H) Cooler | 30 | 0.77724 |
| I) Scrubber | 20 | 0.84535 |

The system reliability is given by:

$$R_{SYS}(t) = \prod_i^n R_i(t)$$

$$= 0.81058 \times 0.65705 \times 0.65705 \times 0.43171 \times 0.84535$$

$$\times 0.65705 \times 0.43171 \times 0.77724 \times 0.84535$$

$$= 0.02380$$

**c)** The operational availability of the system components is given by Eq. (7.49):

$$A_{OPi} = \frac{MTBM_i}{MTBM_i + MDT_i}$$

where

$A_{OPi}$ = operational availability of Component $i$ (–)

$MTBM$ = mean uptime between maintenance actions (CM and PM)

$$= \frac{\text{Uptime}}{\text{Number of system failures} + \text{Number of PM actions}}$$

$MDT$ = mean downtime for all CM and PM procedures and all delays associated with CM and PM actions

$$= \frac{\text{CM downtime} + \text{PM downtime} + \text{All operational delays}}{\text{Number of system failures} + \text{Number of PM actions}}$$

For this problem, there is no planned maintenance and $MTBM = MTTF$. The mean time to failure of the system components is given by Eq. (7.20):

$$MTTF_i = \frac{1}{\lambda_i}$$

The mean downtime $MDT$ is the sum of the $MTTR$ and the 24-hour period for start-up and shut-down and other delays. The calculated operational availability of each Component $i$ is given in Table 8.6. Assuming the system availability can be calculated by the product rule in this case, the system availability is given by Eq. (8.19):

$$A_{OP} = \prod_i^n A_{OPi}$$

$$= 0.99880 \times 0.99820 \times 0.99820 \times 0.99641 \times 0.99912$$

$$\times 0.99820 \times 0.99641 \times 0.99820 \times 0.99713$$

$$= 0.98085$$

**d)** The two most vulnerable pumps are the Pre-scrubber and Scrubber Circulation Pumps, having the highest failure rates. The reliability block diagram with these standby units is shown in Figure 8.24b. After installation of a cold standby pump, the reliability is given by Eq. (8.32):

## Table 8.6

Component availabilities.

| Equipment | Failure rate (failures per $10^6$ h) | MTTF (MTBM) (h) | MTTR (h) | MDT (h) | $A_{OPi}$ |
|---|---|---|---|---|---|
| A) Fan | 25 | 40 000 | 24 | 48 | 0.99880 |
| B) Quench pump | 50 | 20 000 | 12 | 36 | 0.99820 |
| C) Pre-scrubber feedwater pump | 50 | 20 000 | 12 | 36 | 0.99820 |
| D) Pre-scrubber circulation pump | 100 | 10 000 | 12 | 36 | 0.99641 |
| E) Pre-scrubber | 20 | 50 000 | 20 | 44 | 0.99912 |
| F) Scrubber feedwater pump | 50 | 20 000 | 12 | 36 | 0.99820 |
| G) Scrubber circulation pump | 100 | 10 000 | 12 | 36 | 0.99641 |
| H) Cooler | 30 | 33 333 | 36 | 60 | 0.99820 |
| I) Scrubber | 20 | 50 000 | 120 | 144 | 0.99713 |

$$R_i(t) = (1 + \lambda_i t)\exp(-\lambda_i t)$$
$$= \left(1 + \frac{100}{10^6} \times 8400\right)\exp\left(-\frac{100}{10^6} \times 8400\right)$$
$$= 0.79435$$

The system reliability is given by Eq. (8.3):

$$R_{SYS}(t) = \prod_i^n R_i(t)$$
$$= 0.81058 \times 0.65705 \times 0.65705 \times 0.79435 \times 0.84535$$
$$\times 0.65705 \times 0.79435 \times 0.77724 \times 0.84535$$
$$= 0.08058$$

The pumps have a repair time much shorter than the MTTF. Assume that the availability of the pumps with standby is 1.0. Assuming the system availability can be calculated by the product rule in this case, the system availability is given by Eq. (8.19):

$$A_{OP} = \prod_i^n A_{OPi}$$
$$= 0.99880 \times 0.99820 \times 0.99820 \times 1.0 \times 0.99912$$
$$\times 0.99820 \times 1.0 \times 0.99820 \times 0.99713$$
$$= 0.98792$$

e) The reliability block diagram with standby units on all pumps is shown in Figure 8.24c. Reliability for pumps with standby is given by Eq. (8.32):

$$R_i(t) = (1 + \lambda_i t)\exp(-\lambda_i t)$$
$$= (1 + \lambda_i \times 8400)\exp(-\lambda_i \times 8400)$$

Component reliabilities with cold standby for all pumps are given in Table 8.7. The system reliability is given by Eq. (8.3):

## Table 8.7

Component reliabilities with cold standby on all pumps.

| Equipment | Failure rate (failures per $10^6$ h) | $R_i$ (–) |
|---|---|---|
| A) Fan | 25 | 0.81058 |
| B) Quench pump | 50 | 0.93301 |
| C) Pre-scrubber feedwater pump | 50 | 0.93301 |
| D) Pre-scrubber circulation pump | 100 | 0.79435 |
| E) Pre-scrubber | 20 | 0.84535 |
| F) Scrubber feedwater pump | 50 | 0.93301 |
| G) Scrubber circulation pump | 100 | 0.79435 |
| H) Cooler | 30 | 0.77724 |
| I) Scrubber | 20 | 0.84535 |

$$R_{SYS}(t) = \prod_i^n R_i(t)$$
$$= 0.81058 \times 0.93301 \times 0.93301 \times 0.79435 \times 0.84535$$
$$\times 0.93301 \times 0.79435 \times 0.77724 \times 0.84535$$
$$= 0.23073$$

**8**

Assuming the system availability can be calculated by the product rule in this case, the system availability is given by Eq. (8.19):

$$A_{OP} = \prod_{i}^{n} A_{OPi}$$

$$= 0.99880 \times 1.0 \times 1.0 \times 1.0 \times 0.99912 \times 1.0$$
$$\times 1.0 \times 0.99820 \times 0.99713$$
$$= 0.99327$$

**f)** Additional hours of operation per year if cold standby pumps are installed for all of the pumps, instead of the two most vulnerable ones

$$= (0.99327 - 0.98792) \times 8400$$
$$= 45 \text{ h}$$

**g)** The prediction for the additional availability from installing a standby for all pumps, rather than the most vulnerable, brings a benefit of additional production of just under two days. The

economic justification for the additional standby pumps would be based on the value of the lost production. Additional capital cost would be required in order to take advantage of additional production. However, this does not take account of increasing the reliability of the system. The additional standby pumps would make the system significantly more reliable, which would reduce the disruption, environmental damage, and potential safety problems created by additional shutdowns.

This example results in a process with a high availability, but low reliability. This means that although it might achieve the required target for production, it might also be susceptible to frequent breakdowns. As pointed out previously, breakdowns cause disruption and can cause safety and environmental problems. The number of breakdowns could be reduced beyond that possible through the introduction of standby pumps by choosing more reliable equipment in the design, the application of preventive maintenance in the maintenance management, or the introduction of predictive maintenance through condition monitoring of equipment (e.g. vibration monitoring of machines).

## 8.9 Capacity

Whilst availability measures the fraction of time the process is operating, this does not necessarily directly reflect the production capacity. Consider the process in Figure 8.25a, which features a

single production train with a design capacity of 100 and an availability of 0.95. This results in an operating capacity of $100 \times 0.95 = 95$. By contrast, Figure 8.25b shows the same process constructed in two parallel trains, each with a design capacity of 50 and each with an availability of 0.95. The availability of the system of $1 - (1 - A)^2 = 0.9975$. However, this does not relate to a

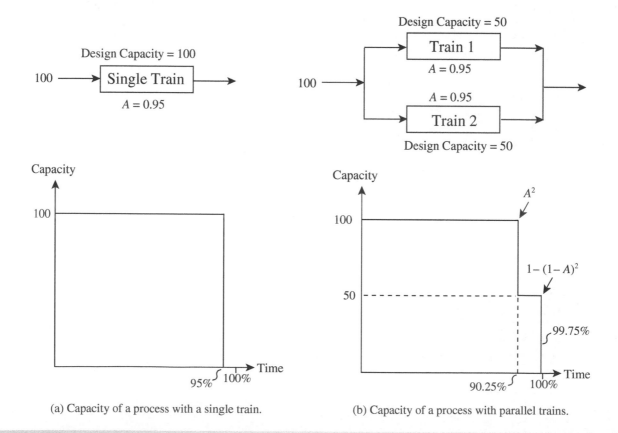

(a) Capacity of a process with a single train.

(b) Capacity of a process with parallel trains.

**Figure 8.25**

An example comparison between a single production train and two parallel trains.

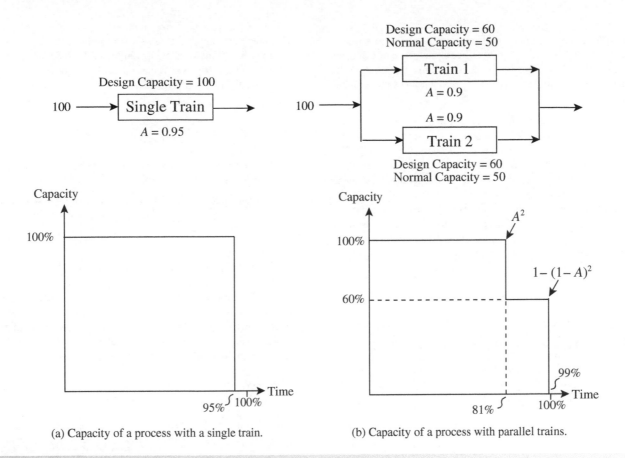

(a) Capacity of a process with a single train.

(b) Capacity of a process with parallel trains.

## Figure 8.26

Comparison between a single production train and two parallel trains with different availabilities and spare design capacity.

capacity of 99.75. Whilst the availability of the system is $1-(1-A)^2$, the availability for both trains operating is $A^2$. The capacity with both trains operating is $100 \times A^2 = 90.25$. The capacity of the parallel system is given by:

Overall capacity $= 100 \times 0.9025 + 50(0.9975 - 0.9025) = 95$

Thus, despite the two designs in Figure 8.25 having different configurations and different availabilities, they both have the same capacity. This is because both designs have the same overall installed design capacity and both designs feature the same availability for all component parts. The same result would be obtained for other designs that also featured the same overall installed design capacity and the same availability for all component parts.

Figure 8.26 shows a slightly different case, but still comparing a single train with two parallel trains. The single train in Figure 8.26a still has an availability of 0.95, but the availability of the parallel trains in Figure 8.26b is now lower at 0.9. It would normally be expected that a lower availability would be associated with a lower capital cost. The parallel trains have design capacities of 60, which would operate at 50 under normal operation. When one of the trains is down, the other train can be operated at the maximum capacity of one of the trains (60). The parallel trains now have a system availability of $1-(1-0.9)^2 = 0.99$. The

### Table 8.8

Alternative availability and capacity for single and parallel trains.

|  | Single train | Parallel trains |
| --- | --- | --- |
| Overall availability | 0.95 | 0.99 |
| Overall capacity | 95 | 91.8 |

availability for both trains operating is $0.9^2 = 0.81$. The overall capacity is now given by:

Overall capacity $= 100 \times 0.81 + 60(0.99 - 0.81) = 91.8$

A comparison of the system performance is now given in Table 8.8.

Thus, in this example, with decreased availability for the individual parallel trains, the increase in system availability brings no benefit in overall capacity for the parallel trains. The system will produce a product for an increased time, but at a lower overall capacity.

Thus, availability on its own is not meaningful in many cases. Both capacity and availability need to be considered together.

8

**Example 8.10**  A liquefied natural gas process has a design capacity of 120 000 t y$^{-1}$. Three design options are to be evaluated involving a single and parallel production trains. In each case, the availability of the single train or each parallel train can be assumed to be 0.95.

a)  For a single production train with a design capacity of 120 000 t y$^{-1}$, calculate the actual production capacity.

b)  For two parallel trains, each with a design capacity of 70 000 t y$^{-1}$, Calculate the actual production capacity assuming that if one of the parallel trains fails, the other operates to its maximum design capacity.

c)  For three parallel trains, each with the design capacity of 45 000 t y$^{-1}$, calculate the actual production capacity assuming that if one or two of the parallel trains fails, each train remaining in operation will operate to its maximum design capacity.

## Solution

a)  For the single train in Figure 8.27a, the capacity is given by:

$$\text{Overall capacity} = 0.95 \times 120\,000 = 114\,000\,\text{t y}^{-1}$$

b)  Two trains in parallel are shown in Figure 8.27b. The availability for both trains operating is $A^2 = 0.9025$ and availability of the system is $1-(1-A)^2 = 0.9975$. The capacity is given by:

$$\text{Overall capacity} = 0.9025 \times 120\,000 + (0.9975 - 0.9025)$$
$$\times 70\,000 = 114\,950\,\text{t y}^{-1}$$

c)  Three trains in parallel are shown in Figure 8.27c. The availability for all three trains operating is $A^3 = 0.8574$. The availability of exactly two trains operating is given by the 2oo3 formula given in Eq. (8.38):

$$A_{SYS} = 3A^2 - 2A^3$$
$$= 3 \times 0.95^2 - 2 \times 0.95^3$$
$$= 0.9928$$

The availability of the complete system is $1-(1-A)^3 = 0.9999$. The capacity is given by:

$$\text{Overall capacity} = 0.8574 \times 120\,000 + (0.9928 - 0.8574)$$
$$\times 90\,000 + (0.9999 - 0.9928)$$
$$\times 45\,000 = 115\,394\,\text{t y}^{-1}$$

The three cases are summarized in Table 8.9. It can be seen that in this case increasing the overall availability by introducing parallel trains only has a small effect on the overall capacity. Parallel trains are more often imposed on the design of large-scale processes when the size of the major items of equipment are increased to the maximum size available. Thereafter, any increase in capacity can only be achieved by introducing parallel trains.

## Table 8.9

Availability and capacity for a single train and two and three and parallel trains.

|  | Single train | Two parallel trains | Three parallel trains |
|---|---|---|---|
| Overall availability | 0.95 | 0.9975 | 0.9999 |
| Overall capacity (t y$^{-1}$) | 114 000 | 114 950 | 115 394 |

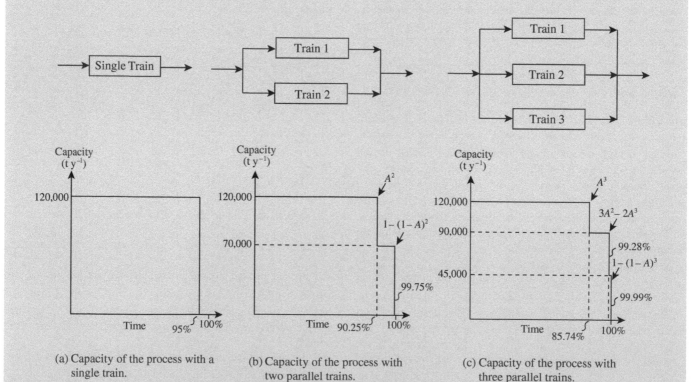

(a) Capacity of the process with a single train.

(b) Capacity of the process with two parallel trains.

(c) Capacity of the process with three parallel trains.

## Figure 8.27

Comparison between a single production train, two trains in parallel, and three trains in parallel.

# 8.10  Reliability, Availability, and Capacity

When assessing a process for its reliability, maintainability, and availability, it is important not to judge the quality of the design by a single criterion. It has been demonstrated in the previous section that availability and capacity need to be considered together. The ultimate aim of the process design is to achieve production targets, safely and with the minimum impact on the environment.

Consider the two alternative designs represented by the operating profiles in Figure 8.28. Design A and Design B in Figure 8.28 both have the same availability. Design A has a lower reliability but a shorter *MTTR* than Design B. If both designs have a single train with the same design capacity, then both will meet the required production capacity. However, Design A can only achieve this by frequent shut-downs and start-ups due to low reliability. As pointed out in Chapter 7, shut-downs and start-ups introduce significant hazards. Most process safety incidents occur during start-ups, shut-downs, and other events that infrequently occur. This is because the start-ups and shut-downs involve many non-routine procedures, and these periods can result in unexpected and unusual situations. Disrupted conditions within the process during shut-down and start-up also result in process waste. The waste created by the need for shut-down for maintenance can be environmentally damaging through waste of materials. For these reasons Design B in Figure 8.28 would normally be preferred to Design A.

Comparison between designs is not always as clear-cut as the options in Figure 8.28, but reliability, maintainability, availability, and capacity must all be weighed together when assessing a design.

# 8.11  Monte Carlo Simulation

The methods presented so far are adequate for relatively simple problems. However, networks might involve a large number of units and significant complexity. The reliability and maintainability functions might be different for different units, and more complex than a simple exponential function. Another factor that has not been considered so far is that when maintaining repairable systems, the system might or might not be restored to the original state. Figure 8.29a shows a repair of a unit to the original state, or as good as new (AGAN). On the other hand, in Figure 8.29b the repair of the unit might be to the same state as when the failure occurred, or as good as old (AGAO). Alternatively, a restoration factor can be applied to bring the system between AGAN and AGAO. When a failure occurs to a repairable system, the remaining components have a current age and the next failure depends on the current age.

The problem might include constraints associated with maintenance, e.g. the maintenance crew might need to repair multiple units, or there may be delays created by the availability of spare parts, and so on. The influence of preventive maintenance and condition monitoring might need to be included. Cost trade-offs might need to be optimized. For example, the amount of redundancy included in the design should be based on cost trade-offs, rather than judgment applied to the predicted reliability and availability of the system. To include all of these features a more automated approach is needed to determine the reliability and availability of complex networks.

To introduce the concept on which the automated approaches are based, consider the failure function for the single unit shown in Figure 8.30. An exponential function is shown for the sake of

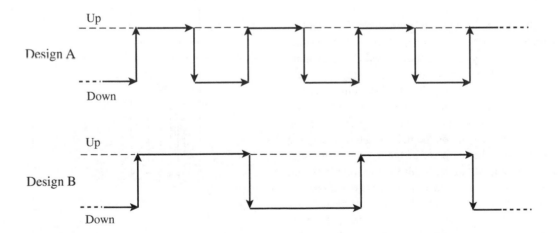

**Figure 8.28**

Two alternative designs with the same availability but different reliability.

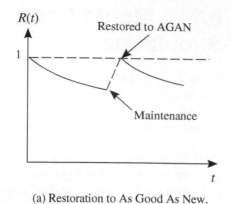

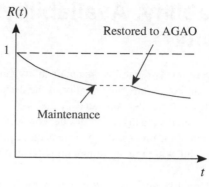

(a) Restoration to As Good As New.    (b) Restoration to As Good As Old.

**Figure 8.29**

Reliability after maintenance.

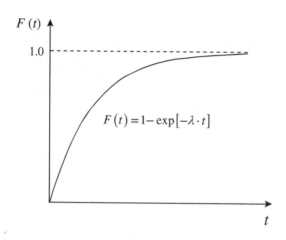

**Figure 8.30**

A single unit characterized by an exponential reliability distribution.

$$R(t) = \frac{\text{Number of simulations in the up mode at mission time}}{\text{Total number of simulations}}$$

(8.39)

Thus, for the results in Figure 8.32, at 5000 hours:

$$R(5000) = \frac{7}{10} = 0.7$$

(8.40)

At 10 000 hours:

$$R(10\,000) = \frac{3}{10} = 0.3$$

(8.41)

The availability can be determined from the mean of the uptime and downtime:

$$A(t) = \frac{\text{Mean uptime}}{\text{Mean uptime} + \text{Mean downtime}}$$

(8.42)

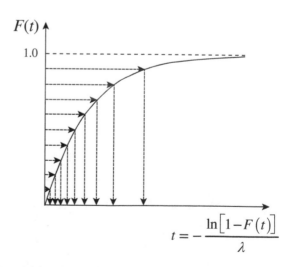

**Figure 8.31**

Random points generated for $F(t)$ over the interval (0, 1) with an equal probability of occurring anywhere in the interval (0, 1) will mostly end up close to zero.

illustration, but the concept is perfectly general. Suppose random points are generated for $F(t)$ over the interval (0, 1) with an equal probability of occurring anywhere in the interval (0, 1), as illustrated in Figure 8.31. Of the random points generated, most will be close to zero, with only a few points having high values. In other words, it is less likely for a failure to occur in a shorter period than a longer period. This is what would be expected from an exponential distribution. If a random point is picked on the failure function and assumed to be the point of failure, this will correspond with a certain time of failure. If this time is compared with the *mission time* (e.g. one year), then, if the time is greater than the mission time, the unit can be considered to have survived. However, if the time is less than the mission time then the unit can be considered to have failed. Repeating the assessment by picking many random points will eventually follow the failure function. Carrying out a number of iterations could lead to the result illustrated in Figure 8.32. Failures are assumed to occur at random points along the failure function in Figure 8.31. Over a number of iterations in Figure 8.32 the reliability can be determined from:

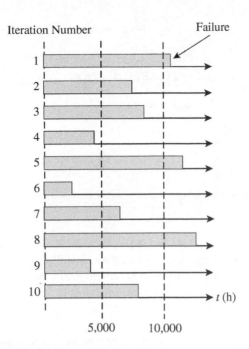

**Figure 8.32**

Monte Carlo simulation for a single unit.

The greater the number of iterations, the closer the agreement with the actual $R(t)$ and $A(t)$. In practice, 1000–5000 iterations might be used, depending on the size and complexity of the problem. This approach is known as a Monte Carlo Simulation.

For a total system, the Monte Carlo Simulation is based on the reliability block diagram for the network. The failure function (reliability function) must be defined for each block. Then, for a single iteration, random points are chosen along the failure function for each block independently, as illustrated in Figure 8.33. The time of failure for each block can be compared with the mission time (e.g. one year). Each block will then have survived if the time associated with the random point is greater than the mission time. However, if the time associated with the random point is less than the mission time, then the block will have failed. An analysis of the paths between the source and sink needs to determine if there is a feasible path from source to sink along which none of the units have failed in the mission time, as illustrated in Figure 8.34. If there is a feasible (non-failed) path along which none of the units have failed, then the network has not failed. If there is no feasible path, the network has failed. However, this is only a single iteration in the calculation. This is then repeated for typically greater than 1000 iterations and the reliability and availability determined from Eqs. (8.39) and (8.42). The more iterations performed, the greater will be the agreement between the distribution of the generated failure times the failure function for each block. Thus, the Monte Carlo Simulation can be formulated in a number of steps:

*Step 1:* Set up the system reliability block diagram.

*Step 2:* For each block define the failure (reliability) and maintainability functions.

*Step 3:* For each block convert a random number from a random number generator into $F$ and $t$ and check whether the blocks have survived to the mission time or failed.

**8**

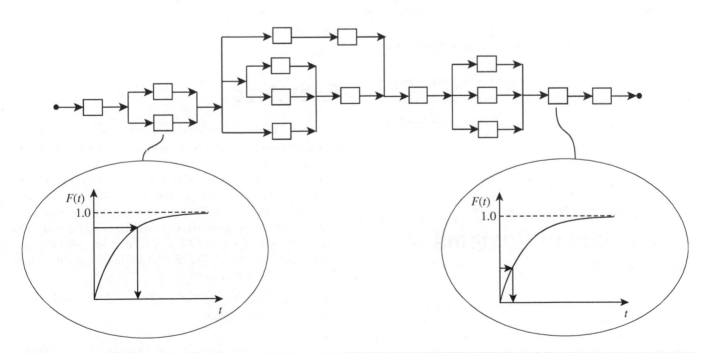

**Figure 8.33**

Monte Carlo simulation for a network.

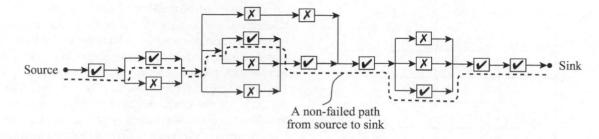

A non-failed path
from source to sink

### Figure 8.34

For each individual simulation the network will not have failed if there is a feasible (non-failed) path from source to sink at the mission time.

*Step 4:* Check whether the network is up or down from the connectivity of the blocks in the network (blocks that are down are effectively removed) for the mission time.

*Step 5:* Repeat for a large number of iterations (typically greater than 1000).

*Step 6:* Calculate the reliability and availability of the network using Eqs. (8.39) and (8.42).

Monte Carlo Simulation is capable of calculating the reliability and availability for the most complex networks and can account for:

- complex reliability functions;
- corrective and preventive maintenance;
- restoration of the repaired equipment to different states (e.g. AGAN);
- operational availability as well as inherent availability;
- constraints on the availability of maintenance crews and spare parts;
- on-line condition monitoring (e.g. measuring vibration on rotating shafts);
- the influence of inspection frequency for off-line condition monitoring;
- sensitivity analysis for standby and maintenance options;
- optimization (e.g. frequency of preventive maintenance).

Given the increased level of detail, adding costs also allows optimization to be carried out.

## 8.12   Reliability, Maintainability, and Availability of Systems – Summary

Reliability of systems can be represented as either fault trees or reliability block diagrams. Reliability block diagrams can feature series and parallel paths. Any failure along a series path causes the entire series path to fail. Parallel paths are redundant and all parallel paths must fail for that parallel network to fail. The reliability of series and parallel systems for independent components can be determined from the probability rule of independent events.

The rules for reliability also give the availability of systems of independent components, but for the availability of series systems the components are not strictly independent. An alternative formula can be used for series systems, but differs little from the product rule for small numbers of components with high availability.

Redundancy uses additional components beyond the minimum number required in order to improve reliability or availability of the system:

- active redundancy shares the load;
- non-active redundancy does not share the load until failure of a component;
- passive redundancy has components that are switched off completely until required.

Common mode failures cause simultaneous failure of two or more components or systems from a common cause. Capacity and availability must be considered together for process systems with parallel trains. Monte Carlo Simulation can be used for the analysis of complex problems.

## Exercises

1. A choice needs to be made between two options for a process pumping step. Option *A* uses a pair of identical pumps in parallel redundancy, both active and each with a failure rate of 200 per $10^6$ h. The probability of failure of each of the parallel units can be assumed to be independent, and if one of the units fails, the other can take up the whole load without the failure rate increasing. As an alternative, Option *B* uses a single pump with a failure rate of 100 per $10^6$ h. Calculate the reliability and mean time to failure (*MTTF*) for the two options and determine which would be preferable.

2. For the reliability block diagram in Figure 8.35, determine the system reliability, given:

$$R_1 = R_2 = R_3 = R_4 = 0.95$$
$$R_5 = R_6 = 0.9$$
$$R_7 = R_8 = 0.92$$

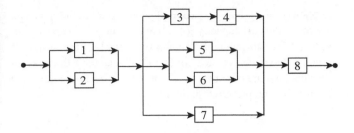

**Figure 8.35**

Reliability block diagram for Exercise 2.

**3.** Figure 8.36 shows a pumping arrangement for a filter. Both pumps are active. Failure and repair data for the components are given in Table 8.10.

a) Represent the system as a reliability block diagram.

b) Calculate the reliability for the system over a one year time horizon, assuming the process operates for $8500 \, \text{h} \, \text{y}^{-1}$.

c) Calculate the inherent availability of the system.

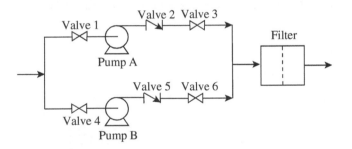

**Figure 8.36**

Pumping system for Exercise 3.

**Table 8.10**

Failure and repair data for Exercise 3.

| Component | Type | Failure rate (failures per $10^6$ h) | *MTTR* (h) |
|---|---|---|---|
| Pump | Centrifugal | 300 | 12 |
| Filter | Liquid | 80 | 24 |
| Valve 1 | Plug | 3 | 4 |
| Valve 2 | Non-return | 4 | 4 |
| Valve 3 | Plug | 3 | 4 |

**4.** A pump can be characterized as having four independent modes of failure, each of which would cause the pump to fail, and each with a constant failure rate. The mechanical seal has a failure rate of 50 failures per $10^6$ hours. The pump also has three bearings, each of which has a failure rate of 10 failures per $10^6$ hours. Calculate the reliability of the pump for a time horizon of 8400 hours.

**5.** In a shell-and-tube heat exchanger the fluid on the tube-side is split and flows in parallel through each tube. The most common cause of failure of heat exchangers is the tubes themselves. Tubes can block through fouling, causing a deterioration in the equipment performance. However, failure is mainly attributed to corrosion of the tubes and leakage from the tubes. Despite the parallel flows in the heat exchanger, as far as reliability is concerned the tubes are functionally in series. Explain why this is the case.

**6.** If the main mode of failure of heat exchangers is by corrosion of individual tubes, then, in principle, the reliability of the heat exchanger can be related to that of the individual tubes. If historical data indicates that in a shell-and-tube heat exchanger with 500 tubes, individual tubes have a mean time to failure (*MTTF*) of two years as a result of corrosion. Assuming a constant failure rate, calculate the reliability of the complete heat exchanger. Does the answer seem reasonable for a corrosion failure mechanism and why? Are any of the assumptions unreasonable?

**7.** Figure 8.37 shows the cooling system for the outlet gas from a chemical reactor. Because of fouling problems, a heat exchanger cannot be used. Instead, the gas from the reactor outlet is quenched by direct contact with water. Some of the water is recycled with heat being removed from the recycle in a cooling water cooler, and some water is purged. There is a continuous make-up of water to compensate for the purge. The pumps are driven by electric motors from a mains power supply. Failure data for the equipment in the process are given in Table 8.11.

a) Draw a reliability block diagram of the system marking the blocks *A* to *D*.

b) Calculate the reliability of the system over a one-year time horizon of 8760 hours.

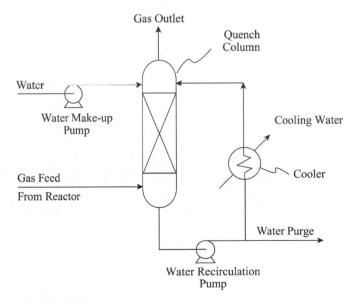

**Figure 8.37**

Direct contact heat transfer system for Exercise 7.

**8**

## Table 8.11

Failure rates for Exercise 7.

| Equipment | Failure rate (failures per $10^6$ h) |
|---|---|
| A) Water make-up pump | 50 |
| B) Water recirculation pump | 100 |
| C) Cooler | 30 |
| D) Absorber | 20 |

c) Calculate the reliability of the system if an identical standby pump is installed with the water recirculation pump on cold standby.

8. Figure 8.38 shows two design options for a pumping system. One option is a single pump with the design capacity of $50 \, m^3 \, h^{-1}$ and an inherent availability of 0.97. The second option is a system of two pumps in parallel, both active, with a design capacity of $30 \, m \, h^{-1}$, normal capacity of $25 \, m^3 \, h^{-1}$, and inherent availability of 0.93. If the system is required to have a capacity of $50 \, m^3 \, h^{-1}$, calculate the inherent availability and overall capacity of the two design options and determine the preferred option.

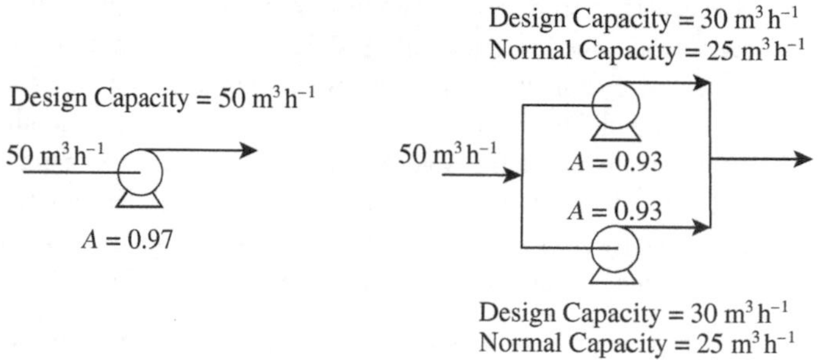

## Figure 8.38

Different pump designs for Exercise 8.

# References

Beasley, M. (1991). *Reliability for Engineers*. Macmillan.

Billinton, R. and Allan, R.N. (1983). *Reliability Evaluation of Engineering Systems*, 2e. Plenum Press.

Chatfield, C. (1999). *Statistics for Technology*, 3e. CRC Press.

Davidson, J. (1994). *The Reliability of Mechanical Systems*. London: The Institution of Mechanical Engineers.

Ebeling, C.E. (1997). *An Introduction to Reliability and Maintainability Engineering*. McGraw-Hill.

Henley, E.J. and Kumamoto, H. (1981). *Reliability Engineering and Risk Assessment*. Prentice-Hall.

Jumonville, P. and Lesso, W.G. (1969). Determining the mean time to failure for certain redundant systems. *IIE Transactions* 1: 81–82.

Sherwin, D.J. and Bossche, A. (2012). *The Reliability, Availability and Productiveness of Systems*. Springer.

Smith, R. (2016). *Chemical Process Design and Integration*, 2e. Wiley.

# Chapter 9

# Storage Tanks

## 9.1 Feed, Product, and Intermediate Storage

Most processes require storage for the feed and product. Storage of feed is required if the delivery of the feed is in batches (e.g. barge, rail car, road truck). Even if the feed is being delivered continuously via pipelines for gases and liquids, or conveyors in the case of solids, there will be no guarantee that feed will be free from interruptions in supply. For example, the upstream plant providing the continuous feed will need to be shut down for various reasons, and there might be unexpected failures of the delivery, for example, as a result of breakdowns.

Whilst solids and liquids are straightforward to store, gases are difficult. Relatively small quantities of gas can be stored in the gaseous state at ambient temperature in pressurized vessels. Larger quantities of gas storage require the gas to be liquefied. This can be achieved by decreasing the temperature using refrigeration, or increasing the pressure, or a combination of both. High-pressure storage has a high capital cost, as it requires thick-walled vessels. Low-temperature storage also has a high capital cost, as it requires capital investment in refrigeration equipment. Low-temperature storage also has a significant operating cost for power to run the refrigeration. The most appropriate method of storage for gases depends on a number of factors and involves safety, as well as capital and operating cost considerations.

If the feed is delivered in batches and used continuously, there is a fluctuating amount of storage, known as the *active stock* (see Figure 9.1). For example, suppose a plant operating with a liquid feed is at steady state. The liquid feed tank after a delivery might be perhaps 80% full. As the plant operates at steady state, the liquid level falls continuously to, say, 20%, when the next delivery arrives and the level returns to 80% as a result of the delivery. The amount of liquid between the 20% and 80% levels is the *active stock* and 20% is *inactive* (Figure 9.1). Storage tanks for liquids

should typically not be designed to operate with less than 10% inactive stack at the minimum, as this would create difficulties in operation. On the other hand, tanks for liquids should not be designed to operate more than typically 90% full at the maximum. An empty space above the liquid (known as *ullage*) is required when the tank is full to allow for safety and expansion. The capital cost of the storage includes the capital cost of storage tanks in the case of gases and liquids, and silos in the case of solids, capital cost of materials handling equipment (e.g. pumps, conveyors, etc.) associated with storage, and the capital cost of refrigeration equipment in the case of storage of liquefied gases. In addition to the capital cost of the equipment, there is also the working capital associated with the value of the material being stored. The greater the value of the material in storage the greater the disincentive to store large quantities of material.

The feed storage:

- provides a supply of feed to the plant between feed deliveries;
- compensates for interruptions in feed delivery due to unforeseen circumstances (e.g. breakdown in the plant manufacturing the feed material);
- allows short-term increases in production if the market for the product is favorable;
- compensates for interruptions in feed delivery due to holiday periods;
- compensates for seasonal variations in feed supply;
- allows feed to be bought under favorable market conditions when it is cheaper and stored for later use;
- dampens out variations in the feed properties.

The amount of feed storage will depend on:

- the frequency of deliveries;
- the size of deliveries;
- the reliability of deliveries;
- the capacity of the plant;
- the phase of the feed (gas, liquid, or solid);
- the hazardous nature of the feed material (the inventory of hazardous feed should be kept to a minimum);

*Process Plant Design*, First Edition. Robin Smith.
© 2024 John Wiley & Sons Ltd. Published 2024 by John Wiley & Sons Ltd.
Companion website: www.wiley.com/go/processplantdesign

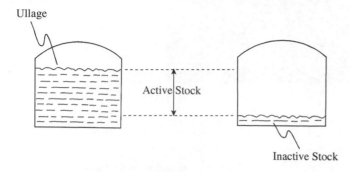

**Figure 9.1**

Capacity of liquid storage tanks.

- the capital and operating costs (e.g. refrigeration system for the storage of liquefied gas) associated with the feed storage equipment;
- the working capital locked up in the stored feed;
- the economic benefit to be gained from being able to take advantage of market fluctuations in the purchase cost of the raw materials.

The product must also be stored, for similar reasons to those for feed storage. The product delivery will often not be continuous. Also, the product will often be delivered to different customers. If the product is being delivered via pipeline in the case of a liquid or gas, or continuous conveyer in the case of a solid, then product storage can be minimized.

The product storage:

- balances the difference between the rates of production and dispatch;
- maintains product delivery during plant shut-down for maintenance;
- maintains product delivery during unforeseen plant shut-down;
- compensates for peak and seasonal demands;
- holds materials when holidays prevent dispatch;
- allows material to be held up for later sale if short-term market conditions are unfavorable, leading to a short-term decrease in the sales price;
- allows variations in product quality to be dampened out.

The amount of product storage will depend on:

- the frequency of product dispatches;
- the size of dispatches;
- the reliability of the dispatches;
- the capacity of the plant;
- the phase of the product (gas, liquid, or solid);
- the hazardous nature of the product (the inventory of a hazardous product should be kept to a minimum);
- the capital and operating costs (e.g. a refrigeration system for the storage of liquefied gas) associated with product storage equipment;

- the working capital locked up in the stored product;
- the economic benefit to be gained from being able to take advantage of market fluctuations in the sales price of the product.

In addition to the storage of feed and product, chemical intermediates are also often stored within the process. *Intermediate storage* for chemical intermediates is required, particularly when the process requires a number of transformation steps between the feed and the product. It creates flexibility in the operation of the plant. For example, consider a process involving a complex reaction system, followed by a complex separation system. The start-up and control of the two sections can be simplified if they can be decoupled. This can be achieved by introducing intermediate storage between the reaction and separation sections (see Figure 9.2). For start-up, the reaction section can be started up independently of the separation section. When starting up, the reaction section produces an intermediate chemical that is accumulated in the intermediate storage. When the reaction section is producing material of a suitable quality and quantity to be fed to the separation section, then the separation section can be started up by feeding from intermediate storage. Off-specification material can be kept separate for reworking at a later time, or disposal. The intermediate storage allows the two sections to have some operational independence from each other. This is not only important for start-up and shut-down, but allows one of the sections to be operated even if the other breaks down for a short period. The intermediate storage also decouples the control of the two sections.

The intermediate storage between the reaction and separation systems of a process can also help dampen out variations in composition, temperature, and flowrate between the two sections (for gases and non-viscous liquids, but not solids). Variations in the outlet properties from the storage are reduced compared with variations in the inlet properties.

The greater the amount of intermediate storage, the greater the flexibility created in the operation of the process, and the simpler will be the control. However, like feed and product storage there are significant costs associated with intermediate storage, involving capital, working, and operating costs. In addition, intermediate storage will bring additional safety problems if the material being stored is of a hazardous nature.

In summary, the amount of feed, product, and intermediate storage will depend on the capital cost, working capital, and operating costs, together with operability, control, and safety considerations.

**Figure 9.2**

Intermediate (buffer) storage.

# 9.2 Intermediate (Buffer) Storage and Process Availability

Intermediate storage can in some situations increase availability (Henley and Hoshino 1977; Henley and Kumamoto 1981). Consider the process in Figure 9.3a. This shows a process partitioned into a reaction section and the separation section with intermediate storage between the two. Figure 9.3b shows that if the upstream unit needs repair, the downstream unit can continue operating by being fed from the intermediate storage. If the downstream unit needs repair, the upstream unit can continue operating to re-fill the tank, as shown in Figure 9.3c. Three possible operating policies can be identified.

1) *Operating Policy A.* Suppose the upstream and downstream units are both operating and the storage tank is full (Figure 9.3a). If the upstream unit needs repair, the downstream unit can continue operating. The downstream unit can be run until the upstream unit is repaired using the inventory in the tank (Figure 9.4a). Now the upstream unit can be started up, the downstream unit shut down, and the tank refilled (Figure 9.4b). Once the tank is refilled the downstream unit can be restarted until the next maintenance (Figure 9.4c). The upstream and downstream operating profile and level in the tank are shown in Figure 9.5. When the upstream unit fails, the tank gradually empties while the downstream unit maintains operation. Figure 9.5 then shows the shut-down of the downstream unit while the tank is refilled. Unfortunately, this operating policy brings no benefit to the overall availability since production is interrupted to refill the tank (Henley and Hoshino 1977).

2) *Operating Policy B.* Figure 9.6a shows the upstream unit failed with the downstream unit fed from the tank inventory until the upstream unit is repaired. Figure 9.6b then shows both

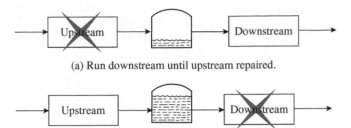

(a) Run downstream until upstream repaired.

(b) Now startup upstream, shutdown downstream and fill the tank.

(c) Now startup downstream and operate until the next maintenance.

## Figure 9.4

Operating Policy A.

upstream and downstream units operating, with no attempt to refill the tank. Only when the downstream unit is down for maintenance is the tank refilled by maintaining the operation of the upstream unit, as shown in Figure 9.6c. One possible scenario for refilling the tank after the scenario in Figure 9.6a has left the tank partially empty. Figure 9.7 shows a possible scenario for the upstream and downstream operating profile to refill the tank. The profiles show a starting point with the tank partially empty. The tank is refilled in steps only when the downstream unit is down for maintenance. This will lead to an increase in the availability as production is not stopped to refill the tank, as the downstream unit is run for a time on the tank inventory when the upstream unit is down. The tank is refilled only when the operating scenario allows it (Henley and Hoshino 1977).

3) *Operating Policy C.* Figure 9.8 shows a third operating policy in which this time the upstream and downstream units have the flexibility to operate at different capacities. In Figure 9.8a both the upstream and downstream units are initially at the same capacity $m$. In the event of the upstream unit being down for maintenance, the intermediate storage can be used to maintain production in the downstream unit using the tank inventory. In Figure 9.8b the upstream unit is now running, but now at a higher capacity than the downstream unit. In Figure 9.8c the intermediate storage has been filled and the upstream unit can be run at the same capacity as the downstream unit until further maintenance is required. Figure 9.9 shows a possible scenario for the upstream and downstream operating profile to refill the tank by running the upstream unit at a higher capacity until the tank is refilled (Henley and Hoshino 1977).

Consider the effect of intermediate storage on availability. If there is no intermediate storage, as shown in Figure 9.10a, with an upstream availability of $A_U$ and a downstream availability of $A_D$, then the availability of the system $A_{SYS}$ is given by:

$$A_{SYS} = A_U A_D \tag{9.1}$$

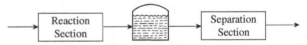

(a) Upstream and downsteam units are both operating with the level in intermediate storage being maintained.

(b) If upstream unit needs repair, downstream unit can continue operating.

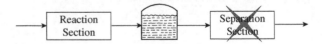

(c) If downstream unit needs repair, the upstream unit can continue operating to re-fill the tank.

## Figure 9.3

Intermediate storage can be used to maintain production in the event of failure of the upstream unit.

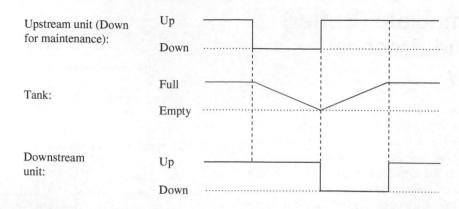

**Figure 9.5**

Operating Policy A tank level.

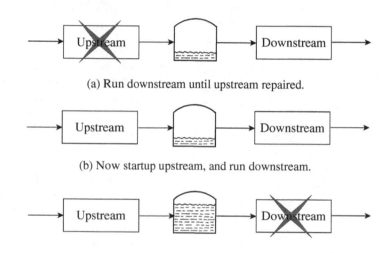

(a) Run downstream until upstream repaired.

(b) Now startup upstream, and run downstream.

(c) Fill the tank only when the downstream unit is down for maintenace.

**Figure 9.6**

Operating Policy B.

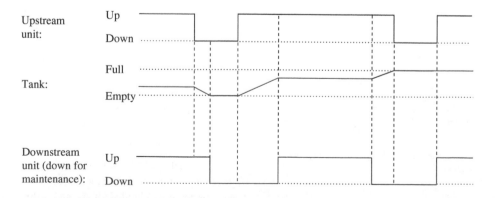

**Figure 9.7**

Operating Policy B tank level.

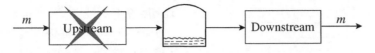

(a) Run downstream unit until upstream unit repaired.

(b) Now startup upstream unit at increased capacity, and run
downstream unit.

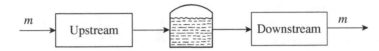

(c) Run upstream unit at increased capacity until tank is refilled.

**Figure 9.8**

Operating Policy C.

9

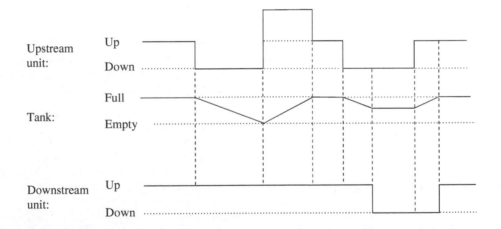

**Figure 9.9**

Operating Policy C tank level.

Having introduced an intermediate storage tank with volume $V$, as shown in Figure 9.10b, there are three cases that can be considered assuming that the capacity of the upstream and downstream units are equal (Henley and Hoshino 1977):

1) *Upstream availability is greater than the downstream availability*. Defining the mean time to repair of the upstream unit $MTTR_U$:

$$MTTR_U \ll \frac{V}{m/\rho} \qquad (9.2)$$

where

$MTTR_U$ = mean time to repair for the upstream unit (h)

$V$ = volume of the intermediate storage (m³)

$m$ = mass flowrate of the process stream (kg h⁻¹)

$\rho$ = density of the process stream (kg m⁻³)

$\dfrac{V}{m/\rho}$ = residence time in the intermediate storage (h)

Then, for this case:

$$A_{SYS} \approx A_D \qquad (9.3)$$

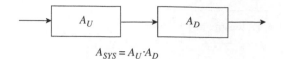

$$A_{SYS} = A_U \cdot A_D$$

(a) Availability without intermediate storage.

(b) Availability with intermediate storage.

(c) Availability with intermediate storage and differing upstream and down stream capacity.

## Figure 9.10

System availability.

Thus, the system availability lies in the range:

$$A_U A_D \leq A_{SYS} \leq A_D \qquad (9.4)$$

2) *Upstream availability is less than the downstream availability*. For:

$$MTTR_U \gg \frac{V}{m/\rho} \qquad (9.5)$$

Then, for this case:

$$A_{SYS} \approx A_U A_D \qquad (9.6)$$

Thus, if the upstream availability is less than the downstream availability then intermediate storage has little effect on availability.

3) *Upstream and downstream availability are equal*. In the unusual eventuality that the availability of the upstream and downstream are equal, then the availability is given by Eq. (9.4).

So far it has been assumed that the upstream and downstream capacities are equal. However, as pointed out previously, varying the upstream capacity if it is possible can be used to operate the storage strategically (Figure 9.10c). Defining the upstream capacity to be $C_U$ and the downstream capacity to be $C_D$, then over a period of time $\Delta t$, the various scenarios are detailed as shown in Table 9.1. Let $V(t)$ be the inventory in the tank at time $t$, assuming constant density (Henley and Kumamoto 1981):

$$V(t + \Delta t) = V(t) + (C_U - C_D)A_U A_D \Delta t - C_D(1 - A_U)A_D \Delta t \qquad (9.7)$$
$$+ C_U A_U (1 - A_D)\Delta t$$

To keep the tank inventory constant at all $t$:

$$V(t + \Delta t) = V(t) \qquad (9.8)$$

## Table 9.1

Upstream and downstream operating scenarios.

| Upstream | Downstream | Rate of tank fill ($m^3\ h^{-1}$) | Time available for tank fill over period $\Delta t$ (h) |
|---|---|---|---|
| Operating | Operating | $C_U - C_D$ | $A_U A_D \Delta t$ |
| Down | Operating | $- C_D$ | $(1 - A_U) A_D \Delta t$ |
| Operating | Down | $+ C_U$ | $A_U (1 - A_D) \Delta t$ |
| Down | Down | 0 | $(1 - A_U)(1 - A_D) \Delta t$ |

Combining Eqs. (9.7) and (9.8) and rearranging (Henley and Kumamoto 1981):

$$\frac{C_U}{C_D} = \frac{A_D}{A_U} \qquad (9.9)$$

Thus, if $A_U \geq A_D$ then $C_U \leq C_D$ and if $A_U \leq A_D$ then $C_U \geq C_D$. This only refers to a mean through time. The actual operating profile needs to be followed through time in a similar way to that illustrated in Figure 9.9.

So far the discussion has related to a process partitioned into two parts only. More complex processes lead to more options, as illustrated in Figure 9.11.

**Example 9.1** Figure 9.12 shows a process partitioned into upstream and downstream units. Also shown in Figure 9.12 are the *MTTF* and *MTTR* of the upstream and downstream units and the capacity of the downstream unit.

a) What is the availability of the system without the storage tank?
b) Is the storage tank likely to increase availability if the capacity of the upstream and downstream units is the same?
c) What are the limits for the availability of the system?
d) What capacity is required in the upstream unit to maintain the level in the tank constant?

### Solution

a) $A_U = \dfrac{100}{110} = 0.91$

$A_D = \dfrac{250}{290} = 0.86$

$A_{SYS} = A_U \times A_D = 0.91 \times 0.86 = 0.78$

b) If both upstream and downstream units have the same capacity, since $A_U > A_D$ the availability will be increased.

c) Limits for the availability

$$A_U A_D \leq A_{SYS} \leq A_D$$

Thus,    $0.78 \leq A_{SYS} \leq 0.86$

d) $C_U = C_D \dfrac{A_D}{A_U} = 10 \times \dfrac{0.86}{0.91}$

$$= 9.45\ m^3\ h^{-1}$$

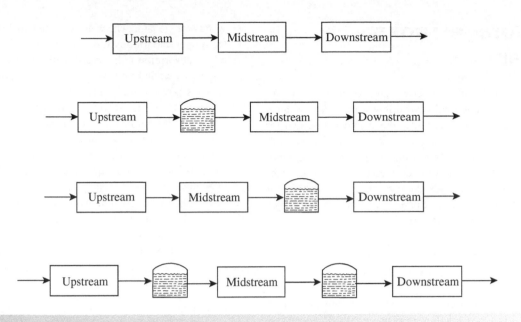

**Figure 9.11**

Processes with multiple units.

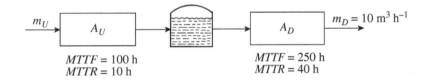

**Figure 9.12**

Capacity of liquid storage tank for Example 9.1.

# 9.3 Optimization of Intermediate Storage

The approximate methods used so far to analyze the influence of intermediate storage on the availability are limited to high-level understanding of the role of storage. To determine the effect of intermediate storage with greater accuracy requires an analysis of the details of the upstream and downstream units and the influence of a given storage volume between them. This needs to be followed through time as in the example in Figure 9.9. This, together with the complexity that more complex process structures and options for intermediate storage (as illustrated in Figure 9.11), calls for a much more detailed analysis (Koolen 2002).

A more detailed analysis can be carried out using Monte Carlo Simulation, as discussed in Chapter 8. Reliability block diagrams for the upstream and downstream units are connected by a volume of storage (Koolen 2002). The resulting capacity of the process will depend on the availability and capacity of the upstream and downstream units and the volume of intermediate storage.

A Monte Carlo Simulation can be carried out for fixed upstream and downstream unit structures and a fixed intermediate storage volume. This allows the detailed performance through time to be evaluated. To optimize the system requires various parameters to be varied to maximize the profit:

- substituting less reliable components for more reliable (and possibly more expensive) components;
- installation of redundancy on components in the upstream and downstream units;
- possible restructure of upstream and downstream units into parallel trains;
- vary the capacity of the upstream and downstream units;
- vary the volume of intermediate storage;
- vary preventive maintenance;
- introduce condition monitoring on critical components.

The parameters can then be varied in order to maximize the profit.

# 9.4 Storage Tanks – Summary

Raw material, product, and intermediate storage are all important to process operations. Intermediate (buffer) storage can be introduced to increase system availability, but it is often needed for many other reasons (operability, reworking material, control, etc.). Different strategies for operation of the storage inventory are possible. The effectiveness of intermediate storage to increase the system availability depends on the relative availability of upstream and downstream units. Spare upstream capacity allows the storage tank to be refilled effectively.

# Exercise

1. Figure 9.13 shows a process to recover acetone solvent from an air stream that is to be vented to atmosphere. The flowrate of air to the process is fixed by the upstream process. The acetone-laden air stream is fed to an absorption column where it is contacted with water. The water dissolves the acetone and the air is sent for further treatment before discharge. The water with the dissolved acetone from the absorption column is fed to a distillation column to recover pure acetone. The distillation column has an internal reboiler serviced by steam and a total condenser serviced by cooling water. The water from the bottom of the distillation column is cooled and recycled to the water storage tank. The water storage tank has a water makeup to compensate for any water losses from the process at the top of the absorption column and with the acetone product. The pumps shown in Figure 9.13 are driven by electric motors from a mains power supply. Failure rates and repair times for the equipment in the solvent recovery process are given in Table 9.2. If equipment failure causes the system to fail, then the process must be shut down temporarily. The shutdown of the process, isolation of the failed equipment for maintenance, issue of the required safety permit, and subsequent start-up after the repair of the equipment require a total of 24 hours in addition to the repair time for the equipment. It is anticipated that there will be no planned maintenance other than during an annual planned shut-down. The process is shut down every year for two weeks to carry out routine maintenance and safety checks. This creates a time horizon for reliability and availability of 8400 hours.

a) Explain briefly why it is a series system.

b) Calculate the reliability of the system for a time horizon of 8400 hours.

c) Calculate the operational availability of the system components and the system operational availability of the process assuming the product rule is accurate enough for series systems.

d) Calculate the reliability and operational availability of the system if cold standby pumps are installed for only the most vulnerable three pumps.

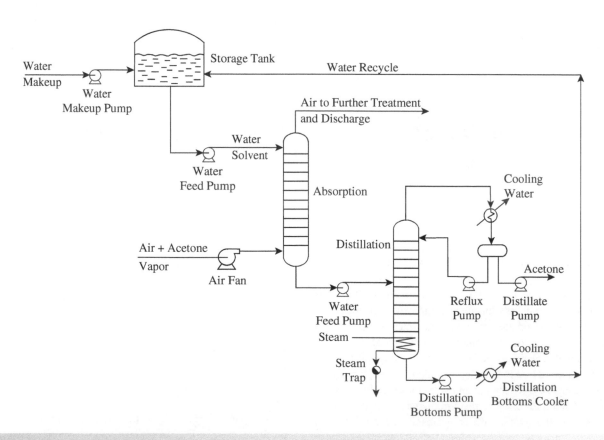

# Figure 9.13

A solvent recovery process.

## Table 9.2

Failure rates and repair times for the equipment in the solvent recovery process.

| Equipment | Failure rate (failures per $10^6$ h) | Repair time (h) |
|---|---|---|
| A) Water Makeup Pump | 50 | 12 |
| B) Water Feed Pump | 50 | 12 |
| C) Distillation Feed Pump | 50 | 12 |
| D) Reflux Pump | 70 | 12 |
| E) Distillate Pump | 70 | 12 |
| F) Distillation Bottoms Pump | 70 | 12 |
| G) Absorption Column | 10 | 120 |
| H) Distillation Column | 15 | 120 |
| I) Distillation Condenser | 15 | 20 |
| J) Cooler | 20 | 20 |
| K) Steam Trap | 15 | 8 |

e) Calculate how many additional hours of operation are anticipated per year if cold standby pumps are installed for the three most vulnerable pumps only.

f) Calculate the reliability and operational availability of the system if all of the pumps have cold standby pumps installed.

g) Calculate how many additional hours of operation are anticipated per year if cold standby pumps are installed for all of the pumps, instead of just the three most vulnerable ones.

h) Describe the advantages of having the additional standby pumps for all pumps and how the extra investment in standby pumps might be justified economically.

i) Determine how incorporation of an additional storage tank could increase the availability of the process.

## References

Henley, E.J. and Hoshino, H. (1977). Effect of storage tanks on plant availability. *Industrial and Engineering Chemistry Fundamentals* 16 (4): 439–443.

Henley, E.J. and Kumamoto, H. (1981). *Reliability Engineering and Risk Assessment*. Prentice-Hall.

Koolen, J.L.A. (2002). *Design of Simple and Robust Process Plants*. Wiley–VCH Verlag GmbH & Co.

**9**

# Chapter 10

# Process Control Concepts

## 10.1 Control Objectives

Once the basic configuration of the chemical process has been determined, including standby equipment, parallel trains, and intermediate storage, then the strategy for control and operation needs to be developed. At the commissioning phase of the project, the control system aims to ensure that the process consistently meets its design requirements, despite the influence of external disturbances, such as changes in feed flowrate, feed conditions, product demand, product specifications, ambient temperature, and so on. Later during the operation phase, process optimization, changes in feedstock and changes in the quantities and specifications of the products will often require process conditions and flowrates different from those in the initial process design. However, the control system still needs to maintain consistently the desired requirements, despite the influence of external disturbances. Consideration here will focus on the conceptual design of the control system. After the conceptual design of the control system has been developed it needs to be evaluated, possibly evolved, and further detail added to finalize the design.

The control system needs to achieve a number of key objectives:

1) *To maintain process variables within safe operating limits.* Ensuring safe operation is the most important task of the control system. This is achieved by monitoring the process conditions and maintaining them within safe operating limits.

2) *To achieve and maintain material and energy balances.* For a continuous process, a control system should ensure that a stable material and energy balance can be achieved and maintained and control both the overall process inventory and individual inventories of the process operations. By contrast batch processes are inherently transient in nature.

3) *To maintain the products to be within specified quality standards.* The products will be constrained by quality specifications. For bulk chemicals and fine chemicals production, this will typically be dictated by the concentration of the main component in the product and the maximum concentrations of impurities allowed. For specialty chemicals additional specifications will apply because of the functional nature of the product.

4) *To maintain the required production rate.* The productivity and revenue from the process will depend on maintaining the required production capacity. The control system being able to maintain the required production rate is critical to the process economics.

5) *To ensure the stability of the process.* A process is stable if a disturbance entering the process produces a bounded response. Figure 10.1a illustrates a stable response to a disturbance affecting the operation, in which the control system gradually overcomes the disturbance. By contrast, Figure 10.1b illustrates examples of different possible unstable responses to the input of a disturbance to the process in which the disturbance becomes amplified and unbounded. It should be noted that the existence of process variations does not necessarily mean instability. Some processes, such as adsorption, vary continually within bounds as part of their operation.

6) *To optimize the performance of the process commensurate with the other objectives.* It is often desirable for the control system to be able to optimize the process (e.g. minimize energy cost) whilst maintaining the other objectives above at the same time.

## 10.2 The Control Loop

Figure 10.2 illustrates a basic control loop. A measurement is carried out of a *process variable* by an *instrument* (*sensor*). This transmits a signal of the measured value of the process variable to a *controller*. The controller compares the measured value of the process variable with that of a *setpoint*. The setpoint is the desired value of the process variable. Using a comparison of the measured value of the process variable and the setpoint, the controller generates a control signal, which is

*Process Plant Design*, First Edition. Robin Smith.
© 2024 John Wiley & Sons Ltd. Published 2024 by John Wiley & Sons Ltd.
Companion website: www.wiley.com/go/processplantdesign

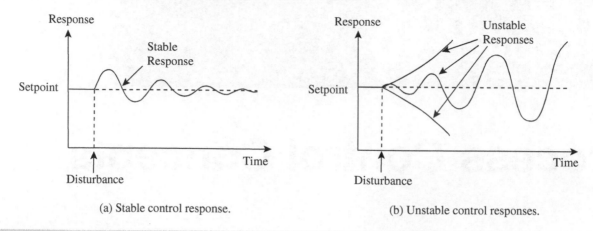

(a) Stable control response.

(b) Unstable control responses.

**Figure 10.1**

Stability of control response.

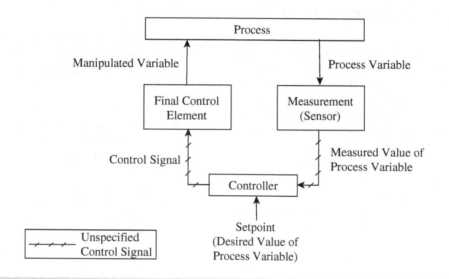

**Figure 10.2**

A basic control loop.

transmitted to the *final control element* (typically a valve). The final control element manipulates a variable to control the original measured variable. Thus, there are four components to a basic control loop:

- measurement (sensor);
- signals;
- controller;
- final control element.

# 10.3  Measurement

Table 10.1 lists the process variables that need to be measured and the commonly used sensors (instruments) used for the measurements. Some of the sensors in Table 10.1 are suited to use in control loops. Others are best used in isolation for process monitoring. The measurements will not be discussed in detail here, but it is important to note that it is necessary to understand exactly what is being measured for the design of the control system. The measurement might not be equally accurate across the whole range. Also, compensation for process conditions may be required. For example, the measurement of differential pressure for a gas might need to be compensated for a variable inlet temperature and pressure.

The speed at which the sensors produce a measurement is also an important issue. Some measurements, such as chemical analysis that require sampling, can involve a significant time delay in creating a signal. This has important implications for the control system. The importance of such time delays will be discussed later.

Not all important variables can be measured in real time. Unmeasured variables can sometimes be *inferred* (calculated) from other measurements to create *inferential control*. This

**Table 10.1**

Measurement sensors used in process control.

| Variable | Sensor |
| --- | --- |
| Flowrate | Differential pressure devices (orifice plate, venturi meter, pitot tube)<br>Turbine meter<br>Rotameter<br>Hot wire anemometer |
| Temperature | Thermocouple<br>Resistance thermometer<br>Thermistor<br>Expansion thermometer<br>Optical pyrometer |
| Pressure | Differential pressure cell (mechanical or semi-conductor)<br>Manometer<br>Bourdon gauge |
| Liquid Level | Differential pressure<br>Float<br>Weight of vessel |
| Liquid-liquid Interface | Differential pressure<br>Float |
| Analysis (Composition) | Gas chromatograph<br>Liquid chromatograph<br>Infrared spectrometer<br>Near-infrared spectrometer<br>Ultraviolet spectrometer |
| Physical Properties | Density<br>Conductivity<br>pH<br>Refractive index |

approach uses *soft sensors* (software sensors), which take input data from other measurements into a model to predict the desired measurement. Models might be physical models, correlations, artificial intelligence, or a combination of these.

## 10.4   Control Signals

Signals can be transferred around the control loop using different mechanisms. There are two broad classes of signal:

1) *Analog signal*. A range is defined for the signal, with the value to be transmitted in proportion to the span. A live zero is used to allow identification of transmission system failure:

a) *Electrical signal*. An electrical signal uses a direct current of 4–20 mA. For example, if a temperature measurement is calibrated to work between 20 °C (4 mA) and 100 °C (20 mA), then a 12 mA signal represents 60 °C.

b) *Pneumatic signal*. A pneumatic signal uses air pressure typically of 0.2–1 barg.

A signal of zero mA or zero barg indicates failure of the signal. Analog signals might need to be *converted*; e.g. a current-to-pressure (I/P) *transducer* converts an electrical analog signal to a pneumatic analog signal.

2) *Digital signal*. A digital signal uses binary code from a computer or microprocessor. Digital signals might need to be converted to analog (D/A) and vice versa (A/D). For example, a measurement to be sent to a computer control system will need A/D conversion.

## 10.5   The Controller

The controller is a device, or more often a control algorithm in a computer-controlled system, that aims to keep the controlled variable at near the *setpoint* for the controlled variable (Buckley 1964; Stephanopoulos 1984; Shinskey 1988; Smith 2009). The controller creates a signal from the difference between the setpoint and measured process variable, which is the *error*. The error can be defined as *reverse acting*:

$$e(t) = SP - PV(t) \qquad (10.1)$$

where

$e(t)$ = control error at time $t$

$SP$ = setpoint

$PV(t)$ = process variable at time $t$

A negative error indicates that the process variable is above the setpoint and a positive error indicates that it is below the setpoint. Figure 10.3 shows a controller *comparator* to calculate the error for a reverse acting controller. Alternatively, the error can be defined to be the other way around as *direct acting*:

$$e'(t) = PV(t) - SP \qquad (10.2)$$

For this case, if the value of the process variable is below the setpoint, the error is negative. The error is positive if the process variable is greater than the setpoint. The control error is used to generate the control actions:

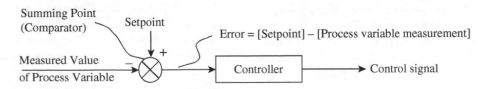

**Figure 10.3**

A reverse acting comparator calculates the difference between the setpoint and the process variable measurement and passes the calculated error to the controller to generate a control signal.

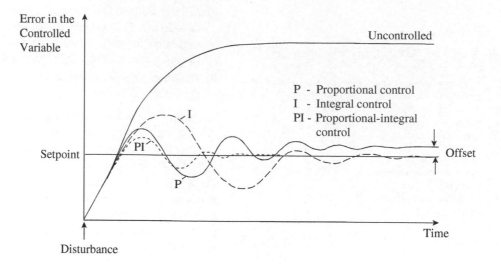

## Figure 10.4

Typical responses from proportional and integral control modes.

- A reverse acting controller action decreases as the process variable increases and increases as the process variable decreases.
- A direct acting controller action increases as the process variable increases and decreases as the process variable decreases.

Both types of control action might be required in different circumstances.

Not only does the error create different directions for the control action, but also the characteristics of the control action for a given direction. Consider now different control characteristics.

1) *Proportional.* The simplest control characteristic is proportional (*P*), which generates the controller output in proportion to the error:

$$CO(t) = K_P e(t) \qquad (10.3)$$

where

$CO(t)$ = controller output at time $t$

$K_P$ = proportional gain

$e(t)$ = control error at time $t$

The controller gain is normally dimensionless, based on input, and the output signals are scaled to be between 0% and 100%. In some controllers *proportional band* (*PB*) is used instead of gain (Shinskey 1988). This can be defined as $PB = 100/K_P$ and thus $K_P = 100/PB$. Controller gain will be used here.

There is a problem if proportional control is applied on its own based on Eq. (10.3). If the controller has acted to remove the error, then $e(t) = 0$, which means that $CO(t) = 0$. This will cause the final control element (e.g. valve) to be fully open or fully closed. This makes no sense. To overcome this problem a bias is introduced:

$$CO(t) = CO_{BIAS} + K_P e(t) \qquad (10.4)$$

where $CO_{BIAS}$ = controller bias.

The controller bias creates a controller output to set the final control element to its desired position when there is a zero error. The values of $CO_{BIAS}$ and $K_P$ can be set independently to give the best control action (Smith 2009). An example of a proportional controller output is shown in Figure 10.4 and compared with a typical uncontrolled response.

Figure 10.4 also illustrates another fundamental problem with proportional control. At steady state there is a long-lived error in the controlled variable, known as the *offset*. This results from the fact that an error is required to generate any controller output signal.

2) *Integral.* If it is important to eliminate the offset associated with proportional control, the controller output can be based on the cumulative error calculated from the integral of the error:

$$CO(t) = CO_{BIAS} + K_I \int_0^{t'} e(t)dt \qquad (10.5)$$

where

$CO_{BIAS}$ = controller bias

$K_I$ = integral gain

In this case, the controller bias is the controller output at time zero. The value $K_I$ can be set to give the best control action (Smith 2009). A typical integral controller output is shown in Figure 10.4. It can be seen that, compared with proportional control, it eliminates the offset. However, because the integral action responds to the cumulated error, the response time is longer than proportional control. Figure 10.4 also shows that integral control tends to be less stable than proportional control.

3) *Proportional-integral.* The relative advantages of proportional only (*P*) and integral only (*I*) control can be combined to give proportional-integral (*PI*) control. From Eq. (10.4) for

proportional control, the controller output bias $CO_{BIAS}$ needs to be adjusted at a rate proportional to the error:

$$\frac{dCO_{BIAS}}{t} = K_I e(t) \tag{10.6}$$

where $K_I$ = integral gain.

Integrating Eq. (10.6):

$$CO_{BIAS} = K_I \int_0^{t'} e(t)dt \tag{10.7}$$

Combining Eqs (10.4) and (10.7):

$$CO(t) = K_P e(t) + K_I \int_0^{t'} e(t)dt \tag{10.8}$$

The integral characteristic can also be expressed using an integral time $\tau_I$:

$$\begin{aligned} CO(t) &= K_P e(t) + \frac{K_P}{\tau_I} \int_0^{t'} e(t)dt \\ &= K_P \left[ e(t) + \frac{1}{\tau_I} \int_0^{t'} e(t)dt \right] \end{aligned} \tag{10.9}$$

where $\tau_I$ = integral time.

The values of $K_P$ and $\tau_I$ can be set independently to give the best control action (Smith 2009). The *PI* controller error can be reduced to zero whilst maintaining a control action because the integral of the error is non-zero. Figure 10.4 shows a typical response for a *PI* control. It can be seen that this eliminates the offset associated with proportional control.

**4)** *Proportional-derivative.* Proportional-derivative (*PD*) control uses gradients to project a value of the process variable $PV(t)$. At a time $\tau_D$ in the future a projected value of $PV(t)$ can be calculated from:

$$\hat{PV}(t) = PV(t) + \tau_D \frac{dPV(t)}{dt} \tag{10.10}$$

where

$\hat{PV}(t)$ = projected value of $PV(t)$ at time $\tau_D$ in the future

$PV(t)$ = process variable at time $t$

$\tau_D$ = derivative time

The control action can be based on the projected error, rather than the current error. For a reverse acting controller:

$$\begin{aligned} \hat{e}(t) &= SP - \hat{PV}(t) \\ &= SP - PV(t) - \tau_D \frac{dPV(t)}{dt} \\ &= e(t) - \tau_D \frac{dPV(t)}{dt} \\ &= e(t) + \tau_D \frac{de(t)}{t} \end{aligned} \tag{10.11}$$

where $\hat{e}(t)$ = projected error at time $\tau_D$ in the future.

Equation (10.4) can now be based on the projected error, instead of the actual error, by substituting Eq. (10.11) into Eq. (10.4):

$$\begin{aligned} CO(t) &= CO_{BIAS} + K_P \hat{e}(t) \\ &= CO_{BIAS} + K_P \left[ e(t) - \tau_D \frac{dPV(t)}{dt} \right] \end{aligned} \tag{10.12}$$

For a reverse acting controller, from Eq. (10.1), $-dPV(t)/dt = de(t)/dt$. Thus, Eq. (10.12) can be written as:

$$CO(t) = CO_{BIAS} + K_P \left[ e(t) + \tau_D \frac{de(t)}{dt} \right] \tag{10.13}$$

This provides the basis for proportional-derivative control. Figure 10.5 shows an example of a response for a *PD* control compared with proportional control. The derivative action creates a more aggressive response to a disturbance, but the derivative action makes the control output sensitive to any noise in the measurements. The values of $CO_{BIAS}$, $K_P$, and $\tau_D$ can be set

**10**

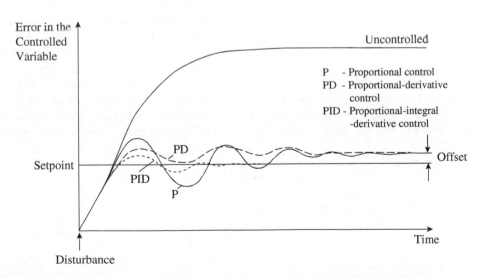

P - Proportional control
PD - Proportional-derivative control
PID - Proportional-integral-derivative control

**Figure 10.5**

Typical responses from derivative control modes.

independently to give the best control action (Smith 2009). Derivative control can dampen the oscillations and increase stability, but this depends on the dynamic behavior of the process, the presence of noise in the measurements, and the appropriate setting of the control parameters. As with proportional only control, the response retains the problem of an offset. It is also possible to write Eq. (10.13) as:

$$CO(t) = CO_{BIAS} + K_P e(t) + K_D \frac{de(t)}{dt} \qquad (10.14)$$

where $K_D$ = derivative gain.

**5)** *Proportional-integral-derivative.*   Proportional-integral-derivative (*PID*) control, also known as *three term control*, is created by basing the error in the *PI* equation (Eq. (10.9)) by the projected error from Eq. (10.11):

$$
\begin{aligned}
CO(t) &= K_P \hat{e}(t) + \frac{K_P}{\tau_I} \int_0^{t'} e(t) dt \\
&= K_P \left[ e(t) + \tau_D \frac{de(t)}{dt} \right] + \frac{K_P}{\tau_I} \int_0^{t'} e(t) dt \qquad (10.15) \\
&= K_P \left[ e(t) + \frac{1}{\tau_I} \int_0^{t'} e(t) dt + \tau_D \frac{de(t)}{t} \right]
\end{aligned}
$$

The values of $K_P$, $\tau_I$, and $\tau_D$ can be set independently to give the best control action (Smith 2009). Figure 10.4 shows an example of a response for a *PID* control. Again, derivative control can dampen the oscillations and increase stability, but again this depends on the dynamic behavior of the process, the presence of noise in the measurements, and the appropriate setting of the control parameters (Shinskey 1988). Figure 10.6 shows the operation of a *PID* controller based on Eq. (10.15). A *comparator* calculates the error from the difference between the setpoint and the measured variable. Then the individual *P*, *I*, and *D* actions are summed and multiplied by the proportional gain. An alternative form of Eq. (10.15) is given by:

$$CO(t) = K_P e(t) + K_I \int_0^{t'} e(t) dt + K_D \frac{de(t)}{dt} \qquad (10.16)$$

where
$K_P$ = proportional gain
$K_I$ = integral gain
$K_D$ = derivative gain

However, it is not advisable to use the form of Eqs (10.13) to (10.16) for derivative control. If a step change is made to the setpoint by a process operator then the value $de(t)/dt$ from

Eq. (10.1) becomes an *impulse function* (*delta function*), which is an *infinite spike*. The same problem will not occur with $dPV(t)/dt$ as a result of a step change in the setpoint. From Eq. (10.1) it is possible to substitute $-dPV(t)/dt$ for $de(t)/dt$ in Eqs (10.13) to (10.16). Equation (10.15) becomes:

$$CO(t) = K_P \left[ e(t) + \frac{1}{\tau_I} \int_0^{t'} e(t) dt - \tau_D \frac{dPV(t)}{dt} \right] \qquad (10.17)$$

Whilst a bias term is not required for PID control under automatic control, it is good practice to add a bias. This is because control loops are often left *open* and the final control element (control valve) is operated manually. Manual operation might be required to follow procedures for plant start-up, to deal with problems associated with a control valve or poor control loop performance, for example resulting from poor tuning of the controller. When a control loop is placed in *manual* or *open loop*, the PID controller no longer determines the position of the final control element, but the final control element is adjusted by the plant operator. While the loop is in manual mode it is unlikely that the process variable will be the same as the setpoint. If the control loop is then switched from manual to automatic mode (from open loop to closed loop), a sudden and potentially large error will be imposed on the control loop. This can lead to abrupt control actions and a major disruption to the process when the switch is made. To avoid this abrupt changeover and achieve a "bumpless" transfer, the setpoint can be temporarily adjusted to the current value of the process variable ($SP = PV(t)$), and a bias equal to the current controller output added ($CO_{BIAS} = CO(t)$):

$$CO(t) = CO_{BIAS} + K_P \left[ e(t) + \frac{1}{\tau_I} \int_0^{t'} e(t) dt - \tau_D \frac{dPV(t)}{dt} \right] \qquad (10.18)$$

where $CO_{BIAS}$ = the controller bias set by a bumpless changeover.

Making $SP = PV(t)$ means there is no error to drive change, and making $CO_{BIAS} = CO(t)$ the current operation is maintained at the changeover. Thus, the final control element is maintained at a fixed position for changeover. The controller is then ramped slowly to the required setpoint and the bumpless bias ramped down. In this way changeover from manual to automatic mode is gradual and the process experiences no abrupt changes. Figure 10.7 shows the operation of a *PID* controller based on Eq. (10.18). A comparator calculates the error

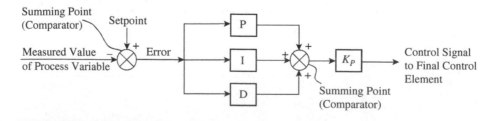

## Figure 10.6

A reverse acting comparator calculates the error, generates P, I, and D signals to create the final control signal and adds them.

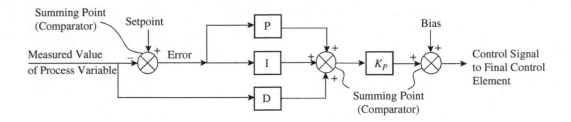

**Figure 10.7**

A reverse acting comparator calculates the error that is used to calculate the P and I components, with the D component calculated from the process measurement before summing and including the gain and bias.

from the difference between the setpoint and the measured variable, which is then used to calculate the proportional and integral components. The change in the process variable is used to calculate the derivative component. Then the individual *P*, *I*, and *D* actions are summed and multiplied by the proportional gain and the bias added. These different options will be discussed again later within the context of digital control.

The units for the control actions can be written in terms of the primary units (e.g. temperature, pressure, etc). However, in practice the signal range from minimum to maximum in the primary units is scaled to the range from 0% to 100% for implementation in the controller. This means that the controller output and bias have no units, as does the proportional gain. The integral gain then has units of reciprocal of time and the derivative gain units of time. The integral time and derivative time both have units of time.

The control equations have been developed based on reverse acting control. For direct acting control, with the error defined by Eq. (10.2), the sign of the derivative term in Eq. (10.18) would be positive.

Different combinations of *P*, *I*, and *D* control actions are suited to different applications:

*P* is sometimes used.

*I* is sometimes used.

*PI* is most common.

*PD* is almost never used.

*PID* is sometimes used.

Once the control action is generated, a signal is sent to the final control element.

# 10.6  Final Control Element

The *final control element* is most often a valve in the process industries. However, it might also be louvres (to control gas flow in a duct) or a variable speed motor drive. Valves and louvres need an *actuator* to adjust the final control element in response to the control signal. An actuator needs a source of energy to function and a control signal:

- *Pneumatic actuator* uses pneumatic air pressure, created by an instrument air compressor, operating on a diaphragm and

responds to an air pressure signal operating typically between 0.2 and 1 barg.

- *Electrical actuator* uses an electric motor and responds to an electrical analog signal operating between 4 and 20 mA.

- *Hydraulic actuator* uses an electrically-driven pump to create pressure in a hydraulic oil, which acts on a piston and cylinder to respond. Whilst commonly used on machines such as packaging machines, these are rarely used in process applications.

1) *Pneumatically actuated control valves*. Figure 10.8 shows schematics of two pneumatically actuated control valves. The design in Figure 10.8a shows an air signal acting on a flexible diagram that compresses a spring to close the valve. If the air signal fails for some reason, then the spring forces the valve open. A small travel indicator gives an approximate indication of the position of the valve in the event of a problem with the valve. In Figure 10.8b the air signal acts on a flexible diagram to compress the spring to open the valve. If the air signal fails, then the spring forces the valve to close.

The shape of the plug and seat arrangement, referred to as the valve *trim*, creates differences in the valve characteristics as it opens and closes. A linear characteristic valve trim is shaped so that the flowrate is directly proportional to the position of the valve stem at a constant pressure difference across the valve. This is achieved by a linear relationship between the position of the valve stem and the orifice area in the valve trim. Another option is an equal percentage characteristic, which creates equal percentage changes in the flowrate from equal changes in the valve stem position. A small change in the valve position at small openings results in a small increase in the flowrate, while the same increment of the valve position at large openings creates a larger increase in the flowrate. A third option is the fast-opening characteristic, which will give a large change in the flowrate for a small change in the position of the valve from the closed position. Determining the most suitable valve trim for an application requires a detailed analysis of the flowrate versus pressure drop characteristics of the system. Control valve actuators can be used with other valve designs than the type illustrated in Figure 10.8. Different valve types used in process industry applications will be discussed in Chapter 15. An example of a control valve with a pneumatic actuator is shown in Figure 10.9.

**10**

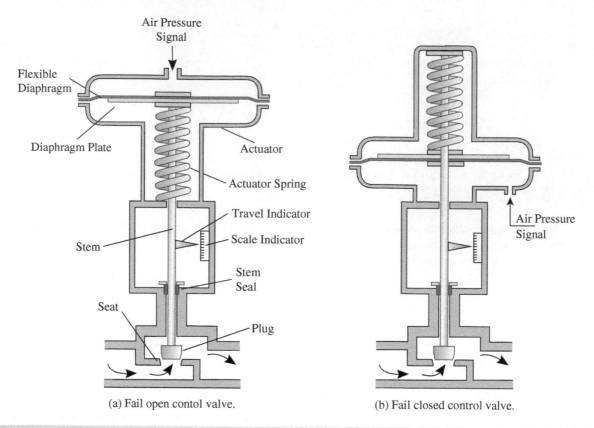

(a) Fail open contol valve.

(b) Fail closed control valve.

**Figure 10.8**

Pneumatically actuated control valves.

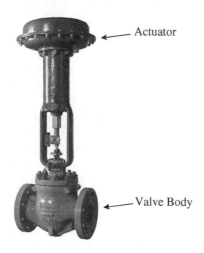

Actuator

Valve Body

**Figure 10.9**

An example of a pneumatically actuated control valve. Source: With permission from Flow Control Equipment, LLC / http://www. flowcontrolequip.com / last accessed March 08, 2023.

For some applications the 0.2–1 barg air pressure in the diaphragm chamber of the control valve might not be enough to cope with friction within the valve mechanism, or with high differential pressures across the valve seat. To overcome this, a *valve positioner* can be used, as illustrated in Figure 10.10. The positioner receives an input pneumatic signal from a control device and increases or decreases the signal if necessary. The positioner is attached to the side of the valve and linked mechanically to the valve stem by a feedback arm to monitor the valve position. The positioner has its own higher pressure instrument air supply to boost the signal to the valve actuator to ensure, if necessary, that the valve achieves the correct position. The positioner compares the actual position of the valve stem with the incoming pneumatic signal. If the valve stem is in the correct position, then the pneumatic signal passes through the positioner to the diaphragm chamber of the control valve unchanged. However, if the valve stem is incorrectly located then the signal is adjusted by the positioner, if necessary, boosting the air pressure to the valve chamber using its supplementary higher pressure instrument air. This ensures that the control valve plug is positioned correctly relative to the valve seat. The positioner can be reverse acting, to send a decreased output signal as the import signal increases, or direct acting to send an increased output signal to the control valve as the input signal to it increases. An example of a valve positioner is shown in Figure 10.11.

2) *Electrically actuated control valves.* An electrical actuator uses an electric motor to provide the force to operate the valve. Fail open or fail closed action is achieved through energy storage. This could be energy input to a spring to open or close the valve in the event of electrical failure, or electrical storage using a capacitor or battery. Electrical actuators if used in

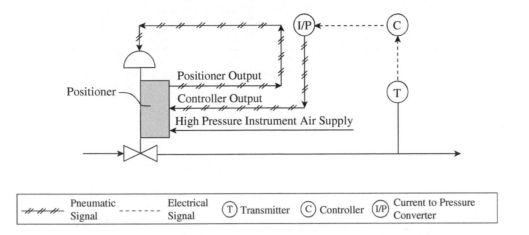

| | Pneumatic Signal | | Electrical Signal | (T) Transmitter | (C) Controller | (I/P) Current to Pressure Converter |

**Figure 10.10**

Pneumatic control valve positioner.

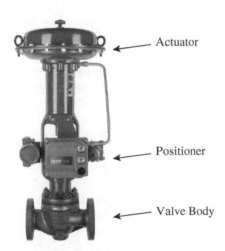

Actuator

Positioner

Valve Body

**Figure 10.11**

An example of a pneumatic control valve positioner. Source: With permission from Flow Control Equipment, LLC / http://www. flowcontrolequip.com / last accessed March 08, 2023.

hazardous environments must have an explosion proof enclosure. Electrical actuators have advantages when working in environments with very low temperatures, where instrument air for pneumatic systems can fail. If the instrument air is not dried adequately, any moisture in the air will freeze in a low temperature environment, blocking the signal. An example of a control valve with an electrical actuator is shown in Figure 10.12.

If a fail-safe condition can be identified for a control valve, and the actuating energy source fails, then the valve must be designed such that it results in a *fail-safe* condition. Figure 10.13 shows the three options. The valve can fail open, fail closed, or fail locked in the last position. To identify the fail-safe option, the whole process needs to be considered when selecting the design.

Consider the example of a distillation column in Figure 10.14. This shows a distillation column with the reboiler operating from

**10**

**Figure 10.12**

An example of an electric control valve actuator. Source: With permission from Westwick-Farrow Pty Ltd / https://www. processonline.com.au / last accessed March 08, 2023.

steam and the condenser operating from cooling water. The flows of both steam and cooling water are controlled by control valves. In the case of each control valve, what would be the most appropriate fail-safe condition? In the case of the steam to the reboiler, suppose it *failed open*. In this event, it would create a high heat

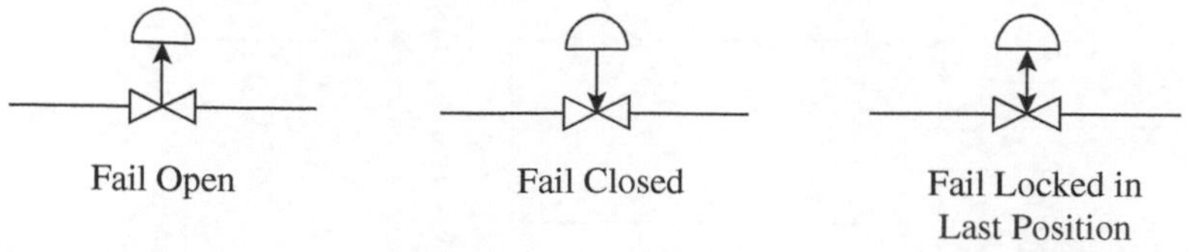

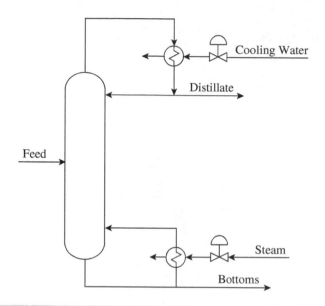

**Fail Open**        **Fail Closed**        **Fail Locked in Last Position**

## Figure 10.13

Fail safe positions for control valves.

Cooling Water

Distillate

Feed

Steam

Bottoms

## Figure 10.14

A distillation column with control valves on the heating and cooling utilities.

input to the reboiler and a high flowrate of vapor up the column. If the condenser could not condense the very high flowrate of vapor, or even worse, if the flow of cooling water to the condenser failed at the same time, then a high flowrate of vapor would need to be vented from the distillation. This could be extremely hazardous. As a result, *fail closed* for the steam to the reboiler would be fail-safe. For the condenser, if the flow of cooling water failed because of failure of the control system, then there would be no condensation of the vapor and the uncondensed vapor would need to be vented. Again, this could be extremely hazardous. As a result, *fail open* for the cooling water to the condenser would be fail-safe. In other words, if something goes wrong it is safest to stop the flow of energy into the system and keep sending energy coming out to a safe place (e.g. the cooling tower).

The choice of fail-safe design for the control valve has important implications for the required control action, which depends on the context. This is illustrated in Figure 10.15, which shows the problem of liquid level control in a tank. A level transmitter measures the level in the tank and transmits this to a level controller. From the previous section:

- A reverse acting controller action decreases as the process variable increases and increases as the process variable decreases.

- A direct acting controller action increases as the process variable increases and decreases as the process variable decreases

Figure 10.15a and b show arrangements for control of the level from the outlet in which an increase in the liquid level above the setpoint requires the outlet flow to increase and vice versa. The outlet control valve in Figure 10.15a is fail open, as in Figure 10.8a. If the liquid level rises above the setpoint, then the valve needs to open wider. A fail open control valve requires a reverse acting controller in this case. The control valve in Figure 10.15b is fail close, as in Figure 10.8b. This time, if the level rises above the setpoint a direct acting controller is required. Figure 10.15c and d shows arrangements for control of the level from the inlet in which an increase in the liquid level above the setpoint requires the inlet flow to decrease and vice versa. The control valve in Figure 10.15c is fail open, as in Figure 10.8a. Now when the level rises above the setpoint, the valve needs to close. A fail open valve requires a direct acting controller in this case. The control valve in Figure 10.15d is fail closed, as in Figure 10.8b. This time, if the level rises above the setpoint a reverse acting controller is required.

**Example 10.1**   A fail-open control valve with a pneumatic actuator receives an output signal from a controller of 0.4 bar. What should be the expected per cent open?

**Solution**

The controller output signal determines the per cent closed:

$$\text{Per cent open} = 100 - \frac{(0.4-0.2)}{(1.0-0.2)} \times 100$$

$$= 75\%$$

**Example 10.2**   A fail-closed control valve with an electrical actuator receives an output signal from a controller of 8 mA. What should be the expected per cent open?

**Solution**

The controller output signal determines the per cent open:

$$\text{Per cent open} = \frac{(8-4)}{(20-4)} \times 100$$

$$= 25\%$$

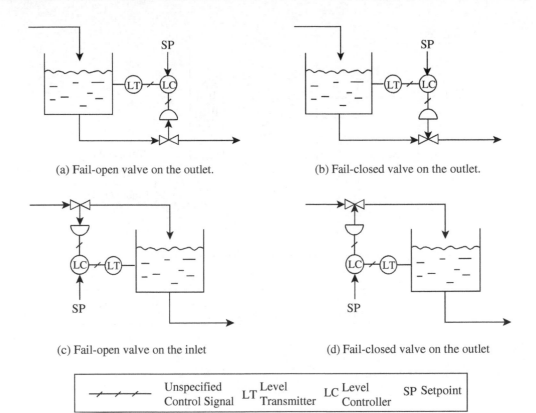

(a) Fail-open valve on the outlet.

(b) Fail-closed valve on the outlet.

(c) Fail-open valve on the inlet

(d) Fail-closed valve on the outlet

| | Unspecified Control Signal | LT | Level Transmitter | LC | Level Controller | SP | Setpoint |
|---|---|---|---|---|---|---|---|

## Figure 10.15

Fail closed or fail open valves requires reverse or direct acting control depending on the application.

**10**

## 10.7   Feedback Control

Figure 10.16 shows a *feedback control* loop (Buckley 1964; Stephanopoulos 1984; Shinskey 1988; Smith 2009). A feedback controller compares the measured control variable to its setpoint and calculates a *compensating move*. The controller responds to an error *after* it has happened.

Figure 10.17 shows an example of a feedback control loop. A process liquid in a vessel is to be heated by a steam coil.

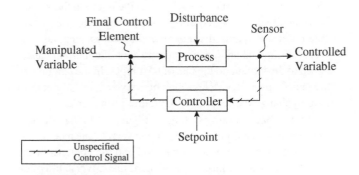

| | Unspecified Control Signal |
|---|---|

## Figure 10.16

A basic feedback control loop.

A temperature sensor after the exit of the vessel transmits a signal to the feedback controller, which compares the measurement with the setpoint. If the temperature measurement is below the setpoint, then the controller will send a signal to the control valve to open. If the temperature measurement is above the setpoint, then the controller will send a signal to the control valve to close.

Ideally, any load change or disturbance entering the process would be followed by an immediate response to bring the process to a new condition of equilibrium. However, this is not possible to achieve in a physical system. The response to a change in load or a disturbance will take time to complete its effect. For example, consider what happens to the system in Figure 10.17 if there is an increase in the flowrate. The increased flow of cold liquid will gradually cause the temperature of the bulk liquid and the vessel to decrease. The temperature of the overflow will then start to decrease. This will flow to the outlet where the temperature of the liquid in the outlet pipe and the outlet pipe itself will gradually decrease. Only then will the temperature sensor detect a decreasing temperature. Then the changed signal sent to the controller will cause the controller to respond. There is a delay between the disturbance entering the system and the disturbance being detected by the temperature sensor.

The ability of the control loop to respond to a disturbance is restrained by *time lags*. Such time lags impede the system performance and can cause instability. Because the controller in feedback control acts *after* the error has happened, time lags are a

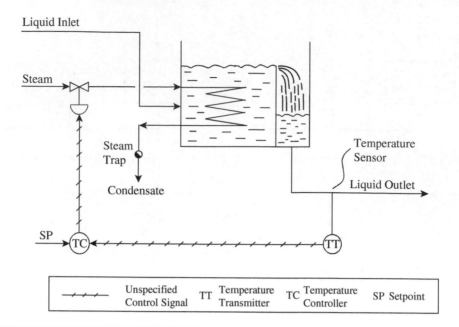

## Figure 10.17

A simple feedback control loop to heat a liquid tank.

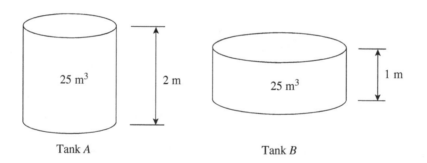

## Figure 10.18

Two tanks full of liquid, each with the same capacity, but different capacitance with respect to level.

major problem for good performance. There are three categories of time lag:

1) *Capacitance*. Any storage property of the process will retard change. Those parts of the process that have the ability to store energy (thermal, chemical, pressure, electrical, or potential energy) or store mass (gas, liquid, or solid) introduce *capacitance*. Capacitance and capacity are not the same. Consider the capacitance of liquid level of two tanks with the same capacity shown in Figure 10.18. Tank $A$ has a depth of 2 m and Tank $B$ a depth of 1 m, but both tanks have a capacity of 25 m$^3$. Tank $A$ has a capacitance with respect to a liquid level of 12.5 m$^3$ m$^{-1}$ and Tank $B$ a capacitance with respect to liquid level of 25 m$^3$ m$^{-1}$. The liquid in the tanks also has *thermal capacitance*, which measures the energy change required to change the temperature by one degree. If the tanks are full of water with a specific heat capacity of 4.2 kJ kg$^{-1}$ K$^{-1}$ and a density of 1000 kg m$^{-3}$, then both tanks have a thermal capacitance of $25 \times 1000 \times 4.2 = 1.05 \times 10^5$ kJ K$^{-1}$. If the liquid in the vessels is heated, it takes a period of time for the liquid to rise in temperature to the required value after the heat is supplied. The period of time taken depends primarily on the thermal capacitance of the liquid and the rate of heat input. In the example of the tank, capacitance delays the corrective action of a control loop to change the level or temperature in the tank.

The feedback control system in Figure 10.17 has thermal capacitance in the liquid contents of the vessel. The metal walls of the vessel and the heat transfer coil and the contents of the coil also introduce capacitance. This has an adverse effect on the ability of the feedback control loop to control the temperature in the tank by retarding any change.

2) *Resistance*. Any part of the process that resists transfer of energy or mass will also retard change. *Resistance is*

opposition to the flow of energy or mass. It is measured in units of the potential required to produce a unit change in flow of energy or mass. Examples are the differential pressure required to cause a quantity of flow through a pipe, valve, or instrument. The pressure loss causes a resistance to flow. Changing the resistance, for example through changing the opening of a control valve, causes a change in the flow. In the case of a shell-and-tube heat exchanger, there is a resistance to the flow of heat caused by the film heat transfer resistances and fouling resistances on the inside and outside of the tubes and the thermal conductivity of the tube walls (Smith 2016).

The feedback control system in Figure 10.17 has resistance to the transfer of heat from the steam to the liquid contents of the vessel from the inside and outside film and fouling resistances for the coil (Smith 2016), together with the resistance caused by the thermal conductivity of the steam coil.

3) *Dead time.* The *dead time* of a process is the time elapsed between the change of an input variable to the first change in the output variable being detected. As the signal passes through the dead time there is no change in the signal. Dead time is also known as *transportation time* or *transportation lag*, because it is often the time delay to transport material from the input to the location where the measurement is made. Dead time is also created by the time the sensor, controller, and final control element take to react. Dead time contrasts with capacitance and resistance, as it is not just a slowing down of the change, but a discrete delay. This introduces more difficulties to control than capacitance and resistance, and should be kept to a minimum.

The control system in Figure 10.17 has dead time that will result from the location of the sensor some distance downstream of the vessel. For example, if liquid with a lower temperature enters the vessel, there is a finite time required for this to travel through the tank and the outlet pipe and reach the temperature sensor. The measurement should be located as close as possible to the vessel, or inside the vessel.

Figures 10.16 and 10.17 show *closed loops*. If there is a problem with the control loop, then the process operator can disconnect the loop and operate the loop manually by adjusting the final control element manually on *open loop*.

## 10.8  Cascade Control

In the feedback control systems discussed so far, a single variable is manipulated to satisfy the specification for a single controlled variable. A single independent setpoint is used. The performance of such systems can be improved in some cases by manipulating more than one variable, but still to satisfy the specification for a single controlled variable.

Figure 10.19 shows a *cascade control* system (Buckley 1964; Stephanopoulos 1984; Shinskey 1988; Smith 2009). A cascade control system uses multiple sensors for measuring process conditions and can often perform better than a single-measurement feedback control. Two controllers are used. The output of one controller is used to drive the setpoint of the other. There is only one independent setpoint. The controller driving the setpoint is the *primary* (or *outer*) controller. The controller receiving the setpoint is the *secondary* (or *inner*) controller. The secondary measurement is located to recognize the disturbance *before* the controlled variable. Effective application of cascade control requires the secondary (inner) control loop to have a significantly faster response than the outer (primary) control loop. This allows the secondary controller enough time to compensate for inner loop disturbances before they can affect the primary process.

As an example of the application of cascade control, consider the fluidized bed dryer shown in Figure 10.20a. A wet granular solid is fed to the dryer and dried by an air flow that has been heated with steam. The fluidized bed maintains good contact between the granular solid and the air. The temperature above

**10**

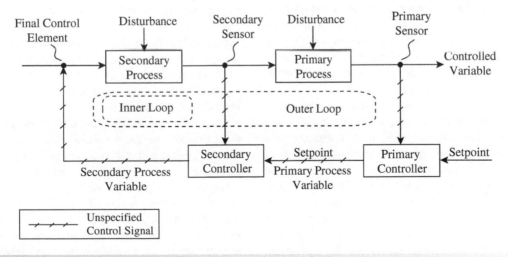

## Figure 10.19

A cascade control system.

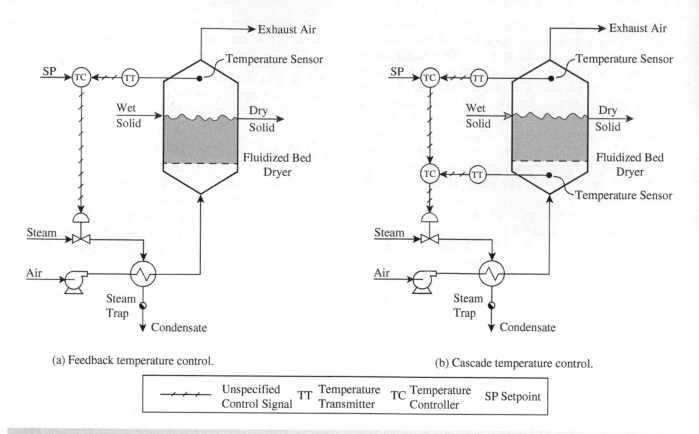

(a) Feedback temperature control.　　　　　　　　(b) Cascade temperature control.

| | Unspecified Control Signal | TT | Temperature Transmitter | TC | Temperature Controller | SP Setpoint |
|---|---|---|---|---|---|---|

## Figure 10.20

Dryer temperature control system.

the fluidized bed is controlled by the flowrate of steam to the air heater. Controlling the temperature at the exit of the fluidized bed ensures that the solids are dried to the required moisture specification at the top of the bed before exiting the dryer. Disturbances can affect the control of the system through changes in the ambient air temperature and variations in the steam conditions. The process is subject to significant time lags. The bed of solids creates a large heat capacitance. The equipment itself also provides some heat capacitance. There is heat transfer resistance in the steam heater. There will be some dead time from the location of the temperature sensor and possibly dead time from the equipment in the control loop. These time delays will act to create a poor performance for the feedback control loop.

An alternative cascade control system is shown in Figure 10.20b. The primary control loop is combined with a secondary control loop. The secondary control loop measures and controls the temperature of the air at the inlet of the fluidized bed by adjusting the steam flow directly. The primary control loop provides the setpoint for the secondary control loop. If a disturbance enters the system through a change in the ambient air temperature or change in the steam conditions, the secondary control loop will detect the disturbance at the inlet to the fluidized bed. Since the steam heater controller setpoint will be unchanged, the steam flowrate controller will respond by changing the flowrate of steam to return the temperature of the inlet air to the desired value at the inlet to the fluidized bed. By responding quickly to the

disturbance, the secondary controller corrects the disturbance before the fluidized bed exit temperature is significantly affected by the disturbance.

## 10.9　Split Range Control

In a *split range control* loop, the output of the controller is split and sent to two or more final control elements, as shown in Figure 10.21a. The splitter defines how each final control element is sequenced as the controller output changes from 0 to 100%. For example, the sequence could follow the control illustrated in Figure 10.21b. As the controller output is increased from 0%, Final Control Element A is gradually closed. At 50% in this example, it is fully closed. For a controller output of greater than 50%, Final Control Element A remains closed and Final Control Element B gradually opens. For a controller output of 100%, Final Control Element B is fully open (with Final Control Element A remaining fully closed). In many split range applications, the controller adjusts the opening of one of the final control elements when its output is in the range of 0 to 50% and the other when its output is in the range of 50 to 100%. However, in principle the split range can be adjusted to other values. The range can be chosen such that the magnitude of the effect of each percentage change in the range is approximately equal whichever final control element is active.

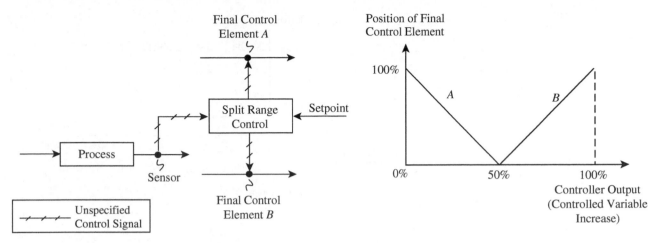

(a) The split range arrangement.

(b) The position of the final control elements.

**Figure 10.21**

Split range control.

As an example of split flow control, Figure 10.22 shows a batch reactor that requires heating to initiate the reaction and then requires cooling. Split range control can be used to maintain the reactor temperature. When the temperature is below the desired setpoint, first the cooling water valve is closed and the steam valve opens. When the temperature is above the desired setpoint, first the steam valve is closed and then the cooling water valve opens.

In another example of split flow control, Figure 10.23a shows a vessel containing a hazardous liquid that requires a nitrogen blanket for safety reasons. The spilt flow control maintains a pressure of nitrogen above the liquid at a setpoint. When the pressure decreases, nitrogen is introduced. When the pressure increases, nitrogen is discharged to vent. In the configuration in Figure 10.23b, if the pressure is low then the controller opens the inlet valve. If the pressure is still too low, the controller closes the outlet valve. Figure 10.23c shows an alternative configuration in which the operation of the control valves overlap at 50% of the controller output. The outlet valve is fully open at a controller output of 0% and fully closed at 100% controller output.

**10**

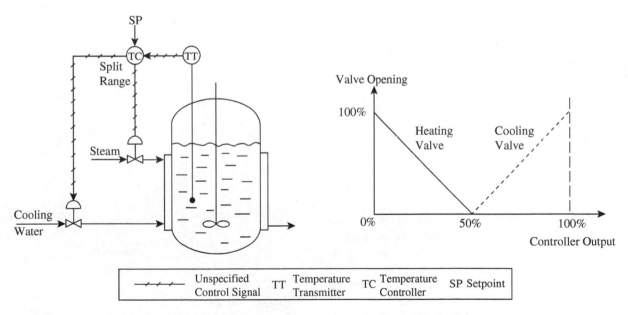

(a) Split range control of a batch reactor heating and cooling.

(b) The position of the control valves.

**Figure 10.22**

Split range control of a batch reactor.

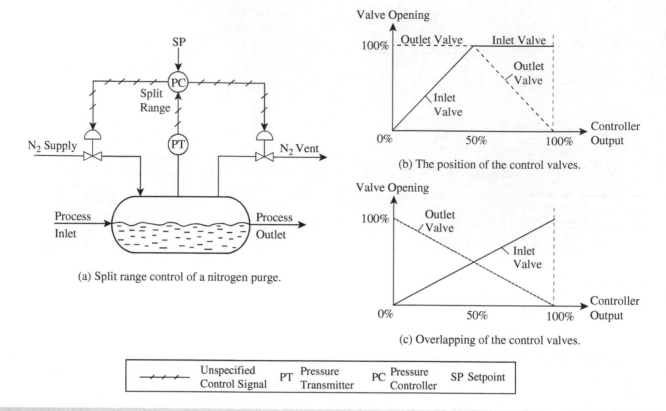

(a) Split range control of a nitrogen purge.

(b) The position of the control valves.

(c) Overlapping of the control valves.

| | | | | |
|---|---|---|---|---|
| ⤙⤙⤙ Unspecified Control Signal | PT | Pressure Transmitter | PC | Pressure Controller | SP Setpoint |

## Figure 10.23

Split range control of a nitrogen purge.

The inlet valve is fully closed at a controller output of 0% and fully open at 100% controller output.

# 10.10   Limit and Selector Control

Limits can be imposed on the magnitude of control signals. *High limit* and *low limit* functions receive a signal and limit the value by a bound. If the bound is not reached the signal is unchanged. In Figure 10.24a the signal is set to be below the limit $X$. In Figure 10.24b the signal is set to be above the limit $X$. In this way, a limit is set on how high or low a signal can go before being sent to another function.

*High selectors* take two or more signals as input, select whichever input signal is the highest, and send that signal as output, as illustrated in Figure 10.24c. *Low selectors* take two or more signals as input, select whichever input signal is the lowest, and send that signal as output, as illustrated in Figure 10.24d. This provides automatic selection amongst multiple measurements.

As an example of the use of a high limit, Figure 10.25a shows the cascade control of the fluidized bed dryer from Figure 10.20a,

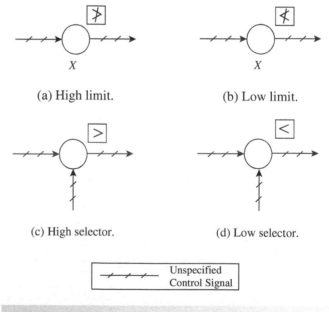

(a) High limit.

(b) Low limit.

(c) High selector.

(d) Low selector.

| | |
|---|---|
| ⤙⤙⤙ | Unspecified Control Signal |

## Figure 10.24

High and low limits and selectors.

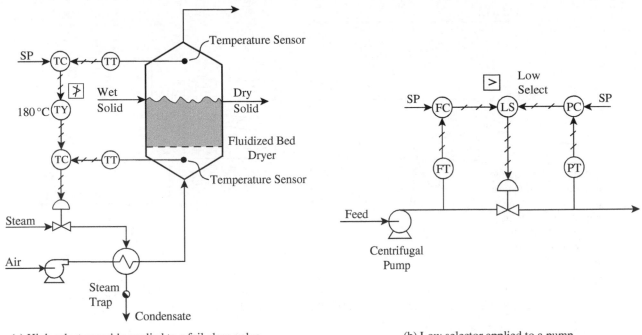

(a) High select override applied to a fail close valve.    (b) Low selector applied to a pump.

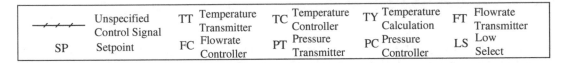

| | | | Temperature | | Temperature | | Temperature | | Flowrate |
|---|---|---|---|---|---|---|---|---|---|
| ⌁⌁⌁ | Unspecified Control Signal | TT | Temperature Transmitter | TC | Temperature Controller | TY | Temperature Calculation | FT | Flowrate Transmitter |
| SP | Setpoint | FC | Flowrate Controller | PT | Pressure Transmitter | PC | Pressure Controller | LS | Low Select |

## Figure 10.25

High and low select overrides for a pumping system.

but with a high limit included. Cascade control systems can in some cases demand a setpoint from the primary controller that is unacceptable for the process materials or equipment or is unsafe to the secondary controller to attain. The limit is set between the controllers. In this example the temperature limit is set to be below 180 °C.

An example of a low select is shown in Figure 10.25b. This shows a centrifugal pump with the outlet flowrate controlled. However, in this example the downstream process has a pressure limitation that must not be exceeded. Thus, pressure control has been implemented downstream of the control valve. In Figure 10.25b the control signals from both the flowrate and pressure controller are sent to a low select override. The lower signal is passed to the valve actuator.

## 10.11  Feedforward Control

The principle of feedforward control is illustrated in Figure 10.26. Feedforward control uses a sensor to measure a disturbance before it affects the controlled variable and passes a signal to the feedforward controller to create an immediate corrective action. This allows a control action to be initiated before the disturbance has affected the process. The controlled variable in Figure 10.26 is not used by the control system, otherwise this would constitute feedback control. Thus, in a feedforward system, the manipulated

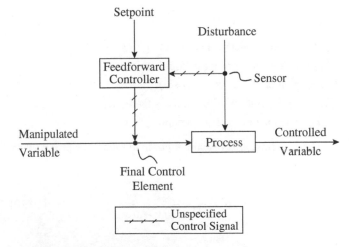

## Figure 10.26

Feedforward control.

variable adjustment is not error-based. Instead it is based on knowledge of the process from a model of the process.

As an example, Figure 10.27 shows a feedforward control scheme for the same operation shown in Figure 10.17 where feedback control was used. In Figure 10.27 a disturbance

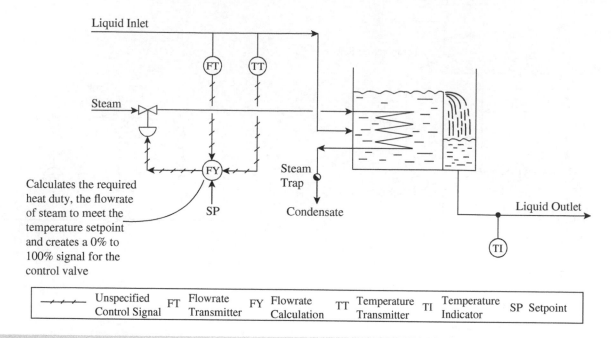

**Figure 10.27**

A pure feedforward control arrangement to heat a liquid tank.

entering the process caused by change of inlet flowrate or temperature is measured and the signals passed, along with the temperature setpoint for the outlet liquid, to a function that calculates the required steam flowrate. The process heat duty is calculated from

$$Q_{PROC} = m_{PROC} C_P (T_{SP} - T_{FEED}) \qquad (10.19)$$

where

$C_P$ = specific heat capacity (kJ kg$^{-1}$ K$^{-1}$)

$m_{PROC}$ = flowrate of process liquid (kg s$^{-1}$)

$Q_{PROC}$ = process heat duty (kJ s$^{-1}$)

$T_{FEED}$ = temperature of the process feed (°C)

$T_{SP}$ = temperature of the setpoint (°C)

Then the flowrate of steam required to satisfy the process duty is calculated from:

$$m_{STEAM} = \frac{Q_{PROC}}{(H_{STEAM} - H_{COND})} \qquad (10.20)$$

where

$H_{COND}$ = specific enthalpy of the saturated steam condensate (kJ kg$^{-1}$)

$H_{STEAM}$ = specific enthalpy of the steam (kJ kg$^{-1}$)

$m_{STEAM}$ = flowrate of steam (kg s$^{-1}$)

The calculated steam flowrate is transformed to a 0 to 100% output to the steam control valve. This provides the required steam to feed flowrate ratio for the temperature control. The feedforward scheme does not wait until the disturbance changes the process conditions and for this to be detected at the outlet before a control action is taken. Note in Figure 10.27 that the liquid outlet-controlled variable is not used by the control system.

Rather than the feedforward controller adjusting the steam control valve directly as shown in Figure 10.27, it can be used to provide the setpoint for a steam flowrate control loop, as shown in Figure 10.28. This has advantages if the steam supply itself is subject to disturbances. For example, if the steam supply pressure varies, this will cause the flowrate of steam to vary, even for the same control valve position. The arrangement in Figure 10.28 will compensate for this before such steam pressure disturbances affect the rest of the system.

Ideally, feedforward control should entirely eliminate the effect of the measured disturbance on the process output. However, to achieve perfect feedforward control requires an accurate model of the process, including accurate knowledge of process conditions that will affect the model (e.g. steam conditions), coupled with accurate measurements. If this is not possible it will result in an offset between the setpoint and the condition of the controlled variable. Because of this, feedforward control is often implemented in conjunction with feedback control. The feedforward controller acts on the major disturbance with feedback control acting on other factors that might cause the process variable to deviate from its setpoint.

Figure 10.29 shows an alternative feedforward control scheme implemented in conjunction with feedback control. The inlet flowrate and temperature are measured and the signals passed, along with the temperature setpoint, to a function to calculate the required steam flowrate. This is passed to another calculation function, which adjusts the 0 to 100% signal from the feedback temperature controller according to the calculated steam flowrate.

It should be noted that feedforward control schemes will still not provide perfect control, as there are still lags in the temperature measurement, control valve response, and so on.

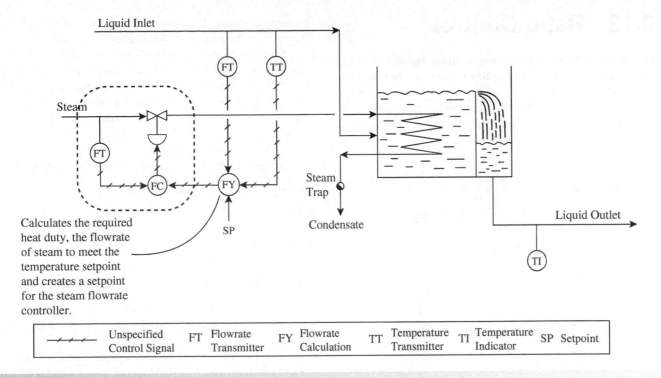

Calculates the required heat duty, the flowrate of steam to meet the temperature setpoint and creates a setpoint for the steam flowrate controller.

| | Unspecified Control Signal | FT | Flowrate Transmitter | FY | Flowrate Calculation | TT | Temperature Transmitter | TI | Temperature Indicator | SP | Setpoint |
|---|---|---|---|---|---|---|---|---|---|---|---|

## Figure 10.28

Rather than the feedforward controller controlling the steam control valve directly, it can provide a setpoint for a steam flowrate controller.

**10**

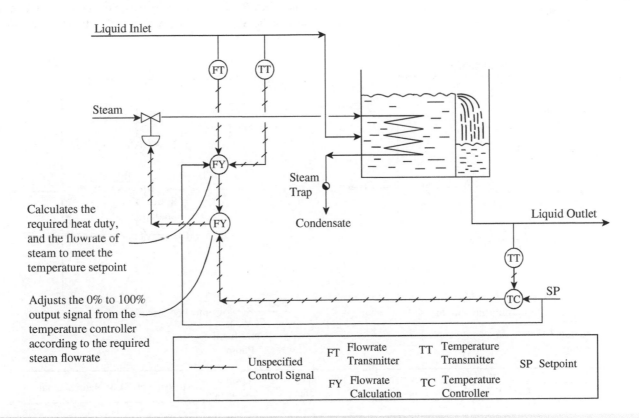

Calculates the required heat duty, and the flowrate of steam to meet the temperature setpoint

Adjusts the 0% to 100% output signal from the temperature controller according to the required steam flowrate

| | Unspecified Control Signal | FT | Flowrate Transmitter | TT | Temperature Transmitter | SP | Setpoint |
|---|---|---|---|---|---|---|---|
| | | FY | Flowrate Calculation | TC | Temperature Controller | | |

## Figure 10.29

A feedforward control arrangement including feedback features to heat a liquid tank.

# 10.12   Ratio Control

Ratio control systems are feedforward control systems in which one variable is controlled to be in a fixed ratio to another variable. Figure 10.30 shows a general *ratio control* system (Buckley 1964; Stephanopoulos 1984; Shinskey 1988; Smith 2009). In Figure 10.30 the *wild* (*disturbance*) stream has its flowrate fixed

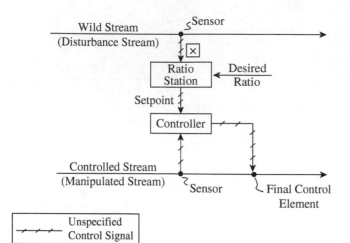

## Figure 10.30

A ratio control system.

elsewhere upstream. The flowrate of the wild stream is measured and sent to a *ratio station* or *signal multiplying relay*, which multiplies this value by the *desired ratio* and sends the signal as a setpoint to a controller on the *controlled* (*manipulated*) stream. This controller maintains the control of the controlled (manipulated) stream at the desired ratio between the two flowrates. It should be noted that the scheme in Figure 10.30 does not control the flowrate of either stream to an absolute value. It simply maintains the ratio of the flowrates between the two streams.

Examples of where ratio control might be used are to control:

- reflux ratio in a distillation column;
- ratio of two reactants entering a reactor;
- ratio of two blended streams (to maintain constant composition);
- ratio of purge stream flow to recycle stream flow;
- ratio of fuel to air flows in a burner;
- ratio of liquid flow to vapor flow in an absorber.

As an example, Figure 10.31a shows a mixing arrangement using two centrifugal pumps. Feed *A* and Feed *B* are to be mixed in a pipe junction such that Feed *B* is maintained at a fixed ratio *k* of the flowrate of Feed *A*. In this example, Feed *A* is the wild stream. The flowrate of the wild stream is measured and sent to a ratio station, where the flowrate is multiplied by the desired ratio for Feed *B*. This provides the setpoint to a feedback control loop controlling the flowrate of Feed *B*. Whilst this arrangement maintains Feed *B* to be a fixed ratio to the flowrate of Feed *A*, the flowrate of the wild stream (Feed *A*) would need to be fixed upstream

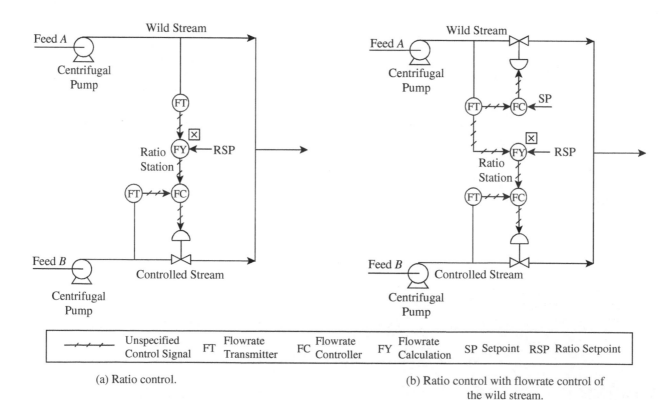

(a) Ratio control.

(b) Ratio control with flowrate control of the wild stream.

## Figure 10.31

Ratio control of a process mixing arrangement.

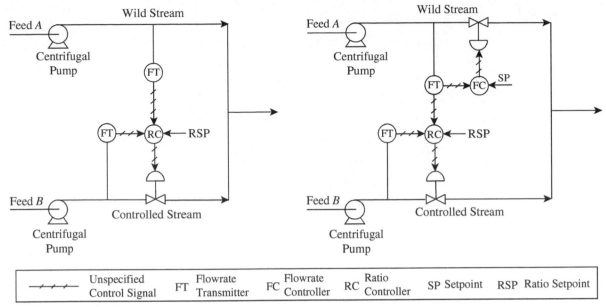

**Figure 10.32**
Ratio control of a process mixing arrangement.

of the mixing arrangement to control the absolute value of the flowrates. Figure 10.31b shows an alternative ratio control system in which the absolute value of the wild stream is controlled before maintaining Feed B to be in a fixed ratio. A ratio controller combines the ratio station and the flowrate controller for the controlled stream in a single unit. Figure 10.32a and b show the equivalent ratio control arrangements using ratio controllers to the ratio stations in Figure 10.31a and b.

Referring back to the control systems used to heat a liquid tank shown in Figures 10.17, 10.27, and 10.29, by contrast Figure 10.33 shows an alternative feedforward control system using a ratio control scheme. The ratio of the flowrates of liquid feed and steam is controlled by a ratio station. The liquid feed is the wild stream and the steam is the controlled stream. The setpoint for the ratio controller is set by the output from a temperature control on the tank outlet. If the liquid feed is subject to a change in flowrate, the ratio controller maintains the required ratio. Because the disturbance has not reached the outlet temperature sensor, the setpoint for the ratio control is also unchanged. Thus, the disturbance is corrected before the tank contents are affected. Again, it should be noted that, whilst feedforward control systems can perform significantly better than feedback systems, they are still not perfect.

## 10.13   Computer Control Systems

Process control is most often implemented using a computer or microprocessor as digital control (Shinskey 1988). The controller would thus be a control algorithm in a computer-controlled system, rather than a separate device. Computer control can be implemented in a number of different ways.

**10**

1) *Direct digital control.* Direct digital control is illustrated in Figure 10.34a. In this arrangement, measurement signals from the process are sent directly to the computer. The computer calculates the control signal from the setpoint and the *PID* control equations, and outputs control signals directly to the final control element. Analog signals coming from the process to the computer need an analog-to-digital (A/D) converter. Commands issued by the computer to the process are binary and need a digital-to-analog (D/A) converter. Figure 10.34a illustrates computer control of a single control loop. In practice there will be many control loops. The many different signals coming from the process are channeled through an electronic *multiplexer*. This is effectively a switch that rotates between the various signals from the process, opening for long enough for the data for each process variable to be digitized and passed to the computer for processing. In this way, the many different signals coming from the process are reduced to a single channel to the computer. Signals coming from the computer to the process go through the reverse arrangement after being converted from digital to analog. A single channel from the computer goes to a *demultiplexer* to channel the appropriate signal to the correct final control element. The operator can change the various setpoints and monitor the process conditions via a control console linked to the computer. Such direct digital control systems have largely been superseded by supervisory computer control.

2) *Supervisory computer control.* Supervisory computer control is illustrated in Figure 10.34b. In this arrangement the process

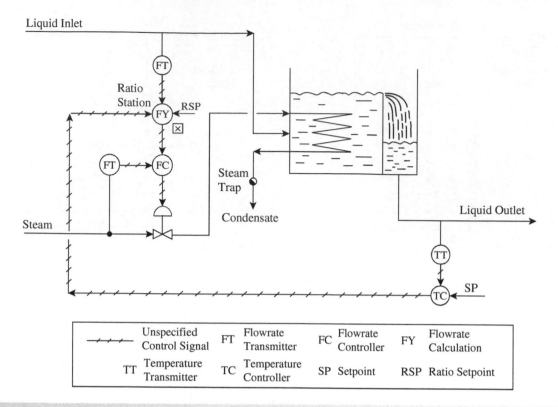

**Figure 10.33**

A ratio control arrangement providing feedforward control to heat a liquid tank.

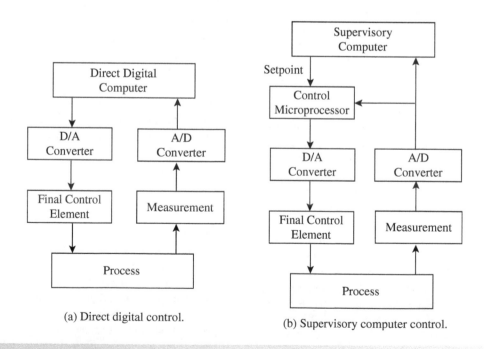

(a) Direct digital control.

(b) Supervisory computer control.

**Figure 10.34**

Computer control systems.

is controlled by a local controller (most often a microprocessor). The main computer now acts in a supervisory role in which it monitors the process variables with the capability of remotely changing the setpoint of the controller. In practice there will be multiple control loops that will require multiplexing devices to channel the signals. Supervisory control creates a buffer between the computer and the process. This allows problematic loops to be disabled and operated manually.

3) *Distributed control system (DCS).* A *distributed control system* (*DCS* extends the concept of supervisory control and involves the placement of controllers throughout a process or site. The controllers (usually microprocessors) are distributed and located close to where they are used, rather than in a *central control room*. Local controllers operate a loop involving its own sensors and final control elements. Each controller can be remotely controlled at a central console. Controllers are networked by a *data highway* (*fieldbus*) to a central console. *Redundancy* (spare equipment) is built into the system for reliability. Plant operations have a centralized location, to allow:

- process control;
- process monitoring;
- recording trends;
- *alarm management*;
- reporting of individual plant components;
- *historian* to record historical information for future reference and analysis.

Figure 10.35 shows the features of a typical DCS. In practice, the architecture of the system will depend on the size and complexity of the facility being controlled and the objectives of the site management. Figure 10.35 shows two processes being controlled by a supervisory computer with the controllers distributed as microprocessors. Process data are routed along a *process data highway* between the control microprocessors, supervisory computer, process console, and data mass storage (historian). Figure 10.35 also shows an *enterprise data highway*, enterprise computer and enterprise console a level up in the hierarchy. This level in the hierarchy has the objective of managing and optimizing the enterprise, rather than controlling individual process variables. Depending on the process control and managerial objectives for the system, there might be several layers of hierarchy above the process control layer. Figure 10.35 simply shows an example of what might be used. The DCS approach allows localized operational supervision and lower exposure to component or subsystem failure by incorporating redundancy, while offering global access to information and control capabilities.

## 10.14 Digital Control

For digital control in a computer-controlled system, the *PID* control algorithms discussed earlier would need to be based on difference equations, rather than differential equations. The *PID* Eq. (10.18) at time $t$ becomes (Shinskey 1988):

$$CO_k = CO_{BIAS} + K_P \left[ e_k + \frac{\Delta t}{\tau_I} \sum_{j=0}^{j=k} e_j - \tau_D \frac{(PV_k - PV_{k-1})}{\Delta t} \right]$$

(10.21)

**10**

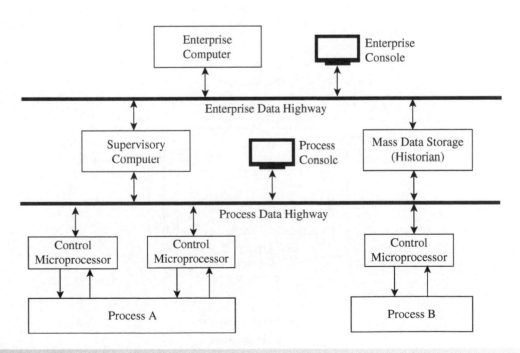

**Figure 10.35**

A typical distributed control system (DCS).

where

$CO_k$ = controller output at the present time $t$

$k$ = interval corresponding with the present time $t$

$CO_{BIAS}$ = controller bias

$e_k$ = error at the present time $t$

$\Delta t$ = sampling time interval

$\tau_I$ = integral time

$\tau_D$ = derivative time

$PV_k$ = process (measured) variable at the present time $t$

$PV_{k-1}$ = process (measured) variable at time $t - \Delta t$

Writing Eq. (10.18) at time $t - \Delta t$ becomes:

$$CO_{k-1} = CO_{BIAS} + K_P\left[e_{k-1} + \frac{\Delta t}{\tau_I}\sum_{j=0}^{j=k-1} e_j - \tau_D \frac{(PV_{k-1} - PV_{k-2})}{\Delta t}\right]$$

$$(10.22)$$

where

$CO_{k-1}$ = controller output at time $t - \Delta t$

$e_{k-1}$ = error at time $t - \Delta t$

$PV_{k-1}$ = process (measured) variable at time $t - \Delta t$

$PV_{k-2}$ = process (measured) variable at time $t - 2\Delta t$

The change in the controller output between $t$ and $t - 2\Delta t$ is given by the difference between Eqs. (10.21) and (10.22):

$$\Delta CO = CO_k - CO_{k-1}$$

$$= K_P\left[(e_k - e_{k-1}) + \frac{\Delta t}{\tau_I}e_k - \tau_D\frac{(PV_k - 2PV_{k-1} + PV_{k-2})}{\Delta t}\right]$$

$$(10.23)$$

Whilst digital control has many advantages (e.g. logic capabilities, ability to optimize, and so on), a disadvantage is introduced relative to analog devices by the need for *sampling*. Sampling at fixed time intervals creates a phase lag, which is not necessarily a feature of analog devices. Successful application of Eq. (10.23) requires an appropriate sampling interval $\Delta t$ to be chosen. If $\Delta t$ is too long then it will create a lag in the system. If $\Delta t$ is too short then it can prevent the integral term from adequately removing offset from the control (Shinskey 1988). However, the sample interval should be short relative to the response of the control loop. In addition, signal noise and process noise create random behavior in the process variable measurements. If a derivative action is used in the controller, such noise in the measured process variable can be amplified to create excessive movement in the controller output. Filtering in the computer software and controller tuning can be used to mitigate the problem (Shinskey 1988). Most often, the application of Eq. (10.23) involves only the proportional and integral control terms.

*PID* controllers using single-input–single-output (SISO) control act independently and do not account for interactions between the various controllers. Interactions between the controllers might is some cases carry serious consequences. Perhaps the major advantage of digital control is the ability to use a more comprehensive model, including all of the control loops and the process, and to solve the whole system simultaneously. Such a model allows for interactions by solving all of the control actions simultaneously in conjunction with a process model. This allows multi-input-multi-output (MIMO) control, as illustrated in Figure 10.36. MIMO systems have more than one input and more than one output and involve multiple control loops. MIMO control systems can be put into practice using *model predictive control* (MPC). One disadvantage of MIMO (MPC) systems is that they may not be able to operate without all of the sensors

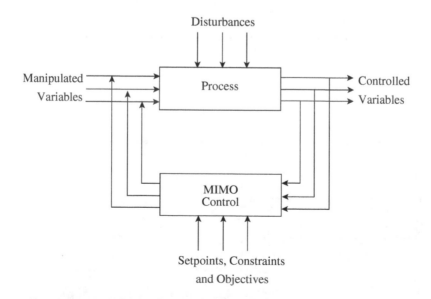

**Figure 10.36**

Multiple-input–multiple-output (MIMO) control.

and final control elements functioning because of component failure or maintenance requirements. SISO systems do not necessarily suffer from this problem and can still often function overall if part of the system is not functioning. A common approach to mitigate this problem with MIMO (MPC) systems is to use conventional controllers to control flowrates and levels, with the setpoints for these cascaded from the higher-level MIMO (MPC) system.

Model predictive control uses a *dynamic model* of the process to control the process. The resulting MIMO system can take into account the interactions between the controller actions. Optimization of the system on-line allows optimal control actions to drive the control action to the target for the manipulated variables. Controller performance is enhanced by using the model to predict future behavior.

The dynamic models used in MPC can be:

1) Physical or empirical (but are often empirical).
2) Linear or nonlinear (but are often linear).

The underlying models for processes are most often nonlinear in nature. However, linear models can be used to approximate the nonlinear behavior of the process over a limited range. Nonlinear models can also be used. These might either have a physical basis or be empirical in nature and based on fitting of data (e.g. fitting an artificial neural network to the data). A common way to develop a model is to carry out a series of step changes to a process variable, measuring how the process responds, and fitting a linear empirical model to the response.

The dynamic model is used to determine optimal control actions based on future predictions of the measured variables, together with current measurements. Figure 10.37 shows the basic structure of model predictive control. Operator personnel provide input objectives and constraints into the model predictive control. Inequality and equality constraints are included in the control calculations, along with the required control objectives. An optimizer is used to minimize the deviations between predicted future control actions and the required trajectory, e.g. to minimize the sum of the squares of the deviations between future predictions of the controlled variables and the required trajectories to the targets of the controlled variables. The optimizer interacts with the dynamic model to create the optimal control actions for the process. The calculated manipulated variables are adjusted by

changing the setpoints for lower-level control loops in supervisory control (although direct control of the final control elements could in principle be used).

The objectives of model predictive control are typically:

- Prevent violations of controller input and output constraints.
- Drive some manipulated variables to their optimal setpoints, while maintaining others within specified ranges.
- Prevent excessive sudden changes to the controller input variables.
- If a sensor or actuator is not available, control as much of the process as possible.

Figure 10.38 shows a typical sampling profile of a model predictive control system. The current time in Figure 10.38 is sampling time $k$. The process is sampled at constant time intervals. Predictions of the measured variables are made for more than one time interval ahead. The number of predictions $P$ ahead at each time interval creates the *prediction horizon*. Future control moves are also calculated at the current time. The number of control moves $M$ ahead at each time interval creates the *control horizon*. Even though $M$ control moves are calculated at sampling time $k$, only the first move is applied. At the next sample point new measurements become available, a new sequence of predictions and control moves is calculated. Again, only the first control move is implemented. At each time step the prediction horizon keeps moving forward. This is known as *receding horizon control*. The principle advantages of MPC include:

- the ability to control large MIMO systems;
- determine optimal control actions, accounting for constraints.

Certain control applications might demand intensive use of logical control actions. Such digital control systems will normally be part of a DCS, as illustrated in Figure 10.35. *Programmable logic control* uses digital control to monitor inputs and outputs and make *logic-based* decisions for process control. It is used for:

- on–off control;
- batch process control;
- timing (e.g. timing of batch operation);
- counting (e.g. production of pharmaceutical or consumer product packages);

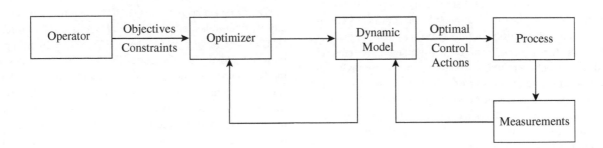

**Figure 10.37**
Model predictive control (MPC).

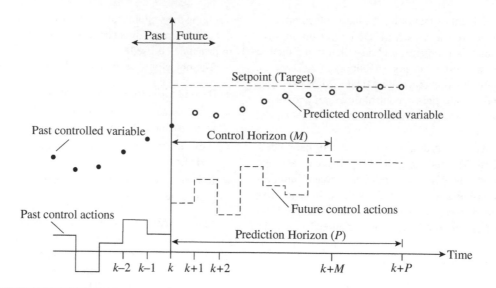

**Figure 10.38**

Model predictive control sampling and horizons.

- sequencing operations, e.g. to control the correct sequence in start-up and shut-down;
- *interlocks* to make functions or operations mutually dependent to prevent undesirable states. It is sometimes necessary to follow a fixed sequence of operations. This might be necessary to control the correct sequence in start-up and shut-down or in batch operations. If this is the case, then interlocks can be included in the control system to prevent operators departing from the required sequence. This can be achieved, for example, by including programmable logic control as part of the DCS.

It is important to distinguish between interlocks that simply avoid undesirable operating practice and interlocks that avoid safety problems. The interlocks to avoid safety problems will normally be part of a *safety instrumented system* (*SIS*) to be considered next.

## 10.15   Safety Instrumented Systems

*Safety instrumented systems* (*SISs*) are strictly not part of the control system. However, they work alongside the control system and in some cases in conjunction with the control system, to maintain the safety of the process. Safety instrumented systems are designed to prevent or mitigate hazardous events by taking the process to a safe state as a result of violations in predetermined conditions (usually at or near safe operating limits). Safety instrumented systems are comprised of a collection of *safety instrumented functions* (*SIFs*) that use a combination of sensors, logic solvers, and final control elements:

1) *Sensors*. Sensors collect data (e.g. temperature, pressure, flow-rate). The sensors used in SIFs are not necessarily of the same type used in process control. High accuracy might be desirable in process control, whereas the sensors used in an SIF need to be more reliable. Separate sensors are most often used for the control and SIF. If the same sensor is used for both, then a high reliability of the sensor must be guaranteed.

2) *Logic solvers*. Logic solvers determine the action to be taken, given the data gathered.

3) *Final control element*. Logic solvers execute pre-programmed actions with the final control element, which is not part of the control system.

The result should be safe *isolation* of the hazardous part of the process. A SIS may also trigger the process control system to move its control valves to a safe state. Separation of the SIS from the control systems reduces the probability that both the control system and a SIS are unavailable at the same time. Also, separation ensures that changes in the control system do not affect the functionality of the SIS. Reliability of SIS systems is paramount. Duplicate or even triplicate components might be used to reduce the risk of failure of the system. Voting systems can be incorporated (e.g. two-out-of-three voting), as discussed for $k$-out-of-$n$ systems in Chapter 8.

In practice, safety instrumented systems are added from a *Layer of Protection Analysis* (*LOPA*) to mitigate potentially hazardous situations identified by the *Hazard and Operability Study* (*HAZOP*) (Crawley and Tyler 2015). Safety instrumented systems are added after the control system has been designed and the P&ID has been developed. It must be emphasized again that safety instrumented systems should be independent of the control system. Safety instrumented systems stay in the background until an abnormal situation requires some action. A failure in the safety instrumented system might not be detected until either there is a demand on the system, in which case a hazardous situation might occur, or the failure is detected in a proof test. The safety instrumented system therefore needs to be tested regularly (see Chapter 15).

# 10.16 Alarms and Trips

*Alarms* are used to alert process operators to abnormal operating situations at the limits of the normal operating envelope. This would be typically high or low levels of process settings, e.g. high or low liquid level in a vessel or an inconsistency in a process parameter. An alarm normally has both visual and sound annunciators. Therefore, if an alarm arises, a light will flash on a screen or alarm panel, accompanied by a horn that sounds intermittently. No action is automated from an alarm, but an operator must decide what action to take. Operator intervention is required to:

- prevent hazards (harm to people, assets, and environment);
- maintain equipment integrity;
- maintain product quality.

Too many alarms can cause confusion if there is a serious problem. Thus, *alarm management* is required to prioritize key alarms to prevent operators being overwhelmed if there is a serious incident. Alarm management involves the integration of instrumentation and ergonomic human factors to increase the usability of an alarm system. Shortcomings of the alarm system can cause a multitude of avoidable incidents, creating unnecessary risks to people, environment, equipment, and can potentially increase operating costs. *Trips* have the function to shut down the system in an orderly fashion when a malfunction occurs. Alternatively, a trip might switch a failed unit to a standby unit. A trip can be part of the control system and initiated as a result of problems with product quality, and other production issues. Alternatively, the trip can be initiated by a safety instrumented system to avoid a hazardous situation occurring.

In principle, there can be three types of alarms:

1) *Process alarms. Process alarms* indicate problems with operation of the plant that might lead to loss of product quality or quantity, process inefficiency or equipment problems. These are normally incorporated into the process control system. They will share the same sensors as the process control system.

2) *Safety related alarms. Safety-related alarms* as part of the control system alert operators to approaching dangerous conditions and should have a high priority. High reliability is necessary, by possibly avoiding shared components or including multiple components and voting systems, to avoid failure of the system or false alarms. Safety-related alarms will often be a warning that action will be taken by an SIS if the condition becomes more severe.

3) *Shutdown alarms. Shutdown alarms* alert operators that an automatic shutdown event has been initiated by a safety instrumented system. This is a monitoring function for a safety instrumented system.

# 10.17 Representation of Control Systems

The conceptual design of process control systems requires the identification of which variables need to be measured and indicated only for process monitoring and which need to be measured and controlled. If control is required, then for conceptual design the measured and manipulated variables need to be initially connected using a single-input–single-output (SISO) feedback control. Later where appropriate, multivariable regulatory control can be applied with single loops cascaded, or ratioed, or split range or feedforward control applied, or MIMO control applied in a model predictive control system.

Figure 10.39 shows the abbreviations used for the representation of control system functions (American National Standard 2022). The three letters within the circle in Figure 10.39 specify the function. The first letter indicates the variable to be measured (e.g. pressure). The following letters designate the function to be

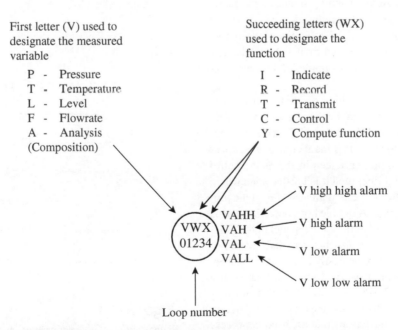

**Figure 10.39**
Abbreviations used for instrumentation and control functions.

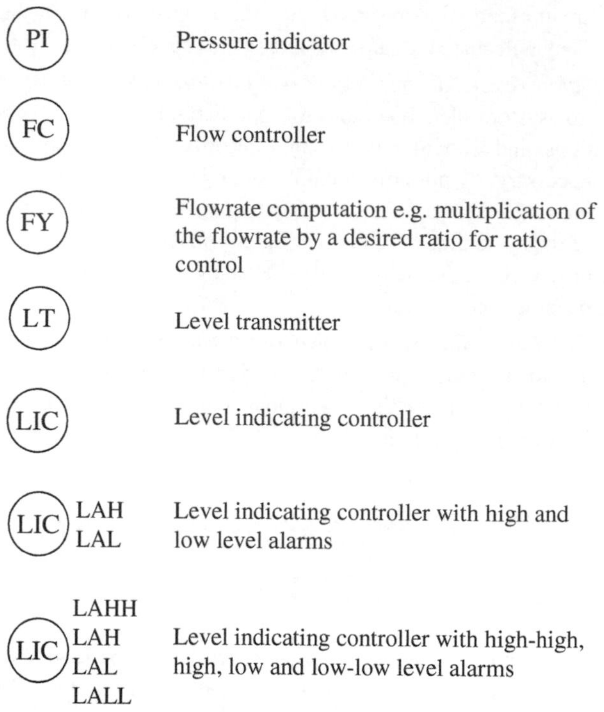

PI    Pressure indicator

FC    Flow controller

FY    Flowrate computation e.g. multiplication of the flowrate by a desired ratio for ratio control

LT    Level transmitter

LIC    Level indicating controller

LIC LAH LAL    Level indicating controller with high and low level alarms

LIC LAHH LAH LAL LALL    Level indicating controller with high-high, high, low and low-low level alarms

**Figure 10.40**

Some examples of abbreviations for instrumentation and control functions.

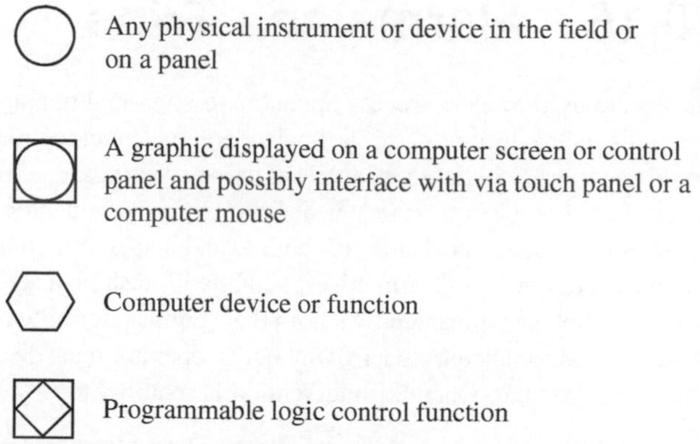

applied to the measurement (e.g. indicate). Alarms can also be added to the function where necessary. These would normally alarm at high or low values of the measured variable, or both. In addition to alarms for high and low levels, alarms can also be high–high and low–low. High–high and low–low conditions are often connected to a shut-down or some other major process change. For example, the high–high setting might be the maximum level before tripping the supply pump to a tank to prevent overflow. The low–low setting might be the minimum level before tripping the pump draining a tank to avoid cavitation and damage to the pump. Figure 10.39 shows some examples of different combinations of functions.

The representation of different kinds of control devices is illustrated in Figure 10.41. Horizontal lines are used to represent the location of the device. A single horizontal line across the representation means that it is located on a main control panel near the control room or some computer screen in the main control room. If there is no horizontal line, then it is located somewhere in the field, probably close to the general area shown on the piping and instrumentation diagram. A double horizontal line means it is located at some secondary (satellite) local panel in the field. A single dashed horizontal line means that it is inaccessible or not generally located where it can be easily accessed or viewed. It may also be used for hidden or password protected areas of a control system. Figure 10.42 shows some examples of commonly used devices. Figure 10.43 illustrates the representation of the different signals and connections that are used in process control.

Any physical instrument or device in the field or on a panel

A graphic displayed on a computer screen or control panel and possibly interface with via touch panel or a computer mouse

Computer device or function

Programmable logic control function

**Figure 10.41**

Representation of different kinds of control devices.

| | Field or Locally Mounted | Located at Central Console |
|---|---|---|
| Discrete device | ○ | ⊖ |
| Function in distributed control system | ⊡ | ⊟ |
| Computer device or function | ⬡ | ⬡ |
| Programmable logic device or function | ◈ | ◈ |

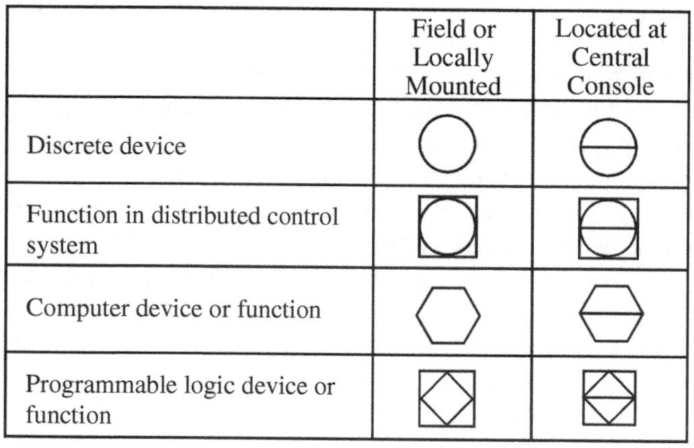

**Figure 10.42**

Examples of different kinds and locations of control devices.

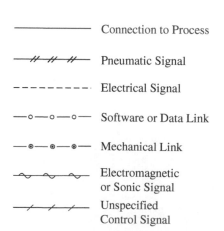

—————— Connection to Process

—#—#—#— Pneumatic Signal

- - - - - - - - - Electrical Signal

—○—○—○— Software or Data Link

—◉—◉—◉— Mechanical Link

〜〜〜 Electromagnetic or Sonic Signal

—/—/—/— Unspecified Control Signal

**Figure 10.43**

Representation of different types of signals used in control.

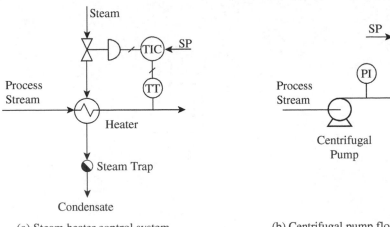

(a) Steam heater control system.          (b) Centrifugal pump flowrate control system.

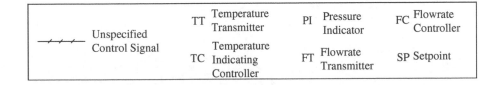

## Figure 10.44

Examples of conceptual process control schemes.

Figure 10.44 shows two examples of representation of conceptual control systems. Figure 10.44a shows temperature control of a process stream being heated by a steam heater. Figure 10.44b shows a flowrate control system for a centrifugal pump. The design shows a pressure indicator on the discharge of the pump in addition to the flowrate control. This is common practice with centrifugal pumps, which allows the operators to see locally whether the pump is operating properly. The conceptual designs in Figure 10.44 require further detail to be added for completion.

**Example 10.3**    Figure 10.45 shows the feed system for a steam reformer reactor. Methane and steam are reacted over a solid catalyst to produce a gas that is mainly hydrogen and carbon monoxide. The ratio of methane to steam needs to be controlled to be a fixed value. The flowrate of the product is to be controlled by controlling the inlet flowrate of methane.

a)  Create a conceptual design for the feed system to maintain a fixed ratio of the two reactants with the product flowrate being controlled by the feed flowrate of methane.

b)  Suggest a fail-safe condition for the two feed control valves.

c)  If the feed control system is to be part of a DCS system and the control valves are to be pneumatically actuated, modify the conceptual design accordingly.

## Solution

a)  A ratio control system is needed to maintain the feed flowrates to be in a fixed ratio. In terms of ratio control from Section 10.12, the methane is the "wild" stream, which needs flow control with a setpoint to maintain an absolute value. The steam is the "controlled stream." The conceptual design is shown in Figure 10.46. The flow transmitter for the methane and ratio station provide the setpoint for the steam flowrate controller. The signals in the conceptual control system are all left undefined.

b)  Where possible, the fail-safe condition of the control valves should be specified. To assess the most appropriate fail-safe condition requires the overall system to be considered. If the methane control valve failed, then a fail open position would allow large flowrates of methane, which is flammable, to enter the process in an uncontrolled fashion. On the other hand, to fail open for the steam control valve would allow large flowrates of steam, which is inert except under reaction conditions, to enter the process. If the steam control valve was to fail closed and the methane feed kept operating, then pure methane would be fed into the process. Taking the two feed control valves together, the safest combination seems to be to specify the methane control valve to be fail closed and the steam control valve to be fail open. However, it would be necessary to ensure in the process design that a large flowrate of steam under pressure in itself did not lead to other hazards.

c)  Figure 10.47 shows the conceptual design modified to be part of a DCS system. Electrical signals are passed from the flowrate transmitters to the controllers. The ratio station and the controllers would be digitally controlled within the computer. The connection between the ratio station and the flowrate controller for the steam would be data transferred within the computer. Given that the control valves are specified to have pneumatic actuators, the electrical signals from the controllers need to be converted from current to pneumatic signals. Also, because digital control is being used, analog to digital and digital to analog converters would be required. However, these are not shown in the control design to simplify in the representation.

**10**

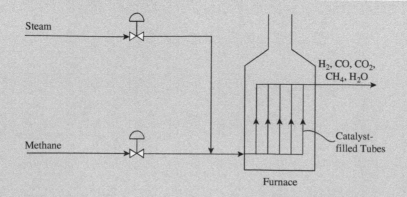

## Figure 10.45

Feed system for a steam reformer to react steam and methane over a catalyst to produce hydrogen.

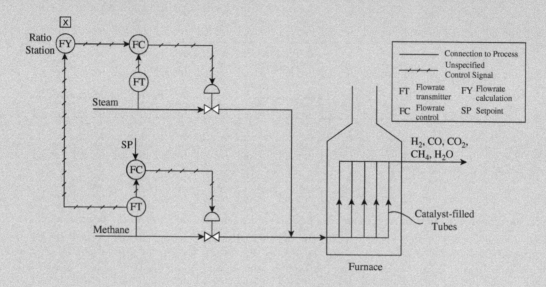

## Figure 10.46

Conceptual design of feed control to a steam reformer.

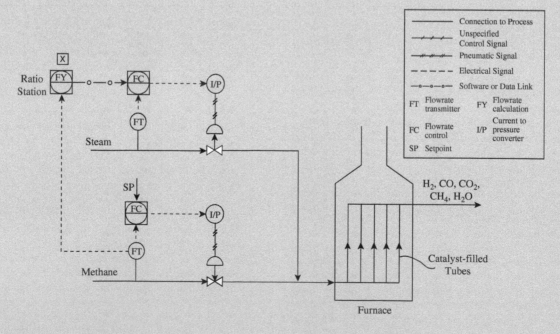

## Figure 10.47

Design of feed control to a steam reformer.

# 10.18 Process Control Concepts – Summary

Process control objectives require production targets to be achieved in a stable operation, while keeping within safe operating limits. A basic control loop involves measurement, signals, a controller, and a final control element (if necessary manipulated by an actuator). There is a hierarchy for process control strategies that increases in sophistication:

- open loop – no feedback, and control implemented manually by a process operator;
- closed-loop feedback control – single-input–single-output (SISO);
- multivariable regulatory control – single loops cascaded, or ratioed, or split range;
- feedforward control – requires a process model;
- model predictive control – MIMO dynamic model capable of optimizing the operation.

Distributed control systems use a network of computers/microprocessors for control.

Safety instrumented systems are added later to mitigate potential hazards, but are independent of the basic process control system.

Alarms are used to alert the process operators to abnormal operating conditions. These can be classified as process, safety-related, and shut-down alarms.

# Exercise

1. Figure 10.48 shows the feed system for an ammonia synthesis reactor. Hydrogen and nitrogen are reacted to produce ammonia. The ratio of hydrogen to nitrogen needs to be controlled to the stoichiometric ratio. The flowrate of the ammonia product is to be controlled by controlling the inlet flowrate of hydrogen.

   a) Create a conceptual design for the feed system to maintain the stoichiometric ratio of the two reactants with product ammonia flowrate being controlled by the feed flowrate of hydrogen.

   b) Suggest a fail-safe condition for the two feed control valves.

   c) If the feed control system is to be part of a DCS system and the control valves are to be pneumatically actuated, modify the conceptual design accordingly.

**10**

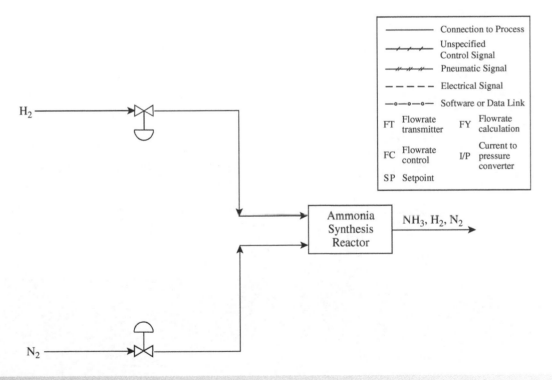

**Figure 10.48**

Feed arrangement for an ammonia synthesis reactor.

# References

American National Standard ANSI/ISA-5.1-2022 (2022). *Instrumentation Symbols and Identification*.

Buckley, P.S. (1964). *Techniques of Process Control*. Wiley.

Crawley, F. and Tyler, B. (2015). *HAZOP: Guide to Best Practice*, 3e. Elsevier Ltd.

Shinskey, F.G. (1988). *Process Control Systems*, 3e. McGraw-Hill Inc.

Smith, C.L. (2009). *Practical Process Control: Tuning and Troubleshooting*. Wiley.

Smith, R. (2016). *Chemical Process Design and Integration*, 2e. Wiley.

Stephanopoulos, G. (1984). *Chemical Process Control: An Introduction to Theory and Practice*. Prentice-Hall.

# Chapter 11

# Process Control – Flowrate and Inventory Control

## 11.1 Flowrate Control

Measuring and controlling flowrate is central to the control of individual operations and overall processes. The most common flowrate to be measured in the process industries is in pipe-flow. Figure 11.1a shows a simple flowrate control loop. The flowrate measurement and controller send a signal to the control valve, which alters its opening to change the pressure drop across the valve and changes the restriction to the flow, thereby changing the flowrate. The flow meters required to measure the flowrate in pipes most often require a well-developed velocity profile across the pipe upstream of the flow meter for accurate measurement. Throttling a control valve causes the velocity profile downstream of the valve to become significantly distorted. Thus, flow meter performance is adversely affected by locating the control valve (and its distorted velocity profile) in a control loop upstream of a flow meter, as illustrated in Figure 11.1a. Hence, locating the control valve downstream of the flow meter, as shown in Figure 11.1b, keeps its distorting effect on the velocity profile downstream of the flow meter. It is normal to specify a minimum length of straight pipe upstream of a flow meter to ensure a reliable reading.

## 11.2 Inventory Control of Individual Operations

Measuring flowrate is also key to control of the *process inventory*. Process inventory measures the *mass of material held up in the process*. Any accumulation of material (positive or negative, total or component) creates a change in the inventory. Control of inventory is a key element in most process control systems.

A material balance for total mass can be written for each operation in a process and the total process, quantified by Eq. (11.1):

$$[\text{Total Mass Out}] = [\text{Total Mass In}] - [\text{Accumulation of Mass}] \quad (11.1)$$

The accumulation of total mass in Eq. (11.1) is the change in inventory. Thus Eq. (11.1) can be written as:

$$\begin{bmatrix} \text{Change of Inventory} \\ \text{for Total Mass} \end{bmatrix} = [\text{Total Mass In}] - [\text{Total Mass Out}] \quad (11.2)$$

For an individual chemical component, material can be generated or consumed by chemical reaction. Thus for each Chemical Component $i$:

$$[\text{Mass Out}]_i = [\text{Mass In}]_i + [\text{Generation}]_i - [\text{Consumption}]_i - [\text{Accumulation}]_i \quad (11.3)$$

The change of inventory for Component $i$ can then be written as:

$$\begin{bmatrix} \text{Change of} \\ \text{Component Inventory} \end{bmatrix}_i = [\text{Mass In}]_i + [\text{Generation}]_i - [\text{Mass Out}]_i - [\text{Consumption}]_i \quad (11.4)$$

To maintain the material balance in an operation, or the overall process, requires the inventory to be kept constant, which means that the change in inventory defined by Eqs (11.2) and (11.4) should be kept as close as possible to zero. In practice, for continuous processes the inventory needs to be regulated to be within acceptable bounds. Inventory control requires inventories of total, component, and phase masses to be maintained between acceptable bounds for both the individual operations in the process and the overall process. This is a key element of the control of both individual operations and overall process control. Once the basic

*Process Plant Design*, First Edition. Robin Smith.
© 2024 John Wiley & Sons Ltd. Published 2024 by John Wiley & Sons Ltd.
Companion website: www.wiley.com/go/processplantdesign

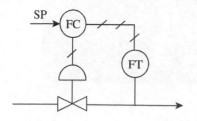
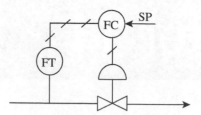

(a) Locating the control valve upstream of the flow meter disturbs flow to the flowmeter and the accuracy of its measurement.

(b) Locating the control valve downstream of the flow meter allows more accurate measurement of flowrate.

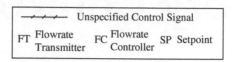

| | Unspecified Control Signal |
|---|---|
| FT Flowrate Transmitter | FC Flowrate Controller SP Setpoint |

## Figure 11.1

Location of the flowmeter relative to the control valve in a control loop.

control mechanisms are in place to maintain the inventory and the mass balance, the rate of change of inventory can then be studied.

A key issue is to understand the purpose of each inventory. If there is no clearly identifiable reason to have an inventory then it can become an unnecessary safety hazard if flammable or toxic. Inherently safer design requires the inventory of hazardous material to be minimized. For a reactor, maintaining the necessary inventory (and the resulting residence time) is critical to ensure the required reactor performance. Thus, the inventory will most often need to be controlled between close bounds. By contrast, some inventories are required simply to allow the process to absorb disturbances in flowrate. If the inventory is there purely

to absorb disturbances in flowrate then the inventory can in principle be allowed to vary significantly in the event of a disturbance. If an inventory is necessary to absorb disturbances in flowrate, and the inventory is set to be controlled between close bounds, then this defeats the object of having the inventory. Inventories should have a clear purpose when designing the control system.

The inventory of an operation must be controlled by only its inflows *or* outflows (Aske and Skogestad 2009). Figure 11.2 shows two operations involving the processing of an assumed incompressible liquid with no reaction or phase change in the operations, with each having a fixed volume. Figure 11.2a shows a closed operation with a fixed volume. Figure 11.2b shows a

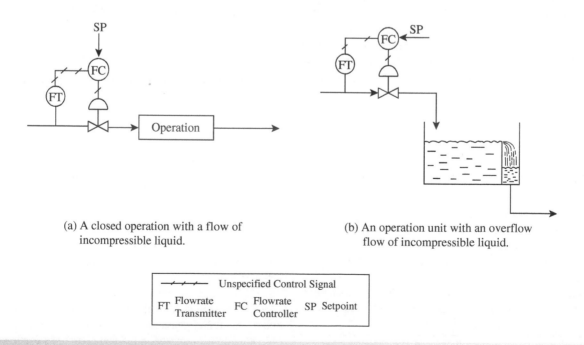

(a) A closed operation with a flow of incompressible liquid.

(b) An operation unit with an overflow flow of incompressible liquid.

| | Unspecified Control Signal |
|---|---|
| FT Flowrate Transmitter | FC Flowrate Controller SP Setpoint |

## Figure 11.2

Incompressible liquid flows in a fixed volume with no reaction or phase change.

vessel with an open surface, but the volume is maintained constant by a weir with an overflow. In both situations in Figure 11.2 the inventory is constant because the volume of liquid is fixed and incompressible. Thus, there is no need for inventory control and the system will *self-regulate*.

Figure 11.3a shows an operation that is processing a gas (or vapor) stream. Now, because the gas is compressible, the inventory *for a fixed volume* depends on the pressure and temperature. If the gas is assumed to follow ideal behavior, then for an ideal gas the inventory is given by:

$$N = \frac{PV}{RT} \tag{11.5}$$

where

$N$ = moles of gas (kmol)

$P$ = pressure (N m$^{-2}$)

$V$ = volume occupied by $N$ kmol of gas (m$^3$)

$R$ = gas constant (8314 N m kmol$^{-1}$ K$^{-1}$ or J kmol$^{-1}$ K$^{-1}$)

$T$ = absolute temperature (K)

For a fixed volume and temperature with no reaction or phase change, the inventory depends on the pressure. Thus, there is no inventory control for the arrangement in Figure 11.3a. The inventory is self-regulating and, for a fixed operation volume and temperature the inventory depends on the inlet pressure, pressure drop through the operation, and the back-pressure created downstream (Aske and Skogestad 2009). Figure 11.3b shows an arrangement attempting to control the inventory by flowrate control on the inlet

and the outlet. Even though it superficially appears to control the inventory by controlling the inlet and outlet, this does not provide inventory control, because both controllers must provide flowrate controls that are identically equal to maintain the inventory. In practice there will always be some difference between the two flowrate controls, even if it is very small. The inventory will change through time, over minutes, hours, days, or weeks. Figure 11.3c shows an arrangement with flowrate control at the inlet and pressure control at the outlet. However, the pressure transmitter has been located downstream of the pressure control valve and is therefore controlling the pressure of the downstream process, rather than the pressure within the operation. This arrangement does not provide any inventory control, as the output does not depend on the inventory. Finally, Figure 11.3d shows an arrangement with flowrate control on the inlet and pressure control on the outlet, with the pressure controller controlling the pressure within the operation. This now provides inventory control (Aske and Skogestad 2009). However, it should be noted that the inventory in Figure 11.3d might be subject to some change with changing flowrate if flowrate changes cause a significant change in the pressure drop through the operation.

Figure 11.4 shows two ways in which the inventory of an operation processing gaseous compressible material might be controlled. In Figure 11.4a the flowrate is controlled upstream at the inlet of the operation and the pressure within the operation controlled from the outlet. This is inventory control *in the direction of flow* (Buckley 1964). Alternatively, the flowrate control can be placed on the outlet and the pressure control placed at the inlet, as shown in Figure 11.4b. This is inventory control *in*

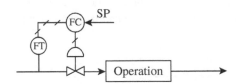

(a) No inventory control, self regulating.

(b) No inventory control, both controllers must be set identical.

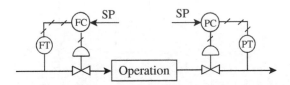

(c) No inventory control, because of the location of the pressure transmitter.

(d) Inventory control with the output depending on the inventory.

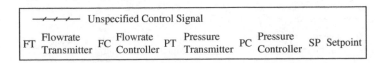

| | — Unspecified Control Signal |
|---|---|

| FT | Flowrate Transmitter | FC | Flowrate Controller | PT | Pressure Transmitter | PC | Pressure Controller | SP | Setpoint |

### Figure 11.3
Compressible gaseous flows in a fixed volume with no reaction or phase change.

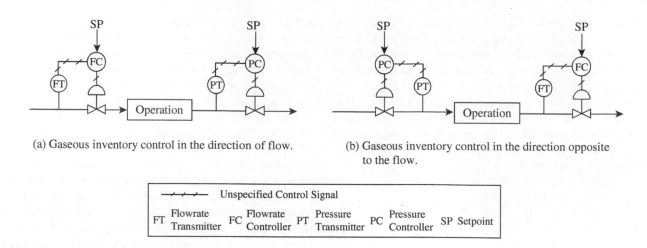

(a) Gaseous inventory control in the direction of flow.

(b) Gaseous inventory control in the direction opposite to the flow.

| | | | | | | | | |
|---|---|---|---|---|---|---|---|---|
| | | | Unspecified Control Signal | | | | | |
| FT | Flowrate Transmitter | FC | Flowrate Controller | PT | Pressure Transmitter | PC | Pressure Controller | SP Setpoint |

### Figure 11.4

Compressible gaseous inventory control in different directions with no reaction or phase change.

*the direction opposite to the flow* (Buckley 1964). Gaseous inventory can thus be controlled in the direction of flow or in the direction opposite to the flow.

Another method of gaseous inventory control was discussed in Chapter 10 using split range control. Figure 10.23 shows the arrangement in which the spilt flow control maintains a pressure of gas above a liquid at a setpoint. When the pressure decreases gas is introduced by one control valve. When the pressure increases, gas is discharged through another control valve.

Figure 11.5 shows inventory control of liquid in an operation in which the volume can vary as the free surface moves up and down. Figure 11.5a shows an arrangement in which there is flowrate control on both the inlet and the outlet. As with the gaseous inventory case in Figure 11.3b, this does not provide inventory control,

because both controllers must provide flowrate controls that are identically equal to maintain the inventory. In practice there will always be some difference. The mismatch, no matter how small, will cause the inventory to change. Alternatively, Figure 11.5b shows an arrangement in which the inlet flowrate is controlled and the outlet is controlled by a level controller for the operation. This now controls the liquid inventory.

Figure 11.6 contrasts two alternative ways in which the level can be controlled for a liquid being processed in an operation in which the volume can vary as the free surface moves up and down. Figure 11.6a shows level control controlling the outlet from the operation. This is *level control in the direction of flow*. Alternatively, the level control can be used to control the inlet, as shown in Figure 11.6b. This is *level control in the direction opposite to*

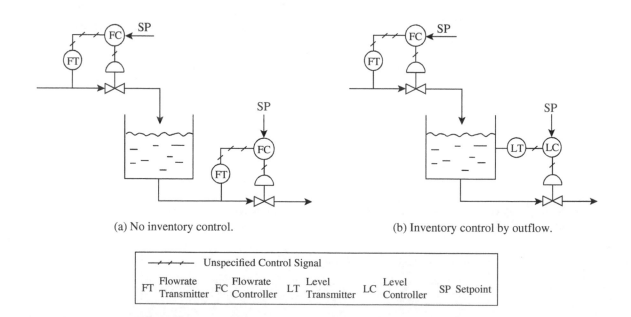

(a) No inventory control.

(b) Inventory control by outflow.

| | | | | | | | | |
|---|---|---|---|---|---|---|---|---|
| | | | Unspecified Control Signal | | | | | |
| FT | Flowrate Transmitter | FC | Flowrate Controller | LT | Level Transmitter | LC | Level Controller | SP Setpoint |

### Figure 11.5

Liquid inventory control with no reaction or phase change.

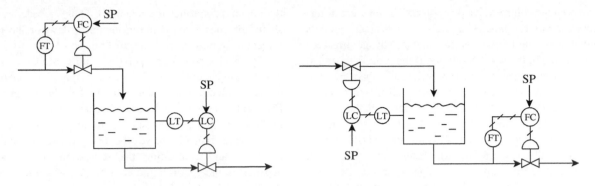

(a) Liquid inventory control in the direction of flow.

(b) Liquid inventory control in the in the direction opposite to the flow.

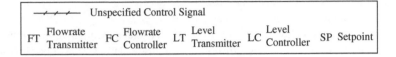

| | Unspecified Control Signal | | | | |
|---|---|---|---|---|---|
| FT Flowrate Transmitter | FC Flowrate Controller | LT Level Transmitter | LC Level Controller | SP Setpoint | |

## Figure 11.6

Direction of liquid inventory control.

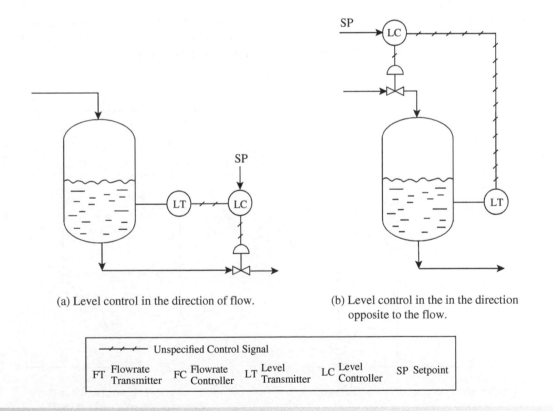

(a) Level control in the direction of flow.

(b) Level control in the in the direction opposite to the flow.

| | Unspecified Control Signal | | | | |
|---|---|---|---|---|---|
| FT Flowrate Transmitter | FC Flowrate Controller | LT Level Transmitter | LC Level Controller | SP Setpoint | |

## Figure 11.7

Direction of level control.

*the flow*. Control in the direction of the flow is more common, but the arrangement is not always possible (Buckley 1964).

Figure 11.6a can give adequate control of the level, as long as the downstream process is not subject to pressure variations.

Consider Figure 11.7a. If the downstream process pressure changes, then the pressure drop across the control valve will change, even if there is no change in the position of the control valve from the level controller. The flowrate through the valve depends on both the control valve position and the pressure difference across the valve. Thus, changing the downstream process pressure will cause the flowrate at the outlet to change, irrespective of any change in the level. Similarly, if the process upstream of Figure 11.7b is subject to pressure variations, then this will change the pressure drop across the valve and the flowrate to the inlet, irrespective of any change in the level. If the control valves in Figure 11.7a and b were under flowrate control, rather than level control, then any such pressure disturbances would be dealt with rapidly, resulting in little disturbance to the level. This can be achieved by applying cascade control to the level. Figure 11.8a shows a cascade level control in the direction of flow. The level controller drives the setpoint of the flowrate controller to keep the level at its setpoint. The level controller is the *primary* (or *outer*) controller. The flowrate controller is the *secondary* (or *inner*) controller. The flowrate controller responds rapidly to pressure changes, but the level controller responds slowly to flowrate changes. Thus, any short-term pressure disturbances downstream of Figure 11.8a are dealt with rapidly by the flowrate controller. Because the level has not changed, the setpoint passed from the level controller to the flowrate controller is unchanged. This allows the system to operate effectively when the downstream process is subject to pressure disturbances. The analogous cascade level control in the direction opposite to the flow is shown in Figure 11.8b. This time, short-term fluctuations in the upstream pressure are corrected by the flowrate controller. Because the level has not changed, the setpoint passed from the level controller to

the flowrate controller is unchanged. Thus, cascade level control should be considered if the upstream or downstream process is subject to significant pressure disturbances.

Figure 11.9 shows an operation with two phases. Vapor and liquid are separated in a flash drum. This time, there are two inventories involved: vapor and liquid. The liquid inventory in Figure 11.9 is controlled by a level controller and the vapor inventory is controlled by a pressure controller. It should be noted that the pressure controller serves the dual purpose of controlling the inventory and controlling the separation, as the separation depends on the vapor–liquid equilibrium, which depends on the pressure of the flash drum.

There are three important principles to be applied for inventory control of individual process operations (Aske and Skogestad 2009):

1)  The total material inventory must be regulated by its inflows *or* outflows, i.e. at least one flow *in or out* must depend on the inventory.

2)  The inventory of each chemical component must be regulated *by its inflows or outflows or by chemical reactions*.

3)  With more than one phase, the inventory of each phase must be regulated *by its inflows or outflows or by phase transition*.

The principles of inventory control for individual operations must now be extended to overall processes. Consider next a number of operations in series.

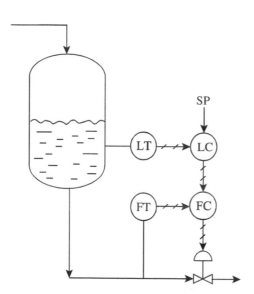

(a) Cascade level control in the direction of flow.

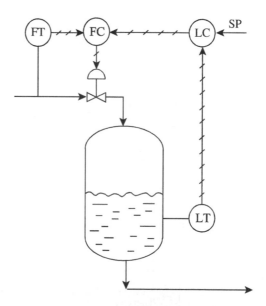

(b) Cascade level control in the in the direction opposite to the flow.

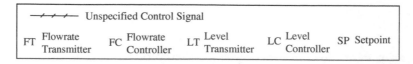

| | | | | |
|---|---|---|---|---|
| ⟶⟋⟋⟋⟵ Unspecified Control Signal | | | | |
| FT Flowrate Transmitter | FC Flowrate Controller | LT Level Transmitter | LC Level Controller | SP Setpoint |

**Figure 11.8**

Cascade level control.

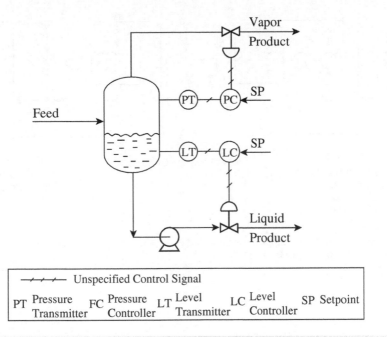

**Figure 11.9**

Two-phase vapor–liquid separator.

## 11.3 Inventory Control of Series Systems

When considering multiple operations, overall inventory control requires that a production rate change allows inventory control throughout the process (Aske and Skogestad 2009). Figure 11.10 shows two operations in series. Inventory control for multiple operations in series must operate consistently in the same direction relative to the point that fixes the flowrate. Liquid inventory control should be in the direction of flow if downstream of flow control, as shown in Figure 11.10a. Liquid inventory control should be in the direction opposite to the flow if upstream of flow control, as shown in Figure 11.10b.

**11**

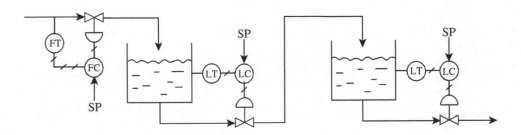

(a) Liquid inventory control in the direction of flow.

(b) Liquid inventory control in the in the direction opposite to the flow.

**Figure 11.10**

Liquid inventory control for series systems.

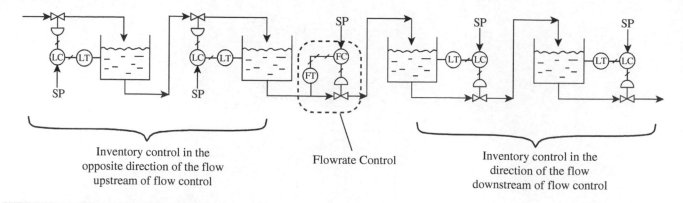

## Figure 11.11

Liquid inventory control for series systems with flowrate control in the middle of the process. Source: L. Bolton and R. Smith, 2022, *Process Integration and Optimization for Sustainability*, DOI: 10.1007/s41660-022-00283-x.

Figure 11.11 shows a more complex process in which the point of flowrate control is in the middle of the process. Upstream of the point of flowrate control the liquid inventory control is in the direction opposite to the flow. Figure 11.11 also shows that downstream of the point of flowrate control the liquid inventory control is in the direction of flow.

**Example 11.1**   Figure 11.12 shows a gaseous chemical reactor using a heterogeneous solid catalyst. The feed flowrate is controlled. Suggest a control scheme to control the inventory of material in the reactor.

### Solution

The feed flowrate is under flowrate control. Inventory depends on the inlet pressure, pressure drop, and the back-pressure from the downstream process. The inventory is self-regulating. Because both feed and product are gaseous, the inventory in the reactor needs to be controlled by controlling the reactor pressure. Figure 11.13 shows a control scheme that controls the reactor pressure. The pressure transmitter can have different locations, as long as it measures the vessel pressure and is upstream of the pressure control valve. It could have been located within the reactor or in the pipework at the exit of the reactor.

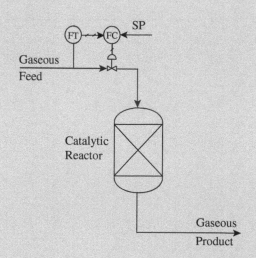

## Figure 11.12

A gaseous catalytic reactor.

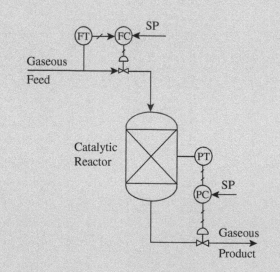

## Figure 11.13

A gaseous catalytic reactor with inventory control.

**Example 11.2**  For the same reactor as in Example 11.1, Figure 11.14 shows the gaseous reactor product being partially condensed and the resulting vapor–liquid mixture separated in a flash drum. Suggest a control scheme to control the process inventory.

**Solution**

Figure 11.15 shows a control scheme in which the gaseous inventory in the reactor is controlled by pressure control, as in Example 11.1. Liquid inventory in the flash drum is controlled by the controlling level. Gaseous inventory in the flash drum is controlled by controlling the pressure of the flash drum. The pressure control arrangement in Figure 11.15 allows the pressure in the reactor and flash drum to be different, in addition to controlling the gaseous inventory. It might be acceptable to control the reactor pressure to control the gaseous inventory in the reactor to be essentially the same (accepting the pressure drop through the system). In that case the pressure control on the flash drum can also control the inventory in the reactor, as in Figure 11.16. However, there might be a significant pressure drop through the system, and that pressure drop might change significantly as a result of flowrate changes. Change in the pressure drop will cause the reactor inventory to change. If this was the case, it would need detailed evaluation.

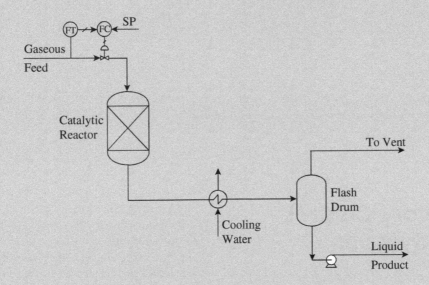

**Figure 11.14**

Partial condensation and separation of the reactor effluent.

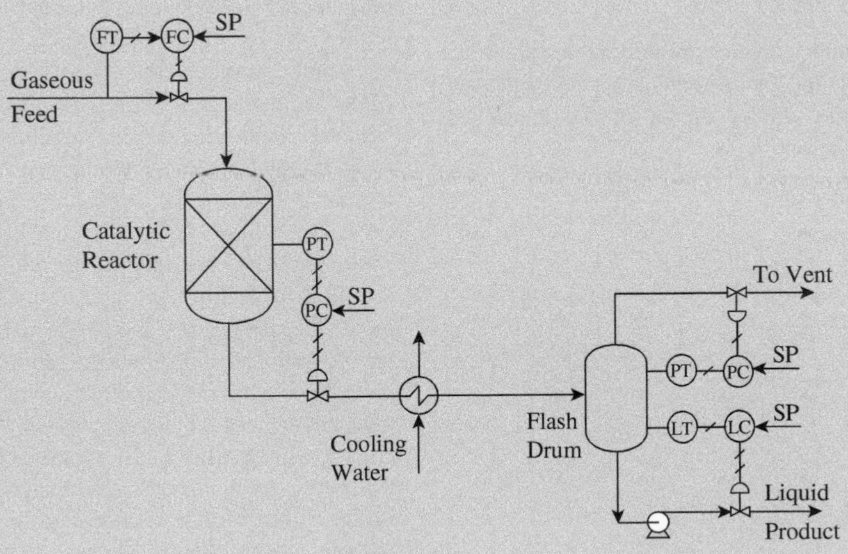

**Figure 11.15**

Inventory control for the reactor and flash drum.

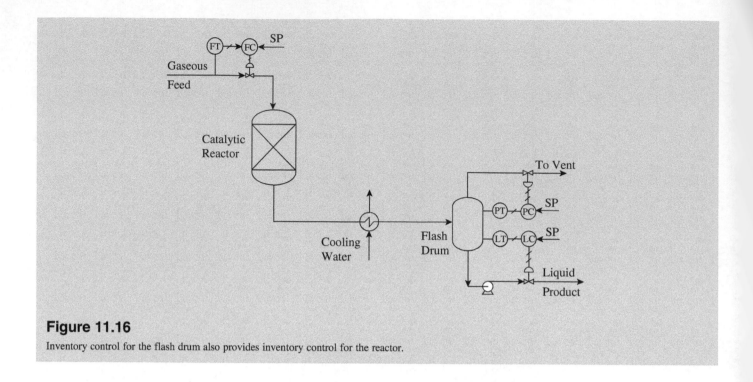

**Figure 11.16**

Inventory control for the flash drum also provides inventory control for the reactor.

# 11.4 Inventory Control of Recycle Systems

An example of a simple recycle system is shown in Figure 11.17. This involves two operations in a recycle system. In a recycle system, inventory control is required for (Aske and Skogestad 2009):

1) The individual operations in the recycle loop.
2) The recycle system as a whole.

In Figure 11.17, inventory control not only requires that the inventory in Operation 1 and Operation 2 are both controlled, but that the inventory in the recycle loop must also be controlled. Inventory control is required both for:

1) The total mass inventory and
2) Each chemical component.

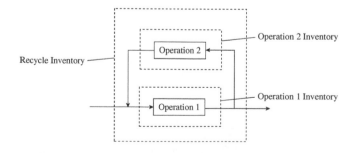

**Figure 11.17**

Inventory control for recycle processes.

Figure 11.18a shows the recycle system with two operations and a control system for the inventory. The inventory control might be level for liquids or pressure for gases. In Figure 11.18a the output of Operation 2 is controlled in the direction of flow. The inventory in the recycle loop is controlled by the input to the recycle loop and inventory control on both operations. Thus, the inventory control system in Figure 11.18a controls the inventory of both the individual operations and that of the recycle loop (Aske and Skogestad 2009). Figure 11.18b shows an alternative inventory control system. This time the output of Operation 2 is controlled in the direction opposite to the flow. The inventory in the recycle loop is controlled by the input to the recycle loop and inventory control on both operations (Aske and Skogestad 2009). Thus, both of the control systems in Figure 11.18 provide inventory control for the recycle system.

Figure 11.19a shows the same recycle loop with another inventory control system. This time, the recycle loop inventory is controlled overall. However, the inventory of Operation 2 is not controlled. The inventory in Operation 2 must be regulated by its inflows or outflows; i.e. at least one flow in or out must depend on the inventory (Aske and Skogestad 2009). Yet another inventory control system is shown in Figure 11.19b. In this case there is inventory control on both of the operations. However, there is no inventory control for the recycle loop. As before, placing flowrate control on the inlet and the outlet does not provide inventory control, because any mismatch, no matter how small, will cause the inventory to change (Aske and Skogestad 2009). Thus, neither of the control systems in Figure 11.19 provide inventory control for both the individual processes and the recycle system.

The inventory control in Figures 11.18 and 11.19 examines the control of the *total* material inventory. However, as with individual process operations, the inventory of each chemical component

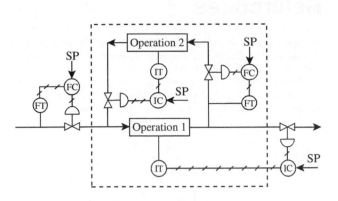

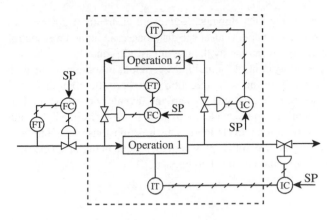

(IT) Inventory transmitter (level for liquids, pressure for gases)

(IC) Inventory controller (level for liquids, pressure for gases)

(a) Output of Operation 2 controlled in the direction of flow with invertory control of the recycle loop.

(b) Output of Operation 2 controlled in the direction opposite to the flow with inventory control of the recycle loop.

## Figure 11.18
Arrangements that control the inventory of the recycle system.

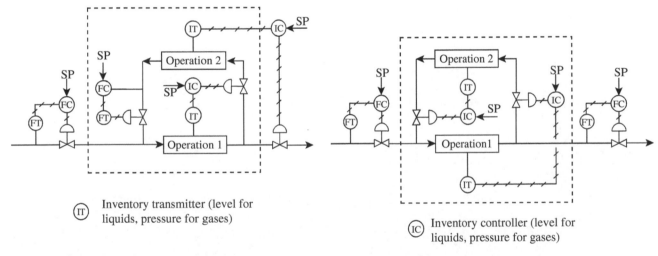

(IT) Inventory transmitter (level for liquids, pressure for gases)

(IC) Inventory controller (level for liquids, pressure for gases)

(a) Recycle loop inventory controlled but Operation 2 inventory not controlled.

(b) Operation 1 and Operation 2 with inventory control, but no inventory control for the reccyle loop.

## Figure 11.19
Arrangements that do not control the inventory of the recycle system.

needs to be controlled, both in the individual operations and the recycle as a whole. Examples of this will be explored in Chapter 14.

# 11.5 Flowrate and Inventory Control – Summary

Flowrate control is best achieved by locating the flowrate measurement and transmitter upstream of the control valve. Control of the material balance requires the process inventory to be controlled.

For an incompressible fluid in a fixed volume with no reaction or phase change, no inventory control is necessary. For gases in a fixed volume, the inventory depends on the pressure. Flowrate control valves on the inlet and outlet of an operation do not control the inventory. Any mismatch, no matter how small, will cause the inventory to change. Inventory control can be in the direction of flow or in the direction opposite to the flow (Buckley 1964).

For incompressible liquid flows in a variable volume, inventory control can be achieved by level control of the outflow. Alternatively, liquid inventory control can be achieved by liquid level

11

control of the inlet flowrate. Significant upstream or downstream pressure disturbances can create disturbances in liquid level control that can be overcome by the use of cascade control.

Inventory control for multiple operations in series must operate consistently in the same direction relative to the point that fixes the flowrate. In a recycle system, the inventory of the individual operations in the recycle loop need to be controlled, but the inventory in the recycle loop as a whole must also be controlled. Inventory control must apply both to the total material balance and each individual chemical component.

# References

Aske, E.M.G. and Skogestad, S. (2009). Consistent inventory control. *Industrial and Engineering Chemistry Research* 48: 10892–10902.

Bolton, L. and Smith, R. (2022). On the role of material balances in the synthesis of overall process control systems. *Process Integration and Optimization for Sustainability*. http://dx.doi.org/10.1007/s41660-022-00283-x.

Buckley, P.S. (1964). *Techniques of Process Control*. Wiley.

# Chapter 12

# Process Control – Degrees of Freedom

## 12.1 Degrees of Freedom and Process Control

One of the basic problems when synthesizing a control system for a process operation, or the process as a whole, is to determine how many control loops are required. Too few control loops and the process is under-controlled. Too many control loops and the process is over-controlled, creating conflicts between the control loops. Either circumstance does not lead to a viable control system. To determine the number of control loops necessary requires knowledge of the *degrees of freedom* for the system.

Consider first the number of degrees of freedom that need to be specified for a process system. If there are $N_{Variables}$ and $N_{Equations}$ *independent* equations in a process model, then if $N_{Variables} = N_{Equations}$ there is a unique solution to the model. The difference between $N_{Variables}$ and $N_{Equations}$ is the number of *degrees of freedom* (Smith 1963; Himmelblau and Riggs 2003):

$$N_{DOF} = N_{Variables} - N_{Equations} \quad (12.1)$$

where

$N_{DOF}$ = number of degrees of freedom

$N_{Variables}$ = number of variables (unknowns)

$N_{Equations}$ = number of *independent* equations describing the process system

There are three cases that need to be identified regarding the number of degrees of freedom:

1) $N_{DOF} = 0$. The problem can be solved and the design completely specified.

2) $N_{DOF} > 0$. The problem is under-specified and ($N_{Variables} - N_{Equations}$) additional independent equations (or independent variable specifications) are required to solve the problem and completely specify the process system.

3) $N_{DOF} < 0$. The problem is over-specified and ($N_{Equations} - N_{Variables}$) independent equations (or independent variable specifications) must be removed to solve the problem and completely specify the process system.

When applying this principle to a process design, if all the variables required to completely define the process are included, then the number of degrees of freedom defines the *design degrees of freedom*. Once the process design has been specified, some of the variables are no longer adjustable (e.g., equipment dimensions, number of distillation trays, and so on). The remaining variables are adjustable and can in principle be controlled. These adjustable variables are:

- temperatures;
- pressures;
- compositions;
- stream flowrates;
- component flowrates.

The number of these adjustable variables that are unspecified defines the *control degrees of freedom*. The control degrees of freedom can be satisfied by adding control loops, one for each control degree of freedom, to allow a unique solution to the model. Thus, the control degrees of freedom define the number of manipulated variables that can in principle be controlled after the non-adjustable variables have been fixed. It defines the number of *single-input–single-outlet control loops* that can be used, and therefore the number of *final control elements* (Luyben 1996). There is a precise number of control loops required for a given process operation in isolation and a precise number for the overall process.

Consider the example of a single-stage equilibrium flash separator, as illustrated in Figure 12.1. The design has been assumed to have fixed the non-adjustable variables (e.g. vessel dimensions) and the system is at steady-state conditions. The feed enters the separator, where it partially vaporizes and is allowed to reach equilibrium between the liquid and vapor. The equations describing the process are listed in Table 12.1. These must include the steady-state material balance, vapor-liquid equilibrium relationships,

*Process Plant Design*, First Edition. Robin Smith.
© 2024 John Wiley & Sons Ltd. Published 2024 by John Wiley & Sons Ltd.
Companion website: www.wiley.com/go/processplantdesign

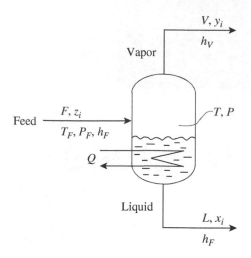

**Figure 12.1**

A flash drum separation.

## Table 12.1

Equations describing a simple flash separator.

| | Number of equations | Type of equation |
|---|---|---|
| $F = V + L$ | 1 | Material balance |
| $Fz_i = Vy_i + Lx_i$ | $N_C - 1$ | Component material balance |
| $y_i = K_i x_i$ | $N_C$ | Vapor-liquid equilibrium (VLE) |
| $K_i = K_i(T, P, x_i, y_i)$ | $N_C$ | VLE correlations |
| $\sum_{i=1}^{Nc} z_i = 1$ | 1 | Mole fraction constraint for feed |
| $\sum_{i=1}^{Nc} y_i = 1$ | 1 | Mole fraction constraint for vapor |
| $\sum_{i=1}^{Nc} x_i = 1$ | 1 | Mole fraction constraint for liquid |
| $FH_F + Q = VH_V + LH_L$ | 1 | Enthalpy balance |
| $H_F = H_F(T_F, P_F, z_i)$ | 1 | Enthalpy correlation for feed |
| $H_V = H_V(T, P, y_i)$ | 1 | Enthalpy correlation for vapor |
| $H_L = H_L(T, P, x_i)$ | 1 | Enthalpy correlation for liquid |
| Total | $3N_C + 7$ | |

where

$F$ = feed flowrate
$V$ = vapor flowrate
$L$ = liquid flowrate
$z_i$ = mole fraction of Component $i$ in the feed
$y_i$ = mole fraction of Component $i$ in the vapor
$x_i$ = mole fraction of Component $i$ in the liquid
$N_C$ = number of components
$K_i$ = vapor-liquid equilibrium constant for Component $i$
$H_F, H_V, H_L$ = specific enthalpy of the feed, vapor and liquid
$Q$ = heat input
$T_F$ = temperature of the feed
$P_F$ = pressure of the feed
$T$ = temperature of the flash drum
$P$ = pressure of the flash drum

vapor-liquid equilibrium correlations, mole fraction constraints, energy balance, and enthalpy correlations. From Table 12.1, the number of equations is $(3N_C + 7)$. The number of variables can be counted to be $(4N_C + 11)$. Thus, the number of *control degrees of freedom* is given by:

$$N_{CDOF} = (4N_C + 11) - (3N_C + 7)$$
$$= N_C + 4 \tag{12.2}$$

where

$N_{CDOF}$ = number of control degrees of freedom
$N_C$ = components

The composition and condition of the feed is likely to be fixed by the upstream process. Assuming $(N_C - 1)$ mole fractions are fixed from the upstream unit ($\sum z_i = 1$ defines the complete composition) and inlet conditions $T_F$, $P_F$, the control degrees of freedom become:

$$N_{CDOF} = (N_C + 4) - (N_C - 1) - 2$$
$$= 3 \tag{12.3}$$

This could mean, for example, specifying the feed flowrate, heat input, and pressure of the flash separator. This would then allow the other variables to be calculated and the system completely defined. In terms of a control system, it means that the feed flowrate, heat input, and pressure of the flash drum would need to be controlled to completely specify the system.

If it is assumed that the flash separator is adiabatic, then $Q = 0$ and there are two control degrees of freedom (e.g. feed flowrate and flash drum pressure). In terms of the control system, this means that the feed flowrate and pressure of the flash drum could be controlled to completely specify the system, as shown in Figure 12.2.

It should be noted for the flash separator material balance in Table 12.1 is specified by an overall material balance of the form:

$$F = V + L \tag{12.4}$$

together with $(N_C - 1)$ component material balance equations of the form:

$$Fz_i = Vy_i + Lx_i \tag{12.5}$$

Had $N_C$ component material balance equations of the form of Eq. (12.5) been included, then there are $(N_C + 1)$ material balance equations and one of these equations is not independent. For example, the $N_C$ component balances can be combined, along with the mole fraction summation equations, to derive the overall material balance Eq. (12.4). The equations when counted for the degrees of freedom must be independent. If an equation can be derived by algebraically combining the other equations, it is not independent. As an alternative for the definition of the material

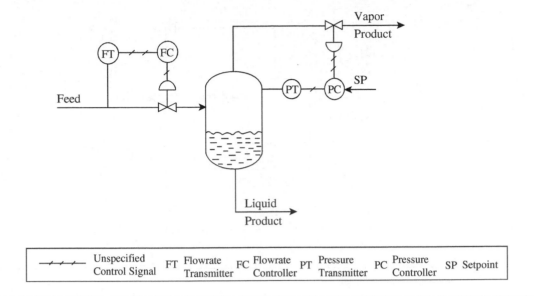

**Figure 12.2**

Control system for the flash drum separation.

balance in Table 12.1, $N_C$ component balances as in Eq. (12.5) could have been specified and the total balance as in Eq. (12.4) not specified.

Whilst feed flowrate and pressure of the flash drum can be controlled to completely specify the system, the whole analysis has assumed steady state. In a dynamic system the material balance would also need to be controlled to maintain steady state. The mass of vapor in the flash drum is maintained by the pressure control, if the liquid level is steady. If it is assumed that the vapor follows ideal gas behavior then:

$$N = \frac{PV}{RT} \qquad (12.6)$$

where

$N$ = moles of gas (kmol)

$P$ = pressure (N m$^{-2}$)

$V$ = volume occupied by $N$ kmol of gas (m$^3$)

$R$ = gas constant (8314 N m kmol$^{-1}$ K$^{-1}$ or J kmol$^{-1}$ K$^{-1}$)

$T$ = absolute temperature (K)

Thus, if the pressure, volume, and temperature are fixed, the moles (mass) of vapor are also fixed by the equation of state. In addition to controlling the pressure, the liquid level also needs to be controlled to control both the mass of the liquid and that of the vapor. Figure 12.3 shows a control scheme that controls the inventory, in addition to the other control variables. The pressure control serves a dual purpose of controlling the conditions for the phase equilibrium and controlling the vapor inventory. The liquid level control maintains the liquid inventory and by controlling the level is also necessary to control the vapor inventory.

Degrees of freedom analysis is normally only performed using steady-state equations; thus adjustments that affect only the inventory are not included in the degrees of freedom analysis, unless

they affect the state variables (e.g. pressure of a vapor). In the flash drum example, there is an inventory of liquid that must be maintained at steady state. This requires an additional single-input–single-output control loop (typically liquid level control), but is not included in a steady-state degrees of freedom analysis. Such inventory control degrees of freedom must be added to the steady-state degrees of freedom. Thus, the total control degrees of freedom are the sum of the steady-state control degrees of freedom and the inventory control degrees of freedom.

Whilst analyzing the number of variables and independent equations can reveal the number of control degrees of freedom, it is tedious and time consuming, but most importantly it is error prone. Even the simple example of a flash drum presented earlier is challenging in order to avoid errors. More complex operations, such as distillation arrangements, become even more challenging. However, the problems start to become insurmountable when considering large complex process flowsheets with many different operations and many connections. It requires a viable and detailed steady-state model of the process to be developed. Equations might be missing or not independent and variables might be omitted. An alternative approach is required (Kwauk 1956; Smith 1963; Ponton 1994; Konda et al. 2006). The approach used here will be to first analyze the steady-state control degrees of freedom and then to add the inventory control degrees of freedom.

## 12.2  Degrees of Freedom for Process Streams

Before considering the control degrees of freedom for process operations, consider first the degrees of freedom for process streams. The variables describing a single-phase stream comprising $N_C$ components are listed in Table 12.2.

**12**

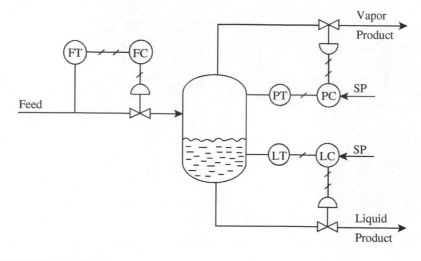

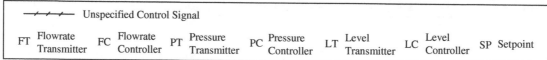

## Figure 12.3

Control system for the flash drum separation with liquid inventory control.

## Table 12.2

Variables describing a single-phase stream.

| Variable | | Number |
|---|---|---|
| Stream flowrate | $F$ | 1 |
| Component mole fractions | $x_i$ | $N_C$ |
| Temperature | $T$ | 1 |
| Pressure | $P$ | 1 |
| Stream specific enthalpy | $H$ | 1 |
| Total | | $N_C + 4$ |

## Table 12.3

Variables describing a single-phase stream using component flowrates.

| Variable | | Number |
|---|---|---|
| Component flowrates | $F_i$ | $N_C$ |
| Temperature | $T$ | 1 |
| Pressure | $P$ | 1 |
| Stream specific enthalpy | $H$ | 1 |
| Total | | $N_C + 3$ |

To define the degrees of freedom for an individual stream requires any relationships between the variables to be defined. The mole fractions must sum to unity:

$$\sum x_i = 1 \tag{12.7}$$

Also, the specific enthalpy can be defined to be a function of temperature, pressure, and composition:

$$H = H(T, P, x_i) \tag{12.8}$$

The *design degrees of freedom* for the single-phase stream are given by deducting the number of equations from the number of variables:

$$\begin{aligned} N_{DOF} &= (N_C + 4) - 2 \\ &= N_C + 2 \end{aligned} \tag{12.9}$$

where

$N_{DOF}$ = design degrees of freedom for a process steam

Thus, it is necessary to specify $(N_C + 2)$ variables to completely define the stream. Alternatively, the variables could have been listed as in Table 12.3. However, defining component flowrates as in Table 12.3 no longer requires the constraint that the mole fractions must sum to unity, as in Eq. (12.7). Including total flowrate in Table 12.3 would not provide an independent source of information. The equation for the specific enthalpy in Eq. (12.8) is still required. Thus:

$$\begin{aligned} N_{DOF} &= (N_C + 3) - 1 \\ &= N_C + 2 \end{aligned} \tag{12.10}$$

This gives the same result of $(N_C + 2)$ design degrees of freedom for an individual single-phase stream. Thus, $(N_C + 2)$ variables need to be specified to define a stream. The same result of $(N_C + 2)$ is obtained if the stream enthalpy is omitted from the variables, and along with it, the equation for the enthalpy (Smith 1963).

Now consider an operation with $N_{in}$ inlet streams and $N_{out}$ outlet streams having the same single-phase. If there are $N_C$ components in the streams, then there are $(N_C + 2)$ design degrees of freedom for each stream. However, if the design is fixed, then only the flowrate of the inlet streams can generally be controlled, with the other variables in the inlet (composition, temperature, pressure) fixed by the upstream unit for that stream. Thus, if each inlet stream contributes only one degree of freedom, and each outlet stream contributes $(N_C + 2)$ degrees of freedom, then the *control degrees of freedom* are given by (Ponton 1994):

$$N_{CDOF} = N_{in} + N_{out}(N_C + 2) \qquad (12.11)$$

where

$N_{CDOF}$ = control degrees of freedom
$\quad N_{in}$ = number of inlet streams
$\quad N_{out}$ = number of outlet streams

If heating or cooling is provided by utility streams, or there is an input or output of mechanical energy (work), then each of these energy streams contributes a degree of freedom. If there are $N_E$ inputs or outputs of energy:

$$N_{CDOF} = N_{in} + N_E + N_{out}(N_C + 2) \qquad (12.12)$$

where

$N_E$ = number of inputs or outputs of energy

It should be noted that the energy inputs or outputs in Eq. (12.12) are assumed to be adjustable.

Now consider the application of Eq. (12.12) to single-phase process operations.

# 12.3  Individual Single-Phase Operations

Consider now the design degrees of freedom for *individual single-phase operations*. It will be assumed that a number of inlet streams are mixed to homogeneity and all outlet streams have the same composition, temperature, and pressure. If all outlet streams are assumed to have the same composition, this imposes $(N_C - 1)$ constraints for each outlet stream, with the remaining mole fraction defined by the mole fractions summing to unity. For $N_{out}$ outlet streams, the total constraints imposed become $(N_C - 1)(N_{out} - 1)$, with the remaining outlet stream defined by the material balance. If all outlet streams are at the same temperature, this imposes a further $(N_{out} - 1)$ constraints. It is $(N_{out} - 1)$ rather than $N_{out}$, as only $(N_{out} - 1)$ independent equalities can be written, with the final equality not being independent. For example, suppose a unit has three outlet streams with temperatures $T_1$, $T_2$, and $T_3$. If the outlet temperatures are constrained to be equal, then $T_1 = T_2$ and $T_1 = T_3$. The third equation $T_2 = T_3$ can be derived by combining the first two equations and is therefore not independent. Similarly, if all outlet streams are at the same pressure,

**Table 12.4**

Constraints for individual single-phase operations.

|  | Number |
|---|:---:|
| Component material balances | $N_C$ |
| Energy balance | 1 |
| Composition constraints | $(N_C - 1)(N_{out} - 1)$ |
| Temperature equality outlet constraints | $(N_{out} - 1)$ |
| Pressure equality outlet constraints | $(N_{out} - 1)$ |
| Total | $N_C N_{out} + N_{out}$ |

this imposes an additional $(N_{out} - 1)$ constraints. The total constraints are given in Table 12.4.

The control degrees of freedom from Eq. (12.12) need to be decreased by the constraints listed in Table 12.4. Thus, for an individual single-phase operation (Ponton 1994):

$$\begin{aligned} N_{CDOF} &= N_{in} + N_E + N_{out}(N_C + 2) - N_C N_{out} - N_{out} \\ &= N_{in} + N_E + N_{out} \end{aligned} \qquad (12.13)$$

It should be noted that the constraints in Table 12.4, which are incorporated in Eq. (12.13), do not cover all cases and additional constraints will need to be added to deal with certain operations. Prominent is the absence of a general expression to reflect the outlet pressure constraints. The constraints in Table 12.4 constrain all outlet temperatures to be equal and the outlet temperature to be fixed by the energy balance. By contrast, the outlet pressures, besides being equal, remain unspecified. There are two general ways the outlet pressure can be specified:

1) Pressure control of the operation.
2) Allowing the outlet pressure to *self-regulate*. Without pressure control, the outlet pressure for a fixed inlet pressure will be dictated by some combination of fluid frictional pressure losses in the operation, pressure drop across valves controlling the flowrate of the outlet streams (or liquid level in the operation), fluid frictional pressure losses in the outlet pipework and vessels, hydraulic head pressure change for liquids, the back-pressure from the process downstream, and the input or output of mechanical energy (work).

Either of these options takes up a degree of freedom. In the case of *self-regulation*, this adds an additional constraint as an equation. For example, the outlet pressure for an operation might be specified to be equal to the inlet pressure for the operation, or the outlet pressure might be specified to be equal to the inlet pressure for the next operation. Alternatively, for example, the outlet pressure might be specified by the fluid frictional losses in the operation and connected downstream operations, and so on. There are many possible scenarios for the specification of the outlet pressure, which is why a general treatment for pressure is difficult and must be considered on a case-by-case basis. Also, whilst outlet pressure is the most prominent constraint missing from Table 12.3, other constraints will need to be included in some cases that will be considered later.

**12**

**Example 12.1**   Figure 12.4 shows a vessel with a single inlet and single outlet, both liquid. No phase change occurs and there is no energy input or output. The inventory of liquid is constant by virtue of the weir and overflow. Determine the number of control degrees of freedom.

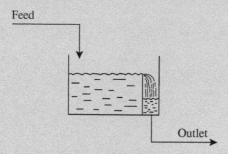

Feed

Outlet

**Figure 12.4**

Control degrees of freedom of a simple vessel with a constant inventory.

**Solution**

Given that there is no energy input or output, the temperature can be assumed to be unchanged across the vessel. From Eq. (12.13), the steady-state control degrees of freedom are given by:

$$N_{CDOF} = N_{in} + N_E + N_{out}$$
$$= 1 + 0 + 1 = 2$$

Before concluding that the number of control degrees of freedom is 2, the outlet pressure needs to be considered. Table 12.4 accounts for the material balance. The energy balance is also included, which in principle fixes the outlet temperature. However, Table 12.4 provides no constraint for the outlet pressure, which must be specified to fully define the outlet stream. The outlet pressure can in principle be controlled, which would take up a degree of freedom. In the case of the vessel in Figure 12.4, pressure control would be difficult, but in general pressure control is an option. Rather than control the pressure, it can be left to self-regulate. In the case of Figure 12.4, it can be assumed that the inlet and outlet pressures are equal. If the outlet pressure is left to self-regulate, this adds another constraint in the form of a pressure drop equation, for example $P_{out} = P_{in}$. If the outlet pressure is left to self-regulate, then the steady-state control degrees of freedom are given by:

$$N_{CDOF} = (1 + 0 + 1) - 1 = 1$$

No inventory control is required in this case, as the inventory is fixed by the overflow arrangement. Thus, with a single control degree of freedom the obvious option to control the system would be to control the feed flowrate.

**Example 12.2**   Figure 12.5 shows two possible ways in which to implement a stream split. The first in Figure 12.5a is simply a junction between two pipes. The second in Figure 12.5b is a single feed to a vessel, with two outlets from the vessel. Having a vessel as shown in Figure 12.5b might be preferred if, for example, a liquid stream needs to be split and a secure head of liquid needs to be ensured for the outlet streams for pumping, or some buffering is desirable. No phase change occurs and there is no energy input or output. Determine the number of control degrees of freedom.

**Solution**

Given that there is no energy input or output, the temperature can be assumed to be unchanged across the operation. From Eq. (12.13), the steady-state control degrees of freedom are given by:

$$N_{CDOF} = N_{in} + N_E + N_{out}$$
$$= 1 + 0 + 2 = 3$$

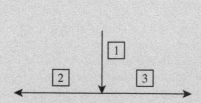

(a) No inventory control required for an incompressible fluid, with pressure self-regulating.

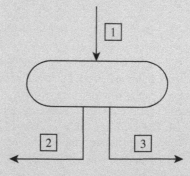

(b) Inlet and outlet flowrates must be matched identically for inventory control.

**Figure 12.5**

Degrees of freedom for splitting operations.

This assumes from the constraints imposed in Table 12.4 that both outlet streams have the same pressure. If the outlet pressure is left to be self-regulating, as argued in Example 12.1, this adds one more constraint (e.g. $P_{out} = P_{in}$):

$$N_{CDOF} = (1 + 0 + 2) - 1$$
$$= 2$$

Thus, with two steady-state control degrees of freedom, the flowrate of two of the three streams needs to be controlled. Figure 12.6 shows possible control schemes. Although the inclusion of the control valves changes the pressures slightly, this does not change the principle. Inventory control will need to be considered later for the design in Figure 12.6b to complete the control system synthesis.

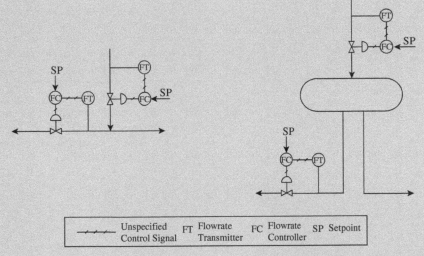

| | | | | |
|---|---|---|---|---|
| ⎯ᴧ⎯ Unspecified Control Signal | FT | Flowrate Transmitter | FC | Flowrate Controller | SP Setpoint |

(a) One option is to control the inlet and one of the outlet streams, allowing the outlet stream pressures to be self-regulating.

(b) One option is to control the inlet and one of the outlet streams, allowing the outlet stream pressures to be self-regulating, but inventory control will also be required later.

**Figure 12.6**

Control of splitting operations.

**Example 12.3**    Figure 12.7 shows two of the different ways to implement a mixer. The first in Figure 12.7a is simply a junction between two pipes. The second in Figure 12.7b is two feeds to a vessel, with a single outlet from the vessel. No phase change occurs and there is no energy input or output. Determine the number of control degrees of freedom.

**12**

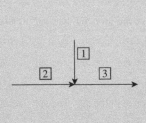

(a) No inventory control required for an incompressible fluid, with outlet pressure self-regulating.

(b) Inlet and outlet flowrates must be matched identically for inventory control.

**Figure 12.7**

Degrees of freedom for mixing operations.

## Solution

Given that there is no energy input or output, the temperature can be assumed to be unchanged across the operation. From Eq. (12.13), the steady-state control degrees of freedom are given by:

$$N_{CDOF} = N_{in} + N_E + N_{out}$$
$$= 2 + 0 + 1 = 3$$

If the outlet pressure is left to be self-regulating, as argued in Example 12.1, this adds one more constraint from the pressure drop equation (e.g. $P_{out} = P_{in}$):

$$N_{CDOF} = (2 + 0 + 1) - 1$$
$$= 2$$

Thus, with two steady-state control degrees of freedom, the flowrate of two of the three streams needs to be controlled.

Figure 12.8 shows possible control schemes in which the flowrates of both inlet streams are controlled. As discussed previously, this would be a case that is likely to benefit from ratio control of the two inlet streams.

An example of a more sophisticated control scheme is shown in Figure 12.9. This shows an analysis controller controlling the composition of the product. The outlet composition control provides the setpoint for one of the inlet flowrates in a cascade control system. The degrees of freedom remain the same as before, since there is one new variable (composition) and one new equation (the analysis controller provides the setpoint for the flowrate controller). In this scheme, in addition to controlling the flowrate of both inlet streams, pressure of the mixer is controlled. The 3 steady state control degrees of freedom are taken up by 3 single-input–single-outlet control loops. The arrangement is suited to cascade control because composition control is much slower than flowrate control. Again, this would be a case that is likely to benefit from ratio control of the two inlet streams.

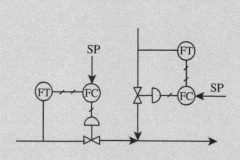

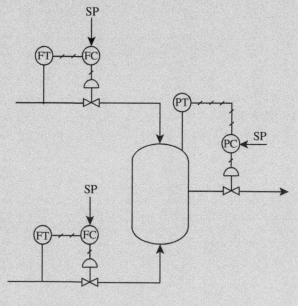

| | | | |
|---|---|---|---|
| ⌇⌇ Unspecified Control Signal | FT Flowrate Transmitter | FC Flowrate Controller | PT Pressure Transmitter | PC Pcessure Controller | SP Setpoint |

(a) One option is to control the flowrate of both inlet streams.

(b) One option is to control both inlet streams, with pressure control on the outlet, but liquid inventory control will also be required later.

## Figure 12.8

Control of splitting operations.

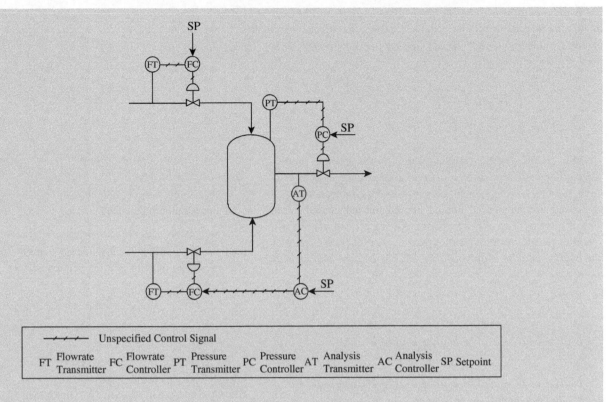

**Figure 12.9**

Mixer control with flowrate control of both inlet streams, pressure control of the outlet, and outlet analysis used for cascade control of one of the inlet flowrates to maintain product concentration.

## 12.4    Heat Transfer Operations with No Phase Change

Consideration of heat exchange will initially be restricted to operations in which there is no change of phase. Figure 12.10 shows a heat exchange operation between two streams. Such operations would most often feature two inlet and two separate outlet streams. Utility heat exchangers employing a hot utility for heating, or cold utility for cooling, can be analyzed for their steady-state control degrees of freedom using Eq. (12.13). Applying Eq. (12.13) to the heat exchanger assumes that there is only one process stream, with the other being accounted for as an energy stream (hot or cold utility):

$$N_{CDOF} = N_{in} + N_E + N_{out}$$
$$= 1 + 1 + 1 = 3 \tag{12.14}$$

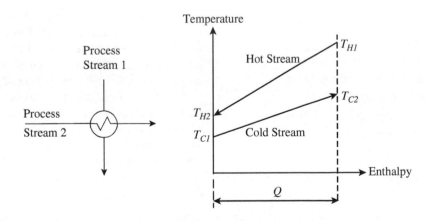

**Figure 12.10**

A process heat exchanger.

If the outlet pressure of the process stream is left to be self-regulating from the fluid frictional losses, this adds one more constraint from the pressure drop equation for the process side of the heat exchanger, giving:

$$N_{CDOF} = 3 - 1$$
$$= 2 \qquad (12.15)$$

This analysis of heat exchangers assumes a single process stream with an energy stream in or out (hot or cold utility). However, the heat exchanger can also be treated as an operation with two process streams, even if one of the streams happens to be a hot or cold utility. Such heat exchange operations considering two streams can be treated as two single stream units connected by an energy stream. Applying Eq. (12.13) to the two separate streams:

$$N_{CDOF} = 2(N_{in} + N_E + N_{out})$$
$$= 2(1 + 1 + 1) = 6 \qquad (12.16)$$

The separate units when connected are constrained by an energy balance across the two units:

$$Q = Q_H = Q_C \qquad (12.17)$$

where

$Q$ = heat exchanger duty

$Q_H$ = heating duty on the hot stream

$Q_C$ = cooling duty on the cold stream

Substituting for $Q_H$ and $Q_C$ in Eq. (12.17) gives the energy balance:

$$m_H C_{PH}(T_{H1} - T_{H2}) = m_C C_{PC}(T_{C2} - T_{C1}) \qquad (12.18)$$

where

$m_H$ = mass flowrate of the hot stream

$m_C$ = mass flowrate of the cold stream

$C_{PH}$ = specific heat capacity flowrate of the hot stream

$C_{PC}$ = specific heat capacity flowrate of the cold stream

$T_{H1}$ = inlet temperature of the hot stream

$T_{H2}$ = outlet temperature of the hot stream

$T_{C1}$ = inlet temperature of the cold stream

$T_{C2}$ = outlet temperature of the cold stream

Also, the heat transfer is constrained by the equation for the rate of heat transfer:

$$Q = UA\Delta T_{LM} F_T$$
$$= UA \frac{(T_{H1} - T_{C2}) - (T_{H2} - T_{C1})}{\ln\left(\dfrac{T_{H1} - T_{C2}}{T_{H2} - T_{C2}}\right)} F_T$$

$$(12.19)$$

where

$Q$ = heat exchanger duty

$U$ = overall heat transfer coefficient

$A$ = heat transfer area

$F_T$ = logarithmic mean temperature difference correction factor for non-countercurrent flow

If both of the outlet pressures are left to be self-regulating from the fluid frictional losses, this adds two more constraints from the pressure drop equations for the heat exchanger, giving:

$$N_{CDOF} = 2(N_{in} + N_E + N_{out}) - 4$$
$$= 2(1 + 1 + 1) - 4 = 2 \qquad (12.20)$$

As a check on this result, if $m_H$, $m_C$, $C_{PH}$, $C_{PC}$, $U$, $A$, and inlet temperatures $T_{H1}$ and $T_{C1}$ are known, then Eqs (12.18) and (12.19) constitute two equations with two unknowns (the outlet temperatures $T_{H2}$ and $T_{C2}$). The equations can be solved to determine the outlet temperatures $T_{H2}$ and $T_{C2}$ without iteration, as detailed in Smith (2016, Chapter 12). Thus, if $m_H$, $m_C$, $C_{PH}$, $C_{PC}$, $U$, $A$, $T_{H1}$, and $T_{C1}$ are specified, there are no degrees of freedom. If the two inlet flowrates are allowed to vary, this introduces two degrees of freedom.

If steam heating is being used, the heat duty of the hot stream is the heat from desuperheating the inlet steam plus the latent heat of condensation. Thus, the heat balance is given by:

$$m_{STEAM}(H_{SH} - H_{COND}) = m_C C_{PC}(T_{C2} - T_{C1}) \qquad (12.21)$$

where

$m_{STEAM}$ = mass flowrate of the hot stream

$H_{SH}$ = specific enthalpy of superheated steam

$H_{COND}$ = specific enthalpy of the saturated condensate

Now the outlet temperature of the hot stream is fixed by steam properties at the steam pressure. Thus, if $H_{SH}$, $H_{COND}$, $m_C$, $C_{PC}$, $T_{C1}$, $U$, and $A$ are known, Eqs (12.21) and (12.19) (with $F_T = 1$ for a steam heater) can be solved for $m_{STEAM}$ and $T_{C2}$. Varying $m_{STEAM}$ allows the outlet temperature $T_{C2}$ to be manipulated.

Figure 12.11 shows two utility heat exchanger control schemes. Figure 12.11a shows a process heater using steam, in which the process flowrate inlet is controlled, with the outlet temperature controlled by the flowrate of steam. Figure 12.11b shows a process cooler using cooling water in which the process flowrate inlet is controlled, with the outlet temperature controlled by the flowrate of cooling water.

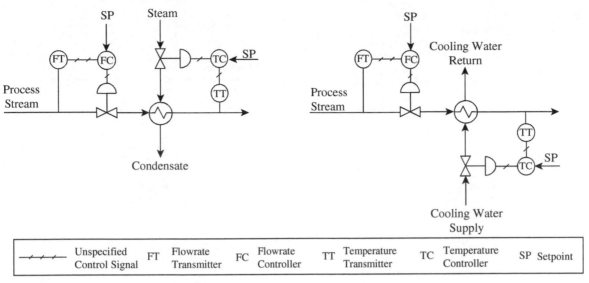

<table>
<tr><td>⌇⌇⌇</td><td>Unspecified<br>Control Signal</td><td>FT</td><td>Flowrate<br>Transmitter</td><td>FC</td><td>Flowrate<br>Controller</td><td>TT</td><td>Temperature<br>Transmitter</td><td>TC</td><td>Temperature<br>Controller</td><td>SP</td><td>Setpoint</td></tr>
</table>

(a) Control of a process heater using steam.    (b) Control of process cooler using cooling water.

### Figure 12.11

Control of utility exchangers.

**Example 12.4**   Figure 12.12 shows a process-to-process heat exchanger. All inlet conditions for both streams are fixed by the upstream operations. This includes flowrate, composition, temperature, and pressure:

a) Determine the number of control degrees of freedom.
b) Devise alternative schemes to control the outlet temperature of the hot stream and select the most appropriate.
c) Devise alternative schemes to control the outlet temperature of the cold stream and select the most appropriate.

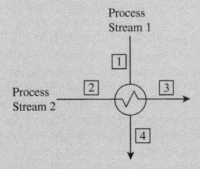

### Figure 12.12

A process to process a heat exchanger.

**Solution**

a) The number of steady-state control degrees of freedom for a heat exchanger with the stream outlet pressures left to self-regulate is given by Eq. (12.20). If the inlet process flowrate for both streams is fixed, this removes two degrees of freedom. The result is that there are no control degrees of freedom.

$$N_{CDOF} = 0$$

b) If the temperature of the hot stream outlet is to be controlled, there are no degrees of freedom. This requires a new degree of freedom to be created. Introducing a partial bypass around the heat exchanger introduces a new degree of freedom. Figure 12.13a shows control of the hot stream with a hot stream partial bypass. Figure 12.13b shows control of the hot stream with a cold stream partial bypass. Of the two schemes, partial bypass on the same side as the control is required is favored. This minimizes the effect of heat exchanger dynamics in the control loop and time lags. Thus, the control scheme in Figure 12.13a is preferred.

c) If the temperature of the cold stream outlet is to be controlled, this also requires a partial bypass around the heat exchanger to be introduced. Figure 12.13c shows control of the cold stream with a cold stream partial bypass. Figure 12.13d shows control of the cold stream with a hot stream partial bypass. Again, of the two schemes, partial bypass on the same side as the control is required is favored, minimizing the effect of heat exchanger dynamics in the control loop and time lags. Thus, the control scheme in Figure 12.13c is preferred.

**12**

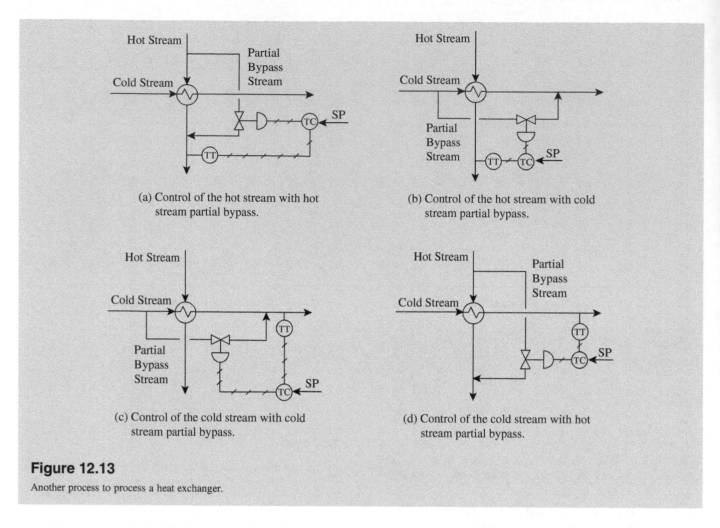

**Figure 12.13**

Another process to process a heat exchanger.

In some situations, process heat needs to be supplied by radiant heat transfer from the combustion of fuel in a *fired heater* or *furnace*, as discussed in Chapter 4. Figure 12.14 shows an example of a process fired heater. A process feed stream is assumed heated without change of phase in the furnace coil. Again, this can be represented by two separate units linked together. For the furnace combustion zone there are two process streams in (air and fuel, which are assumed to have fixed temperature, pressure, and composition), one energy stream out (including the heat released from combustion), and one process stream out (the flue gas), which from Eq. (12.13) gives $N_{CDOF} = 4$. Similarly, for the furnace coil there is one process stream in, one energy stream in, and one process stream out, giving $N_{CDOF} = 3$. There is an energy balance across both units (that includes the heat released from combustion), which creates a constraint. Also, only part of the heat is transferred from the combustion gases to the furnace coil, with the remainder lost to the stack. The ratio of the heat transferred to the coil relative to the total heat input to the combustion zone can be defined by a furnace efficiency (see Chapter 4). This creates an additional constraint and is analogous to the rate equation constraint in the degrees of freedom for heat exchangers, discussed above. Finally, if the outlet pressures from the stack and the furnace coil are left to self-regulate, this adds 2 further constraints.

$$N_{CDOF} = 4 + 3 - 1 - 1 - 2$$
$$= 3 \qquad (12.22)$$

Thus, the fired heater has 3 control degrees of freedom. These would typically be taken up by applying flowrate control to the combustion air, fuel, and process feed in Figure 12.14. Control of fired heaters will be discussed in Chapter 13.

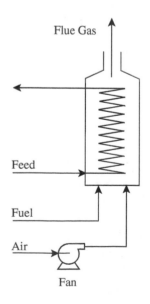

**Figure 12.14**

A process fired heater.

# 12.5  Pumps and Compressors

Whereas heat transfer operations involve an input or output of energy in the form of heat, pumps and compressors involve an input of energy that is mechanical in nature. Figure 12.15 shows a centrifugal pump and its operating characteristics. Centrifugal pumps normally work at constant rotational speed and deliver an output pressure or *head* that is a function of flowrate (Figure 12.15). The *pump head curve* is fixed by the design of the pump. As the flowrate increases, the head (outlet pressure) decreases according to the pump operating curve, which is a fixed characteristic of the design of the machine and cannot be manipulated without changing the design of the pump. There is one degree of freedom *inherent* to a fixed speed centrifugal pump. If the outlet pressure is specified, this fixes the flowrate. If the outlet flowrate is specified, this fixes the pressure. When the pump is located in the process, any upstream pressure changes the location of the pump head curve, but still constrained by the design of the pump. The pumping action is against a *system pressure head* created by the design of the outlet system. The pumping action is against a pressure created by the frictional pressure losses in pipes, pipe fittings, and vessels, together with any hydraulic head from change in elevation for liquids and backpressure from the downstream process (Smith 2016). As the flowrate through the system increases, so does the pressure drop, due to increased fluid frictional losses. The actual performance of the pump cannot be specified until it operates within the process. The flowrate and pressure created by the pump for the combination of pump and outlet system is where the two head curves cross (Figure 12.15). Note that the system head in Figure 12.15 does not in this case go through the origin of the graph, indicating that a difference in elevation between the feed vessel and pump or back pressure from the downstream process creates a static head at zero flowrate. Thus, for a fixed design of pump and system, the flowrate and outlet pressure of the pump are both fixed by the conditions where the pump head curve and system head curves cross.

The number of control degrees of freedom is given by Eq. (12.13). Whilst there is an energy balance across the pump, as included in Eq. (12.13), the energy (mechanical) input is not free to be adjusted independently. For a given pump, the energy input is fixed to be a function of the flowrate, inlet conditions, pressure difference across the pump, liquid physical properties, and pump efficiency. This creates an additional constraint (equation) for the control degrees of freedom. In addition, the pump head curve and system head curve are two additional constraints (equations) for the control degrees of freedom. Thus, Eq. (12.13) gives:

$$N_{CDOF} = N_{in} + N_E + N_{out} - 3$$
$$= 1 + 1 + 1 - 3 = 0 \tag{12.23}$$

The outlet pressure and flowrate are thus fixed where the pump head and system head curves cross, as shown in Figure 12.15. If the outlet pressure or flowrate are to be controlled, then the shape of the system head curve needs to be manipulated such that it crosses the pump head curve at the required point of operation. The point at which the curves cross can be manipulated by introducing a control valve to adjust the system head curve by changing the pressure drop in the system head and relaxing the constraint of the fixed system head curve. This will be considered in Chapter 13.

If the pump is a positive displacement pump, rather than centrifugal, then at a fixed rotational speed the pump delivers a fixed volume of liquid, irrespective of the outlet pressure (assuming no slippage of liquid in the clearances in the pump). Firstly, this means the flowrate is fixed, creating a constraint. Secondly, unlike centrifugal pumps, which produce pressure, positive displacement pumps do not produce pressure. The pressure is created by the system itself, which creates a back-pressure. The back-pressure is dependent on the pressure in the discharge vessel and the outlet system pressure drop. The system head introduces an additional constraint. Thirdly, the energy input cannot be adjusted independently but for a given pump is fixed to be a function of the flowrate, pressure difference across the

**12**

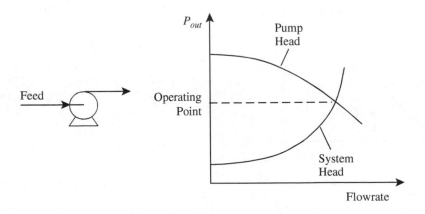

**Figure 12.15**

A fixed speed centrifugal pump.

pump, liquid properties, and pump efficiency. This adds a third additional constraint. Thus:

$$N_{CDOF} = N_{in} + N_E + N_{out} - 3$$
$$= 1 + 1 + 1 - 3 = 0 \qquad (12.24)$$

Thus, constant speed-positive displacement pumps have no degrees of freedom. Their control will be considered in the Chapter 13.

Turning now to compressors, Figure 12.16a shows the operating characteristics of a fixed speed centrifugal compressor. It should be noted that the shape of the operating curve is different from that of a centrifugal pump. The operating curve is a feature of the individual compressor design and cannot be manipulated. The maximum in the curve has important implications, as will be discussed in Chapter 13. If the rotational speed is fixed, there is *inherently* one degree of freedom if the compressor is considered in isolation. For a fixed inlet pressure, if the outlet pressure is specified, this fixes the flowrate. If the outlet flowrate is specified, this fixes the outlet pressure if the inlet pressure is fixed. However, like centrifugal pumps, centrifugal compressors operate against a system head and the operation of the compressor self-regulates to match the system head where the two curves cross. The flowrate and pressure created by the compressor for the combination of compressor and system is where the two head curves cross in Figure 12.16a.

The number of control degrees of freedom is again given by Eq. (12.13). Like centrifugal pumps, whilst there is an energy balance across the compressor, as included in Eq. (12.13), the energy

(mechanical) input is not free to be adjusted independently in this case. For a given compressor, the energy input is a function of the flowrate, inlet conditions, pressure difference across the compressor, gas properties, and compressor efficiency. This creates an additional constraint (equation) for the control degrees of freedom. The compressor operating profile and the system head curve create two additional constraints. Thus, Eq. (12.13) gives:

$$N_{CDOF} = N_{in} + N_E + N_{out} - 3$$
$$= 1 + 1 + 1 - 3 = 0 \qquad (12.25)$$

For a fixed inlet pressure, the outlet pressure and flowrate are thus fixed where the compressor performance and system head curves cross, as shown in Figure 12.16a.

Whilst centrifugal pumps normally have a fixed rotational speed, many centrifugal compressors can have a variable speed. Figure 12.16b shows a centrifugal *compressor map* in which the speed can be varied. Each fixed speed has its own operating curve. Once the compressor is placed in the process context it must match the system pressure and flowrate characteristics of the system head curve, as shown in Figure 12.16b. The system head curve crosses the compressor operating curves for different speeds at different points. The compressor is constrained to operate along the system head curve. If the speed is fixed, then the compressor will self-regulate to the conditions where the system head curve crosses the compressor operating curve for that speed. This means that the outlet pressure, flowrate, and rotational speed cannot be chosen independently. Equation (12.25) applies at each compressor speed and the freedom to change the speed introduces

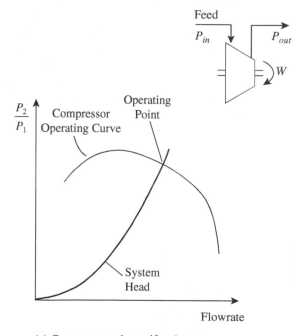

(a) Constant speed centrifugal compressor characterisitics.

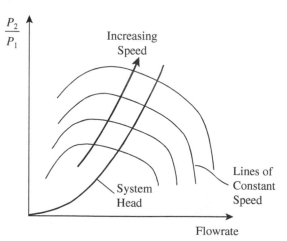

(b) Variable speed centrifugal compressor characterisitics.

**Figure 12.16**

Centrifugal compressor operating curves matched against system head curves.

a single degree of freedom. For example, if the compressor speed is specified, this will fix both the pressure and flowrate where the system head curve crosses the compressor operating curve at that speed. Thus, the speed can be used to control either the flowrate *or* outlet pressure.

If the compressor is positive displacement, rather than centrifugal, then, as with positive displacement pumps, at a fixed rotational speed the compressor delivers a fixed volume of gas, irrespective of the outlet pressure (assuming no slippage of gas in the compressor clearances or internal compressor valves). For a fixed speed, the flowrate is fixed and the outlet pressure is fixed by the fixed pressure difference across the compressor, plus any backpressure from the downstream process and frictional pressure losses. For a fixed speed the input of energy cannot be adjusted independently, but is a function of the flowrate, inlet conditions, pressure difference across the compressor, gas properties, and compressor efficiency. This requires three additional constraints to be subtracted from the control degrees of freedom, and as with positive displacement pumps, for a fixed speed there are no degrees of freedom as Eq. (12.23). Control of positive displacement compressors will be considered in Chapter 13.

# 12.6 Equilibrated Multiphase Operations

There are many process operations that involve the contacting of multicomponent systems with multiple phases in which there is mass transfer between the phases. The contacting might involve existing multiple phases being fed, or the creation of multiple phases through, for example, the partial condensation of a vapor, or partial vaporization of a liquid. These process operations are most often designed by assuming that the phases are allowed to come to phase equilibrium, or are *equilibrated*. For such processes, the control degrees of freedom need to account for both the additional variables and additional constraints.

Consider an operation involving $N_C$ components that contain $N_P$ phases, in which the phases are equilibrated in a *single operation*. If the operation contains $N_P$ phases in equilibrium, this introduces $(N_C - 1)$ composition variables for each additional phase, giving a total of $N_P (N_C - 1)$ composition variables for $N_P$ phases. This adds $N_P (N_C - 1)$ degrees of freedom to Eq. (12.12), giving (Ponton 1994):

$$N_{CDOF} = N_{in} + N_E + N_{out}(N_C + 2) + N_P(N_C - 1) \quad (12.26)$$

where

  $N_C$ = number of components
  $N_P$ = number of phases in equilibrium

For such an equilibrated operation, the following assumptions are made (Ponton 1994):

1) Each outlet stream is assumed to connect with one phase only. This relates to the objective of many multiphase equilibrated operations. However, even though the assumption is made here, it should be noted that it does not apply to all operations with multiple phases.
2) Each outlet stream has the same composition as a phase.
3) The temperatures of all outlet streams are equal.
4) The pressures of all outlet streams are equal.
5) Phase equilibrium relationships constrain the concentrations within each phase.

The constraints for multiphase equilibrated operations are listed and summed in Table 12.5. Subtracting the constraints total given in Table 12.5 from Eq. (12.26) gives the steady-state control degrees of freedom for an individual multiphase operation (Ponton 1994):

$$\begin{aligned} N_{CDOF} &= N_{in} + N_E + N_{out}(N_C + 2) + N_P(N_C - 1) \\ &\quad - N_{out} - N_C N_{out} - N_C N_P + 1 \qquad (12.27) \\ &= N_{in} + N_E + N_{out} - N_P + 1 \end{aligned}$$

Care must be exercised when applying Eq. (12.27) to ensure that the operation complies with the assumptions made in its derivation. For example, an operation for partial condensation of a vapor or partial vaporization of a liquid would comply if each of the two phases leave in a separate outlet. In another example, Eq. (12.27) would also apply to an individual equilibrium stage in a distillation operation. However, it would not apply overall to a cascade of distillation equilibrium stages. This results from the outlet streams from the cascade not all being at the same temperature and all outlet streams not being at the same pressure, as assumed in Table 12.5. Also, the equilibrium relationship between the phases on each stage of the cascade will be different. Even if the same correlation is used for the phase equilibrium on each stage, the parameters in the correlation will be different on each stage because of the different conditions in each stage. If a number of equilibrated operations are linked together, then whilst each operation might comply with the assumptions made in the derivation of Eq. (12.27), this does not necessarily mean that when combined together, it will comply overall.

**12**

## Table 12.5

Constraints for multiphase equilibrated systems.

|  | Number of constraints |
|---|---|
| Component material balances | $N_C$ |
| Energy balance | 1 |
| Temperature equality outlet constraints | $(N_{out} - 1)$ |
| Pressure equality outlet constraints | $(N_{out} - 1)$ |
| Each outlet stream has the same composition as a phase | $N_{out}(N_C - 1)$ |
| Equilibrium phase relationships | $N_C(N_P - 1)$ |
| Total | $N_{out} + N_C N_{out} + N_C N_P - 1$ |

**Example 12.5**   Consider again the example of a single-stage equilibrium flash separator from Figure 12.1 with the stream count shown in Figure 12.17. Determine the steady-state control degrees of freedom using Eq. (12.27).

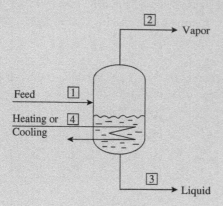

**Figure 12.17**

Streams for the flash drum separation.

**Solution**

Assuming an energy stream in or out of the separator, from Eq. (12.27):

$$N_{CDOF} = N_{in} + N_E + N_{out} - N_P + 1$$
$$= 1 + 1 + 2 - 2 + 1 = 3$$

Note that steams leaving the operation are already constrained to be at the same pressure and temperature. For an adiabatic flash $N_E = 0$:

$$N_{CDOF} = 1 + 0 + 2 - 2 + 1 = 2$$

This is the same result as obtained earlier from the number of variables and number of independent equations, but is much simpler. Again, control could be control of the feed flowrate and flash drum pressure. To maintain steady state it will also require inventory control. This will be considered again in Chapter 13.

**Example 12.6**   Consider the vapor-liquid-liquid separator shown in Figure 12.18.

a) Determine the steady-state control degrees of freedom using Eq. (12.27) and suggest parameters to be controlled.
b) Determine what inventory control would be required.

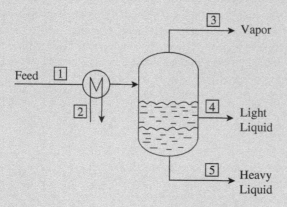

**Figure 12.18**

Vapor–liquid–liquid separator with energy input/output.

**Solution**

a) Assuming an energy stream in or out of the separator, from Eq. (12.27) the steady-state degrees of freedom are given by:

$$N_{CDOF} = N_{in} + N_E + N_{out} - N_P + 1$$
$$= 1 + 1 + 3 - 3 + 1 = 3$$

Again, note that steams leaving the operation are already constrained to be at the same pressure and temperature. For an adiabatic flash $N_E = 0$:

$$N_{CDOF} = 1 + 0 + 3 - 3 + 1 = 2$$

As with the single-stage equilibrium flash, this could be control of the feed flowrate and flash drum pressure.

b) To maintain steady state will require control of the inventory of both liquids in the unit. Vapor inventory is controlled by pressure control, together with the combined level control of both liquids. For inventory control of the liquids, control of the level of the light liquid and control of the level of the liquid-liquid interface would be required. Figure 12.19 shows a possible control scheme for an adiabatic vapor-liquid-liquid separator.

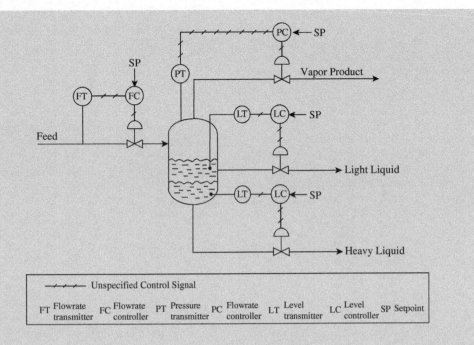

**Figure 12.19**

Control of an adiabatic vapor–liquid–liquid separator.

**Example 12.7**   Figure 12.20a shows an operation for the partial vaporization of the process stream using steam heating. Figure 12.20b shows a partial condensation of the process stream using cooling water cooling. Determine the number of steady-state control degrees of freedom.

**Solution**

Assuming an energy stream in or out of the partial vaporization or partial condensation, from Eq. (12.27):

$$N_{CDOF} = N_{in} + N_E + N_{out} - N_P + 1$$
$$= 1 + 1 + 2 - 2 + 1$$
$$= 3$$

Note again from the assumptions for Eq. (12.27) that each outlet stream is assumed to connect with one phase only. The partial vaporization is the same as the non-adiabatic flash drum from Example 12.5. The process flowrate needs to be controlled, then two from the outlet temperature, outlet pressure, or vapor-liquid split fraction (e.g., via control of the vapor outlet flowrate). Note that temperature control cannot be applied for single component systems, as the temperature of the liquid does not change with vaporization of the liquid. Again, steams leaving the operation are already constrained to be at the same pressure and temperature.

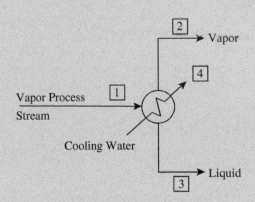

(a) Streams for a partial vaporization          (b) Streams for a partial condensation.

**Figure 12.20**

Partial vaporization and condensation.

12

**Example 12.8**  Figure 12.21a shows an operation for the total vaporization of the process stream using steam heating. Figure 12.21b shows a total condensation of the process stream using cooling water cooling. Determine the number of control degrees of freedom.

**Solution**

Total vaporization and total condensation are limiting cases from the partial vaporization and partial condensation considered in Example 12.7. If equilibrium is assumed throughout the process operation, then it is limited by phase equilibrium (which is not necessarily always the case). As the final vaporization or condensation occurs, the number of outlet streams changes from 2 to 1. Thus, from Eq. (12.27):

$$N_{CDOF} = N_{in} + N_E + N_{out} - N_P + 1$$
$$= 1 + 1 + 1 - 2 + 1$$
$$= 2$$

Process flowrate, together with temperature *or* pressure can be controlled. Note that the count is still correct if the product stream is not assumed to be in equilibrium with the feed. In this case there is still a relationship between the compositions of the two phases, i.e., equality of phase compositions, rather than an equilibrium relation as assumed in Table 12.5 (Ponton 1994).

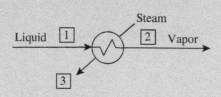

(a) Streams for a total vaporization

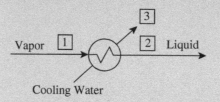

(b) Streams for a total condensation.

**Figure 12.21**

Total vaporization and condensation.

## 12.7  Control Degrees of Freedom for Overall Processes

Connected operations sharing a stream lose one degree of freedom for each connecting stream from the sum of those for the individual operations. The control degrees of freedom for connected operations can be obtained by summing the degrees of freedom for each operation and subtracting the number of shared streams (Ponton 1994).

Consider the cascade of countercurrent equilibrium stages shown in Figure 12.22. This is the arrangement internal to absorption or stripping columns and the rectifying or stripping sections of distillation columns. The number of steady-state control degrees of freedom for each equilibrium stage is given by Eq. (12.27):

$$N_{CDOF} = N_{in} + N_E + N_{out} - N_P + 1$$
$$= 2 + 0 + 2 - 2 + 1 \tag{12.28}$$
$$= 3$$

For a cascade of *n* stages, deducting $2(n-1)$ shared streams:

$$N_{CDOF} = 3n - 2(n-1)$$
$$= n + 2 \tag{12.29}$$

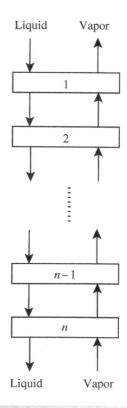

**Figure 12.22**

Cascade of equilibrium stages.

In principle, it is possible to maintain each stage at a different pressure by a valve arrangement on the vapor between the trays. However, this would not be exploited in practice. If the pressures of the stages are left to be self-regulating then the pressure drop equation for each stage adds a constraint for each stage (even if $\Delta P = 0$). Given that liquid and vapor streams are already constrained to leave each stage at the same pressure, this imposes an additional $(n - 1)$ constraints:

$$\begin{aligned} N_{CDOF} &= (n + 2) - (n - 1) \\ &= 3 \end{aligned} \tag{12.30}$$

Figure 12.23 shows a cascade of equilibrium stages with an intermediate feed. Above the feed stage there are $n_R$ stages in the rectifying section. Below the feed stage there are $n_S$ stages in the stripping section. Thus, from Eq. (12.30) there are 3 steady-state control degrees of freedom for the rectifying section above the feed stage and 3 for the stripping section below the feed stage. For the feed stage the steady-state control degrees of freedom are given by Eq. (12.27):

$$\begin{aligned} N_{CDOF} &= N_{in} + N_E + N_{out} - N_P + 1 \\ &= 3 + 0 + 2 - 2 + 1 \\ &= 4 \end{aligned} \tag{12.31}$$

Adding the number of control degrees of freedom together for the rectifying section, the feed stage, and the stripping section, and subtracting the 4 connecting streams gives:

$$N_{CDOF} = 3 + 4 + 3 - 4 = 6 \tag{12.32}$$

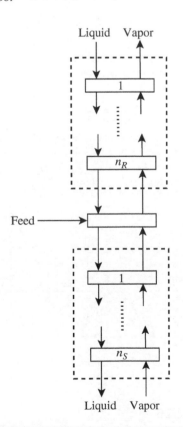

**Figure 12.23**

Cascade of equilibrium stages with intermediate feed.

However, additional constraints dictate that the three parts should be at the same pressure (or the pressure is self-regulating), which adds two constraints, giving:

$$N_{CDOF} = 3 + 4 + 3 - 4 - 2 = 4 \tag{12.33}$$

Two alternative but related approaches can be used to determine the control degrees of freedom for an overall process:

1) Sum the control degrees of freedom for the individual operations, subtract the number of shared (connecting) streams and subtract any additional constraints for the outlet pressure of the individual operations.

2) Rather than add the control degrees of freedom for each operation from Eq. (12.27) and subtracting the number of shared streams, as implied by Option 1 above, Eq. (12.27) implies that the number of streams can simply be added across the whole process, irrespective of the operations, subtracting $(N_P - 1)$ for each occurrence of phases greater than 1 that are *constrained by mass transfer equilibrium* and subtracting the number of shared streams. This can be applied by simply adding the number of streams across the process directly (which accounts for the shared streams) and subtracting the number of occurrences of $(N_P - 1)$ where there are mass transfer equilibrium constrained operations. This equates to adding up the streams and subtracting the number of *mass transfer equilibrium constrained phase interfaces* in the process units around which the stream count occurs (Ponton 1994):

$$\begin{aligned} N_{CDOF} &= N_{in} + N_E + N_{out} - N_P + 1 \\ &= N_{Streams} - (N_P - 1) \\ &= N_{Streams} - N_{INT} \end{aligned} \tag{12.34}$$

where

$N_{Streams}$ = number of streams

$N_P$ = number of mass transfer equilibrium constrained phases

$N_{INT}$ = number of mass transfer equilibrium constrained phase interfaces

Any additional constraints for the outlet pressure of the individual operations where pressure control is not applied must also be subtracted from Eq. (12.34).

Once the steady-state control degrees of freedom have been established, allowance can be made for inventory control. This will require additional single-input–single-output control loops to maintain the material balance in both individual operations and the overall process. For liquids, vessels in which the liquid level can change will require level control to control the inventory (Buckley 1964; Aske and Skogestad 2009). Pressure control will also be required in some cases for vapor inventory control. These inventory controllers will need to be added after control of the other process variables counted as steady-state control degrees of freedom. Thus, in addition to the control degrees of freedom identified to control flowrate, temperature, pressure, and composition in the steady-state control degrees of freedom analysis, additional controllers will be required to control the process inventory where necessary.

**12**

**Example 12.9** Return to the three-phase equilibrium flash separator analyzed previously, shown in Figure 12.18 and analyzed in Example 12.6. Determine the steady-state control degrees of freedom from the number of streams and mass transfer equilibrium constrained phase interfaces.

## Solution

Number of streams = 5
Number of mass transfer equilibrium constrained phase interfaces ($N_{INT} = N_P - 1$) = 2
From Eq. (12.34):

$$N_{CDOF} = N_{Streams} - N_{INT}$$
$$= 5 - 2 = 3$$

For an adiabatic flash:

$$N_{CDOF} = N_{Streams} - N_{INT}$$
$$= 4 - 2 = 2$$

This is the same result as obtained previously. This could be control of the feed flowrate and flash drum pressure. To maintain the steady state will later require control of the inventory of both liquids in the unit, as discussed previously.

**Example 12.10** Figure 12.24 shows a liquid process stream being heated without vaporization from the supply temperature of 20 to 200 °C by low-pressure steam and then high-pressure steam. Determine the steady-state control degrees of freedom and the number of control loops required.

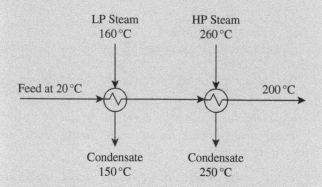

**Figure 12.24**

A liquid feed needs to be preheated by low-pressure steam and high-pressure steam.

## Solution

Figure 12.25 shows the stream count for the system, counting the steam streams as energy streams:

Number of streams = 5
Number of mass transfer equilibrium constrained phase interfaces: ($N_{INT} = N_P - 1$) = 0

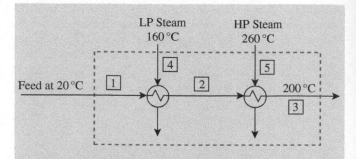

**Figure 12.25**

Stream numbering for the feed preheat system with the steam streams counted only as energy streams.

From Eq. (12.34):

$$N_{CDOF} = 5 - 0 = 5$$

The degrees of freedom that fix the pressures of Streams 2 and 3 would not normally be exploited. Instead, the pressures would be self-regulating and two constraints from the equations for the process-side pressure drop equations for each heater need to be subtracted:

$$N_{CDOF} = 5 - 2 = 3$$

Alternatively, consider both sides of the heaters as shown in Figure 12.26 and treat the steam heaters as condensers. Even though the heat exchangers are now considered as condensers they are not constrained by mass transfer equilibrium constrained interfaces. Hence:

$$N_{CDOF} = N_{Streams} - N_{INT}$$
$$= 7 - 0 = 7$$

Two constraints from the equations for the process-side and steam-side pressure drop equations for each of the two heaters need to be subtracted:

$$N_{CDOF} = 7 - 4 = 3$$

This is the same result, but a slightly different analysis of the system. However, in practice the outlet pressure on the steam side would be constrained by a steam trap. The control system requires 3 control loops, as shown in Figure 12.27.

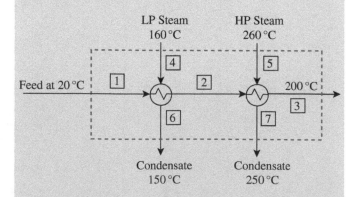

**Figure 12.26**

Stream numbering for the feed preheat system with all streams counted.

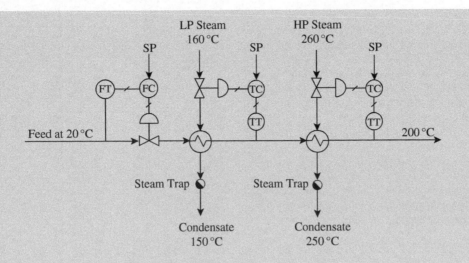

**Figure 12.27**

Control system for the preheat.

**Example 12.11**    Determine the control degrees of freedom for a distillation column with a total condenser and with:

a) A kettle reboiler.
b) A thermosyphon reboiler.
c) An integral reboiler.

**Solution**

a) Figure 12.28 shows a distillation with a kettle reboiler and the resulting stream count:

$$N_{Streams} = 10$$

There are mass transfer equilibrium constrained interfaces in the distillation column itself, the condenser, and the kettle reboiler. Note that the distillate receiver is not counted as having a constrained interface, as it is just collecting condensate from the condenser, acting as a splitter and providing an inventory to assist operability, and is not constrained by vapor–liquid equilibrium:

$$N_{INT} = 3$$

Thus:

$$N_{CDOF} = 10 - 3 = 7$$

However, it would not be usual to control the outlet pressure for the condenser, reflux drum, or reboiler and these would be left to self-regulate, requiring 3 further constraints to be subtracted. Thus:

$$N_{CDOF} = 7 - 3 = 4$$

An alternative analysis is to count the control degrees of freedom for each of the operations that make up the distillation system and subtract the number of connecting streams. As calculated earlier, the control degrees of freedom for the column itself with only the vapor and liquid streams feeding and leaving at the top and bottom is 4. As shown previously the condenser

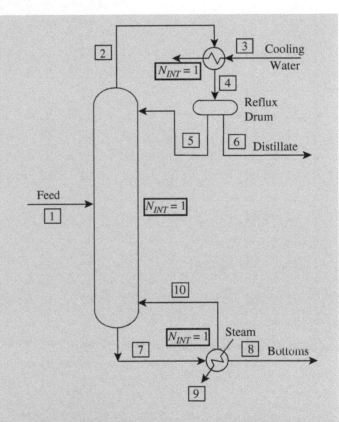

**Figure 12.28**

Degrees of freedom for distillation with a total condenser and a kettle reboiler.

adds 2 control degrees of freedom, the reflux drum (as a splitter) adds 3, and the kettle reboiler (as a partial vaporization with two outlets) adds 3. This gives a total of 12 and there are 5 connecting streams, giving the number of control degrees of freedom to be 7. Then subtracting the three self-regulating pressure constraints for the condenser, reflux drum, and reboiler gives a total control degrees of freedom of 4, which is in agreement with the above analysis.

**12**

Analysis of the steady-state control degrees of freedom shows that there will be 4 single-input–single-output control loops required. For inventory control, there are liquid levels in the distillation reflux drum and distillation column bottoms that need to be controlled to control the material balance. When inventory control is included, the number of single-input–single-output control loops will need to increase from 4 to 6.

An alternative arrangement for splitting the reflux is shown in Figure 12.29. The stream count is now 11 with 3 mass transfer equilibrium constrained interfaces as before, resulting in 8 steady-state control degrees of freedom. Then subtracting 4 self-regulating pressure constraints for the condenser, reflux drum, splitter at the outlet of the reflux drum, and reboiler gives 4 steady-state degrees of freedom, as before.

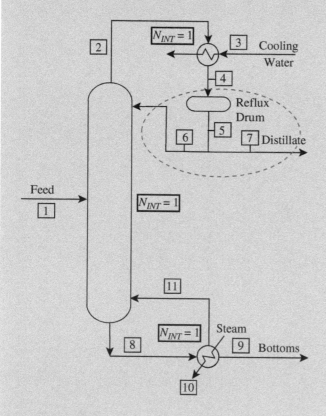

**Figure 12.29**

Degrees of freedom for an alternative arrangement around the reflux for distillation with a total condenser and a kettle reboiler.

**b)** Figure 12.30 shows a distillation with a thermosyphon reboiler and the resulting stream count:

$$N_{Streams} = 11$$

Again, there are mass transfer equilibrium constrained interfaces in the distillation column itself, the condenser, and the thermosyphon reboiler:

$$N_{INT} = 3$$

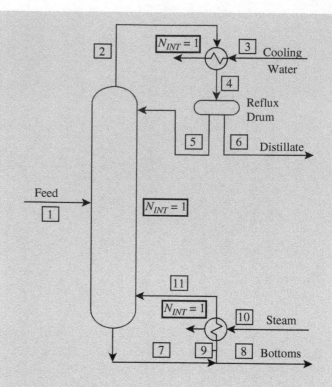

**Figure 12.30**

Degrees of freedom for distillation with a total condenser and a thermosyphon reboiler.

Thus:

$$N_{CDOF} = 11 - 3 = 8$$

However, it would not be usual to control the outlet pressure for the condenser, reflux drum, splitter at the base of the column, or reboiler. These would be left to self-regulate, requiring 4 further constraints to be subtracted. Thus:

$$N_{CDOF} = 8 - 4 = 4$$

Again, there are four control degrees of freedom. An alternative analysis is to count the control degrees of freedom for each of the operations that make up the distillation and subtract the number of connecting streams. The control degrees of freedom for the column itself are 4, as given previously, the condenser adds 2, the reflux drum (as a splitter) adds 3, the splitter at the base of the column adds 3, and the thermosyphon reboiler adds 2. This gives a total of 14 and there are 6 connecting streams, giving the number of control degrees of freedom to be 8. Then, subtracting the 4 outlet pressure constraints for the condenser, reflux drum, splitter at the base of the column, and thermosyphon reboiler gives a total control degrees of freedom of 4, which is in agreement with the above analysis.

As with kettle reboilers, for thermosyphon reboilers analysis of the control degrees of freedom shows that there will be 4 single input–single-output-control loops required. For inventory control, there are liquid levels in the distillation reflux drum and distillation column bottoms that need to be controlled to control the material balance. When inventory control is included, the number of single-input–single-output control loops will again need to increase from 4 to 6.

**c)** Figure 12.31 shows a distillation with an integral reboiler and the resulting stream count:

$$N_{Streams} = 10$$

There are mass transfer equilibrium constrained interfaces in the distillation column itself, the condenser, and the integral reboiler:

$$N_{INT} = 3$$

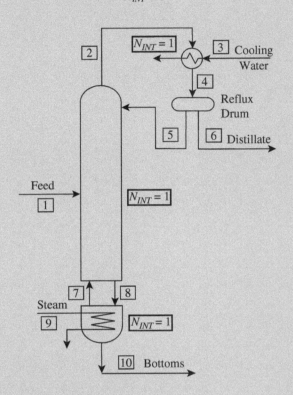

**Figure 12.31**

Degrees of freedom for distillation with a total condenser and an integral reboiler.

Thus:

$$N_{CDOF} = 10 - 3 = 7$$

If the outlet pressure for the condenser, reflux drum, and integral reboiler are left to self-regulate, this introduces 3 further constraints to be subtracted. Thus:

$$N_{CDOF} = 7 - 3 = 4$$

Again, there are four control degrees of freedom. It should be noted that if the distillation and integral reboiler are combined in Figure 12.31, the stream count decreases from 10 to 8, the number of mass transfer equilibrium constrained interfaces decreases from 3 to 2 and there is one fewer outlet pressure constraint from the integral reboiler. This gives the same result of 4 control degrees of freedom.

An alternative analysis for Figure 12.31 is to count the control degrees of freedom for each of the operations that make up the distillation and subtract the number of connecting streams. In Figure 12.31 the control degrees of freedom for the column itself are 4, as shown previously, the condenser adds 2, the reflux drum (as a splitter) adds 3, and the integral reboiler as a partial vaporizer adds 3. This gives a total of 12 and there are 5 connecting streams, giving the number of control degrees of freedom as 7. Then subtracting the 3 outlet pressure constraints for the condenser, reflux drum, and integral reboiler gives a total control degrees of freedom of 4, which is in agreement with the above analysis.

As with the kettle and thermosyphon reboilers, analysis of the control degrees of freedom shows that there will be 4 single input–single-output control loops required. For inventory control, there are liquid levels in the distillation reflux drum and distillation column bottoms that need to be controlled to control the material balance. When inventory control is included, the number of single-input–single-output control loops will again need to increase from 4 to 6.

The control of distillation columns will be discussed in Chapter 13.

**12**

**Example 12.12**   Figure 12.32 shows a process for the recovery of acetone from an air stream that is to be vented to atmosphere. The acetone laden air stream is fed to an absorption column where it is contacted with water. The water dissolves the acetone and the air is sent for further treatment before discharge. The water with the dissolved acetone from the absorption column is fed to a distillation column to recover pure acetone. The distillation column has an *internal reboiler* fed by steam and a total condenser serviced by cooling water. The water from the bottom of the distillation column is cooled and recycled to the water storage tank. The water storage tank has a water make up to compensate for any water losses from the process at the top of the absorption column and with the acetone product. Determine the number of control degrees of freedom.

**Solution**

Figure 12.33 shows the process with the stream count:

$$N_{Streams} = 14$$

There are mass transfer equilibrium constrained interfaces in the absorption column, the distillation column, and the distillation column condenser:

$$N_{INT} = 3$$

Thus:

$$N_{CDOF} = 14 - 3 = 11$$

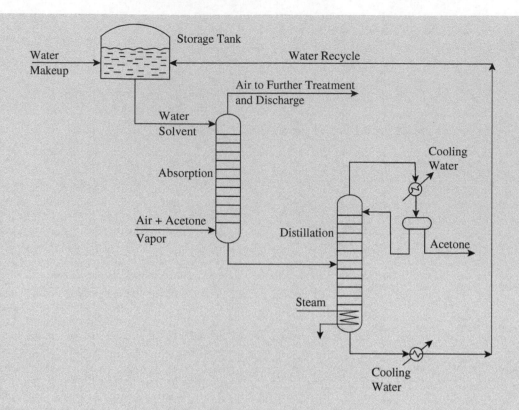

**Figure 12.32**

A solvent recovery process.

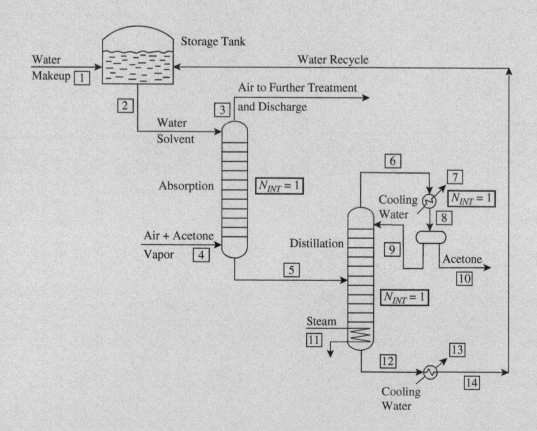

**Figure 12.33**

Stream count for the solvent recovery process.

However, additional constraints not accounted for in Table 12.4 and Eqs (12.13) and (12.34) need to be included, in particular, any outlet pressures that are left to self-regulate. The degree of freedom to fix the outlet pressure from the storage tank will be left to self-regulate. Also, the degrees of freedom to operate the distillation column condenser and the distillate reflux drum at different pressures from the distillation column would not normally be exploited, removing 2 additional degrees of freedom through self-regulation of the pressure outlet. Finally, the outlet of the recycle cooler will not have a pressure control, but outlet pressure will be self-regulating. Thus:

$$N_{CDOF} = 11 - 4 = 7$$

Thus, 7 single-input–single-output control loops will be required.

An alternative analysis is to count the control degrees of freedom for each of the operations that make up the process and subtract the number of connecting streams. For the operations:

| Operation | $N_{CDOF}$ |
|---|---|
| Storage Tank (represented as a mixer) | 3 |
| Absorption Column | 3 |
| Distillation Column | 4 |
| Cooler | 3 |
| Total | 13 |

This gives a total of 13 control degrees of freedom and there are 4 connecting streams, giving the number of control degrees of freedom to be 9. Then subtracting the 2 self-regulating pressure constraints for the storage tank and the distillation bottoms cooler gives a total control degrees of freedom of 7, which is in agreement with the above analysis. Note that the outlet pressure constraints for the distillation condenser and reflux drum have already been accounted for in the distillation column degrees of freedom.

For inventory control, there are liquid levels in the storage tank, absorption column bottoms, distillation reflux drum, and distillation column bottoms that need to be controlled, requiring 4 inventory controllers. When inventory control is included, the number of single-input–single-output control loops will need to increase from 7 to 11.

However, one final point is that the flowrate of air to the process will be fixed by the upstream process, removing the degree of freedom to set the feed flowrate. This decreases the control degrees of freedom and the number of single-input–single-output control loops from 11 to 10. The analysis will be validated in Chapter 14 when the control system for this process is synthesized.

**Example 12.13** Figure 12.34 shows the process for the reaction between Feed $A$ and Feed $B$ to produce Product $C$ according to the reaction:

$$A + B \rightarrow C + \text{light byproducts}$$

Whilst the main reaction is to Product $C$, some light (high-volatility) byproducts are also formed. There is some vaporization of components from the reactor. As much volatile material as possible is condensed using cooling water and returned to the reactor. The uncondensed light (non-condensable) byproducts are vented. The flowrate of the vented non-condensable byproducts can be used to control the pressure of the reactor. The reactor effluent is a mixture of Components $A$, $B$, and $C$, which is first cooled using cooling water and then separated in a vacuum distillation column with a thermosyphon reboiler. The order of relative volatility of the three components is $\alpha_A > \alpha_B > \alpha_C$. The unreacted Components $A$ and $B$ are removed as the distillation overhead and recycled to the reactor. Product $C$ is taken from the distillation column bottoms. The flowrates of Feed $A$ and Feed $B$ to the process are not in this case fixed by the outlet of upstream processes and can be varied to meet the flowrate demands of Product $C$. Determine the number of control degrees of freedom.

**Solution**

For this more complex case, the degrees of freedom for the process will be determined by dividing the process into operations and the

degrees of freedom determined for each operation. These will then be combined to give the degrees of freedom for the whole process.

Starting with the reactor, it is assumed that the reactor design and reaction conditions are fixed, thereby freezing the degrees of freedom associated with the reaction system itself. Figure 12.35 shows the reactor with the stream count:

$$N_{Streams} = 10$$

There are mass transfer equilibrium constrained interfaces in the reactor and the condenser:

$$N_{INT} = 2$$

Thus:

$$N_{CDOF} = 10 - 2 = 8$$

The outlet pressure for the condenser and reflux drum will not be controlled to be a different pressure from the reactor and will be left to self-regulate, requiring 2 further constraints to be subtracted. Thus:

$$N_{CDOF} = 8 - 2 = 6$$

For inventory control, there are liquid levels in the reactor and the reflux drum, requiring 2 level controllers. Thus:

$$N_{CDOF} = 6 + 2 = 8$$

**12**

$$A + B \rightarrow C$$
$$\alpha_A > \alpha_B > \alpha_C$$

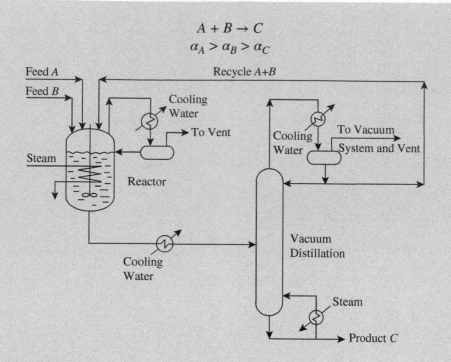

## Figure 12.34

A process involving a reaction, separation, and recycle system.

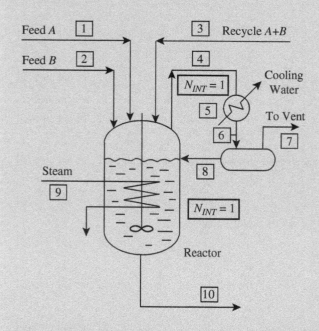

## Figure 12.35

Degrees of freedom for the process reactor

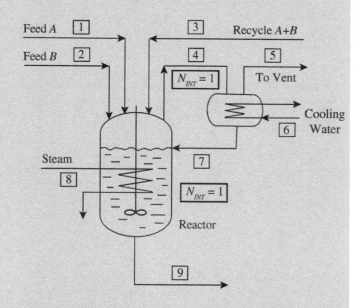

## Figure 12.36

An alternative representation for the degrees of freedom for the reactor.

To check the degrees of freedom for the reactor condenser and reflux arrangement, an alternative representation is shown in Figure 12.36. Conceptually, the reflux condenser and drum together are essentially the same as the non-adiabatic flash drum shown in Figure 12.17, in this case using cooling water to carry out a partial condensation. The stream count from Figure 12.36 is given by:

$$N_{Streams} = 9$$

There are mass transfer equilibrium constrained interfaces in the reactor and the non-adiabatic flash drum:

$$N_{INT} = 2$$

Thus:

$$N_{CDOF} = 9 - 2 = 7$$

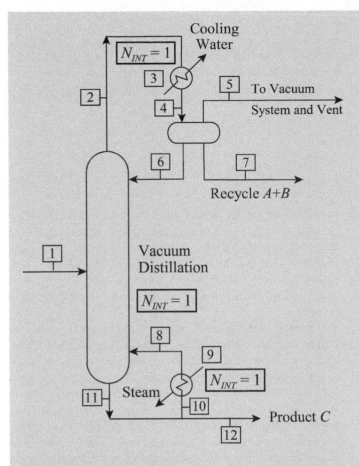

The outlet pressure for the non-adiabatic flash drum will not be controlled to be a different pressure from the reactor and will be left to self-regulate, requiring 1 further constraint to be subtracted. Thus:

$$N_{CDOF} = 7 - 1 = 6$$

Adding the 2 inventory controls for the reactor and the non-adiabatic flash drum gives the same result of 8 degrees of freedom.

The degrees of freedom for the product cooler have been analyzed previously and are 2. Figure 12.37 shows a vacuum distillation with a thermosyphon reboiler and the resulting stream count:

$$N_{Streams} = 12$$

There are mass transfer equilibrium constrained interfaces in the distillation column itself, the condenser, and the thermosyphon reboiler:

$$N_{INT} = 3$$

Thus:

$$N_{CDOF} = 12 - 3 = 9$$

However, it would not be usual to control the outlet pressure for the condenser, reflux drum, splitter at the base of the column, or reboiler and these would be left to self-regulate, requiring 4 further constraints to be subtracted. Thus:

$$N_{CDOF} = 9 - 4 = 5$$

For inventory control, there are liquid levels in the reactor and the reflux drum. Thus:

$$N_{CDOF} = 5 + 2 = 7$$

**Figure 12.37**

Degrees of freedom for the vacuum distillation with a thermosyphon reboiler.

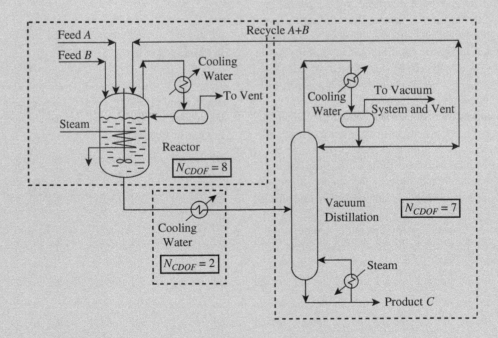

**Figure 12.38**

The nodes can be combined to give the overall process control degrees of freedom.

12

Again, the combination of condenser and reflux drum is conceptually the same as the non-adiabatic flash drum shown in Figure 12.17 using cooling water to carry out a partial condensation. An alternative analysis assuming a non-adiabatic flash drum to represent the combination of condenser and reflux drum gives the same result.

Connecting the 3 operations together in Figure 12.38 gives a total of $(8 + 2 + 7) = 17$ control degrees of freedom and there are 3 connecting streams, giving the number of control degrees of freedom to be $17 - 3 = 14$. The analysis will be validated in Chapter 14 when the control system for this process is synthesized in Chapter 14. In particular, the control of the material balance needs careful consideration, as will be discussed in Chapter 14.

# 12.8 Degrees of Freedom – Summary

Once the process design has specified the non-adjustable variables (e.g., equipment dimensions), the remaining degrees of freedom are the control degrees of freedom. The number of control degrees of freedom dictates the number of single-input–single-output control loops, and therefore the number of final control elements (Luyben 1996).

Degrees of freedom analysis is performed using steady-state equations. Thus, adjustments that affect only the inventory are not included in the control degrees of freedom analysis, unless they affect state variables. This means that the approach starts with the steady state analysis to determine the number of control degrees of freedom. The control of inventory is considered in a second stage.

For single phase operations, the number of degrees of freedom can be determined by counting the degrees of freedom for the inlet and outlet streams.

For equilibrated multiphase operations there are additional variables and additional constraints that need to be included in the analysis. Each outlet stream is assumed to connect to one phase only. The temperatures and pressures of all outlet streams are assumed to be equal and phase equilibrium relationships constrain the concentrations of each phase. The degrees of freedom from counting the inlet and outlet streams are then adjusted by these additional variables and constraints.

In an overall process, each connected operation sharing a stream loses one degree of freedom from the sum of those for the individual operations. A simple approach to determine the control degrees of freedom for an overall process is to add up all the streams in the process and subtract the number of mass transfer equilibrium constrained phase interfaces in the process units.

Care must be taken to account for the outlet pressures of operations. The outlet pressure from an operation might be controlled or left to self-regulate. Without pressure control the outlet pressure will be dictated by a combination of fluid frictional pressure losses, the input or output of mechanical energy (work), and the back pressure from the process downstream. The effect of self-regulation of the back pressure is to add another constraint in the form of a pressure drop equation.

The basis of this analysis of control degrees of freedom is to assume that the material balance is controlled separately. However, to maintain the material balance requires the process inventory to be controlled. Inventory control may be needed for gases and vapors through the control of pressure. For liquids, vessels in which the liquid level can change will require level control to control the inventory (Buckley 1964; Aske and Skogestad 2009). These inventory controllers will need to be added after control of the of the other process variables counted as steady-state control degrees of freedom in this analysis. Thus, in addition to the degrees of freedom identified to control flowrate, temperature, pressure, and composition in the degrees of freedom analysis, additional controllers will be required to control the process inventory where necessary.

# Exercises

1. Figure 12.39 shows three different absorber systems. In each case water is used as a solvent to dissolve components in a low composition from a gas feed. Determine the control degrees of freedom for:

   a) The absorber in Figure 12.39a with the solvent water supplied only as fresh feed.

   b) The absorber with a local recycle and purge shown in Figure 12.39b.

   c) The absorber with a local recycle, purge, and cooler in the recycle stream shown in Figure 12.39c.

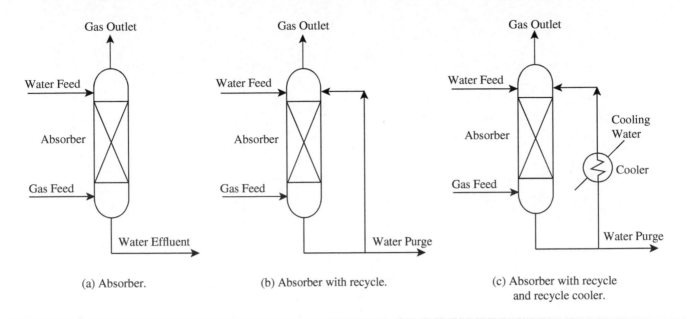

**Figure 12.39**

Control degrees of freedom for absorption systems.

2. A distillation operation is illustrated in Figure 12.40. The distillation has an internal reboiler and a partial condenser. Determine the control degrees of freedom.

# References

Aske, E.M.G. and Skogestad, S. (2009). Consistent inventory control. *Industrial and Engineering Chemistry Research* 48: 10892–10902.

Buckley, P.S. (1964). *Techniques of Process Control*. Wiley.

Himmelblau, D.M. and Riggs, J.B. (2003). *Basic Principles and Calculations in Chemical Engineering*, 7e. Prentice-Hall.

Konda, N.V.S.N., Rangaiah, G.P., and Krishnaswamy, P.R. (2006). A simple and effective procedure for control degrees of freedom. *Chemical Engineering Science* 61: 1184–1194.

Kwauk, M. (1956). A system for counting variables in separation processes. *AIChEJ* 2: 240–248.

Luyben, W.L. (1996). Design and control degrees of freedom. *Industrial and Engineering Chemistry Research* 35: 2204–2214.

Ponton, J.W. (1994). Degree of freedom analysis in process control. *Chemical Engineering Science* 49: 2089–2095.

Smith, B.D. (1963). *Design of Equilibrium Stage Processes*, Chapter 3. McGraw-Hill.

Smith, R. (2016). *Chemical Process Design and Integration*, 2e. Inc.

**12**

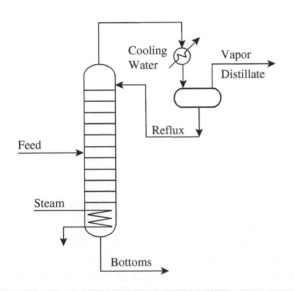

**Figure 12.40**

Distillation with an internal reboiler and external partial condenser.

# Chapter 13

# Process Control – Control of Process Operations

The control degrees of freedom specify the number of process variables that may be manipulated once design specifications are fixed. This dictates the number of single-input–single-output control loops and the number of final control elements. This dictates the number of control loops required for regulating flowrate, temperature, pressure, composition, pH, and so on. However, inventory control is also required to maintain the material balance in a pseudo steady state through time. For liquids, vessels in which the liquid level can change will require level control to control the inventory. For gases, pressure control might be required for inventory control (see Chapter 11). Once the total number of single-input–single-output control loops has been established for all variables, including inventory, then the control configuration needs to be synthesized. The approach to be adopted here will be to decompose the process into *process operations* or *process nodes*. Each node consists of one or more processing units, which together have a common goal. For example, a distillation node would consist of the column, reboiler, condenser, and distillate receiver (reflux drum). A reactor node would consist of the reactor, together with any heating or cooling system required to service the heat of reaction. A control configuration will then be developed for each process node in isolation, before considering the overall process control configuration, which will be considered in Chapter 14.

Before developing the control configuration for a process node, it is important to understand the control objectives. Sometimes the control objectives are clear and unambiguous. For example, a process heater has a clearly defined objective of input of heat to a process stream to increase the temperature to a desired value. In other cases, the control objectives might not be so straightforward. For example, a distillation column control configuration can give priority to the control of the recovery of components, or product composition (or both), of either the distillate or column bottoms. Even further for distillation, there might be a mix of objectives, possibly involving both the distillate and column bottoms. It is also important to understand the types of disturbance

to be expected in order to design the most effective system to overcome the disturbances. For example, a distillation column might be subjected to disturbances in the feed flowrate, but not perhaps the feed composition.

Once the process nodes are combined together to give the complete process, the overall process control configuration must then satisfy the overall process control objectives when subjected to disturbances. In many cases the flowrate of the inlet streams to individual nodes will not need to be controlled when placed in context of the overall process, but will be fixed by the process upstream. However, when considering the process nodes in isolation it will be assumed that inlet streams will need to be controlled. The control configuration of individual process nodes will be synthesized to satisfy the control degrees of freedom. Only when the process nodes are located within the overall process will the unnecessary inlet flowrate controls be removed. The conceptual design of the control configuration needs first to be developed. Later, the conceptual design of the control configuration of the process nodes will be combined to develop the corresponding conceptual design for the overall process. This later provides the basis for the design of the distributed control system (DCS), possibly using model predictive control. First the conceptual design of the control configuration of the most common process operations will be considered.

## 13.1 Pump Control

Figure 13.1 shows a constant speed centrifugal pump. Centrifugal pumps deliver an output pressure or *head* that is a function of flowrate (Figure 13.1). The pump head curve is fixed by the design of the pump and represents the performance of the pump under different conditions. The performance of the pump when installed is a consequence of both the pump design and the surrounding process. In Chapter 12 it was concluded that there is one degree of freedom *inherent* to a fixed speed centrifugal pump. If the outlet pressure is specified (or fixed by the surrounding process), this fixes the flowrate. If the outlet flowrate is specified (or fixed by the surrounding process), this fixes the pressure.

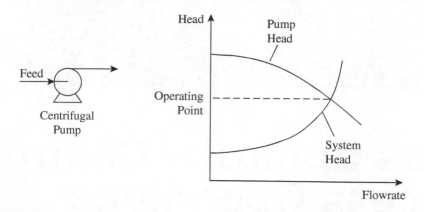

## Figure 13.1

Centrifugal pump flowrate control.

When the pump is located in the process, the pumping action is against a pressure head created by frictional losses in the outlet system of pipes, pipe fittings and vessels, the hydraulic head created by differences in elevation, and the back-pressure from vessels in the system. The pressure drop for the outlet system, or *system head*, increases as the flowrate through the system increases from increased fluid frictional losses (Figure 13.1). The *system head curve* takes up a degree of freedom when combined with the pump, as it introduces another equation. The pump therefore self-regulates to match the system head curve, and there are no remaining degrees of freedom. The flowrate created by the pump when located in the process is where the two head curves cross (Figure 13.1).

Figure 13.2 shows flowrate control on the outlet of a centrifugal pump. As the control valve opens or closes, it decreases or increases the pressure drop across the valve. When the pressure drop across the valve is combined with the system head, this changes the shape of the curve for the outlet system. Thus, the control valve adjusts the shape of the system head curve so that it intersects the pump head curve at the desired point. In this case, the control valve releases the constraint created by the fixed system head curve and the control loop takes up the resulting degree of freedom. The control valve has a limited range over which the outlet pressure or flowrate can be varied. The flowrate can be reduced to zero by closing the valve completely, but the maximum flowrate will be dictated by the intersection of the pump curve with the system head when the valve is fully open and the pressure drop across the valve is effectively zero (in practice there will also be some pressure drop even when the valve is fully open). It is often important to avoid zero discharge flow caused by the outlet flow being blocked either by the control valve, or another downstream valve, or by a blockage in the outlet pipe, known as *deadheading*. This can cause damage to the pump in some cases, or create undesirable heating of the process liquid from the pump energy input. If this is the case, a small percentage of the total flow through the centrifugal pump is routed back to the suction (source tank), known as a *spill-back* or *minimum-flow protection*. These issues will be

discussed again later in the context of piping and instrumentation diagrams. Such spill-backs become even more important for positive displacement pumps.

For fixed speed positive displacement pumps, it was concluded in Chapter 12 that there are no degrees of freedom. The pump delivers a fixed volume of liquid, irrespective of the outlet pressure (assuming no slippage of liquid in the clearances in the pump). The flowrate is fixed but positive displacement pumps do not produce pressure. The pressure is created by the system itself, which creates a back pressure. The back pressure is dependent on the pressure in the discharge vessel and the system pressure drop. Thus, if the flowrate from a fixed speed positive displacement pump is to be controlled, a new degree of freedom needs to be introduced, as shown in Figure 13.3a. This shows a *recycle* or *spill-back* control configuration. Because the flow resistance of the outlet system does not change the flowrate through the pump, the flowrate can only be controlled by controlling the flowrate of the recycle back to the source tank. The control valve in Figure 13.3a is designated to fail open to ensure that control configuration failure does not create a hazardous overpressure in the pump discharge in the event of the outlet system being blocked by closure of a valve or an accidental blockage. Figure 13.3b also shows a recycle control arrangement that features a *high-pressure override* in which the outlet pressure is controlled in addition to the flowrate control. The flowrate control and pressure control signals are passed to a *low selector* to choose the lower of the two signals, so that the desired flowrate is delivered subject to not exceeding some maximum acceptable pressure. Since the valve is fail open, the lower of the two signals drives the valve to the safe state (Driedger 1996). It should be noted that if there is a high recycle flowrate in the schemes in Figure 13.3, then the input of energy to the liquid from pumping will build up in the recycle, causing a significant temperature increase in the liquid. If this is a concern, then a cooler (e.g., cooling water cooler) needs to be located in the recycle to remove the energy of pumping.

An often favorable alternative to using the recycle arrangement shown in Figure 13.3 is to use a positive displacement pump with a variable speed drive, e.g., a variable speed motor. In this case,

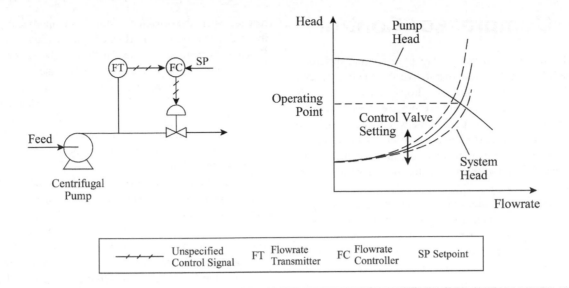

## Figure 13.2

Centrifugal pump flowrate control.

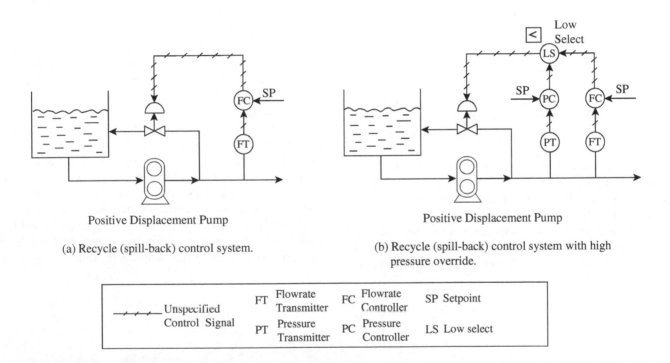

## Figure 13.3

Constant speed positive displacement pump flowrate control.

the flowrate controller controls the speed of the pump, and therefore the volume delivered. Whilst a variable speed motor will be more expensive than a constant speed motor, it can save energy when operating at low throughput and might also avoid adverse effects on the process fluid from the unnecessary input of pumping energy at a low flowrate. There is a trade-off between capital and energy costs.

In some situations, a positive displacement pump is used to dose the process with a small flowrate of a chemical. For example, the process might need an acid or alkali dose to the process to control pH. In such situations, on/off control might be used to start the pump motor automatically when dosing is required. Alternatively, some piston and cylinder positive displacement pumps can control the flowrate by varying the piston stroke (cylinder loading).

# 13.2   Compressor Control

Centrifugal and axial compressors create a pressure difference between the inlet and the outlet that delivers a flowrate depending on the back-pressure and pressure drop characteristics of the upstream and downstream parts of the process. Centrifugal and axial compressors must be operated between safe limits both at low and high flowrates. If there are restrictions in the suction or discharge of a centrifugal or axial compressor that reduce the flow, it can be driven to an unstable region known as the *surge region*.

Figure 13.4 illustrates a simplified operating curve for a typical fixed speed centrifugal compressor. If the compressor is operating at Point A, then the flow can be reduced to Point B by restricting the outlet flow. However, reducing the flow further beyond the peak at Point C results in an outlet pipe pressure higher than the compressor pressure. The flow will then reverse, with the operating point moving to zero flow at Point D in Figure 13.4. Once the outlet pipe clears, the operating point moves to Point E. The excessive flow then moves the operation back along the curve to Point C and the cycle repeats. The resulting compressor *surge* causes a violent oscillating flow, which can be extremely hazardous and can damage the machine.

In Chapter 12 it was noted that *if considered in isolation* there is a single control degree of freedom *inherent* to centrifugal compressors with a fixed rotational speed. Similar to fixed speed centrifugal pumps, if the ratio of outlet to inlet pressure across the compressor is fixed, then the outlet flowrate is fixed. The pressure ratio and flowrate cannot be chosen independently. Figure 13.5a shows the *compressor operating curve* for a constant rotational speed centrifugal compressor. The compressor operating curve is fixed by the design of the compressor and represents the performance under a range of conditions. The actual compressor performance cannot be defined until it operates within the process. The operation must somehow be maintained with a flowrate higher than the *surge point*, which corresponds with the maximum point on the operating curve. For high flowrates, the performance curve becomes very steep as the *choke point* is reached. This limit is reached where the flow in the compressor approaches sonic velocity (Mach 1) and no more flow can pass through the compressor. When the compressor is placed in the process context, the pressure drop for the system increases as the flowrate increases, as represented by the system head curve indicated by a solid line in Figure 13.5a. The fixed system head curve takes up a degree of freedom through the introduction of another equation. The compressor then self-regulates to match the system head curve where the two head curves cross in the same way as for a centrifugal pump (Figure 13.5a).

Introduction of a control valve relaxes the constraint of the fixed system head curve and allows the shape of the system head curve to be changed through adjusting the pressure drop across the valve. The resulting degree of freedom is taken up by a control loop on the discharge in Figure 13.5a, known as *discharge throttling*. A flowrate controller on the outlet of the compressor is used to restrict the flow out of the compressor. Figure 13.5b shows an alternative flowrate control using *suction throttling*. In this case, a flowrate controller on the inlet of the compressor is used to restrict the flow into the compressor. Rather than using a control valve, as shown in Figure 13.5b, the final control element could have been the adjustment of louvers in the compressor inlet duct. In this way, the operating point can be moved along the operating curve, as shown in Figure 13.5a.

In terms of process design, discharge throttling (Figure 13.5b) will require a lower power demand than suction throttling (Figure 13.5c) for the same mass flowrate, overall inlet and outlet pressure and the same pressure drop across the control valve. Under the same overall conditions, discharge throttling will have a lower pressure ratio across the compressor and a lower volume at the compressor inlet, both of which will reduce the power demand for a given mass flowrate (Smith 2016). However despite this, throttling if used is more often implemented on the suction side of the compressor to maintain a greater distance from the surge limit. Figure 13.6 shows the effect of changing the suction pressure on the discharge pressure. It can be seen in Figure 13.6 that for a fixed flowrate, the lower the suction pressure, the further away the operating point is from the surge limit. Thus, using suction

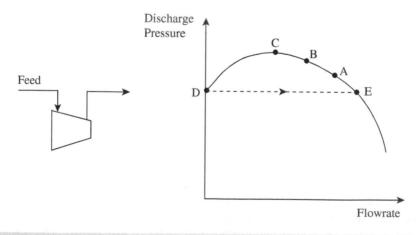

**Figure 13.4**

Centrifugal compressor surge.

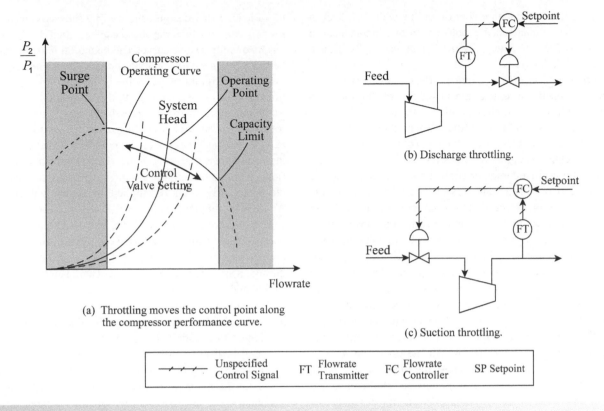

(a) Throttling moves the control point along the compressor performance curve.

(b) Discharge throttling.

(c) Suction throttling.

| | Unspecified Control Signal | FT | Flowrate Transmitter | FC | Flowrate Controller | SP Setpoint |

### Figure 13.5

Centrifugal compressor throttling at constant compressor speed.

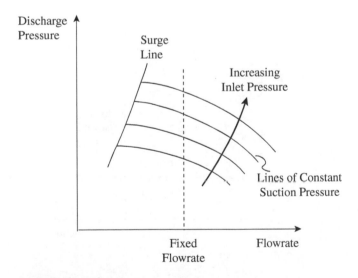

### Figure 13.6

The effect of changing inlet pressure on centrifugal compressor discharge pressure.

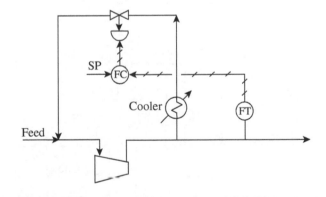

### Figure 13.7

Fixed speed centrifugal compressor recycle (spill-back) control.

throttling is safer for the avoidance of surge. The main problem with both arrangements in Figure 13.5 is the absence of any control to avoid surge, to which both arrangements are vulnerable.

Figure 13.7 shows a centrifugal compressor with a *recycle* or *spill-back* control. In this arrangement, neither the flowrate

through the compressor itself, nor the outlet pressure from the compressor, are controlled. The flowrate to be delivered is varied by varying the flowrate in the recycle. This maintains an essentially constant flowrate through the compressor itself for a fixed pressure difference across the compressor and avoids the low flowrates that would lead to surge. A cooler in the recycle removes the heat of compression. The cooler might be a cooling water cooler, air cooling, refrigeration, or heat rejection to heat recovery, depending on the application. In practice, the cooler might need a temperature control configuration to control the outlet temperature

**13**

of the gas. Cooler temperature control will be addressed later in this chapter. The fundamental problem with the arrangement in Figure 13.7 is that it is potentially inefficient, especially if there is a large recycle flowrate.

An *anti-surge control* arrangement offers a much better solution than spill-back control to avoid surge. Figure 13.8 shows an anti-surge control configuration applied to the suction throttling control configuration in Figure 13.5c. Measurement of the flowrate through the compressor itself, inlet and outlet pressures, and inlet and outlet temperatures for the compressor, allows a model of the compressor in a computer control configuration to locate the operating point on the compressor curve. If the anti-surge control configuration detects that the compressor is approaching the surge region then it takes action to reverse the movement of the operating point toward the surge point by opening the *anti-surge valve*, and increasing the flowrate through the compressor by creating a recycle around the compressor. The anti-surge valve is a fast opening and relatively slow closing control valve that can be pneumatically or electrically actuated. The anti-surge valve should be closed under normal operation, but should fail open in the event of a control configuration failure. Anti-surge control configurations typically prevent the operation moving to within 10–15% of the surge flowrate (the *surge control limit*). It should be noted that it is during start-up, shut-down, and abnormal operation that the compressor is most vulnerable to surge.

The discussion of centrifugal compressors so far has assumed a constant rotational speed. However, the compressor speed can

be varied in many designs. Figure 13.9 shows a centrifugal *compressor map* in which the speed can be varied. Each fixed speed has its own operating curve. The operation must be maintained between the *surge line* and the *capacity limit*. The surge line follows the maximum point on the operating curve for each rotational speed. From Figure 13.9, the surge limit depends on the rotational speed of the compressor, decreasing with decreasing rotational speed. For high flowrates, the performance curves become very steep as the choke point is reached where the flow in the compressor reaches sonic velocity (Mach 1) and no more flow can pass through the compressor. Different types of variable speed drive can be applied to centrifugal compressors. A variable compressor drive will be more expensive in capital cost than a constant speed drive, but it can save energy when operating at low throughput by avoiding the flow resistance from a throttling control valve, or flow resistance from recycle through a spill-back control valve. There is a trade-off between capital and energy costs.

In Chapter 12 it was noted that if the rotational speed can vary there are *inherently* two degrees of freedom if the compressor is considered in isolation. In principle two from the speed, outlet pressure, and outlet flowrate can be specified. For example, if the flowrate through the compressor and rotational speed are specified, the discharge pressure is fixed or if the outlet pressure and rotational speed are specified, the flowrate is fixed. Figure 13.9 shows the *compressor operating map* for a variable speed compressor. The operating map shows compressor operating curves for different discrete speeds. Whilst discrete curves are shown

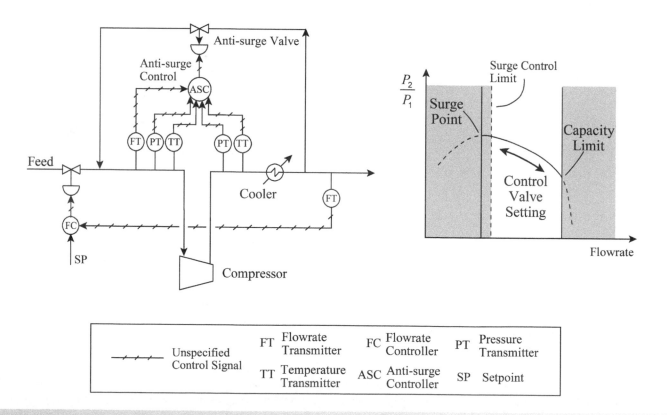

**Figure 13.8**

Fixed speed centrifugal compressor using suction throttling with anti-surge control.

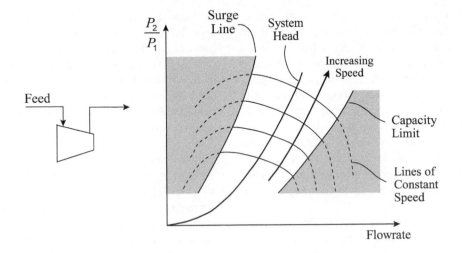

**Figure 13.9**

Variable speed centrifugal compressor operating map.

in Figure 13.9, in principle the compressor taken in isolation can operate anywhere between the surge line and capacity limit, but between the maximum and minimum speeds for the compressor. Once the compressor is placed in the process context it must match the system pressure and flowrate characteristics. Also shown in Figure 13.9 is the system head curve. The system head curve takes up one of the degrees of freedom by introducing another equation. The system head curve crosses the compressor operating curves

for different speeds at different points. The compressor is now constrained to operate along the system head curve between the maximum and minimum speeds for the compressor. For example, if the compressor speed is specified, both the pressure ratio and flowrate are fixed.

As an example, Figure 13.10a shows a centrifugal compressor with a steam turbine driver. The rotational speed of the steam turbine (and hence the compressor) can be varied by varying

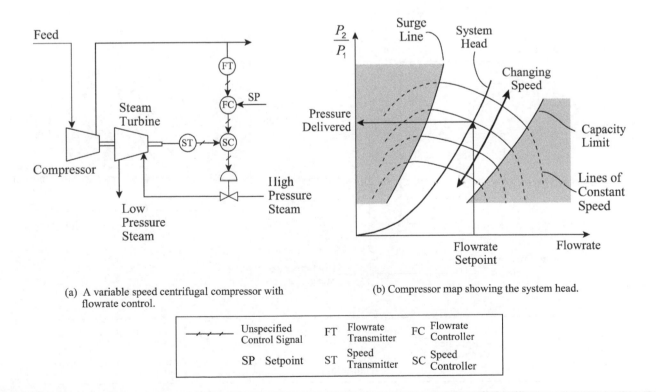

(a) A variable speed centrifugal compressor with flowrate control.

(b) Compressor map showing the system head.

| | | | | | |
|---|---|---|---|---|---|
| ⟋⟋⟋ | Unspecified Control Signal | FT | Flowrate Transmitter | FC | Flowrate Controller |
| SP | Setpoint | ST | Speed Transmitter | SC | Speed Controller |

**Figure 13.10**

Centrifugal compressor with speed control of the flowrate.

the flowrate of high-pressure steam into the steam turbine. A speed transmitter on the steam turbine sends a signal to a speed controller that controls the flowrate of steam into the steam turbine. In the example in Figure 13.10a control of flowrate is cascaded to the speed controller to adjust the setpoint for speed. Figure 13.10b shows the corresponding compressor map with the system head curve superimposed. The setpoint for the flowrate controller fixes the intersection between the system head curve and the compressor curve for the appropriate speed. This in turn fixes the outlet pressure. Rather than use a steam turbine to drive the compressor, an alternative could have been to use a variable speed electric motor. However, the arrangement in Figure 13.10 has no control to avoid surge.

Figure 13.11 shows a centrifugal compressor control configuration with an anti-surge control configuration. Measurements of flowrate, pressures, and temperatures around the compressor, together with the operating speed, allows a model in a computer control configuration to locate the operating point on the compressor map. If surge is approached, the anti-surge valve opens in order to move the operating point away from the surge region. Again, anti-surge control configurations typically create a surge control limit to prevent the operation moving to within 10–15% of the surge flowrate. The design should maintain the anti-surge valve to be closed under normal operation. Rather than use a steam turbine in Figure 13.9, a variable speed electric motor could be used.

Positive displacement compressors deliver a fixed volumetric flowrate of gas. The inlet and outlet pressures adjust to allow the flow into the compressor and flow out of the compressor to balance the fixed volumetric flowrate. Positive displacement compressors, unlike centrifugal compressors, cannot self-regulate their capacity against a discharge pressure. Although there are many different designs of positive displacement compressor, reciprocating machines are the most common (Smith 2016). It was noted in Chapter 12 that, as with positive displacement pumps, there are no control degrees of freedom for a constant speed positive displacement compressor. A constant volume is displaced and the pressure is created by the system itself, which creates a back-pressure. Figure 13.12 shows a control arrangement for a reciprocating compressor that uses a recycle (spill-back) control arrangement around the compressor to introduce a control degree of freedom. A cooler is located in the recycle in order to remove the heat of compression. The cooler might require temperature control. Rather than use a recycle arrangement, the performance of the machine can be changed to provide control. Suction throttling can in principle be used, in which the pressure drop across the inlet control valve (or inlet louvers) decreases the suction pressure. This decreases the volumetric efficiency of the compressor (see Smith 2016, Chapter 13) and lowers capacity as the inlet pressure to the compression cylinders decreases. However, discharge throttling has no effect and cannot be used. A further option within the machine is to control the *cylinder loading*. The inlet and discharge valves within the compressor that allow gas to enter or leave the compression cylinder can be controlled to regulate the volume of gas compressed in each stroke of the piston. Varying the speed of the compressor with a variable speed motor is also possible. If the

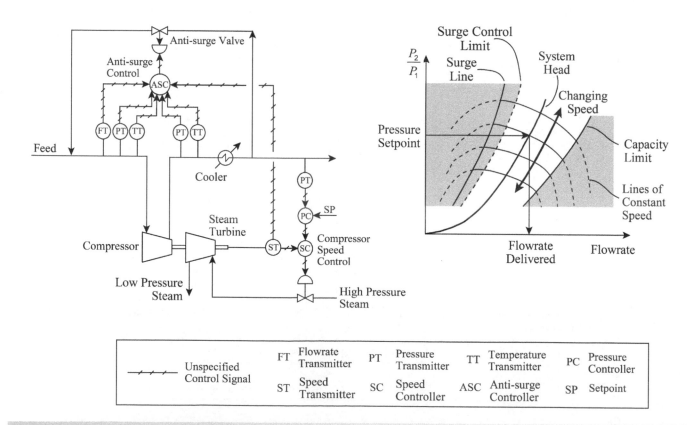

| | | FT | Flowrate Transmitter | PT | Pressure Transmitter | TT | Temperature Transmitter | PC | Pressure Controller |
|---|---|---|---|---|---|---|---|---|---|
| ⌁⌁⌁ | Unspecified Control Signal | ST | Speed Transmitter | SC | Speed Controller | ASC | Anti-surge Controller | SP | Setpoint |

**Figure 13.11**

Pressure control of variable speed centrifugal compressor with anti-surge control.

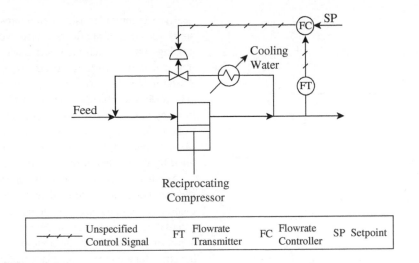

## Figure 13.12

Reciprocating compressor with recycle (spill-back) control.

demand for compressed gas is intermittent, then the capacity can be controlled by starting and stopping the motor drive to the compressor by means of *on–off control*.

Compressors are major components of refrigeration systems and dominate the operating costs. Examples of refrigeration systems are shown in Figures 5.10 to 5.15. In this case, the suction pressure controls the vaporization temperature of the refrigerant fluid. The principles discussed here for discharge pressure control are readily adapted to control the suction pressure.

# 13.3   Heat Exchange Control

In Chapter 12 it was concluded that heat exchangers using a hot or cold utility involve three control degrees of freedom. If the outlet pressure of the process stream is left to self-regulate from the fluid frictional losses, this adds one more constraint from the pressure drop equation for the process side of the heat exchanger, reducing the control degrees of freedom to two. One of these remaining degrees of freedom would be typically taken up to control the flowrate of the process feed. Figure 13.13 shows a commonly used control arrangement for a steam heater using a steam trap to allow the condensate to pass, but not steam. Steam traps were discussed in Chapter 4. In Figure 13.13, if the steam pressure changes for some reason, then the pressure drop across the control valve will change, even if there is no change in the position of the steam control valve from the temperature controller. The flowrate through the valve depends on both the control valve position and the pressure difference across the valve. Thus, changes in the steam pressure will cause the flowrate of steam to change, irrespective of any change in the process temperature. If the steam control valve in Figure 13.13 was under flowrate control, rather than temperature control, then any such pressure disturbances would be dealt with rapidly, resulting in little disturbance to the temperature. This can be achieved by applying cascade control to the temperature, as shown in Figure 13.14. The temperature controller drives the setpoint of the flowrate controller to keep

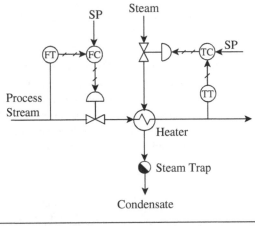

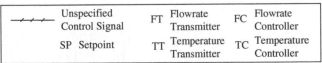

## Figure 13.13

Steam heater control arrangement with steam trap.

the temperature at its setpoint. The temperature controller is the *primary* (or *outer*) controller. The flowrate controller is the *secondary* (or *inner*) controller. The flowrate controller responds rapidly to pressure changes, but the temperature controller responds slowly because of the time required for the heat transfer to take place through the heat exchanger.

Rather than use a steam trap as in Figures 13.13 and 13.14, another method that can be used is to collect the steam condensate in an external *condensate drum* after the heat exchanger, as illustrated in Figure 13.15. The same control arrangement is used as in Figure 13.13, but the steam trap is substituted by a condensate drum with level control. The condensate level control simply maintains an inventory of condensate in the drum. In this way,

**13**

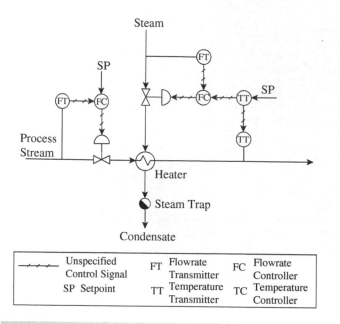

**Figure 13.14**

Steam heater control using cascade control.

the condensate provides a liquid seal that prevents uncondensed steam from passing directly to the condensate return system. A vapor balance line is provided to connect the steam inlet to the heater or the outlet steam vapor space of the heater (as in Figure 13.15). This allows the condensate to flow freely from the heater to the condensate drum under gravity. This arrangement is not

nearly as common as using steam traps, but is better when processing large condensate flows. It also avoids the possibility of steam passing to the condensate system, which can happen with a steam trap if it is not working efficiently, or fails in an extreme case. A significant flow of steam passing to the condensate system can create safety hazards. The control scheme in Figure 13.15 will in many cases benefit from implementing the same cascade control configuration as used in Figure 13.14. The steam is put under flowrate control, with the setpoint for the flowrate control supplied by the temperature controller.

Another option is shown in Figure 13.16, which allows the steam condensate to accumulate in the bottom of the heater. Level control of the condensate in the heater acts in a similar way to the external condensate drum in Figure 13.15. However, the accumulation of condensate at the bottom of the heater also reduces the heat transfer area for steam condensation in the heater. This allows additional control of the heat transfer. It also allows the condensate to be subcooled. Figure 13.16 shows the temperature controlled by the condensate level in the heater. The process outlet temperature controller determines the condensate level control valve setting as long as the condensate level exceeds its set point. If condensate level drops below its set point, then the condensate level controller determines the valve setting. This is achieved by using a low select to choose between the outputs of the liquid outlet temperature controller and the condensate level controller. In this way, a condensate level is always maintained in the heat exchanger.

Like utility heaters, cooling water coolers involve two degrees of freedom if the process stream pressure is left to self-regulate. For such coolers, the cooling water can be controlled with the control valve on the cooling water supply or return side of the cooler.

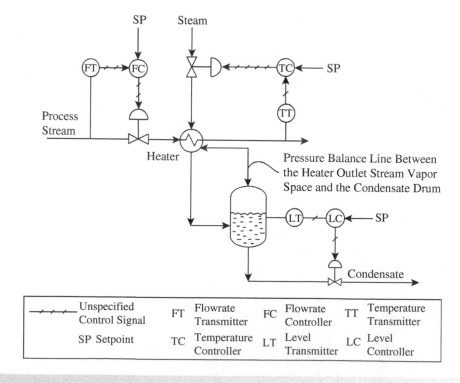

**Figure 13.15**

Steam heater control arrangement with condensate pot.

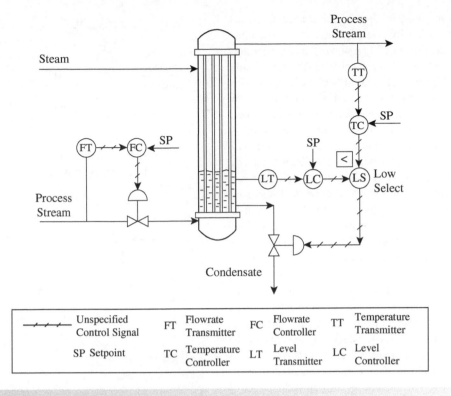

| | Unspecified Control Signal | FT | Flowrate Transmitter | FC | Flowrate Controller | TT | Temperature Transmitter |
|---|---|---|---|---|---|---|---|
| SP | Setpoint | TC | Temperature Controller | LT | Level Transmitter | LC | Level Controller |

## Figure 13.16

Steam heater control arrangement with condensate control valve.

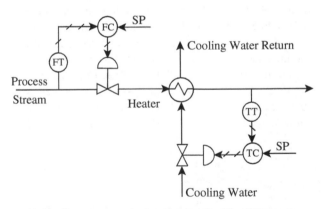

(a) Cooling water control on the supply side of the cooler.

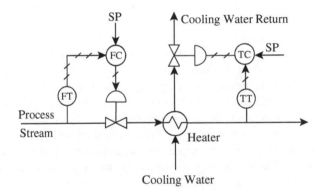

(b) Cooling water control on the return side of the cooler.

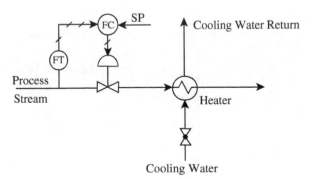

(c) Cooling water cooler with manual control on the supply side of the cooler.

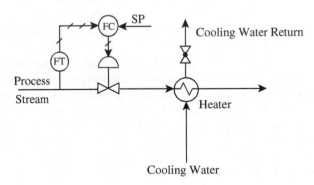

(d) Cooling water cooler with manual control on the return side of the cooler.

13

## Figure 13.17

Cooling water cooler control.

Figure 13.17a shows the temperature control arrangement for a process stream being cooled by cooling water with the control valve on the supply side. Figure 13.17b shows the corresponding control arrangement with the control valve on the return side. The arrangements in Figure 13.17a and b are almost equivalent in control terms. However, there are practical issues that might suggest one of the schemes to be preferable. The arrangement in Figure 13.17a will result in the cooling water being at a lower pressure in the heat exchanger than is the case with Figure 13.17b. The cooling water will contain dissolved gases. As the temperature of the cooling water is increased in the heat exchanger, the solubility of the dissolved gases will decrease. The lower pressure of the cooling water in the heat exchanger in Figure 13.17a will result in greater release of the dissolved gases than the arrangement in Figure 13.17b, possibly reducing the heat transfer coefficient and possibly resulting in an accumulation of released gases inside the heat exchanger. An upward flow of cooling water, rather than a downward flow, in such a heat exchanger has the advantage of facilitating the escape of any released gases. The choice between the schemes in Figure 13.17a and b can also have safety implications in the event of failure of tubes from corrosion leading to leakage across the tube walls. Figure 13.17a would create a potential scenario of the process fluid leaking into the cooling water in the event of tube failure from corrosion, whereas Figure 13.17b would create a greater potential for the cooling water to leak into the process fluid in the event of failure of a heat exchanger tube. Thus, the designer might choose between the arrangements in Figure 13.17a and b to ensure that any potential leak occurs in the less hazardous direction in the event of tube failure.

For cooling water coolers, it is often the case that installing temperature control cannot be justified, as accurate control of the process outlet temperature is often not necessary, for example, in the case of the process fluid going to storage with a storage temperature that is not critical. If this is the case, then the cooler will be left on manual control by adjustment of a valve to regulate the flow of cooling water through the cooler. Once set up, the valve setting the flowrate of cooling water would not be adjusted, unless there was a significant change in the process operation. Figure 13.17c and d show the manual equivalents to the automatic control arrangements.

In Chapter 5 it was discussed how heat from a process fluid can be rejected directly to the environment by the use of *air-cooled heat exchangers*. The fluid to be cooled passes through the inside of tubes and ambient air is blown across the outside by the use of fans. The outside surface generally features extended surfaces to enhance the heat transfer. Figure 13.18a shows a control arrangement for an air-cooled heat exchanger in which the temperature controller changes the speed of a variable speed electric motor driving the cooler fan. This changes the flowrate of air and the rate of heat transfer. Rather than changing the speed of the motor, the pitch (angle) of the fan blades can be altered in the control loop. Altering the pitch of the fan blades changes the flowrate of air across the outside of the tubes. If the speed of the motor driving the fan is fixed and the blade angle of the fan is fixed, then there are no control degrees of freedom. However, Figure 13.18b shows a control arrangement in which the speed of the motor and pitch of

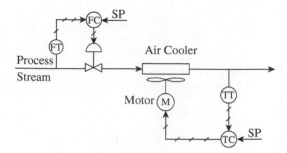

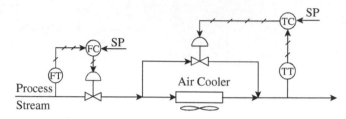

(a) Air cooler control with variable speed motor.

(b) Air cooler control with bypass.

**Figure 13.18**

Air cooler control.

the fan blades are fixed but introduces a new degree of freedom with a partial bypass of the process fluid around the air cooler.

It should be noted that Figures 13.13 to 13.18 show flowrate control on the process stream inlet. Whilst this satisfies the degrees of freedom of the individual operations, when these operations are considered in the context of an overall process, it will commonly be the case that the process stream flowrate will be fixed by an upstream operation and will thus not need flowrate control.

Figure 13.19 shows a process-to-process heat exchanger between two process fluids. Such heat exchangers are common in heat exchanger networks (Smith 2016). A common situation

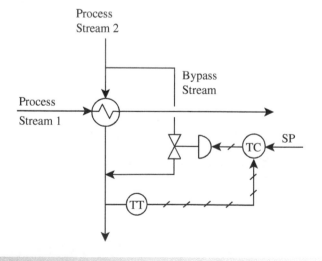

**Figure 13.19**

Process to process heat exchanger control with both stream inlet streams fixed.

is that the inlet conditions of both process streams are fixed, including the flowrate. If this is the case, then, as noted in Chapter 12, there are no control degrees of freedom. Thus, if the outlet temperature of one of the streams needs to be controlled, an additional degree of freedom needs to be introduced. Figure 13.19 shows the introduction of a new degree of freedom by creating a partial bypass around the heat exchanger for one of the process streams. Manipulating the flow in the bypass allows the temperature of one of the streams to be controlled. Alternatively, if one of the streams undergoes a change of phase (vaporization or condensation), then instead of using a partial bypass, another way to introduce a degree of freedom is to put the stream undergoing the phase change on the shell side of the heat exchanger and create a liquid level for the partially vaporized liquid (for vaporization) or condensed liquid (for condensation) in the heat exchanger. Manipulating the outlet flow to control the liquid level creates a degree of freedom in the heat exchanger that changes the effective heat transfer area, and hence the outlet temperatures in a similar way to the control system in Figure 13.16. Then one of the outlet temperatures can be controlled by manipulating the liquid level.

## 13.4   Furnace Control

In some situations, process heat needs to be supplied by radiant heat transfer from the combustion of fuel in a *fired heater* or *furnace*, as discussed in Chapter 4. In Chapter 12 it was concluded that there are five degrees of freedom. If the process stream outlet and the flue gas stream have outlet pressures that are self-regulating, this removes two control degrees of freedom, giving three control degrees of freedom. Figure 13.20 shows an example of a basic furnace control configuration for the process fired heater from Figure 12.14. The first control degree of freedom has been

taken up by control of the feed flowrate. The control objective in Figure 13.20 is control of the process outlet temperature. The control objective dictates the *firing demand*. In a different context, for a steam boiler the control objective (firing demand) could be the steam generation flowrate or steam pressure. The firing demand is used to control the flowrate of fuel to the combustion. The corresponding flowrate of air to the combustion is under ratio control, with the final control element in this case being a variable speed motor controlling the fan. An alternative final control element might have been adjustment of louvers in the air duct for a fixed speed fan delivering the combustion air. Yet another option for the final control element might have been adjustment of louvers in an air duct for the combustion air driven by natural draft. In the context of ratio control, the fuel flowrate is the "wild" stream and the combustion air flowrate the "controlled" stream. As discussed in Chapter 10 for ratio control, the ratio station and air flowrate controller in Figure 13.20 can be combined as a *ratio controller*. It is important to ensure that there is excess oxygen in the flue gas. Figure 13.20 shows the oxygen concentration in the flue gas being monitored by analysis to ensure complete combustion of the fuel and avoid carbon monoxide formation. Oxygen in the flue gas would typically be in excess by 3%, or possibly less. The excess air might be as low as 1.5% if designed to give efficient mixing and combustion. Figure 13.20 shows a *damper* in the furnace stack. This allows the size of opening in the stack to be adjusted manually to control the pressure in the combustion chamber. Note that pressure in the combustion chamber was not included in the degrees of freedom analysis and is left to manual adjustment for efficient operation of the furnace.

For the control configuration in Figure 13.20, if both the fuel control loop and the air control loop respond rapidly to a change in the firing demand, then the required air to fuel ratio will be maintained. However, it is likely that the air flowrate control will respond more slowly than the fuel flowrate control. In this case, a rapid increase in the firing demand will temporarily lead to a

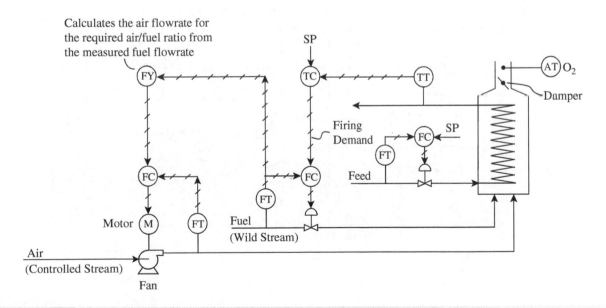

**Figure 13.20**

Ratio control of a process fired heater.

fuel-rich environment, possibly causing carbon monoxide and unburnt fuel to be emitted from the stack. This problem can be overcome to some extent by appropriate tuning of the controllers for the fuel flow and air flow, effectively to slow the response of the fuel control relative to the air control. Alternatively, a rapid decrease in the firing demand will lead to too much excess air, decreasing the furnace efficiency through unnecessary oxygen and nitrogen lowering the combustion temperature and increasing the stack losses. Whilst appropriate tuning of the controllers can overcome these problems to some extent, a more robust solution can be obtained by introducing a *low select override* and a *high select override*, as shown in Figure 13.21 (Calabrese et al. 2006). If the firing demand increases rapidly:

- The HS override will pass the firing demand signal to the air flow controller, rather than the signal from the fuel flow controller, causing air flow to increase immediately, following the increase in firing demand.
- The LS override will pass the air flow signal to the fuel flow controller, rather than the signal from the firing demand, preventing the fuel flow from increasing until the air flow has increased proportionally.

If the firing demand decreases rapidly:

- The LS override will pass the firing demand signal to the fuel flow controller, rather than the air flow signal, causing fuel flow to decrease immediately, following the decrease in firing demand.
- The HS override will pass the fuel flow signal to the air flow controller, rather than the signal from the firing demand, preventing the air flow from decreasing until the fuel flow has done so.

Overall, for the LS and HS overrides, if the firing demand increases or decreases rapidly the control strategy ensures that the air flow is increased before the fuel flow is increased, and that the fuel flow is decreased before the air flow is decreased.

The furnace control configurations discussed so far do not have the capacity to change the air to fuel ratio. This may be required, for example, if there are changes in the quality of the fuel, which can happen when using waste gases as fuel. Figure 13.22 shows an *oxygen trim* control configuration in which the oxygen in the stack is measured and used to automatically adjust the air to fuel ratio. The oxygen measurement is compared with a setpoint (e.g., 3% $O_2$) and a signal generated such that the fuel to air ratio required to correct the excess oxygen can be calculated before calculating the required air flowrate from the measured fuel flowrate. The calculated required air to fuel ratio scales the 0–100% output from the composition controller to calculate an actual working range of physical air to fuel ratio. This scheme, like the one in Figure 13.20, can still result in a temporary deficit of excess air resulting from rapid changes in the firing load if, as is likely, the control of air flowrate is slower than that of the fuel flowrate.

Figure 13.23 shows an oxygen trim control configuration with a low select override and a high select override. The control scheme is the same as the scheme in Figure 13.21, but the signal from the oxygen analysis control is compared with a setpoint (e.g., 3% $O_2$) and a signal generated such that the fuel to air ratio required to correct the excess oxygen can be calculated before calculating the required air flowrate from the firing demand and calculating the fuel flowrate required from the measured air flowrate.

The oxygen measurement for the schemes in Figures 13.22 and 13.23 will be slow and the controller parameters will need to be tuned accordingly. Also, the oxygen measurement might not have

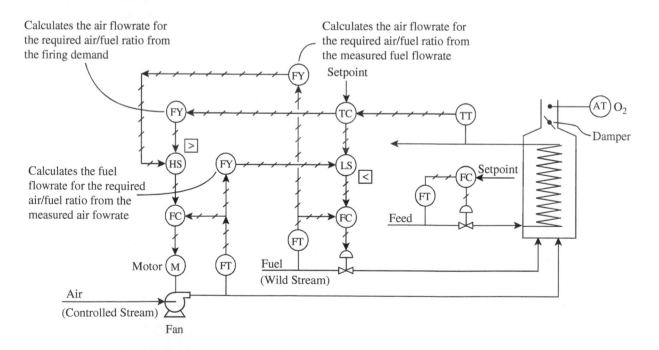

**Figure 13.21**

Ratio control of a process fired heater with high and low select to maintain excess air.

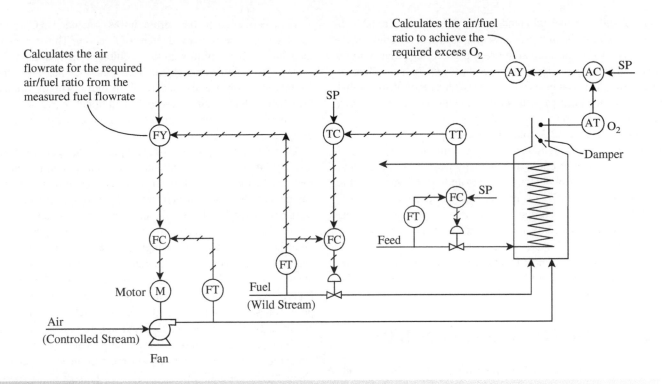

**Figure 13.22**

Fired process heater with oxygen trim control.

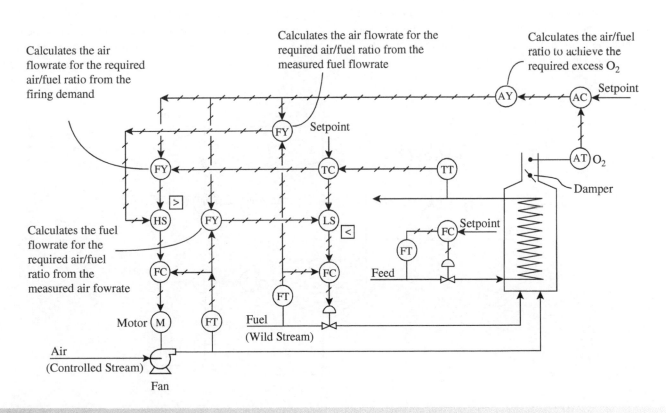

**Figure 13.23**

Fired process heater with oxygen trim control and high and low select to maintain excess air.

a high reliability, which will render the whole system unreliable. If this is the case, while the system is being repaired, it would be better to have a system that can be reconfigured temporarily to be without the oxygen trim control in the event of the failure of the analysis. This can be accomplished with a ratio calculator that can be switched locally to either following the oxygen control or a local automatic mode with a fixed ratio.

Figures 13.20 to 13.23 only show examples of the many different control configurations possible for furnace control. Control configurations can be configured in many ways. It is important to understand the control objectives and to understand what possible disturbances need to be overcome. Many other, more sophisticated, furnace control configurations can be applied to overcome these problems, especially for fuel with a variable composition and calorific value (Calabrese et al. 2006).

Finally, it should be noted that the ratio stations and air flow-rate controllers in Figures 13.20 to 13.23 can be combined as ratio controllers, as discussed in Chapter 10.

# 13.5   Flash Drum Control

In Chapter 12 it was noted that a flash drum without input or output of energy has two control degrees of freedom. Feed flowrate and flash drum pressure are the obvious control choices, and it will also normally be necessary to apply a liquid level control to maintain the steady-state liquid mass balance. Figure 13.24 shows a typical arrangement where the feed flowrate is controlled. The amount and composition of liquid and vapor produced by flashing depends on the pressure. The vapor inventory is maintained by controlling the pressure through manipulation of the vapor outlet flow, while the liquid inventory is maintained by controlling its level by manipulating the liquid outlet.

For a non-adiabatic flash drum, it was concluded in Chapter 12 that there are three control degrees of freedom. This is similar to the adiabatic flash drum with the addition of the heat input as an extra degree of freedom. Figure 13.25 shows a possible control configuration for a non-adiabatic flash drum. This is similar to the adiabatic flash drum, but with the additional degree of freedom to control the separation through the input or output of energy using temperature control.

However, it is important that the input or output of energy does actually influence the temperature. The control configuration in Figure 13.25 could function for a feed containing a mixture of components, but not for a feed that is a pure component. For a pure component, varying the heat input changes the amount of vaporization, but not the temperature (for a constant pressure). On the other hand, for a mixture, changing the amount of vaporization changes the composition of the liquid and vapor and the temperature. For an adiabatic flash, the control configuration in Figure 13.24 will potentially function with either a mixture or a pure component. An alternative control scheme for a non-adiabatic flash is shown in Figure 13.26. In this arrangement, the amount of vaporization controls the pressure, rather than the temperature. This does not rely on the amount of vaporization changing the temperature.

# 13.6   Absorber and Stripper Control

Figure 13.27 shows a basic absorber configuration in which a gas or vapor mixture is contacted with a liquid solvent. The solvent preferentially dissolves or reacts with one or more components in the gas. The absorption process might be *physical absorption*

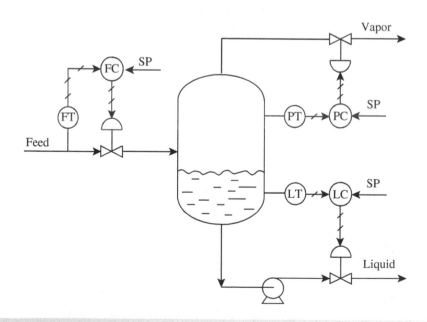

**Figure 13.24**
Adiabatic flash drum separator control.

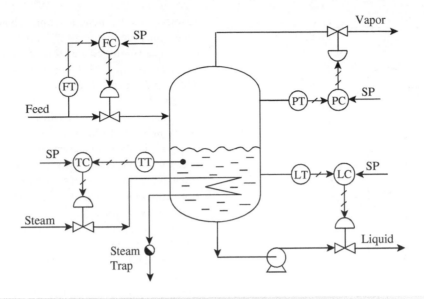

**Figure 13.25**

Non-adiabatic flash drum separator control.

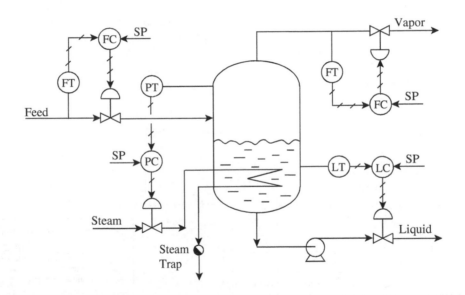

**Figure 13.26**

Alternative control scheme for a non-adiabatic flash drum.

or it might be *chemical absorption* in which a reaction takes place (Smith 2016). In physical absorption the mass transfer is most often enhanced by a low temperature and a high pressure. In Chapter 12, the control degrees of freedom for a cascade of non-reacting equilibrium stages was analyzed and it was concluded that there were three control degrees of freedom. Thus, in the absorption process in Figure 13.27 there will be three control degrees of freedom plus inventory control for the liquid in the base of the absorption column. One of the control degrees of freedom would normally be taken up by controlling the pressure, which also controls the inventory of the gas. The pressure is shown to be controlled in Figure 13.28 by the flowrate of the outlet gas. The other two control degrees of freedom would normally be taken up by the inlet flowrate for the gas and the liquid solvent. However, given that the design and operation of such an absorption process is based on the ratio of the liquid to gas flowrate (Smith 2016), it would seem logical in many cases to apply ratio control to the two feed streams. An example is shown in Figure 13.28. The choice of which streams are wild and controlled depends on the process context. Also, in Figure 13.28, inventory control is used at the base of the absorption column to control the inventory of the liquid solvent. It should also be noted that

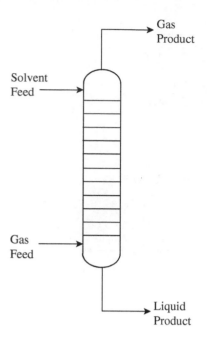

**Figure 13.27**

An absorber with gas and solvent feeds.

temperature control might need to be added to the absorber in some cases. As already noted, physical absorption is generally favored by a lower temperature. The heat of absorption might need to be removed at intermediate points. The control of the

absorber might benefit from changing the level control arrangement in Figure 13.28 to one in which the bottom stream is flowrate controlled with the setpoint for the flowrate controller provided by the level controller in a cascade arrangement. This kind of level control arrangement was discussed in Chapter 11.

A more sophisticated control configuration for the absorber is shown in Figure 13.29. This is similar to the control configuration in Figure 13.28, but is based on a chemical analysis of the outlet gas. The chemical analysis is sent to a controller that provides the setpoint for the control of the liquid to gas ratio in a cascade control configuration. This kind of control configuration would be useful if the control objective was to control the chemical composition of the outlet gas. If the chemical composition of the outlet liquid was the control objective, then the corresponding scheme could be applied to the liquid outlet. However, it should be noted that the analysis is likely to involve a significant time delay. The analysis instrument is also likely to be unreliable and not available for a significant time.

Stripping is the reverse of absorption and involves the transfer of solute from the liquid to the gas (or vapor) phase. For stripping, the mass transfer is most often enhanced by a high temperature and a low pressure. The design and operation of such a stripping process is based on the ratio of the gas to liquid flowrate (Smith 2016). Like absorption, it would seem logical in many cases to apply ratio control to the two feed streams. An example is shown in Figure 13.30. The choice of which stream is wild and which is controlled depends on the process context. As with absorber control, a chemical analysis of one of the outlet streams can be used to

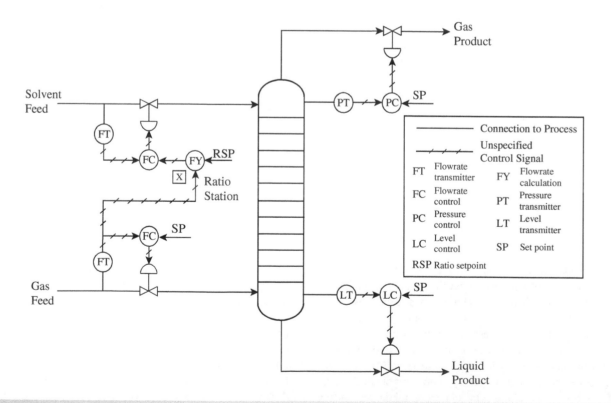

**Figure 13.28**

An absorber control system.

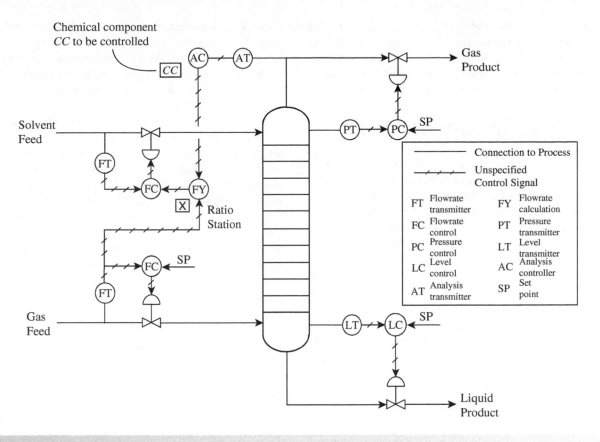

## Figure 13.29

An absorber control system with composition control of the gas product.

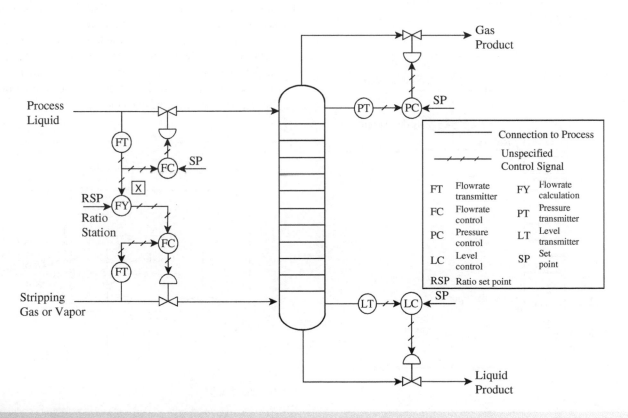

## Figure 13.30

A stripper control system.

13

provide the setpoint for the control of the gas to liquid ratio in a cascade control configuration. However, again the analysis is likely to introduce a significant time delay and to be unreliable. It should be noted that, as with absorber control, temperature control of the stripper might need to be added in some cases.

# 13.7 Distillation Control

Distillation is the most common method used for the separation of homogeneous fluid mixtures. The principal advantages of distillation are that it is capable of separating mixtures with a wide range of throughputs, with a wide range of feed concentrations, and can produce high purity products. The conceptual design of distillation columns has been dealt with elsewhere (Smith 2016). Different column configurations are possible (Smith 2016), but here consideration will be restricted to *simple* distillation columns in which a single feed is split between two products using a column with a reboiler and condenser, as illustrated in Figure 13.31.

The control objectives might be related to recovery of one or more components in the distillate or bottoms, or control of the composition of the distillate or bottoms, or a combination of these. *Single-sided* composition specifications require a component to be at or below a certain value, or at or above a certain value. *Two-sided* composition specifications require a component to be retained within a certain range. The separation control can be used to control the distillate or the bottoms product compositions, or both. The discussion here will be restricted to *single-point* composition control in which the composition of only one end of the distillation column is controlled directly. The choice of whether to control the distillate or bottoms product will normally be based on which of the two distillation products is most critical. However,

it must be emphasized that changes to a distillation column operation will tend to affect both the overheads and bottoms, no matter which is being given the greatest emphasis in control. When designing the control configuration, it is particularly important to understand the nature of the disturbances that are likely to affect the column operation.

For the simple distillation system as a whole, it was concluded in Chapter 12 that there are four control degrees of freedom. One sets the column pressure. Another sets the feed flowrate when the column is considered in isolation. Then two control degrees of freedom are used to control the separation (product purity or recovery). However, there are two additional degrees of freedom to control the liquid inventory in the reflux drum and the column sump in the base of the column (Figure 13.31), giving six degrees of freedom for a simple column and requiring six single-input–single-output control loops in total.

The primary controlled variables are distillate and bottom product compositions, liquid levels in the reflux drum and the column sump, and column pressure. Manipulated variables to achieve control object directives are feed flowrate (for a column in isolation), distillate flowrate, bottoms flowrate, reflux flowrate, heat input to the reboiler, and heat removal from the condenser.

### 13.7.1 Distillation Column Pressure Control

One of the first and most important decisions to be made for the design of a distillation column is the operating pressure. This is also a critical control parameter, as it affects the separation directly. There are a number of ways in which the pressure can be controlled (Luyben 1992; Sloley 2001; Smith 2012)

1) *Rate of condensation in the condenser*. A common method used for controlling column pressure is through control of the heat removal in the condenser, as illustrated in Figure 13.32. If the control valve opens in Figure 13.32, then the flowrate of cooling water increases, increasing the amount of the vapor condensed. This decreases the column pressure. Closing the control valve does the reverse.

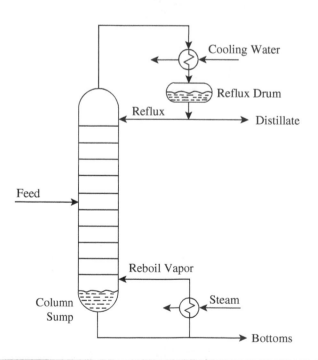

**Figure 13.31**

A simple distillation.

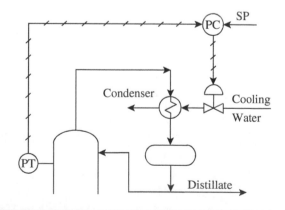

**Figure 13.32**

Distillation column pressure control using total condenser cooling water flowrate.

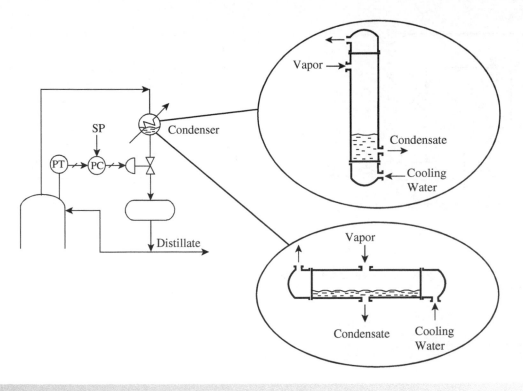

## Figure 13.33

Distillation column pressure control using a flooded condenser.

2) *Flooded condenser.* Flooded condenser arrangements are shown in Figure 13.33. Whilst the total heat transfer area in the condenser is fixed, retaining condensate within the condenser varies the heat transfer area exposed to the condensing vapor. Thus, controlling the level of condensate inside the condenser can be used to control the pressure by varying the amount of condensation occurring. If the control valve closes, condensate builds up, decreasing the condensation. This decreases the amount of vapor being condensed and increases the column pressure. Opening the control valve does the reverse. The major drawback of this option is a slow response for the control of pressure to changes in the control valve setting.

3) *Hot vapor bypass.* Hot vapor bypass also uses a flooded condenser, as shown in Figure 13.34, but the level in the condenser is controlled by the liquid level in the reflux drum. The condenser is located at an elevation slightly below that of the reflux drum. The pressure in the condenser is the same as that in the distillation column, but the pressure in the reflux drum is lower than that in the distillation column, as a result of the pressure drop across the control valve. When the control valve opens, it equalizes the pressure between the vapor line and the reflux drum, causing the condenser to become flooded because of the difference in level. The flooding of the condenser decreases the rate of condensation, causing the pressure to increase. Closing the control valve does the reverse. Like flooded condenser pressure control, this option gives a slow response for the control of pressure.

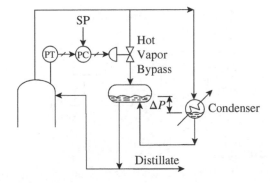

## Figure 13.34

Distillation column pressure control using condenser hot vapor bypass.

4) *Partial condensers.* If the overhead product is required as a vapor, a partial condenser can be used. Figure 13.35a shows the column pressure being controlled by manipulating the flowrate of cooling water to the condenser. This controls how much reflux will be returned to the column. Alternatively, the flowrate of vapor distillate can be used to control the column pressure, as illustrated in Figure 13.35b.

5) *Venting non-condensibles.* It might be the case that there are components in the overhead vapor that cannot be condensed in the condenser because the condenser temperature is not low enough. Figure 13.36 shows distillation column pressure control using the flowrate of non-condensed material in the

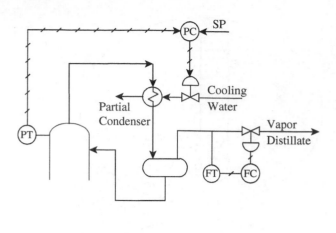

(a) Pressure control using partial condenser cooling water flowrate.

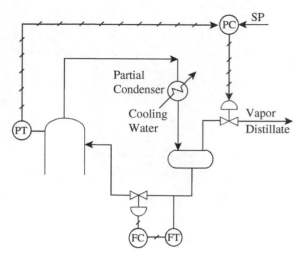

(b) Pressure control for a partial condenser using vapor distillate flowrate.

## Figure 13.35

Distillation column pressure control with partial condenser.

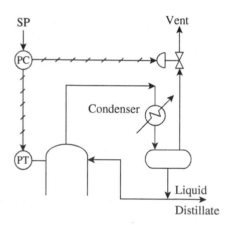

## Figure 13.36

Distillation column pressure control using the flow of non-condensed material in the distillate.

distillate. This method of control might require the condensate from the condenser to be subcooled in order to avoid losses of vapor to the vent, along with the non-condensibles.

6) *Vacuum pressure control.* If the column needs to work under vacuum, the overhead vapor from the column is condensed, largely leaving non-condensibles that are removed by the vacuum system. To ensure as little of the volatile materials as possible are lost to the vacuum system, the condensate from the condenser should be *subcooled* to below saturation. Both ejectors and vacuum pumps are used to remove the non-condensibles. Figure 13.37 shows examples of a vacuum pressure control configuration in which a steam ejector is used to create the vacuum. In Figure 13.37a a bleed of inert gas (e.g., nitrogen or steam) into the vacuum system allows pressure

control by adding additional load to the steam ejector. Figure 13.37b shows an alternative arrangement with a recycle around the steam ejector, which avoids the use of large amounts of inert gas. Other control configurations for ejectors are possible (Luyben 1992; Sloley 2001; Smith 2012). If the vacuum is created by a vacuum pump (e.g., liquid ring pump), then pressure control can be implemented using a variable speed drive on the vacuum pump, or recycle around the vacuum pump.

### 13.7.2 Distillation Column Inventory Control

There are two liquid inventories that need to be managed in the distillation column. These are in the reflux drum and the column sump (Figure 13.31). There are two options to manage the inventory for each of these. The inventory in the reflux drum can be controlled by varying either the distillate product flowrate or the reflux flowrate. The inventory in the column sump can be controlled by varying either the bottoms product flowrate or the reboiler duty (boil up).

Figure 13.38 shows inventory control options in which the inventory in the reflux drum is controlled by varying the distillate product flowrate, with the reflux flowrate fixed. Figure 13.38a shows the option of combining this with control of the sump inventory by manipulating the bottoms flowrate with the reboiler duty (boil up) fixed. Figure 13.38b shows the alternative option of control of the sump inventory by manipulating the reboiler duty (boil up) with the bottoms flowrate fixed.

Figure 13.39 shows inventory control options in which the inventory in the reflux drum is controlled by varying the reflux flowrate, with the distillate product flowrate fixed. Figure 13.39a shows the option of combining this with control of the sump inventory by manipulating the bottoms flowrate with the reboiler duty (boil up) fixed. Figure 13.39b shows the

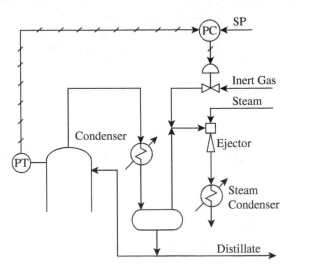

(a) Pressure control using a bleed of inert gas.    (b) Pressure control using s reccle around the ejector.

**Figure 13.37**

Distillation column pressure control for a vacuum column using a steam ejector for vacuum creation.

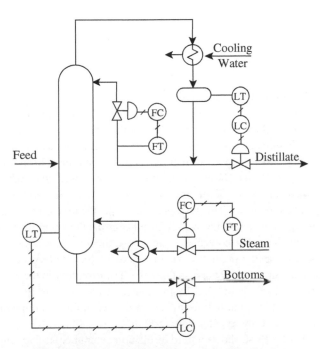

(a) Distillate flowrate inventory control with column sump inventory control from bottoms flowrate.    (b) Distillate flowrate inventory control with column sump inventory control from reboiler duty (boilup).

**Figure 13.38**

Distillate inventory control using the distillate flowrate.

alternative option of control of the sump inventory by manipulating the reboiler duty (boil up) with the bottoms flowrate fixed.

Of these four combinations, not all are viable. If the feed flowrate is fixed either by feed flowrate control or an upstream process, then the material balance can be maintained by fixing either the distillate flowrate *or* the bottoms flowrate, but not both. Thus, the inventory control in Figure 13.39b is not viable. Neither inventory controller manipulates the flow out of the column for a fixed feed. In Chapter 11 it was concluded that the total material inventory must be regulated by its inflows or outflows, i.e., at least one

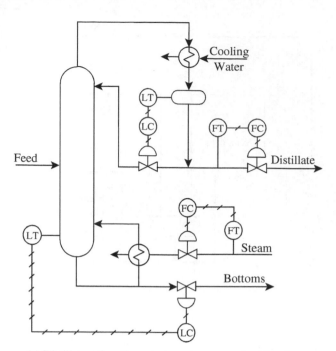

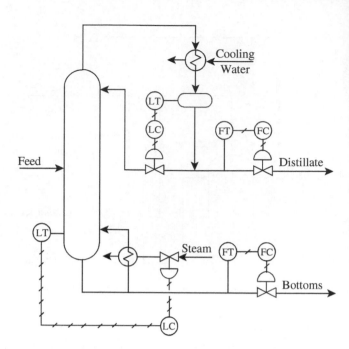

(a) Distillate reflux flowrate inventory control with column sump inventory control from bottoms flowrate.

(b) Distillate reflux flowrate inventory control with column sump inventory control from reboiler duty (boilup) is problematic.

## Figure 13.39

Distillate inventory control using the reflux flowrate.

flow in or out must depend on the inventory. In Figure 13.39b the level controllers for the reflux drum and the column sump will be in conflict with each other.

The three inventory control arrangements in Figures 13.38a and b and 13.39a can form the basis for different arrangements for the control of simple columns.

### 13.7.3   Distillation Product Composition Control

Control of the separation can be achieved by measuring the composition of the products directly using online chemical analysis of the composition of the products. The methods of chemical analysis that can be used include:

- gas chromatography;
- liquid chromatography;
- infrared spectrometry;
- near-infrared spectrometry;
- ultraviolet spectrometry.

Figure 13.40 shows distillation column control by *indirect control of the distillate with constant boil up using chemical analysis*. The distillation throughput is controlled by controlling the flowrate of feed. This arrangement adopts the inventory control arrangement from Figure 13.38a and controls the composition of the distillate. The inventory in the reflux drum is controlled by manipulating the flowrate from the reflux to the column and the inventory in the column sump is controlled by the bottom flowrate. Chemical analysis of the distillate is cascaded with the

reflux flowrate control. This provides the setpoint for the reflux flowrate control loop. Constant boil-up is maintained through control of the flowrate of steam to the reboiler. Pressure control in this case is through control of the rate of condensation, as discussed earlier. Other methods of pressure control could have been used.

Figure 13.41 shows distillation column control by *indirect control of the distillate with constant bottoms flowrate using chemical analysis*. This arrangement adopts the inventory control arrangement from Figure 13.38b and controls the composition of the distillate. The control arrangement at the top of the column is the same as the previous case in Figure 13.40. The feed control and pressure control are also the same. The inventory in the reflux drum is again controlled by manipulating the flowrate from the reflux to the column, but now the inventory in the column sump is controlled by reboiler duty (boil up).

Figure 13.42 shows a third distillation column control arrangement by *direct control of the distillate with constant boil up using chemical analysis*. Again, the distillation throughput is controlled by controlling the flowrate of feed but this time the arrangement adopts the inventory control arrangement from Figure 13.39a and controls the composition of the distillate. The inventory in the reflux drum is this time controlled by manipulating the flowrate of the reflux to the column and the inventory in the column sump is controlled by the bottoms flowrate. Chemical analysis of the distillate is cascaded with the distillate flowrate control. This provides the setpoint for the distillate flowrate control loop. Distillate flowrate is manipulated directly to control the distillate composition or recovery. The feed control and pressure control are the same as before.

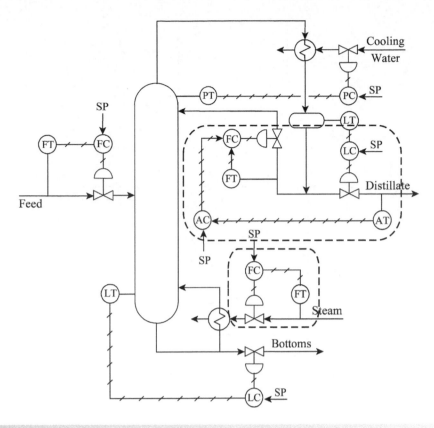

**Figure 13.40**

Distillation column control by indirect control of the distillate using analysis of composition with constant boil up.

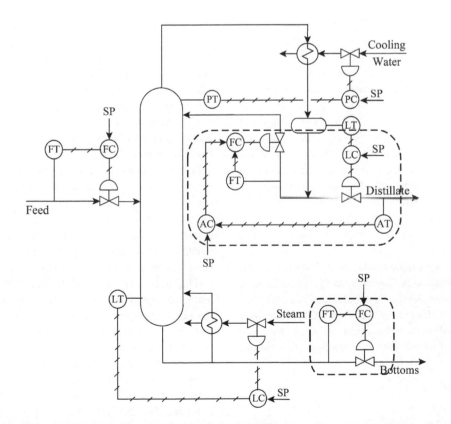

**Figure 13.41**

Distillation column control by direct control of the distillate using analysis of composition with constant bottoms flowrate.

13

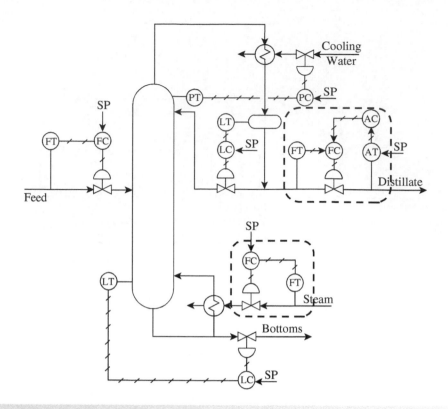

## Figure 13.42

Distillation column control by direct control of the distillate using analysis of composition with constant boil up.

Equivalent arrangements to those in Figures 13.40–13.42 used for control of the distillate can be used to control the bottoms. Figure 13.43 shows distillation column control by *indirect control of the bottoms with a constant reflux flowrate using chemical analysis*. This adopts the inventory control arrangement from Figure 13.38a and is analogous to the distillate control in Figure 13.40. Figure 13.44 shows distillation column control by *direct control of the bottoms with a constant reflux flowrate using chemical analysis*. This adopts the inventory control arrangement from Figure 13.38b and is analogous to the distillate control in Figure 13.41. Figure 13.45 shows a third distillation bottoms column control arrangement by *indirect control of the bottoms with a constant distillate flowrate using chemical analysis*. This adopts the inventory control arrangement from Figure 13.39a and is analogous to the distillate control in Figure 13.42.

Whilst chemical analysis can be used to control the separation, there are a number of drawbacks. Chemical analysis is expensive, introduces significant time delays into the control, is unreliable and might not be available for extended periods. Rather than measuring the composition directly, temperature measurement inside the column can be used to *infer* the composition. Temperature measurements are:

- cheaper than composition;
- more reliable;
- require less maintenance;
- avoid the significant time lag introduced by chemical analyzers.

Equivalent temperature control arrangements to those using chemical analysis can be developed. Figures 13.46 to 13.48 show the temperature control equivalents for distillate control using chemical analysis in Figures 13.40 to 13.42. Figures 13.49 to 13.51 show the temperature control equivalents for bottoms control using chemical analysis in Figures 13.43 to 13.45.

The location of the temperature measurement in the distillation column for control is critical. Figure 13.52 illustrates how an optimum location of the temperature measurement can be identified. This would normally be determined by carrying out a rigorous simulation of the column at the base case conditions to calculate the temperature profile through the column. The simulation is then repeated, varying the manipulated variable up and down. For example, for direct control of the distillate, the reboiler heat load would be fixed and the distillate flowrate varied. Figure 13.52 shows a typical result after varying the parameter from base case conditions by, say, −1% and +1%. A suitable location is where the temperature change varies significantly and is nearly equal when the manipulated variable is increased or decreased for the composition to be controlled. If it is possible to control the temperature close to the top or bottom of the column, then a position closest to the most important product should be chosen. Rather than use a single temperature measurement, it might be better to take the average of more than one measurement or take variations of the difference between two temperatures. This can be explored in the first instance by perturbing a steady-state simulation of the column.

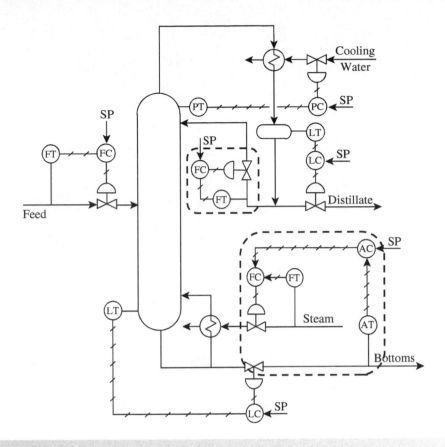

**Figure 13.43**

Distillation column control by indirect control of the bottoms using analysis of composition with constant reflux flowrate.

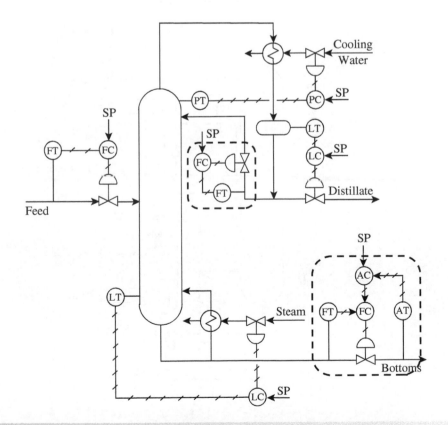

**Figure 13.44**

Distillation column control by direct control of the bottoms using analysis of composition with constant reflux flowrate.

13

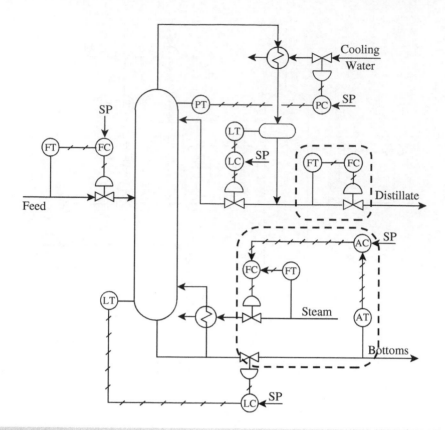

**Figure 13.45**

Distillation column control by indirect control of the bottoms using analysis of composition with constant distillate flowrate.

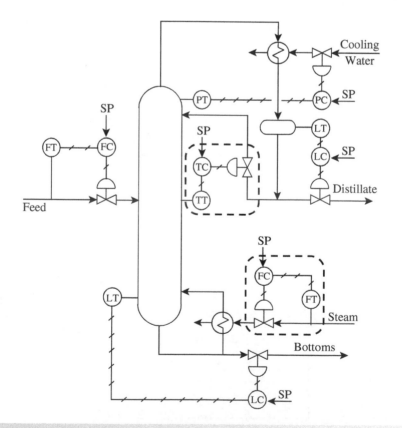

**Figure 13.46**

Distillation column control by indirect control of the distillate using temperature with constant boil-up.

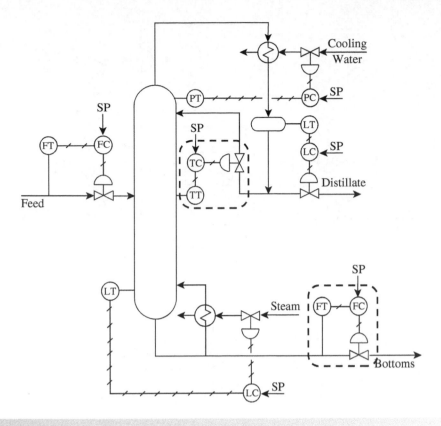

**Figure 13.47**
Distillation column control by indirect control of the distillate using temperature with constant bottoms flowrate.

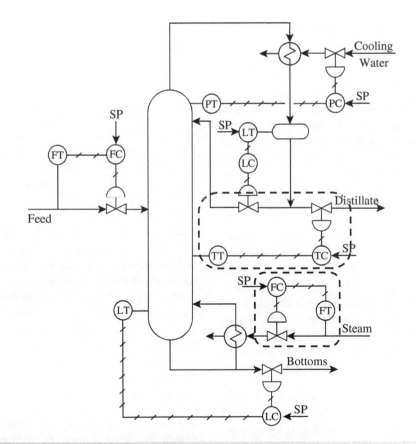

**Figure 13.48**
Distillation column control by direct control of the distillate using temperature with constant boil-up.

13

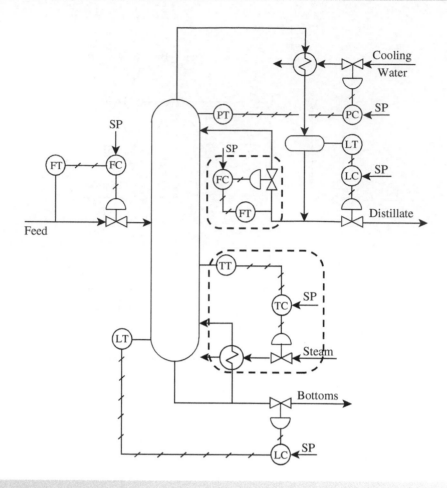

## Figure 13.49

Distillation column control by indirect control of the bottoms using temperature with constant reflux flowrate.

It is clear that control of the composition by chemical analysis and by temperature measurement both have advantages and disadvantages. It is possible to combine measurement of chemical analysis and temperature in cascade systems. Figure 13.53 shows distillation column control by indirect control of the distillate using chemical analysis cascaded to temperature control with constant boil-up. In this scheme the slow-acting chemical analyzer controls the setpoint for the faster-acting inferential temperature control. Another advantage of this scheme is that it helps to overcome the problems associated with the poor reliability of the chemical analyzer. If the analyzer needs to be taken off-line for repair or maintenance, the temperature control allows the operation of the column to continue until the analyzer becomes available again. Figure 13.54 shows the corresponding distillation column control by indirect control of the bottoms using chemical analysis cascaded to temperature control with a constant reflux flowrate. Similar schemes can be readily developed for the other cases considered above.

Many other arrangements can be used for the control of distillation columns (Luyben 1992; Smith 2012). Different designs of columns (e.g., columns with side streams) and complex columns (e.g., dividing wall columns) require more specialized solutions.

### 13.7.4   Choice of Distillation Control Configuration

The most appropriate choice of control configuration, even for simple columns, will depend on the specific characteristics of the separation. The relative flowrates in the distillation system can in the first instance guide the choice of the control configuration.

The split of the flowrate of the feed between the distillate and bottoms will vary according to the separation. For example, it may be that the feed splits to give 10% as distillate or it might be 90% as distillate, depending on the separation. Uneven splits will create significant differences in flowrates between the streams in the distillation. If the split is, say, 10% in the overheads, controlling the level in the reflux drum using a stream with a low flowrate is likely to be problematic. It might therefore be preferable to control the reflux drum inventory by controlling the reflux flowrate in these circumstances. Similarly, if the split is, say, 90% of the feed to the overheads, then controlling the sump inventory with the bottoms flowrate is likely to be more problematic than controlling the sump inventory with the reboiler duty (boil up). The bottoms flowrate will be small, and again controlling the inventory with a small flowrate is likely to be problematic.

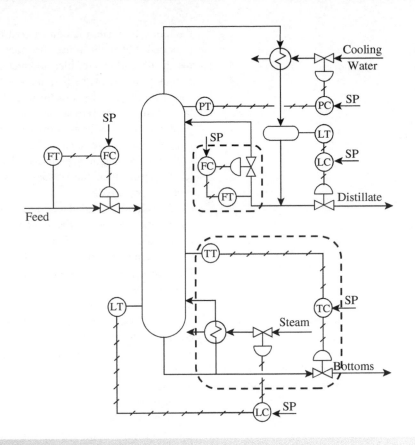

**Figure 13.50**
Distillation column control by direct control of the bottoms using temperature with constant reflux flowrate.

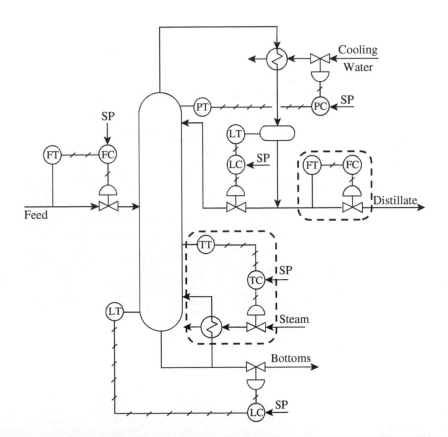

**Figure 13.51**
Distillation column control by indirect control of the bottoms using temperature with constant distillate flowrate.

13

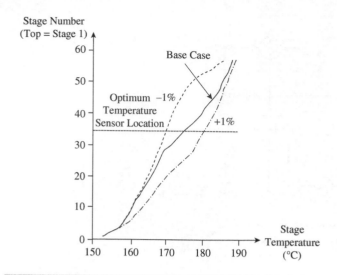

**Figure 13.52**

Optimum location of the distillation temperature sensor.

Even if the split of the feed between the distillate and bottoms is more even, significant differences in flowrate between different streams in the distillation can also influence the choice of control configuration. For example, in the distillate, if one of the flowrates of the distillate and reflux is significantly larger (e.g., by a factor of 5–10), then the composition or temperature measurement should in the first instance be used to control the larger flowrate. This means that if the distillate flowrate is significantly larger than

the reflux flowrate, then direct control of the distillate might be preferred, but if the reflux flowrate is significantly larger than the distillate flowrate, then indirect control might be preferred. Similarly, if one of the bottoms and boil-up flowrates is significantly larger (e.g., by a factor of 5–10), then the composition or temperature measurement should again in the first instance be used to control the larger flowrate. Thus, if the bottoms flowrate is significantly larger than the boil-up flowrate, then direct control of the bottoms might be preferred. If the boil-up flowrate is significantly larger than the bottoms flowrate, then indirect control might be preferred.

Unless there are good reasons to do otherwise, temperature and composition control of distillation should adopt indirect control, rather than direct control. Temperature and composition control of the overheads using the reflux flowrate, rather than the distillate flowrate, generally gives a faster response. Similarly, temperature and composition control of the bottoms using the reboil, rather than the bottoms flowrate, generally gives a faster response.

The control schemes discussed above can benefit from cascade control of the steam flowrate to the reboiler, as well as cascade control of levels in the column sump and reflux drum. If the steam to the reboiler is being controlled by temperature or composition, then flowrate control of the steam can be used with the setpoint for the steam flowrate controller provided by the temperature or composition controller, as used in heater control in Figure 13.14. The control of level in the column sump by the bottoms flowrate can benefit from using a bottoms flowrate controller, with the setpoint supplied by the level controller, as discussed in Chapter 11. Similarly, the control of level in the reflux drum by the flowrate of

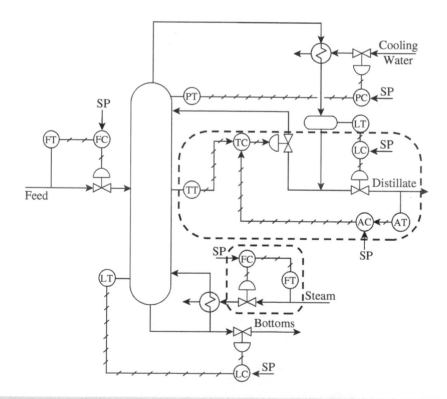

**Figure 13.53**

Distillation column control by indirect control of the distillate using concentration–temperature cascade control with constant boil-up.

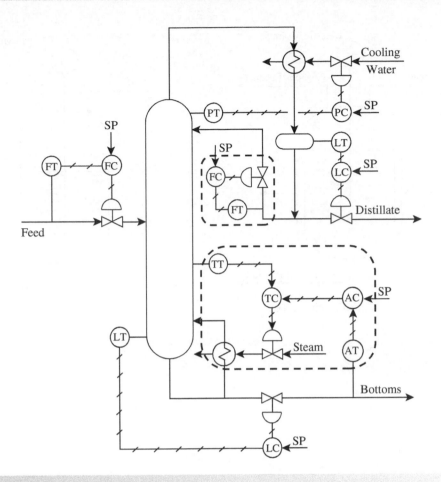

## Figure 13.54

Distillation column control by indirect control of the bottoms using concentration–temperature cascade control with constant reflux.

reflux or the overhead product can benefit from using flowrate control of the stream with the setpoint for the flowrate controller supplied by the reflux drum level controller. The distillation control schemes in Figures 14.40 to 14.51 do not show such features for the sake of clarity, but also such features are not always included in distillation control schemes.

It should be emphasized that these are only general guidelines and more detailed examination of the process dynamics of the design might change the decision. Rigorous dynamic simulation is a more robust way to evaluate the process response to a disturbance and the dynamic performance of the different control structures (Luyben 2014). *Relative gain array (RGA)* analysis can also be helpful in selecting the most appropriate control scheme (Bristol 1966; Shinskey 1988). RGA provides guidance for pairing measured and manipulated variables so as to reduce interactions between the control loops.

## 13.8  Reactor Control

In the case of reactors, reactor conversion is a key parameter both for design and control, defined by (Smith 2016):

$$\text{Conversion} = \frac{(\text{Reactant consumed in the reactor})}{(\text{Reactant } fed \text{ to the reactor})} \quad (13.1)$$

Residence time is a key variable to control reaction conversion. Here, the reactor design is considered to be fixed, along with the required inventory. Temperature and pressure need to be controlled.

### 13.8.1  Reactor Stability

Poor design of the heat transfer can lead to an inherent instability that will prove to be a major problem for control. Figure 13.55 shows two different designs for a non-adiabatic continuous stirred tank (mixed flow) reactor. Figure 13.55a shows a design with heating or cooling supplied by a jacket. Figure 13.55b shows the corresponding design with heating or cooling supplied by a heat transfer coil.

Figure 13.56 shows the characteristic curve for reactor conversion versus temperature for a single irreversible exothermic reaction. At a low temperature there will be a low rate of reaction. As the temperature is increased, the reaction rate will increase exponentially according to the Arrhenius Law (Smith 2016). At high conversion, depletion of the reactant will cause the reaction rate to decrease again, even as the temperature increases. The result is the characteristic S-shaped curve in Figure 13.56. The actual operation point in the reactor is established by matching the heat generated by the reaction against the reactor cooling.

Figure 13.57 shows the *heat evolved* ($Q_R$) for a single irreversible exothermic reaction as a function of temperature. This follows

**13**

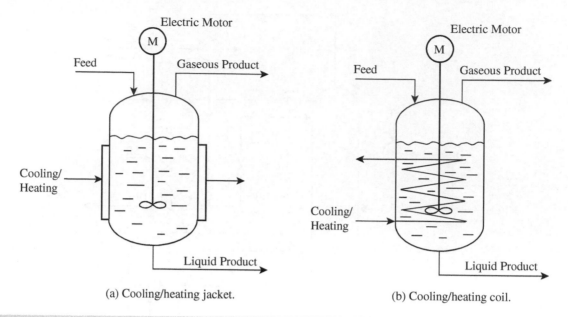

(a) Cooling/heating jacket.  (b) Cooling/heating coil.

**Figure 13.55**

Cooling/heating for a continuous stirred tank reactor.

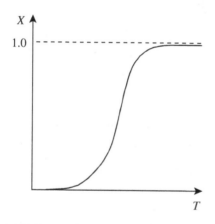

**Figure 13.56**

Variation of reactor conversion with temperature for a single irreversible reaction.

the same profile as the reactor conversion shown in Figure 13.56. Heat is removed from the continuous stirred tank reactors in Figure 13.55 by the sensible heat in the product leaving the reactor and the heat removed from the cooling system. An approximate expression for the reactor cooling can be developed by neglecting the cooling for the sensible heat in the product and assuming both the temperature of the reactor contents and the coolant temperature in the jacket or coil are both constant. In this case, the heat removal can be approximated by:

$$Q_C = UA(T - T_C) \qquad (13.2)$$

where

$Q_C$ = cooling heat duty (J s$^{-1}$ = W)

$U$ = overall heat transfer coefficient (W m$^{-2}$ K$^{-1}$)

$A$ = heat transfer area (m$^2$)

$T$ = temperature of the reactor (K)

$T_C$ = temperature of the coolant (K)

This is shown in Figure 13.57a to be a straight line with slope $UA$ and intercept on the temperature axis at $T_C$. The operating point for the reactor occurs where the curves for $Q_R$ and $Q_C$ cross ($Q_R = Q_C$). If the reactor has been designed and the $UA$ fixed, then the performance of the cooling is determined by:

$$Q_C = mC_P(T - T_C) \qquad (13.3)$$

where

$Q_C$ = cooling heat duty (J s$^{-1}$ = W)

$m$ = mass flowrate of the coolant (kg s$^{-1}$)

$C_P$ = specific heat capacity of the coolant (J kg$^{-1}$ K$^{-1}$)

$T$ = outlet temperature of the coolant (K, °C)

$T_C$ = inlet temperature of the coolant (K, °C)

If $C_P$ is constant, this is the equation of a straight line with a slope of $mC_P$ and an intercept on the temperature axis at $T_C$. Thus, once the system has been defined, the intersection between the characteristic curve for the reactor and the cooling system profile is controlled by the flowrate of the coolant.

Figure 13.57a shows a possible cooling system with a high rate of cooling. The high rate of cooling results in a low rate of reaction

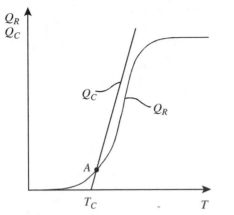

(a) A high cooling rate leads to a low conversion.

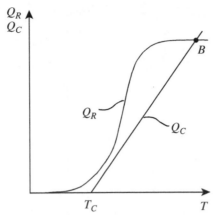

(b) A low cooling rate leads to a high conversion.

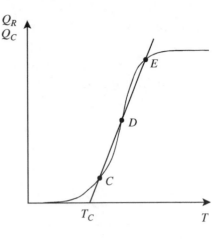

(c) An intermediate cooling rate gives an unstable operating point.

**Figure 13.57**

Multiple steady states for a single exothermic irreversible reaction.

and low reactor conversion. By contrast, Figure 13.57b shows a cooling system with a low rate of cooling. The low rate of cooling results in a higher rate of reaction and higher reactor conversion. The S-shape of the heat generation curve means that the intersection between the heat generation and cooling curves is most likely to occur at low or high conversion. However, Figure 13.57c shows a cooling curve that intersects the heat generation curve at three points. Point $C$ in Figure 13.57c is similar to Point $A$ in Figure 13.57a. Point $E$ in Figure 13.57c is similar to Point $B$ in Figure 13.57b. However, Point $D$ in Figure 13.57c cannot be operated in a stable manner, even though the rates of heat generation and removal are equal. If a disturbance causes the temperature to increase slightly above Point $D$, then the heat generation is higher than the heat removal and the temperature would continue to increase until Point $E$ is reached. On the other hand, if a disturbance causes the temperature to decrease slightly below Point $D$, then the heat generation is lower than the heat removal and the temperature would continue to decrease until Point $C$ is reached. Thus, Point $D$ is not a stable operating point for the reactor design. Systems characterized by the behavior in Figure 13.57c are extremely challenging to control and effective control requires application of advanced control methods. If possible, it is better to design the process such that the potential for instabilities of this kind is avoided in the basic process design.

Figure 13.58 shows the characteristic curve for reactor conversion versus temperature for a different reaction, this time involving a single reversible exothermic reaction. Given enough time the reaction will reach equilibrium for a given temperature. For a single reversible exothermic reaction, the equilibrium conversion shown in Figure 13.58 will decrease with an increase in temperature according to *Le Châtelier's Principle* (Smith 2016). For example:

$$A + B \rightleftharpoons C + \text{Heat}$$

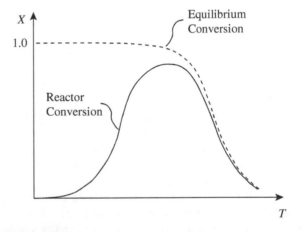

**Figure 13.58**

Variation of reactor conversion with temperature for a single exothermic reversible reaction.

**13**

If the temperature of an exothermic reversible reaction is increased, the equilibrium will be displaced in a direction to oppose the effect of the change; i.e., decrease the conversion to decrease the heat evolution. In terms of reaction kinetics, the reverse reaction is accelerated more than the forward reaction as the temperature increases. However, in a practical reactor the actual conversion will not reach equilibrium, but will be kinetically limited as shown in Figure 13.58. As with irreversible reactions, at a low temperature there will be a low rate of reaction far from equilibrium. As the temperature is increased, the reaction rate will increase exponentially. At high conversion, there will be a decrease in the rate of reaction caused by depletion of the reactant, but the reaction will also slow as it reaches the equilibrium conversion. Thus, the reactor conversion will increase to a maximum and then decrease as the temperature increases, as

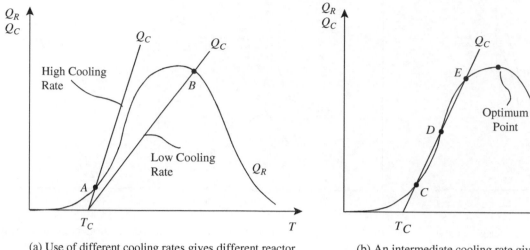

(a) Use of different cooling rates gives different reactor conversions.

(b) An intermediate cooling rate gives an unstable operting point.

## Figure 13.59

Multiple steady states for s single exothermic reversible reaction.

shown in Figure 13.58. The rate of heat evolution will follow the same basic shape.

Figure 13.59a shows the rate of heat generation for a single reversible exothermic reaction matched against two different cooling rates. This shows that at high rates of cooling there is a low reactor conversion where the generation and cooling curves cross at Point $A$. At low rates of cooling there is a high reactor conversion where the curves cross at Point $B$. Figure 13.59b shows a cooling curve that intersects the heat generation curve at three points. Point $D$ in Figure 13.59b cannot be operated in a stable manner, for the same reasoning as Point $D$ in Figure 13.57c. For the reaction shown in Figure 13.59b, the optimum cooling rate should intersect the heat generation curve at its maximum point.

Figures 13.56 and 13.58 show two simple examples of single reactions. More complex systems involving multiple reactions will create different and more complex characteristic shapes for the rate of heat generation with temperature. These more complex shapes can create their own unstable operating points from the match between heat generation and cooling curves (Denbigh and Turner 1984). For the different shapes it is generally possible in principle to design the cooling system to have a line steeper than all parts of the $Q_R$ curve and thus be inherently stable. Capital cost can be saved by designing for a smaller $UA$, requiring the control configuration to overcome a potential instability. However, the instability of the reactor will not allow open-loop control by the operator and the controller will create significant problems for controller tuning (Shinskey 1988). The design of the reactor should, where possible, avoid the inherent instability illustrated in Figures 13.57c and 13.59b.

These stability issues are by no means restricted to the liquid phase reactors such as the ones illustrated in Figure 13.55a and b. The characteristic curves illustrated in Figures 13.56 and 13.58 do not depend on the reactor configuration or whether

the reaction is liquid or gas phase. Probably the worst instability problem occurs when an exothermic reaction generates heat at a faster rate than the cooling system can remove it. Such *runaway reactions* happen because the rate of reaction, and hence the rate of heat generation, increases exponentially with temperature, whereas the rate of cooling increases only linearly with temperature. Once heat generation exceeds available cooling capacity, the rate of temperature rise becomes progressively faster. If the energy release is large enough, liquids will vaporize and over-pressurization of the reactor can occur.

### 13.8.2 Liquid Phase Reactors

Figure 13.60 shows a continuous stirred tank reactor for an exothermic liquid phase reaction with a cooling jacket to remove the heat of reaction. In terms of degrees of freedom, from Eq. (12.12) there are two inlet streams and two outlet streams. However, outlet pressure of the coolant will be left to self-regulate, resulting in three control degrees of freedom. Additionally, inventory control for the liquid in the reactor will be needed through liquid level control. Thus, four single-input–single-output control loops and four final control elements will be required.

The feed in Figure 13.60 is flowrate controlled. The temperature in the reactor is controlled by controlling the flowrate of coolant from a temperature measurement in the bulk of the reactor liquid. In this case, gases are released as a byproduct of the reaction. Also in this case, the bulk of the reactor contents have a low volatility and pressure control is by control of the venting of non-condensables. Liquid inventory is controlled by level control. One shortcoming of this scheme is that it responds slowly to variations in coolant temperature or coolant flowrate. This is because the reactor temperature changes only slowly when there is a disturbance in the coolant temperature or coolant flowrate. There is a

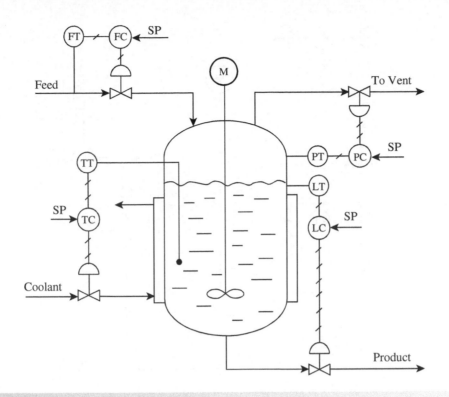

## Figure 13.60

Control of a continuous stirred tank reactor with a cooling jacket and pressure control by venting inerts.

large thermal inertia from the jacket and the volume of reactants. Furthermore, the temperature measurement may also lag the actual temperature significantly due to thermal inertia. The temperature measuring device is commonly enclosed in a *thermowell*, which is a tube with a closed end that protrudes into the process vessel. The other end of the thermowell is open, which allows the measuring device to be inserted into the process vessel whilst being protected by the thermowell. Lags for the cooling jacket, reactor contents, and thermocouple temperature measurement are all in series. This will lead to a slow control response. Eliminating the thermowell for the thermocouple reduces the time lag associated with the measurement. The speed of the control response can also be increased by designing the reactor for a high rate of heat transfer. However, the discussion earlier related to Figures 13.57 and 13.59 shows that it is important to take into account the effect of the cooling rate on reactor conversion.

Whilst design of the reactor can have an important influence on the speed of the control configuration response to disturbances, the design of the control configuration itself can make significant improvements in the response to disturbances. Figure 13.61 shows an alternative temperature control configuration. This uses a cascade control configuration in which the primary controller for the bulk reactor temperature provides the setpoint for the secondary controller for the coolant flowrate. If there is a disturbance in the cooling temperature or flowrate, this disturbance can be quickly detected by the coolant sensor. Since the coolant controller setpoint will be unchanged, the coolant controller will respond

to change the flowrate to its desired value for temperature control. By responding quickly to the coolant disturbance, the secondary controller corrects the disturbance before the reactor temperature is significantly affected by the disturbance. The remainder of the control configuration is the same as that in Figure 13.60.

If the contents of the reactor are volatile, then it might be preferred to condense the volatile material and return it to the reactor. Any non-condensibles can in principle be vented and used to control the pressure. Figure 13.62 shows a reactor with a reflux condenser and reflux drum. The reflux drum requires inventory control, which introduces an additional degree of freedom. Condensed material is returned to the reactor from the reflux drum under level control and non-condensibles are vented from the reflux drum. The flowrate of non-condensibles is in this case used to control the pressure of the reactor. The remaining part of the control configuration is the same as in Figure 13.61.

The arrangement in Figure 13.62 can be exploited to remove large quantities of heat from the reactor for an exothermic reaction that might otherwise be difficult to remove using a cooling jacket or cooling coil. Allowing vaporization of the contents followed by condensation (e.g., using cooling water) exploits the latent heat of vaporization to remove the heat of reaction, rather than just exploiting sensible heat. The vaporized material is condensed and returned to the reactor. The vaporized material might be feed or product materials, or a solvent used as a reaction medium. This allows quantities of heat to be removed from the reactor that might otherwise be difficult using conventional heat exchange. Vaporization of reactor contents can in many cases be used to remove the

**13**

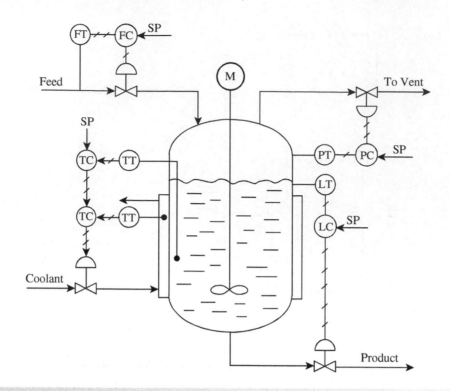

**Figure 13.61**

Cascade control of a continuous stirred tank reactor with a cooling jacket and pressure control by venting inerts.

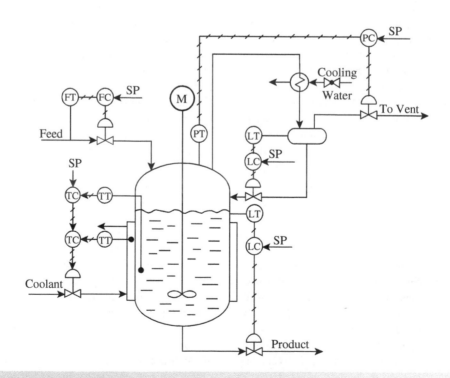

**Figure 13.62**

Cascade control of a continuous stirred tank reactor with a cooling jacket and pressure control by venting non-condensibles after refluxing condensibles.

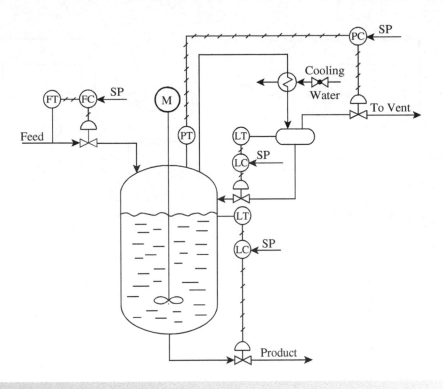

**Figure 13.63**

Cooling of the reactor through vaporization and refluxing of the reactor contents with pressure control by venting inerts.

heat of reaction without the need for a cooling jacket or cooling coils. This is shown in Figure 13.63, where the pressure is again controlled by venting of non-condensibles.

In some cases there might be no significant non-condensibles. If this is the case, then the pressure of the reactor will be the vapor pressure of the reactor contents at the reactor temperature. Thus, controlling the reactor pressure will control the reactor temperature. Figure 13.64a shows a control scheme in which all of the heat of reaction is removed by vaporization of the reactor contents. Pressure is controlled by manipulating the flowrate of cooling water to the condenser, which controls the amount of condensation taking place. This controls the reactor temperature indirectly. Because the reactor temperature and pressure are no longer independent of each other and therefore cannot be controlled separately, this removes a degree of freedom and the number of control loops is decreased by one. Indeed, the pressure controller could even be replaced by a temperature controller manipulating the same valve. If the composition in the reactor varies significantly, then the relationship between pressure and temperature also varies. In this situation the temperature in the reactor can be measured and used to control the setpoint of the pressure controller in a cascade system, as shown in Figure 13.64b. This allows for greater variability in the reactor concentrations. Also, the pressure controller will tend to act faster to disturbances in the cooling water conditions than the temperature controller. In both schemes in Figure 13.64 provision should be made for the venting of any non-condensibles that might build up during operation, particularly at start-up.

Many other control arrangements are possible with stirred tank reactors, but only some prominent examples have been discussed here. For example, if the reaction involves two feeds, then the control of one of the feeds can be used to control the rate of reaction, and hence the rate of heat evolution and the temperature. Rather than stirred tank reactors, many other different configurations can also be encountered for liquid phase reactions.

### 13.8.3  Gas Phase Reactors

Gas phase reactions using a heterogeneous catalyst are common in chemical processes. Figure 13.65 shows an adiabatic gas phase catalytic reactor in which the catalyst is loaded into a packed bed. The feed flowrate is controlled and then the temperature increased to the required inlet temperature for the reaction, in this example by steam heating, but could also have been furnace heating or heat recovery. It might be necessary for the inlet temperature to be increased slowly through time as the catalyst deactivates, but only by a few tens of degrees at most. The adiabatic reactor needs to be designed such that any temperature increase for exothermic reactions or temperature decrease for endothermic reactions is within acceptable bounds. Figure 13.65 also features a pressure control. This might be necessary to maintain the required process conditions for the reaction or to maintain inventory control in the reactor. When the reactor is placed in the context of the process as a whole, the feed flowrate and pressure control in Figure 13.65 might not be necessary. However, initially at least, it should be assumed that they are required.

**13**

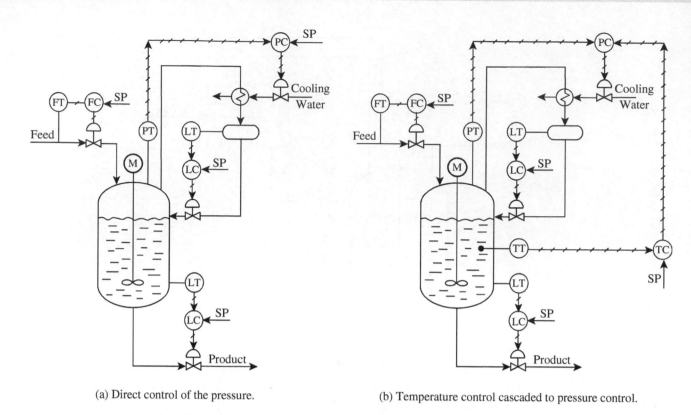

(a) Direct control of the pressure.          (b) Temperature control cascaded to pressure control.

## Figure 13.64

Cooling of the reactor through vaporization and refluxing of the reactor contents in the absence of non-condensibles.

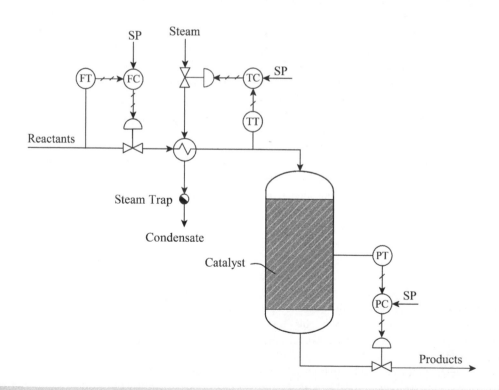

## Figure 13.65

Control of an adiabatic gas-phase catalytic reactor.

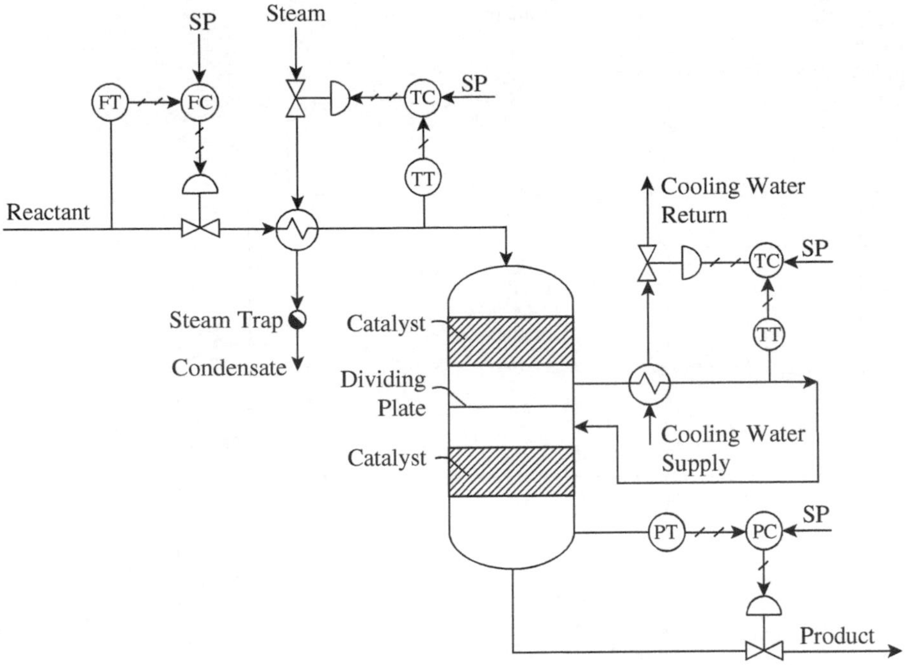

## Figure 13.66

Control of a catalytic gaseous reactor with intermediate external cooling.

If adiabatic operation of a gaseous catalytic reactor is not feasible and would lead to an unacceptable temperature increase (if exothermic) or decrease (if endothermic) in the reactor, then temperature control of the reactor can be maintained in different ways. Figure 13.66 shows an example of a gaseous catalytic reactor for an exothermic reaction with the catalyst split between two beds with intermediate cooling between the beds. In principle, the cooling could have been achieved using coils within the reactor. However, the example shown in Figure 13.66 shows the reacting stream taken out of the reactor after the first bed, cooled externally, and returned to the reactor to complete the reaction. Again, the feed flowrate and pressure control in Figure 13.66 might not be necessary when the reactor is placed in the context of the process as a whole. This approach can be extended to reactions with greater exothermic or endothermic heat of reaction by increasing the number of beds with intercooling or heating between each of the beds. Typically, up to eight beds at most are used.

An alternative to intermediate cooling with heat exchange for an exothermic reaction, as shown in Figure 13.66, is to use *cold shot cooling*, as shown in Figure 13.67. The reactor in Figure 13.67 involves two feeds. Reactant 1 and Reactant 2 are mixed under flowrate control, preheated to the required reactor temperature (in this example by steam heating), and fed to the first catalyst bed. The reacting stream from the first reactor bed is cooled by feeding further cold Reactant 2 as *cold shot* under temperature control. There are two mechanisms for temperature control involved with cold shot arrangements. The first is the cooling caused by the injection of cold reactant. However, the staged feed in the reactant also controls the rate of reaction by changing the concentration of the reactants. The feed arrangement for the

two feeds to the first bed shown in Figure 13.67 might benefit from *ratio control* of the feeds. This, along with the need for pressure control, would be best determined when the reactor is placed in the context of the process as a whole. Cold shot arrangements can be used with a single feed as well as the example with two feeds shown in Figure 13.67. In that arrangement the single feed is split between the two beds. As with intermediate cooling between the beds, cold shot can be extended to reactions with greater exothermic heat of reaction by increasing the number of beds with cold shot between each of the beds. Again, typically up to eight beds at most are used.

However, there is a limit to the amount of temperature control that can be achieved with the packed bed arrangements in Figures 13.66 and 13.67. In such situations, the reaction can be carried out inside tubes packed with catalyst. A cooling medium (or heating medium for endothermic reactions) on the outside of the tubes would remove (or supply) the heat of reaction. Figure 13.68 illustrates a reactor that is carrying out an exothermic gas phase reaction. The reactor feed is first preheated to an optimum inlet temperature. The preheating might be carried out by steam, fired heating, or heat recovery. It would be common for the inlet temperature to be increased slowly through time as the catalyst deactivates. The feed in Figure 13.68 then enters the reactor tubes, where the reaction is initiated by the catalyst. Cooling of the reactor tubes can be carried out using a flow of heat transfer oil, molten salt, or boiling of boiler feedwater to generate steam on the outside of the tubes. For heat transfer oil or molten salt, the temperature control configuration would control the flowrate of coolant, as shown in Figure 13.68. If steam is to be generated, temperature control would adjust the pressure of the steam generation. For

**13**

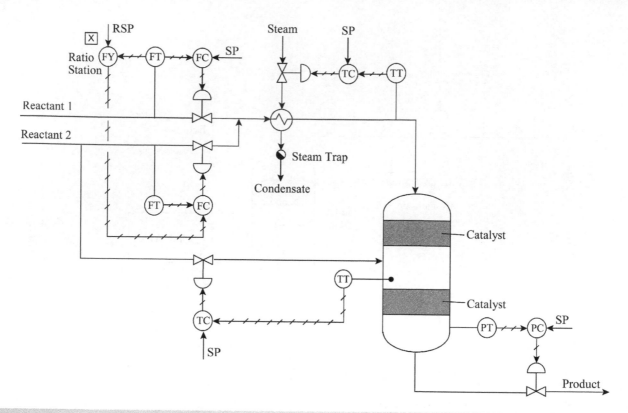

**Figure 13.67**

Control of a catalytic gaseous reactor with cold shot cooling.

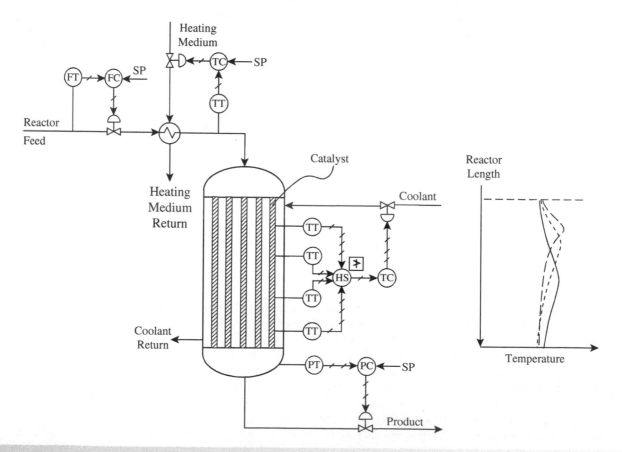

**Figure 13.68**

Control of a catalytic tubular reactor for an exothermic gaseous reaction.

cooling by steam generation, the boiler feedwater is normally fed to the bottom of the reactor in a counter-current arrangement. If molten salt or heat transfer oil are used, then the coolant is normally fed to the top of the reactor in a co-current arrangement, which maintains the lowest temperature at the top of the reactor and highest cooling temperature at the bottom.

For a highly exothermic reaction, a high rate of reaction would be expected close to the inlet of the tubes, where there is a high concentration of reactants. If the catalyst is a uniform design along the tubes, this would initially create an increase in temperature in the tubes that would then be expected to decrease along the tubes as the rate of the reaction decreased and the coolant removed the heat of reaction. As shown in Figure 13.68, any peak in temperature can be in different locations along the tubes. The location and size of any peak in temperature would depend on the feed flowrate, composition of the reactants, inlet temperature, design of the catalyst, catalyst activity, and coolant performance (Singh et al. 2008).

However, the reactor design in most cases would strive to minimize the size of a peak in temperature through changing the catalyst design in different layers along the tubes. The size, shape, and porosity of the catalyst pellets and the distribution of active sites within the pellets can be changed in different layers along the tubes in order to control the rate of reaction along the tubes. If the catalyst within the tubes is different in different layers, then there might in principle be multiple peaks in temperature.

Multipoint thermocouple probes contained in a metal sleeve are used in such reactors to measure the temperature at different points along the tube length. Such tubular reactors might contain many hundreds (even thousands) of tubes. Some of the tubes will be fitted with probes. The tubes fitted with such probes will not necessarily behave in exactly the same way as the tubes without probes. A high selector system can be used to detect the location and magnitude of temperature peaks and control the coolant to keep any temperature peaks below an upper limit. Again, the feed flowrate and pressure control in Figure 13.68 might not be necessary when the reactor is placed in the context of the process as a whole.

Further discussion of reactor control can be found in Luyben (2007).

# 13.9   Control of Process Operations – Summary

Having determined the control degrees of freedom for individual operations, the conceptual control configuration can be synthesized. Additional control loops will be required where inventory control is required. The number of degrees of freedom defines the number of single-input-single-output control loops required. At this stage, the control degrees of freedom and inventory control are satisfied by the control configuration with no consideration given to the overall process context. This chapter has developed control configurations for individual process operations:

- pumps;
- compressors;
- heat exchangers;
- furnaces;
- flash drums;
- distillation;
- reactors.

When put together in an overall process, these individual process operation control schemes are likely to interact with each other. This will be considered in the next chapter.

# Exercises

1. A process cooling duty is to be serviced by cooling water. It is important to as much as possible avoid process fluid leaking into the cooling water in the event of failure of one of the heat exchanger tubes. Devise a control scheme to control the outlet temperature of the cooler.

2. Figure 13.69 shows the control configuration for a constant speed positive displacement pump. A recycle system is used to control the flowrate with a high-pressure override. A low selector passes the correct signal to the pneumatic control valve. The pneumatic control valve uses a pressure range

**13**

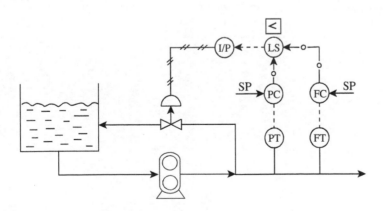
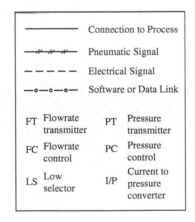

| | Connection to Process |
| --- | --- |
| | Pneumatic Signal |
| | Electrical Signal |
| | Software or Data Link |

| FT | Flowrate transmitter | PT | Pressure transmitter |
| --- | --- | --- | --- |
| FC | Flowrate control | PC | Pressure control |
| LS | Low selector | I/P | Current to pressure converter |

## Figure 13.69

Control system for a constant speed positive displacement pump.

between 0.2 and 1 barg. At a certain set of conditions the flow-rate controller creates a signal of 0.4 barg and the pressure controller creates a signal of 0.5 barg. Calculate the valve percent open.

**3.** Figure 13.70 shows a flash drum that has been fitted with a steam coil to control the vaporization. The steam supply is sub-ject to variations in pressure. Devise a control scheme for the flash drum that can respond quickly to variations in the steam conditions.

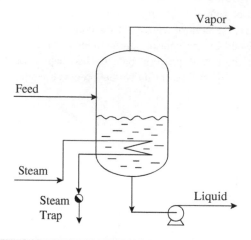

**Figure 13.70**

Flash drum with a heating coil.

# References

Bristol, E.H. (1966). On a new measure of interaction for multivariable process control. *IEEE Transactions on Automatic Control* 1: 133–134.

Calabrese, A.M., Jensen, B.A., Gaertner, S.E., and Liptak, B.G. (2006). Furnace and reformer controls. In: *Liptak BG Process Control and Optimization*, 4e, vol. II. CRC Press.

Denbigh, K.G. and Turner, J.C.R. (1984). *Chemical Reactor Theory*, 3e. Cambridge University Press.

Driedger, W.C. (1996). Controlling positive displacement pumps. *Hydrocarbon Processing* 47–54.

Luyben, W.L. (1992). *Practical Distillation Control*. Van Nostrand Reinhold.

Luyben, W.L. (2007). *Chemical Reactor Design and Control*. Wiley.

Luyben, W.L. (2014). *Plantwide Dynamic Simulators in Chemical Processing and Control*. CRC Press.

Shinskey, F.G. (1988). *Process Control Systems*, 3e. McGraw-Hill Inc.

Singh, S., Lal, S., and Kaistha, N. (2008). Case study on tubular reactor hot-spot temperature control for throughput maximization. *Industrial and Engineering Chemistry Research* 47 (19): 7257–7263.

Sloley, A.W. (2001). Effectively control column pressure. *Chemical Engineering Progress* 97 (1): 38–48.

Smith, C.L. (2012). *Distillation Control - an Engineering Perspective*. Wiley.

Smith, R. (2016). *Chemical Process Design and Integration*, 2e. Wiley.

# Chapter 14

# Process Control – Overall Process Control

The process flow diagram provides the basis for the development of the overall process control configuration, but it does not provide a specification for the control configuration. To design the overall process control system it is first necessary to understand the process control objectives for the overall process (Luyben et al. 1999; Larsson and Skogestad 2000). The first objective to be clarified is whether the process is required to accept a fixed amount of feed from an upstream process or is required to produce a fixed amount of product by varying the feed. Less commonly, the production rate might be fixed at an intermediate point within the flowsheet. Other control objectives typically include temperature, pressure, composition, and so on, in various parts of the process and the products. Any constraints related to process yields or waste byproducts need, as much as possible, to be accommodated by the control system. Whilst safety systems will be added later, all decisions on process control must be cognizant of the safety implications and, where possible, hazards minimized at each stage as the design progresses.

In addition to understanding the control objectives, it is also necessary as much as possible to understand the types of disturbances that will affect the control system. The types of disturbance anticipated can have a major influence on the decisions made for the control system design. External disturbances could be variations in feed flowrate, composition, temperature, or pressure. Other external disturbances can come from variations in the condition of utilities, such as changes in cooling water temperature, steam pressure, steam temperature, or fuel gas composition, pressure, and flowrate. Disturbances can also be created internally and passed between operations in the process resulting from poor control systems on other operations (e.g., control valve malfunction, or poor controller tuning giving a poor control response to disturbances), failure of control equipment on other operations (e.g., instrument failure), process equipment malfunctions (e.g., pumps not performing to

specifications), or process equipment failures (e.g., switching to standby equipment).

Given an understanding of the control objectives and the likely disturbances, the control configuration can be synthesized. Consideration is first restricted to the conceptual design of the control configuration. Once the conceptual configuration has been developed and evaluated, details of the control signals, location of the controllers, integration with model predictive control algorithms, and the design of the distributed control system (DCS) can be developed. When synthesizing the conceptual control configuration for the overall process, the problem has certain characteristics that do not necessarily feature in the design of control configurations for single operations discussed in Chapter 13 (Stephanopoulos and Ng 2000):

1) The variables to be controlled by an overall process are not as clearly or as easily defined as for single operations.
2) Local control decisions, made within the context of single units, may have long-range effects throughout the process.
3) The size of the control problem for the overall process is significantly larger than that for the individual operations, making its solution considerably more difficult.

The following illustrative example highlights a number of issues for the design of overall process control systems that do not feature when considering the control of individual operations.

## 14.1 Illustrative Example of Overall Process Control Systems

Figure 14.1 illustrates a simple example showing a process flow diagram for a reaction, separation, and recycle system (Bolton and Smith 2022). Fresh Feed $A$ is assumed to be pure and reacted to Product $B$ via the reaction:

$$A \longrightarrow B \qquad (14.1)$$

*Process Plant Design*, First Edition. Robin Smith.
© 2024 John Wiley & Sons Ltd. Published 2024 by John Wiley & Sons Ltd.
Companion website: www.wiley.com/go/processplantdesign

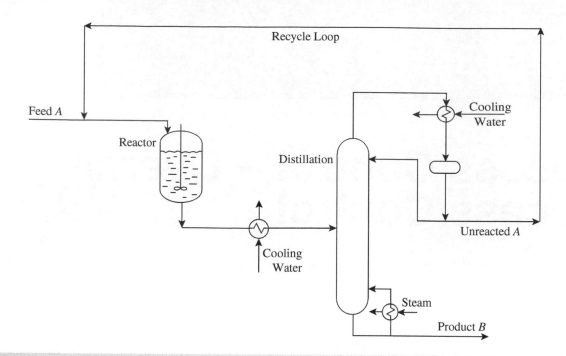

## Figure 14.1

A simple process with reaction, separation, and recycle. Source: Bolton and Smith (2022).

The rate of reaction is assumed to be given by (Rase 1977; Denbigh and Turner 1984; Levenspiel 1999):

$$r_A = kC_A^a \qquad (14.2)$$

where

$r_A$ = rate of reaction of Component $A$ (kmol m$^{-3}$ s$^{-1}$)
$k$ = reaction rate constant
$C_A$ = molar concentration of Component $A$ (kmol m$^{-3}$)
$a$ = order of reaction

To keep the example simple, the reactor is assumed to be operated adiabatically. The reaction rate constant $k$ is assumed to be characterized by the Arrhenius Law (Rase 1977; Denbigh and Turner 1984; Levenspiel 1999):

$$k = k_0 \exp\left[-\frac{E}{RT}\right] \qquad (14.3)$$

where

$k_0$ = frequency factor
$E$ = activation energy (kJ kmol$^{-1}$)
$R$ = universal gas constant (kJ kmol$^{-1}$ K$^{-1}$)
$T$ = absolute temperature (K)

In Figure 14.1 (Bolton and Smith 2022) the fresh Feed $A$ is mixed with recycled Component $A$ and enters an adiabatic mixed flow reactor (continuous stirred tank reactor, CSTR), where it is partially converted to Product $B$. The reactor is assumed to have a fixed volume. The reactor temperature is assumed to be constant and thus the reaction rate constant $k$ is also fixed. The reactor effluent is cooled and separated in a distillation column. The relative volatility of Component $A$ is greater than that of Component $B$. Thus, unconverted Component $A$ is separated to the overhead of the distillation column and recycled to the reactor with Component $B$ taken as product from the distillation bottoms.

Figure 14.2 shows one possible control system for the whole process. Before developing an approach for the synthesis of such control schemes, it is necessary to understand some important problems that can arise with overall process control strategies, especially when the process features recycles (Bolton and Smith 2022). Figure 14.2 highlights the recycle path. The recycle stream between the reactor and distillation in Figure 14.2 potentially creates significant instability problems for the control arrangement when the control for the material balance is considered. A problem can potentially arise if the flowrate of the fresh Feed $A$ increases for some reason. Then the level in the reactor increases temporarily until the reactor level controller increases the flowrate from the reactor. The increased flowrate from the reactor increases the flowrate to the distillation, which increases the level in the

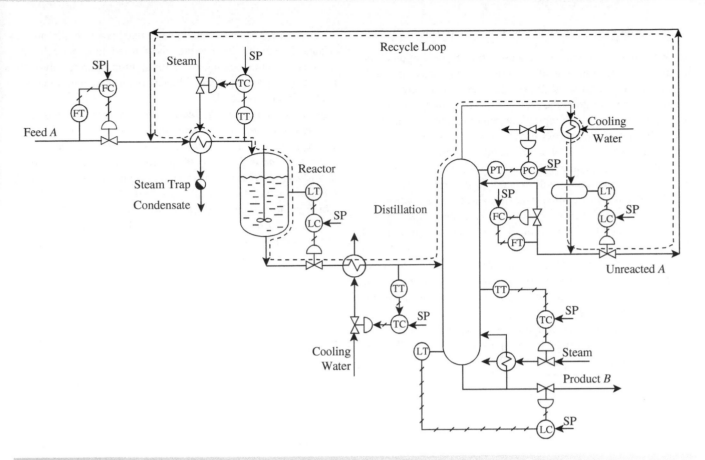

**Figure 14.2**

An overall control system for the simple process. Source: Bolton and Smith (2022).

distillation reflux drum. The reflux level control increases the recycle flow, which in turn increases the flowrate to the reactor, which again temporarily increases the level in the reactor. This creates a cycle in which the flows throughout the recycle loop keep increasing. The phenomenon is known as the "snowball" effect (Luyben 1994; Luyben et al. 1999). The basic problem with the control arrangement in Figure 14.2 is that all flows in the recycle are set by level control, but the material balance and total inventory in the recycle loop is not controlled (Luyben 1994; Luyben et al. 1999). A circuit of level controllers each attempt to pass on unwanted extra inventory to each other. A small change in the feed can therefore in principle create a large change in the recycle flowrate, as there is no control on the overall inventory. Any disturbance that tends to increase the total inventory (e.g. increase in the flowrate of fresh feed) in principle can create a large increase in all flowrates around the recycle, leading to an unstable operation (Luyben 1994; Luyben et al. 1999). Conversely, a disturbance that decreases the total inventory, such as a decrease in the fresh feed, can create a large decrease in the flowrates around the recycle, though this is generally less problematic in practice.

Accepting that there are potentially serious problems associated with control of the total material inventory in

Figure 14.2, consider now the inventory of the chemical components in the process. If there are no impurities in the feed and no byproducts formed in the reactor, then only the inventory of Components A and B need to be controlled. For the recycle system in Figure 14.2, Component B once formed can leave from the bottom of the distillation column as a product under level control. The inventory of Component B in the process is therefore controlled. However, unreacted Component A cannot leave the process unless it reacts to Component B. The feed to the process in Figure 14.2 is under flowrate control and not linked to the inventory of Component A in the process. There is thus no control of the inventory of Component A in the overall process.

Generally, there are three ways these problems can be resolved (Bolton and Smith 2022):

- allow the process to self-regulate if this feasible;

- modify the process design and/or operating conditions to allow self-regulation if the existing process cannot self-regulate to the extent required;

- modify the control system to remove any potential for the snowball effect.

**14**

### 14.1.1 Allow the Process to Self-Regulate If This Is Feasible

In the example in Figure 14.2 at least some self-regulation of the inventory of Component $A$ is possible in principle. Self-regulation at constant reactor temperature can be created by change in the flowrate of the recycle changing the concentration in the reactor and thus changing the rate of reaction. If the flowrate of Feed $A$ to the process increases, then this will lead to an increased concentration of Component $A$ in the reactor, both from the fresh feed and then the recycle. The effect of this increase in concentration will depend on the parameters in the kinetic equation. If the process design and process control system are unchanged then two cases can be distinguished, which can be illustrated by a simple quantitative analysis (Bolton and Smith 2022).

**Case 1**  *If an increase in the concentration of reactant in the reactor has no influence on the conversion to product, then there will be no self-regulation and no mitigation of the potential for the snowball effect.*

Starting with the basic equation for the reactor conversion in the mixed-flow reactor (CSTR) in Figure 14.2 (Rase 1977; Denbigh and Turner 1984; Levenspiel 1999):

$$X_A = \frac{-r_A V}{F_{A,in}} = \frac{-r_A V}{Q C_{A,in}} \tag{14.4}$$

where

$X_A$ = conversion of Component $A$ (–)

$r_A$ = rate of reaction of Component $A$ (kmol m$^{-3}$ s$^{-1}$)

$V$ = volume (m$^3$)

$F_{A,in}$ = molar flowrate of Component $A$ in the reactor inlet (kmol s$^{-1}$)

$Q$ = volumetric flowrate of the feed (m$^3$ s$^{-1}$)

$C_{A,in}$ = molar concentration of Component $A$ at the reactor inlet (kmol m$^{-3}$)

In an extreme case, somewhat unusual in practice, the kinetic equation in Eq. (14.2) might have $a = 0$, a zero-order reaction, giving $r_A = k$ and Eq. (14.4) becomes:

$$X_A = \frac{kV}{Q C_{A,in}} \tag{14.5}$$

For a zero-order reaction in which the feed concentration is fixed, the reaction rate is independent of the concentration of reactant in the reactor. Changing concentration in the reactor therefore has no effect on the rate of the reaction and therefore cannot prevent the potential for the snowball effect. The recycle rate would therefore increase (or, in the case of decreasing feed, reduce) inexorably until it reaches the limit of the process equipment in the recycle loop, typically either the maximum flow which can be delivered by a pump or admitted by a valve (or zero flow for reducing feed).

In summary, such a scenario has a fixed reaction rate, and any mismatch between the feed rate and this reaction rate will inevitably cause an accumulation or diminution of inventory of the reactant in the entire process, which will not cease until some other process limit is reached.

**Case 2**  *If increasing the recycle flowrate increases the rate of reaction, even at constant reactor temperature, this can create self-regulation and prevent the snowball effect. However, any self-regulation is limited by the maximum possible reaction rate that allows the reactor conversion to be increased to match the increase in the flowrate of fresh feed at steady state, beyond which point the snowball effect cannot be prevented.*

It is clear from the above arguments that a zero-order reaction cannot self-regulate to prevent the snowball effect. Consider now the example of a first-order reaction:

$$-r_A = kC_{A,out} \tag{14.6}$$

If it is assumed that the density is constant, then combining Eqs (14.4) and (14.6) gives:

$$X_A = \frac{C_{A,out} V}{Q C_{A,in}} = \frac{k C_{A,in}(1 - X_A)V}{Q C_{A,in}} = \frac{k(1 - X_A)V}{Q} \tag{14.7}$$

Rearranging:

$$X_A = \frac{1}{\frac{Q}{kV} + 1} \tag{14.8}$$

For the process in Figure 14.2 a material balance around the reactor at steady state gives, assuming that both the fresh feed and the recycle are pure:

$$F_B = (F_A + R_A)X_A \tag{14.9}$$

where

$F_B$ = molar flowrate of Component $B$ in the reactor outlet (kmol s$^{-1}$)

$F_A$ = molar flowrate of Component $A$ in the fresh feed (kmol s$^{-1}$)

$R_A$ = molar flowrate of Component $A$ in the recycle (kmol s$^{-1}$)

Substituting for $X_A$ from Eq. (14.8) into Eq. (14.9) gives:

$$F_B = \frac{F_A + R_A}{\frac{Q}{kV} + 1} \tag{14.10}$$

The volumetric flowrate to the reactor can be written as:

$$Q = (F_A + R_A)\frac{\rho_A}{M_A} \tag{14.11}$$

where

$\rho_A$ = density of Component $A$ (kg m$^{-3}$)

$M_A$ = molar mass of Component $A$ (kg kmol$^{-1}$)

Recognizing that an overall material balance for the process in Figure 14.2 requires that at steady state $F_B = F_A$ and substituting Eq. (14.11) into Eq. (14.10) gives:

$$F_B = F_A = \frac{F_A + R_A}{\dfrac{(F_A + R_A)\rho_A}{kVM_A} + 1}$$

$$= \frac{1}{\dfrac{\rho_A}{kVM_A} + \dfrac{1}{(F_A + R_A)}} \qquad (14.12)$$

Rearranging Eq. (14.12) gives:

$$\frac{F_A \rho_A}{kVM_A} + \frac{F_A}{F_A + R_A} = 1 \qquad (14.13)$$

Defining the recycle ratio $RR$:

$$RR = \frac{R_A}{F_A} \qquad (14.14)$$

Substituting Eq. (14.14) into Eq. (14.13) and rearranging finally gives:

$$RR = \frac{\dfrac{F_A \rho_A}{kVM_A}}{1 - \dfrac{F_A \rho_A}{kVM_A}} \qquad (14.15)$$

Consider now the consequence of Eq. (14.15) in terms of the control system in Figure 14.2 and its ability to self-regulate. For a reactor with constant volume $V$ and constant temperature, if the flowrate of fresh feed $F_A$ increases, then the group $F_A \rho_A / kVM_A$ increases. This means that from Eq. (14.15) the recycle ratio must increase to maintain the material balance and allow the process to self-regulate. As the flowrate of fresh feed $F_A$ increases, the concentration of the reactant in the reactor increases, and the concentration of the product decreases to maintain a constant reactor conversion. However, the increase in the concentration of Component $A$ in the reactor is less than proportional to the increase in the feed rate, because of the increase in the recycle flowrate. This means that the increase in the rate of reaction is also less than proportional to the increase in the feed rate. If $F_A \rho_A \ll kVM_A$, a material balance can be maintained and the process will self-regulate with a modest increase in the recycle rate, preventing the snowball effect. However, as $F_A \rho_A$ increases, the extent of the increase in the recycle rate needed becomes much larger, and as $F_A \rho_A$ approaches $kVM_A$, the group $F_A \rho_A / kVM_A$ approaches unity and the recycle ratio from Eq. (14.15) tends to infinity.

Thus, as the flowrate of fresh Feed $A$ in Figure 14.2 increases, the process can initially self-regulate by increasing the recycle flowrate, which prevents the snowball effect. However, as $F_A \rho_A$ approaches $kVM_A$ the change in recycle, which follows a change in feed, becomes larger and tends towards infinity. Such a system therefore can accommodate some change in feed rate, but for larger changes will typically be constrained by some other process limit, as in Case 1. The ratio $F_A \rho_A / kVM_A$ is essentially a ratio of the feed rate to the maximum possible reaction rate (Bolton and Smith 2022). This maximum possible reaction rate is the reaction rate as the reactor composition approaches pure Component $A$.

In summary, provided it is possible to increase the recycle sufficiently for the resulting change in reactor composition to deliver an increase in conversion of Component $A$ to Component $B$ to match the increase in fresh Feed $A$ flowrate at steady-state, then the snowball effect will be prevented. However, as the flowrate of Feed $A$ starts to approach the maximum possible reaction rate, greater increases in the recycle ratio result, such that ultimately the conversion of Component $A$ to Component $B$ cannot match the increase in fresh Feed $A$ at steady-state and the snowball effect cannot be prevented.

### 14.1.2 Modify the Process Design and/or Operating Conditions to Allow Self-regulation if the Existing Process Cannot Self-regulate to the Extent Required

If increasing the recycle flowrate reaches the maximum possible reaction rate and does not allow the reactor conversion to match the increase in fresh feed at steady-state, then the snowball effect can be prevented by increasing the reactor volume or increasing the reaction rate (for example by increasing the reactor temperature).

The limit of increased reaction rate and therefore the limit for self-regulation can be increased if the volume in the reactor is increased by allowing the level in the reactor to increase, or by changing to a larger reactor, or by increasing the reactor temperature. Increasing the reactor volume $V$ or increasing the reaction rate constant $k$ (for example by increasing the reaction temperature) allows the group $F_A \rho_A / kVM_A$ to decrease. This allows an increase in the range over which the fresh feed flowrate can be increased and the process remain self-regulating.

### 14.1.3 Modify the Control System to Remove any Potential for the Snowball Effect

The potential problems associated with the control system in Figure 14.2 have been presented as a *control problem*, which is how it typically manifests itself. However, it reflects a more fundamental *material balance problem*. The only way out of the process for the feed Component $A$ is by reaction to Component $B$. The snowball effect arises because there is insufficient reaction capacity available to convert the feed to the product,

14

a situation that should be apparent from analysis using a steady-state model. Indeed, the denominator term in Eq. (14.15), $kVM_A$, can be seen as a representation of the maximum reaction capability and the snowball effect results when the feed $F_A\rho_A$ approaches that maximum.

If the process cannot self-regulate, it is also possible to change the design of the control scheme to remove the potential for such instability. This can, in principle, be achieved by ensuring there is a flowrate controller somewhere in the recycle (Luyben 1994; Luyben et al. 1999). Figure 14.3 shows an alternative control arrangement for the process in Figure 14.2. In Figure 14.3, the reactor effluent control has been changed to flowrate control. Inventory control for the reactor is now from the flowrate of fresh Feed $A$. The inclusion of reactor effluent flowrate control in the recycle loop prevents the snowball effect since the inventory of Component $A$ in the process is now being controlled by the flowrate of fresh Feed $A$ to the process under level control into the reactor. This breaks the cycle by changing the overall material balance in the process so as only to accept as much fresh feed as the reactor is capable of processing. Therefore the feed to the process can no longer be set upstream. It should be noted that in the control scheme in Figure 14.3 the rate of production is now

controlled by a combination of the reactor outlet flow controller, the level control in the reactor, and the reactor temperature. How this works in practice depends on the degree of influence that concentration has on the reaction rate.

1) If there is no influence of concentration on the reaction rate (as in Case 1 above), then the production rate of Component $B$ can only be influenced by changing the volume or temperature in the reactor.

2) If there is an influence of concentration on the reaction rate (as in Case 2 above), then increasing the flowrate by changing the setpoint for the flowrate controller at the exit of the reactor will cause the reactor level to lower and the flowrate of fresh Feed $A$ to increase to maintain the level. This will cause the concentration of Component $A$ in the reactor to increase, both from the increased flowrate of fresh Feed $A$ and unconverted Component $A$ in the recycle, and the production of Component $B$ will increase. Following the definitions used under Case 2 above, if $F_A\rho_A \ll kVM_A$ the increase in production will be almost proportional to the increase in flowrate, but as $F_A\rho_A$ approaches $kVM_A$ smaller increases in production result from increases in the reactor exit flowrate. In addition to the flowrate control

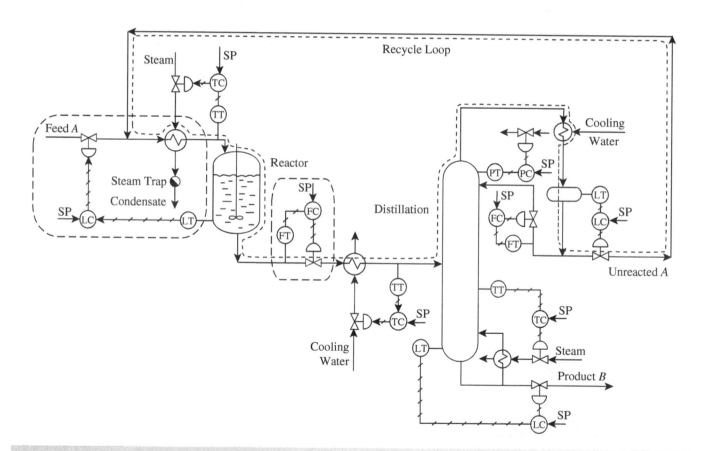

**Figure 14.3**

Placing a flowrate controller on the reactor effluent in the recycle loop can avoid the snowball effect and maintain the inventory of the Feed $A$. Source: Bolton and Smith (2022).

of the reactor outlet, the reactor level can also be changed to adjust the residence time in the reactor and adjust the reactor conversion.

Yet another control option to the one shown in Figure 14.2 is shown in Figure 14.4. This time the snowball effect is countered by measuring the combined feed to the reactor (fresh feed plus recycle) and using this to control the flowrate of the fresh Feed *A*. This means that there is flowrate control in the recycle, even though the control valve is outside the recycle loop, thus avoiding the snowball effect by admitting only as much feed to the process as can be reacted. Controlling the combined flowrate to the reactor within the recycle loop maintains the inventory of Component *A* in the process. The rate of production can be controlled by a combination of the combined flowrate of Component *A* to the reactor and the level in the reactor.

It should be noted that just placing a flowrate controller somewhere in the recycle does not necessarily solve the control problems. An alternative location for flowrate control in the recycle to attempt to solve the control problems for this flowsheet could be changing the distillation control to indirect

control of the bottoms with a constant distillate flowrate, as shown in Figure 14.5. This introduces a flowrate controller into the recycle loop. However, there will still be no effective inventory control of the lighter unreacted feed chemical component, which will accumulate in the distillation column instead of the entire recycle loop. For example, if disturbances in the reactor result in a lower conversion, this will cause the level in the reflux drum to increase from the increased flow of unreacted Component *A*, resulting in increased reflux. Then the temperature control of the column bottoms will respond by driving the lighter chemical component back up the column. This will lead to a conflict between the overhead and bottoms control for the distillation as the two level controllers will both try to reject the additional inventory. The root cause of the problem is that there is no inventory control of the unreacted Component *A*. For the unreacted feed there is only one way to leave the process and that is by reaction to the product.

It should also be noted that the schemes in Figures 14.3 to 14.5 would not be viable options if the flowrate of fresh Feed *A* was fixed by the outlet from an upstream process and the process was subject to disturbances created by the upstream process.

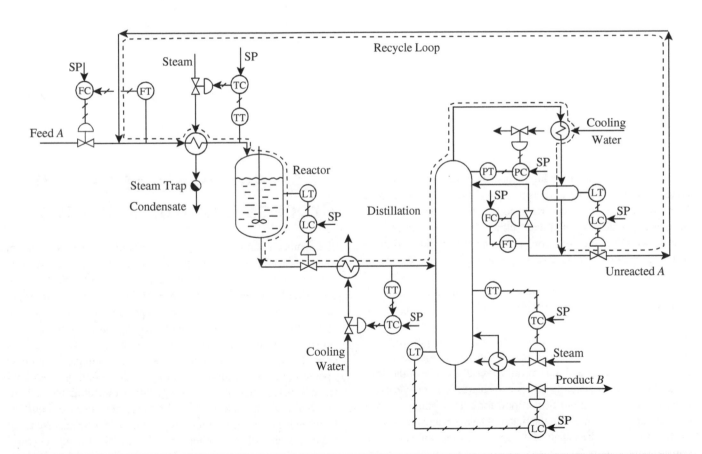

### Figure 14.4

Controlling the combined flowrate to the reactor within the recycle loop can also avoid the snowball effect and maintains the inventory of the Feed *A*.
Source: Bolton and Smith (2022).

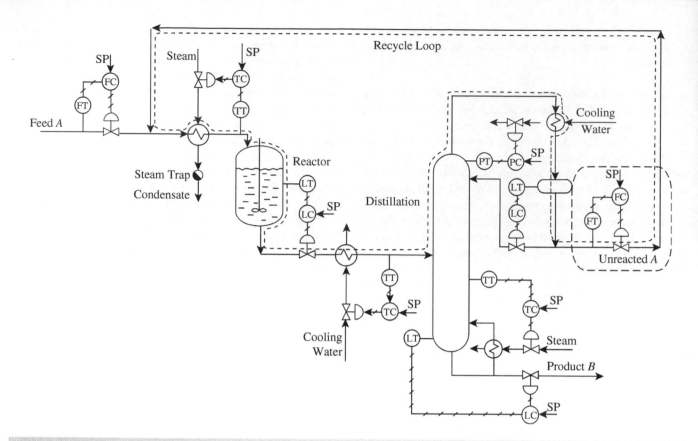

## Figure 14.5

Change of the distillation column control features a flowrate control in the recycle to avoid the snowball effect, but creates problems with the control of the distillation. Source: Bolton and Smith (2022).

In this situation, some mechanism must be found to vary the overall reaction rate in order to consume the required amount of feed. This would typically require a composition measurement on the reactor outlet to manipulate either the reactor temperature (and hence the rate of reaction) or the level setpoint (and hence residence time in the reactor) to achieve the required reactor conversion.

Whilst the above simplified analysis is specific to a simple first-order single component reaction and process configuration, it nevertheless illustrates several important principles for the design of overall process control configurations and the link with the material balance of the process. The key is for the control designer is to understand the overall material balance of the process and what can practically be done to change it. For any recycle, there must be a route out of the loop for every component with enough capacity to maintain the material balance, and the capacity of that route must be capable of adapting to the amount of that component present. Fundamentally, the overall process control configuration should ensure that the process delivers the intended material balance. It should also be noted that if self-regulation is not feasible, simply putting a flowrate controller somewhere in the recycle will not necessarily lead to a stable control system.

## 14.2   Synthesis of Overall Process Control Schemes

The synthesis of process control systems requires the specification of the measured and manipulated variables and their connection with controllers in order to achieve the control objectives for the individual operations and those for the overall process. Many different approaches have been suggested for this. A simple approach to the synthesis of overall process control configurations is to divide the process into a collection of operations, synthesize a control scheme for each operation, and then combine the process operation control configurations to give the overall process control design (Stephanopoulos 1983, 1984). Connecting the individual control schemes for the operations together can create conflicts between the connecting control schemes. Conflicts can arise when the output control of one operation clashes with the input control of the downstream operation. This typically manifests itself as two control valves in the same connecting stream, one for the upstream control system and one for the downstream control system. Such conflicts can often be readily resolved by simply removing one of the controllers. However, this does nothing to reconcile long-range interactions throughout the process, especially when the process has recycles.

For the synthesis of the overall process control strategy, it is first necessary to understand the structure and purpose of the process to ensure that the control scheme is structured to deliver the intent of the process design. The above illustrative example demonstrates that controlling the material balance of the process is central to achieving this. To deliver the intended process design material balance requires both the flowrate and inventory to be controlled in both the individual operations and the overall process. In Chapter 11 three important principles were introduced for inventory control of individual process operations that cannot self-regulate their inventory (Aske and Skogestad 2009):

1) The total material inventory must be regulated by its inflows *or* outflows, i.e. at least one flow *in or out* must depend on the inventory.

2) The inventory of each chemical component must be regulated by *its inflows or outflows or by chemical reactions*.

3) With more than one phase, the inventory of each phase must be regulated by its *inflows or outflows or by phase transition*.

For many operations, these three issues overlap with each other to a greater or lesser extent. Inventory control of process recycles was also discussed in Chapter 11 (Tyreus and Luyben 1993; Aske and Skogestad 2009), in which inventory control is required for both the individual units in the recycle loop and the recycle system as a whole. Just as the inventory of individual operations and recycle systems needs to be controlled, so must the inventory of the overall process. This requires inventory control for both the total material inventory and each chemical component.

A simple procedure will now be developed for the synthesis of overall process control schemes (Bolton and Smith 2022). The procedure starts from the overall control objectives. The first objective to be clarified is whether the process is required to accept a fixed flowrate of feed from an upstream process or is required to produce a fixed flowrate of product by varying the feed flowrate. Less commonly, the production rate might be fixed at an intermediate point within the flowsheet. Other control objectives typically include temperature, pressure, composition, and so on, in various parts of the process and the products. The process flow diagram is then decomposed into *process nodes*. Each process node consists of one or more processing units, which together have a common goal, e.g., a reactor with its heating/cooling system to deal with the heat of reaction or a distillation column with its reboiler, condenser, and reflux drum, or heat exchange, and so on. The control configuration of each of the process nodes is first addressed in relation only to adjacent nodes, but in a specific sequence. Unless control of the flowrate and material inventory can be achieved, the process will not be able to achieve the intent of the process design, even before other control objectives are addressed. Importantly, the procedure is initiated where the overall process capacity is fixed. This is typically fixed by the feed to the process, or the need to deliver a fixed product rate from the process. It is the point where the

material balance is most constrained. From the process node that sets the process capacity, the approach works node-by-node through the process, synthesizing a control system for each node, as discussed in Chapter 13. If the feed flowrate to the process is fixed by an upstream process, then the synthesis should start at the node where the feed enters the process. Conversely, if there are nodes upstream of the node that sets the process capacity, the procedure requires working backwards upstream node-by-node through the process from where the flowrate is fixed. If the product flowrate is to be fixed, then the synthesis should start at the node where the product leaves the process, working backwards upstream node-by-node. Flows from previously assessed nodes should be treated as having been fixed. Recycles are most straightforward to deal with if assumed to be only able to be manipulated at their origin, although this is not strictly the only option. Initially, the control of each node is applied in isolation, but in a fixed order in relation to adjacent nodes. This allows a first overall control scheme to be established. This then needs to be assessed in terms of whether it is capable of delivering the performance of the process design, particularly in relation to process recycles, and evolved if appropriate.

# 14.3 Procedure for the Synthesis of Overall Process Control Schemes

The arguments developed so far can be used to create a procedure for the synthesis of process control schemes for continuous processes in six steps (Bolton and Smith 2022):

1) First determine the overall objectives of the process. There are two common cases for process capacity. In one, the feed to the process is fixed by an upstream process and the process must receive a flowrate of a feed that is determined elsewhere. In the second, the process is required to deliver a fixed amount of a product to a downstream process without downstream storage to manage the downstream inventory. Less commonly, the production rate might be fixed at an intermediate point within the flowsheet. Although the capacity objectives are important to establish, there may be other overall objectives to impose.

2) Divide the process flow diagram into process nodes. Each process node consists of one or more processing units, which together have a common goal.

3) Starting from the node where the overall process flowrate (process capacity) is fixed, determine process control options by first assessing how flowrate and inventory control is to be achieved for each node of the process in isolation. If the process feed is fixed, start with the node where the feed enters the process and work forward, downstream node-by-node through the process. Conversely, if the product rate is fixed, start with the node where the product leaves the process and work backwards upstream node-by-node through the process. In both cases, treat flows from previously assessed nodes as having

been fixed, and treat recycles at this stage as normally being manipulable only at their origin. For each node:

**a)** Establish how the total material inventory can be regulated. The total material inventory must either self-regulate (e.g., incompressible liquid flow through a closed vessel with no liquid interface or gas flow through a vessel driven by pressure difference, and so on) or be controlled by manipulating at least one of its inflows or outflows. Inventory control should be in the direction of flow if downstream of the location flow control. Inventory control should be in the direction opposite to the flow if upstream of the location of flow control (see Chapter 11). Inventory control for multiple operations in series must operate consistently in the same direction relative to the point that fixes the flowrate (Buckley 1964; Aske and Skogestad 2009). This is illustrated in Figure 11.11.

**b)** Establish how the inventory of each chemical component can be regulated. The inventory of each chemical component must either self-regulate or be controlled by its inflows or outflows or chemical reactions.

**c)** If more than one phase is present, establish how the inventory of each phase can be regulated. The inventory of each phase must also either self-regulate or be regulated by its inflows or outflows or phase transition (e.g., having both a pressure controller and a level controller in a gas–liquid system or an interface controller for two immiscible liquid phases).

**d)** Once the flowrate and inventory control have been achieved, modify the node control scheme to achieve the other control objectives. These would include temperature, pressure, pH, and so on. For quality control, such as

composition, the overall material balance and the chemical component material balances must be capable of being manipulated to achieve this (e.g., for distillation, increasing the reflux ratio either by increasing the reflux flowrate or reducing the top product flowrate).

**4)** Where a process contains recycles, for each recycle repeat the substeps from Step 3 for all of the nodes connected together in the recycle *as though it were a single node.*

**5)** If appropriate, check the overall number of control loops against the control degrees of freedom, as discussed in Chapter 12. It should be noted that just because a control configuration agrees with the target for the number of control loops, this does not necessarily mean that the design will work efficiently, or will even work at all. It is simply a check to ensure that the design is not under-controlled or over-controlled.

**6)** Once Steps 1 to 5 have been completed, a relatively small number of plausible control structures will be seen to be viable. Then dynamic simulation of parts, or all, of the process can then be carried out for the candidate structures to allow the controller settings to be tuned, the stability and response of the process to disturbances to be checked and, if necessary, the design evolved to ensure that it achieves the overall process control objectives in a stable and optimal way (Luyben 2014). *Relative gain array* (*RGA*) analysis can also be helpful in selecting the most appropriate control scheme (Bristol 1966; Shinskey 1988). RGA provides guidance for pairing measured and manipulated variables so as to reduce interactions between the control loops.

Consider now the application of the procedure to three examples to illustrate the main features of the approach.

## Example 14.1    Acetone Recovery System

Consider again the process from Example 12.12 shown in Figure 14.6. Air laden with acetone needs the acetone to be separated from the air and the acetone recycled to another process. The air flows directly from the upstream process and its flowrate is fixed by the upstream process. The separation of acetone from the air is achieved by feeding the air to an absorption column and the acetone absorbed in water. The air is given further treatment using an activated carbon bed downstream of the process before discharge to the atmosphere. The acetone–water mixture from the absorber is separated in a distillation column and the water is recycled to the absorption. A makeup water stream is required to make up for water losses from the air discharge and the acetone product. In Chapter 12 it was concluded that there were 10 degrees of freedom, 4 of which were inventory control. Synthesize a control configuration for the process and check that it achieves the target for the number of control loops required.

### Solution

#### Step 1

The objective of the process is to recover a fixed fraction of the acetone carried by an air stream from an upstream process prior

to the air being vented. The flowrate of air, and acetone carried by the air, are fixed by the upstream process.

#### Step 2

The process flow diagram in Figure 14.6 is shown decomposed into four process nodes: storage tank, absorber, distillation, and cooler.

#### Step 3

Given that the feed is fixed by the flowrate of air upstream of the absorber, the procedure is started from the absorber node. The procedure could then follow downstream or upstream from the absorber. In this case, initially the procedure will follow upstream from the absorber and then downstream from the absorber.

*Absorber*

The absorber control scheme is shown in Figure 14.7.

**(3a)** Given that the flowrate of air is fixed by the upstream process and is connected directly to the upstream process, the flowrate of the inlet air cannot be regulated. The total material inventory for the air entering the absorber as feed can be regulated by pressure control of the outlet air from the absorber. The total material inventory of the water solvent feed can be regulated by level control of the absorber sump.

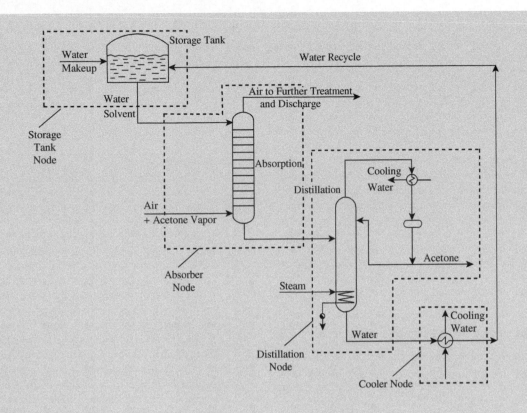

**Figure 14.6**

Process flow diagram for a solvent recovery process.

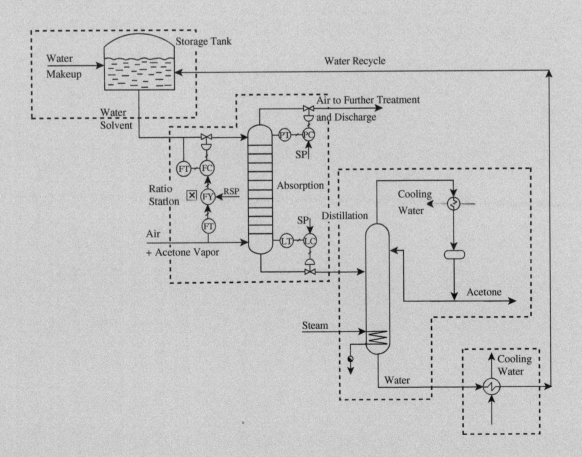

14

**Figure 14.7**

Absorber control using ratio control from the air flowrate.

**(3b)** Neglecting any air dissolved in the solvent water, the inventory of the air component will be regulated by absorber pressure control. The water component entering the absorber will leave in the absorber bottoms under level control. A small amount of water will be lost with the air vent. The acetone component entering the absorber will leave in the bottoms under level control or will leave unrecovered with the air stream. This allows chemical component inventories to be controlled.

**(3c)** The liquid phase inventory balance is achieved by level control of the absorber bottoms and the vapor phase inventory balance is achieved by absorber pressure control.

**(3d)** The absorber must achieve a specified separation of the acetone from the air feed. Once the number of plates for the absorber has been fixed in the equipment design, the performance depends on the ratio of the liquid to vapor flowrate. For a flowrate of air to the absorber set by the upstream process that cannot be controlled and can fluctuate, the flowrate of water solvent should be ratioed to the inlet air flowrate.

*Storage Tank*

The storage tank control scheme is shown in Figure 14.8.

**(3a)** The flowrate from the storage tank to the absorber is set by the absorber controls. The storage tank is supplied by recycled water. Any imbalance between the water fed to the absorber and that fed by the recycle must be supplemented by makeup water under level control to maintain the total material inventory.

**(3b)** Water from both the recycle and makeup and any acetone entering with the recycle water will leave and be fed to the absorber under flowrate control. Hence component inventories will be controlled.

**(3c)** Only the balance for the liquid phase inventory needs to be considered. This will be maintained by the level control as discussed above.

**(3d)** No other control objectives.

*Distillation Column*

The distillation control is shown in Figure 14.9.

**(3a)** Since the feed to the distillation column is fixed by the upstream absorber, one or both of the main liquid product stream flows must be manipulated by an inventory controller to achieve regulation of the total material inventory across the distillation, as discussed in Chapter 13. There are three available options: (i) the overhead level manipulates the reflux flowrate and the sump level manipulates the flowrate of the water recycle; (ii) the overhead level manipulates the overhead acetone product flowrate and the sump level manipulates the reboil; and (iii) the overhead level manipulates the overhead acetone product flowrate and the sump level manipulates the recycle water flowrate. As discussed in Chapter 13, a control scheme where the overhead level manipulates reflux and sump level manipulates reboil is not viable because there is no control of the column inventory.

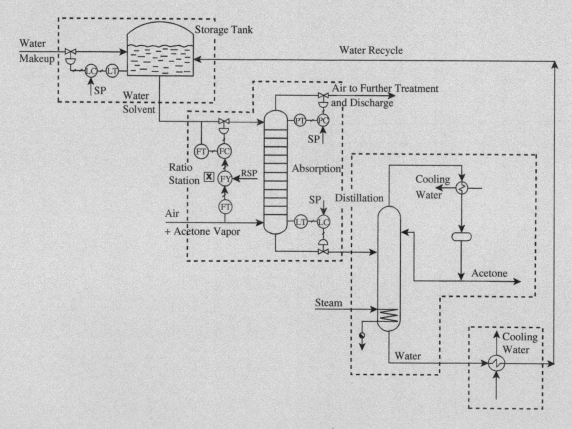

**Figure 14.8**

Control of the storage tank inventory.

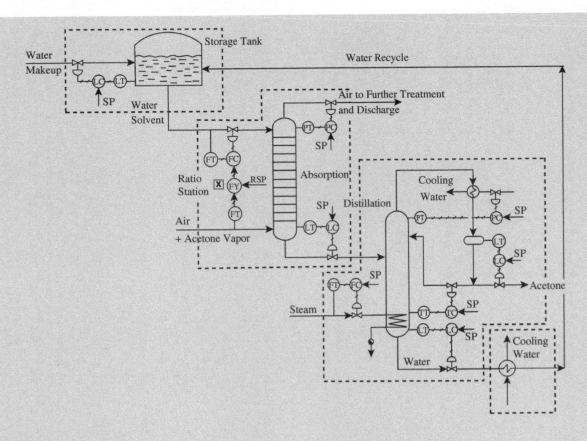

**Figure 14.9**

Distillation control uses direct control of the overhead.

**(3b)** Considering the individual chemical components, noting that the purpose of the distillation column is to separate the acetone (which leaves in the top product) from water (which leaves in the bottom stream to recycle), there must therefore be an inventory control manipulating both product streams, and Option (iii) from Step (3a) above is the most straightforward way to achieve this.

**(3c)** The liquid inventory is maintained by Option (iii) from Step (3a) above and the vapor inventory is maintained by column pressure control.

**(3d)** The control of the separation can be based on chemical analysis, temperature measurement of the column profile, or a combination of the two, as discussed in Chapter 13. The simplest and most robust option of temperature control will be adopted. The distillation temperature profile can be controlled by manipulating either the steam flow to the reboiler or the reflux flowrate. Temperature control of the reflux flowrate will be adopted in this case, given that the main product is the acetone leaving the process and the bottom product is recycled.

Figure 14.9 shows the distillation control using inventory control of the overhead and bottoms product and temperature control of the reflux flowrate.

*Cooler*
The cooler control is shown in Figure 14.10.

**(3a)** The total material inventory across the cooler will self-regulate by continuity.

**(3b)** Component inventories across the cooler will be maintained by continuity.

**(3c)** With no change of phase, phase inventories will be maintained by self-regulation.

**(3d)** The heat exchanger outlet temperature is controlled by manipulating the flowrate of cooling water.

**Step 4**
For the process recycle, considering the storage tank, absorber, distillation, cooler, and recycle as a single entity:

**(4a)** The total inventory of gas through the process is maintained by pressure control of the absorber. Liquid is recycled in the process with water losses made up by makeup water to the storage tank under level control, thereby maintaining the total inventory. The recycle does not feature only level controllers and includes a flowrate controller. This will prevent the snowball effect for the recycle.

**(4b)** Consider the individual chemical components. If the quality controls for the distillation column function as required, then acetone entering the process will leave the process from the distillation overhead, with small losses of acetone to the air vent. Water is recycled, but small losses to the air vent and to the acetone product will necessitate water makeup under level control in the storage tank, as discussed above. The inventory of the chemical components is maintained across the process.

**(4c)** The recycle is liquid-only and the liquid inventory is maintained by level control of makeup in the storage tank. Vapor

**14**

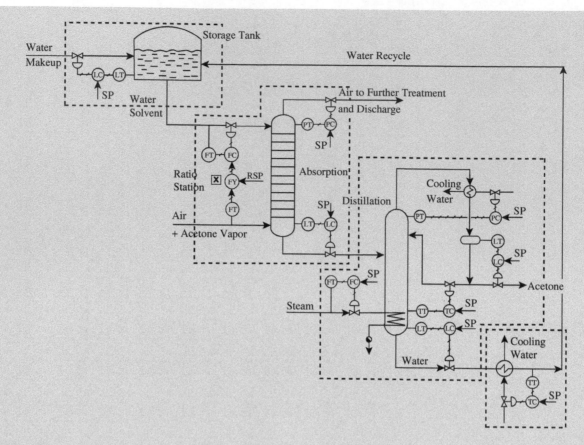

## Figure 14.10

Cooler control of the distillation bottoms for recycle.

inventory is maintained by pressure controls in the absorber and distillation columns.

**(4d)** Cooling of the recycle water to the storage tank in the cooler maintains a low temperature in the storage tank, which helps the absorption process to recover acetone.

## Step 5

Considering the degrees of freedom, Figure 14.10 features ten control valves, four of which are for inventory control. Example 12.12 concluded that there were 11 degrees of freedom, but then fixing the feed flowrate by the upstream process removes one degree of freedom. Thus, the control system synthesized in Figure 14.10 is in agreement with the degrees of freedom target.

## Step 6

As already noted, the recycle includes a flowrate controller, which will prevent the snowball effect. Having identified a potentially viable control scheme for the overall process, dynamic simulation can now focus on resolving the remaining key questions. Which is the more effective option for controlling the distillation separation, adjusting reflux flow or reboiler duty? Figure 14.10 shows reflux flowrate adjustment. Dynamic simulation might also be used to determine controller tuning for all of the controllers and evaluate how effective the controls are against potential disturbances. Relative gain array (RGA) analysis might also be helpful in selecting the most appropriate control scheme to reduce interactions between the control loops (Bristol 1966; Shinskey 1988).

## Example 14.2    Process Involving Reaction Between Two Feeds with One of the Feed Flowrates Fixed

Figure 14.11 shows a process for the reaction between Feed $A$ and Feed $B$ to produce Product $C$ according to the reaction (Bolton and Smith 2022):

$$A + B \longrightarrow C + \text{Light Byproducts}$$

Whilst the main reaction is to Product $C$, some light (high-volatility) byproducts are also formed. The rate of reaction can be assumed to be represented by the rate equation:

$$r = kC_A^a C_B^b \qquad (14.16)$$

where

$r$ = rate of reaction
$k$ = reaction rate constant
$C_A$ = molar concentration of Component $A$
$C_B$ = molar concentration of Component $B$
$a, b$ = orders of reaction

The reaction rate constant $k$ can be assumed to follow the Arrhenius Law given by Eq. (14.2). There is some vaporization

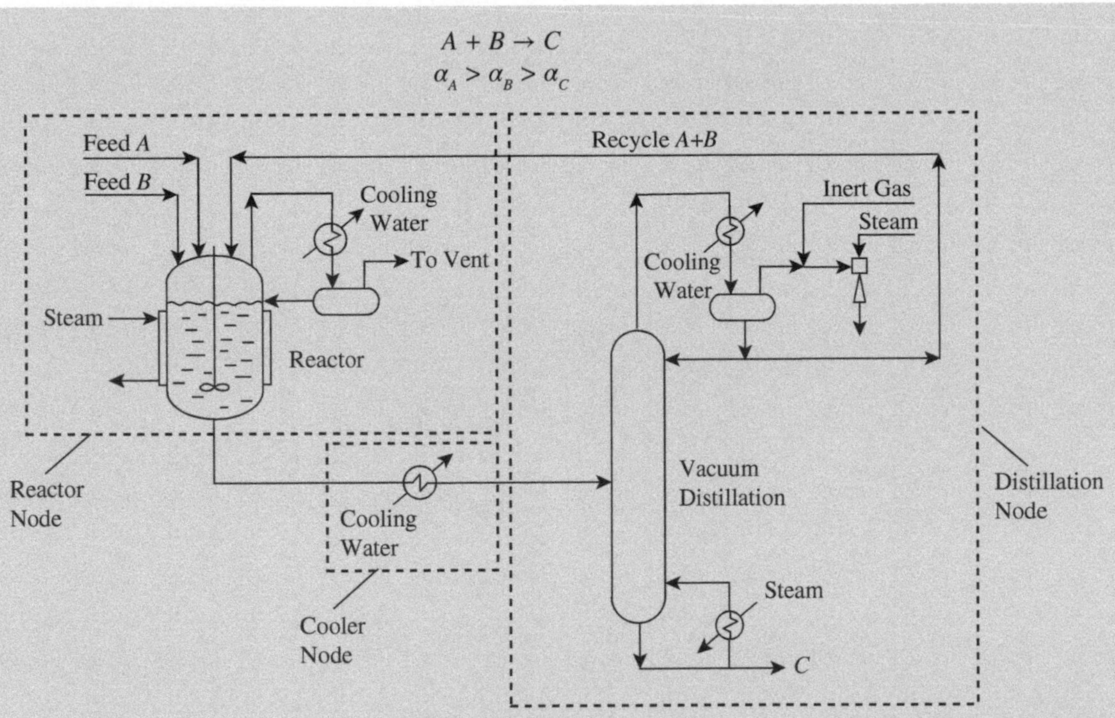

$$A + B \rightarrow C$$
$$\alpha_A > \alpha_B > \alpha_C$$

## Figure 14.11

The process flow diagram for reaction of two feeds. Source: Bolton and Smith (2022).

of chemical components from the reactor. As much volatile material as possible is condensed using cooling water and is returned to the reactor. The uncondensed light (non-condensable) byproducts are vented. The flowrate of the vented non-condensable byproducts can be used to control the pressure of the reactor. The reactor effluent is a mixture of Components $A$, $B$, and $C$, which is first cooled using cooling water and then separated in a vacuum distillation column. The vacuum is to be created by a steam ejector similar to the scheme in Figure 13.37a. The order of relative volatility of the three components is $\alpha_A > \alpha_B > \alpha_C$. The unreacted Components $A$ and $B$ are removed as the distillation overhead and recycled to the reactor. The Product $C$ is taken from the distillation column bottoms. The flowrate of Feed $A$ is fixed and the process must be operated so as to convert this into Product $C$. Synthesize a control configuration for the process and check that the control system achieves the target for the number of control loops required, set in Example 12.14 to be 14.

## Solution

### Step 1

The process flowrate is intended to be operated to consume a fixed amount of Feed $A$.

### Step 2

The process flow diagram in Figure 14.11 is shown decomposed into three process nodes: the reactor, cooler, and distillation nodes. Given that the feed is fixed in this case, the procedure is started from the reactor node and continues forward downstream (Bolton and Smith 2022).

### Step 3

*Reactor*

(3a) Figure 14.12 shows a reactor control system in isolation. The flowrate of Feed $A$ is fixed. However, the flowrate of Feed $B$ should be fixed by the flowrate of Feed $A$ in order to maintain

a stoichiometric ratio of the two feeds. The recycle flowrate can only be manipulated at its origin. The feed and recycle flowrates thus cannot be used to control the inventory in the reactor. Vaporized material from the reactor is condensed using cooling water and the condensed liquid and uncondensed gases separated in a reflux drum. The condensed liquid is returned to the reactor under level control in the reflux drum. The flowrate of uncondensed gases from the reflux drum is used to control the pressure of the reactor. In principle an additional control loop could be added in Figure 14.12 to control the temperature of the condensate, creating a separation by the partial condensation. However, a temperature control loop is not always necessary if:

- There is a substantial difference between the volatility of condensable Components $A$, $B$, and $C$ and the light non-condensable byproducts, with essentially all of Components $A$, $B$, and $C$ being condensed and essentially all of the light byproducts being vented.

- The heat duty on the condenser is small relative to the heat input (or heat output in the case of an exothermic reaction) from the reactor.

The key to deciding whether a temperature control loop is necessary is whether the condensation is sensitive to the flowrate of cooling water. If not sensitive, then the additional control loop is unnecessary and the cooling water can be left to manual control using for example a globe valve.

The only practical option to control the overall material balance is to use level control to control the liquid outlet from the reactor.

(3b) If the two feeds are manipulated to be in a stoichiometric ratio and the ratio is assumed at this stage to be maintained in the recycle stream, then the level controller will maintain a material balance for Components $A$, $B$, and $C$.

**14**

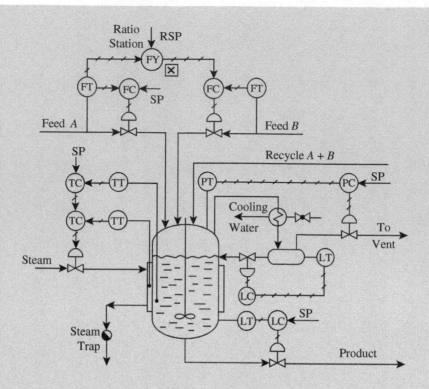

## Figure 14.12

Control of the reactor overall material balance for a fixed flowrate of Feed *A*.

**(3c)** The vapor material balance is maintained by pressure control from the flowrate of non-condensable gases.

**(3d)** The reactor temperature is controlled from the flowrate of steam to the reactor jacket. Cascade control has been adopted for the reasons discussed in Chapter 13. Adjusting the reactor temperature setpoint will change the reactor conversion by changing the rate of reaction.

*Cooler*

**(3a)** Figure 14.13 shows a cooler control system in isolation. The-flowrate is fixed by the reactor inventory control upstream. The inventory in the cooler is liquid phase and naturally self-regulates.

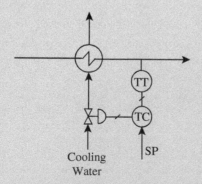

## Figure 14.13

Control of the cooler.

**(3b)** The component material balances will also self-regulate.

**(3c)** There is no change of phase in the cooler and thus the phases will also self-regulate.

**(3d)** The cooler outlet temperature is controlled by the flowrate of cooling water.

*Distillation*

**(3a)** Figure 14.14 shows a control system for the vacuum distillation in isolation for a feed flowrate fixed upstream. A distillation vacuum control system was shown in Figure 13.36a. In the current problem, the vacuum is created by a steam ejector and the pressure controlled by a bleed of inert gas (nitrogen). Conceptually the condenser and reflux drum arrangement is similar to the non-adiabatic flash drum shown in Figure 13.25, but instead using cooling water to carry out a partial condensation. In principle an additional control loop could be added in Figure 14.14 to control the temperature of the condensate. However, as with the reactor condenser, it is desirable to condense as much of Components *A* and *B* as possible, leaving only non-condensable gases to be removed by the vacuum system. To ensure as little as possible of the volatile Components *A* and *B* are lost to the vacuum system, the condensate from the condenser should be *subcooled* to below saturation. As with the reactor reflux condenser, an additional control loop for the cooling water is considered not necessary in this case and the cooling water is left to manual control using a globe valve. Since the feed to the distillation column is fixed by the upstream parts of this process, one or both of the main product stream flows must be manipulated by an inventory controller to achieve a liquid material balance. The vapor material

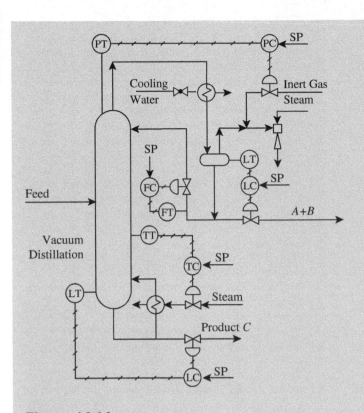

**Figure 14.14**

Control of the vacuum distillation for fixed feed flowrate.

balance is managed by the pressure control. There are three available options to control the overall material balance: (i) the overhead level manipulates the reflux flowrate and the sump level manipulates the flowrate of Product $C$; (ii) the overhead level manipulates the overhead product flowrate and the sump level manipulates the reboil; and (iii) the overhead level manipulates the overhead product flowrate and the sump level manipulates Product $C$ flowrate. A control scheme where the overhead level manipulates reflux and sump level manipulates reboil is not viable because there is no control of the column inventory.

**(3b)** Considering the individual chemical components, noting that the purpose of the column is to separate the Product $C$ (which leaves in the bottoms stream) from unreacted Components $A$ and $B$ (which leave in the overhead stream), there must therefore be an inventory control manipulating both product streams, and Option (iii) from Step (3a) above is the most straightforward means of achieving this.

**(3c)** The vapor material balance is achieved by using a suitable pressure control. The liquid material balance is discussed in Step (3a) above.

**(3d)** The distillation temperature profile can be controlled by manipulating either the steam flow to the reboiler or the reflux flowrate. Figure 14.14 shows the option of manipulating the steam flow to the reboiler.

**Step 4**

For the process recycle, considering the reactor, cooler, distillation, and recycle as a single entity, Figure 14.15 shows the three process nodes together (Bolton and Smith 2022):

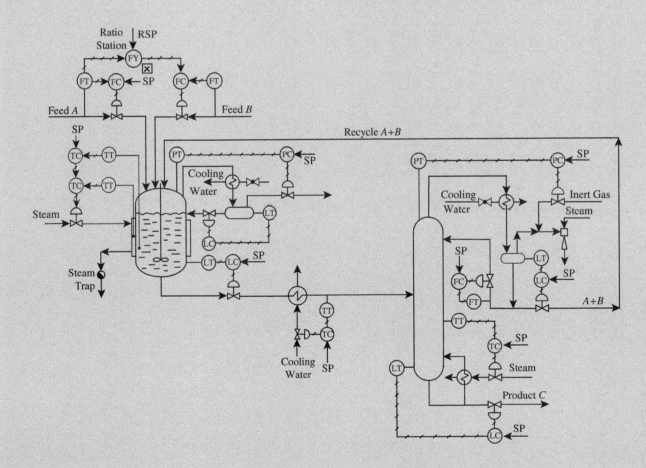

**Figure 14.15**

An initial control system for the whole process has a number of potential problems.

**(4a)** For the overall process material balance, the only significant route out for the process is Product $C$, which is taken from the distillation column bottoms. Components $A$ and $B$ cannot leave the process without first reacting to Component $C$. Option (iii) from Step (3a) for the distillation column control allows an overall material balance to be maintained by controlling the only significant route out of the process. However, when considering the total material inventory, the recycle loop from the reactor outlet through the distillation column overhead and back to the reactor features only level controllers. This may suffer from the snowball effect.

**(4b)** For the individual component material balances, there are a number of potential issues. Product $C$ from the distillation bottoms can deliver an effective material balance for Component $C$, but not for either of the Components $A$ and $B$. Since the feed flowrate of Feed $A$ to the process is fixed, the only effective means of achieving a material balance for Component $A$ is to vary the consumption of Feed $A$ in the reactor. The most straightforward way to achieve this is to analyze the reactor composition and manipulate either the reactor level setpoint (thereby varying the residence time and conversion) or the temperature setpoint (thereby varying the rate of reaction). Figure 14.16 shows an analysis used to control the temperature of the reactor. However, it is usually desirable to constrain the temperature range within relatively narrow limits either for safety reasons or to avoid loss of reaction selectivity. The material balance for Component $B$ needs also to be controlled. Like Component $A$, Component $B$ can only

leave the process by reacting to Component $C$. The consumption of Feed $B$ is essentially equal to that of Feed $A$ and cannot be varied independently. In principle, matching the feed rate of Feed $B$ to be exactly that for reaction with Feed $A$ should be possible. However, in practice this approach is not feasible as the measurements will not be exact and any difference in calibration between the two would lead to an accumulation of Component $A$ or Component $B$. Depending on the size of the mismatch in flow, this may take place over hours, days, or weeks (Luyben et al. 1999). One way to overcome this problem, shown in Figure 14.16, is to feed Feed $B$ in the ratio to Feed $A$ but to adjust the molar ratio (which should always be close to one) using a controller based on an analysis of composition in the reactor (such as the mole fraction of Component $B$).

**(4c)** Since the recycle is liquid only, there are no additional vapor inventory considerations across the process recycle.

**(4d)** There are no additional temperature or quality control considerations around the process recycle. One possible concern with this scheme is that there is a closed loop of controlled inventories, comprising the reactor level, the distillation column reflux drum, and the recycle stream. Such a closed loop can in principle suffer from the snowball effect. Whilst the adjustment of conversion in the reactor should ensure that neither Feed $A$ nor Feed $B$ can accumulate and thus limit the extent of this, it will depend on the performance of the various control loops. The chemical analysis control of the reactor is likely to be slow

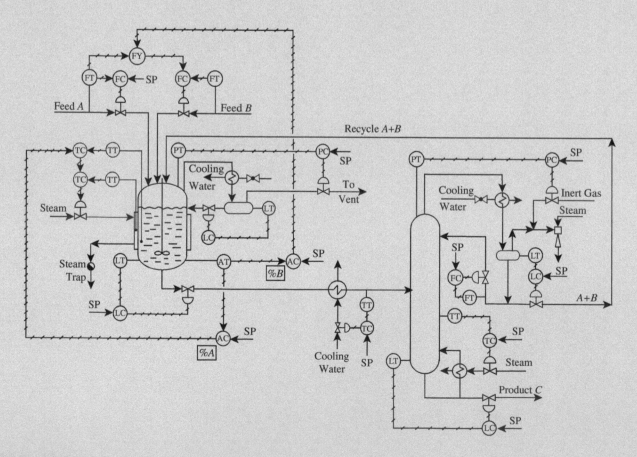

**Figure 14.16**

Control system for a fixed process feed flowrate. Source: Bolton and Smith (2022).

compared with the level controllers, which might still contribute to the snowball effect.

## Step 5

In Figure 14.16 there are 14 control loops including the two condensers on manual control, which agrees with the target for the degrees of freedom determined in Example 12.14. It should again be emphasized that just because a control system agrees with the target for the number of control loops, this does not necessarily mean that the design will work efficiently or even work at all. It is simply a check to ensure that the process has the correct number of control loops, avoiding under-control or over-control of the process.

## Step 6

Having identified a potentially viable control scheme for the overall process, dynamic simulation can now focus on resolving the remaining key questions.

1) Which is the more effective and safe option for controlling the reactor conversion (analysis of Feed *A* in the reactor), adjusting the level, or adjusting temperature? Figure 14.16 shows temperature adjustment.

2) Which is the more effective option for controlling the distillation separation, adjusting reflux flow or reboiler duty? Figure 14.16 shows reboiler duty adjustment.

3) How should the inventory controllers in the closed recycle loop be tuned to avoid a problematic snowball effect?

Dynamic simulation might also be used to determine controller tuning more generally and evaluate how effective the control system is when subjected to potential disturbances. Relative gain array (RGA) analysis might also be helpful in selecting the most appropriate control scheme to reduce interactions between the control loops (Bristol 1966; Shinskey 1988).

## Example 14.3   Process Involving Reaction Between Two Feeds with the Product Flowrate Fixed

Consider now Example 14.2 for the scenario where the process is required to deliver a fixed amount of Product *C* to a downstream process without downstream storage to manage the downstream inventory (Bolton and Smith 2022).

### Solution

#### Step 1

The process flowrate is intended to be operated to produce a fixed amount of Product *C*.

#### Step 2

Again, the process flow diagram in Figure 14.11 shows the process decomposed into three process nodes: the reactor, cooler, and distillation nodes. This time the flowrate is fixed by the

product flowrate from the distillation node. The procedure starts from the distillation and moves backwards upstream (Bolton and Smith 2022).

#### Step 3
*Distillation*

(3a) Figure 14.17 shows a control system for the vacuum distillation in isolation for a fixed product flowrate from the process. The amount of Product *C* is fixed; therefore in principle either the feed to the column or the overhead liquid product could be manipulated to achieve an overall material balance.

(3b) The feed composition entering the column is not being controlled. Varying the feed flowrate to the column is the only means available to control the inventory of Component *C*

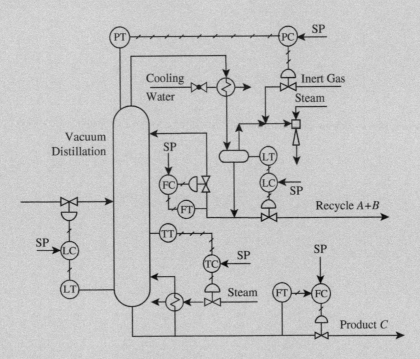

## Figure 14.17

Control of the vacuum distillation for fixed product flowrate.

(the level in the column sump). That leaves the overhead product flowrate as the only means available to control the inventory of Components *A* and *B* (the level in the reflux drum), which forms the process recycle.

**(3c)** A vapor material balance can be achieved by using a suitable pressure control.

**(3d)** The distillation temperature profile can be controlled by manipulating either the steam flow to the reboiler or the reflux flowrate. Figure 14.17 shows the steam flow to the reboiler option.

### Cooler

**(3a)** The cooler control system shown in isolation in Figure 14.13 can also be used for this case. The flowrate will be fixed by the reactor inventory control upstream. The inventory in the cooler is liquid phase and naturally self-regulates.

**(3b)** The component material balances will also self-regulate.

**(3c)** There is no change of phase in the cooler and thus the phases will also self-regulate.

**(3d)** The cooler outlet temperature is controlled by the flowrate of cooling water.

### Reactor

**(3a)** Figure 14.18 shows a reactor control system in isolation for the fixed product flowrate case. Since the outlet flowrate from the reactor, and the recycle flowrate to the reactor, are now fixed by the distillation controls, one of the feeds must be manipulated to maintain an overall material balance.

**(3b)** Following the same logic as for the fixed feed case in Example 14.2, it is desirable to ensure that the two feeds are maintained in ratio to each other, and that the ratio is adjusted in order to ensure that the feed rates of Feeds *A* and *B* are

precisely matched. The simplest way to achieve this is to use a level controller to manipulate the feed flowrate of Feed *A*, manipulate the feed rate of Feed *B* in a ratio to the feed flowrate of Component *A*, and adjust the ratio by composition analysis so as to achieve a target composition in the reactor (Figure 14.18).

**(3c)** The vapor material balance is achieved by using a suitable pressure control.

**(3d)** The reactor temperature is controlled by manipulating steam flow to the reactor jacket. Adjusting the reactor temperature setpoint will change the reactor conversion by changing the rate of reaction.

### Step 4

For the process recycle, considering the reactor, cooler, distillation, and recycle as a single entity. Figure (14.19) shows the three process nodes together (Bolton and Smith, 2022):

**(4a)** For the overall process material balance considering the reactor, heat exchanger, distillation column and recycle as a single entity, from Figure 14.19 the reactor level control delivers an overall process material balance by manipulating the feeds into the process.

**(4b)** Similarly, the reactor composition control ensures that Feeds *A* and *B* are fed in the required proportions to deliver a material balance of both. In this case, the variation in the outlet from the reactor will cause a change in the reactor composition, which automatically adjusts the composition to match the reaction rate with the removal of Component *C*.

**(4c)** Since the recycle is liquid-only, there are no additional vapor inventory considerations across the process with recycle.

**(4d)** There are no additional temperature or quality control considerations across the process with recycle.

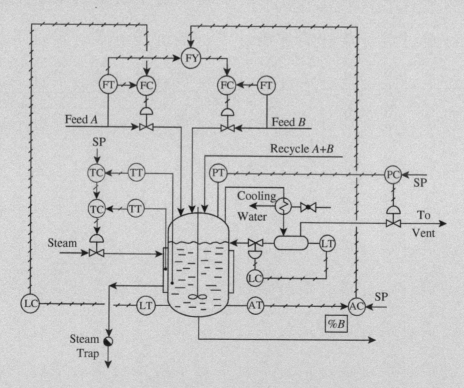

**Figure 14.18**

Control of the reactor overall material balance for a fixed process product flowrate.

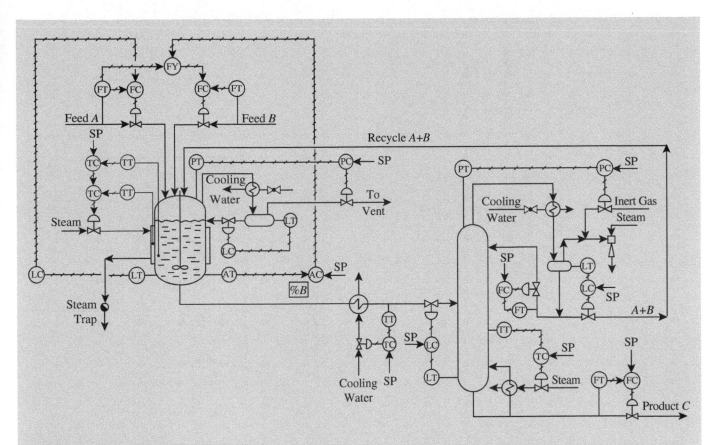

**Figure 14.19**

Control system with a fixed product flowrate. Source: Bolton and Smith (2022).

**Step 5**

In Figure 14.19 there are again 14 control loops including the two condensers on manual control, which agrees with the target for the degrees of freedom determined in Example 12.14.

**Step 6**

Dynamic simulation can now be focused on determining which is the more effective option for controlling the distillation separation, either adjusting the reflux flowrate or reboiler duty. It may also be used to determine controller tuning and evaluate how effective the controls are against potential disturbances. There could be concerns relating to the control of the distillation feed from level control of the bottoms, since this features a potentially significant time lag (depending on the number of trays and the type of column internals in the lower section of the column). It may be necessary or desirable to increase the capacity of the column sump to ensure that there is sufficient capacity to allow this level controller to work well despite the time lag. Again, relative gain array (RGA) analysis might also be helpful in selecting the most appropriate control scheme (Bristol 1966; Shinskey 1988).

# 14.4   Evolution of the Control Design

Examples 14.2 and 14.3 investigated overall process control for the two cases of fixed feed flowrate and fixed product flowrate. Should the final step in the procedure highlight problems with the performance of the control system, it might then be necessary to evolve the control configuration. It is important that the inventory of each node in isolation *and* the inventory of the whole flowsheet is managed. In the overall control scheme, there may be different ways in which this can be achieved. As an example, Figure 14.20 shows an evolution of the control configuration design that avoids the potential for the snowball effect for the fixed feed case by placing a flow controller in the recycle. The reactor inventory is now controlled by the flowrate of fresh Feed A with the ratio of Feed A and Feed B controlled by the analysis of composition in the reactor. Figure 14.20 also avoids the potential problem with the time lag for the control of the distillation in Figure 14.19 for the fixed product case (Bolton and Smith 2022). However, now neither the feed flowrate nor the product flowrate can be controlled directly, but is fixed implicitly by the feed rate to the distillation column. Intermediate storage could be used downstream to dampen out differences between the required flowrates and those achieved by the process control system.

**14**

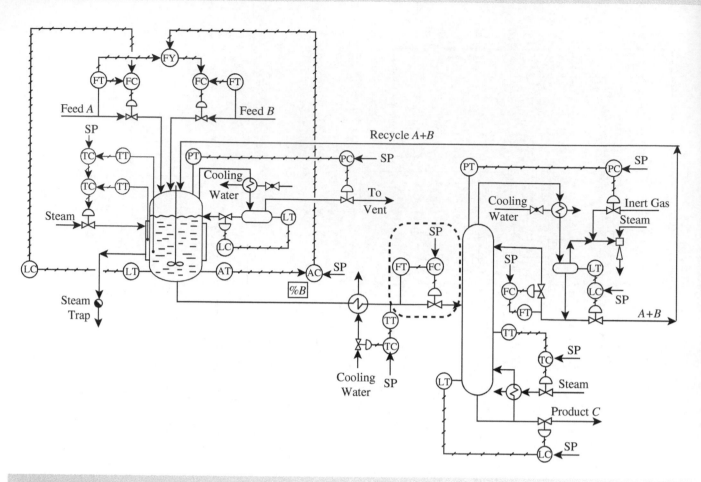

**Figure 14.20**

An alternative control system that does not allow the feed or product flowrate to be fixed directly. Source: Bolton and Smith (2022).

# 14.5 Process Dynamics

Ensuring that the material balance and inventory control for the process can in principle be maintained, together with control of the other process variables (temperature, pressure, pH, etc.), does not remove the need to study the process dynamics. Even when a process can in principle be self-regulating, as in some cases of the process in Figure 14.2, the resulting dynamics may still lead to an inoperable system. For example, if the changes in the material balance caused by the changes in reactor conditions act more slowly than the level controllers in the recycle loop, then the snowball effect may still occur.

A key issue is to understand the purpose of each inventory. For a reactor, the inventory provides volume for the reaction to occur. By contrast, reflux drums and column sumps simply allow the process to absorb disturbances. Inventories should have a clear purpose. For example, in Figure 14.19 a change in the demand for Product C will result in a lowering in the level in the distillation column sump, which will cause the level control to demand an increased flow from the reactor and affect its performance.

Instead, the inventory in the column sump can be used to smooth the impact of that change in demand on the rest of the process by allowing the sump level to vary considerably in the event of a disturbance. In this way the other levels can be kept relatively constant.

Once a basic control configuration has been established to develop the conceptual design of the control system, further evaluation is necessary. This can be achieved using dynamic simulation of parts of the process, or preferably the whole process, tuning the controller settings, and checking the stability and response of the process to disturbances (Luyben 2014). Relative gain array (RGA) analysis can also be helpful in selecting the most appropriate control scheme (Bristol 1966; Shinskey 1988). RGA analysis uses a matrix of interaction measures for all possible single-input–single-output pairings of the variables considered. Guidance is provided for pairing measured and manipulated variables so as to reduce interactions between the control loops. If necessary, the design can be evolved to other control structures to ensure that it achieves the overall process control objectives in a stable and optimal way. Finally, the design provides the basis for the design of the DCS system, possibly using model predictive control. Details are beyond the scope of this text.

# 14.6  Overall Process Control – Summary

The synthesis procedure for the overall process control scheme must start from the overall control objectives. The first objective to be clarified is whether the process is required to accept a fixed flowrate of feed from an upstream process or is required to produce a fixed flowrate of product by varying the feed flowrate. Other control objectives typically include process conditions and compositions in various parts of the process and the products. The likely disturbances entering the process must also be appreciated. The conceptual design of the control configuration can then be developed.

The process flow diagram is divided into *process nodes*. Each process node consists of one or more processing units, which together have a common goal. The control scheme of each of the process nodes is first addressed in relation only to adjacent nodes, but in a specific sequence. The purpose of most controllers in chemical processes is to control the material balance. Unless control of the material balance can be achieved, the process will not be able to achieve the intent of the process design, even though other control objectives need to be addressed. Importantly, the procedure is initiated where the overall process capacity is fixed. This is the point where the material balance is most constrained. From the process node that sets the process capacity the approach works node-by-node through the process synthesizing a control configuration for each node. If the feed flowrate to the process is fixed by an upstream process, then the synthesis should start at the node where the feed enters the process and works forward downstream. Conversely, if there are nodes upstream of the node that sets the process capacity, the procedure requires working backwards upstream node-by-node through the process from

where the flowrate is fixed. If the product flowrate is to be fixed, then the synthesis should start at the node where the product leaves the process, working backwards upstream node-by-node. Flows from previously assessed nodes should be treated as having been fixed. Recycles are most straightforward to deal with if assumed to be only manipulable at their origin.

Initially, the control of the material balance for each node is applied in isolation, but in relation to adjacent nodes. This approach typically identifies a small number of candidate control structures that can be compared using dynamic simulation. A flowsheet might require markedly different control structures depending on whether the feed or product flowrate is to be fixed. Dynamic simulation of the process can be used to tune the controller settings and check the stability and response of the process to disturbances and the design evolved to ensure that it achieves the overall process control objectives in a stable and optimal way (Luyben 2014).

# Exercises

1. Figure 14.21 shows a process that reacts two liquids, Feeds *A* and *B*, in a stirred tank reactor with a steam heating jacket, producing small amounts of volatile byproducts with no commercial value. Neither of the two feeds is fixed by an upstream process. The reactor effluent is cooled using cooling water and then volatiles separated in a flash drum. Reactor pressure can be controlled by restricting the venting of the gaseous byproducts of reaction. Develop a conceptual control configuration for the process:

   a) Divide the process into process nodes.

   b) Synthesize control schemes for each process node.

   c) Combine the process nodes to form a complete process with its control scheme.

   d) Eliminate conflicts between the process node control schemes.

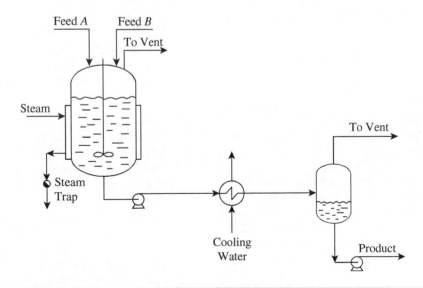

**Figure 14.21**

A reactor and separation system for Exercise 1.

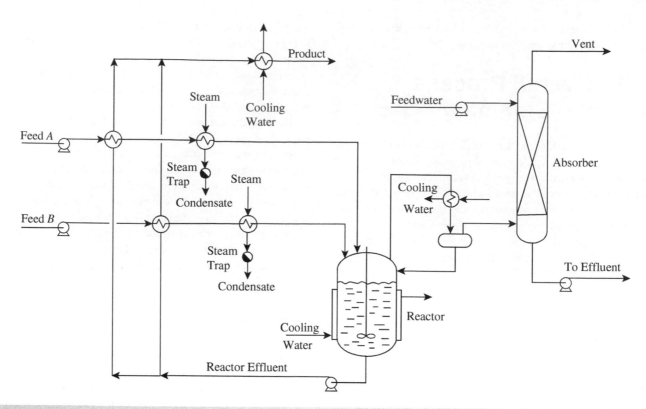

## Figure 14.22
A reactor and separation system for Exercise 2.

2. Figure 14.22 shows a stirred tank reactor in which two Feeds *A* and *B* are reacted to a Product *C*. Neither of the two feeds is fixed by an upstream process. A small amount of gaseous byproducts are produced in the reactor. The reaction is exothermic and the heat of reaction is removed by the cooling jacket serviced by cooling water. This is to be used to control the temperature in the reactor. The pressure of the reactor needs to be controlled. Of the material vaporized from the reactor, as much as possible is condensed by a cooling water condenser and refluxed to the reactor. It is not considered necessary to provide temperature control for the reflux condenser for the reactor. The uncondensed gases are passed to an absorber where as much as possible is absorbed in water. Any gases not absorbed are vented. Liquid from the absorber is sent to effluent. The feeds to the reactor are to be temperature controlled. The feeds are first heated by heat recovery and then by steam before entering the reactor. The product is to be temperature controlled by cooling water.

a) Develop a conceptual design for the control schemes for the reactor as a process node. The production flowrate should be controlled by the flowrate of Feed *A*. The two feeds are to be controlled to be in their stoichiometric ratio. The flowrate split ratio of the reactor effluent needs to be controlled to be the same as the ratio of the flowrates of the two feed streams.

b) Develop a conceptual design for the control schemes for the absorber as a process node.

c) Combine the two process nodes to form a complete process. The pressure of the absorber needs to be controlled to a different pressure from that in the reactor.

3. Figure 14.23 shows an environmental protection system for a process to safely vent a tail gas (Henley and Kumamoto 1981). The flowrate of the tail gas is fixed by the upstream process. The temperature of the tail gas is first decreased to the required temperature by quenching with water. The cooled tail gas is then passed to a scrubber that removes solid particles using a spray of water. A mesh pad at the top of the scrubber prevents entrainment of particles and droplets from the scrubber. The tail gas is then passed to an absorber with a fixed number of absorber plates where the gaseous contaminants are removed by absorption in water. The tail gas is then vented.

a) Develop a conceptual design for the control configuration for the environmental protection system in Figure 14.23. Because the scrubber and absorber are open to the atmosphere, there is no need for pressure control.

b) If it is essential for the vent stream from the outlet of the absorber to be constrained to be below certain concentration limits, then a process analyzer needs to be added to the vent stream. Modify the design of the absorber control scheme from Part (a) above to allow the analysis to control the performance of the absorber.

c) Concern over the amount of water being consumed by the absorber and subsequent volume of effluent created has led to the proposal to introduce a recycle of water around the

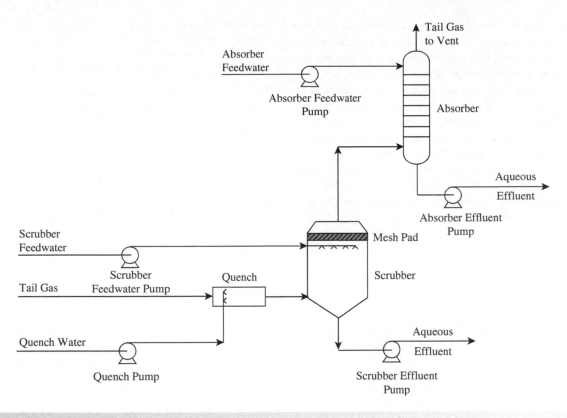

**Figure 14.23**

Process environmental protection system for Exercise 3. Source: Adapted from Henley and Kumamoto (1981).

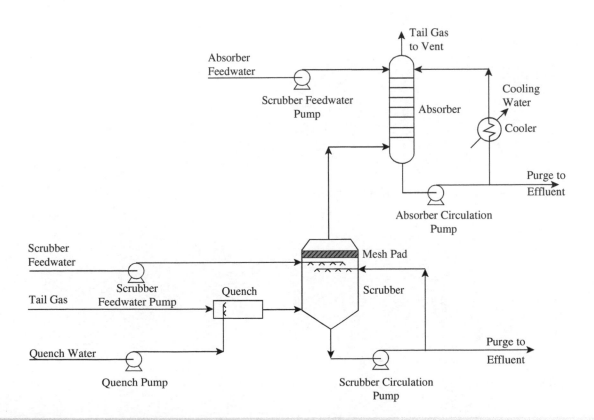

**Figure 14.24**

Process environmental protection system with water recycles for Exercise 3. Source: Adapted from Henley and Kumamoto (1981).

14

scrubber and the absorber. Figure 14.24 shows a modified design to decrease the water consumption of the system. The water is being recycled around both the scrubber and the absorber, with a purge in the recycle to remove the build-up of solid particles and dissolved contaminants from the recycle. Also in the recycle, a cooler serviced by cooling water has been incorporated in the recycle around the absorber to remove heat transferred to the water from the process gas and energy input from the energy of pumping water around the recycle. It is not necessary to add an analysis to control the concentration of the vent in this case. Develop a conceptual design for the control configuration for the environmental protection system in Figure 14.24.

# References

Aske, E.M.G. and Skogestad, S. (2009). Consistent inventory control. *Industrial and Engineering Chemistry Research* 48: 10892–10902.

Bolton, L. and Smith, R. (2022). On the role of material balances in the synthesis of overall process control systems. *Process Integration and Optimization for Sustainability* http://dx.doi.org/10.1007/s41660-022-00283-x.

Bristol, E.H. (1966). On a new measure of interaction for multivariable process control. *IEEE Transactions on Automatic Control* 1: 133–134.

Buckley, P.S. (1964). *Techniques of Process Control*. Wiley.

Denbigh, K.G. and Turner, J.C.R. (1984). *Chemical Reactor Theory*, 3e. Cambridge University Press.

Henley, E.J. and Kumamoto, H. (1981). *Reliability Engineering and Risk Assessment*. Prentice-Hall.

Larsson, T. and Skogestad, S. (2000). Plantwide control – A review and a new design procedure. *Modelling, Identification and Control* 21 (4): 209–240.

Levenspiel, O. (1999). *Chemical Reaction Engineering*, 3e. Wiley.

Luyben, W.L. (1994). Snowball effect in reactor/separator processes with recycle. *Industrial and Engineering Chemistry Research* 33: 299–305.

Luyben, W.L. (2014). *Plantwide Dynamic Simulators in Chemical Processing and Control*. CRC Press.

Luyben, W.L., Tyreus, B.D., and Luyben, M.L. (1999). *Plantwide Process Control*. McGraw-Hill.

Rase, H.F. (1977). *Chemical Reactor Design for Process Plants*, vol. 1. Wiley.

Shinskey, F.G. (1988). *Process Control Systems*, 3e.

Smith, C.L. (2009). *Practical Process Control: Tuning and Troubleshooting*. Wiley.

Smith, R. (2016). *Chemical Process Design and Integration*, 2e. Wiley.

Stephanopoulos, G. (1983). Synthesis of control systems for chemical plants – a challenge for creativity. *Computers and Chemical Engineering* 7 (1): 331–365.

Stephanopoulos, G. (1984). *Chemical Process Control - An Introduction to Theory and Practice*. Prentice-Hall.

Stephanopoulos, G. and Ng, C. (2000). Perspectives on the synthesis of plantwide control structures. *Journal of Process Control* 10: 97–111.

Tyreus, B.D. and Luyben, W.L. (1993). Dynamics and control of recycle systems. 4. Ternary systems with one or two recycle streams. *Industrial and Engineering Chemistry Research* 32: 1154–1162.

# Chapter 15

# Piping and Instrumentation Diagrams – Piping and Pressure Relief

## 15.1 Piping and Instrumentation Diagrams

Development of Piping and Instrumentation Diagrams (P&IDs) starts after the process flow diagram (PFD) has been developed and expands to include:

- the entire process, logically ordered, from feed inlets to product outlets;
- multiple items of equipment that are represented in the process flow diagram as a single item (e.g., a single reactor in a PFD that in practice is a multiple reactor in parallel and needs to be included as such in a P&ID);
- pumps and compressors, with drivers;
- standby equipment (redundancy);
- all material storage linked directly to the process (centralized storage systems will have their own P&IDs);
- utility, effluent treatment and waste disposal equipment directly linked to the process (centralized utility, effluent treatment, and waste disposal systems will have their own P&IDs);
- steam traps and steam condensate drums;
- all instrumentation and control, including location of control devices (or algorithms);
- control valves, with actuator type and failure mode;
- safety instrumented functions;
- emergency shutdown valves, including the type of actuator and the failure mode as part of the safety instrumented functions;
- signal types between control functions and safety functions;
- pressure relief valves, bursting discs, and venting and blow-down systems;
- all process and utility pipes, including vents, drains, bypass lines, and sampling points (including those required for start-up, shut-down, maintenance, and abnormal operation);

- pipe dimensions and specifications;
- heat tracing on pipes;
- slopes on pipes to allow liquids to flow by gravity during normal operation or draining when the process is shut down;
- all valves, together with the valve type, and isolations, including those for both operation and maintenance;
- locking requirements for valves required for safety reasons (e.g., valves locked open to only be closed in special circumstances under a safety permit);
- inspection access;
- piping connections at the edges of sheets to neighboring P&ID sheets;
- all vessels, equipment, instruments, control loops, and piping should ultimately have a unique identification number. For example, in the case of pipes, the identification number would include a code to specify uniquely which pipe it refers to, the pipe service, diameter, schedule (wall thickness), material of construction, tracing type, and insulation.

Processes normally require a number of P&IDs to represent the whole process. Much of the detailed information regarding the process is incorporated graphically in the P&ID to guide the engineering of the plant, analyze safety, operate, troubleshoot, maintain and revamp the process. Different disciplines input to the development of P&IDs. These include process engineers, instrumentation and control engineers, mechanical engineers, operations personnel and maintenance engineers, together with the owner of the process. The P&IDs are amongst the most important documents relating to a plant. It is necessary to keep them up to date as the plant is modified through its working life.

Although standard symbols are used for the construction of P&IDs, there is no universal standard and different companies use their own preferences (American National Standard ANSI/ISA-5.1 2022; BS EN ISO 10628-2 2012; BS EN ISO 10628-1 2015a; BS ISO 15519-2 2015b; DIN 2429-2 1988; DIN 19227-1 1993; Process Industry Practices (PIP) PIC001 2008). For example, Figure 15.1 shows two possible symbols for centrifugal pumps. Because of this, P&IDs should start with a legend of the process symbols to be used.

*Process Plant Design*, First Edition. Robin Smith.
© 2024 John Wiley & Sons Ltd. Published 2024 by John Wiley & Sons Ltd.
Companion website: www.wiley.com/go/processplantdesign

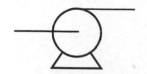
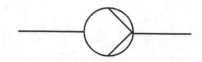

**Figure 15.1**

Alternative symbols for centrifugal pumps.

After the preliminary P&IDs are developed, they need to evolve with further detail gradually added as the P&IDs evolve through the Project Definition (Front End Engineering Design) phase of the project, including changes resulting from Hazard and Operability (HAZOP) and Layer of Protection Analysis (LOPA) studies. Consideration here will focus on the preliminary development of P&IDs.

## 15.2  Piping Systems

In *preliminary design*, the pipes can be sized on the basis of an assumed acceptable fluid velocity or pressure drop or both. For non-viscous liquids ($\mu < 10\,\text{mN s m}^{-2}$ or $10\,\text{cP}$) a velocity of $1–2\,\text{m s}^{-1}$ is normally used. An excessively high liquid velocity leads to a high pressure drop, water hammer, vibration, and erosion–corrosion. For viscous fluids, the velocity may be constrained by the allowable pressure drop or shear degradation of the fluid (e.g., high shear rates breaking large molecules down into smaller molecules). Typical values are given in Table 15.1. For gases and vapors, typical fluid velocities are in the range of $15–30\,\text{m s}^{-1}$.

Although preliminary sizing of pipes can be based on the velocity, it is also necessary to ensure that the pressure drop is acceptable. The pressure drop per 100 m pipe is often used as a criterion for an acceptable pressure drop. Different companies have different criteria for the pressure drop per 100 m pipe, but typical values are:

- for low-viscosity subcooled liquids 0.25 bar/100 m;
- for dry gases 0.1 bar/100 m.

However, pressure drop in pipes is caused by a combination of frictional pressure loss in straight pipes, frictional pressure loss in pipe fittings, and for liquids the static head caused by change in elevation. If the fluid is a gas, the change in elevation can normally be neglected. Pipe fittings include bends, isolation valves, control valves, orifice plates, expansions, and reductions. For short transmission pipes, the frictional pressure loss is often dominated by the pressure loss in pipe fittings. It is important to ensure that the dominant pressure loss is not across the control valve to ensure that the control valve can maintain effective control.

The optimum pressure drop is ultimately a trade-off between capital cost and operating cost for pumping or compression. A larger pipe diameter increases the capital cost of the pipe, but decreases the velocity and pressure drop, resulting in decreased pumping or compression costs and decreased capital cost of pumps/compressors. A smaller pipe diameter decreases the capital cost of the pipe, but increases the velocity and pressure drop, resulting in increased pumping or compression costs and increased capital cost of pumps/compressors. However, the trade-off should also take account of the effects of fluid velocity on erosion and corrosion.

The fluid velocity and pressure drop must take account of standard pipe sizes. Appendix H gives dimensions for commonly used pipe sizes. Pipe sizes are specified by a nominal pipe size (NPS), or nominal diameter (DN), and a schedule. The larger the schedule number, the thicker the pipe wall. As the schedule increases, the outside diameter stays constant, but wall thickness increases. The schedule of pipe required depends on the maximum internal pressure and the appropriate corrosion allowance. It is common to specify a minimum diameter for mechanical strength or to avoid blockages (e.g., minimum DN 50). For steam systems, the smallest schedule normally used is Schedule 40.

The choice of pipe diameter and schedule depends on:

- the internal diameter required for fluid velocity or pressure drop requirements;
- the wall thickness required to withstand the maximum working pressure at the design temperature;
- the corrosion allowance (and possibly erosion allowance) required;
- mechanical supports (increased mechanical strength of a pipe allows a greater distance between pipe supports).

Pipe support requirements depend on diameter, schedule, and number of pipes being supported (pipe supports are not necessarily for a single line). A thicker wall than necessary to retain the internal pressure might be specified to avoid too many pipe supports or to guard against excessive pipe vibration, especially for smaller diameter pipes. Pipe vibration can cause failure of the pipe from metal fatigue and can be avoided by extra clamping of the pipe to support structures.

Figure 15.2 shows the symbols for different piping options. Pipes might need a slope for gravity flow in normal operation

**Table 15.1**

Typical fluid velocities for viscous liquids.

| Viscosity (mN s m$^{-2}$ = cP) | Velocity (m s$^{-1}$) |
|---|---|
| 50 | 0.5–1.0 |
| 100 | 0.3–0.6 |
| 1000 | 0.1–0.3 |

| | Pipe |
| | Pipe With Slope for Draining |
| | Pipe Flanged Joint |
| | Blank (Blind) Flange |
| | Reducer (Change in Pipe Diameter) |
| | Traced Pipe |

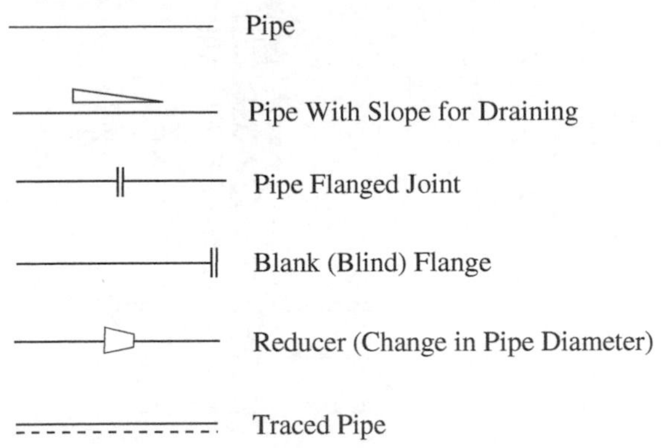

**Figure 15.2**

Symbols for different piping options.

or for draining after shut-down. Piping systems are normally constructed in sections using joints created by bolting two flanges together with a gasket between them to provide a seal. Flanged joints are used to connect pipe sections, valves, pumps, and other process equipment. Sections can be welded together for safety or structural integrity reasons, but flanged joints retain many advantages. Flanged joints provide easy access and dismantling for maintenance, cleaning, inspection, or modification of pipe sections or process equipment. Flanged joints can be shown on the P&ID and reduction or expansion sections for change in the pipe diameter. Tracing is required when pipes are transporting fluids that have characteristics that cause them to freeze, become too viscous, or condense undesirably at ambient temperature, even if the pipes are insulated. Tracing provides additional heat underneath a layer of insulation from an electrical heating tape, or steam heating, or hot water heating from a tube or small-diameter pipe on the exterior of the pipe. Tracing might also be required for the small-diameter pipes used as *impulse lines* to connect the process in which the pressure is measured and transmitted to an instrument. Tracing should be shown on the P&ID.

**Example 15.1**   Steam with a flowrate of $5\,\text{t}\,\text{h}^{-1}$ is required to flow through a schedule 40 pipe. The steam is at 15 bar and 250 °C. Assuming a velocity of steam in the pipe of $20\,\text{m}\,\text{s}^{-1}$, carry out a preliminary sizing for the pipe.

**Solution**

From steam tables for the given steam conditions, the specific volume $v$ of the steam is given by:

$$v = 0.1520\,\text{m}^3\,\text{kg}^{-1}$$

The density $\rho$ is given by:

$$\rho = \frac{1}{0.1520}$$
$$= 6.58\,\text{kg}^3\,\text{m}^{-3}$$

For a velocity of $20\,\text{m}\,\text{s}^{-1}$, the internal area of the pipe $A$ is given by:

$$A = \frac{5 \times 10^3}{3600} \times \frac{1}{6.58} \times \frac{1}{20}$$
$$= 0.0106\,\text{m}^2$$

The internal diameter $D$ is given by:

$$D = \sqrt{\frac{4A}{\pi}}$$
$$= 0.116\,\text{m}$$
$$= 116\,\text{mm}$$

From Appendix H choose DN 150 (OD = 168.3 mm). The mechanical strength of the pipe and pipe support arrangements would need to be considered later, and might change the choice.

Valves are also required for piping systems. A *valve* regulates, directs, or controls the flow of a fluid by opening, closing, or partially obstructing the flow in a piping system. *Block valves* turn the flow *on or off*, but are not intended to regulate. The P&ID needs to show not only where valves are to be used, but what type of valve is to be used. Figure 15.3 shows some commonly used valve types:

1) *Globe valve*. A typical globe valve is illustrated in Figure 15.3a. A screw mechanism is used to move a *plug* or *disc* against a *seat*. Figure 13.3a shows both a *plug disc* and a *composite disc* design. The composite disc design uses a seating material incorporated into the disc. Many different designs of seating arrangements are available. Globe valves can be used to regulate the flow, as well as starting and stopping the flow.

2) *Butterfly valve*. A typical butterfly valve is illustrated in Figure 15.3b. The angle of a disc is manipulated relative to the flow. When the disc is perpendicular to the flow, a seal is created between the disc and the lining of the valve to prevent flow. When the disc is rotated through a quarter turn, the disc is in line with the flow and the valve is open. Butterfly valves can be used to regulate the flow, as well as starting and stopping the flow. Their use is normally restricted to larger diameter pipelines.

3) *Gate valve*. A typical gate valve is illustrated in Figure 15.3c. This operates by a screw mechanism raising and lowering a circular gate in the path of the fluid. The gate faces can form a wedge shape or can be parallel. Gate valves are primarily used for starting and stopping flow, rather than regulating the flow.

4) *Plug valve*. A typical plug valve is illustrated in Figure 15.3d. A cylindrical or more often conically tapered plug with a hollow port is rotated through a quarter turn to allow or prevent flow. The installation space is smaller than a gate valve and operation to open and close is simple and rapid. The tapered plug allows a tight shut-off.

5) *Ball valve*. A typical ball valve is illustrated in Figure 15.3e. A sphere with a cylindrical hollow port is rotated through a quarter turn to allow or prevent flow. Conventional ball valves

**15**

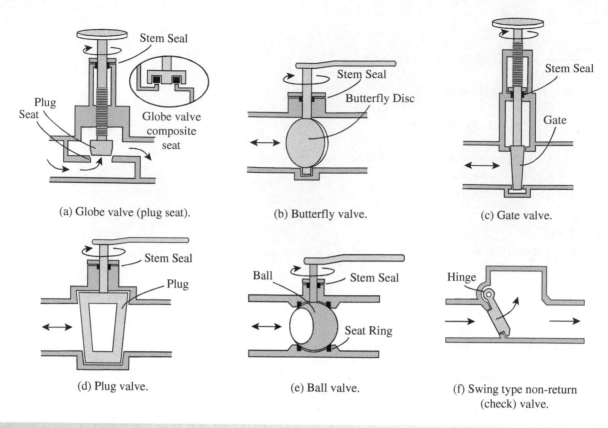

(a) Globe valve (plug seat).    (b) Butterfly valve.    (c) Gate valve.

(d) Plug valve.    (e) Ball valve.    (f) Swing type non-return (check) valve.

### Figure 15.3

Examples of different valve types. Source: Smith (2016).

have a port smaller than the pipe diameter. Full port valves with the same diameter of the pipe are also available, but are less common. Tight shut-off can be achieved by virtue of the seat rings, but the seat material creates temperature limitations. As with plug valves, ball valves require an installation space smaller than a gate valve and operation to open and close is simple and rapid.

6) *Non-return (check) valve*. Non-return valves are used purely to prevent reversal of flow, for example to prevent liquid from siphoning back from a tank if a pump is switched off. A *swing* type non-return valve is illustrated in Figure 15.3f. The disc swings on a hinge either onto the seat to prevent reverse flow or swings to allow forward flow. Other designs of non-return valves use mechanisms such as forward flow moving a ball mounted in a slot away from a circular seat to open the valve, with reversal of flow moving the ball in the reverse direction to seal with the seat to prevent reverse flow.

Solenoid valves are used in some small-scale applications. A solenoid valve is an electrically operated valve that is either fully closed or fully open. The valve features a solenoid, which is an electric coil with a movable ferromagnetic core (plunger) at its center. An electric current through the coil creates a magnetic field that causes the core to open or close the valve. They are used for low flowrates and applications with a low pressure difference across the valve.

Three-way valves offer multiple outlets from the same valve body. For example, Figure 15.4 shows the function of an L-type three-way valve. In Figure 15.4, as the valve plug is rotated the inlet flow is directed to a different outlet. Further rotation closes the flow. Other three-way valve designs offer different outlet arrangements. Solenoids can also be used to operate three-way valves.

Some valves need to be *locked open* (*LO*) or *locked closed* (*LC*) for safety reasons. These can only be operated with a key. Figure 15.5a shows an example of a locking valve. Many different designs are available. Rather than valves being locked, *valve car seals* can be used. These are cable seals that might need to be physically cut to operate the valve, but reusable car seals are also available. Figure 15.5b shows a typical car seal arrangement. Thus, valves might be *car seal open* (*CSO*) or *car seal closed* (*CSC*). Whether a valve is locked or car sealed, operation of the valve would normally not be allowed except under a safety permit.

Figure 15.6a shows a blank flange used to seal the end of a pipe. In general, pipes should not be left open to the environment, even if downstream of a closed valve, and even for a short period. If a pipe section or equipment is removed from a pipe system, then an opening to the environment should be sealed with a blank flange. A valve alone cannot guarantee a safe isolation between

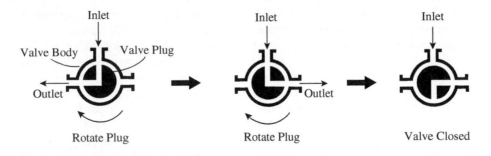

## Figure 15.4

Operation of an L-type three-way valve.

(a) Example of a locking valve.
(Reproduced with permission from IPF Online)

(b) Example of a car seal lock.
(Reproduced by permission of Industrial
Total Lockout (Safety) Ltd)

## Figure 15.5

Examples of valve locking arrangements.

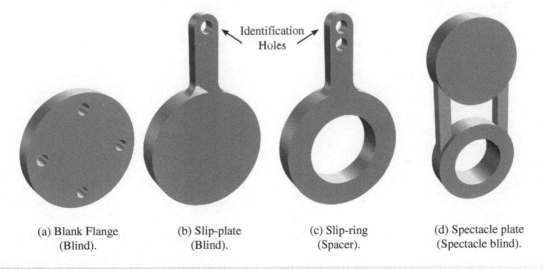

(a) Blank Flange
(Blind).

(b) Slip-plate
(Blind).

(c) Slip-ring
(Spacer).

(d) Spectacle plate
(Spectacle blind).

15

## Figure 15.6

Pipe fittings used for positive isolation of equipment. Source: Reproduced from Health and Safety Executive HSG 253 (2006), The Safe Isolation of Plant and Equipment, by permission of HSE. HSE has not reviewed this product and does not endorse its use.

two sections of piping and cannot guarantee isolation of equipment for maintenance. *Positive isolation* (complete separation) can only be guaranteed by removing a section of pipe (referred to as a *spool*) or inserting a *slip plate* or *spectacle plate* in a flange joint. A slip plate is shown in Figure 15.6b. Inserting a slip plate into a flanged joint between two sections of piping, or between a section of pipe and a vessel or process equipment, can achieve positive isolation. Once the slip plate is removed to allow flow it is replaced by a *slip ring* to fill the gap in the pipe allowed for the slip plate (Figure 15.6c). Exterior markers indicate whether the plate inserted is a slip plate or a slip ring. Figure 15.6d shows a spectacle plate that can be rotated to act as either a slip plate or a slip ring.

Figure 15.7 shows examples of symbols used to represent valves and isolation devices in P&IDs.

Safe isolation between a fluid and equipment to be isolated and maintained might require more than a valve or even two valves in series, as valves alone cannot guarantee isolation. Figure 15.8a shows an example of *positive isolation* of an item of equipment from a hazardous fluid with a *single block and bleed and slip plate*. A block valve and slip plate are located in the pipe between

the hazardous fluid and the equipment to be isolated. The block valve isolates the fluid to install the slip plate. The slip plate ensures positive isolation. An open drain valve for a liquid, or vent valve for a gas, is used to depressurize and drain/vent the pipe for installation and removal of slip or spectacle plate. A pressure indicator ensures that there is no build-up of pressure between the valve and the slip/spectacle plate. Figure 15.8b shows an example of *proved isolation* using *double block and bleed*. Two block valves are located in the pipe between the hazardous fluid and the equipment to be isolated. Double block and bleed requires draining or venting of fluid to *prove* isolation. Each block valve is *proved* separately by a drain/vent valve. The drain/vent valves are normally open when the block valves are closed. Pressure indicators ensure there is no build-up in pressure. Other arrangements are possible for both positive and proved isolation (Health and Safety Executive 2006).

As discussed in Chapter 10, control valves need to adopt a fail-safe position in the event of failure of the control system (e.g., in the event of loss of instrument air), providing a fail-safe position can be identified. The fail-safe position for a control valve can be open, closed, or kept in the current position, as discussed in

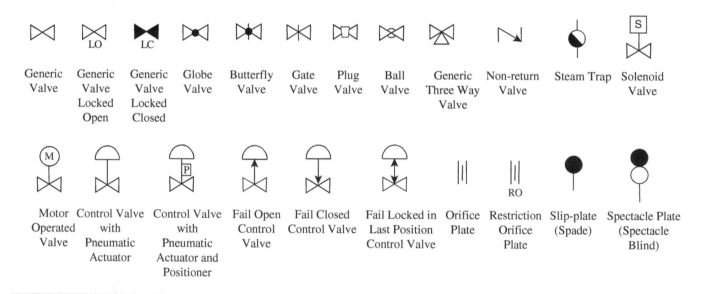

**Figure 15.7**

Examples of valve and isolation device symbols.

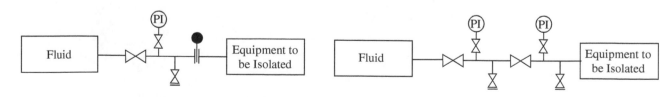

(a) Positive isolation with single block and bleed and slip plate.

(b) Proved isolation with double block and bleed

**Figure 15.8**

Positive and proved isolation of equipment.

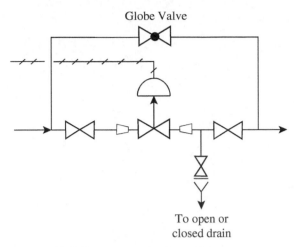

Globe Valve

(a) Fail open control valve with bypass.

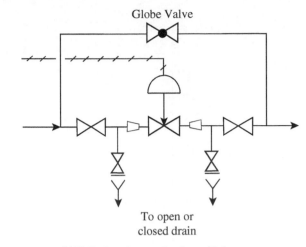

Globe Valve

To open or
closed drain

(b) Fail closed control valve with bypass.

**Figure 15.9**

Piping arrangements for control valves.

Chapter 10. Figure 15.9a shows the P&ID for a fail-open control valve and Figure 15.9b the P&ID for a fail-closed control valve. The symbols for the different types of control signal are given in Figure 10.43. The control valve size is often smaller than the corresponding line size, requiring a pipe reducer and enlarger on either side of the control valve. If the failure of the control valve causes the plant to shut down, this causes lost production, and the shut-down and subsequent start-up brings additional safety, product quality, and environmental issues. A bypass line is often installed around the control valve to allow control valve maintenance without shutting down the plant. A manual control valve in the bypass line allows the plant to keep operating until the control valve is repaired or replaced. Exceptions to having a bypass are when the plant can be stopped quickly (e.g., a batch operation), or if it is hazardous to operate on manual control. Isolation valves are located upstream and downstream of the control valve. Whether these additional isolation valves are required depends on the availability of other valves in the same line that would allow the control valve to be isolated. The control valves in Figure 15.9 are for liquid service. Drain valves allow the system to be depressurized and drained for maintenance. If the control valve fails to open, as shown in Figure 15.9b, then a single drain valve downstream of the control valve is often adequate to allow the system to be drained for maintenance, but sometimes both upstream and downstream drain valves are included, even for fail-open valves (especially if the valve is prone to plugging). If the control valve fails closed, as shown in Figure 15.9b (or in the current position), then drain valves upstream and downstream of the control valve are needed for the system to be depressurized and drained for maintenance, because of the isolating effect of the fail-closed control valve. If the control valves are for a gaseous service, then vent valves would be required instead of drain valves. Venting must be to a safe location (Figure 15.9).

## 15.3  Pressure Relief

Reactor and separator vessels, storage tanks, heat exchangers, and piping are designed to contain a certain pressure. Process design specifies the *operating pressure* for a specified temperature under normal operating conditions. The *maximum operating pressure* extends beyond normal operating conditions to include start-up, shut-down, and other off-design conditions. The *maximum allowable working pressure* (*MAWP*) is the maximum gauge pressure that the weakest component of the system can safely withstand *for a designated temperature*. It must be emphasized that the weakest component needs to be assessed at the maximum working temperature. The reason for this is illustrated in Figure 15.10, which shows the variation of the maximum working stress for carbon steel sheet with temperature. Figure 15.10 shows a steep decline in the maximum working stress with increasing temperature. Based on the applicable standards, *MAWP* can vary from 10% to 25% above the maximum operating pressure of the system.

A variety of causes may lead to pressure in excess of the maximum allowable working pressure in the equipment. This can result from:

- Heat input from heating utilities, heat recovery, heat of reaction, external fire or solar gain, which increases pressure through vaporization of liquid or thermal expansion. The problem of the heat input might be associated with loss of cooling.
- Direct pressure input from higher-pressure sources.

Common causes of overpressure include:

1) Exposure to external fire of a closed volume of process or utility fluid, often referred to as the "fire case". The fire case must be considered if there is flammable material in the proximity

**15**

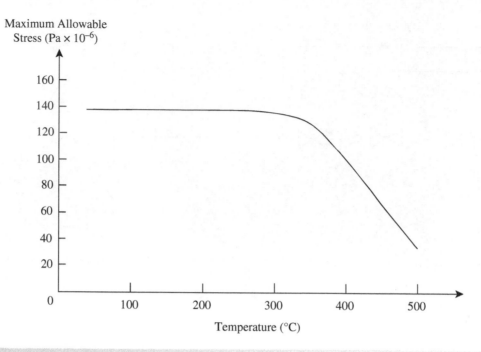

**Figure 15.10**

Variation of maximum allowable stress of carbon steel sheet with temperature.

that can cause an external fire in the event of a safety incident. Water sprinkler and deluge systems can provide fire protection to spray water onto a fire to mitigate the problem.

2) Blocked outlet, resulting from operator error, maintenance error, or blockage from process material formed under abnormal conditions.

3) Pipe or vessel blocked in by valves with continued heat input from utility or tracing or solar gain.

4) Equipment failure from heat exchanger tube rupture, control or non-return valve failure, or air cooler failure.

5) Utility failure from:

- steam failure causing failure of steam turbines driving equipment or steam ejectors;
- cooling water failure preventing the output of energy when energy input continues;
- power failure causing failure of electrical equipment;
- instrument air failure causing failure of control equipment.

6) Absorbent failure in an absorber causing accumulation of gases not removed by the failed absorbent.

7) Runaway reaction where exothermic reaction generates heat at a faster rate than the cooling system can remove it.

8) Other possibilities.

Any system that has the potential to have its pressure increased beyond safe levels requires safety devices (and/or safety instrumented functions) to protect the equipment and any resulting consequences for safety and the environment. Figure 15.11 illustrates two pressure relief devices.

1) *Pressure relief valves.* An example of a pressure relief valve is illustrated in Figure 15.11a. An adjustable spring is used to counter the internal pressure and maintain it closed. The *set pressure* is the pressure at which the force imposed by the spring is no longer able to keep the valve closed. If the internal pressure is greater than the set pressure, the valve opens to relieve any overpressure and then closes. Different valve seating designs and different mechanisms from the simple spring arrangement in Figure 15.11a can be used. There are different designs of pressure relief valves, depending on the application.

a) *Safety valve.* A safety valve is characterized by a rapid opening or "pop" action and is normally used on gas and steam applications. Safety valves are designed to open completely when the internal pressure exceeds the setting for which the spring has been set.

b) *Relief valve.* A relief valve has a gradual lift, generally proportional to the increase in pressure over the opening pressure. The opening depends on pressure, but is normally fully open at around 10% over the set pressure. It is primarily used on liquid service applications.

c) *Safety relief valve.* A safety relief valve can be used for compressible fluids or liquids depending on its application. It functions as a safety valve for compressible fluids with rapid opening and as a relief valve for liquid applications with opening in proportion to the increase in pressure.

2) *Bursting (rupture) discs.* A bursting disc is illustrated in Figure 15.11b. A bursting disc is a thin membrane (usually metal) that fails and ruptures at a critical pressure. The bursting disc has a one-time use and, unlike a pressure relief valve,

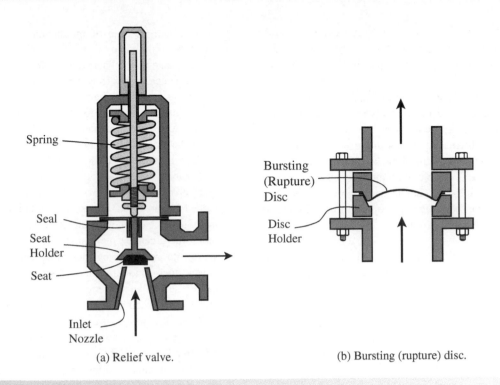

(a) Relief valve.

(b) Bursting (rupture) disc.

**Figure 15.11**

Pressure relief devices. Source: Smith (2016).

cannot close after the pressure has been relieved. There are different classes of bursting discs:

a) *Forward acting*. Figure 15.11b illustrates a forward acting disc in which the internal pressure acts against the concave surface. As the pressure increases beyond the allowable pressure the disc bulges and fails in tension. Some designs use scoring on the convex surface of the disc to allow a more controlled failure.

b) *Reverse acting*. In a reverse acting disc, the internal pressure acts against the convex side of the disc. Compression loading causes it to reverse, snapping through the neutral position and causing it to fail by a scoring pattern on the surface of the disc on the low-pressure side, or a knife blade penetration located on the low-pressure side.

Figure 15.12 shows examples of P&ID symbols for pressure relief devices. Figure 15.12a shows an example of a symbol for

pressure relief valve. Figure 15.12b shows an example of a symbol for a bursting disc.

The main distinction between bursting discs and pressure relief valves is that pressure relief valves will reclose, whereas bursting discs cannot and need to be replaced if activated, which might mean a plant shut-down. This would suggest that pressure relief valves are the best solution, and they are more commonly used. However:

a) If pressure relief valves are exposed to process material that corrodes the device, it might prevent it from operating properly after some time in service.

b) Process material might cause plugging of the pressure relief valve when activated. This is particularly a problem if the process material contains solids or can polymerize.

c) Pressure relief valves can leak and sometimes not re-seat properly after opening, causing continued discharge of potentially hazardous material. It is possible to monitor the opening of pressure relief valves and to detect when they have not reseated properly.

d) Pressure relief valves will need regular testing, inspection, and repair, according to the service, applicable standards, and the maintenance history. Inspection and repair are typically every 1 to 5 years.

e) In some cases, the transient pressures mean that the pressure relief valve cannot act fast enough and a bursting disc must be used. Bursting disc opening times are typically 1–10 ms (almost instantaneous). Pressure relief valve opening times are typically 50–100 ms. Actual opening times depend on

**15**

(a) Pressure relief valve.

(b) Bursting disc.

**Figure 15.12**

Symbols for pressure relief devices.

the size and design of the device and the process pressure. By comparison, control valve opening times are typically 0.5–10 s (but can be much longer, for example, in the case of large motor operated valves).

**f)** Pressure relief devices have a set pressure at which they open. However, it is important to note that it is actually the difference in pressure across the device that causes it to operate. A 1 bar back-pressure will increase the pressure at which it opens by 1 bar above its normal setpoint. A back-pressure might be created on the valve, for example if a major process upset causes multiple discharges into the relief header at the same time. Pressure relief valves can be designed to open at a specified absolute pressure to avoid this problem.

Relief devices are often used in conjunction with other layers of protection. For example, a safety instrumented function can be used to trip the process on detection of pressure that has increased beyond safe levels before the safety device opens. In this way, the relief device will only open if the safety instrumented function fails on demand or does not prevent further increase in pressure.

# 15.4 Relief Device Arrangements

Relief devices can be arranged in different ways. It is sometimes advisable or necessary to use multiple pressure relief devices. Start with a single pressure relief valve.

**1)** *Single pressure relief valve.* Figure 15.13a shows a single pressure relief valve both with and without isolation valves. Isolation valves on the inlet and the outlet of the pressure relief valve facilitate ease of maintenance on the pressure relief valve when the process is shut down. If isolation valves are installed, then they must be locked or car sealed open until the process is shut down. The isolation valves should be full-port ball or gate block valves to avoid restriction of the flow. The isolation

valves facilitate ease of maintenance of the pressure relief valve, but only when the process is shut down for the maintenance to take place. There are no circumstances under which it would be acceptable to isolate the pressure relief valve while the process is in operation.

An alternative arrangement is shown in Figure 15.13b, which does allow maintenance of the pressure relief valve while the process is in operation. Two valves are shown in parallel, both sized for full flow for the relief. In one arrangement both valves have isolation valves on the inlet and the outlet, one valve locked (or car sealed) open, and the other valve locked (or car sealed) closed. The isolation valves should again be full-port ball or gate block valves. The on-line pressure relief valve can be maintained by first bringing the off-line valve into operation by unlocking and opening its isolation valves, then unlocking and closing the isolation valves for the pressure relief valve that was on-line. The isolation valves are then re-locked in their new position. An alternative shown in Figure 15.13b is to use a three-way valve to switch between the two pressure relief valves. The off-line valve can then in principle be removed for maintenance.

A fundamental problem to be avoided is that of *chattering* of the pressure relief valve. Chattering is high-frequency opening and closing of the pressure relief valve. If the valve is oversized then, for a low demand rate for the relief device relative to the device capacity, the valve opens and the build-up of back-pressure resulting from the discharge flow causes the valve to close. The valve opens again when the discharge flow has stopped. The cycle repeats at high frequency. This high-frequency opening and closing causes vibration that can result in misalignment of the valve internals, valve seat damage, or mechanical failure of the valve internals and associated piping. To avoid chattering, pressure relief valves should be sized to be less than 140% of the required capacity (Hellemans 2009). The problem is exacerbated if loads on the relief system vary significantly. Whilst it seems instinctively correct to oversize relief systems, significant oversizing can also bring hazards.

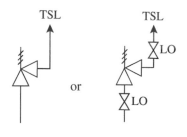

(a) Single pressure relief valve.

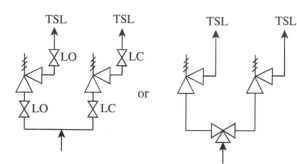

(b) Two full-flow pressure relief valves in parallel with one open to the system.

| TSL | To Safe Location |
| --- | --- |

# Figure 15.13

Examples of different arrangements for a single pressure relief valve.

**2)** *Pressure relief valves in parallel.* Multiple pressure relief valves are required when the relief load exceeds the capacity of the largest available pressure relief valve. It is also good practice to install multiple pressure relief valves for loads that might vary significantly in order to minimize chattering on small discharges. When capacity variations are frequently encountered in normal operation, multiple smaller pressure relief valves with set pressure staggered slightly can help to avoid chattering. Figure 15.14a shows two pressure relief valves in parallel, both sized for part load and both open to the system with staggered set pressures. In this arrangement the pressure relief valve with the lowest setting will be capable of handling minor upsets, and the other valve opening at a slightly higher set pressure as the capacity requirement increases. Arrangements both with and without isolation valves are shown in Figure 15.14a. If isolation valves are installed, then they must be locked or the car sealed open until the process is shut down. The isolation valves facilitate ease of maintenance for the pressure relief valves, but only when the process is shut down for the maintenance. Again, full-port ball or gate block valves should be used.

An alternative arrangement is shown in Figure 15.14b allows maintenance of the pressure relief valves while the process is in operation. Four valves are shown in parallel, all sized for part flow for the relief. Each valve pair in Figure 15.14b work like the pair in Figure 15.14a, with one of each pair having a slightly higher set pressure than the other.

In one arrangement all pressure relief valves have isolation valves on the inlet and the outlet. One pair are locked (or car sealed) open, and the other pair locked (or car sealed) closed. The on-line pressure relief valve pair can be maintained by first bringing the off-line valve pair into operation, then unlocking and closing the isolation valves for the valve pair that was on-line. The valves should then be locked in the new position. An alternative shown in Figure 15.14b is to use a three-way valve to switch between the two pairs of pressure relief valves.

**3)** *Pressure relief valve and bursting disc in parallel.* Figure 15.15a shows an arrangement with a pressure relief valve and bursting disc in parallel. This attempts to combine the advantages of both devices. As an example, consider a duty in which a pressure relief valve might be in danger of plugging from process solids formed due to upset conditions such as polymerization reactions. In such circumstances, a bursting disc would be less susceptible to blockage from solids. In Figure 15.15a the two devices are used in parallel with the set pressures such that the pressure relief valve opens first and the bursting disc operates only if the relief valve fails to relieve the overpressure. Thus, if the pressure relief valve operates normally to relieve any overpressure, it then resets, allowing the process to continue in operation. However, if a serious problem occurs blocking the pressure relief valve, then the bursting disk can act as a backup, being less susceptible to plugging. Of course, if the bursting disc does operate, then the process will need to be shut down.

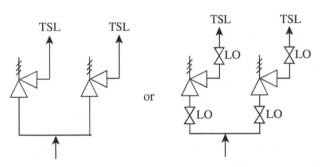

(a) Two pressure relief valves in parallel open to the system.

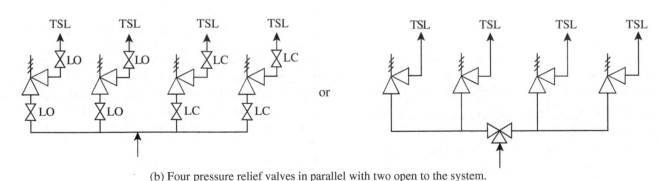

(b) Four pressure relief valves in parallel with two open to the system.

| TSL | To Safe Location |
|-----|------------------|

**Figure 15.14**

Examples of different arrangements for two pressure relief valves in parallel.

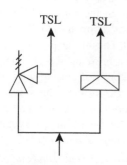
TSL    TSL

(a) Relief valve and bursting disc in parallel.

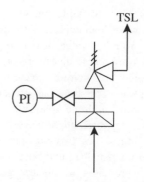
TSL

(b) Relief valve and bursting disc in series.

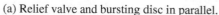

| TSL | To Safe Location |
|-----|------------------|

## Figure 15.15

Combinations of relief valves and bursting discs.

4) *Relief valve and bursting disc in series*. If the process stream that needs pressure relief is particularly corrosive, such that it could damage the relief valve, then a bursting disc and pressure relief valve can be used is series, as shown in Figure 15.15b. The bursting disc, constructed from appropriate materials of construction, protects the valve from aggressive process material. If pressure relief is required, then both open, but the pressure relief valve then closes, allowing the process operation to continue. The replacement of the bursting disc can then be deferred. It is important to have a pressure indicator between the bursting disc and pressure relief valve. If a leak occurs in the bursting disc, for example from a pinhole developing from corrosion, then the process material will leak into the gap between the bursting disc and the pressure relief valve. This will cause a pressure build-up in the gap and decreases the pressure difference across the bursting disc. This decreased pressure difference will cause the bursting disc to not burst at the correct pressure and the system can overpressurize. The pressure indicator allows the problem to be identified.

When designing a relief arrangement, it is important not to create a significant pressure drop in the inlet to pressure relief valves or create a significant back-pressure on the discharge. Either of these can lead to chattering of pressure relief valves (Hellemans 2009).

Different standards apply for relief devices, for example:

- API RP 520 Sizing, Selection, and Installation of Pressure Relieving Devices in Refineries;
- API RP 521 Guide for Pressure Relieving and Depressurizing Systems Petroleum, Petrochemical and Natural Gas Industries – Pressure Relieving and Depressurizing systems;
- ASME Boiler and Pressure Vessel Code, Section VIII;
- EN 764-7 Pressure Equipment. Safety Systems for Unfired Pressure Vessels.

Figure 15.16 summarizes the important pressures for relief systems:

1) *Operating pressure*. The gauge pressure during normal service.
2) *Set pressure*. The pressure at which the relief device begins to activate. The expected maximum operating pressure should be low enough to prevent the relief device activating during normal operation of the process.
3) *Maximum allowable working pressure (MAWP)*. The maximum gauge pressure (allowing for any hydrostatic head) that the weakest component of the system can safely withstand for a designated temperature.
4) *Accumulation*. The pressure increase over the MAWP during the relief process, expressed as a percentage of the MAWP.
5) *Overpressure*. The pressure increase above the set pressure during the relieving process expressed as a percentage of the set pressure.

It should be noted from Figure 15.16 that relief devices are normally set to open at the MAWP. If multiple relief devices are used, then one relief device must be set no higher than the MAWP, but the others can be set as high as 105% of the MAWP. The allowable accumulation for a single relief device is 110%. The exception to this is fire exposure, for which the allowable accumulation is 121% of the MAWP. When multiple relief devices are used for non-fire scenarios, the allowable accumulation is 116%.

Once an emergency discharge has been created from pressure relief, it must be safely dealt with. Emergency discharge from pressure relief valves and bursting discs can be dealt with in a number of ways:

1) Direct discharge to an atmospheric safe location (ASL) far away from people and ignition sources, under conditions leading to rapid and safe dilution. For example, this would be

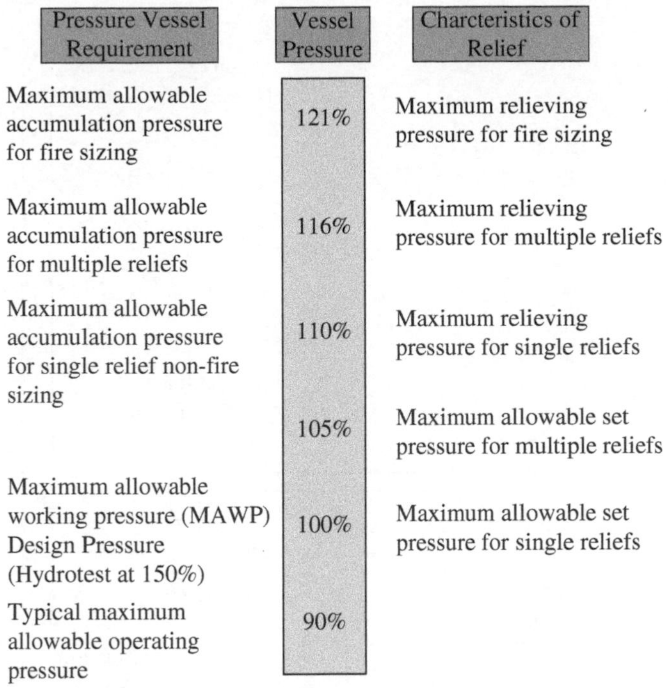

| Pressure Vessel Requirement | Vessel Pressure | Charcteristics of Relief |
|---|---|---|
| Maximum allowable accumulation pressure for fire sizing | 121% | Maximum relieving pressure for fire sizing |
| Maximum allowable accumulation pressure for multiple reliefs | 116% | Maximum relieving pressure for multiple reliefs |
| Maximum allowable accumulation pressure for single relief non-fire sizing | 110% | Maximum relieving pressure for single reliefs |
| | 105% | Maximum allowable set pressure for multiple reliefs |
| Maximum allowable working pressure (MAWP) Design Pressure (Hydrotest at 150%) | 100% | Maximum allowable set pressure for single reliefs |
| Typical maximum allowable operating pressure | 90% | |

**Figure 15.16**

Relief pressures. Source: Reproduced by permission from Crowl and Tipler (2013).

appropriate for utility system discharges that tend to be less hazardous.

2) Total containment in a connected vessel, with ultimate disposal being deferred. For example, this would be appropriate for small-scale chemical operations.

3) Partial containment, in which some of the discharge is separated either physically (e.g., gravitational or centrifugal means) or chemically using, for example, absorption. For example, this would be appropriate for larger scale chemicals operations.

4) Combustion in a flare after separation of any liquid. This would be appropriate for petrochemicals and petroleum refining operations. It would be typical to have many relief valves routed to the same flare header.

Gaseous emergency discharges of hydrocarbons are most often dealt with by routing the discharge through a *flare header* to a central *elevated flare*. Figure 15.17 shows an elevated steam-assisted flare. The discharge is combusted in an open combustion process with oxygen for the combustion provided by air around the frame. Good combustion depends on the flame temperature, residence time in the combustion zone, and good mixing to complete the combustion. Destruction efficiency can be up to 98%. Figure 15.17 shows steam being used to assist the mixing in the flame tip for efficient combustion. Before the discharge stream enters the flare stack a knockout is required to remove any entrained liquid. Liquids must be removed as they can extinguish the flame or lead to irregular combustion in the flame. There is also a danger that liquid might not be completely combusted, which can result in liquid reaching the ground and creating hazards. The flare header needs to be protected against the flame propagating back into the header. A *gas barrier* prevents the flame

passing down into the flare header. A flashback seal drum containing water provides additional protection. In some designs the flashback seal drum is incorporated into the base of the flare stack. Halogenated and sulfur-containing compounds should not be flared.

Figure 15.18 shows an example of a relief system for a process vessel. If the vessel contains any liquid, the release of pressure in the process vessel will result in both vapor and liquid being vented from the vessel. A *blowdown drum* is used to separate the liquid from the vapor. Vapor carried from the blowdown drum can then be sent to a scrubber, flare stack, or vented to a safe location.

In high-pressure gas systems, the outlet is often sent to an atmospheric safe location. If the outlet is connected to a header, the operation of a relief device will cause a pressure build-up in the header, which will create a back-pressure on other relief devices and interfere with their safe function.

# 15.5 Reliability of Pressure Relief Devices

Like other items of process equipment, relief devices have reliability issues that need to be considered. Failure rates for pressure relief devices depend on the mode of failure. Failure rates for pressure relief valves are typically (Mannan 2012):

- failure to open 0.01 $y^{-1}$;
- premature open 0.02 $y^{-1}$;
- failure to close 0.02 $y^{-1}$.

**15**

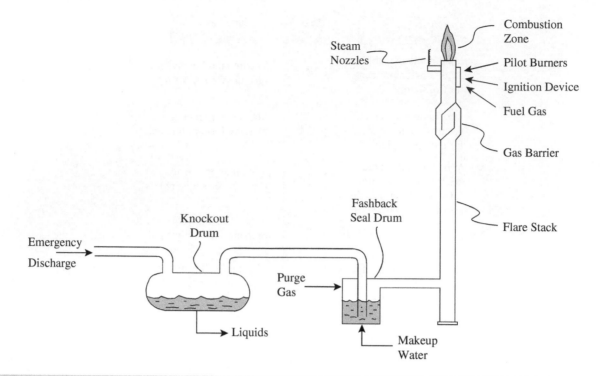

**Figure 15.17**

Typical arrangement for an elevated steam-assisted flare.

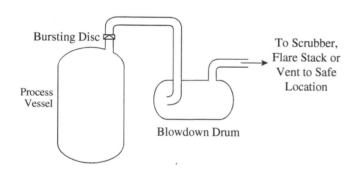

**Figure 15.18**

Example of a relief system for a process vessel.

Failure rates for bursting discs are typically (Mannan 2012):

- failure to rupture at normal pressure 0.2 y$^{-1}$.

Care needs to be exercised interpreting failure rates for relief devices. The failure mode of most concern is failure to open for pressure relief valves or failure to rupture for bursting discs, at the required set pressure. Premature opening of pressure relief valves or premature rupture for bursting discs is likely to cause less of a hazard. Premature rupture of bursting discs can occur due to corrosion or metal fatigue of the disc from cyclic stress on the disc.

However, to assess the reliability of a safety system requires the failure to respond to a hazardous situation over a time horizon to be considered. Safety systems stay in the background until an abnormal situation requires some action to keep the system safe. A dangerous undetected failure will not be detected until either there is a demand on the system (from overpressurization) or it is detected in a *proof test*. Proof testing can be carried out for pressure relief valves by either removing for bench testing in a workshop, which can be a time-consuming and expensive operation, or they can be tested in situ. In situ testing involves attaching a hydraulic ram to the valve. The hydraulic ram assists the valve to overcome the spring tension, slightly opening the valve for a short period in order to determine the valve set point. Proof testing of safety instrumented systems can be carried out by manipulation of the process pressure without driving the process into the demand condition for the safety system.

The *probability of failure on demand* (*PFD*) measures the *unavailability* or *fractional dead time* (i.e., the fraction of total time that the protective device is in a failed state), which is a measure of the probability of its failure to provide the protection required. Assuming an exponential distribution (see Chapter 7), the probability of failure of a safety device or system during the interval between proof tests $\tau_P$ is given by:

$$F(\tau_P) = 1 - \exp(-\lambda\tau_P) \qquad (15.1)$$

where

$F(\tau_P)$ = probability of failure at time $\tau_P$ (−)

$\quad \lambda$ = failure rate $(y^{-1})$

$\quad \tau_P$ = interval between proof tests (y)

The safety device is only required on demand. The demand rate is the number of times per year at which the protective system is required to act (normally obtained from historical records). The probability of a demand occurring during the dead time after failure can be defined, again assuming an exponential distribution:

$$Pr_D(\tau_D) = 1 - \exp(-D\tau_D) \qquad (15.2)$$

where

$Pr_D(\tau_D)$ = probability that a demand will occur during the dead time $(y^{-1})$

$\quad \tau_D$ = total dead time when the safety device is unavailable (y)

$\quad D$ = demand rate $(y^{-1})$

The hazard rate can be defined as the number of times per year a hazard can occur, given by:

$\quad H$ = failure rate × probability that a demand

$\quad\quad$ will occur during the dead time

$$= \lambda[1 - \exp(-D\tau_D)] \qquad (15.3)$$

where

$H$ = hazard rate $(y^{-1})$

For a *random failure*, on average:

$$\tau_D = \frac{\tau_P}{2} \qquad (15.4)$$

The hazard rate becomes:

$$H = \lambda\left[1 - \exp\left(-\frac{D\tau_P}{2}\right)\right] \qquad (15.5)$$

If the safety device never failed $\lambda = 0$, then the hazard rate would be zero. If $D\tau_P/2$ is large, then $H = \lambda$ and the hazard rate would equal the demand rate.

The probability of failure on demand is the fraction of time that the safety device is unavailable:

$$PFD = \frac{1}{\tau_P}\int_0^{\tau_P} F(t)\,dt$$

$$= \frac{1}{\tau_P}\int_0^{\tau_P} 1 - \exp(-\lambda t)\,dt$$

$$= \frac{1}{\tau_P}\int_0^{\tau_P} dt - \frac{1}{\tau_P}\int_0^{\tau_P} \exp(-\lambda t)\,dt \qquad (15.6)$$

$$= \frac{1}{\tau_P}[t]_0^{\tau_P} - \frac{1}{\tau_P}\left[\frac{1}{-\lambda}\exp(-\lambda t)\right]_0^{\tau_P}$$

$$= \frac{\lambda\tau_P + \exp(-\lambda\tau_P) - 1}{\lambda\tau_P}$$

Approximate expressions for $F(\tau_P)$, $P_D(\tau_D)$, $H$, and $PFD$ can be developed by approximating the exponential term as a Maclaurin Series:

$$1 - \exp(-\lambda t) = 1 - \left[1 + (-\lambda t) + \frac{1}{2!}(-\lambda t)^2 + \frac{1}{3!}(-\lambda t)^3 + \cdots\right] \qquad (15.7)$$

If $\lambda \ll 1$ then the exponential can be approximated by the first term in the series. Thus:

$$1 - \exp(-\lambda t) \approx \lambda t \qquad (15.8)$$

Similarly, if $D \ll 1$ then:

$$1 - \exp(-D\tau_D) \approx D\tau_D = \frac{D\tau_P}{2} \qquad (15.9)$$

Using these approximations, the following expressions can be developed:

$$Pr_D(\tau_D) \approx \frac{D\tau_P}{2} \qquad (15.10)$$

$$H \approx \frac{\lambda D\tau_P}{2} \qquad (15.11)$$

$$PFD = \frac{1}{\tau_P}\int_0^{\tau_P} 1 - \exp(-\lambda t)\,dt$$

$$\qquad (15.12)$$

$$\approx \frac{1}{\tau_P}\int_0^{\tau_P} \lambda t\,dt = \frac{1}{\tau_P}\left[\frac{\lambda t^2}{2}\right]_0^{\tau_P} = \frac{\lambda\tau_P}{2}$$

The approximate expression for $H$ can diverge significantly from the exponential expression as $D$ or $\tau_P$ increase. The approximate expression for the $PFD$ agrees quite closely with the exponential expression. Differences for $PFD$ emerge when $\lambda$ and $\tau_P$ become larger, but, even then, the errors tend to remain small. Also, the equation will be used with data that has significant uncertainty. This analysis assumes that any failure will remain undetected until the next proof test. For logic solver computer systems, such as trip systems, diagnostic tests can be used to identify failures as they occur and can be repaired directly. However, diagnostic cover is never 100% and there is always the need to conduct complete proof tests, but at a reduced rate. By contrast, failure of relief valves will remain undetected until there is a proof test or a demand on the system.

**15**

**Example 15.2**    A relief valve has a failure rate of 0.01 failures $y^{-1}$. Calculate the probability of failure on demand and the hazard rate assuming proof testing every one year and two years with demand rates of 0.1, 1, and 10 per year.

**Solution**

For $\tau_P = 1$ y and $D = 0.1$ $y^{-1}$ the $PFD$ is given by Eq. (15.12):

$$PFD = \frac{\lambda\tau_P}{2} = \frac{0.01 \times 1}{2} = 0.005$$

For comparison, the more accurate formula in Eq. (15.6) yields $PDF = 0.004983$. The error is negligible and unimportant compared with the other assumptions made. Using the more accurate formula for the hazard rate, from Eq. (15.5):

$$H = \lambda\left[1 - \exp\left(-\frac{D\tau_P}{2}\right)\right]$$

$$= 0.01\left[1 - \exp\left(-\frac{0.1 \times 1}{2}\right)\right]$$

$$= 0.000488 \text{ y}^{-1}$$

Period over which hazard is expected

$$= 1/H = 2050 \text{ y}$$

For $\tau_P = 1$ y:

| $D$ (y$^{-1}$) | PFD | $H$ (y$^{-1}$) | Period over which hazard expected (y) |
|---|---|---|---|
| 0.1 | 0.005 | 0.000488 | 2050 |
| 1 | 0.005 | 0.003935 | 254 |
| 10 | 0.005 | 0.009933 | 101 |

For $\tau_P = 2$ y:

| $D$ (y$^{-1}$) | PFD | $H$ (y$^{-1}$) | Period over which hazard expected (y) |
|---|---|---|---|
| 0.1 | 0.01 | 0.000952 | 1051 |
| 1 | 0.01 | 0.006321 | 158 |
| 10 | 0.01 | 0.010000 | 100 |

It can be seen that the *PFD* increases significantly with an increase in the proof testing period, as does the hazard rate. Increasing the demand rate increases the hazard rate significantly.

**Example 15.3** A pressure relief valve has a failure rate of $0.01 \text{ y}^{-1}$. Historical records indicate a demand rate on the system of $0.5 \text{ y}^{-1}$.

a) For a hazard rate of $1 \times 10^{-3} \text{ y}^{-1}$, calculate the required interval of proof testing.
b) In this case, the minimum acceptable proof testing interval is considered to be 6 months. If the calculated interval is less than 6 months, what might be changed to give an acceptable interval?

**Solution**

a) The hazard rate is given by:

$$H = \lambda\left[1 - \exp\left(-\frac{D\tau_P}{2}\right)\right]$$

Rearranging:

$$\tau_P = -\frac{2}{D}\ln\left[1 - \frac{H}{\lambda}\right]$$

$$= -\frac{2}{0.5}\ln\left[1 - \frac{1 \times 10^{-3}}{0.01}\right]$$

$$= 0.42 \text{ y}$$

b) The proof testing interval is less than 6 months. To increase the interval for a fixed hazard rate, then changes must be made to either the failure rate $\lambda$ or the demand rate $D$. Measures that could be taken include:

- Replace the current relief valve with one that has a lower failure rate.
- Decrease the demand rate through changes in operating practice.
- Change the control system in the process to safeguard against increases in pressure.
- Introduce safety instrumented systems to shut off pressure-producing equipment in the event of overpressure being created.

**Example 15.4** A trip system with a failure rate of $0.5 \text{ y}^{-1}$ has a demand rate of $100 \text{ y}^{-1}$ and an interval of proof testing of 1 month. Calculate the probability of failure on demand and the hazard rate and assess the viability of the system (Kletz 1999).

**Solution**

For $\lambda = 0.5 \text{ y}^{-1}$ and $\tau_P = 1/12$ y the *PFD* is given by Eq. (15.12):

$$PFD = \frac{\lambda\tau_P}{2} = \frac{0.5 \times 1/12}{2} = 0.0208$$

For $D = 100 \text{ y}^{-1}$ the hazard rate is given by Eq. (15.5):

$$H = \lambda\left[1 - \exp\left(-\frac{D\tau_P}{2}\right)\right]$$

$$= 0.5\left[1 - \exp\left(-\frac{100 \times 1/12}{2}\right)\right]$$

$$= 0.492 \text{ y}^{-1}$$

The hazard rate is almost the same as the failure rate. The *PFD* measures the unavailability or fractional dead time (i.e. the fraction of total time that the protective device is in a failed state). In this case the trip is unavailable for 7.6 days per year. There is demand every 3.65 days and there will be a high probability of a demand in the dead period. Proof testing in this case is not meaningful as failures will most likely be followed by a demand before the next proof test is due. The fundamental problem is that the demand rate is extremely high and the root causes of this need to be investigated.

**Example 15.5**   A relief valve when used in a clean environment is expected to have a failure rate of 0.01 $y^{-1}$. However, it is to be used in an environment that is fouling and corrosive, leading to an expected failure rate of 0.5 $y^{-1}$. To overcome this problem, a bursting disc with a failure rate of 0.2 $y^{-1}$ is to be installed before the relief valve. This is expected to decrease the failure rate of the relief valve to that for clean conditions of 0.01 $y^{-1}$. The demand rate on the system from historical records is expected to be 0.1 $y^{-1}$.

a) Calculate the failure rate of the combined bursting disc and relief valve.

b) For an acceptable hazard rate of 0.01 $y^{-1}$, calculate the required period for proof testing.

**Solution**

a) The reliability of a series system is given by the product rule (Chapter 8):

$$R_{SYS}(t) = R_{DISC}(t)R_{PRV}(t)$$

For exponential distributions:

$$R_{SYS}(t) = \exp(-\lambda_{SYS}t) = \exp(-\lambda_{DISC}t) \times \exp(-\lambda_{PRV}t)$$
$$= \exp[-(\lambda_{DISC} + \lambda_{PRV})t]$$

Thus:

$$\lambda_{SYS} = \lambda_{DISC} + \lambda_{PRV}$$
$$= 0.2 + 0.01$$
$$= 0.21 \ y^{-1}$$

b)

$$H = \lambda_{SYS}\left[1 - \exp\left(-\frac{D\tau_P}{2}\right)\right]$$

Rearranging:

$$\tau_P = -\frac{2}{D}\ln\left[1 - \frac{H}{\lambda_{SYS}}\right]$$
$$= -\frac{2}{0.1}\ln\left[1 - \frac{0.01}{0.21}\right]$$
$$= 0.98 \ y$$

# 15.6  Location of Relief Devices

To locate relief devices requires potential overpressure scenarios to be identified. External fire, heat transfer causing overpressure through fluid expansion or liquid vaporization, interstream leakage resulting from equipment failure, control system failure, and many other scenarios can lead to overpressure events. After studying such scenarios, relief devices are commonly found at the following locations:

● All pressure vessels for the containment of pressure require a pressure relief device (ASME 2021). A pressure vessel is a vessel in which the pressure is obtained from an indirect source (e.g. a pump or compressor) or by the application of heat from an indirect source or a direct source, with a MAWP exceeding 1 barg for the ASME code (ASME 2021) or a MAWP exceeding 0.5 barg for the European Union code (EU Directive PED 2014/68/EU 2014). The relief device should be connected as directly and close as possible to the vessel being protected to minimize the pressure drop between the vessel and the pressure relief device. However, a pressure vessel does not require a pressure relief device if it is protected from overpressure by the system design (ASME 2021). A pressure vessel does not require a pressure relief device if the pressure is self-limiting (e.g., the maximum discharge pressure of a pump or compressor) and this pressure is less than or equal to the MAWP of the vessel at the working temperature (ASME 2021).

● At the discharge of positive displacement pumps and positive displacement compressors before the next block valve.

● At the discharge of centrifugal pumps (e.g., gear pumps) and centrifugal compressors (e.g., reciprocating compressors) before the next block valve if the maximum head delivered exceeds the pressure rating of the outlet pipe. It is important to consider not just the maximum pressure added by the pump or compressor in isolation, but consider the total pressure of the combined inlet pressure and the maximum pressure added by the pump or compressor.

● On the low-pressure side of heat exchangers before the next block valve to protect against tube failure. In the worst case there might be a sudden catastrophic failure of the tube. Although uncommon, sudden catastrophic tube failure causes a transient pressure peak much higher than the steady-state pressure and might necessitate the use of a bursting disc, rather than a relief valve, if the relief valve cannot relieve the overpressure quickly enough.

● On the outlet of the cold side of heat exchangers before the next block valve to protect against heating causing expansion of fluid or vaporization of liquid that is blocked in by valves or blockages, resulting in overpressure.

● On the outlet of furnace coils before the next block valve to protect against heating causing expansion of fluid or vaporization of liquid that is blocked in by valves or blockages, resulting in overpressure.

● Steam side of steam heaters, vessel steam jackets and vessel heating coils to prevent overpressure if there is a risk of failure of upstream steam pressure control.

● Outlet of turbines and expanders. This is to safeguard against the case when the turbine/expander is shut down and the turbine/expander outlet approaches the inlet pressure. If the inlet pressure of the turbine or expander exceeds the pressure rating of the outlet piping and equipment, then pressure relief will be required.

● Piping sections and vessels that can be blocked in by valves or blockages and exposed to external heat from external fire, tracing or solar gain.

● Storage vessels without a vent to atmosphere. This will be considered in more detail in the next chapter.

**15**

This is not a comprehensive list. Studying overpressure scenarios might lead to the requirement for pressure relief for other scenarios than listed above. However, all of the locations in the above list might not need pressure relief, except those mandated by the applicable standards. Individual locations must be studied in the context of the design and operation of the rest of the process for their pressure relief requirements. Formal study is conducted by a Hazard and Operability Study (HAZOP) followed by a Layer of Protection Analysis (LOPA) to determine the required layers of protection.

## 15.7  P&ID Piping and Pressure Relief – Summary

Piping and Instrumentation Diagrams (P&IDs) take the conceptual aspects of the Process Flow Diagram (PFD) and expands to include:

- the entire process from raw materials in to products out;
- multiple items of equipment shown in the PFD as a single item;
- all material storage linked directly to the process (centralized storage facilities will have their own P&IDs);
- utility and effluent treatment equipment directly linked to the process (centralized utility and effluent treatment systems will have their own P&IDs);
- steam traps and steam condensate drums;
- pumps and compressors with drivers;
- standby equipment;
- all process and utility pipes, including vents, drains, bypass lines, sampling points, tracing and slopes required for both operation and maintenance;
- pipe dimensions and specifications;
- all valves or isolations for operation or maintenance and their type;
- control valves, with actuator type and failure mode;
- instrumentation and control, including location of control devices (algorithms);
- safety instrumented functions;
- emergency shut-down valves, including the type of actuator and the failure mode as part of the safety instrumented functions;
- signal types between control functions and safety functions;
- pressure relief valves, bursting discs, and venting and blow-down systems;
- piping connections at the edges of sheets to neighboring P&ID sheets;
- preliminary pipe sizing can be carried out on the basis of an assumed velocity or pressure drop per unit length. Valve types need to be chosen and specified for a P&ID. Valves need to be locked or car sealed in some cases. Positive isolation for maintenance might require the use of slip plates or spectacle plates.

Any system that has the potential to have its pressure increased beyond safe levels requires safety devices. Pressure relief can be from pressure relief valves or bursting discs. Pressure relief valves are preferred, but bursting discs are sometimes necessary. Various arrangements of relief devices can be developed. Once an emergency discharge has been created from pressure relief, it must be safely dealt with. To assess the reliability of a safety system requires the failure to respond to a hazardous situation over a time horizon to be considered.

# Exercises

1. A pressure relief valve has a failure rate of 0.01 $y^{-1}$. Historical records indicate a demand rate on the system of 0.1 $y^{-1}$. Calculate:

   a) The hazard rate for an assumed interval of proof testing of 2 years.

   b) The interval of proof testing necessary to achieve a probability of failure on demand *PFD* of $2.5 \times 10^{-3}$. Use the approximate formula.

2. Figure 15.19 shows a relief valve arrangement with two valves operating in parallel, both open to the process.

   a) Why are relief valves preferred to bursting discs?

   b) Why do the relief valve settings need to be different for both valves?

   c) What criterion normally determines the set pressure of the primary relief valve in such parallel arrangements?

   d) A single relief valve with a failure rate of 0.01 failures per year is mounted on the top of the pressure vessel to protect against overpressure. The acceptable hazard rate is 0.001 $y^{-1}$. If the demand rate is expected to be once every 3 years, calculate the required proof testing period in years.

   e) If the inspection rate in Part (d) is considered to be too frequent, what options can allow a longer proof testing period, given that it is necessary to use a relief valve rather than a bursting disc and mechanical design of the vessel on which it is mounted restricts pressure relief to be with a single relief valve.

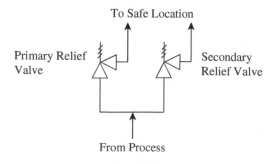

## Figure 15.19

Arrangement of two relief valves in parallel.

**3.** Figure 15.20 shows a vertical thermosiphon reboiler at the base of a distillation column. Vaporization of the column bottoms takes place on the inside of the tubes. Heating is provided by steam condensing on the shell-side. Steam is the high-pressure side of the heat exchanger. Two relief valves have been included in the design. Briefly explain:

**a)** Why a relief valve has been placed on the tube-side through which the process fluid is flowing and being vaporized.

**b)** Why a relief valve has been placed on the shell-side through which the steam is flowing and providing the heat of vaporization.

**4.** Figure 15.21 shows a batch polymerization reactor.

**a)** The reactor is protected from overpressure using a pressure relief valve mounted on top of a bursting disc. Explain two advantages and two disadvantages of such arrangements.

**b)** A pressure indicator between the bursting disc and the pressure relief valve checks for any leakage from the bursting disc. Briefly explain what will happen if there is a small *undetected* leak past the bursting disc, e.g., from a pin hole on the bursting disc, followed by an overpressure in the vessel.

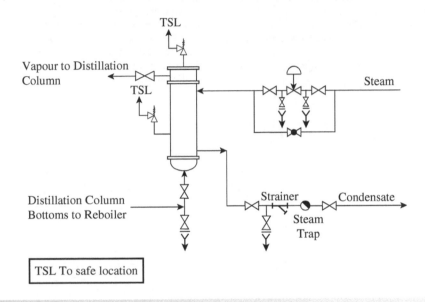

**Figure 15.20**

A vertical thermosyphon reboiler.

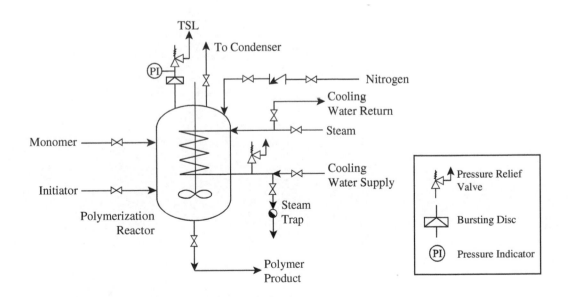

**Figure 15.21**

A batch polymerization process.

**c)** Discuss briefly the appropriateness or otherwise of using an arrangement of a combination of a bursting disc and pressure relief valve in the polymerization process in Figure 15.21.

**d)** Suggest alternative arrangements for the pressure relief on the polymerization reactor.

# References

American National Standard (2022). ANSI/ISA-5.1-2022. *Instrumentation Symbols and Identification*.

ASME Boiler and Pressure Vessel Code (2021). *Section VIII Division 1*.

BS EN ISO 10628-1 (2015a). *Diagrams for the chemical and petrochemical industry. Part 1: Specification of diagrams*.

BS EN ISO 10628-2 (2012). *Diagrams for the chemical and petrochemical industry. Part 2: Graphical symbols*.

BS ISO 15519-2 (2015b). *Specifications for diagrams for process industry. Part 2: Measurement and control*.

Crowl, D.A. and Tipler, S.A. (2013). Sizing pressure-relief devices. *Chemical Engineering Progress* 109 (10): 68–76.

DIN 19227-1 (1993). *Graphical Symbols and Identifying Letters for Process Measurement and Control Functions*. Symbols for Basic Functions.

DIN 2429-2 (1988). *Symbolic Representation of Pipework Components on Engineering Drawings; Functional Representation*.

EU Directive PED 2014/68/EU (2014) *Pressure Equipment*.

Health and Safety Executive HSG 253 (2006). *The Safe Isolation of Plant and Equipment*.

Hellemans, M. (2009). *The Safety Relief Valve Handbook*. Butterworth-Heinemann Elsevier Ltd.

Kletz, T. (1999). *Hazop and Hazan*, 4e. IChemE.

Mannan, S. (2012). *Lees Loss Prevention in the Process Industries*, 4e. Butterworth-Heinemann.

Process Industry Practices PIP PIC001 (2008). *Piping and Instrumentation Diagram Documentation Criteria*.

Smith, R. (2016). *Chemical Process Design and Integration*, 2e. Wiley.

Smith, D.J. (2017). *Reliability, Maintainability and Risk*, 9e. Butterworth-Heinemann Elsevier.

# Chapter 16

# Piping and Instrumentation Diagrams – Process Operations

## 16.1 Pumps

Many different types of pump are used in chemical processes. These can be generally classified as centrifugal and positive displacement types. In the chemical process industry as a whole, centrifugal pumps are the most common type used. Figure 16.1a shows examples of symbols used for the representation of centrifugal pumps in piping and instrumentation diagrams (P&IDs). Also shown in Figure 16.1 are examples of symbols used for the representation of positive displacement pumps. Figure 16.1b shows examples of symbols for rotary gear pumps, Figure 16.1c examples of symbols for screw pumps, and Figure 16.1d examples of symbols for piston pumps. Whilst the pump types represented in Figure 16.1 are the most common, other types are used for specialist applications.

Start by considering the development of P&IDs for centrifugal pumps. Before looking into the detail of a P&ID arrangement for centrifugal pumps, it is good practice to consider a preliminary layout arrangement. Figure 16.2a shows an arrangement in which a centrifugal pump is pumping liquid from a lower level to a higher level. This arrangement will not function for most centrifugal pumps. At start-up the inlet piping and the pump casing will be empty of liquid and will contain only gases. Until the suction pipe and pump casing are full of liquid it will fail to perform its required function. Standard centrifugal pumps are designed to pump liquids and not gases. Therefore the inlet piping and pump casing will remain empty and the pump will not function. There are different ways to solve this problem. If the pump and tanks are kept in the same location and elevation, then some feature needs to be added to *prime* the pump and fill casing with liquid for start-up. This could be an auxiliary tank of liquid from which liquid could be pumped using a *priming pump* or flow by gravity of liquid from a tank at a higher elevation than the pump into the centrifugal pump, purely for start-up. Another option would be to use a *self-priming* centrifugal pump. Self-priming pumps generally work by retaining liquid in the pump casing that is mixed with the gases on start-up of the pump to form a two-phase mixture that can be pumped, albeit inefficiently. The liquid and gas are then separated at the exit of the casing and the liquid returned to the pump casing. As the self-priming process continues, more gas is drawn into the pump casing and mixed with liquid. This process continues until all of the gas in the inlet piping and the pump casing is expelled. Different mechanisms can be used to facilitate this self-priming. Self-priming pumps are less efficient, and more expensive than standard centrifugal pumps, and are not commonly used. Rather than use a self-priming pump, other options could involve using a pump design in which the centrifugal casing is permanently submerged below the level of liquid in the feed tank.

Figure 16.2b shows an alternative layout. In this arrangement, the centrifugal pump is located at an elevation below the level of liquid in the feed tank. For start-up the centrifugal pump is primed simply by allowing liquid to flow under gravity from the feed tank to expel gases from the suction pipe and the pump casing. However, there must be a route to allow the gases to be vented. The arrangement in Figure 16.2b features a discharge tank at a higher elevation than the feed tank. Figure 16.2c shows the arrangement from Figure 16.2b after liquid has been pumped to a higher elevation. There is now a problem if the pump is stopped. There is nothing to stop the liquid from the discharge tank flowing back through the centrifugal pump to the feed tank under gravity if the pump is stopped. This could result in the feed tank overflowing.

Once some basic decisions have been made about the layout of the centrifugal pump arrangement, then details of the P&ID can start to be developed. Figure 16.3 shows a typical P&ID arrangement for standard centrifugal pumps. It should be recalled that the P&ID needs to show all the instrumentation and control, and all valves, including those for both operation and maintenance. Figure 16.3 shows inlet and outlet block valves to isolate the pump for maintenance. A non-return valve prevents backflow from discharge to suction through the pump. Drain valves are used to drain liquid from the pump and pipework prior to maintenance. This includes a drain valve between the non-return valve and the outlet

*Process Plant Design*, First Edition. Robin Smith.
© 2024 John Wiley & Sons Ltd. Published 2024 by John Wiley & Sons Ltd.
Companion website: www.wiley.com/go/processplantdesign

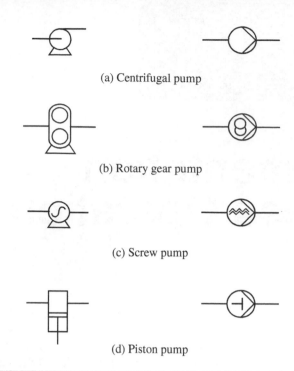

(a) Centrifugal pump

(b) Rotary gear pump

(c) Screw pump

(d) Piston pump

**Figure 16.1**

Common P&ID symbols for pumps.

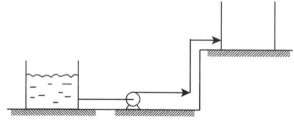

(a) Pump without priming.

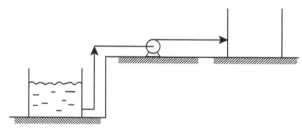

(b) Pump priming from the feed tank.

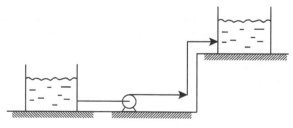

(c) Pump layout susceptable to backflow.

**Figure 16.2**

Layout arrangements for centrifugal pumps.

block valve, as liquid might be trapped under pressure when the pump is shut down. The liquid drained from the system might flow to a *closed drain* (not open to the atmosphere) if the liquid is hazardous and can give off hazardous vapor. Alternatively, if the liquid is not hazardous it might flow to an *open drain* (open to the atmosphere). A *vent* or *priming* valve is required to vent off gases from the pump and pipework for start-up. The venting procedure might involve the venting of gases that are hazardous. Thus, venting should be to a safe location. It is also common practice to include a pressure measurements on the pump suction and discharge. The discharge pressure measurement is particularly important, allowing the mechanical health of the pump to be monitored and allow hydraulic problems to be identified, such as blockages. Pressure measurement on the suction side is also useful to check for issues such as net positive suction head. The pressure measurement might be a local pressure indicator (pressure gauge) or a pressure transmitter sending a pressure measurement to the distributed control system (DCS), or both. A local pressure indicator allows process operators to tell at a glance whether the pump is working properly on not. Both are included in Figure 16.3.

A relief valve will be required at the pump discharge if the shut-in discharge pressure for the pump exceeds the pressure rating for the piping or that of any downstream equipment, as shown in Figure 16.4. For a centrifugal pump the maximum pressure delivered is when the discharge is shut in and there is no flow. It is important to consider not just the pressure added by the pump in isolation, but also the total pressure of the combined pump inlet pressure and the pressure added by the pump. The relief valve is located before the non-return valve and the discharge block valve. Valves should not be located in the relief path in case a block valve is closed inadvertently or there is a failure of a non-return valve. Most centrifugal pumps do not require a relief valve on the exit, as the discharge piping is normally designed to be capable of withstanding the shut-in head capable of being delivered by the pump.

Figure 16.5 shows the same basic centrifugal pump arrangement with flowrate control added. In this case, the control valve can be isolated for maintenance and features a bypass capable of operating the system for short periods under manual control. Drain valves have been included both upstream and downstream of the control valve. It might be possible to feature a single drain valve only if the control valve is fail-open.

Suspended particles can cause obstruction of the pump impellor, as well as wear on the impellor, shaft, bearings, and casing. A *strainer* provides a means of removing large solids from the flowing liquid using a removable perforated metal or mesh straining element. The strainer elements are designed with different mesh sizes to remove particles of a specified size. The strainer element retains all particles that are equal to, or larger than, the size of the strainer element openings. Figure 16.6 shows a strainer added to the inlet of the pump. The three most common types of strainer are shown in Figure 16.7. *Basket* or *simplex* strainers are shown in Figure 16.7a, like the one shown at the pump inlet in Figure 16.6. This type of strainer is a closed vessel with a removable and cleanable strainer element. The major disadvantage of this type of strainer is that the pump needs to be shut down for cleaning of the strainer. Figure 16.7b shows another type of basket strainer, termed a *duplex strainer*, that features two screening elements, each in their own vessel with one on-line and the other off-line. The flow can be switched between the straining elements using

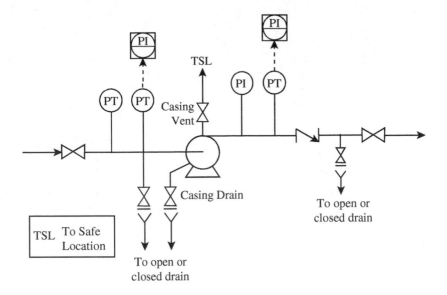

## Figure 16.3

Example of a typical P&ID arrangement for centrifugal pumps.

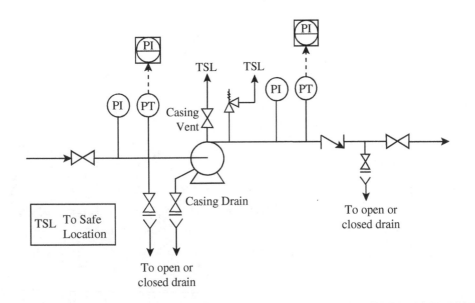

## Figure 16.4

A relief valve might be required for the centrifugal pumps discharge.

a three-way valve mechanism. This means that the pump does not need to be shut down for cleaning of the strainer. One basket can be cleaned while the other is still in operation. These are normally used where larger amounts of solids need to be removed. A third type of strainer is shown in Figure 16.7c, known as a Y-strainer. The screening element is contained in the Y-branch of the strainer. These are normally used in applications where there is only a small amount of solids to be separated.

It is very common to have a cold standby for centrifugal pumps, as shown in Figure 16.8. In the example in Figure 16.8

each pump is protected by a basket strainer, and a non-return valve. Relief valves are not required in this case. Operations can continue in the event of failure of the on-line pump by opening the isolations for the standby pump, starting the standby pump, and isolating the failed pump. Maintenance can then be carried out on the off-line pump whilst maintaining operation.

It should be noted that it might not be necessary to include all of the features shown in Figures 16.6 and 16.8. For example, the strainers might not be necessary and some of the drains and vents might not be necessary. Drainage and venting might be possible

**16**

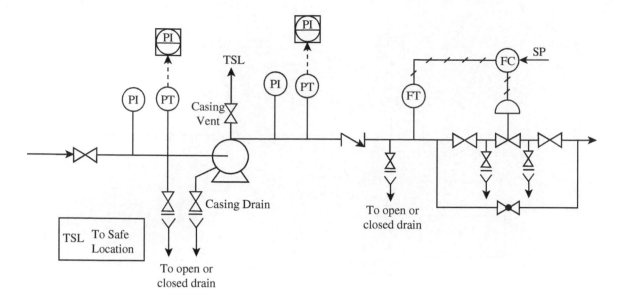

**Figure 16.5**

A centrifugal pump with flowrate control.

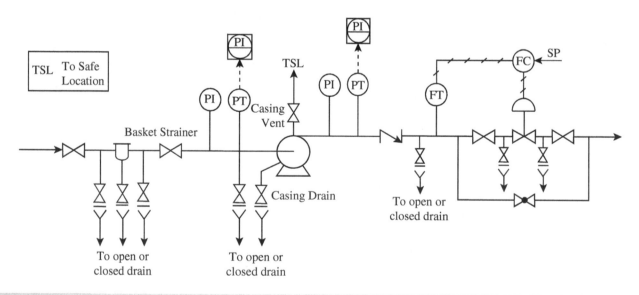

**Figure 16.6**

A centrifugal pump with an inlet strainer.

via connected adjacent equipment. Each case must be examined according to the individual circumstances and in the context of the surrounding process.

In some circumstances it is necessary to maintain a minimum flow through the pump to avoid overheating of liquid if there is a closed outlet or to keep solids in suspension. Figure 16.9 shows an example of a pumping arrangement in which the flowrate is controlled by a recycle to avoid overheating of liquid, or to keep solids in suspension, or to avoid shear damage of the product from the pump, if there is a closed or significantly restricted outlet. Fail-safe for the control valve is fail-open. The example in Figure 16.9 also shows a duplex

strainer in the suction for the pump to sieve out the larger particles that would otherwise damage the pump or lead to blockage of the system.

Alternatively, continuous flow through the centrifugal pump can be maintained by a recycle with a restriction orifice, as shown in Figure 16.10. The restriction orifice is a plate with a hole smaller than the pipe diameter designed to achieve a restricted flow. This again maintains a minimum flowrate through the pump to avoid overheating of the liquid, or to keep solids in suspension, or to avoid shear damage of the product from the pump, if there is a closed or significantly restricted outlet. However, the continuous flow in the recycle creates a continuous energy waste. Rather

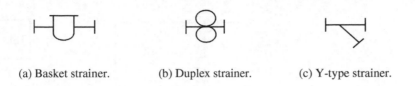

(a) Basket strainer.    (b) Duplex strainer.    (c) Y-type strainer.

## Figure 16.7

Pumps can require strainers to remove solids that might damage the pump.

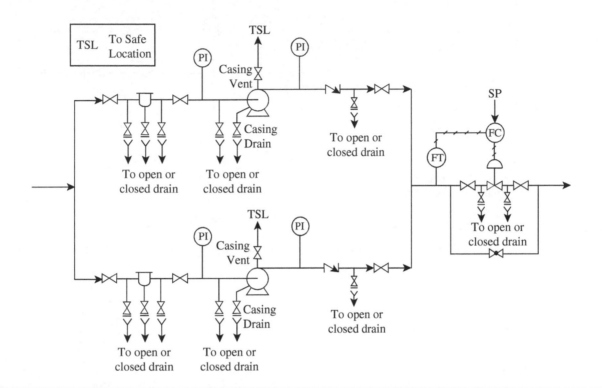

## Figure 16.8

An example of a preliminary P&ID for a centrifugal pump arrangement with standby.

than use a restriction orifice to maintain a recycle flow, an automatic spring-loaded three-way valve can be used to split the flow at the pump outlet and maintain a minimum recycle flowrate.

Figure 16.11 shows an example of a P&ID for a positive displacement pump. Unlike most centrifugal pump designs, positive displacement pumps are self-priming because of the close tolerances within the pump. The control arrangement in Figure 16.11 follows the *recycle* or *spill-back* control system from Figure 13.3b. Because the flow resistance of the outlet system does not change the flowrate through the pump, the flowrate is controlled by controlling the flowrate of the recycle back to the source tank. The control valve is designated to be fail-open to ensure that control system failure does not create a hazardous overpressure in the pump discharge in the event of the outlet system being blocked by closure of a valve or an accidental blockage. Figure 16.11 features a *high-pressure override* in which the outlet pressure is controlled in addition to the flowrate control. The

flowrate control and pressure control signals are passed to a *low selector* to choose the lower of the two signals. Since the valve is fail open, the lower of the two signals drives the valve to the safe state (Driedger 1996). A relief valve is also included in the example shown in Figure 16.11 to prevent overpressurization in the event of the high-pressure override not functioning. In principle, inclusion of *both* the high-pressure override *and* the relief valve is not necessary. The control system from Figure 13.3a with relief valve protection could have been used. However, if this is adopted, there might be a high demand rate on the relief valve, the implications of which were discussed in Chapter 15. It should be noted that there are no valves in the relief path, other than the locked open valves on the relief valve that have been included for maintenance. As pointed out in Chapter 13, if there is a high recycle flowrate in the schemes in Figure 16.11, then the energy of pumping input to the liquid will build up in the recycle, causing a significant temperature increase in the liquid. If this is the case,

**16**

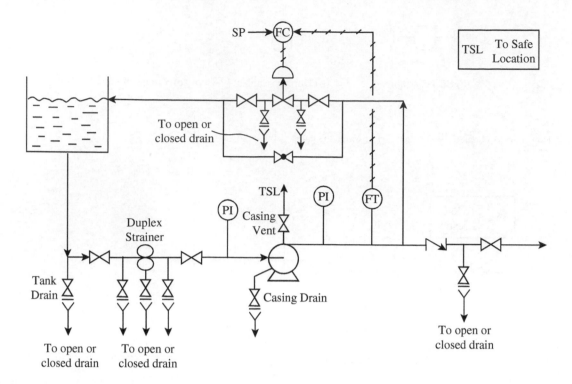

**Figure 16.9**

Recycle flowrate control maintains a minimum flowrate through the pump to avoid overheating of liquid if there is a closed outlet or to keep solids in suspension.

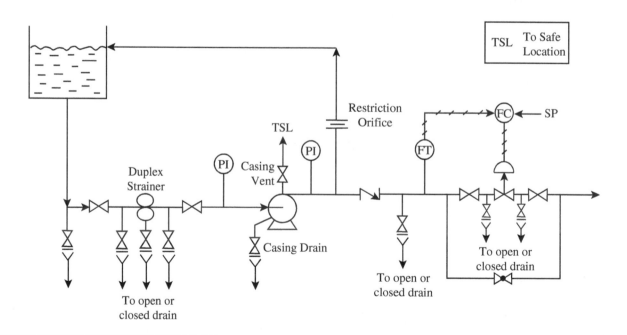

**Figure 16.10**

A continuous recycle using a restriction orifice maintains a minimum flowrate through the pump to avoid overheating of liquid if there is a closed outlet or to keep solids in suspension.

then a cooler (e.g. cooling water cooler) needs to be located in the recycle to remove the energy of pumping. Inlet and outlet block valves for maintenance are included in Figure 16.11. A non-return valve prevents back-flow when the pump is stopped. Even though there might be close tolerances within the positive displacement pump to prevent backflow through the pump, backflow can still occur through the recycle system. Drain valves are included to drain the system for maintenance, similar to the arrangements for centrifugal pumps.

An often favorable alternative to use of the recycle arrangement shown in Figure 16.11 is to use a positive displacement pump with a variable speed drive, e.g., a variable speed motor drive. In this

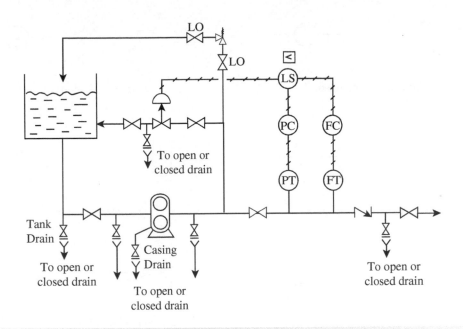

## Figure 16.11

An example of a preliminary P&ID arrangement for positive displacement pumps using recycle flowrate control.

case, the flowrate controller controls the speed of the pump, and therefore the volume delivered.

## 16.2 Compressors

Gas compressors can generally be classified as:

**a)** *Dynamic compressors* or *turbo-compressors* that transfer energy to the gas by dynamic means from a rotating impeller or blades. The kinetic energy of the gas is increased, which is then converted to pressure energy. The direction of gas flow in a *centrifugal compressor* is radial with respect to the axis of rotation. In an *axial flow compressor*, the gas flow is parallel to the axis of rotation. Figure 16.12a shows examples of symbols used for centrifugal compressors. A dynamic compressor with a low-pressure ratio is normally termed a fan or blower. Figure 16.12b shows examples of symbols used for fans or blowers.

**b)** *Positive displacement compressors* confine successive volumes of fluid within a closed space in which the pressure of the fluid is increased as the volume of the closed space is decreased. Figure 16.12c shows examples of symbols used for reciprocating compressors, Figure 16.12d examples of symbols used for rotary compressors, and Figure 16.12e examples of symbols used for liquid ring pumps.

**c)** *Ejectors* in which the kinetic energy of a high-velocity *working fluid* or *motive fluid* (steam or a gas) entrains and compresses a second fluid stream. The device has no moving parts. They are inefficient devices and use is normally for vacuum service, where small quantities of gas are handled. Figure 16.12f shows examples of symbols used for ejectors.

Consider first centrifugal compressors. As discussed in Chapter 13, for variable speed compressors for a given inlet pressure and flowrate through the compressor, the discharge pressure can be controlled by controlling the rotational speed. Alternatively, for a given inlet and discharge pressure, the flowrate can be controlled by controlling the rotational speed. Figure 16.13 shows an example of a preliminary P&ID for a centrifugal compressor based on the control system from Figure 13.11 using a steam turbine to drive the compressor with speed control. Figure 16.13 uses speed control of the steam turbine to control the outlet pressure. Recycling around the compressor is used for *anti-surge* control in the event of the flowrate approaching surge conditions. The anti-surge valve is a fast-opening and relatively slow closing control valve controlling the recycle. Other control arrangements have been discussed in Chapter 13. For the example in Figure 16.13:

- A cooler has been included on the compressor discharge to remove the heat of compression. Rather than use a cooling water cooler, an air cooler might have been used, or another cooling medium, as discussed in Chapter 13. If the compressor is part of a low-temperature process, where the compressor might be, for example, part of a refrigeration cycle, then rather than using cooling water or air cooler cooling, a refrigerated fluid might be required.

- The cooling water cooler has been shown to be under manual control in this case. Alternatively, automatic temperature control might have been included.

- A knockout drum in the compressor suction has been provided to give protection to the compressor against any liquid droplets that might be carried through to the compressor with the gas and otherwise damage the compressor. It is not intended for use as a gas–liquid separator to separate a significant amount of liquid from the gas before the compressor. If that is necessary, an additional separation drum should be used upstream of the compressor. The knockout drum in Figure 16.13 shows a mesh pad at the top of the drum. This allows any droplets

**16**

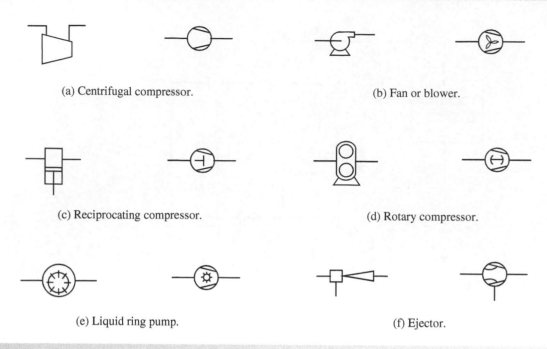

(a) Centrifugal compressor.

(b) Fan or blower.

(c) Reciprocating compressor.

(d) Rotary compressor.

(e) Liquid ring pump.

(f) Ejector.

**Figure 16.12**

Common P&ID symbols for compressors.

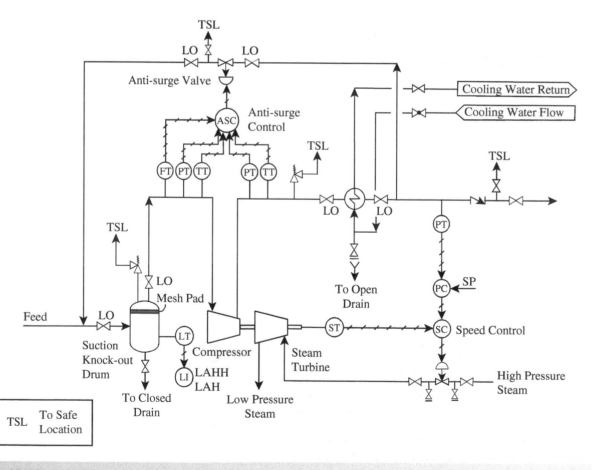

**Figure 16.13**

Example of a preliminary P&ID for a centrifugal compressor.

carried from the drum to coalescence on the surface of the mesh and fall to the base of the drum to increase the efficiency of separation of the settling drum. In other applications, if there is no danger of any droplets being carried through to the compressor, then, rather than using a knockout drum, a simple strainer similar to those used in centrifugal pumps might have been used on the compressor suction to avoid the danger of any solids being drawn into the compressor. If the gas is free of liquid droplets or solid particles, then it might not be necessary to provide protection.

- Isolation valves have been included around the recycle loop to allow maintenance on the various components after shutdown. However, the isolation valves in this case are shown as locked open to prevent the recycle anti-surge system from being disabled by inadvertent closure of the valves.

- No bypass is shown on the anti-surge valve to avoid the possible use of manual operation of the anti-surge recycle. In the event of failure of the anti-surge valve, the system should be shut down.

- A non-return valve has been included on the outlet of the compressor to prevent any danger of back-flow. Such non-return valves are different in design from those described for liquid service in Chapter 15.

- Maintenance after shut-down requires vents at various points to relieve the pressure and clear the equipment of hazardous gases. Venting to a safe location depends on the application.

For oil and gas applications this might be to a flare system. For other applications, the vent might be to a vent header system feeding a thermal oxidizer or high-level discharge to the environment. For less hazardous compression duties, such as air or nitrogen compression, venting directly to atmosphere should be acceptable.

- The level control in the knock-out drum is shown in the example to be under manual control, but automatic level control could have been incorporated.

- Pressure relief valves are shown on both the knockout drum and the compressor discharge. The relief valve on the compressor discharge might not be required if the maximum discharge pressure of the compressor is lower than the pressure rating of the piping and downstream equipment.

It should be emphasized that the arrangement in Figure 16.13 is only an example and the specific details will depend on the application and process context. Not all features shown in Figure 16.13 might be necessary. Axial compressors require similar control and P&ID arrangements.

For applications in oil and gas processing, emergency shutdown valves (ESD or ESDV) are often included in compressor systems. Figure 16.14 shows a preliminary P&ID for a centrifugal compressor with some elements of relief and blowdown shown for this case. Emergency shutdown valves have been included at the inlet and outlet of the compressor. These valves are part of the

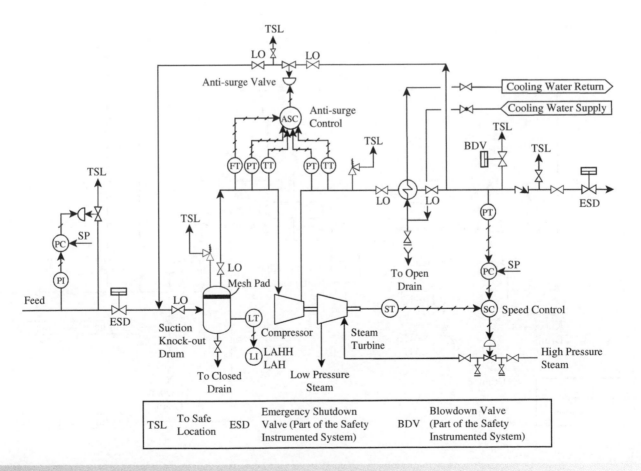

**Figure 16.14**

Example of a preliminary P&ID for a centrifugal compressor with emergency shutdown valves.

safety instrumented system and are not designed to control the flow, but to stop the flow of gases on detection of a dangerous event. The emergency shutdown valves are typically ball valve designs, are fail-safe, and should close in the event of any system failure. Either pneumatic or electric actuators can be used, but pneumatic actuators are generally preferred. Figure 16.14 also features a blowdown valve (BDV). This is used during an emergency shutdown to release the pressure, but at a controlled flowrate from a restriction orifice at the outlet of the blowdown valve. It should again be fail-safe and fail-open in the event of any system failure. During start-up, when the emergency shutdown valves are closed, there might be a very high pressure difference across the valve, requiring the actuator to produce a very high force to open the valve. Rather than provide a large actuator to deal with this condition, a bypass line can be installed around the emergency shutdown valve with its own small shut-down valve and a throttling valve across the main emergency shutdown valve to equalize the pressure across the main emergency shutdown valve and assist the valve opening for start-up.

Consider now positive displacement compressors. The most common type are reciprocating compressors. An example of a preliminary P&ID for a reciprocating compressor is shown in Figure 16.15:

- Pressure pulses are created during the compression cycle because suction and discharge valves in the compressor are only open for part of the cycle (Smith 2016). Pulsation

dampener vessels are used on the suction and discharge to reduce the pressure and flowrate pulsations. The volumetric pulsation entering the vessel is partly absorbed by the volume (capacitance) of the vessel.

- The flowrate control uses a recycle around the compressor. Other control arrangements are possible from within the machine, as discussed in Chapter 13.

- As with the centrifugal compressor example, a knockout drum has been included to provide protection for the compressor from a small amount of liquid droplets that might be carried through with the gas and damage the machine. For other applications, a simple strainer might be adequate. Alternatively, if the gas is free from liquid droplets and solid particles, then protection might not be required.

- A cooling water cooler has been included at the compressor discharge to remove the heat of compression.

- The flowrate of cooling water for the cooler and the level in the knockout drum are shown in the example to be under manual control. Both could also be on automatic control.

- Isolation valves in the recycle loop are shown locked open for safety reasons.

- In this case, a bypass around the recycle control valve has not been included for safety reasons, preventing manual control. In the event of failure of the recycle control valve, the compressor will need to be shut down.

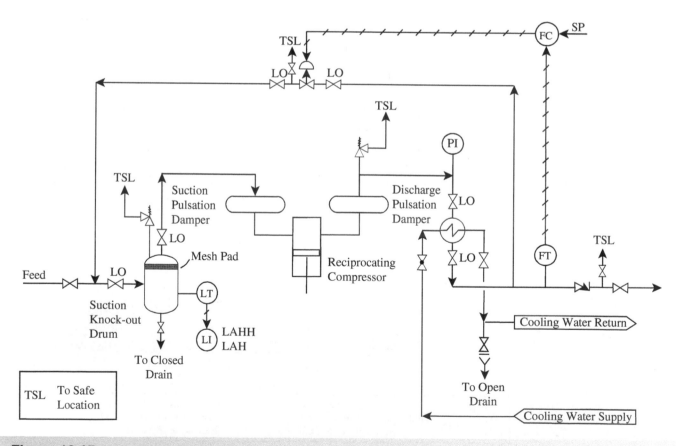

**Figure 16.15**

Example of a preliminary P&ID for a reciprocating compressor.

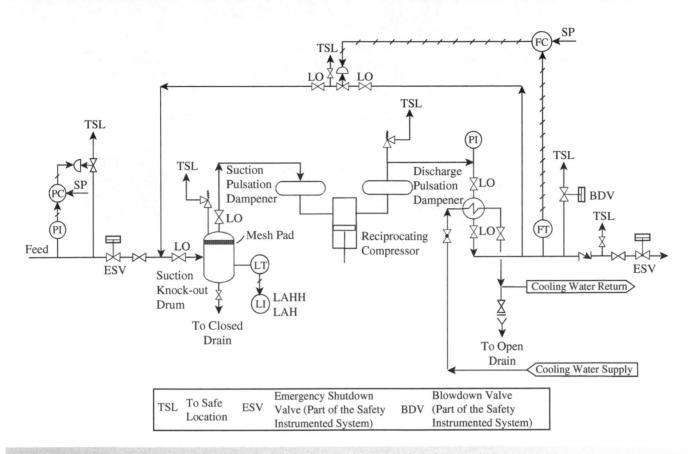

**Figure 16.16**

Example of a preliminary P&ID for a reciprocating compressor with emergency shut-down valves.

- Maintenance requires vents at various points to relieve the pressure and clear the equipment of hazardous gases. Venting to a safe location depends on the application.
- Pressure relief valves are shown on both the knockout drum and the compressor discharge.

For applications in oil and gas processing, as with centrifugal compressors, emergency shutdown valves are often included at the inlet and outlet of the system, controlled by the safety instrumented system, as shown in Figure 16.16. The emergency shutdown valves again require blowdown of the gas to a safe location in the event of their closure. Also, as before, the emergency shut-down valves might require pressure equalization for start-up.

It should again be emphasized that the examples shown in Figures 16.13 to 16.16 will not be typical of all compressor arrangements. The details will be specific to the application and the process context.

## 16.3   Heat Exchangers

Many different types of heat exchanger are used in the process industries. Figure 16.17a shows examples of symbols to represent generic heat exchangers in P&IDs. Overall, by far the most

commonly used type used is the shell-and-tube design. Within shell-and-tube designs there are a number of different flowrate arrangements for the tube side and the shell side (Smith 2016). Figure 16.17b shows an example of a symbol for a shell-and-tube design with a fixed tube sheet, two passes on the tube side and one pass on the shell side. Figure 16.17c shows an example of a symbol for a shell-and-tube design with the floating head tube sheet, with two passes on the tube side and one pass on the shell side. Figure 16.17d shows an example of a symbol for a U-tube heat exchanger. Figure 16.17e shows an example of a symbol for a kettle reboiler. Finally, Figure 16.17f shows an example of a symbol for a plate heat exchanger.

Figure 16.18 shows an example of a preliminary P&ID for a two-pass shell-and-tube heat exchanger with a fixed tube sheet exchanging heat between two liquid streams. The cold fluid is on the tube side and the hot fluid is on the shell side, where the high-pressure side is the tube side in this example:

- Inlet and outlet block valves for both streams have been included for maintenance isolation.
- Drain valves have been included to drain liquids for maintenance. In other cases, it might be possible to drain the system from other equipment around the heat exchanger.
- Vent valves have been included on both the tube side and shell side to allow gases and vapors to be vented to a safe location at

**16**

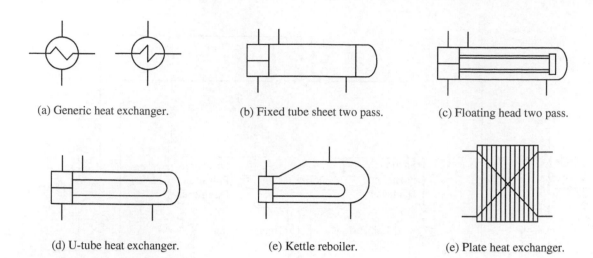

(a) Generic heat exchanger.  (b) Fixed tube sheet two pass.  (c) Floating head two pass.

(d) U-tube heat exchanger.  (e) Kettle reboiler.  (e) Plate heat exchanger.

**Figure 16.17**

Common P&ID symbols for heat exchangers.

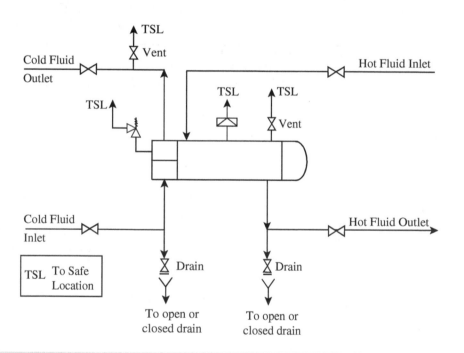

**Figure 16.18**

Example of a preliminary P&ID for a two-pass shell-and-tube heat exchanger for liquid streams.

start-up. Trapped vapors and gases can prevent the heat exchanger from functioning properly. In other cases, it may be possible to vent from other equipment around the heat exchanger.

- A pressure relief valve has been included on the cold side (tube side) to prevent excess pressure from expansion of shut-in fluid from the hot fluid. Although shut-in liquid expansion will cause an increase in pressure, the pressure relief valve required will only be small.

- Given that the tube side is the high-pressure side, pressure relief has been included on the shell side to prevent excess pressure from a tube failure. Tubes can fail in different ways. Although uncommon, the worst case is shear of the tube across the complete cross-section. Such sudden catastrophic tube failure causes a transient pressure peak much higher than the steady-state pressure. In such circumstances it is often necessary to use a bursting disc, rather than a relief valve, if the relief valve cannot relieve the overpressure quickly enough. If a

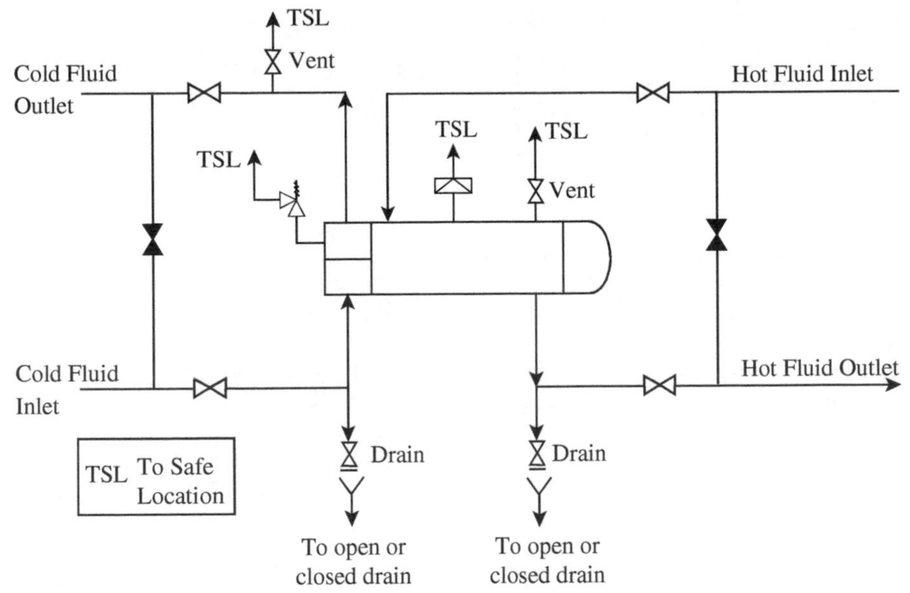

TSL

Cold Fluid Outlet

Vent

Hot Fluid Inlet

TSL

TSL      TSL

Vent

TSL

To Safe Location

Drain

Drain

Cold Fluid Inlet

Hot Fluid Outlet

To open or closed drain

To open or closed drain

## Figure 16.19

Example of a preliminary P&ID for a two-pass shell-and-tube heat exchanger for liquid streams with bypass for both streams for maintenance.

quantitative analysis of the peak pressure resulting from tube failure indicates a maximum pressure at or below the MAWP at the working temperature, then a pressure relief device may not be required. But a device might still be required for the fire case.

Figure 16.19 shows the same example from Figure 16.18, but with bypasses for both streams. This allows maintenance or cleaning of the heat exchanger without shutting down the plant in some circumstances. However, the process must be able to operate on a temporary basis without the heat exchanger (e.g., with heat exchangers in parallel).

It should be emphasized again that it might not be necessary to include all of the features in Figures 16.18 and 16.19. For example, it might be possible to vent and drain from other connected adjacent equipment.

If the heat exchanger is a steam heater, then provision must be made for the removal of the steam condensate from the heat exchanger. Figure 16.20 shows a generic steam heater with the steam outlet being controlled by a steam trap (see Chapter 4). The steam trap is protected by a strainer at its inlet to prevent solids (particularly products of corrosion) from entering the steam trap and preventing it functioning efficiently. The inlet and the outlet of the steam trap feature isolation valves for maintenance on shut-down of the heat exchanger. A drain valve is also included to allow relief of the internal pressure and draining of liquid prior to maintenance.

Although steam traps are one of the most common means for controlling the condensate outlet from steam heaters, a condensate drum can also be used, particularly for large condensate flows (see Chapter 13). Figure 16.21 shows the preliminary P&ID for a condensate drum on a steam heater. The condensate level control in the drum provides a liquid seal that prevents uncondensed steam

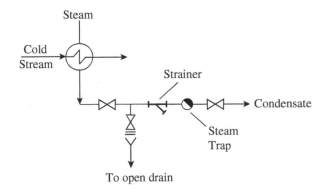

Steam

Cold Stream

Strainer

Condensate

Steam Trap

To open drain

## Figure 16.20

Example P&ID for a heater steam trap.

from passing directly to the condensate return system. A vapor balance line is provided to connect the inlet steam to the heater or the heater outlet steam vapor space (as shown in Figure 16.21) to the condensate drum. This allows the condensate to flow freely to the condensate drum under gravity.

## 16.4 Distillation

Distillation operations require vapor and liquid streams to be contacted countercurrently using either trays or packing in the distillation column. Figure 16.22a shows a distillation column fitted with *trays* or *plates*. Liquid reflux enters the first tray at the top of the column and flows across what is shown in Figure 16.22a as a perforated plate (*sieve* tray). Liquid is prevented from

**16**

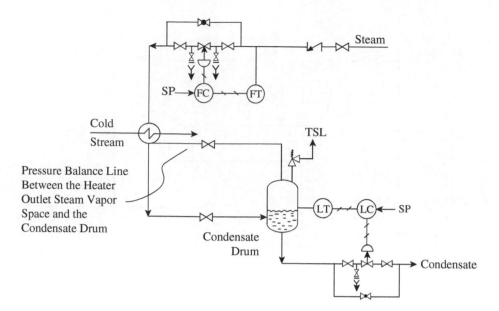

## Figure 16.21

Large steam flows can require a condensate drum for a steam heater.

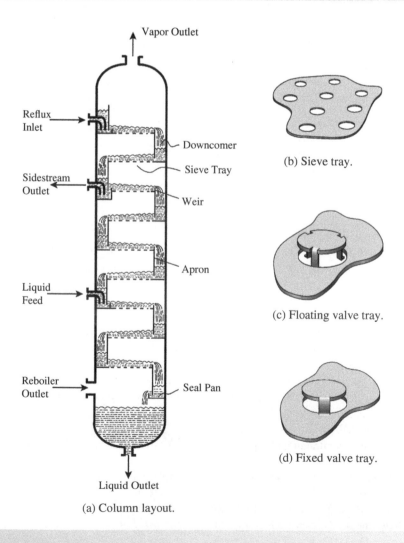

(a) Column layout.

(b) Sieve tray.

(c) Floating valve tray.

(d) Fixed valve tray.

## Figure 16.22

A distillation column fitted with trays. Source: Smith (2016), *Chemical Process Design and Integration*, John Wiley & Sons Ltd.

*weeping* through the holes in the tray by the upcoming vapor. The liquid from each tray flows over a *weir* and down a *downcomer* to the next tray, and so on. Different downcomer designs are possible (Smith 2016). The simplest and most commonly used type of tray is the sieve tray, as illustrated in Figure 16.22b. One major disadvantage of sieve trays is their lack of flexibility in operation. Figure 16.22c shows a *floating valve* tray. These use small movable metal flaps to adjust the size of the openings of the holes in the tray. The flap rises or falls in the hole as the vapor rate increases or decreases. The design of floating valve in Figure 16.22c allows movement guided and restrained by three metal legs. Other designs mount the metal flap in a cage to guide and restrain movement. The principal advantage of floating valve trays is that they improve the flexibility of operation. Figure 16.22d shows fixed valve trays. These use metal flaps punched from the tray at a fixed distance from the tray. The principal advantage of fixed valves is that the horizontal vapor velocity created by the valve promotes intense radial mixing on the tray, and also reduces fouling. If the weir loading is too great for a single-pass tray, then multiple tray passes can be used, as illustrated in Figure 16.23 (Smith 2016). Figure 16.23a shows a single-pass tray, Figure 16.23b a design with two passes, and Figure 16.23c a design with four passes.

Figure 16.24 shows a distillation column fitted with packing. The column is filled with a solid material that has a high surface area. The liquid flows across the surfaces of the packing and vapor flows upward through the voids in the packing, contracting the liquid on its way up the column. Many different designs of packing are available. Figure 16.24a illustrates a column arrangement that uses structured packing. Structured packing is most often manufactured from corrugated metal sheets with the corrugations inclined to the horizontal in alternating layers, as illustrated in Figure 16.24b. These pre-formed sheets of metal are joined together to produce preformed cylindrical packing slabs with inclined flow channels and built up in alternating slabs within the column. Designs can feature additional crimping of the metal sheets and holes in the sheets. The structured packing slabs are often fitted with *wall wipers* to redirect liquid away from the column wall and prevent bypassing of the liquid from the packing

down the column wall. An alternative design of packing is *random* or *dumped* packing. The random packing is pieces of pre-formed metal, plastic, or ceramic, which when dumped in the column produce a body with a high surface area. Figure 16.24c shows the simplest form of random packing, a Raschig Ring. Figure 16.24d shows a metal Pall Ring©, Figure 16.24e shows a metal Intallox Saddle©, and Figure 16.24f a Berl Saddle. Many other designs of random packing are available. The pieces of packing are dumped in beds onto a support plate. Packed columns require column internals for the distribution of liquid and vapor and the collection of liquid. A liquid distribution device is required for the reflux. A liquid collector is required at the base of the rectifying section. The liquid cascading down from the rectifying section then needs to be redistributed for the stripping section, along with the liquid feed. Vapor distribution devices are required at the base of the rectifying and stripping sections. Tall packed columns might be divided into a number of beds, rather than just two beds above and below the feed. Liquid collection and redistribution would be carried out between the beds.

Figure 16.25a shows the representation of some of the features for tray columns used in P&IDs. Generic trays are represented by a dashed line running across the column. Figure 16.25a also shows single-pass trays and two-pass trays. The chimney tray shown in Figure 16.25a is a device where no vapor and liquid contact is created, but are commonly used to take an intermediate stream from the distillation column. Rather than use a chimney tray for a side draw, a draw-off tray can be used, as shown in Figure 16.25a. This takes a liquid side draw from a sump, while the remaining liquid continues down the column.

Figure 16.25b shows the representation of some of the features for packed columns used in P&IDs. At the top of the column a metal wire mesh *demister pad* prevents carryover of liquid droplets with the vapor by causing the liquid droplets to coalesce on the surface wire mesh. Liquid reflux or liquid feed requires a liquid distribution system to distribute liquid over the entire diameter of the packed section. Many different designs of liquid distributer are available. Figure 16.25b shows a spray nozzle design. As discussed above, packed columns require devices for liquid collection and redistribution. For example, chimney trays

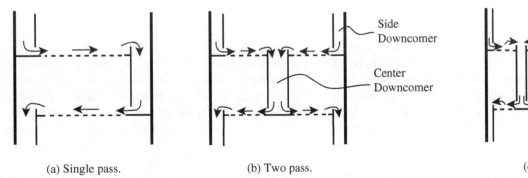

(a) Single pass.  (b) Two pass.  (c) Four pass.

Side Downcomer

Center Downcomer

**16**

**Figure 16.23**

Multi-pass tray layouts. Source: Smith (2016), *Chemical Process Design and Integration*, John Wiley & Sons Ltd.

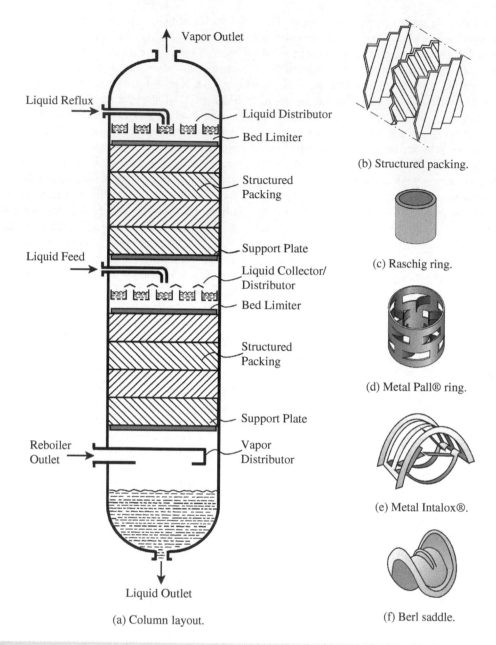

Vapor Outlet

Liquid Reflux

Liquid Distributor

Bed Limiter

Structured Packing

Support Plate

Liquid Feed

Liquid Collector/ Distributor

Bed Limiter

Structured Packing

Support Plate

Reboiler Outlet

Vapor Distributor

Liquid Outlet

(a) Column layout.

(b) Structured packing.

(c) Raschig ring.

(d) Metal Pall® ring.

(e) Metal Intalox®.

(f) Berl saddle.

**Figure 16.24**

A packed distillation column. Source: Smith (2016), *Chemical Process Design and Integration*, John Wiley & Sons Ltd.

can be used for liquid collection and redistribution. Also shown in Figure 16.25 are access ways for maintenance and inspection. Both tray and packed columns would feature access ways.

Figure 16.26 shows an example of a preliminary P&ID for a distillation column overhead. The pressure of the column is being controlled by the flowrate of non-condensibles from the reflux drum. The flowrate of cooling water to the condenser is under manual control in this example. Temperature control of the cooling water could have been used as an alternative. The reflux is under flowrate control. The level in the distillate drum is being controlled by the flow of distillate. Other control systems might use the flowrate of cooling water to control the column pressure. The centrifugal pumps for the reflux and distillate both have cold

standbys. Both the distillation column and the reflux drum are classified as pressure vessels and each has a pressure relief valve. The condenser has pressure relief valves on both the tube side and the shell side. The tube side is the cold side and the cooling water as the cold fluid is capable of being blocked in while heat is being added from the vapor, potentially creating excess pressure from expansion of the cooling water. The cooling water is the high-pressure side and might over-pressurize the shell in the event of a tube failure. As discussed in Section 16.3, tube failure can result in a transient pressure peak significantly higher than the steady-state pressure. Such a transient pressure might require a bursting disc, rather than a pressure relief valve, as a bursting disc will be much faster in operation. If the peak pressure is lower than the

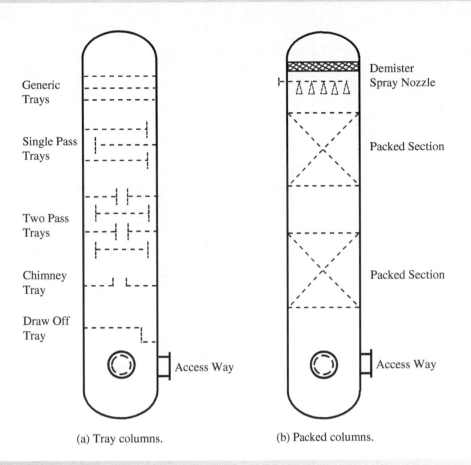

(a) Tray columns.       (b) Packed columns.

**Figure 16.25**

Representation of distillation columns.

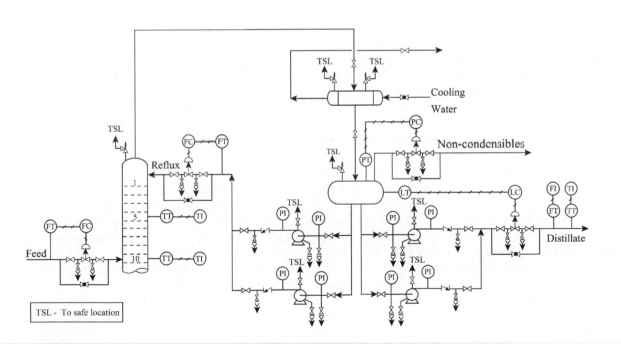

**Figure 16.26**

Example of a preliminary P&ID for a distillation column overheads.

16

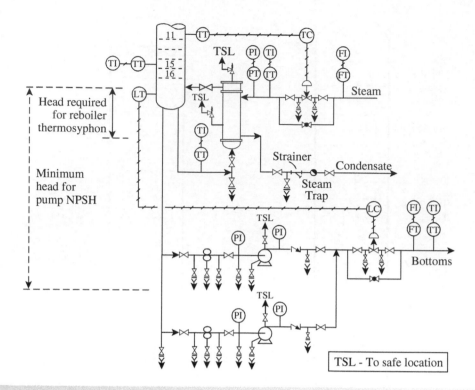

**Figure 16.27**

Example of a preliminary P&ID for a distillation column bottoms.

MAWP at the working temperature, then a relief device might not be required. However, the fire case might still require a safety device.

Figure 16.27 shows an example of a preliminary P&ID for a distillation column bottom. The design features a vertical thermosiphon reboiler under natural circulation. The physical layout will require the relative height of the column sump and the reboiler to be such that it will create the head necessary for the natural circulation system. The bottom product is pumped from the distillation sump with a centrifugal pump that has a cold standby. The physical layout will also demand that the elevation of the column sump above the bottoms pumps is high enough to maintain enough head for the *net positive suction head (NPSH)* requirement for the pumps (Smith 2016). The thermosiphon reboiler uses steam on the shell side of the reboiler and as the high-pressure side of the exchanger, with condensate flow from the reboiler controlled by a steam trap. A pressure relief valve has been included on the steam side of the reboiler. This will safeguard against any problems associated with control of the steam pressure to the reboiler. A pressure relief valve has also been included on the tube side of the reboiler. This will safeguard against expansion of liquid in the tubes blocked in by valves with the continued addition of heat from steam. Also, with the tubes being on the low-pressure side, failure of a tube could lead to a pressure peak on the tube side.

As noted previously, the preliminary P&IDs in Figures 16.26 and 16.27 will need to evolve through time as more detail of the

project becomes available. Additional features might need to be added and some of the features in Figures 16.26 and 16.27 might prove not to be necessary.

# 16.5   Liquid Storage

Liquid storage tanks are very often required for raw materials, products, and intermediate storage. Most large liquid storage tanks are cylindrical with flat bottoms and operate at atmospheric pressure, or close to atmospheric pressure. Atmospheric storage tanks are defined to operate at pressures between atmospheric and 6.9 kPa gauge. Figure 16.28 illustrates three designs of liquid storage tank. Figure 16.28a and b illustrate *fixed roof designs*. Figure 16.28c shows an *external floating roof* design. The floating roof is buoyant and rises and falls with the level of liquid in the tank. Vapor seals are necessary at the edges of the floating roof to prevent vapor losses. Such designs must feature drains to allow rainwater to run away from the surface of the roof. However, snowfall can present significant problems to such external floating roof designs. The design of cylindrical storage tanks will be considered again in more detail regarding their mechanical design in Chapter 19.

If a fixed roof tank is closed, then filling will pressurize the tank and emptying will create a vacuum. Either can damage the tank. If the tank is freely ventilated, emptying and filling storage tanks causes the tank to "breathe" to atmosphere. Figure 16.29 illustrates the problem. As the tank is emptied, as shown in

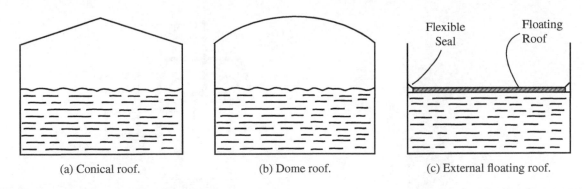

(a) Conical roof.    (b) Dome roof.    (c) External floating roof.

**Figure 16.28**

Different designs of cylindrical flat bottom storage tanks.

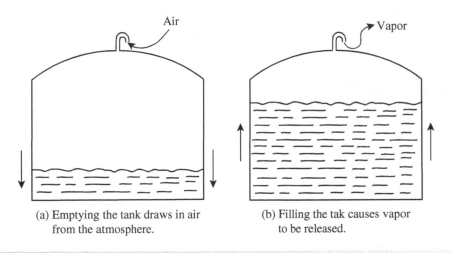

(a) Emptying the tank draws in air
from the atmosphere.

(b) Filling the tak causes vapor
to be released.

**Figure 16.29**

Emptying and filling storage tanks causes them to "breathe" to the atmosphere.

Figure 16.29a, air is drawn in through the vent to prevent a vacuum condition that could damage the tank. When the tank is filled, as shown in Figure 16.29b, air is expelled to the atmosphere through the vent to prevent a pressure condition that could also damage the tank. The expelled air will most likely carry with it vapor from the tank contents, possibly releasing unacceptable emissions (e.g., volatile organic compounds – VOCs). It is not just emptying and filling the tank that can cause the tank to breathe to atmosphere. Changes in ambient temperature (e.g., diurnal temperature cycles) can cause the contents of the tank to contract and expand, again causing the tank to breathe to atmosphere and possibly creating unacceptable emissions. In addition to the environmental consequences of the tank breathing to atmosphere, there is also the possibility of formation of a flammable mixture in the vapor space, especially when the tank is "breathing in" air. If a flammable mixture can form then an open vent should be protected by a flame arrestor. The flame arrestor prevents ignition sources from traveling into the vessel. The flame arrestor functions by absorbing the heat from an external flame front and cooling the burning mixture below its auto-ignition temperature. The

heat is absorbed by metal channels typically comprising crimped metal ribbons or wire mesh. The storage temperature should be kept below the flash point (for further criteria, see API 2014). It is often considered to be safe to use a freely ventilated fixed-roof tank if the material is not toxic and has a saturated liquid vapor pressure below 0.1 bara. However, ventilation to atmosphere is generally restricted.

Storage tanks for less hazardous materials (non-toxic, low vapor pressure) can be fitted with pressure-vacuum relief valves to reduce evaporation losses to atmosphere, as shown in Figure 16.30. Figure 16.30a shows a *pressure-vacuum relief valve* and *flame arrestor*. Figure 16.30b illustrates how the pressure-vacuum valve functions. As the tank is emptied, the vacuum *pallet* lifts to allow air to enter the tank and protect against a vacuum. At the same time, the overpressure pallet remains on its seat. As the tank is filled, the overpressure pallet lifts to allow air to exit to the atmosphere, while the vacuum pallet remains on its seat. The flame arrestor prevents ignition sources from traveling into the vessel. The pressure-vacuum relief valve reduces, but does not eliminate breathing to atmosphere.

**16**

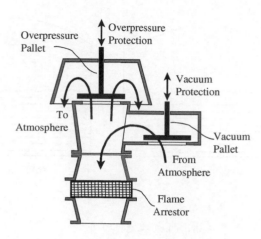

(a) Pressure-vacuum relief and flame arrestor.
(Reproduced with permission from Groth
Corporation)

(b) Pressure-vacuum relief protects the tank from
damge resulting from overpressure when the
tank is filled and vacuum when the tank is
emptied.

## Figure 16.30

Pressure-vacuum relief and flame arrestor for storage tanks.

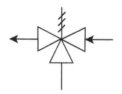

(a) Pressure-vacuum
relief valve.

(b) Vacuum relief valve.

## Figure 16.31

Symbols for vacuum-pressure relief devices.

Figure 16.31a shows an example of a P&ID symbol for a pressure-vacuum relief valve. This protects against both external and internal pressure. Figure 16.31b shows an example of a symbol for vacuum relief valve. This protects against external pressure only. Figure 16.32 shows an example of a P&ID representation of a storage tank fitted with a flame arrestor and a pressure-vacuum relief valve.

Storage tanks also need protection from overfilling (API 2020). If the stored liquid is less hazardous (non-toxic, low vapor pressure), a physical overflow system can be used. An example is illustrated in Figure 16.33. This shows an overflow line connected to a liquid seal. The liquid seal is necessary in this example to ensure that there is no connection of the tank vapor to the atmosphere under normal conditions. The liquid in the seal should not be hazardous in terms of safety or the environment, and must not be prone to freezing in cold conditions. The liquid level in the seal pot will fall and rise slightly during tank emptying and filling. It is important to ensure that the liquid seal is not allowed to run dry. Facilities should allow for refilling and topping up the liquid

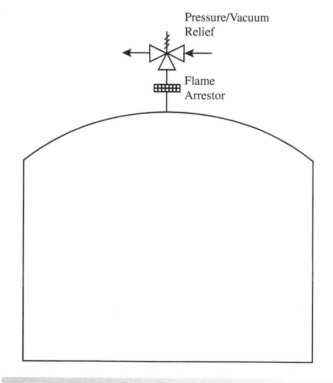

## Figure 16.32

An example of an atmospheric liquid storage tank fitted with a flame arrestor and pressure-vacuum relief to reduce vapor losses to atmosphere.

level and draining the seal pot. Protection against overfilling the tank is provided by a level indicator and high-level alarm in this example. Other physical overflow arrangements are possible depending on whether the overflow should feed from the top or

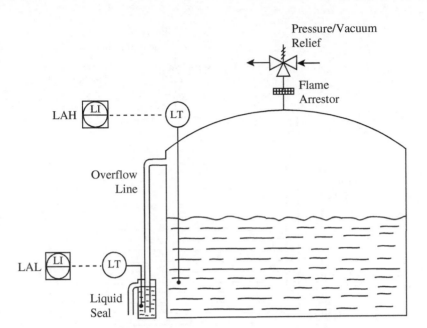

## Figure 16.33

Storage tanks can be protected from overfilling by an overflow line through a liquid seal.

bottom of the tank in the case of stratification of the stored liquid in the tank (Verma and Self 2014). Note that the liquid contents are shown in Figure 16.33 for the sake of illustration, but it is not common practice to show the liquid contents in P&IDs.

Protection from overfilling for hazardous material requires a safety instrumented system, as shown in Figure 16.34. An emergency shutdown valve shuts off the feed to the tank for a dangerously high level or loss of the air signal. The air supply to the emergency shutdown valve is supplied through a solenoid valve with a fail-safe system that shuts off the air supply to the emergency shutdown valve. Two- or three-way solenoid valves can be used. Figure 16.34 shows a three-way valve. When diverting

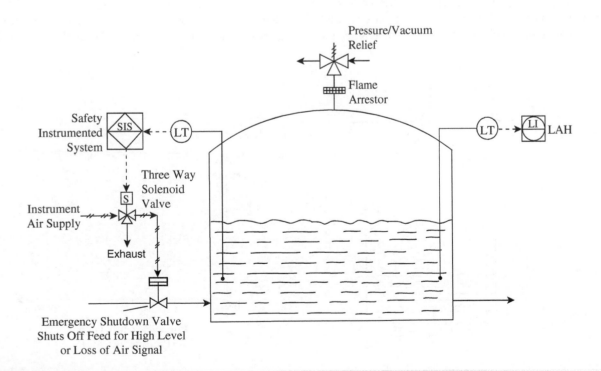

## Figure 16.34

An emergency shut-down valve can be used to prevent overfilling of the tank.

16

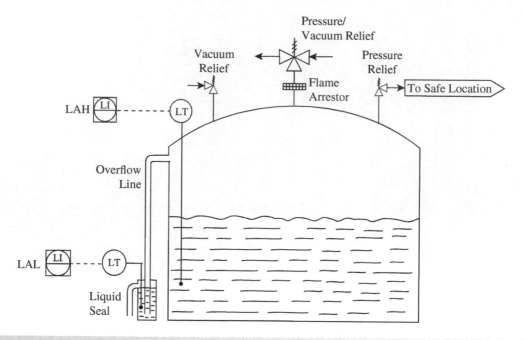

**Figure 16.35**

Separate vacuum and pressure relief valves might be required in addition to pressure-vacuum relief.

to atmosphere this vents all of the air stored in the actuating mechanism, ensuring rapid closure of the emergency shutdown valve. The safety instrumented system and control system should be independent. Hence a separate level indicator is shown for control.

Pressure-vacuum relief valves are designed for pressure control when emptying and filling the tank and thermal breathing. These flows are usually small in comparison to emergency relief for a fire. Relief valves should be fitted for fire relief when much

larger flows might be expected or for failure of the vacuum-pressure relief valve, as shown in Figure 16.35.

Storage of more hazardous liquids demands an inert gas blanket, typically nitrogen, above the liquid in the storage tank. The nitrogen blanket is most often controlled using split range control, which was introduced in Chapter 10 (Mannan 2012). Figure 16.36 shows a nitrogen blanket controlled by maintaining the pressure at a setpoint using split ratio control. When the pressure decreases,

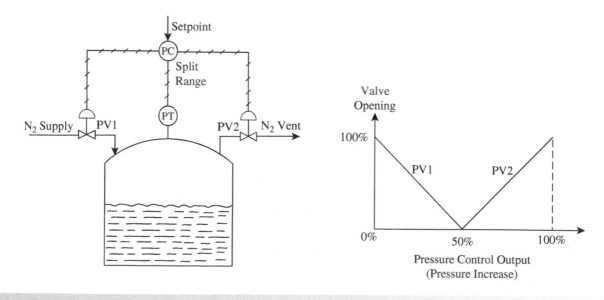

**Figure 16.36**

Control of a storage tank nitrogen blanket using slit range control.

nitrogen is introduced. When the pressure increases, nitrogen is discharged to vent. In Figure 16.36 PV1 needs to close in response to increasing pressure. When the pressure increases beyond the setpoint in the range 0–50%, PV1 is closed. Then, PV2 needs to open when the pressure increases beyond the setpoint. When the pressure increases beyond the setpoint in the range 50–100%, the controller opens PV2 with PV1 closed.

Figure 16.37 shows an example of a storage tank P&ID with a nitrogen blanket and safety instrumented system to protect against overfilling the tank.

Vaporization of liquid in storage tanks can be suppressed by the use of floating roofs. Such floating roofs might be open to the atmosphere with no fixed roof or internally with a fixed roof and vent. Figure 16.38 shows a storage tank with an internal

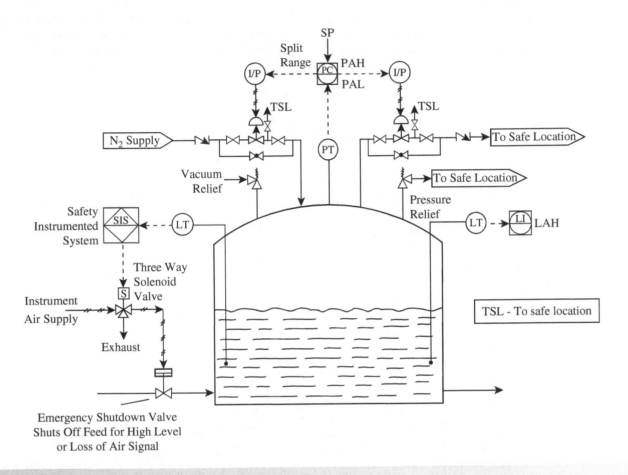

**Figure 16.37**

An inert gas purge is required for storage of more hazardous liquids.

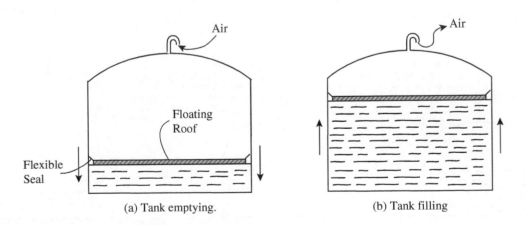

(a) Tank emptying.          (b) Tank filling

**Figure 16.38**

A floating roof can be used to prevent vapor losses.

floating roof vented directly to the atmosphere. The consequences of leaks from the seals at the edges of the floating roof need to be accommodated. For particularly hazardous material, an internal floating roof with a nitrogen purge can in principle be used. Figure 16.39 shows an example of a P&ID for a liquid storage tank with particularly hazardous contents using an internal floating roof, a nitrogen blanket, and a safety instrumented system to protect against overfilling the tank.

Gases such as anhydrous ammonia, propane, and liquefied natural gas can be stored as liquids using a low temperature and/or a high pressure. Refrigerant fluids for refrigeration systems might use the stored gas itself as the refrigerant, or another fluid in a closed cycle. There are generally three methods for storing liquefied gases. These are illustrated in Figure 13.40.

1) *Cylindrical tank with dome roof.* Figure 16.40a shows a dome roof tank for storing liquefied gases. These can be large in volume (typically up to 200 000 m$^3$) with a low pressure (typically 0.3 barg).

2) *Spherical tank.* Figure 16.40b shows a spherical storage tank. These tend to be used for medium volumes (typically up to 40 000 m$^3$). The mechanical construction offers uniform stress

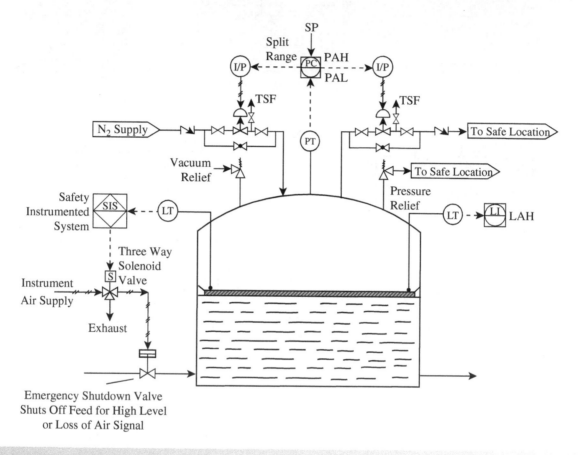

### Figure 16.39

A floating roof can be used in conjunction with an inert gas purge for particularly hazardous liquids.

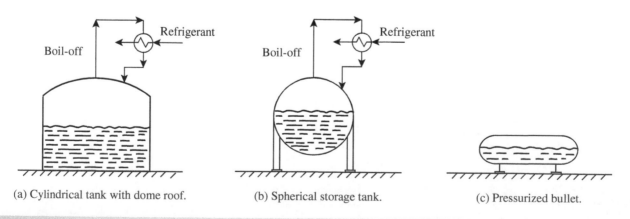

(a) Cylindrical tank with dome roof.     (b) Spherical storage tank.     (c) Pressurized bullet.

### Figure 16.40

Liquefied gases can be stored in different ways.

distribution under internal loading, resulting in highly efficient pressurized storage. This will be discussed in more detail in Chapter 19. Spherical tanks can operate at low pressure with refrigeration or full-pressure storage (vapor pressure of the liquid at ambient temperature) typically up to 28 bar.

3) *Pressurized bullet*. Figure 16.40c shows a pressurized "bullet" tank. These are restricted to small volumes (typically up to 7500 m$^3$). The tank is a full-pressure storage at the vapor pressure of the liquid at ambient temperature.

# 16.6  P&ID Process Operations – Summary

To develop a P&ID for process operations requires:

- control systems;
- vents for start-up;
- valves to isolate for maintenance;
- drain valves for maintenance;
- pressure relief devices.

Liquid storage can be close to atmospheric pressure. Breathing to atmosphere needs to be controlled or eliminated. Vacuum-pressure control valves allow a reduction in the breathing losses to atmosphere. Liquid storage tanks need protection against overfilling. This can be an overflow for less hazardous liquids or by using a safety instrumented system in conjunction with an emergency shutdown valve. Pressure relief devices are required for protection against vacuum and overpressure.

Liquid storage can be above atmospheric pressure when storing liquefied gases. Storage can use refrigeration to condense boil-off or a combination of pressure and refrigeration.

# Exercises

1. Figure 16.41 shows part of a P&ID for a centrifugal pump. The pump does not need a safety relief device or a strainer. Figure 16.41 features seven valves. Explain briefly the function of each of the valves V1–V7.
2. Figure 16.42 shows a steam heater is to be used to pre-heat a reactor feed. Develop a P&ID for the steam heater. Assume:

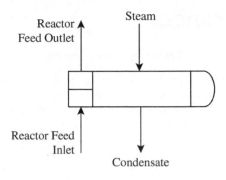

**Figure 16.42**

A steam heater for a reactor feed.

- A two-pass shell-and-tube heat exchanger is to be used with the steam on the shell side.
- The tube side is the low-pressure side of the heater.
- Control of the reactor feed outlet temperature is required.
- Pressure relief needs to be applied where necessary.

Mark the fail-safe condition for any control valves.

3. Figure 16.43 shows a flash drum to be used to separate a multicomponent vapor and liquid feed with a steam coil used to control the vaporization. Develop a P&ID for the flash drum with its pump. Synthesize a conceptual control system. It is not necessary to represent the details of the location of the controllers or the type of control signals. It can be assumed that the steam supply is stable and temperature control of the flash drum would not need to be cascaded in this case. Include pressure relief devices where it is appropriate. It is not necessary to include locking isolation valves for the pressure relief devices in this case. It can be assumed that the pump will not require a relief valve, as the discharge pressure is below the pressure rating of the discharge pipe. Give a first indication of the fail-safe condition for any control valves. This will not be able to be confirmed until the surrounding process context is understood.

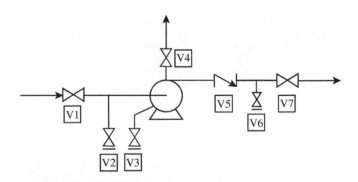

**Figure 16.41**

Valves required for a centrifugal pump.

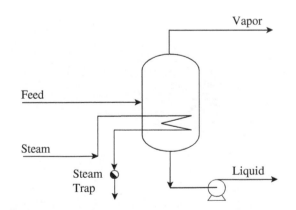

**Figure 16.43**

A non-adiabatic vapor–liquid separator with a steam coil.

# References

API STANDARD 2000 (2014) *Venting Atmospheric and Low-Pressure Storage Tanks*.

API RECOMMENDED PRACTICE 2350 (2020) *Overfill Protection for Storage Tanks in Petroleum Facilities*.

Driedger, W.C. (1996). Controlling positive displacement pumps, *Hydrocarbon Processing*, May 47–54.

Mannan, S. (2012). *Lees Loss Prevention in the Process Industries*, 4e. Butterworth-Heinemann.

Smith, R. (2016). *Chemical Process Design and Integration*. Wiley.

Verma, S. and Self, F. (2014). Design options for overfill prevention for aboveground storage tanks. American Institute of Chemical Engineers 2014 Spring Meeting, 10th Global Congress on *Process Safety*, New Orleans, 30 March–2 April.

# Chapter 17

# Piping and Instrumentation Diagrams – Construction

## 17.1 Development of Piping and Instrumentation Diagrams

The development of piping and instrumentation diagrams (P&IDs) for a complete process starts from the process flow diagram (PFD) and includes all of the details listed in Chapter 15 to allow the process to function safely and to be maintained safely. The P&ID gives a complete graphical documentation of the process. In practice, different disciplines input to the development of the P&IDs. These include process engineers, instrumentation and control engineers, piping engineers, operations personnel, and maintenance engineers. The P&IDs will evolve through the duration of the definition (front end engineering design) phase of the project.

Development of the P&IDs can be carried out in a series of steps:

*Step 1: Create a legend of the symbols to be used.*

Although standard symbols are used for the construction of P&IDs, there is no universal standard and different companies adopt their own preferences. Hence P&IDs should start with a legend of symbols.

*Step 2: From the basic process flow diagram for the process, add any missing steps and equipment from raw material inlets to product outlets.*

The basic process flow diagram (PFD) is expanded to add missing equipment to create a *comprehensive process flow diagram*. This takes the basic PFD and adds the missing steps and equipment from raw material inlets to product outlets, and includes:

i) Multiple items of equipment that are shown in the basic PFD as a single step (e.g., a single reactor represented in a PFD as a single item that in practice is multiple reactors in parallel needs to be included as such in a P&ID).

ii) Pumps and compressors.

iii) Utility and effluent treatment equipment linked directly to the process (centralized utility and effluent treatment systems will have their own P&IDs).

iv) Standby equipment (redundancy).

v) Storage linked directly to the process (centralized storage systems will have their own P&IDs).

*Step 3: Divide the resulting comprehensive flow diagram into process nodes.*

Each node consists of one or more processing units, which together have a common goal.

*Step 4: Synthesize a conceptual instrumentation and control system for the process.*

As discussed in Chapter 14, first set the control objectives. Then, starting from the node where the overall process flowrate (process capacity) is fixed, determine process control options by first assessing how flowrate and inventory control is to be achieved for each node of the process in isolation. Once the flowrate and inventory control has been achieved, modify the node control scheme to achieve the other control objectives of temperature, pressure, pH, and so on. If the process feed is fixed, start with the node where the feed enters the process and work forward, downstream node-by-node through the process. Conversely, if the product rate is fixed, start with the node where the product leaves the process and work backwards upstream node-by-node through the process. Where a process contains recycles, consider the control design with the nodes connected together in the recycle as though it were a single node. Finally, check the stability and response of the process to disturbances.

*Step 5: For each node in isolation add further details for the instrumentation and control and add additional pipes, valves, vents, drains, bypass lines, and sampling points required for operation and maintenance.*

Once the overall conceptual process control system has been developed, the same process nodes can be used to develop

*Process Plant Design*, First Edition. Robin Smith.
© 2024 John Wiley & Sons Ltd. Published 2024 by John Wiley & Sons Ltd.
Companion website: www.wiley.com/go/processplantdesign

preliminary P&IDs for each of the process nodes. Instrumentation and control locations and signaling arrangements need to be added. Also, for each node in isolation add:

**i)** Process and utility pipes required for start-up, shut-down, maintenance, and abnormal operation.

**ii)** Vents, drains, bypass lines, and sampling points required (including those required for start-up, shut-down, maintenance, and abnormal operation).

**iii)** All valves, together with the valve type, for both operation and maintenance, and isolations for maintenance (e.g., slip plates).

**iv)** Inspection access.

*Step 6: Combine the P&IDs for the nodes to form a complete process and remove conflicts or unnecessary duplication of valves, vents, and drains between the nodes.*

The interactions between adjacent nodes in the P&IDs can then be explored and the node P&IDs evolved as necessary. For example, connecting lines between adjacent nodes might have isolation valves from both the upstream and downstream nodes when both are not needed. Isolation valves, vents, and drains might be shared between nodes. If vents and drains are to be shared, then they would normally be located in the downstream node. This avoids unnecessary duplication of features and is best analyzed working from upstream to downstream.

*Step 7: Add pressure relief devices and relief vent and blowdown systems.*

Pressure relief devices, vacuum relief, safety instrumented functions, relief vent and blowdown systems can then be added, as discussed in Chapter 15. It is important for there to be no valves in the relief path (unless full bore valves that are locked or car sealed open).

In practice, the overall P&ID for a process will constitute a number of sheets with the connections between the sheets marked. There would normally be separate P&IDs for the relief and blowdown, centralized utilities (steam, inert gas, compressed air, and so on), centralized waste treatment, and all ancillary process equipment, with the connections between the diagrams marked. It is bad practice to try to squeeze too many items onto a single P&ID sheet. If too many items are placed on a single sheet at the beginning, then the final version will be too congested. Once overall P&IDs have been developed, they can be subjected to a hazard and operability (HAZOP) study and layers of protection analysis (LOPA). The P&ID will evolve from the first iteration of its creation with more and more detail being added as it is developed.

# 17.2   A Case Study

Consider the PFD for a simple reaction process shown in Figure 17.1. This takes raw materials Feed *A* and Feed *B* from storage and reacts to crude Product *C*. The reactor effluent is cooled using cooling water, which is sent to product storage before being

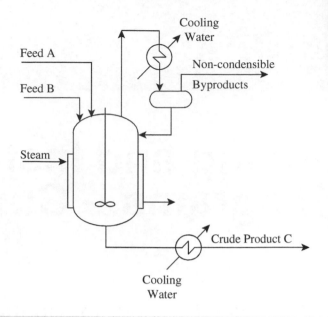

**Figure 17.1**

A process flow diagram.

processed further in another process. Following the steps presented above:

*Step 1: Create a legend of the symbols to be used.*

Figure 17.2 shows a legend of the symbols to be used.

*Step 2: From the basic PFD for the process, add any missing steps and equipment from raw material inlets to product outlets.*

Figure 17.3 shows the PFD with additional equipment added. The raw materials and product storage tanks have been added. Also added are the centrifugal pumps for the reactor feeds and for the reactor effluent, including standby pumps. Figure 17.3 shows the resulting *comprehensive process flow diagram* after all items of equipment have been included.

*Step 3: Divide the resulting comprehensive flow diagram into process nodes.*

Figure 17.4 shows the resulting comprehensive process flow diagram divided into process nodes. Each node consists of one or more processing units, which together have a common goal.

*Step 4: Synthesize a conceptual instrumentation and control system for the process.*

The procedure follows the approach detailed in Chapter 14. In this example the process throughput is fixed by the feed flowrate and the procedure starts with the feed storage and delivery to the process. Figure 17.5 shows a conceptual instrumentation and control configuration for raw materials storage. This shows an overflow line with a liquid seal in this case. Alternatively, a safety instrumented system might have been used to protect against over-filling, as discussed in Chapter 16. The liquid seal ensures that there is no connection of the tank vapor to the atmosphere under normal conditions. The tank is provided with a level indicator and high-level alarm. Figure 17.6 shows a conceptual instrumentation and

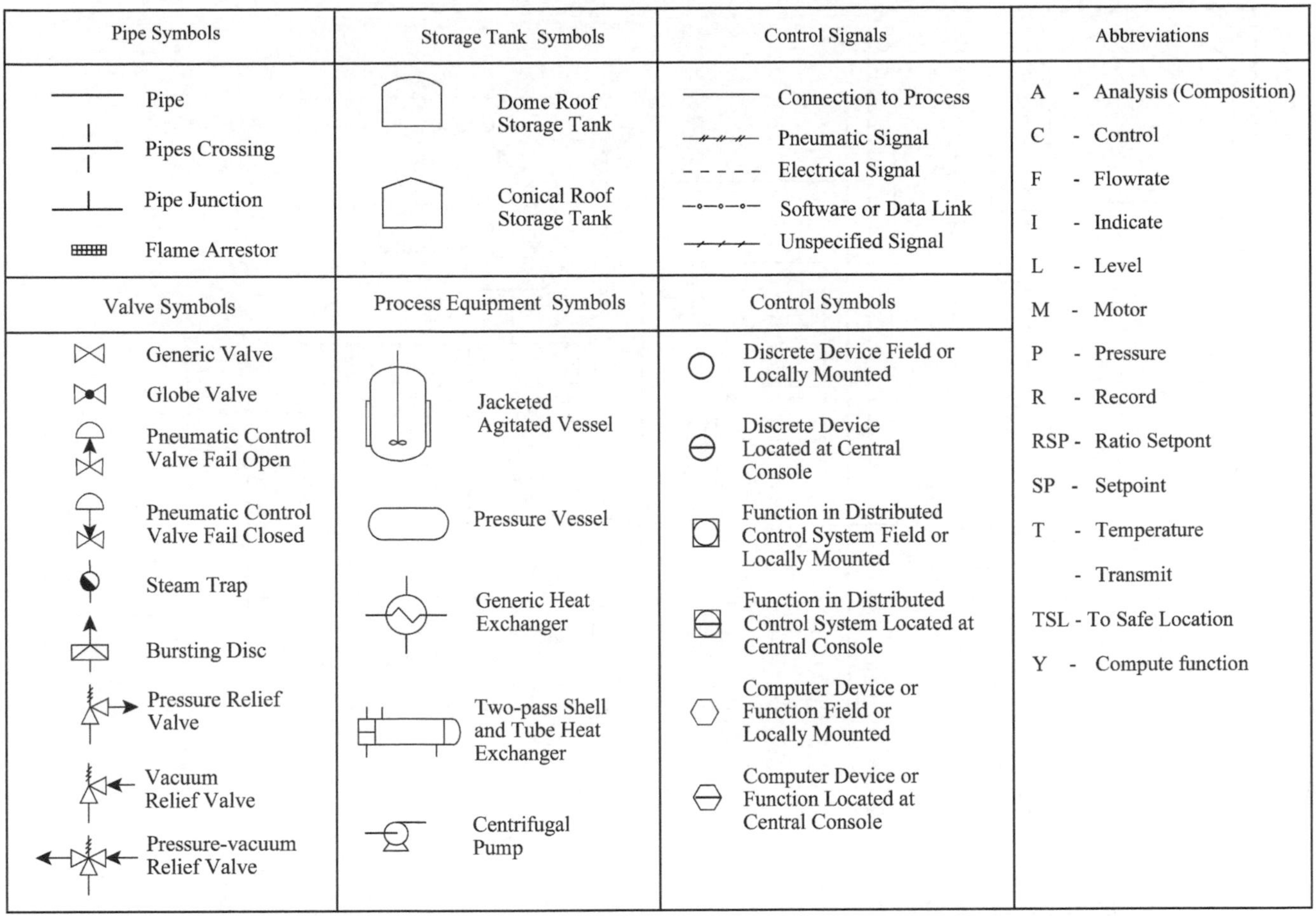

Pipe Symbols
Pipe
Pipes Crossing
Pipe Junction
Flame Arrestor
Storage Tank Symbols
Dome Roof Storage Tank
Conical Roof Storage Tank
Control Signals
Connection to Process
Pneumatic Signal
Electrical Signal
Software or Data Link
Unspecified Signal
Abbreviations
A - Analysis (Composition)
C - Control
F - Flowrate
I - Indicate
L - Level
M - Motor
P - Pressure
R - Record
RSP - Ratio Setpont
SP - Setpoint
T - Temperature
- Transmit
TSL - To Safe Location
Y - Compute function
Valve Symbols
Generic Valve
Globe Valve
Pneumatic Control Valve Fail Open
Pneumatic Control Valve Fail Closed
Steam Trap
Bursting Disc
Pressure Relief Valve
Vacuum Relief Valve
Pressure-vacuum Relief Valve
Process Equipment Symbols
Jacketed Agitated Vessel
Pressure Vessel
Generic Heat Exchanger
Two-pass Shell and Tube Heat Exchanger
Centrifugal Pump
Control Symbols
Discrete Device Field or Locally Mounted
Discrete Device Located at Central Console
Function in Distributed Control System Field or Locally Mounted
Function in Distributed Control System Located at Central Console
Computer Device or Function Field or Locally Mounted
Computer Device or Function Located at Central Console

**Figure 17.2**
A legend of the P&ID symbols to be used.

Feed A
Cooling Water
Non-condensible
Byproducts
Steam
Feed B
Crude Product C
Cooling Water

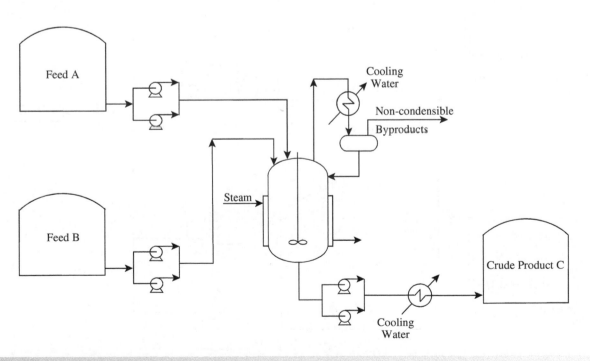

**Figure 17.3**
Add any missing steps and equipment from the PFD from raw materials receipt to product delivery to produce a comprehensive PDF.

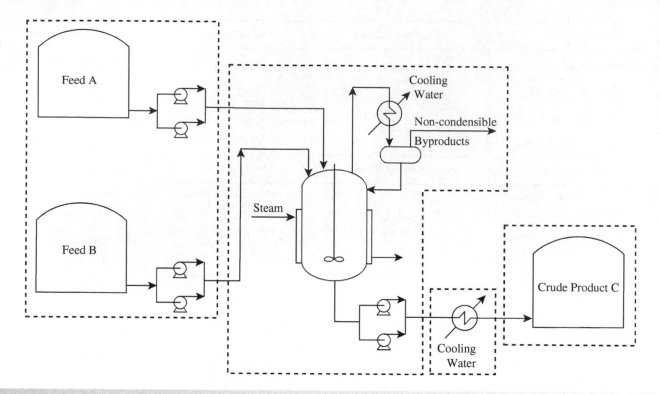

**Figure 17.4**

Divide the resulting comprehensive flow diagram into nodes.

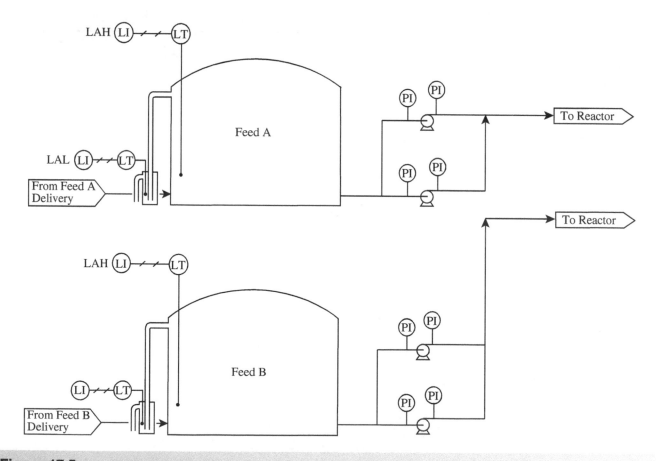

**Figure 17.5**

Create a conceptual instrumentation and control system for raw materials storage.

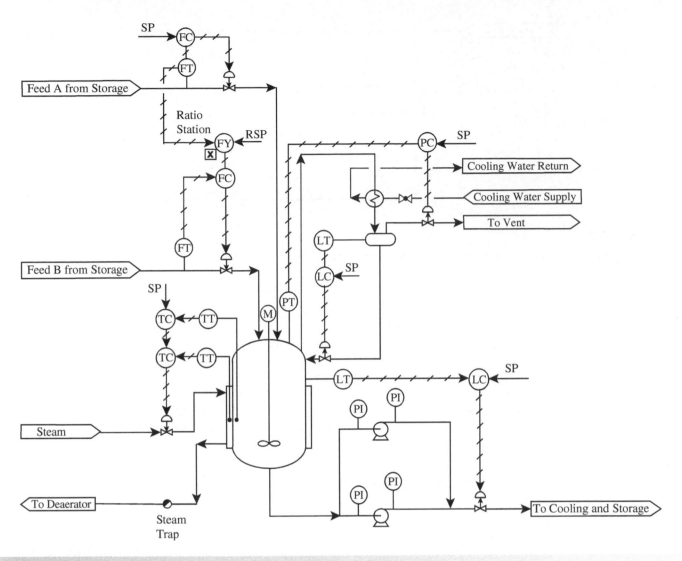

## Figure 17.6

Create a conceptual instrumentation and control system for the reactor.

control system for the reactor. The two feeds are under ratio control to maintain the stoichiometric ratio of the feeds. The reactor is heated by steam and the reactor temperature is controlled by cascade control, as discussed in Chapter 13. As much as possible of the reactor contents that are vaporized are condensed with a cooling water condenser, and the condensed liquid refluxed to the reactor. The flow of non-condensible material from the condensate drum is used to control the pressure of the reactor. No temperature control has been applied to the condenser in this case, but it has been left under manual control. The reactor and the condensate drum both have level control. Figure 17.7 shows a conceptual instrumentation and control system for the reactor cooler. The reactor effluent temperature is not critical in this case, as it is being fed to storage. Thus, temperature control has not been included and the cooling water has been left on manual control. A temperature indicator for the cooler outlet has been included. Figure 17.8 shows a conceptual instrumentation

and control system for the product storage. As with feed storage, an overflow and level indicator have been included. There is no recycle in the example and in this case no conflicts between the node control systems created by recycles.

*Step 5: For each node in isolation add further details for the instrumentation and control and add additional pipes, valves, vents, drains, bypass lines, and sampling points required for operation and maintenance.*

Figure 17.9 shows the P&ID for the feed storage with details of the instrumentation and control added, and also with additional valves for operation and maintenance added. A pressure-vacuum relief valve has been added with a flame arrestor to reduce tank breathing to atmosphere. Figure 17.10 shows the P&ID for the reactor with additional details added. Figure 17.11 shows the P&ID for the cooler with additional details for control and operation. Figure 17.12 shows the P&ID for the product storage with additional details for control and operation.

**17**

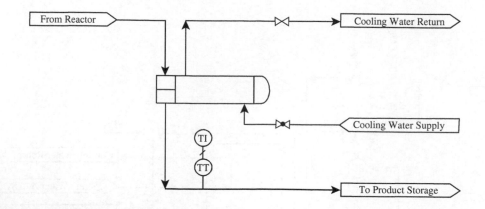

## Figure 17.7

Create a conceptual instrumentation and control system for the reactor cooler.

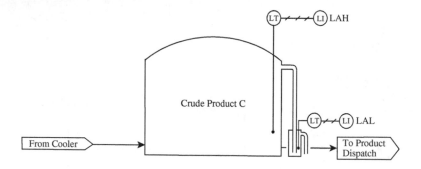

## Figure 17.8

Create a conceptual instrumentation and control system for the product storage.

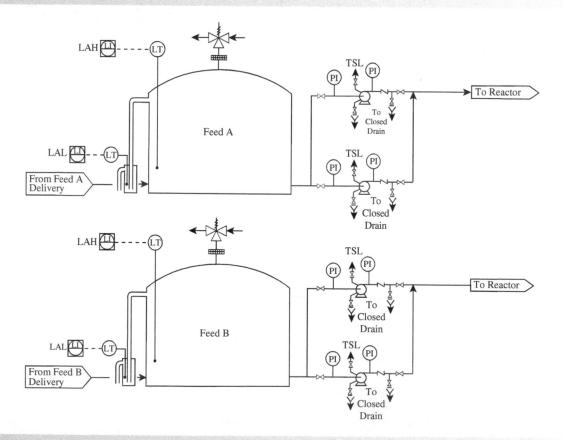

## Figure 17.9

For the raw materials storage add additional details for control and operation.

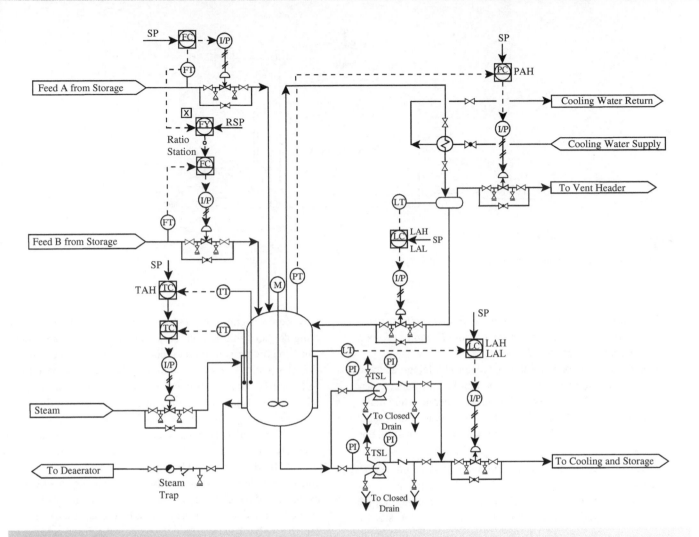

## Figure 17.10

For the reactor add additional details for control and operation.

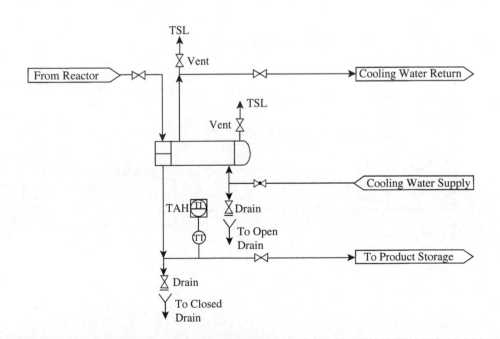

## Figure 17.11

For the cooler add additional details for control and operation.

17

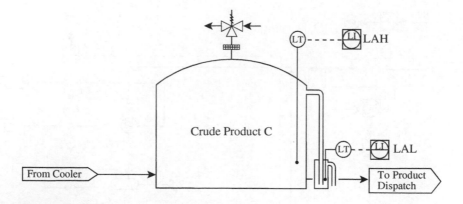

## Figure 17.12

For the product storage add additional details for control and operation.

*Step 6*: *Combine the P&IDs for the nodes to form a complete process and remove conflicts or unnecessary duplication of valves, vents, and drains between the nodes.*

Combining the nodes for the raw materials storage, reactor, cooler and product storage brings no conflicts or unnecessary duplication of valves, vents, and drains for operation and maintenance in this case. Figure 17.13 shows the cooler and product storage nodes combined, as they will be shown together on one P&ID sheet.

*Step 7*: *Add pressure relief devices and relief vent and blowdown systems.*

Figure 17.14 shows the P&ID for the raw materials storage with pressure and vacuum relief valves added to the storage tank. As pointed out in Chapter 16, pressure-vacuum relief

valves are designed for pressure control when filling and emptying the tank and thermal breathing. These flows are usually small in comparison to emergency relief for an emergency or fire. Hence pressure and vacuum relief valves have been fitted.

Figure 17.15 shows the P&ID for the reactor with pressure relief valves. Both the reactor and the condensate reflux drum are classified as pressure vessels and each have a pressure relief valve. The condenser has a pressure relief valve on the tube side. The tube side is the cold side and the cooling water as the cold fluid is capable of being blocked in while heat is being added from the vapor, possibly creating overpressure from expansion of the cooling water. A pressure relief valve is considered not necessary on the shell side in this case. A pressure relief valve has been added to the reactor steam jacket to protect

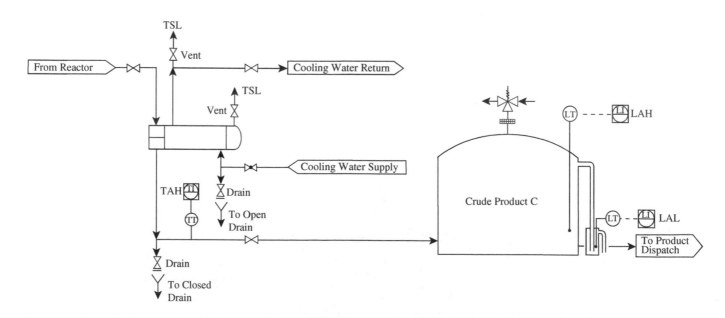

## Figure 17.13

Combine the cooler and product storage nodes.

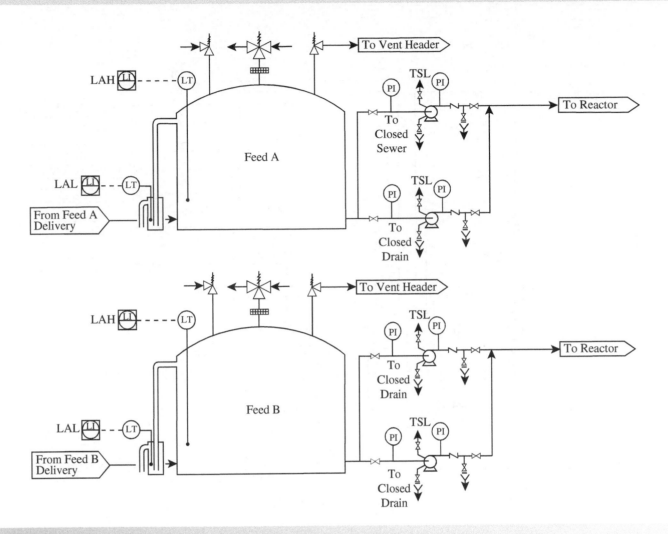

## Figure 17.14

Add the relief and blowdown system to the raw materials storage.

the reactor vessel against the possibility of the steam pressure going out of control and overpressuring the steam jacket.

Figure 17.16 shows the P&ID for the cooler and product storage with pressure relief devices. The cooler has a bursting disc on the shell side. The shell side is the cold side and the cooling water as the cold fluid is capable of being blocked in while heat is being added from the reactor effluent, possibly creating overpressure from expansion of the cooling water. Also, tube failure might be a cause of overpressure. Given the possibility of rapid pressure transients from tube failure, a bursting disc would be preferred to a relief valve, as discussed in Chapter 16. The product storage in Figure 17.16 shows pressure and vacuum relief valves added to the tank, as discussed in Chapter 16.

Finalizing the P&IDs for this stage in their development, Figure 17.17 shows the legend to clarify the symbols and abbreviations to be used in the P&IDs. Figure 17.18 shows the P&ID for raw materials storage, Figure 17.19 the P&ID for the reactor, and Figure 17.20 the P&ID for the cooler and product storage. It should be emphasized that these P&IDs are preliminary only

and must be subject to a process of evolution. A HAZOP study needs to be carried out to check operability and safety (Kletz 1999; Crawley and Tyler 2015). Corrections can then be made to the P&IDs after the HAZOP. Following the HAZOP, a LOPA studies layers of protection, modifies pressure relief if necessary, adds safety instrumented systems, and adds/modifies alarms (Mannan 2012; Smith 2017). Alarm management is specialized and requires a detailed study. This will result in changes to the P&IDs. More detail needs to be added, such as the detail of the destination for streams marked going to a safe location. All equipment and piping need to have a unique identification number (*TAG number*), consisting of a series of letters and numbers. For example, in the case of pipes the TAG number would include a code to specify uniquely which pipe it refers to, the service, diameter, schedule (wall thickness), material of construction, tracing type, and insulation. All instruments and controllers will need a TAG number. Notes are added to the P&ID for non-standard information e.g., flowmeters should be installed with at least 20 straight pipe diameters upstream and 10 straight pipe diameters downstream.

**17**

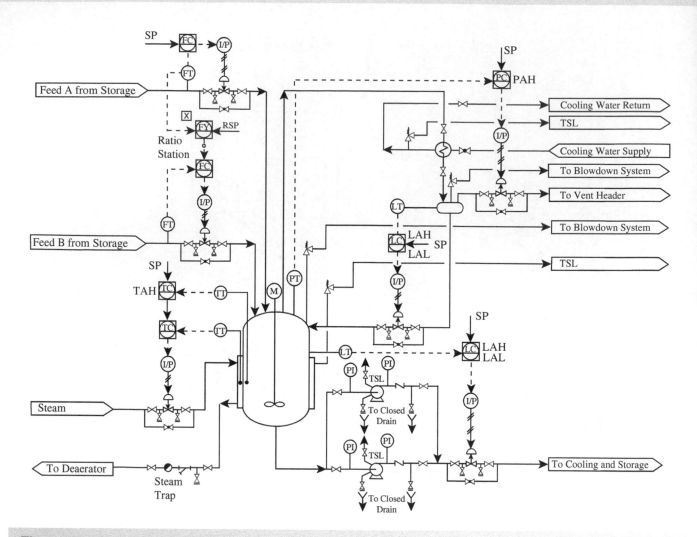

**Figure 17.15**

Add the relief and blowdown system to the reactor.

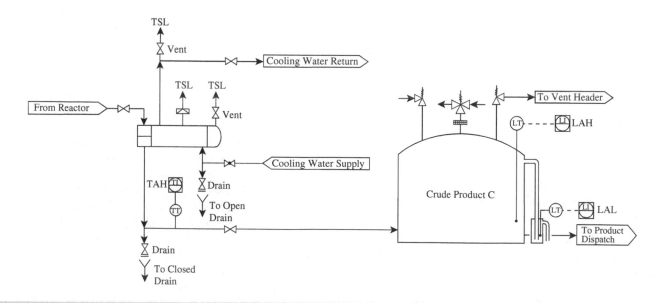

**Figure 17.16**

Add the relief and blowdown system to the cooler and product storage.

| Pipe Symbols | Storage Tank Symbols | Control Signals | Abbreviations | NOTES |
|---|---|---|---|---|
| ——— Pipe | Dome Roof Storage Tank | ——— Connection to Process | A - Analysis (Composition) | |
| ┬ Pipes Crossing | | ┄┄ Pneumatic Signal | C - Control | |
| | Conical Roof Storage Tank | ----- Electrical Signal | F - Flowrate | |
| ┴ Pipe Junction | | ─o─o─ Software or Data Link | I - Indicate | |
| ▦ Flame Arrestor | | ─/─/─ Undefined Signal | L - Level | |

| Valve Symbols | Process Equipment Symbols | Control Symbols | | |
|---|---|---|---|---|
| ⋈ Generic Valve | | ○ Discrete Device Field or Locally Mounted | M - Motor | |
| ⋈ Globe Valve | Jacketed Agitated Vessel | ⊖ Discrete Device Located at Central Console | P - Pressure | |
| Pneumatic Control Valve Fail Open | | | R - Record | |
| Pneumatic Control Valve Fail Closed | Pressure Vessel | ▢ Function in Distributed Control System Field or Locally Mounted | RSP - Ratio Setpont | |
| Steam Trap | Generic Heat Exchanger | ▢ Function in Distributed Control System Located at Central Console | SP - Setpoint | |
| Bursting Disc | | | T - Temperature | |
| Pressure Relief Valve | Two-pass Shell and Tube Heat Exchanger | ⬡ Computer Device or Function Field or Locally Mounted | - Transmit | |
| Vacuum Relief Valve | | ⬡ Computer Device or Function Located at Central Console | TSL - To Safe Location | |
| Pressure-vacuum Relief Valve | Centrifugal Pump | | Y - Compute function | |

**EX Chemical Company**

| Drawing No | PID 01-01 |
|---|---|
| Title | Legend |
| Made By | |
| Checked By | |
| Approved By | |
| Date | |

## Figure 17.17

Preliminary P&ID legend.

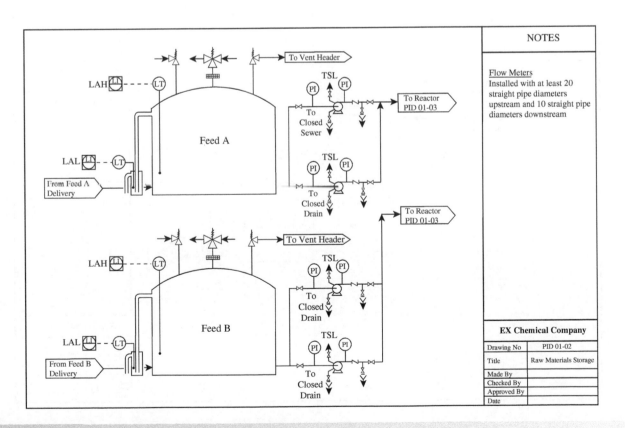

## Figure 17.18

Preliminary P&ID for raw material storage.

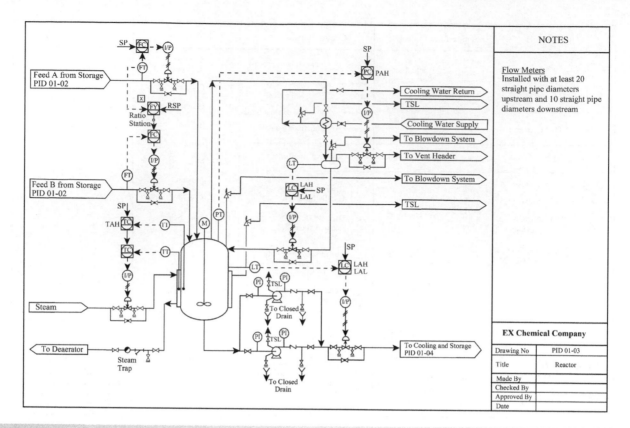

## Figure 17.19

Preliminary P&ID for the reactor.

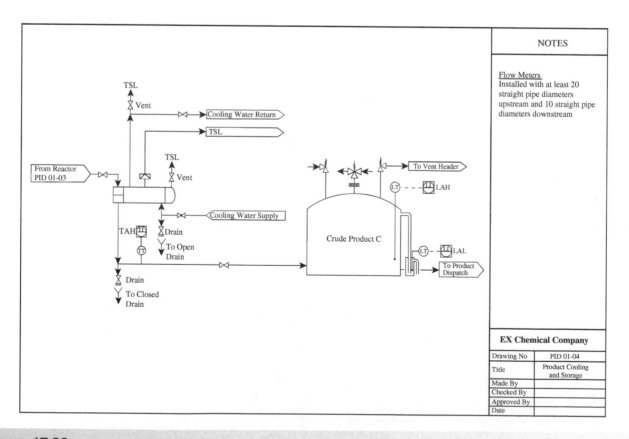

## Figure 17.20

Preliminary P&ID for product cooling and storage.

## 17.3 P&ID Construction – Summary

To develop a P&ID, starting for the PFD:

1) Create a legend of the symbols to be used.

2) From the basic PFD for the process, add any missing steps and equipment from raw materials in to products out.

3) Divide the resulting comprehensive flow diagram into process nodes.

4) Synthesize a conceptual instrumentation and control system for the process.

5) For each node in isolation add further details for the instrumentation and control and add additional pipes, valves, vents, drains, bypass lines, and sampling points required for operation and maintenance.

6) Combine the P&IDs for the nodes to form a complete process and remove conflicts or unnecessary duplication of valves, vents, and drains between the nodes.

7) Add pressure relief devices and relief vent and blowdown systems.

## References

Crawley, F. and Tyler, B. (2015). *HAZOP: Guide to Best Practice*, 3e. IChemE Elsevier.

Kletz, T. (1999). *Hazop and Hazan*, 4e. IChemE.

Mannan, S. (2012). *Lees' Loss Prevention in the Process Industries*, 4e. Butterworth-Heinemann.

Smith, D.J. (2017). *Reliability, Maintainability and Risk*, 9e. Butterworth-Heinemann Elsevier.

# Chapter 18

# Materials of Construction

Choice of materials of construction affects both the mechanical design and the capital cost of equipment. Estimation of the capital cost and specification of equipment requires decisions to be made regarding the materials of construction. Consider first the mechanical properties of materials.

## 18.1  Mechanical Properties

1) *Tensile and compressive strength.* When solid material is subjected to opposing forces acting through the center of the cross-section of the solid, this is referred to as normal (axial) stress, as illustrated in Figure 18.1. Numerically, normal stress is the force divided by the *original* cross-sectional area. The force can be *tensile*, as in Figure 18.1a, or *compressive*, as in Figure 18.1b. Thus:

$$\sigma = \frac{F}{A_0} \qquad (18.1)$$

where

$\sigma$ = stress (N m$^{-2}$)

$F$ = force (N)

$A_0$ = original cross-sectional area (m$^2$)

This assumes that the stress is evenly distributed over the cross-section, which might not be strictly true depending on how the solid is anchored.

Normal (axial) strain, as illustrated in Figure 18.2, is the ratio of the change of length to the original length due to the deformation of the solid:

$$\varepsilon = \frac{\Delta L}{L_0} \qquad (18.2)$$

where

$\varepsilon$ = strain (−)

$\Delta L$ = change in length (m)

$L_0$ = original length (m)

Figure 18.4 shows a stress–strain profile for a metal solid under tensile stress, typical of low carbon steel. Initially, stress is proportional to strain, as expressed by Hooke's law. Within this linear range, deformation is *elastic* and not permanent and the material will return to its original shape when the stress is removed. Within the limits for which Hooke's Law is obeyed, the ratio of stress to strain produced is known as Young's Modulus or Modulus of Elasticity:

$$E = \frac{\sigma}{\varepsilon} \qquad (18.3)$$

where

$E$ = ratio of stress to strain or Young's Modulus or Modulus of Elasticity (N m$^{-2}$)

$\sigma$ = stress (N m$^{-2}$)

$\varepsilon$ = strain (−)

$E$ is constant for a given material and is often assumed to be the same in tension and compression. If stress is increased beyond the linear range up to the *yield strength*, deformation continues to be elastic and deformation is not permanent, even though stress is no longer proportional to strain. The yield strength marks the elastic limit beyond which the material will no longer go back to its original shape when the stress is removed. The yield strength is not a sharply defined point and is often defined as the stress that will cause a permanent deformation of 0.2% of the original dimension, as illustrated in Figure 18.3. As stress is increased beyond the yield strength, *plastic deformation* occurs and strains are only partially recoverable. The material is permanently deformed if the stress is completely released. Increased stress causes the material to elongate uniformly along its length to the point where the stress–strain curve in Figure 18.3 reaches a maximum. The maximum point is known

*Process Plant Design*, First Edition. Robin Smith.
© 2024 John Wiley & Sons Ltd. Published 2024 by John Wiley & Sons Ltd.
Companion website: www.wiley.com/go/processplantdesign

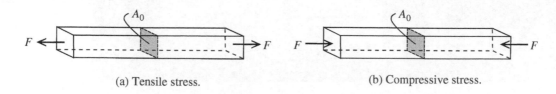

(a) Tensile stress.                  (b) Compressive stress.

**Figure 18.1**

Normal stresses.

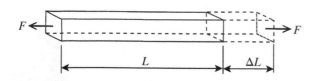

**Figure 18.2**

Normal strain is measured by the change in length.

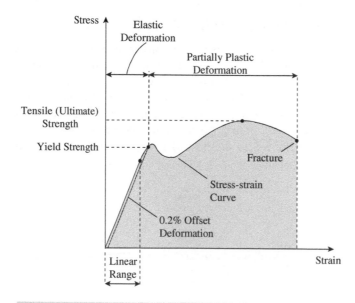

**Figure 18.3**

Typical stress–strain curve for low carbon steel.

which do not exhibit a yield point. The design basis depends on the material of construction and the applicable design code being followed.

The shape of the stress–strain curve varies according to the material. Figure 18.4a illustrates a curve typical for materials such as steel alloys and aluminum alloys. Figures 18.3 and 18.4a illustrate the behavior of *ductile* materials where the strain is greater than 5% before fracture. Figure 18.4b illustrates a curve typical of *brittle* materials, such as cast iron. Brittle materials typically fail while the deformation is elastic and where the strain is less than 5% before fracture. The tensile strength and fracture strength are the same in this case. If the stress is applied normal to the stressed area, as illustrated in Figure 18.1b, but to compress (shorten) the material, it is a *compressive stress*. Compressive strength is not necessarily equal to tensile strength.

If a specimen is loaded in tension, as shown in Figure 18.5a, then the axial elongation is accompanied by a lateral contraction. If the specimen is loaded in compression, as shown in Figure 18.5b, then the axial contraction is accompanied by a lateral elongation. The lateral strain $\varepsilon'$ at any point is proportional to the axial strain $\varepsilon$ at that same point if the material is linearly elastic. The ratio of these strains is a property of the material known as Poisson's Ratio. This is a dimensionless ratio that can be expressed as:

$$\nu = -\frac{\text{Lateral strain}}{\text{Axial strain}} = -\frac{\varepsilon'}{\varepsilon} \qquad (18.4)$$

where

$\nu$ = Poisson's ratio (–)

$\varepsilon$ = axial strain (m m$^{-1}$)

$\varepsilon'$ = lateral strain (m m$^{-1}$)

The minus sign is inserted to compensate for the lateral and axial strains having opposite signs. If the axial strain in the specimen is in tension, it is positive and the lateral strain is negative, since the width of the bar decreases. If the axial strain in the specimen is in compression, it is negative and the lateral strain is positive, since the width of the specimen increases. Poisson's Ratio will have a value that depends both on the composition of the metal and the method of formation. For most metals, Poisson's Ratio lies in the range 0.25–0.35 and for the metals commonly used for the construction of process vessels and equipment is usually around 0.3.

as the *tensile strength* or *ultimate strength*. As stress is increased beyond this point further, plastic deformation is concentrated in the region of the weakest point and the material begins to "neck" or thin down locally until fracture occurs.

For mechanical design of equipment, the allowable stress should be limited to be below the yield strength. However, since the yield strength is difficult to determine accurately, the allowable stress is taken as either the yield strength or tensile (ultimate) strength *at the working temperature* divided by a *safety factor*. For pressure vessels (vessels operating with pressures above 1 barg), the safety factor is typically 3.5 or 3, depending on the design code. The use of tensile strength as the design basis is necessary for materials such as cast iron,

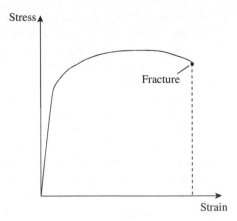

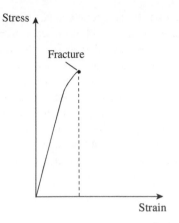

(a) Stress-strain curve typical of steel alloys and aluminium alloys.

(b) Stress-strain curve typical of brittle materials.

**Figure 18.4**

Stress–strain curves vary significantly according to the material.

(a) A uniaxial tensile force creates an axial strain that is positive and a lateral strain that is negative.

(b) A uniaxial compressive force creates an axial strain that is negative and a lateral strain that is positive.

**Figure 18.5**

A uniaxial tensile or compressive force creates both axial and lateral strain.

**2)** *Ductility and malleability. Ductility* is the ability of a material to deform plastically under tensile stress before fracturing. On the other hand, *malleability* measures the ability of a material to deform under compressive stress. Both of these mechanical properties measure the extent to which a solid material can be deformed plastically without fracture. A material with high ductility will be able to be drawn into long, thin wires without breaking. Ductility and malleability are especially important in metal-working, as materials that crack or break under stress cannot be manipulated using metal-forming processes, such as rolling and drawing. Brittle materials cannot be formed by processes such as rolling and drawing and must be cast or thermoformed.

**3)** *Toughness. Toughness* is the ability of a metal to deform plastically and to absorb energy before fracture. Brittleness implies sudden failure; toughness is the opposite. Of prime importance is the ability to absorb energy before fracture. Toughness requires a combination of strength and ductility. One measure of toughness is the area under the stress–strain curve from a tensile test (see Figure 18.3). This value is termed *material toughness* and has units of energy per volume. In order to be tough, a material must have both high strength and ductility. Brittle materials that are strong but with limited ductility are not tough. Ductile materials with low strengths are also not tough. To be tough, a material should withstand both high stresses and high strains.

**4)** *Hardness.* Hardness is a characteristic of material that determines the resistance of a material to wear and abrasion. This is important for materials of construction in duties that are exposed to abrasive material, particularly solids and slurries. Various tests can be used for hardness, leading to different measures of hardness. Hardness can be increased in many metals by heat treatment or cold working of the metals.

**5)** *Fatigue.* Fatigue is progressive and localized structural damage that occurs when a material is subjected to cyclic loading. The stress that causes such damage can be much less than the yield or tensile strength of the material. If the loads are above a certain threshold, microscopic cracks begin to form. Eventually a crack can reach a critical size and propagate suddenly, causing structural fracture. The shape of the structure significantly affects the fatigue life. Square holes or sharp corners can lead to elevated local stresses where fatigue cracks can initiate. Round holes and smooth shapes increase the fatigue strength. *Fatigue limit* quantifies the amplitude of cyclic stress that can be applied to the material without causing fatigue failure. Ferrous alloys and titanium alloys have a distinct limit, i.e., a stress amplitude below which there appears to be no number of cycles that will cause failure. Other metals such as aluminum and copper do not have a distinct limit and will eventually fail even from small stress amplitudes.

**6)** *Creep.* Creep is the tendency of a material to deform slowly under the influence of mechanical stresses. The deformation may become large enough for a component to no longer be

**18**

able to perform its function. It can occur as a result of long-term exposure to high levels of stress below the yield strength. Creep is more severe when the material under stress is exposed to elevated temperature for long periods, and generally increases near its melting point. Creep is of particular concern when materials are subjected to a combination of high stress and high temperature. As a general guideline, for metals the effects of creep deformation generally become noticeable at approximately 30% of the melting point (measured in absolute temperature). Creep creates initially a relatively high level of strain that slows with increasing time.

# 18.2  Corrosion

In chemical manufacture, corrosion resistance is the prime (but not only) consideration in materials selection. Corrosion is the deterioration of materials (usually metals) as a result of chemical reactions between the materials and the surrounding environment, e.g., the oxidation of iron in the presence of water by an electrolytic process. Whilst the main concern is corrosion caused by process fluids, external corrosion in the surrounding environment can also be an issue. There are various mechanisms for corrosion that depend on the material of construction and the corrosive environment. The rate of corrosion is also highly temperature-dependent. In broad terms, corrosion can be categorized as either *uniform* or *localized*. The main types of corrosion are:

1) *Uniform corrosion. Uniform* or *general* corrosion is attack to the material that proceeds more or less uniformly over a surface at a uniform rate. It should be noted that uniform corrosion can also be accompanied by localized corrosion. An acceptable rate of corrosion of relatively low-cost material, such as carbon steel, is of the order of 0.25 mm $y^{-1}$ or less (Hunt 2014). For more expensive materials, such as stainless steel, an acceptable rate of corrosion is of the order of 0.1 mm $y^{-1}$ (Hunt 2014).

2) *Galvanic corrosion. Galvanic* corrosion occurs when two different metals have contact, either physically or electrically, in a common electrolyte. A *galvanic couple* can be created in which the more *active metal* (the anode) corrodes at an accelerated rate and the more *noble metal* (the cathode) corrodes at a lower rate. The surface area of the two different metals affects the corrosion rates. By contrast, if each material was immersed separately in the electrolyte, each metal would corrode at its own uniform rate. The same principle as galvanic corrosion can be applied for corrosion protection through the use of *sacrificial electrodes*. For example, a zinc electrode can be connected electrically as a sacrificial anode to protect steel.

3) *Erosion corrosion. Erosion* corrosion is a corrosion process enhanced by the erosion action of flowing fluids. The erosion action might result from high velocity and turbulence of a fluid, or the presence of impinging solid particles or bubbles in a liquid stream, or impinging solid particles or droplets in a gas stream. *Cavitation*, which is the formation and sudden collapse of vapor bubbles in a liquid, can also facilitate erosion corrosion. The erosion action removes protective (or passive) oxide layers from the surface of the material, enhancing the corrosion. The erosion corrosion may also be accompanied by mechanical erosion. Erosion corrosion can be mitigated through avoiding sudden changes in the direction of fluid flow, avoiding the formation of jets directed at surfaces in the flow, installation of impingement plates to protect vulnerable surfaces, or the use of more resistant materials and surface coatings.

4) *Crevice corrosion. Crevice* corrosion is localized corrosion in small crevices where fluid can stagnate. Such crevices can occur in a metal surface or at joints, screw threads, and so on. To function as a corrosion site, a crevice has to be wide enough to allow entry of the corrosive fluid, but narrow enough to ensure the corrosive fluid remains stagnant. Crevices can then develop a local chemistry that is different from the bulk fluid. Net anodic reactions can occur within the crevice and net cathodic reactions exterior to the crevice, causing corrosion similar to galvanic corrosion. Changing the design to avoid crevices and potential corrosion sites can mitigate crevice corrosion.

5) *Pitting corrosion. Pitting* corrosion is localized corrosion that leads to the creation of small holes in the metal surface. Pitting corrosion is usually found on metals and alloys such as aluminum alloys, stainless steels, and stainless alloys, when an ultra-thin passive film (oxide film) protects the metal surface from corrosion. If this layer is chemically or mechanically damaged and does not immediately repassivate, it can become anodic, with the main area becoming cathodic, leading to localized galvanic corrosion.

6) *Stress corrosion cracking.* Stress corrosion cracking is a failure mechanism caused by a combination of a susceptible material (normally a ductile metal), tensile stress, and a specific corrosive environment. Tensile stress is required to open up cracks in the material. The stress can be either directly applied or can be present in the form of residual stresses frozen into the material from fabrication. Cracks produced by the stress can grow rapidly in a specific corrosive environment. Stress corrosion cracking is highly chemically specific and only occurs when certain alloys are exposed to very specific chemical environments. For example, copper and its alloys are susceptible to ammonia compounds, mild steels are susceptible to alkalis, and stainless steels are susceptible to chlorides. Temperature is also an important factor. Stress corrosion cracking can be mitigated by selecting combinations of materials and process environments that avoid the problem and stress-relieving components by heat treatment after fabrication and welding.

7) *Intergranular corrosion.* The microstructure of a metal is made up of individual crystalline areas known as *grains*. The structure of the grains results from the material composition and the way the material was manufactured. The outside area of a grain that separates it from the other grains is known as the *grain boundary* and is a region of misfit between the grains. Intergranular corrosion is a localized attack along the grain boundaries, or immediately adjacent to grain boundaries.

However, the bulk of the grains remain largely unaffected. The corrosion usually progresses along a narrow path along the grain boundary. In a severe case of intergranular corrosion, entire grains can be dislodged due to complete deterioration of their boundaries. The resulting mechanical properties of the structure can be seriously affected. An example is stainless steels, where chromium-rich grain boundary precipitates can lead to a local depletion of chromium immediately adjacent to these precipitates, leaving these areas vulnerable to corrosive attack.

**8)** *Hydrogen embrittlement.* Embrittlement is a phenomenon that causes loss of ductility in a material, making it brittle and reducing its load-bearing capacity. Hydrogen embrittlement involves the diffusion of hydrogen atoms into the metal. If steel is exposed to hydrogen at high temperatures, hydrogen atoms diffuse into the alloy and react with carbon in the alloy to form methane or combine with other hydrogen atoms to form hydrogen molecules in tiny pockets. Since these molecules are too large to diffuse through the metal, pressure builds at grain boundaries and voids within the metal, causing minute cracks to form. Copper alloys can be embrittled if exposed to hydrogen. The hydrogen diffuses through the copper and reacts with inclusions of $Cu_2O$ forming water, which then forms pressurized bubbles at the grain boundaries and voids. Hydrogen embrittlement can be avoided by choosing a metal alloy that is resistant.

## 18.3  Corrosion Allowance

When equipment is mechanically designed, after determining the wall thickness needed to meet mechanical requirements, an extra thickness is added to compensate for the reduction in the wall thickness as a result of corrosion over the life of the equipment. This is referred to as the *corrosion allowance*. Thus, a corrosion allowance must be added to the wall thickness to meet the mechanical requirements over the life of the equipment. As noted above, an acceptable rate of uniform corrosion for relatively low-cost material, such as carbon steel, is of the order of $0.25$ mm y$^{-1}$ and of the order of $0.1$ mm y$^{-1}$ for more expensive materials, such as stainless steel (Hunt 2014). However, this should be considered to be the maximum. The actual rate of uniform corrosion must be determined and used as the basis of the corrosion allowance. Then, because the penetration depth for the corrosion can vary across the surface, a corrosion allowance is normally assigned a safety factor of two. The allowance is normally a minimum of 1 mm, but usually in the range of 2–4 mm.

## 18.4  Commonly Used Materials of Construction

The most commonly used materials of construction used in chemical plant are:

**1)** *Carbon steel.* Carbon steel is an alloy formed from iron and carbon with nominal amounts of other elements. Steel is considered to be carbon steel when the percentages of other trace elements do not exceed certain values. The maximum content of manganese is 1.65%, silicon 0.60%, and copper 0.60%. Steel that also contains higher or specified quantities of other elements such as nickel, chromium, or vanadium is referred to as *alloy steel*. Changing the amount of carbon changes the properties of the steel. Lower carbon steels are softer and more easily formed and steels with higher carbon content are harder and stronger, but less ductile and more difficult to machine and weld. Carbon steels can be classified as:

- Low carbon steel, also known as mild steel, which has a carbon content of typically 0.05–0.25%, with up to 0.4% manganese. It is a low-cost material that is easy to fabricate. It is not as hard as higher-carbon steels.

- Medium carbon steel has a carbon content of typically 0.29–0.54%, with 0.60–1.65% manganese. It is ductile and strong with good wear properties.

- High carbon steel has a carbon content of typically 0.55–0.95%, with 0.30–0.90% manganese. It is very strong and holds its shape well, making it ideal for springs and wire.

- Very high carbon steel has a carbon content of typically 0.96–2.1%. The high carbon content makes it an extremely strong but brittle material.

An alloy of iron with more than 2.1% carbon as the main alloying element is classed as cast iron. In addition to carbon, cast irons must also contain from 1% to 3% silicon. Cast iron cannot be used for pressure containment because of its brittleness, but can be used for some equipment components, such as pump casings. High silicon irons with carbon up to 1.1% and silicon up to 15% have excellent resistance to attack by sulfuric acid and most organic acids, but still suffer from brittleness.

**2)** *Stainless steel.* Stainless steels are the most commonly used materials of construction for corrosion resistance (Pitcher 1976). Stainless steel differs from carbon steel mainly by the amount of chromium present, with a minimum of 10.5% chromium. Other alloying elements are added to enhance properties such as strength at high or cryogenic temperatures, ease of fabrication, and weldability (ability to be welded). These additional elements include nickel, molybdenum, titanium, copper, carbon, and nitrogen. Stainless steels protect against corrosion as they contain enough chromium to form a passive film of chromium oxide, which prevents further surface corrosion. The oxide bonds strongly to the metal surface and blocks corrosion from spreading to the internal structure of the metal. Thus, stainless steels are most effective in oxidizing environments.

There are many grades of stainless steels. The American Iron and Steel Institute (AISI) and the American Society of Testing and Materials (ASTM) have developed designations such as 304, 430, etc. These are not specifications, but relate to steel grade composition ranges only. Stainless steels have different metallurgical structures and can be divided into five types:

a) *Austenitic grades*.

Austenitic grades are the most commonly used types in chemical plants. The most common austenitic grades are iron-chromium-nickel steels that form the 300 series. Austenitic grades cannot be hardened by heat treatment, but can be hardened by cold-working. The letter "L" after a stainless steel type indicates low carbon (e.g., 304L), in which the carbon is kept to 0.03% or below. These are used to provide extra corrosion resistance after welding. However, L-grades are more expensive and more carbon at high temperatures imparts greater physical strength. High carbon H-grades contain between 0.04% and 0.10% carbon and retain strength at extreme temperatures. Some of the more commonly used grades are given in the table below (Pitcher 1976).

| Type | Typical composition | Typical applications |
|---|---|---|
| 304 | 18–20% Cr, 8–11% Ni, 0.08% C, balance Fe | The most common of the austenitic grades. Widely used for chemical processing equipment in lesser corrosive environments. It is also widely used for process equipment in the food and beverage industries. |
| 304L | 18–20% Cr, 8–11% Ni, 0.03% C, balance Fe | Low carbon version of Type 304. Used when welding and heat treatment are a problem with Type 304. |
| 316 | 16–18% Cr, 10–14% Ni, 2–3% Mo, 0.08% C, balance Fe | Molybdenum used to control pitting corrosion. Widely used for chemical processing equipment in more corrosive environments. Also used for process equipment in the food and beverage and pulp and paper industries. |
| 316L | 16–18% Cr, 10–14% Ni, 2–3% Mo, 0.03% C, balance Fe | Low carbon version of Type 316. Used when welding and heat treatment are a problem with Type 316. |
| 317 | 18–20% Cr, 11–15% Ni, 3–4% Mo, 0.08% C, balance Fe | Higher percentage of molybdenum than Type 316 for extra corrosion resistance. Suitable for highly corrosive environments. |
| 317L | 18–20% Cr, 11–15% Ni, 3–4% Mo, 0.03% C, balance Fe | Low carbon version of Type 317 when welding and heat treatment are a problem with Type 317. |

b) *Martensitic grades*.

Martensitic grades are stainless alloys that are both corrosion resistant and can be hardened by heat treatment. These grades are chromium steels with no nickel. The martensitic grades are used where strength, hardness, and wear resistance are required. Some of the more commonly used grades are given in the table below (Pitcher 1976).

| Type | Typical composition | Typical applications |
|---|---|---|
| 410 | 11.5–13.5% Cr, 0.6% Ni, 1% Mn, 0.15% C, balance Fe | Basic martensitic grade. Low cost, general purpose, stainless steel that can be heat treated and is used when corrosion is not severe. Typical applications are for highly stressed parts needing a combination of strength and corrosion resistance. |
| 420 | 12–14% Cr, 0.6% Ni, 0.2–0.4% C, balance Fe | Increased carbon improves mechanical properties. |
| 431 | 15–17% Cr, 1.25–2.5% Ni, 0.2% carbon maximum, balance Fe | Better corrosion resistance than type 410 and good mechanical properties. Typical applications are for high strength parts, such as valves and pumps. |
| 440 | 16–18% Cr, 0.6% Ni, 0.75% Mo maximum, 0.6–1.2% C, balance Fe | Further increases the chromium and carbon to improve strength and corrosion resistance. |

c) *Ferritic grades*.

Ferritic grades are suited to applications that need to resist corrosion and oxidation, but also require a high resistance to stress corrosion cracking. These grades cannot be heat treated. Ferritic grades are more corrosive resistant than martensitic grades, but generally not as good as the austenitic grades. Like martensitic grades, these grades are chromium steels with no nickel. Some of the more commonly used grades are given in the table below (Pitcher 1976):

| Type | Typical composition | Typical applications |
|---|---|---|
| 430 | 16–18% Cr, 0.12% C maximum, balance Fe | The basic ferritic grade, with slightly less corrosion resistance than Type 304. High resistance to nitric acid, sulfur gases, and many organic and food acids. |
| 405 | 11.5–14.5% Cr, 0.8% C maximum, 0.1–0.3% Al, balance Fe | Lower chromium, but aluminum added to prevent hardening when cooled from high temperatures. Typical applications include heat exchangers. |
| 442 | 18–23% Cr, 1% Mn, 0.2% carbon, balance Fe | Increased chromium to improve resistance to oxidation scaling. Typical applications include furnace and heater parts. |

| 446 | 23–27% Cr, 1.5% Mn maximum, 0.2% carbon maximum, balance Fe | Contains even more chromium than Type 442 to offer high resistance to oxidation scaling at high temperatures and when sulfur may be present. |

**d)** *Duplex grades.*

These grades are a combination of austenitic and ferritic material. They have higher strength and better resistance to stress corrosion cracking.

**e)** *Precipitation hardening grades.*

Precipitation hardening grades offer a combination of strength, ease of fabrication, ease of heat treatment, and corrosion resistance not found in the other classes. These grades are used for bar, rod, wire, forgings, sheet. and strip products.

**3)** *Nickel and nickel alloys.* Nickel and its alloys have good resistance to many chloride-bearing and reducing environments that attack stainless steels. The resistance of nickel to reducing environments is enhanced by molybdenum and copper. Pure nickel is generally only used for a sodium and potassium hydroxide services, with its alloys being preferred for most applications. The most common grades of nickel alloys are (Hughson 1976):

**a)** *Hastelloy.*

Hastelloy alloys offer high resistance to uniform corrosion attack, localized corrosion, stress corrosion cracking, and ease of welding and fabrication. Some of the more commonly used grades are given in the table below.

| Hastelloy | Typical composition | Typical applications |
|---|---|---|
| Type B | 28% Mo, 2% Fe, 1% Cr, balance Ni | Hydrochloric acid, hydrogen chloride gas, and sulfuric, acetic, and phosphoric acids. Thermal oxidation of chlorinated wastes. |
| Type C | 16% Mo, 5% Fe, 16% Cr, balance Ni | Mineral (inorganic) acids (e.g. sulfuric, nitric, phosphoric, etc.), formic and acetic acids, acetic anhydride, chlorine, chlorine contaminated solutions (organic and inorganic), hypochlorite, brine solutions, and seawater. Paper pulp processes. |
| Type G | 3% Mo, 30% Fe, 21% Cr, 2% Cu, balance Ni | Hot sulfuric and phosphoric acids, oxidizing and reducing agents, acid and alkaline solutions, mixed acids, sulfate compounds, contaminated nitric acid, and hydrofluoric acid. |

**b)** *Inconel.*

Inconel is a family of austenitic nickel-chromium-based alloys. When heated, Inconel forms a stable, passive oxide layer protecting the surface from further attack. Inconel retains strength over a wide temperature range and is attractive for high-temperature applications where aluminum and steel would be vulnerable to creep. Some of the more commonly used grades are given in the table below.

| Inconel | Typical composition | Typical applications |
|---|---|---|
| Type 600 | 7% Fe, 16% Cr, balance Ni | Applications that require resistance to corrosion and heat. Reducing conditions of alkaline solutions. Resistant to chloride-ion stress corrosion cracking. |
| Type 625 | 9% Mo, 2.5% Fe, 22% Cr, balance Ni | Sulfuric acid at low temperatures. Phosphoric acid and sodium hydroxide solutions at high temperatures, but low concentrations. |
| Type 825 | 3% Mo, 30% Fe, 21% Cr, balance Ni | Reducing environments such as sulfuric and phosphoric acids. Oxidizing environments such as nitric acid solutions, nitrates, and oxidizing salts. Most organic acids including boiling concentrated acetic acid, acetic-formic acid mixtures, maleic and phthalic acids. Most alkaline solutions. |

**c)** *Monel.*

Monel is a group of nickel alloys, primarily composed of nickel (up to 67%) and copper (typically 32%), iron (typically 2%), with small amounts of manganese, carbon, and silicon. It is used for sulfuric acid, hydrochloric acid, and reducing conditions. It can be used for many alkalis, but not at all concentrations. It has less resistance to alkalis than pure nickel. Monel can be used in seawater and brackish water services. It is not vulnerable to stress corrosion cracking by any of the chloride salts. Compared with steel, Monel is difficult to fabricate as it work-hardens very quickly.

**4)** *Aluminum and aluminum alloys.* Aluminum is principally used in heat-transfer applications because of its high thermal conductivity. Aluminum plate-fin heat exchangers are used extensively in cryogenic applications. Aluminum fins are widely used to enhance the heat transfer in high transverse fins for air-cooled heat exchangers. One of the main limitations of aluminum is that its strength declines significantly above 150 °C. The highest working temperature is normally taken to be 200 °C. However, it has excellent low-temperature properties and can be used down to −250 °C. Many mineral acids attack aluminum, but it can be used with concentrated nitric acid above 82%, concentrated sulfuric acid, and most organic acids. Aluminum cannot be used with strong alkali solutions. As with stainless steels and nickel alloys, corrosion resistance is a result of the formation of a thin oxide layer, which means that it is most suited to use in oxidizing conditions.

**18**

A number of aluminum alloys are available. The alloys are mainly to improve the mechanical properties and most have lower corrosion resistance than the pure metal.

5) *Copper and copper alloys*: Copper has the outstanding potential attraction that it has an extremely high thermal conductivity for use in heat-transfer equipment. However, pure copper is rarely used in chemical plant equipment. Copper is generally attacked by mineral acids, but is generally resistant to caustic alkalis, organic acids, and salts. Cupronickels containing 10–30% nickel have good corrosion resistance and are used for heat exchanger tubing. Resistance to seawater is particularly good for cupronickels and they are often used for heat exchangers where seawater is used as the coolant.

6) *Lead and lead alloys*. Lead was one of the traditional materials of construction in chemical plant. It was particularly important in the construction of sulfuric acid plants. Lead relies for its high resistance to corrosion on layers of insoluble lead salts that form on the surface. However, the use of lead and lead alloys has declined as other metals and plastics have replaced it.

7) *Titanium*. Titanium has good corrosion resistance in oxidizing and mild reducing environments. It is resistant to nitric acid at almost all concentrations. It is resistant to hot chloride solutions and performs better than stainless steel for seawater duties. However, fabrication with titanium is difficult.

8) *Zirconium*. Zirconium is similar to the titanium in its corrosion properties. However, zirconium is more resistant to hydrochloric acid and resistant to chlorides, except ferric and cupric. Zirconium resembles titanium in its difficulty of fabrication. All welding must be carried out in an inert medium.

9) *Tantalum*. Tantalum is practically inert to many oxidizing and reducing acids. It is attacked by hot alkalis and hydrofluoric acid. The mechanical properties of tantalum are similar to mild steel, but it has a higher melting point.

10) *Glass*. Glass chemical plant equipment is available from specialist equipment manufacturers. Pipe joints typically use polytetrafluoroethylene (PTFE) gaskets. Glass has excellent resistance to all acids except hydrofluoric acid. There are two major drawbacks to its use. Firstly, it is brittle and not suited to high pressures and can also be damaged by thermal shock. Secondly, the availability of equipment is limited in scope and restricted to relatively small scales. However, glass lined steel combines the corrosion resistance of glass with the strength of steel. Thus, glass-lined stirred reactor vessels are quite common.

11) *Plastic materials*. Commonly used plastic materials of construction are polyvinyl chloride (PVC), acrylonitrile–butadiene–styrene (ABS), and polyethylene (PE). Generally, plastics have high resistance to weak mineral acids and inorganic salt solutions when metals might not be suited. By comparison with metals, plastics are limited to use at relatively low temperatures. The use of plastic materials is generally restricted to tanks, pipes, and valves. However, specialist designs of heat exchangers and pumps are also available in plastic materials. The most commonly used plastic materials are shown in the following table.

| PVC | Most frequently specified of all thermoplastic-piping materials. Resistant to corrosion and chemical attack by acids, alkalis, salt solutions, and many other chemicals. However, it is attacked by polar solvents such as ketones and aromatics. The maximum service temperature for PVC is 60 °C under pressure and 80 °C in drainage. |
|-----|-----|
| ABS | Temperature range is –40 °C to 82 °C. ABS is resistant to a wide variety of process materials. |
| PE | Of the different grades of PE, process equipment is usually manufactured from high-density PE for strength and hardness. High-density PE tanks can be used up to 55 °C. Pipe is usually made from medium or high density PE. PE is generally used for gas distribution, water lines, and slurry lines. |

12) *Composite materials*. A composite material combines two or more materials, often ones that have very different properties. The properties of the two materials combine together to give the composite properties. The most commonly used composite material is fiberglass. Glass fiber-reinforced plastics (GRP) or fiberglass-reinforced plastics (FRP) use polyester resins reinforced with glass fiber. On its own, glass is strong but brittle and it will break if bent. The matrix holds the glass fibers together and protects them from damage by spreading the forces acting on them. Glass reinforced plastics are relatively strong and have a good resistance to dilute mineral acids, inorganic salts, and many solvents, but are less resistant to alkalis. The use of glass-reinforced plastics is mostly restricted to tanks and vessels that can be fabricated in sizes up to 20 m.

Some advanced composites are now made using carbon fibers instead of glass. These materials are lighter and stronger than fiberglass but more expensive to produce and therefore do not find wide application in a chemical plant.

13) *Linings*. Many corrosion-resistant materials are not suitable for fabrication into large-scale equipment because they are difficult to weld, too brittle, too soft, or too expensive. Rather than use an expensive alloy for the construction of large-scale equipment, a cheaper material such as carbon steel can be used, with a lining of corrosion-resistant material to prevent damage. Glass can be used to line vessels and pipes. Plastic linings can be used. The most chemically resistant plastic for such applications is polytetrafluoroethylene (PTFE), which is resistant to all alkalis and acids except fluorine and chlorine gas at elevated temperatures. Rubber has been used extensively in the past to line tanks. Both natural and synthetic rubbers can be used. Natural rubber has good resistance to most alkalis and most acids, but not nitric acid. Synthetic rubbers can be used with nitric acid and strongly oxidizing environments, but are generally not suited to chlorinated solvents. Metal linings can also be used (e.g. titanium lining of a carbon steel pressure vessel).

In addition to the use of linings for corrosion resistance, linings are also necessary in high-temperature equipment to protect the metal shells of furnaces. Refractory bricks and cements are used as lining for fired heaters, boilers, and furnace reactors. Carbon steel shells must be maintained below 500 °C in operation.

# 18.5   Criteria for Selection of Materials of Construction

Many factors go into the selection of materials of construction for a particular application:

1) *Corrosion*. The primary consideration in many chemical process applications is corrosion. The general characteristics of corrosion and the use of different materials have been discussed earlier. Corrosion charts are available to give more detail in the selection process (Appendix D; Green and Perry 2007). However, such corrosion charts should be used with great caution. Corrosion is very difficult to predict with certainty. In addition to the bulk fluid major components, it is dependent on temperature, on low levels of components in the process materials, and on the preparation of the material of construction. It is also dependent on localized phenomena, as discussed earlier. The most reliable way to predict corrosion resistance is to use experience from working processes operating with the same, or a similar, process. Laboratory tests are useful, but are not as effective as tests carried out on working plant processing the same chemicals. Test samples can be located in working plant and tested for corrosion resistance. However, again the temperature is important in the tests, as well as the composition of the process materials.

   The primary choice for material for most applications is carbon steel, if corrosion permits. An alternative is to use plastic materials, but these are limited by their inability to withstand elevated temperatures and by their low strength. Thus, lack of corrosion resistance for carbon steel, or adverse temperature and pressure in the case of plastic material, force the use of more expensive materials. Most corrosion problems encountered in process design can be solved by using one of the various grades of stainless steel discussed earlier (Pitcher 1976). However, sometimes the corrosion problems are too severe for stainless steels to handle. In such cases, it might be necessary to resort to more expensive metals, such as nickel alloys. As a cheaper alternative, it is sometimes possible to line equipment constructed with cheaper material, such as carbon steel, with glass, PTFE, or some other corrosion-resistant material or metal.

2) *Temperature*. In addition to the corrosion phenomena being highly temperature dependent, temperature significantly affects the mechanical properties of the material. Plastic materials are generally restricted for use below 80 °C. Copper can be used up to 260 °C and aluminum up to 460 °C. Carbon steel is limited to use below 500 °C. Stainless steel grades 304 and 316 and nickel can be used up to 760 °C. Some other stainless steel grades, such as 321 and 347, and Monel, can be used up to 815 °C. Type 446 stainless steel, inconel, and titanium can be used up to 1100 °C. Creep resistance is an important consideration for materials being exposed to high temperatures, especially when in combination with high stress.

   However, a high temperature is not the only problem. Metals can fail catastrophically when exposed to very low temperatures through *brittle fracture*. Carbon steel can be used down to −45 °C. Stainless steels can be engineered for lower temperatures, but this depends on the grade. Some austenitic grades of stainless steel can be used as low as −270 °C. Aluminum can be used down to −250 °C.

3) *Pressure*. For processes working under significant pressure, the main consideration is the strength of the material of construction, which will be temperature dependent.

4) *Abrasive environments*. Processing solids and slurries can create an abrasive environment for materials of construction. Under such conditions, materials of construction without the necessary hardness will quickly deteriorate.

5) *Ease of the fabrication*: Fabrication involves cutting, bending, machining, joining, and assembling to produce functional equipment. The mechanical properties of materials and their ease of fabrication are intimately linked. The mechanical properties can change as a result of the fabrication. Properties such as brittleness, ductility, malleability, and the work hardening and heat-treatment properties of materials are all important factors in the ease of fabrication.

   Joining of materials is another important consideration. *Welding* joins metals by causing coalescence of the metals being joined. This is most often achieved by melting the metals to be joined and adding a filler material to form a pool of molten material that cools to become a strong joint. Many different energy sources can be used for welding. However, not all metals are easily welded and welding different metals together can present problems. Brazing is an alternative joining process for metals in which a filler metal is melted between the metals to be joined. The filler metals used in brazing usually have melting points between 450 °C and 1000 °C, but must have a lower melting point than the metals being joined. Unlike welding, the metals to be joined are not melted. The filler metal melts and reacts metallurgically with the metals to be joined, forming strong, permanent joints. Brazing has the advantage of producing fewer thermal stresses than welding.

   Various methods are available for joining plastics and composites. These can be classed into mechanical fastening, adhesive and solvent bonding, and welding. Welding is an effective method of permanently joining plastic components, but can only be applied to thermoplastics and thermoplastic elastomers.

6) *Availability of standard equipment*. When designing a chemical process, it is necessary to choose equipment (e.g. pumps, compressors, valves, and pipes) from suppliers that have a limited range of specifications, including materials of construction. The choice of material of construction for such equipment might be dictated by the options available from equipment suppliers.

7) *Cost*. Cost effectiveness is one of the main considerations in process design. The choice of material of construction has major cost implications, as illustrated in Tables 2.5 to 2.7, which shows that equipment constructed from stainless steels will be between 2.4 and 3.4 times more expensive than if constructed from carbon steel, between 3.6 and 4.4 times more expensive if constructed from nickel alloys, and for titanium 5.8 times as expensive as carbon steel. There is therefore a great incentive to avoid exotic materials of construction. An alternative discussed above is to use low-cost material with a corrosion-resistant lining.

**18**

## 18.6  Materials of Construction – Summary

Many factors enter into the choice of the materials of construction. Among the most important are:

- corrosion resistance;
- mechanical properties (particularly tensile and yield strength, compressive strength, ductility, toughness, hardness, fatigue limit, and creep resistance);
- effect of temperature on mechanical properties (both low and high temperatures);
- ease of fabrication (machining, welding, and so on);
- availability of standard equipment in the material;
- cost.

A much more detailed discussion of selection of materials of construction has been given by Kirby (1980).

## References

Green, D.W. and Perry, R.H. (2007). *Perry's Chemical Engineer's Handbook*. McGraw-Hill.

Hughson, R.V. (1976). High-nickel alloys for corrosion resistance. *Chemical Engineering* 125.

Hunt, M.W. (2014). Develop a strategy for material selection. *Chemical Engineering Progress* 42.

Kirby, G.N. (1980). How to select materials. *Chemical Engineering* 86.

Pitcher, J.H. (1976). Stainless steels: CPI workhorses. *Chemical Engineering* 119.

# Chapter 19

# Mechanical Design

Whilst the final mechanical design of vessels and equipment is normally carried out by specialist mechanical engineers, it is still necessary for process designers to have an understanding of mechanical design for a variety of reasons:

a) The process design of equipment and vessels must be cognizant of the difficulties that the process design might impose on the consequent mechanical design.

b) The size and dimensions of process vessels, and arrangements for their support, often constrain the design to avoid difficulties in the mechanical design that would lead to excessive fabrication costs (e.g., diameter or height of cylindrical pressure vessels).

c) Even preliminary capital costing can require a preliminary mechanical design (e.g., the cost of the vessel is related to the weight of the material of construction used, which depends on the thickness of the vessel).

d) Systems for the protection of vessels and equipment against excessive pressure will be influenced by the mechanical design.

e) The layout of process equipment in process plots will be influenced by mechanical design issues (e.g., the option of a distillation column being self-supporting or supported by a structure will lead to different mechanical designs, different capital costs, and different space requirements).

f) Poor mechanical design can create major hazards, especially when installed equipment and piping is modified (e.g., the root cause of the Flixborough disaster in 1974 was poor mechanical design for a piping modification).

## 19.1 Stress, Strain, and Deformation

There are a number of different types of force that can induce deformation and ultimately failure of materials. Figure 19.1 illustrates the types of *mechanical forces* that can induce different types of deformation. Starting from the original body in Figure 19.1a, the various forces could be:

- *tensile*, as illustrated in Figure 19.1b, results from axial forces that tend to stretch or lengthen the body;
- *compressive*, as illustrated in Figure 19.1c, results from axial forces that tend to compress or shorten the body;
- *shearing*, as illustrated in Figure 19.1d, results from forces acting parallel but not on the same line that tends to shear the body;
- *torsional*, as illustrated in Figure 19.1e, tends to twist the body
- *bending*, as illustrated in Figure 19.1f, results from a transverse load that tends to bend the body;
- *buckling*, as illustrated in Figure 19.1g, results from axial compression forces that tend to bend the body, governed by the stiffness of the body.

In addition to mechanical forces, differences in temperature in a body can also create forces through differential expansion or contraction in different locations to induce deformation and possibly failure.

### 19.1.1 Normal Stress

Normal (axial) stress, as illustrated in Figure 18.1, occurs when equal and opposite collinear opposing forces (i.e. in the same line) act through the solid. The forces can be *tensile*, as in Figure 18.1a, or *compressive*, as in Figure 18.1b.

### 19.1.2 Shear Stress

Shear stresses are caused by equal and opposite forces acting in parallel, but are not collinear (i.e. not on the same line), as illustrated in Figure 19.2, creating a tendency for one part of the body to slide over another. In Figure 19.3 a shear stress $\tau$ acts on parallel sides of the body. These stresses create opposing forces on each parallel side of magnitude $\tau \cdot x \cdot z$ and form a couple with magnitude $(\tau \cdot x \cdot z)y$, which can only be balanced by *complementary* shear stresses acting normally, as shown in Figure 19.3.

*Process Plant Design*, First Edition. Robin Smith.
© 2024 John Wiley & Sons Ltd. Published 2024 by John Wiley & Sons Ltd.
Companion website: www.wiley.com/go/processplantdesign

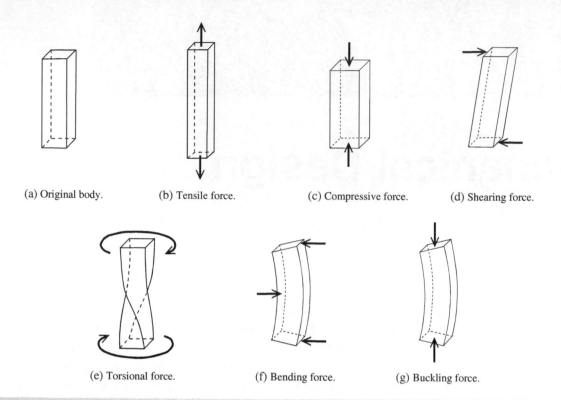

(a) Original body.        (b) Tensile force.        (c) Compressive force.        (d) Shearing force.

(e) Torsional force.        (f) Bending force.        (g) Buckling force.

**Figure 19.1**

Mechanical forces that induce deformation.

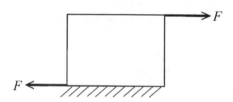

**Figure 19.2**

Parallel forces not in line create shear stress.

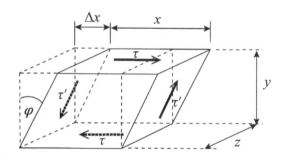

**Figure 19.4**

Shear stress is measured by the angle of deformation.

The complementary shear stress in Figure 19.3 has magnitude $\tau'$. For the forces to be in equilibrium:

$$(\tau \cdot x \cdot z)y = (\tau' \cdot y \cdot z)x \qquad (19.1)$$

Thus:

$$\tau = \tau' \qquad (19.2)$$

Therefore, every shear stress in equilibrium is accompanied by an equal complementary shear stress.

Figure 19.4 shows the distortion produced by the shear stress and is given by:

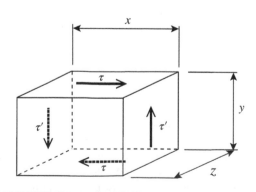

**Figure 19.3**

A shear stress sets up a complementary shear stress to resist rotation.

$$\gamma = \frac{\Delta x}{y} = \tan \varphi \approx \varphi \qquad (19.3)$$

($\tan \varphi \approx \varphi$ providing $\varphi$ is small and measured in radians)

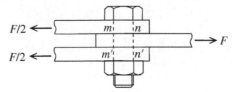

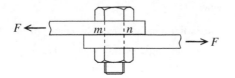

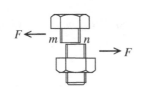

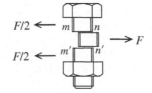

(a) A bolt subject to single shear.          (b) A bolt subject to double shear.

**Figure 19.5**

Shearing force acting on a bolt.

where

$\gamma$ = shear strain (–)

$\Delta x$ = distortion parallel to the shear stress (m)

$y$ = separation of the shear forces (m)

$\varphi$ = change in angle caused by the shear (radians)

For elastic materials shear strain is proportional to the shear stress producing it. The ratio is defined as the modulus of rigidity $G$ (N mm$^{-2}$):

$$G = \frac{\text{Shear stress}}{\text{Shear strain}}$$

$$= \frac{\tau}{\varphi} \qquad (19.4)$$

As an example, Figure 19.5 shows bolted joints in which the bolts are under shear stress. Figure 19.5a shows a bolt subject to single shear across the plane *mn*. Figure 19.5b shows a bolt subject to double shear across the planes *mn* and *m'n'*. If it is assumed that the shearing stresses are uniformly distributed and there are no bending stresses, then for single shear in Figure 19.5a:

$$\tau = \frac{F}{A} = \frac{F}{\pi D^2/4} \qquad (19.5)$$

where

$\tau$ = shear stress (N mm$^{-2}$, N m$^{-2}$)

$F$ = force (N)

$A$ = area under shear (mm, m$^2$)

$D$ = diameter of bolt (mm, m)

For the double shear case in Figure 19.5b:

$$\tau = \frac{F}{A} = \frac{F}{2 \times \pi D^2/4} = \frac{2F}{\pi D^2} \qquad (19.6)$$

Bolted joints are common in structures. When using bolts under shear load, it is important to recognize that the area available to resist a shear stress depends on whether the part of the bolt in the shear plane is threaded or unthreaded. If the part of the bolt in the shear plane is threaded then the area available to resist the shear is smaller than the size of the bolt. For example, a 20 mm diameter bolt might have a diameter between the roots of the threads of 17.7 mm. This *minor diameter* should be used for the area to resist the shear force. When multiple bolts are used, the minimum distance between the bolts is typically 3*D*.

**Example 19.1**    Figure 19.6 shows a beam supported by three bolts attached to a fin plate, which is in turn welded to the supporting beam. The three bolts are in single shear and need to resist the shearing force *F*. The bolts are 20 mm in diameter with threads in the shearing plane, each giving an area to resist the shearing of 245 mm$^2$ and allowable stress of 160 N mm$^2$. Determine the force that the three bolts can support if the failure of the bolts is assumed to be shearing only.

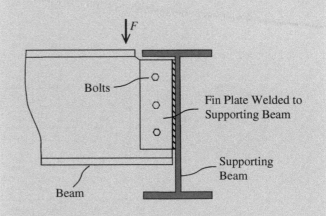

**Figure 19.6**

A beam supported by three bolts subject to shear.

**Solution**

For bolts in single shear each bolt can resist a shearing force of $\tau \times$ area of bolt. Thus, the three bolts can support a force $F$ given by:

$$F = 3 \times \tau \times A$$
$$= 3 \times 160 \times 245$$
$$= 118 \times 10^3 \text{ N}$$

It should be noted that such a bolted joint can fail by other mechanisms than just shear of the bolts. For example, if the holes are significantly larger than the diameter of the bolt, then bending can occur. The plates can also fail (Timoshenko 1956; Ryder 1969; Willems et al. 1981; Gere 2004).

### 19.1.3   Torsion

*Torque* is the moment of a force that causes a body to rotate about an axis and is measured by the product of the force and the distance from the axis of rotation. *Torsion* occurs when two torques of equal magnitude act on the same body in opposite directions. The most important torsion problem in process plant design applies to circular shafts, for example, pumps, compressors, and mixers. Figure 19.7 shows that when torsion is applied to a circular shaft, shear stresses distort the shaft through a small angle. This distortion occurs whether the shaft is fixed at a particular section or is free to rotate transmitting power from the shaft. If the shaft is imagined to be composed of a series of circular discs, it can be assumed that the discs remain perpendicular to the longitudinal axis and that displacement of a point on the cross-section is proportional to the distance from the center.

In Figure 19.7 two equal and opposite torques $T$ are applied to a circular shaft. A straight line $AB$ along the length of the shaft at any radius before the application of the torsion becomes a helix $AC$ under the action of the torsion. If $BC$ is small, then $AC$ can be approximated as a straight line. Thus, at a radius $r$ in Figure 19.7:

$$\text{Arc } BC = r\theta \approx L\varphi \tag{19.7}$$

where

$r$ = radial distance from the center of the shaft (m)
$\theta$ = radial angle of the twist (radians)
$L$ = length of the shaft (m)
$\varphi$ = longitudinal angle of the shear along the shaft (radians)
   = shear strain (−)

From Eq. (19.4) $\varphi = \tau/G$ and substituting in Eq. (19.7) gives:

$$\frac{\tau}{G} = \varphi = \frac{r\theta}{L} \tag{19.8}$$

where

$\tau$ = shear stress (N m$^{-2}$)
$G$ = modulus of rigidity (N m$^{-2}$)

Rearranging:

$$\tau = \frac{G\theta}{L} r \tag{19.9}$$

This shows that for a given torque, the shear stress is proportional to the radius. In Figure 19.7 the torque in an element of radius $r$ can be equated to the sum of the moments of the tangential stresses on the element:

$$dT = \tau (2\pi r \, dr) \, r \tag{19.10}$$

where

$T$ = torque (N m)

$$T = \int_0^R 2\pi r^2 \tau \, dr \tag{19.11}$$

Substituting Eq. (19.9):

$$T = \int_0^R 2\pi r^2 \frac{G\theta r}{L} \, dr$$
$$= \frac{2\pi G\theta}{L} \int_0^R r^3 \, dr$$
$$= \frac{2\pi G\theta}{L} \times \frac{R^4}{4} \tag{19.12}$$
$$= \frac{G\theta}{L} \left( \frac{\pi R^4}{2} \right)$$
$$= \frac{G\theta}{L} \left( \frac{\pi D^4}{32} \right)$$

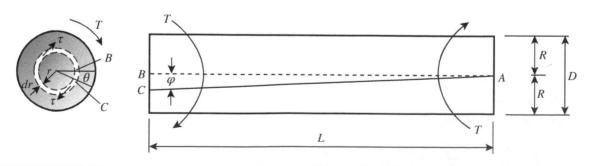

### Figure 19.7
Torsional stress acting on a solid circular cylinder.

where

$R$ = radius of shaft (m)
$D$ = diameter of shaft (m)
$L$ = length of the shaft (m)

Combining Eqs (19.11) and (19.12):

$$\frac{T}{J} = \frac{G\theta}{L} = \frac{\tau}{r} \qquad (19.13)$$

where

$J$ = polar moment of inertia (m$^4$)

$$= \left(\frac{\pi D^4}{32}\right) \text{ for a solid circular shaft}$$

**Example 19.2**   A stainless steel shaft is transmitting power of 1.5 MW to a process machine operating at a speed of 3000 rpm. The yield strength of the material is $215 \times 10^6$ Pa. Assume a safety factor of 2. What should be the minimum diameter of the shaft?

**Solution**

First establish the relationship between shaft power delivered by the shaft, torque, and speed of rotation:

$$\text{Power} = \frac{\text{Work}}{\text{Time}} = \frac{\text{Force} \times \text{Distance}}{\text{Time}}$$

If the distance moved in one revolution is $2\pi r$ and the time taken for 1 revolution is $1/N$, where $N$ is the rotational speed, then the shaft power is given by:

$$\text{Power} = \text{Force} \times 2\pi r N$$

The torque $T$ on the shaft is given by:

$$T = \text{Force} \times r$$

Combining the equations gives:

$$\text{Power} = 2\pi N T$$

Thus:

$$T = \frac{\text{Power}}{2\pi N}$$

$$= \frac{1.5 \times 10^6}{2\pi \times 3000/60}$$

$$= 4775 \text{ N m}$$

From Eq. (19.13):

$$\frac{T}{J} = \frac{\tau}{r}$$

$$\frac{4775}{\pi D^4/32} = \frac{215 \times 10^6/2}{D/2}$$

$$D^3 = \frac{4775 \times 32 \times 2}{\pi \times 215 \times 10^6 \times 2}$$

$$D = 0.061 \text{ m}$$

### 19.1.4   Bending

Consider a beam subjected to a transverse load from a point source $W$, as shown in Figure 19.8a. The deflection occurs in the same plane as the load. The beam in Figure 19.8a features simply supported ends. The points of support $A$ and $B$ allow the ends of the beam to rotate freely during bending. It is also assumed that the beam is supported in such a way that lateral movement is unrestricted. In this way, any reaction on the supports $R_1$ and $R_2$ as a result of the load $W$ can be considered to be vertical. The magnitude of $R_1$ and $R_2$ can be determined by the equations of moments. If the beam is in equilibrium, the sum of the moments of all forces on Point $B$ will be zero. Thus:

$$R_1 L - Wb = 0 \qquad (19.14)$$

where

$R_1$ = reaction force from Support 1 (N)
$W$ = load (N)

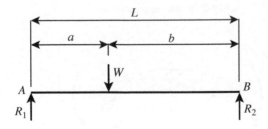

(a) Location of the concentrated load creates a reactive force from the supports.

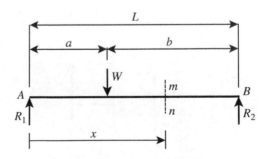

(b) Separation of the forces acting in different parts of the beam.

**Figure 19.8**

Concentrated load acting on a beam with simple supports.

Rearranging:

$$R_1 = \frac{Wb}{L} \qquad (19.15)$$

Similarly, for Support B:

$$R_2 = \frac{Wa}{L} \qquad (19.16)$$

where

$R_2$ = reaction force from Support 2 (N)

Now consider the beam is cut into two parts at $mn$ at a distance $x$ from $A$ in Figure 19.8b. If the portion of the beam between $mn$ and $B$ is assumed to be removed, then the forces at $mn$ must represent the action of the portion between $mn$ and $B$. The forces acting on the portion of the beam between $A$ and $mn$ (where $x$ is greater than $a$ in Figure 19.8b) can be replaced by a vertical force $F$:

$$F = R_1 - W \qquad (19.17)$$

where

$F$ = force (N)

The magnitude of the couple $M$ is given by:

$$M = R_1 x - W(x-a)$$
$$= \frac{Wbx}{L} - W(x-a) \qquad (19.18)$$

where

$M$ = bending moment (N m)

For $x < a$ in Figure 19.8b, Eq. (19.17) would become $F = R_1$ and Eq. (19.18) would become $M = R_1 x = Wbx/L$. The force $F$ is known as the *shearing force* and $M$ is the *bending moment* at the cross-section $mn$. Whereas $F$ is constant for this case, $M$ varies with $x$. $M = 0$ at $x = 0$ and $M = Wba/L$ at $x = a$. By convention the shear force is considered positive when it acts clockwise, as shown in Figure 19.9a, and negative when it acts anticlockwise, as shown in Figure 19.9b. A bending moment is considered positive if the forces result in a bending that is convex downwards, as shown in Figure 19.9c. Forces that result in a bending that is convex upwards are considered negative, as shown in Figure 19.9d. For the Section $mn$ to $B$ in Figure 19.8b:

$$F = R_1 - W = -R_2 = -\frac{Wa}{L} \qquad (19.19)$$

$$M = R_2(L-x)$$
$$= \frac{Wa}{L}(L-x) \qquad (19.20)$$

where $M = 0$ at $x = L$ and $M = Wab/L$ at $x = a$. If the load is at the center of the span, then at $a = b = L/2$ and $M = WL/4$. Figure 19.10 shows the shear force and bending moment diagram for the beam with a point load of $W$.

Consider now a beam with a uniform loading of $w$ per unit length, as shown in Figure 19.11a. The reactions in this case are:

$$R_1 = R_2 = \frac{wL}{2} \qquad (19.21)$$

The shearing force in the portion of the beam between $A$ and $mn$ can be represented by one force acting at the center, as shown in Figure 19.11b.

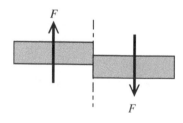

(a) Positive shear force.

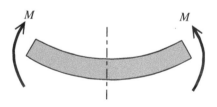

(c) Positive bending moment.

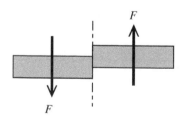

(b) Negative shear force.

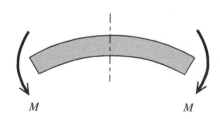

(d) Negative bending moment.

**Figure 19.9**

Sign convention for shear force and bending moment diagrams.

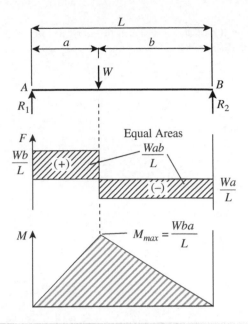

**Figure 19.10**

Shear force and bending moment diagram for a beam with a concentrated load on simple supports.

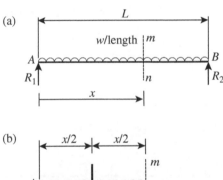

**Figure 19.11**

Distributed load on a beam with simple supports.

$$F = R_1 - wx$$

$$= w\left(\frac{L}{2} - x\right) \tag{19.22}$$

The algebraic sum of the moments for the beam between $A$ and $mn$ is given by:

$$M = R_1 x - wx \times \frac{x}{2}$$

$$= \frac{wL}{2}x - \frac{wx^2}{2} \tag{19.23}$$

$$= \frac{wx}{2}(L - x)$$

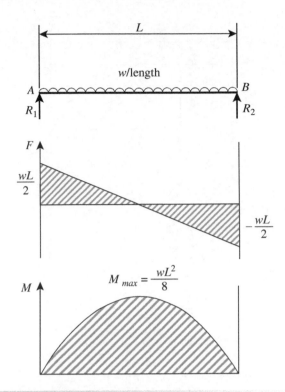

**Figure 19.12**

Shear force and bending moment diagram for a beam with a distributed load on simple supports.

This is a parabolic curve, as shown in Figure 19.12 with a value of $M = 0$ at $x = 0$ and at $x = L$. The maximum $M$ is at $L/2$ (taking $\mathrm{d}M/\mathrm{d}x = 0$ at $L/2$):

$$M_{\text{max}} = \frac{wL^2}{8} \tag{19.24}$$

Figure 19.12 shows the shear force and bending moment diagram for the beam with a distributed load of $w$ per unit length. The maximum shear force acts at the beam supports, whereas the maximum bending moment acts at the center of the beam.

Figures 19.10 and 19.12 illustrate only two of the many possible cases for types of beam, beam supports, and loading. The details of more complex arrangements, such as different supports, cantilevers, and combinations of point and distributed loading, can be found elsewhere (Timoshenko 1956; Ryder 1969; Willems et al. 1981; Gere 2004).

**Example 19.3**    Figure 19.13 shows a simply supported beam with a combined point loading of $W$ and uniform loading of $w$ per unit length. Sketch the shear force and bending moment diagrams.

**Solution**

Taking moments around Point $C$ in Figure 19.13:

$$R_1 \times L = W \times b + wL \times \left(\frac{L}{2}\right)$$

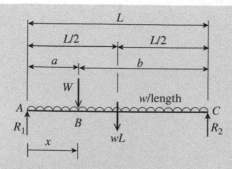

**Figure 19.13**

Combined point and distributed loads.

Solving for $R_1$:

$$R_1 = \frac{Wb}{L} + \frac{wL}{2}$$

Taking moments around Point $A$ in Figure 19.13:

$$R_2 \times L = W \times a + wL \times \left(\frac{L}{2}\right)$$

Solving for $R_2$:

$$R_2 = \frac{Wa}{L} + \frac{wL}{2}$$

For the Section $AB$ in Figure 19.13, the shear force $F_{AB}$ at a distance $x$ from Point $A$ is given by:

$$F_{AB} = R_1 - wx$$

$$= \left(\frac{Wb}{L} + \frac{wL}{2}\right) - wx$$

For the Section $BC$ in Figure 19.13, the shear force $F_{BC}$ at a distance $x$ from Point $A$ is given by:

$$F_{BC} = R_1 - W - wx$$

$$= \left(\frac{Wb}{L} + \frac{wL}{2}\right) - W - wx$$

The bending moment in the Section $AB$ in Figure 19.13 is given by:

$$M_{AB} = R_1 \times x - wx \times \frac{x}{2}$$

$$= \left(\frac{Wb}{L} + \frac{wL}{2}\right)x - \frac{wx^2}{2}$$

The bending moment in the Section $BC$ in Figure 19.13 is given by:

$$M_{BC} = R_1 \times x - W \times (x-a) - wx \times \frac{x}{2}$$

$$= \left(\frac{Wb}{L} + \frac{wL}{2}\right)x - W(x-a) - \frac{wx^2}{2}$$

Figure 19.14 shows a sketch of the shear force and bending moment diagrams. The shear force becomes zero when:

$$F_{BC} = 0 = R_1 - W - wx$$

$$= \left(\frac{Wb}{L} + \frac{wL}{2}\right) - W - wx$$

Solving for $x$:

$$x = \frac{R_1 - W}{w} = \frac{\left(\dfrac{Wb}{L} + \dfrac{wL}{2}\right) - W}{w}$$

The maximum in the bending moment occurs when:

$$\frac{dM_{BC}}{dx} = 0 = \left(\frac{Wb}{L} + \frac{wL}{2}\right) - W - wx$$

Solving for $x$:

$$x = \frac{\left(\dfrac{Wb}{L} + \dfrac{wL}{2}\right) - W}{w}$$

This is the same point where the shear force goes to zero. These features are sketched in Figure 19.14. A particular case occurs when the point load is in the center of the span when $a = b = L/2$. The maximum shear force exists at the supports when:

$$R_1 = R_2 = F_{max} = \frac{W}{2} + \frac{wL}{2}$$

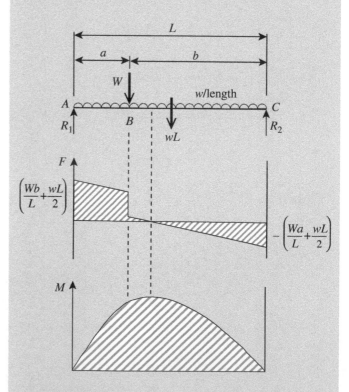

**Figure 19.14**

Shear force and bending moment diagram for a beam with combined point and distributed loads.

For a point load in the center of the span, the maximum bending moment exists at the center of the span when:

$$M_{\text{max}} = \frac{WL}{4} + \frac{wL^2}{8}$$

This is the sum of the individual maximum bending stresses for the separate cases of point load in the center of the span and distributed load.

Consider now the relationship between the shearing force and bending moment. Figure 19.15 shows a section of a beam subjected to a distributed load $w$. Let the shearing force at $x$ be $F$ and be $F + dF$ at $x + dx$. Similarly, let the bending moment at $x$

be $M$ and be $M + dM$ at $x + dx$. If $w$ is evenly loaded, then the total load acting on the section is $w\,dx$ and this acts through the center of the incremental section $C$. Analyzing the couple around point $C$ of Figure 19.15 for the incremental length:

$$M + \frac{F\,dx}{2} + (F + dF)\frac{dx}{2} = M + dM \tag{19.25}$$

If the term $dF\,dx$ is considered to be negligible, then Eq. (19.25) simplifies to:

$$F = \frac{dM}{dx} \tag{19.26}$$

Figure 19.16a shows a beam after being subjected to a bending moment $M$. It is assumed that before bending the beam is straight and the beam is bent into circular arcs with a common center of curvature. It is also assumed initially in this simplified analysis that only the bending moment acts and there is no shearing force. Figure 19.16b shows a cross-section of the beam. In Figure 19.16c it is clear that the inside, concave edge is in compression, whereas the outside convex edge is in tension. There is a *neutral surface* $NN$ between the compressive and tensile forces that does not undergo strain during bending. The intersection between the neutral surface and any cross-section at $XX$ is called the *neutral axis*. In Figure 19.16 the strain at Arc $ab$ is:

$$\begin{aligned}
\varepsilon_{ab} &= \frac{ab - NN}{NN} \\
&= \frac{(R + y)\theta - R\theta}{R\theta} \tag{19.27} \\
&= \frac{y}{R}
\end{aligned}$$

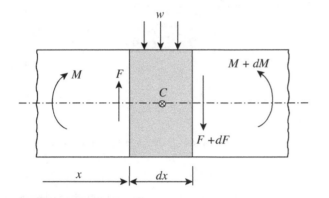

## Figure 19.15

Relationship between shearing force and bending moment.

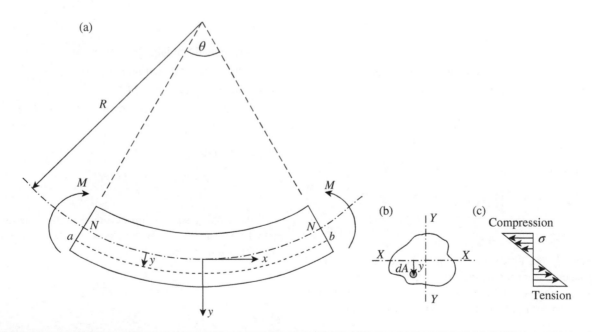

## Figure 19.16

Pure bending with no shearing force for a beam acted on by a constant bending moment.

where

$\varepsilon_{ab}$ = strain at Arc $ab$ (−)

$ab$ = length of Arc $ab$ (m)

$NN$ = length of Arc $NN$ (m)

$R$ = radius of curvature (m)

$y$ = distance from the neutral axis (m)

$\theta$ = angle of curvature (radians)

Combining with Eq. (18.3):

$$\sigma_{ab} = \frac{yE}{R} \tag{19.28}$$

where

$\sigma_{ab}$ = longitudinal stress at Arc $ab$ (N m$^{-2}$)

$E$ = ratio of stress to strain, Young's Modulus (N m$^{-2}$)

Given that $E/R$ is constant, Eq. (19.28) shows that the stress is proportional to the distance from the neutral axis $XX$. Thus, economy and weight reduction would benefit from being concentrated at the greatest distance from the neutral surface. This is why structural steel beams adopt an I-cross-section.

For different shapes of cross-section, it is important to understand the location of the neutral axis $XX$. It can be shown that the location corresponds with the location of the *centroid* of the shape of the cross-section. The centroid of the two-dimensional shape corresponds to the mean position of all the points in the shape. It also corresponds with the *center of gravity* of a two-dimensional shape. Dividing a two-dimensional shape into a series of $n$ areas, the location of the centroid can be determined by:

$$\bar{y} = \frac{\sum\limits_{i=1}^{n} y_i A_i}{\sum\limits_{i=1}^{n} A_i} = \frac{\sum\limits_{i=1}^{n} y_i A_i}{A} \tag{19.29}$$

where

$\bar{y}$ = distance of the centroid from an axis of reference (m)

$y_i$ = distance of the incremental area $i$ from an axis of reference (m)

$A_i$ = area of incremental area $i$ (m$^2$)

$A$ = area of the two-dimensional shape (m$^2$)

More generally:

$$\bar{y} = \frac{\int\limits_A y \, dA}{A} \tag{19.30}$$

In Figure 19.16b, let $dA$ denote an elemental area of cross-section distance $y$ from the neutral axis $XX$. The force acting on this elemental area is $\sigma \, dA$. Given that the forces acting over the cross-section form a couple, the resulting forces must equal zero. Thus:

$$\int\limits_A \sigma A = 0 \tag{19.31}$$

Combining with Eq. (19.28):

$$\frac{E}{R} \int\limits_A y \, dA = 0 \tag{19.32}$$

Given that $E/R$ in Eq. (19.32) is a constant, then the integral must have a value of zero. The point where the first moment of area equals zero corresponds with the centroid of the area. Thus, the neutral axis passes through the centroid of the area.

The bending moment in Figure 19.16 is balanced by the moment of normal forces over the cross-section around the neutral axis $XX$. Thus:

$$M = \int\limits_A y \sigma \, dA \tag{19.33}$$

Combining with Eq. (19.28) gives:

$$\begin{aligned} M &= \frac{E}{R} \int\limits_A y^2 \, dA \\ &= \frac{EI}{R} \end{aligned} \tag{19.34}$$

where

$I$ = *moment of inertia* or *second moment of area* (m$^4$)

$= \int\limits_A y^2 \, dA$

Recognizing that the neutral axis passes through the centroid of the area, Eqs (19.28) and (19.34) can be combined to give:

$$\frac{\sigma_B}{y} = \frac{M}{I} = \frac{E}{R} \tag{19.35}$$

where

$\sigma_B$ = longitudinal bending stress (N m$^{-2}$)

$y$ = distance from the neutral axis (m)

$M$ = bending moment (N m)

$E$ = ratio of stress to strain, Young's Modulus (N m$^{-2}$)

$R$ = radius of curvature (m)

Rearranging Eq. (19.35):

$$\sigma_B = \frac{My}{I} \tag{19.36}$$

Thus, bending stress is proportional to the distance from the neutral axis. The maximum stress occurs at the maximum distance from the neutral axis:

$$\sigma_{B\,max} = \frac{M y_{max}}{I} \tag{19.37}$$

where

$\sigma_{Bmax}$ = maximum longitudinal bending stress (N m$^{-2}$)

$y_{max}$ = distance to the most extreme fiber from the neutral axis (m)

A section modulus $Z$ can be defined as:

$$Z = \frac{I}{y_{max}} \quad (19.38)$$

where

$Z$ = section modulus (m$^3$)

$y_{max}$ = distance of the most extreme fiber from the neutral axis (m)

Combining Eqs (19.37) and (19.38):

$$\sigma_{B\,max} = \frac{M}{Z} \quad (19.39)$$

If the moment of inertia of an area with respect to the centroid is known, then the moment of inertia with respect to any parallel axis can be calculated using the *parallel axes theorem*, which will now be developed. Figure 19.17 shows a general cross-section where $XX$ is an axis through the centroid of the cross-section. In Figure 19.17 the axis $ZZ$ is parallel to the $XX$ axis. Writing the moment of inertia from the $ZZ$ axis gives:

$$I_{ZZ} = \int_A (y + h)^2 \, dA$$
$$= \int_A y^2 \, dA + 2h \int_A y \, dA + h^2 \int_A dA \quad (19.40)$$

where

$I_{ZZ}$ = moment of inertia (second moment of area) from the $ZZ$ axis (m$^4$)

Given that $\int_A y \, dA = 0$ for an axis through the centroid, then:

$$I_{ZZ} = \int_A y^2 \, dA + Ah^2$$
$$= I_{XX} + Ah^2 \quad (19.41)$$

where

$I_{XX}$ = moment of inertia (second moment of area) from the $XX$ axis (m$^4$)

$A$ = area of cross-section (m$^2$)

$h$ = perpendicular distance between the $ZZ$ and $XX$ axes (m)

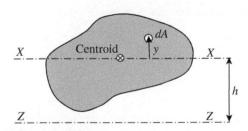

**Figure 19.17**

The parallel axis theorem.

This is the parallel axes theorem, which states that *the moment of inertia (second moment of area) about any axis is equal to the moment of inertia about a parallel axis through the centroid plus the area multiplied by the square of the distance between the axes.* It should be noted that the moment of inertia around an axis that passes through the centroid is a minimum value.

It is often necessary to calculate the moment of inertia of complex cross-sections. If this is the case, the area of the cross-section can be divided into sub-areas, then the moment of inertia of the complete area is the sum of the moments of inertia of the sub-areas if all the sub-areas refer to the same axis. If the area and centroid of the sub-areas are known, then the moment of inertia of the sub-areas can be summed, rather than integrated. The parallel axis theorem can be applied to sum the moments of inertia of the sub-areas as follows:

$$I = \sum_{i=1}^{n} (\bar{I}_i + A_i h_i^2) \quad (19.42)$$

where

$\bar{I}_i$ = moment of inertia of an individual sub-area about its own centroid (m$^4$)

$A_i$ = area of Sub-area $i$ (m$^2$)

$h_i$ = vertical distance from the centroid of Sub-area $i$ to the neutral axis (m)

Consider now a number of examples of different cross-sections. For a solid rectangular cross-section, as shown in Figure 19.18a (see Appendix E):

$$I = \frac{bd^3}{12} \quad (19.43)$$

where

$b$ = breadth of the beam (m)

$d$ = depth of the beam (m)

$$Z = \frac{bd^2}{6} \quad (19.44)$$

$$\sigma_{B\,max} = \frac{6M}{bd^2} \quad (19.45)$$

For a hollow rectangular cross-section, as shown in Figure 19.18b (see Appendix E):

$$I = \frac{b_O d_O^3}{12} - \frac{b_I d_I^3}{12} \quad (19.46)$$

where

$b_O$ = outer breadth of the beam (m)

$d_O$ = outer depth of the beam (m)

$b_I$ = inner breadth of the beam (m)

$d_I$ = inner depth of the beam (m)

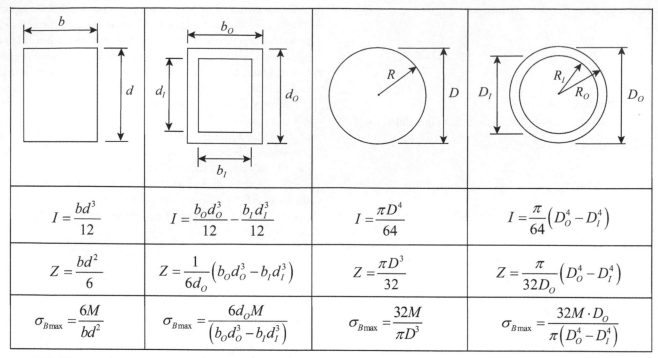

(a) Solid rectangle.   (b) Hollow rectangle.   (c) Solid cylinder.   (d) Hollow cylinder.

### Figure 19.18

Moment of inertia and bending stress for common cross-sections.

$$Z = \frac{1}{6d_O}\left(b_O d_O^3 - b_I d_I^3\right) \qquad (19.47)$$

$$\sigma_{B\max} = \frac{6d_O M}{\left(b_O d_O^3 - b_I d_I^3\right)} \qquad (19.48)$$

Note that the inner and outer rectangles must be concentric for these equations to apply. For a solid circular cross-section, as shown in Figure 19.18c (see Appendix E):

$$I = \frac{\pi D^4}{64} \qquad (19.49)$$

where

$D$ = diameter of the beam (m)

$$Z = \frac{\pi D^3}{32} \qquad (19.50)$$

$$\sigma_{B\max} = \frac{32M}{\pi D^3} \qquad (19.51)$$

For a hollow circular cross-section as shown in Figure 19.18d (see Appendix E):

$$I = \frac{\pi}{64}\left(D_O^4 - D_I^4\right) \qquad (19.52)$$

where

$D_O$ = outer diameter of the beam (m)

$D_I$ = inner diameter of the beam (m)

$$Z = \frac{\pi}{32D_O}\left(D_O^4 - D_I^4\right) \qquad (19.53)$$

$$\sigma_{B\max} = \frac{32M D_O}{\pi\left(D_O^4 - D_I^4\right)} \qquad (19.54)$$

Note that the inner and outer circles must be concentric for these equations to apply. For thin-walled cylinders, these equations can be approximated to be (see Appendix E):

$$I = \frac{\pi D_M^3 t}{8} = \pi R_M^3 t \qquad (19.55)$$

where

$D_M$ = mean diameter of the beam (m)

$R_M$ = mean radius of the beam (m)

$t$ = thickness of the thin cylinder wall (m)

$$Z = \frac{\pi D_M^2 t}{4} = \pi R_M^2 t \qquad (19.56)$$

$$\sigma_{B\max} = \frac{4M}{\pi D_M^2 t} = \frac{M}{\pi R_M^2 t} \qquad (19.57)$$

Consider now a more complex example where the cross-section is the standard I-section of structural steelwork. The actual cross-section of a rolled structural beam can be approximated as three rectangles, as shown in Figure 19.19. The second moment of area for the I-section can be determined from the three rectangles

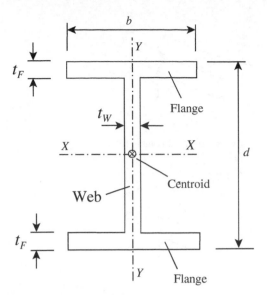

**Figure 19.19**

An I-section beam.

using the parallel axes theorem. Given the distance from the centroid of the section to the centroid of the two flanges is $(d - t_F)/2$ in Figure 19.19:

$$I_{XX} = 2\left[\frac{bt_F^3}{12} + bt_F\left(\frac{d - t_F}{2}\right)^2\right] + \frac{t_W(d - 2t_F)^3}{12} \qquad (19.58)$$

An alternative way to determine the moment of inertia would be to decompose the I-section into a "positive" outer rectangle with area $bd$ and two "negative" rectangles $(b - t_W)/2$ wide and $(d - 2t_F)$ deep. Subtracting the moment of inertia for the two "negative" rectangles from the "positive" moment of inertia for the outer rectangle according to the parallel axes theorem would yield the same result.

---

**Example 19.4**    A beam with a cross-section corresponding to Figure 19.19 is simply supported over a span of 10 m. The dimensions of the beam are:

$d = 300$ mm
$b = 150$ mm
$t_F = 9$ mm
$t_W = 6.5$ mm

The yield strength of the structural steel is $250\,\text{N mm}^{-2}$ with a safety factor of 1.67 to be used. Calculate the maximum concentrated load that can be supported by the bending stress at a distance of 4 m from one of the supports.

**Solution**

Figure 19.20 illustrates the bending moment diagram. The maximum bending moment is given by Eq. (19.20):

$$M_{max} = \frac{4 \times 6 \times W}{10}$$

$$= \frac{12W}{5}\,\text{N m}$$

$$= \frac{12W}{5} \times 10^3\,\text{N mm}$$

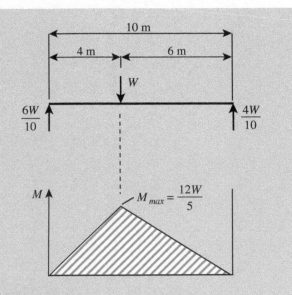

**Figure 19.20**

Bending moment diagram for Example 19.4.

From Eq. (19.58):

$$I_{XX} = 2\left[\frac{bt_F^3}{12} + bt_F\left(\frac{d - t_F}{2}\right)^2\right] + \frac{t_W(d - 2t_F)^3}{12}$$

$$= 2\left[\frac{150 \times 9^3}{12} + 150 \times 9\left(\frac{300 - 9}{2}\right)^2\right] + \frac{6.5(300 - 2 \times 9)^3}{12}$$

$$= 69.33 \times 10^6\,\text{mm}^4$$

From Eq. (19.37):

$$\sigma_{B\,max} = \frac{M\,y_{max}}{I}$$

The maximum stress occurs at the maximum distance from the centroid of 150 mm:

$$\frac{250}{1.67} = \frac{(12W \times 10^3/5) \times 150}{69.33 \times 10^6}$$

$$W = \frac{250 \times 69.33 \times 10^6 \times 5}{1.67 \times 150 \times 12 \times 10^3}$$

$$= 28.8 \times 10^3\,\text{N}$$

This neglects any effect of shear stress and the weight of the beam itself has not been accounted for. In practice, the beam would most likely not be simply supported, but fixed (e.g., by bolting a structural steel beam to a vertical steel stanchion). The distribution of load might be more complex than a simple point load. Also, commercially available beams, whilst retaining the basic I-section, are formed in a single piece with rounded corners, rather than the shape of three simple rectangles.

---

The analysis of bending so far neglects shear stress. In practice, the shearing force at any cross-section of the beam will create shear stresses on transverse sections. The existence of horizontal shear stresses in bending can be demonstrated as in Figure 19.21. In Figure 19.21a two horizontal rectangular beams with the same dimensions and of depth $d$ on simple supports are subjected to a

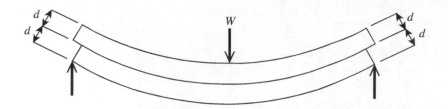

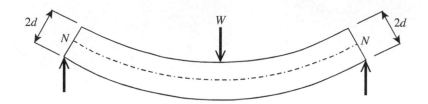

(a) Two identical beams with simple supports subject to a concentrated central load.

(b) Bonding the two beams illustrates that there must of a shear stress to prevent the two beams sliding relative to each other.

### Figure 19.21

Shear stress acting on a beam.

concentrated load $W$. If there is no friction between the beams, each beam will bend independently of the other. Each will be subject to compression at the upper surface and tension at the lower. The beams will slide relative to each other. Figure 19.21b illustrates what happens when the two beams are bonded together to give a single beam with depth $2d$ and subjected to the same load. There will be a shearing force along the neutral axis $NN$ to prevent the sliding of the upper and lower beams relative to each other.

Figure 19.22 shows a beam of rectangular cross-section. A shearing force $F$ acts on a cross-section. This creates shear stresses $\tau$ that can be assumed to act parallel to the shear force $F$. It can be assumed that the distribution of shear stresses is uniform across the width of the beam. Shear stresses on one side of an element are accompanied by complementary shear stresses $\tau'$ of equal magnitude acting on perpendicular faces of the element. Thus, there will be horizontal shear stresses between horizontal layers of the beam

as well as the transverse shear stresses across the vertical cross-section.

Figure 19.23 shows a section of a beam subject to a transverse load. Across the incremental length $dx$ there are bending moments $M$ and $M + dM$ and shearing forces $F$ and $F + dF$. The shear stress on the transverse section will create a complementary shear stress on longitudinal planes parallel to the neutral axis $NN$. Let $\tau$ be the complementary shear stress at a distance $y_0$ from the neutral axis $NN$ and $A$ the shaded area of the cross-section at distances greater than $y_0$ from the neutral axis. The difference in the longitudinal force across the horizontal length $dx$ is $d\sigma\, dA$. This force summed over the whole area $A$ is in equilibrium with the transverse shear stress $\tau$ acting over the longitudinal plane with area $z\, dx$. Thus:

$$\tau z\, dx = \int_A d\sigma\, dA \qquad (19.59)$$

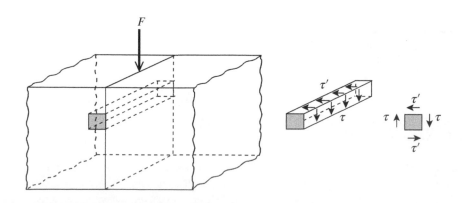

### Figure 19.22

A shearing force acting on a beam creates complementary shear stresses.

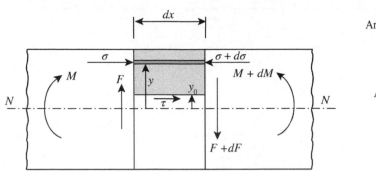
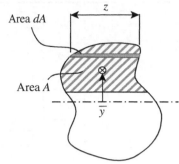

## Figure 19.23

A section of a beam subject to a transverse load.

From Eq. (19.36):

$$\sigma = \frac{M\,y}{I} \quad \text{and} \quad \sigma + \mathrm{d}\sigma = \frac{(M + \mathrm{d}M)y}{I} \qquad (19.60)$$

Taking the difference of these two equations:

$$\mathrm{d}\sigma = \frac{y\,\mathrm{d}M}{I} \qquad (19.61)$$

Substituting Eq. (19.61) into Eq. (19.59) gives:

$$\tau z\,\mathrm{d}x = \int_A \frac{y\,\mathrm{d}M}{I}\,\mathrm{d}A = \frac{\mathrm{d}M}{I} \int_A y\,\mathrm{d}A \qquad (19.62)$$

Rearranging Eq. (19.62):

$$\tau = \frac{1}{z\,I}\frac{\mathrm{d}M}{\mathrm{d}x} \int_A y\,\mathrm{d}A \qquad (19.63)$$

Substituting Eq. (19.26):

$$\tau = \frac{F}{z\,I} \int_A y\,\mathrm{d}A \qquad (19.64)$$

For a given shape of section, a *first moment of area* or *statical moment of area* can be defined as:

$$Q = \int_A y\,\mathrm{d}A \qquad (19.65)$$

where

$Q$ = first moment of area about the neutral axis for the entire section (m³)

$\mathrm{d}A$ = elemental area (m²)

$y$ = perpendicular distance to the element $\mathrm{d}A$ from the neutral axis (m)

The first moment of area is used to calculate the centroid of the section, as discussed earlier.

Combining Eqs (19.64) and (19.65):

$$\tau_B = \frac{FQ}{z\,I} \qquad (19.66)$$

where

$\tau_B$ = transverse bending shear stress acting over the longitudinal plane (N m⁻²)

$F$ = force acting on the beam (N)

$Q$ = first moment of area (m³)

$I$ = second moment of area (m⁴)

$z$ = length of the cutting plane (m)

The maximum value of $Q$ defines the maximum shear stress for the section:

$$\tau_{B\,\max} = \frac{FQ_{\max}}{z\,I} \qquad (19.67)$$

where

$\tau_{B\max}$ = maximum transverse bending shear stress acting over the longitudinal plane (N m⁻²)

$Q_{\max}$ = maximum first moment of area (m³)

It is often necessary to calculate the first moment of area of complex cross-sections. As with the moment of inertia for complex cross-sections, the area of the cross-section can be divided into sub-areas; then the first moment of area of the complete area is the sum of the first moment of area of the sub-areas if all the sub-areas refer to the same axis. As long as the area and centroid of the sub-areas are known, then the first moment of area of the sub-areas can be summed, rather than integrated:

$$Q_{XX} = \sum_{i=1}^{n} A_i y_i \qquad (19.68)$$

where

$Q_{XX}$ = first moment of area about the neutral axis (m³)

$A_i$ = area of sub-area $i$ (m²)

$y_i$ = vertical distance from the centroid of sub-area $i$ to the neutral axis (m)

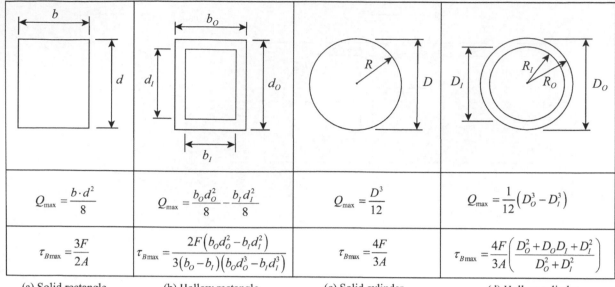

|  |  |  |  |
|---|---|---|---|
| $Q_{max} = \dfrac{b \cdot d^2}{8}$ | $Q_{max} = \dfrac{b_o d_o^2}{8} - \dfrac{b_I d_I^2}{8}$ | $Q_{max} = \dfrac{D^3}{12}$ | $Q_{max} = \dfrac{1}{12}\left(D_O^3 - D_I^3\right)$ |
| $\tau_{B\,max} = \dfrac{3F}{2A}$ | $\tau_{B\,max} = \dfrac{2F\left(b_o d_o^2 - b_I d_I^2\right)}{3\left(b_o - b_I\right)\left(b_o d_o^3 - b_I d_I^3\right)}$ | $\tau_{B\,max} = \dfrac{4F}{3A}$ | $\tau_{B\,max} = \dfrac{4F}{3A}\left(\dfrac{D_O^2 + D_O D_I + D_I^2}{D_O^2 + D_I^2}\right)$ |
| (a) Solid rectangle. | (b) Hollow rectangle. | (c) Solid cylinder. | (d) Hollow cylinder. |

## Figure 19.24

First moment of area and shear stress for common cross-sections.

As with the moment of inertia, the first moment of area can be determined by summing "positive" and "negative" sub-areas.

Consider a number of examples of different cross-sections. For a solid rectangular cross-section in Figure 19.24a (see Appendix F):

$$Q_{max} = \frac{bd^2}{8} \tag{19.69}$$

$$\tau_{B\,max} = \frac{3F}{2A} \tag{19.70}$$

where

$b$ = width of the rectangular section (m)
$d$ = depth of the rectangular section (m)
$A$ = area of the rectangular cross-section (m²)
$\quad = bd$

For a hollow rectangular cross-section as shown in Figure 19.24b (see Appendix F):

$$Q_{max} = \frac{b_o d_o^2}{8} - \frac{b_I d_I^2}{8} \tag{19.71}$$

where

$b_O$ = outer breadth of the beam (m)
$d_O$ = outer depth of the beam (m)
$b_I$ = inner breadth of the beam (m)
$d_I$ = inner depth of the beam (m)

$$\tau_{B\,max} = \frac{2F\left(b_o d_o^2 - b_I d_I^2\right)}{3\left(b_O - b_I\right)\left(b_o d_o^3 - b_I d_I^3\right)} \tag{19.72}$$

Note that the inner and outer rectangles must be concentric, and the centroids coincident, for these equations to apply. For a solid circular cross-section, as shown in Figure 19.24c (see Appendix F):

$$Q_{max} = \frac{2R^3}{3} = \frac{D^3}{12} \tag{19.73}$$

$$\tau_{B\,max} = \frac{4F}{3A} \tag{19.74}$$

where

$R$ = radius of the cylinder (m)
$D$ = diameter of the cylinder (m)
$A$ = area of the cylinder cross-section (m²)
$\quad = \pi R^2 = \dfrac{\pi D^2}{4}$

For a hollow circular cross-section, as shown in Figure 19.24d (see Appendix F):

$$Q_{max} = \frac{2}{3}\left(R_O^3 - R_I^3\right) = \frac{1}{12}\left(D_O^3 - D_I^3\right) \tag{19.75}$$

$$\tau_{B\,max} = \frac{4F}{3A}\left(\frac{R_O^2 + R_O R_I + R_I^2}{R_O^2 + R_I^2}\right) = \frac{4F}{3A}\left(\frac{D_O^2 + D_O D_I + D_I^2}{D_O^2 + D_I^2}\right) \tag{19.76}$$

where

$R_O$ = outside radius of the cylinder (m)
$R_I$ = inside radius of the cylinder (m)
$D_O$ = outside diameter of the cylinder (m)
$D_I$ = inside diameter of the cylinder (m)
$A$ = area of the hollow cylinder cross-section (m²)
$\quad = \pi\left(R_O^2 - R_I^2\right) = \dfrac{\pi}{4}\left(D_O^2 - D_I^2\right)$

Note again that the inner and outer circles must be concentric and the centroids coincident for these equations to apply.

The maximum shear stress occurs at the neutral axis. The analysis of bending shear stress in other cross-section shapes can be found elsewhere (Timoshenko 1956; Ryder 1969; Willems et al. 1981; Gere 2004).

---

**Example 19.5**  A rectangular steel beam with a span of 1 m is to be used to support a pipe. The loading from the pipe can be considered to be concentrated at the center of the beam and the beam can be assumed to be simply supported. The beam is 3 cm wide and 6 cm deep. The yield strength of the structural steel is 250 N mm$^{-2}$ with a safety factor of 1.67 to be used, and the density of the steel can be assumed to be 7850 kg m$^{-3}$.

a) Neglecting the self-weight created by the weight of the beam itself, calculate the maximum bending and shear stresses and determine the maximum load from the pipe that can be supported.
b) Check whether the self-weight from the weight of the beam has a significant effect.

**Solution**

a) The maximum bending moment for a concentrated central load is given by Eq. (19.20):

$$M_{max} = \frac{WL}{4} = \frac{W \times 1 \times 10^3}{4} = 250\,W \text{ N mm}$$

where

$M_{max}$ = maximum bending moment (N mm)
$W$ = beam loading (N)
$L$ = length of the beam (mm)

For a rectangular beam, from Eq. (19.43):

$$I = \frac{bd^3}{12} = \frac{30 \times 60^3}{12} = 540 \times 10^3 \text{ mm}^4$$

where

$I$ = moment of inertia of a rectangular beam (mm$^4$)
$b$ = breadth of the beam (mm)
$d$ = depth of the beam (mm)

From Eq. (19.37):

$$\sigma_{B\,max} = \frac{M\,y_{max}}{I}$$

where

$\sigma_{B\,max}$ = maximum bending normal stress (N mm$^{-2}$)
$y_{max}$ = distance from the neutral axis (mm)
$d$ = depth of the beam (mm)

Thus, for $y_{max}$ = 30 mm:

$$\sigma_{B\,max} = \frac{250 \times W \times 30}{540 \times 10^3} = \frac{250}{1.67}$$

Solving for $W$:

$$W = \frac{250}{1.67} \times \frac{540 \times 10^3}{250 \times 30}$$

$$= 10\,778 \text{ N}$$

$$= \frac{10\,778}{9.81} \text{ kg}$$

$$= 1099 \text{ kg}$$

Maximum shear stress from Eq. (19.70):

$$\tau_{B\,max} = \frac{3F}{2bd}$$

Given the shearing force $F = W/2$:

$$\tau_{B\,max} = \frac{3W}{2 \times 2 \times 30 \times 60} = \frac{250}{1.67}$$

Solving for $W$:

$$W = \frac{250 \times 2 \times 2 \times 30 \times 60}{1.67 \times 3}$$

$$= 359\,281 \text{ N}$$

$$= \frac{359\,281}{9.81} \text{ kg}$$

$$= 36\,624 \text{ kg}$$

b) Weight of the beam

$$= 0.03 \times 0.06 \times 1.0 \times 7850 \times 9.81$$

$$= 138.6 \text{ N}$$

$$= 0.1386 \text{ N mm}^{-1}$$

From Example 19.3, the maximum bending moment is given by:

$$M_{max} = \frac{WL}{4} + \frac{wL^2}{8}$$

$$= \frac{W \times 1.0 \times 10^3}{4} + \frac{0.1386 \times \left(1.0 \times 10^3\right)^2}{8}$$

$$= 250W + 17\,325$$

$$\sigma_{B\,max} = \frac{(250W + 17\,325) \times 30}{540 \times 10^3} = \frac{250}{1.67}$$

Solving for $W$:

$$W = 10\,709 \text{ N}$$

$$= \frac{10\,709}{9.81} \text{ kg}$$

$$= 1092 \text{ kg}$$

Given the shearing force from Example 19.3:

$$F_{max} = \frac{W}{2} + \frac{wL}{2}$$

$$= \frac{W}{2} + \frac{0.1386 \times 1.0 \times 10^3}{2}$$

$$= \frac{W}{2} + 69.3$$

$$\tau_{B\,max} = \frac{3\left(\dfrac{W}{2} + 69.3\right)}{2 \times 30 \times 60} = \frac{250}{1.67}$$

Solving for $W$:

$$W = \frac{250 \times 2 \times 2 \times 30 \times 60}{1.67 \times 3}$$

$$= 359\,143 \text{ N}$$

$$= \frac{359\,143}{9.81} \text{ kg}$$

$$= 36\,610 \text{ kg}$$

Thus, in this case accounting for the self-weight of the beam makes little difference.

The permissible load is thus 1092 kg. There will be a number of pipe supports supporting the pipe length and the weight of the whole pipe length will be distributed across them. The example assumes that the presence of a shear stress does not affect the distribution of bending stress. The combination of stresses will be considered later in this chapter.

There are two commonly encountered situations when bending stresses in process vessels needs to be analyzed. Consider, first, tall free-standing cylindrical vessels, such as distillation and absorption towers, which are subject to bending stresses imposed from wind load. Figure 19.25a illustrates a self-supporting distillation or absorption column subject to a wind load. The column will be supported on a cylindrical *skirt*, assumed to have the same diameter as the column, which is anchored to the ground. Consider the wind load to be uniform along the length of the column as $w$ per unit length. The column can be considered to be a *cantilever* with a *built-in* or *encastre* support, as shown in Figure 19.25b. The shear force diagram is shown in Figure 19.25c. The support must exert a reactive force of $wL$ to counteract the shear force. Figure 19.25d shows the bending moment diagram. In Figure 19.25b, taking moments at Point $x$ from the top of the tower, to maintain equilibrium:

$$M_x = wx\,\frac{x}{2} = \frac{wx^2}{2} \tag{19.77}$$

where

$M_x$ = moment around point $x$ (N m)

$w$ = force per unit length (N m$^{-1}$)

$x$ = distance from the top of the tower (m)

This is a parabolic relationship, as illustrated in Figure 19.25d, with a maximum when $x = L$ at the base of the column (base of the skirt):

$$M_{\max} = \frac{wL^2}{2} \tag{19.78}$$

where

$L$ = length (height) of the tower and skirt (m)

The support must exert a reactive force of $wL$ and a fixing moment of $wL^2/2$ to oppose rotation caused by the wind load. The force $w$ per unit length from the wind load is determined

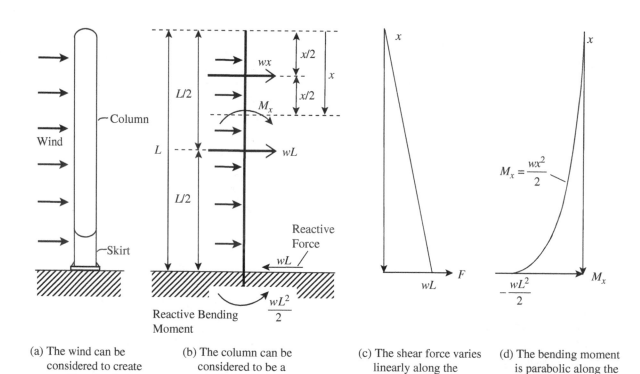

(a) The wind can be considered to create a uniform load on the column.

(b) The column can be considered to be a cantilever with a built-in support.

(c) The shear force varies linearly along the column.

(d) The bending moment is parabolic along the column.

**Figure 19.25**

A tall cylindrical vessel, such as a distillation or absorption tower, is subject to bending stresses from wind load.

by the drag as the wind flows across the cylindrical tower. The force per unit area exerted by the wind can be quantified in terms of the drag coefficient for the cylinder (Knudsen and Katz 1958):

$$\frac{F}{A_P} = \frac{F}{LD} = c_D \frac{\rho_{AIR} v^2}{2} \tag{19.79}$$

where

$F$ = wind load (N)

$A_P$ = projected area on a plane normal to the flow (m$^2$)

$L$ = length (height) of the tower (m)

$D$ = diameter of the tower (m)

$c_D$ = drag coefficient (–)

$\rho_{AIR}$ = air density (kg m$^3$)

$v$ = wind velocity (m s$^{-1}$)

Thus, from Eq. (19.79):

$$w = \frac{F}{L} = \frac{c_D \rho_{AIR} v^2 D}{2} \tag{19.80}$$

Table 19.1 shows the drag coefficient for smooth cylinders of different length-to-diameter ratios in turbulent flow (Knudsen and Katz 1958).

Whilst the application of Eq. (19.80) seems straightforward in principle, there are a number of complications:

a) The values of $c_D$ in Table 19.1 relate to smooth cylinders. In practice, the surface is unlikely to be smooth. Even worse, there will be pipes running axially along the outside of the tower for feed, overhead vapor, and possibly side streams. Also, there will be nozzles and access hatchways projecting from the surface of the column, together with platforms and ladders for access. All of these features will add to the drag on the tower.

b) It is not possible for the designer to evaluate the projected area to the wind accurately because of the pipes, nozzles, access hatchways, platforms, and ladders attached. These will create a different wind resistance from that of the cylinder and, unlike the cylinder, the additional resistance will also depend on the orientation to the wind. The diameter of the column can be increased to allow for these features. Table 19.2 presents

### Table 19.1

Effect of the length-to-diameter ratio on the drag coefficient for a turbulent flow past smooth cylinders.

| $L/D$ | $c_D$ |
|---|---|
| $\infty$ | 1.2 |
| 40 | 0.98 |
| 20 | 0.92 |
| 10 | 0.82 |
| 5 | 0.74 |
| 3 | 0.74 |
| 2 | 0.68 |
| 1 | 0.63 |

### Table 19.2

Factor to increase the outside diameter of the column (including insulation) to allow for pipes, nozzles, access hatchways, platforms, and ladders (Data from Bednar 1991).

| Column outside diameter (m) | Factor to increase diameter |
|---|---|
| Less than 0.91 m | 1.50 |
| 0.91–1.52 m | 1.40 |
| 1.52–2.13 m | 1.30 |
| 2.13–2.74 m | 1.20 |
| Greater than 2.74 m | 1.18 |

factors that can be applied to the outside diameter of the column, including the thickness of insulation, to allow for the additional resistance of the ancillary features (Bednar 1991).

c) The wind velocity in Eq. (19.80) implies a constant value along the length (height) of the column. In practice, the wind velocity varies with height from zero at the surface of the ground, increasing with height (eventually to the geostrophic wind speed at high altitude). This variation in wind velocity with height will depend on the meteorological conditions, surrounding terrain, and surrounding structures (EN 1991-1-4: 2005 +A1 2010). Assuming that the highest wind speed applied along the length (height) of the tower is the safest approach, although this will be conservative. For large structures, the wind speed can be broken down into bands from the top to the bottom of the column with Eq. (19.80) applied in each band. The wind speed must take account of extreme conditions such as hurricanes and typhoons. Also, the tendency of the wind to *gust* for short periods can be accounted for (Bednar 1991).

d) Another problem caused by wind load arises when the wind creates shedding of eddies downwind of a cylindrical structure. If the frequency of the shedding eddies coincides with the natural frequency of the structure, this can give rise to critical conditions. These effects need to be investigated when height-to-diameter ratios are greater than 10. This is outside the scope of this text (Brownell and Young 1959; Bednar 1991; Escoe 1994).

**Example 19.6** A self-supporting distillation column has been sized to have a height of 36.8 m, an inside diameter of 2.5 m, and a shell thickness of 20 mm. Assume that the distillation column is mounted on a cylindrical skirt support with a height of 3 m, the same outside diameter as the column with a thickness of 20 mm. Assume that the distillation column is insulated for its entire length (not the skirt) with 75 mm of mineral wool covered in aluminum cladding, increasing the outside diameter exposed to the wind to 2.69 m. Meteorological records for the location indicate that a maximum wind speed of 50 m s$^{-1}$ can be assumed. This wind speed can be assumed to apply at all heights. Air density can be assumed to be 1.3 kg m$^{-3}$.

a) Calculate the bending moment and maximum bending stress at the base of the skirt.

b) Calculate the maximum shear stress at the base of the skirt.

## Solution

**a)** First determine the diameter of the column subjected to the wind load. From Table 19.2, for a 2.69 m diameter column (including insulation), the diameter exposed to the wind to allow for platforms, ladders, and external pipes is given by:

$$D = 2.69 \times 1.20$$
$$= 3.23 \text{ m}$$

Height of the column with the skirt:

$$L = 36.8 + 3$$
$$= 39.8 \text{ m}$$

Length-to-diameter ratio:

$$\frac{L}{D} = \frac{39.8}{3.23}$$
$$= 12.3$$

From Table 19.1:

$$c_D = 0.84$$

From Eq. (19.80):

$$w = \frac{c_D \rho_{AIR} v^2 D}{2}$$
$$= \frac{0.84 \times 1.3 \times 50^2 \times 3.23}{2}$$
$$= 4409 \text{ N m}^{-1}$$

$$M_x = \frac{w x^2}{2}$$
$$= \frac{4409 \times 39.8^2}{2}$$
$$= 3.492 \times 10^6 \text{ N m}$$

From Eq. (19.54):

$$\sigma_{B \max} = \pm \frac{32 M D_O}{\pi \left( D_O^4 - D_I^4 \right)}$$
$$= \pm \frac{32 \times 3.492 \times 10^6 \times 2.54}{\pi \left( 2.54^4 - 2.5^4 \right)}$$
$$= \pm 35.283 \times 10^6 \text{ N m}^{-2}$$
$$= \pm 35.283 \text{ N mm}^{-2}$$

On the upwind side of the column the bending stress is tensile and $\sigma_{\max} = +35.282 \text{ N mm}^{-2}$. On the downwind side of the column the bending stress is compressive and $\sigma_{\max} = -35.283 \text{ N mm}^{-2}$.

**b)** The maximum shear stress occurs at the base of the skirt and perpendicular to the wind direction:

$$F = wL$$
$$= 4409 \times 39.8$$
$$= 1.755 \times 10^5 \text{ N m}^{-2}$$

The area of the cross-section is given by:

$$A = \frac{\pi}{4} \left( D_O^2 - D_I^2 \right)$$
$$= \frac{\pi}{4} \left( 2.54^2 - 2.5^2 \right)$$
$$= 0.1583 \text{ m}^2$$

From Eq. (19.76):

$$\tau_{B \max} = \frac{4F}{3A} \left( \frac{D_O^2 + D_O D_I + D_I^2}{D_O^2 + D_I^2} \right)$$
$$= \frac{4 \times 1.755 \times 10^5}{3 \times 0.1583} \left( \frac{2.54^2 + 2.54 \times 2.5 + 2.5^2}{2.54^2 + 2.5^2} \right)$$
$$= 2.217 \times 10^6 \text{ N m}^{-2}$$

The normal bending stress is clearly dominant. It should also be noted that these are not the only stresses acting on the skirt. There is also a compressive force created by the weight of the distillation column, ladders and pipework attached to the column, the column contents, and the skirt itself. Combining these stresses will be considered later. The most significant approximation is the assumption that the wind velocity is the same at all heights. In practice the wind velocity increases with height from zero at ground level. Allowance can be made for increase of wind velocity with height at the expense of increased complexity in the calculations (Brownell and Young 1959, Escoe 1986, Bednar 1991). However, there is always uncertainty in both the wind velocity and its variation with height, even at a fixed location.

Another commonly encountered situation when bending stresses in process vessels need to be analyzed is illustrated in Figure 19.26, in which a horizontal cylindrical vessel is supported on *saddles*. Figure 19.27 shows the corresponding shear force and bending moment diagrams. The weight of the vessel and any liquid contents will create bending stresses around the supports. In Figure 19.27, the distance between the center of the beam and support $R_2$ is $(L/2 - b)$. Taking moments about $R_2$:

$$R_1(L - a - b) = Lw \left( \frac{L}{2} - b \right) \tag{19.81}$$

where

$R_1$ = reaction force (N)

$L$ = length of beam (m)

$a$ = distance from support $R_1$ to the end of the beam (m)

$b$ = distance from support $R_2$ to the end of the beam (m)

$w$ = force per unit length from the weight of the vessel and its contents (N m$^{-1}$)

Rearranging Eq. (19.81):

$$R_1 = \frac{Lw \left( \frac{L}{2} - b \right)}{(L - a - b)} \tag{19.82}$$

For the left-hand end overhang from $R_1$:

$$M = -wx \frac{x}{2} = -\frac{w x^2}{2} \tag{19.83}$$

where

$M$ = bending moment (N m)

$x$ = distance from the left-hand end of the beam (m)

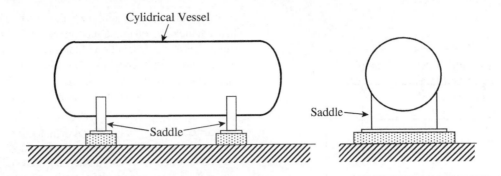

## Figure 19.26

Horizontal cylindrical vessels are subject to bending stresses around their supports from the weight of the vessel and contents.

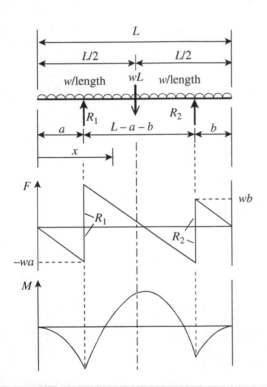

## Figure 19.27

A beam supported at two asymmetric intermediate points.

The maximum value of $M$ for the overhang occurs when $x = a$:

$$M_{max} = -\frac{wa^2}{2} \tag{19.84}$$

Between $R_1$ and $R_2$ the moment is given by:

$$M = -\frac{wx^2}{2} + R_1(x-a) \tag{19.85}$$

The maximum bending moment between $R_1$ and $R_2$ occurs when:

$$\frac{dM}{dx} = 0 = -wx + R_1 \tag{19.86}$$

Rearranging Eq. (19.86):

$$
\begin{aligned}
x &= \frac{R_1}{w} \\
&= \frac{1}{w} \times \frac{Lw\left(\dfrac{L}{2}-b\right)}{(L-a-b)} \\
&= \frac{L\left(\dfrac{L}{2}-b\right)}{(L-a-b)}
\end{aligned} \tag{19.87}
$$

Combining Eqs (19.82), (19.85), and (19.87):

$$
\begin{aligned}
M_{max} &= -\frac{w}{2}\left[\frac{L^2\left(\dfrac{L}{2}-b\right)^2}{(L-a-b)^2}\right] + \frac{wL\left(\dfrac{L}{2}-b\right)}{(L-a-b)}\left[\frac{L\left(\dfrac{L}{2}-b\right)}{(L-a-b)}-a\right] \\[2mm]
&= \frac{wL^2\left(\dfrac{L}{2}-b\right)^2}{(L-a-b)^2}\left[\frac{1}{2}-\frac{a(L-a-b)}{L\left(\dfrac{L}{2}-b\right)}\right] \\[2mm]
&= \frac{wL\left(\dfrac{L}{2}-b\right)}{(L-a-b)^2}\left[\frac{L^2}{4}-aL-\frac{bL}{2}+ab+a^2\right]
\end{aligned} \tag{19.88}
$$

The variation of shear force and bending moment is illustrated in Figure 19.27. If the supports are symmetric, as shown in Figure 19.28, then $a = b$ in Eq. (19.88), which becomes:

$$
\begin{aligned}
M_{max} &= \frac{wL\left(\dfrac{L}{2}-a\right)}{(L-2a)^2}\left[\frac{L^2}{4}-aL-\frac{aL}{2}+a^2+a^2\right] \\[2mm]
&= \frac{wL}{2(L-2a)}\left[\frac{L^2}{4}-\frac{3aL}{2}+2a^2\right] \\[2mm]
&= \frac{wL}{2(L-2a)}\left[(L-2a)\left(\frac{L}{4}-a\right)\right] \\[2mm]
&= \frac{wL}{2}\left[\frac{L}{4}-a\right]
\end{aligned} \tag{19.89}
$$

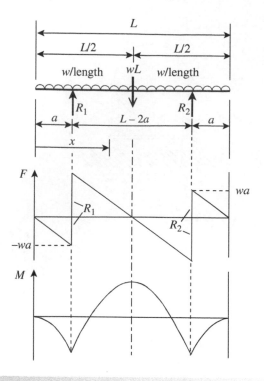

## Figure 19.28

A beam supported at two symmetric intermediate points.

Figure 19.28 shows the variation in shear force and bending moment for symmetric supports. The maximum bending moment in the beam can be minimized by adjusting the location of the supports. This is achieved when the maximum bending moment between the supports is numerically equal to the moment for the overhang (Ryder 1969):

$$\frac{wa^2}{2} = \frac{wL}{2}\left(\frac{L}{4} - a\right) \tag{19.90}$$

Rearranging Eq. (19.90) results in the quadratic equation:

$$0 = 4a^2 + 4La - L^2 \tag{19.91}$$

Solving the quadratic for $a$:

$$a = \frac{-4L \pm \sqrt{(4L)^2 - 4(4)\left(-L^2\right)}}{2(4)} \tag{19.92}$$

$$= 0.207L$$

Thus, the minimum overall bending moment occurs when the distance between the ends of the beam and the supports is $0.207L$. This equates to a distance between the supports of $0.586L$. Moreover, the overall bending moment can be decreased further if the distances between the supports and the ends of the beam are allowed to be different at each end and the supports are allowed to be asymmetric. The location of the two supports can in principle be adjusted independently to give the minimum possible overall bending moment (Ryder 1969). However, symmetric supports are almost always used for horizontal cylindrical vessels and the position of the supports for horizontal cylindrical vessels is

rarely positioned to minimize the overall bending moment. Instead, the position of the supports is chosen to minimize the detrimental effects of distortion of the shell resulting from the weight of the vessel and its contents. This will be discussed in more detail later. The location of the supports will also be influenced by the location of piping at the bottom of the vessel.

**Example 19.7**　A horizontal cylindrical vessel is to be fabricated with an internal diameter of 2 m and an external length of 6 m. The vessel is to have a shell thickness of 12 mm fabricated from flat plate in sections rolled and welded. Assume initially that the vessel has flat ends also with a thickness of 12 mm. It is supported on two saddles each 0.4 m from the end of the vessel. The vessel is to be completely full of water with a density of 957.9 kg m$^{-3}$. The vessel is to be fabricated from carbon steel SA 516-70 with a density of 7850 kg m$^{-3}$ with a corrosion allowance of 2 mm. Assume that the vessel acts effectively as a beam with the weight of the vessel and its contents evenly distributed along the length of the vessel:

a) Calculate the weight of the vessel and its contents.
b) Calculate the bending normal stress halfway along the vessel on its upper and lower surfaces before corrosion.
c) Calculate the bending shear stress on the sides of the vessel above the supports midway between the top and bottom, corresponding with the neutral axis.
d) Repeat the calculation assuming the shell thickness is decreased by the corrosion allowance.

**Solution**

a) The weight of the vessel and contents will be slightly higher before any corrosion occurs.

Weight of the vessel

$$= \left[\frac{\pi}{4}\left(D_O^2 - D_I^2\right)L + \frac{\pi}{4} \times D_I^2 \times t \times 2\right]\rho_{MOC} \times g$$

$$= \left[\frac{\pi}{4}\left(2.024^2 - 2.0^2\right)6 + \frac{\pi}{4} \times 2^2 \times 0.012 \times 2\right]7850 \times 9.81$$

$$= 40\,853\ \text{N}$$

where

$D_O$ = outside diameter of the vessel (m)
$D_I$ = inside diameter of the vessel (m)
$L$ = external length of the vessel (m)
$t$ = thickness of the end plate (m)
$\rho_{MOC}$ = density of material of construction (kg m$^{-3}$)
$g$ = acceleration due to gravity (m s$^{-2}$)

Weight of the contents

$$= \left[\frac{\pi}{4} \times D_I^2 \times (L - 2t)\right]\rho_L \times g$$

$$= \left[\frac{\pi}{4} \times 2^2 \times (6.0 - 2 \times 0.012)\right]957.9 \times 9.81$$

$$= 176\,421\ \text{N}$$

where

$\rho_L$ = density of liquid contents (kg m$^{-3}$)

Total weight $= 40\,853 + 176\,421 = 217\,274\ \text{N}$

**b)** Weight per unit length $w$:

$$w = \frac{217\,274}{6.0}$$

$$= 36\,212\,\mathrm{N\,m^{-1}}$$

The bending moment is given by Eq. (19.89):

$$M_{max} = \frac{wL}{2}\left(\frac{L}{4} - a\right)$$

$$= \frac{36\,212 \times 6.0}{2}\left(\frac{6.0}{4} - 0.4\right)$$

$$= 119\,501\,\mathrm{N\,m}$$

It should be noted that there is an additional bending moment resulting from the difference in the hydraulic head of liquid between the top and bottom of the vessel that has been neglected. This will be discussed later. The maximum normal bending stress is given by Eqs (19.37) and (19.54):

$$\sigma_{B\,max} = \frac{M\,y_{max}}{I} = \pm\frac{32M\,D_O}{\pi\left(D_O^4 - D_I^4\right)}$$

$$= \pm\frac{32 \times 119\,501 \times 2.024}{\pi\left(2.024^4 - 2^4\right)}$$

$$= \pm 3.151 \times 10^6\,\mathrm{N\,m^{-2}}$$

Thus, the bending normal stress at the bottom of the vessel is $+3.151 \times 10^6\,\mathrm{N\;m^{-2}}$ tensile and at the top of the vessel $-3.151 \times 10^6\,\mathrm{N\cdot m^{-2}}$ compressive.

**c)** The bending shear stress on the sides of the vessel above the supports midway between the top and bottom, corresponding with the neutral axis, is given by Eq. (19.76). The shear force is given by (Figure 19.28):

$$F = R_1 - wa$$

$$= \frac{Lw}{2} - wa$$

$$= \frac{6.0 \times 36\,212}{2} - 36\,212 \times 0.4$$

$$= 94\,152\,\mathrm{N}$$

The area of the cross-section is given by:

$$A = \frac{\pi}{4}\left(D_O^2 - D_I^2\right)$$

$$= \frac{\pi}{4}\left(2.024^2 - 2^2\right)$$

$$= 0.07585\,\mathrm{m^2}$$

From Eq. (19.76):

$$\tau_{B\,max} = \frac{4F}{3A}\left(\frac{D_O^2 + D_O D_I + D_I^2}{D_O^2 + D_I^2}\right)$$

$$= \frac{4 \times 94\,152}{3 \times 0.07585}\left(\frac{2.024^2 + 2.024 \times 2.0 + 2.0^2}{2.024^2 + 2.0^2}\right)$$

$$= 2.483 \times 10^6\,\mathrm{N\,m^{-2}}$$

The bending normal stress is significantly higher than the bending shear stress.

**d)** Repeat the calculation after corrosion with the thickness of the vessel walls decreased from 12 mm to 10 mm.

Weight of the vessel:

$$= 34\,093\,\mathrm{N}$$

Weight of the contents:

$$= 177\,246\,\mathrm{N}$$

Total weight:

$$= 211\,338\,\mathrm{N}$$

Weight per unit length:

$$w = 35\,223\,\mathrm{N\,m^{-1}}$$

Total weight of the vessel and weight per unit length decrease as a result of corrosion.

Bending moment:

$$M_{max} = 116\,152\,\mathrm{N\,m}$$

Maximum normal bending stress:

$$\sigma_{B\,max} = \pm 3.664 \times 10^6\,\mathrm{N\,m^{-2}}$$

Thus, the bending normal stress at the bottom of the vessel increases to $+3.664 \times 10^6\,\mathrm{N\;m^{-2}}$ tensile and at the top of the vessel increases to $-3.664 \times 10^6\,\mathrm{N\;m^{-2}}$ compressive.

Shear force:

$$F = 91\,552\,\mathrm{N}$$

Area of the cross-section:

$$A = 0.06327\,\mathrm{m^2}$$

$$\tau_{B\,max} = 2.894 \times 10^6\,\mathrm{N\,m^{-2}}$$

Both the bending normal stress and the bending shear stress increase as a result of corrosion. The bending stress still dominates.

It should be noted that such a horizontal vessel would be unlikely to have flat ends. Other types of heads will be discussed later. Also, the stresses in such a horizontal vessel have been simplified in this calculation. A more detailed discussion of such vessels will be given later in this chapter.

### 19.1.5 Buckling

If a body is subjected to a compressive axial stress, buckling may occur. Buckling is characterized by a sudden sideways deflection, as shown in Figure 19.29. The deflection depends on the type of support at the ends. The deflection may occur well below the yield stress and is not a failure of the material. Buckling is instead governed by the stiffness of the body. Within the limits that Hooke's Law is obeyed, the ratio of stress to strain produced is the Modulus of Elasticity $E$, which is constant for a given material. *Flexural rigidity* is the stiffness of a body when subjected to bending and is a function of both the Modulus of Elasticity and shape of the cross-section, defined by:

$$\text{Flexural rigidity} = EI \tag{19.93}$$

where

$E$ = modulus of elasticity (N m$^{-2}$)

$I$ = moment of inertia or second moment of area (m$^4$)

If all of the cross-sectional area of a member is focused at a point, then at a distance from the neutral axis that gives the same

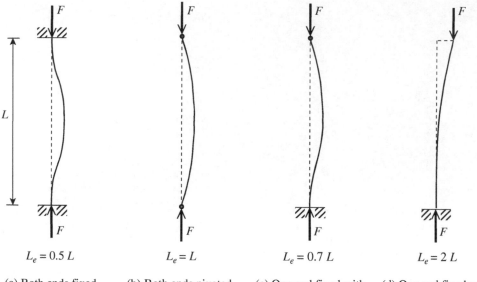

$$L_e = 0.5\,L \qquad L_e = L \qquad L_e = 0.7\,L \qquad L_e = 2\,L$$

(a) Both ends fixed.  (b) Both ends pivoted.  (c) One end fixed with the other pivoted.  (d) One end fixed with the other free.

**Figure 19.29**

Buckling of an object depends on how the ends are anchored.

moment of inertia as the real cross-section, the distance is termed the *radius of gyration k* and is given by:

$$k_{XX} = \left(\frac{I_{XX}}{A}\right)^{1/2} \text{ and } k_{YY} = \left(\frac{I_{YY}}{A}\right)^{1/2} \quad (19.94)$$

where

$k_{XX}$, $k_{YY}$ = radius of gyration around the *XX* and *YY* axis (m)

$I_{XX}$, $I_{YY}$ = moment of inertia around the *XX* and *YY* axis (m$^4$)

$A$ = area of cross-section (m$^2$)

The smaller radius of gyration is used to define the *slenderness ratio*:

$$\text{Slenderness ratio} = \frac{\text{Effective length of member}}{\text{Least radius of gyration}} = \frac{L_e}{k} \quad (19.95)$$

where

$L_e$ = effective length (m)

$k$ = radius of gyration (m)

The column length free to buckle depends on its end support conditions. Restraining the ends of a column with a fixed support increases the load-carrying capacity of a column. Allowing rotation at the end of a column generally reduces its load-carrying capacity. The effective length $L_e$ reflects this (Figure 19.29). The slenderness ratio can be used to predict the mode of failure of a member under compression. For steel members the ranges of failure are given approximately by:

- slenderness ratios $L_e/k < 40$ failure is expected to be yielding;
- slenderness ratios $40 < L_e/k < 120$ failure is expected to be a combination of yielding and buckling;
- slenderness ratio of $120 < L_e/k < 200$ failure is expected to be buckling.

Slenderness ratios of greater than 200 are normally not acceptable for steel members. The most efficient column sections for axial loads are those with almost equal $k_{XX}$ and $k_{YY}$ values.

**Example 19.8**  A structural steel column has a section given in Figure 19.30. The column is 5 m long and fixed at both ends. Calculate the slenderness ratio to determine if it is within acceptable bounds and the most likely mode of failure.

**Solution**

$$\text{Area of the section} = 350 \times 350 - 320 \times 250$$

$$= 42\,500 \text{ mm}^2$$

From Eq. (19.58):

$$I_{XX} = 2\left[\frac{bt_F^3}{12} + bt_F\left(\frac{d - t_F}{2}\right)^2\right] + \frac{t_W(d - 2t_F)^3}{12}$$

$$= 2\left[\frac{350 \times 50^3}{12} + 350 \times 50\left(\frac{350 - 50}{2}\right)^2\right] + \frac{30(350 - 2 \times 50)^3}{12}$$

$$= 833.85 \times 10^6 \text{ mm}^4$$

From Eq. (19.94):

$$k_{XX} = \left(\frac{I_{XX}}{A}\right)^{1/2}$$

$$= \left(\frac{833.85 \times 10^6}{42\,500}\right)^{1/2}$$

$$= 140.1 \text{ mm}$$

For the moment of inertia for the *Y*-axis, the breadth is parallel to the *Y*-axis and the depth is parallel to the *X*-axis. From Eq. (19.43):

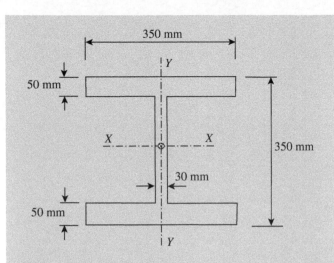

**Figure 19.30**

Cross-section of a structural member for Example 19.8.

$$k_{YY} = 2 \times \frac{50 \times 350^3}{12} + \frac{250 \times 30^3}{12}$$
$$= 357.85 \times 10^6 \text{ mm}^4$$

From Eq. (19.94):

$$k_{YY} = \left(\frac{357.85 \times 10^6}{42\,500}\right)^{1/2}$$
$$= 91.7 \text{ mm}$$

Take $k_{YY}$ to be the least for the calculation of the slenderness ratio:

$$\text{Slenderness ratio} = \frac{L_e}{k}$$
$$= \frac{0.5 \times 5 \times 10^3}{91.7}$$
$$= 27.2$$

The slenderness ratio is within acceptable bounds and the failure mode is expected to be yielding.

### 19.1.6   Thermal Stresses

In addition to mechanical forces, stresses can also be created by changes in temperature. Solid materials expand when the temperature is increased and vice versa. There are a number of ways in which such expansion or contraction can induce thermal stresses:

- Temperature gradients across a solid create different rates of expansion or contraction throughout the solid and result in thermal stresses.
- If the temperature of a solid body is increased or decreased and the body is not allowed to expand or contract freely, then thermal stresses will be induced.
- If a compound component made up from different materials is subjected to a change in temperature, then the change in dimension of the component parts will tend to be unequal. If the

component parts are restrained to remain together, then this will induce thermal stresses.

## 19.2   Combined Stresses

So far, consideration has been restricted to the application of either pure normal stress or pure shear stress to a body. However, situations will be encountered when both normal and shear stresses are acting. Figure 19.31 shows an element from a body being subjected to normal and shear stresses. Faces of the element are defined as positive or negative. Faces oriented toward the positive directions of the axes are positive, whereas opposite faces are negative, as illustrated in Figure 19.31.

1)  The shear stresses are denoted by $\tau_{ij}$, where Subscript $i$ indicates the face of the element on which the stress is acting, which is perpendicular to the $i$-axis. Subscript $j$ indicates the direction in which the stress acts. Stresses are positive if they act on a positive face in a positive direction or act on a negative face in a negative direction. Stresses are negative if they act on a positive face in a negative direction or act on a negative face in a positive direction. An example of a two-dimensional system is shown in Figure 19.32 to illustrate positive and negative shear stresses. It should be noted that to maintain equilibrium requires:

$$\tau_{xy} = \tau_{yx}, \quad \tau_{xz} = \tau_{zx}, \quad \tau_{yz} = \tau_{zy} \tag{19.96}$$

where

$\tau_{xy}$ = shear stress acting on the $x$-plane in the direction of the $y$-axis (N mm$^{-2}$, N m$^{-2}$)

2)  For the normal stresses $\sigma_{ii}$, the face of the element on which the stress is acting and the direction are the same. The normal stresses $\sigma_{xx}$, $\sigma_{yy}$, and $\sigma_{zz}$ are most often simply written as $\sigma_x$, $\sigma_y$, and $\sigma_z$. The sign convention is positive for tension and negative for compression.

If normal and shear stresses are both acting on a body, the resultant stress across any section might be neither normal nor

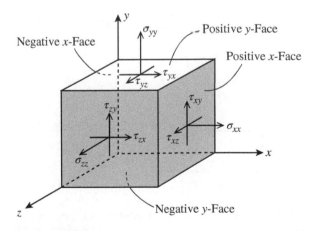

**Figure 19.31**

Sign convention for stresses.

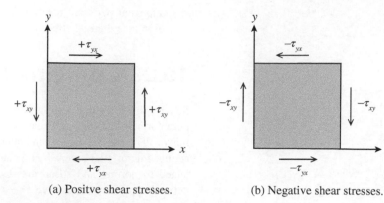

(a) Positive shear stresses.    (b) Negative shear stresses.

## Figure 19.32

Example of a sign convention for shear stresses. (a) Positive shear stresses. (b) Negative shear stresses.

tangential to a given plane. Normal stress and shear stress may vary with respect to different inclined planes. For design, it is necessary to find which planes are subject to extreme values of normal and shear stress components.

First, consider a simple situation in which a body is subjected to only normal forces in tension. Figure 19.33a shows a body subject to a normal tensile force $F$. Also shown is an *imaginary* inclined plane across the body. In Figure 19.33b the imaginary

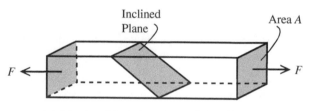

(a) A body subjected to a normal (tensile) force can be cut by an inclined plane.

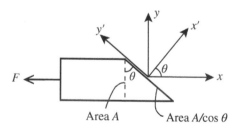

(b) Defining the angle of the inclined plane allows new coordinates to be defined.

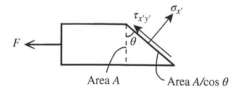

(c) Balancing the forces on the inclided plane requires a normal stress and shear stress to maintain the equilibrium

## Figure 19.33

Stresses on an inclined plane.

inclined plane is shown relative to the axes $x$ and $y$ for the body and $x'$ and $y'$ for the inclined plane. In Figure 19.33c for this plane, a normal stress $\sigma_{x'}$ and shear stress $\tau_{x'y'}$ are necessary to maintain the equilibrium. Note that the shear stress $\tau_{x'y'}$ can in principle be shown to act in either direction (i.e. either up or down the inclined plane in Figure 19.33c). If the direction is reversed, the shear stress would change sign in the analysis. If the cross-sectional area of the body is $A$, then the area of the imaginary inclined plane is $A/\cos \theta$.

$$\sigma_{x'} = \frac{\text{Force}}{\text{Area}} = \frac{F \cos \theta}{A / \cos \theta} = \frac{F}{A} \cos^2 \theta = \sigma_x \cos^2 \theta \quad (19.97)$$

where

$F$ = tensile force (N)

$A$ = cross-sectional area (m$^2$)

$\sigma_x$ = normal stress acting on the original $x$-plane (N m$^{-2}$)

$\sigma_{x'}$ = normal stress acting on the inclined $x'$-plane (N m$^{-2}$)

Given the basic trigonometric relation $\cos^2\theta = (1 + \cos 2\theta)/2$, Eq. (19.97) becomes:

$$\sigma_{x'} = \frac{\sigma_x}{2}(1 + \cos 2\theta) \quad (19.98)$$

The maximum and minimum for this expression are given by:

$$\sigma_{max} = \sigma_x \quad \text{when } \theta = 0° \\ \sigma_{min} = 0 \quad \text{when } \theta = 90° \quad (19.99)$$

Also, for the shear stress:

$$\tau_{x'y'} = \frac{\text{Force}}{\text{Area}} = \frac{-F \sin \theta}{A / \cos \theta} = -\frac{F}{A} \sin \theta \cos \theta = -\sigma_x \sin \theta \cos \theta \quad (19.100)$$

where

$\tau_{x'y'}$ = shear stress acting on the inclined $x'$-plane in the direction of the inclined $y'$-axis (N m$^{-2}$)

Given the basic trigonometric relation $\sin\theta \cos \theta = \sin 2\theta/2$, Eq. (19.100) becomes:

$$\tau_{x'y'} = -\frac{\sigma_x}{2} \sin 2\theta \quad (19.101)$$

The maximum shear stress $\tau_{max}$ and minimum shear stress $\tau_{min}$ from this expression are given by:

$$\tau_{max} = \pm \frac{\sigma_x}{2} \quad \text{when } \theta = \pm 45°$$
$$\tau_{min} = 0 \quad\quad \text{when } \theta = 0° \tag{19.102}$$

It is interesting to note that when a sample of brittle material fails in a tensile test, the fracture occurs on a plane across the sample where the normal stress is a maximum. For ductile materials fracture occurs with angles at the edge of the fracture of 45° to the axis of the tensile force. *This indicates that ductile materials fail in shear, even in a tensile test.*

The extreme values of normal stresses (maximum and minimum) are called the *principal stresses*. The mutually perpendicular planes on which the principal stresses act are called *principal planes*. In two-dimensional problems there are two principal normal stresses, the *major principal stress* (which is the maximum) and the *minor principal stress* (which is the minimum value). Shear stress is zero on the principal plane. The values of the principal stresses are taken algebraically. For example, if the principal stresses are 200 N mm$^2$ tensile and 500 N mm$^2$ compressive, the major principal stress will be taken to be 200 N mm$^2$ and the compressive stress of 500 N mm$^2$ will be taken to be the minor principal stress, as it is algebraically −500 N mm$^2$.

A more general analysis of the principal stresses in two dimensions is developed in Appendix G. From this, the maximum and minimum principal stresses are given by (see Appendix G):

$$\sigma_{max} = \sigma_1 = \frac{\sigma_x + \sigma_y}{2} + \sqrt{\left(\frac{\sigma_x - \sigma_y}{2}\right)^2 + \tau_{xy}^2}$$

$$\sigma_{min} = \sigma_2 = \frac{\sigma_x + \sigma_y}{2} - \sqrt{\left(\frac{\sigma_x - \sigma_y}{2}\right)^2 + \tau_{xy}^2} \tag{19.103}$$

where

$\sigma_x$ = normal stress acting on the x-plane (N m$^{-2}$)

$\sigma_y$ = normal stress acting on the y-plane (N m$^{-2}$)

$\tau_{xy}$ = shear stress acting on the x-plane in the direction of the y-axis (N m$^{-2}$)

The direction of the stresses is shown in Figure 19.31. If the shear stresses are zero ($\tau_{xy} = 0$), then the principal stresses reduce to $\sigma_1 = \sigma_x$ and $\sigma_2 = \sigma_y$. The corresponding maximum and minimum shear stresses are given by (see Appendix G):

$$\tau_{max} = \sqrt{\left(\frac{\sigma_x - \sigma_y}{2}\right)^2 + \tau_{xy}^2}$$

$$\tau_{min} = -\sqrt{\left(\frac{\sigma_x - \sigma_y}{2}\right)^2 + \tau_{xy}^2} \tag{19.104}$$

If $\tau_{xy} = 0$, then this reduces to $\tau_{max} = (\sigma_x - \sigma_y)/2$ and $\tau_{min} = -(\sigma_x - \sigma_y)/2$. The principal stresses and maximum shear stresses are related by (see Appendix G):

$$\tau_{max} = \frac{\sigma_1 - \sigma_2}{2} \tag{19.105}$$

The theory can be extended to triaxial stresses. In a coordinate system of x, y, z, there are three normal stresses $\sigma_x$, $\sigma_y$, and $\sigma_z$ and three shear stresses $\tau_{xy}$, $\tau_{xz}$, and $\tau_{yz}$. One set of axes $x'$, $y'$, and $z'$ exist at a point in a loaded body for which the shear stress components $\tau_{x'y'}$, $\tau_{x'z'}$, and $\tau_{y'z'}$ are all zero. The cutting planes normal to this set of $x'$, $y'$, and $z'$ axes are the principal planes, with the

principal stresses denoted to be $\sigma_1$, $\sigma_2$, and $\sigma_3$. If the order of these three principal stresses is $\sigma_1 > \sigma_2 > \sigma_3$, then:

$$\tau_{max} = \frac{\sigma_1 - \sigma_3}{2} \tag{19.106}$$

It is also necessary to understand the mechanisms of failure in order to apply the appropriate criteria to the design of components and vessels. A number of theories have been proposed to identify the cause of material failure. Of these theories, two are commonly used to quantify failure:

1) *Maximum Normal Stress Theory*. The maximum normal stress theory states that material failure will occur when the greatest principal stress reaches a critical value. The failure stress is measured on a sample of the material in a simple unit axial test. Alternatively, failure occurs when the greatest negative (compressive) principal stress exceeds a critical value. Maximum normal stress theory can be used as a failure criterion for *brittle* materials in static loading when the maximum principal stress is equal to the *ultimate stress* of the material (see Chapter 18). Thus, the maximum normal stress is given by:

$$\sigma_{FS} = \sigma_{US} \tag{19.107}$$

where

$\sigma_{FS}$ = maximum normal stress at failure (N m$^{-2}$)

$\sigma_{US}$ = ultimate tensile stress (N m$^{-2}$)

2) *Maximum Shear Stress Theory*. Maximum shear stress theory states that material failure by yielding will occur for any state of stress when the absolute maximum shear stress reaches a critical value. The critical value of the maximum shear stress is that in the tensile test when yielding of the specimen commences. The maximum shear stress theory applies for *ductile* materials. Maximum shear stress at yield acts at an angle of 45° to the axis of the tensile force. At yielding in a uniaxial test, the principal stress is $\sigma_{YS}$, with the other principal stresses zero. Thus, the shear stress at the yield point is given by:

$$\tau_{YS} = \frac{\sigma_{YS}}{2} \tag{19.108}$$

where

$\tau_{YS}$ = shear stress at the yield point (N m$^{-2}$)

$\sigma_{YS}$ = tensile stress at the yield point (N m$^{-2}$)

Thus, for a biaxial stress in which $\sigma_1 > \sigma_2$, yielding will occur when:

$$\frac{\sigma_1 - \sigma_2}{2} = \frac{\tau_{YS}}{2} \tag{19.109}$$

or

$$\sigma_1 - \sigma_2 = \tau_{YS} \tag{19.110}$$

For a triaxial stress in which $\sigma_1 > \sigma_2 > \sigma_3$, yielding will occur when:

$$\sigma_1 - \sigma_3 = \tau_{YS} \tag{19.111}$$

For combined stresses, maximum shear stress theory is approximately correct, but slightly conservative. It is most often used in mechanical design applications because it is relatively straightforward to apply and slightly conservative.

These relationships are necessary to understand the effect of combined stresses on process vessels, which will be considered later. The elastic strength in tension and in compression for ductile steels is usually considered to be the same. This assumption is sufficiently accurate for many purposes. However, this is not necessarily true of all materials, especially non-metals (e.g. concrete).

# 19.3 · Spherical Vessels Under Internal Pressure

An important mechanical design problem is that of vessels under internal pressure. Both spherical and cylindrical vessels need to be considered. Three important assumptions need to be made:

1) The material is isotropic.
2) The strain resulting from the pressure is small.
3) The thickness of the wall is much smaller than the diameter.

A consequence of the third assumption is that it can be assumed that there is no variation in the stress across the wall, as illustrated in Figure 19.34. This is a reasonable assumption if the ratio of thickness to internal diameter is less than 1/20.

Figure 19.35 illustrates spherical vessel under internal pressure. If the effect of the curvature is neglected because of the thin wall, the balance of forces in Figure 19.35 gives:

$$\sigma_C \, \pi D_I \, t_{SPH} = P \frac{\pi D_I^2}{4} \tag{19.112}$$

where

$\sigma_C$ = circumferential (hoop) stress (N m$^{-2}$)
$D_I$ = internal diameter (m)
$t_{SPH}$ = thickness of the sphere wall (m)
$P$ = internal pressure (N m$^{-2}$)

Rearranging gives:

$$\sigma_C = \frac{PD_I}{4t_{SPH}} \tag{19.113}$$

The accuracy of this equation depends on the thin-walled assumption with $D_I \gg t_{SPH}$. There must be a radial stress at the inside surface equal to the pressure $P$ and zero at the outside surface, as it is a free surface. However, the circumferential (tangential) stress in Eq. (19.113) is the product of $P$ times $D_I/4t_{SPH}$. When this ratio $D_I/4t_{SPH}$ is large, it can be assumed that the radial stress is small compared with the tangential stress and it can be neglected. Equation (19.113) can be rearranged to obtain the wall thickness for a maximum allowable specified stress $\sigma_{ALL}$:

$$t_{SPH} = \frac{PD_I}{4\sigma_{ALL}} \tag{19.114}$$

This thin-wall theory assumes that there is no variation of stress across the wall. Theory can be developed to account for the variation in stress in thick-wall theory (Ryder 1969). Thick-wall theory brings a significant increase in the complexity of the analysis. Alternatively, the thin-wall equation can be modified empirically to account for some variation in stress. For a

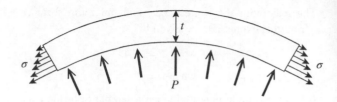

(a) Actual stress across the wall.

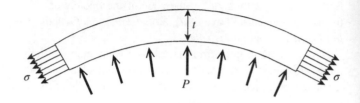

(b) Approximate stress across the wall for a thin wall.

**Figure 19.34**

Thin wall pressure vessel.

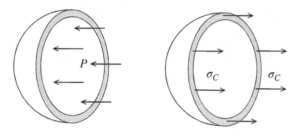

**Figure 19.35**

Stress for a spherical vessel under internal pressure.

moderately thick shell employing thin-shell theory, use the mean diameter instead of the internal diameter:

$$t_{SPH} = \frac{P(D_I + t_{SPH})}{4\sigma_{ALL}} \tag{19.115}$$

Rearranging Eq. (19.115):

$$t_{SPH} = \frac{PD_I}{4\sigma_{ALL} - P} \tag{19.116}$$

This equation is modified empirically in the design codes to be (ASME 2019; BSI Standards Publications 2019):

$$t_{SPH} = \frac{PD_I}{4\sigma_{ALL} - 0.4P} \tag{19.117}$$

Pressure vessels are fabricated from preformed sections welded together. The efficiency of the joints needs to be accounted for by a *joint efficiency* $\eta_J$. Thus Eq. (19.117) becomes:

$$t_{SPH} = \frac{PD_I}{4\sigma_{ALL}\eta_J - 0.4P} \tag{19.118}$$

## Table 19.3

Efficiency of welded joints.

| Type of joint | Extent of radiographic inspection | | |
| --- | --- | --- | --- |
| | **Full** | **Partial** | **None** |
| Butt joints fully welded from both sides | 1.0 | 0.85 | 0.7 |
| Single-welded butt joint with backing strips that remain in place after welding | 0.9 | 0.8 | 0.65 |

where

$\eta_J$ = joint efficiency (–)

The joint efficiency depends on both the welding technique used and the extent of radiographic inspection of the completed weld. Radiographic inspection uses penetrating radiation from an X-ray or gamma-ray source to view the internal structure of the weld and identify any defects. Remedial action required for any defects identified will normally require removal of the affected area of the weld by gouging or grinding to sound metal, followed by re-welding. Table 19.3 presents the efficiencies used for welded joints. Welded joints will be discussed in more detail

later in this chapter. Finally, an allowance for corrosion can be added (see Chapter 18):

$$t_{SPH} = \frac{PD_I}{4\sigma_{ALL}\eta_J - 0.4P} + C \qquad (19.119)$$

where

$C$ = corrosion allowance (m)

This equation is applicable when the thickness of the shell of a wholly spherical vessel does not exceed $0.178D_I$ or $P$ does not exceed $0.665\sigma_{ALL}\eta_J$ (ASME 2019; BSI Standards Publications 2019).

In order to apply Eq. (19.119) requires the maximum allowable stress to be specified. Various materials are used for the fabrication of pressure vessels. Table 19.4 presents the mechanical properties of a number of commonly used materials of construction (ASME 2017a). Values of tensile strength and yield strength are shown, along with allowable stresses, which vary according to temperature. A number of points should be noted:

1) The values quoted in Table 19.4 are for plate materials.
2) Minimum values for tensile strength and yield strength are shown. Mechanical properties of a given material show variability due to the variability of the microstructure and

## Table 19.4

Maximum allowable stresses for a range of materials.

| Material | UNS code | Minimum tensile strength (MPa) | Minimum yield strength (MPa) | Maximum allowable stress (MPa) for sheet metal temperature (°C) | | | | | | | | | | |
| --- | --- | --- | --- | --- | --- | --- | --- | --- | --- | --- | --- | --- | --- | --- |
| | | | | **40** | **65** | **100** | **150** | **200** | **250** | **300** | **350** | **400** | **450** | **500** |
| Carbon Steel SA 516-60 | K02100 | 415 | 220 | 118 | 118 | 118 | 118 | 118 | 118 | 115 | 108 | 88.9 | 62.7 | 31.6 |
| Carbon Steel SA 516-70 | K02700 | 485 | 260 | 138 | 138 | 138 | 138 | 138 | 138 | 136 | 128 | 101 | 67.1 | 33.6 |
| Stainless Steel 304 | S30400 | 515 | 205 | 138 | 138 | 137 | 130 | 126 | 122 | 116 | 111 | 107 | 103 | 99.3 |
| Stainless Steel 304L | S30403 | 485 | 170 | 115 | 105 | 97.0 | 88.1 | 81.2 | 76.0 | 72.3 | 69.7 | 67.6 | 65.7 | 63.4 |
| Stainless Steel 316 | S31600 | 515 | 205 | 138 | 128 | 118 | 107 | 99.2 | 92.8 | 88.1 | 84.1 | 82.0 | 80.6 | 79.2 |
| Stainless Steel 316L | S31603 | 485 | 170 | 115 | 106 | 96.3 | 87.4 | 81.2 | 76.0 | 72.5 | 70.0 | 67.5 | 65.0 | ... |
| Hastelloy B3 | N10675 | 760 | 350 | 216 | 216 | 216 | 209 | 199 | 191 | 183 | 177 | 173 | 169 | ... |
| Hastelloy C-276 | N10276 | 690 | 285 | 188 | 180 | 170 | 158 | 148 | 139 | 131 | 125 | 120 | 117 | 115 |
| Titanium | R50400 | 345 | 275 | 98.6 | 94.6 | 83.5 | 70.8 | 61.4 | 53.9 | 46.7 | ... | ... | ... | ... |

manufacturing process. Thus, the properties can be defined by either a mean or a minimum value, determined statistically from a number of tests. Designing according to the minimum value decreases the probability of failure.

3) ASME (2017a) tables for a number of materials show two sets of allowable stresses with different values. If the design of the vessel allows some permanent deformation, then the highest values can be used. Table 19.4 shows the lower values from the ASME code.

There will be a minimum wall thickness required to ensure that any vessel is sufficiently rigid to withstand its own weight, and any incidental loads. The ASME BPV Code Sec. VIII D.1 (with certain exceptions) specifies a minimum wall thickness of 1.5 mm not including corrosion allowance, and regardless of vessel dimensions and material of construction. The minimum thickness of shells and heads used in compressed air service, steam service, and water service should be 2.5 mm, exclusive of any corrosion allowance. The minimum thickness of shells and heads of unfired steam boilers should be 6 mm, exclusive of any corrosion allowance.

Equation (19.119) can be rearranged for $C = 0$ to give the stress for a given spherical shell thickness:

$$\sigma_{SPH} = \frac{PD_I + 0.4Pt_{SPH}}{4\eta_J t_{SPH}} \tag{19.120}$$

Equation (19.120) refers to the uncorroded state. After corrosion, $D_I$ will increase by $2C$ and $t_{SPH}$ will decrease by $C$. When $\eta_J = 1$, Eqs (19.113) and (19.120) give very similar results for thin-walled pressure vessels.

# 19.4 Cylindrical Vessels Under Internal Pressure

Now consider the corresponding stresses for a cylindrical vessel under internal pressure. Figure 19.36 shows a half cylinder of internal diameter $D_I$, thickness $t$, and length $L$. The pressure $P$ creates a vertical force on the area $D_I L$. This creates a *circumferential stress* $\sigma_C$, or *hoop stress,* which acts on an area $2tL$. Equating the forces:

$$PD_I L = 2t_C L\sigma_C \tag{19.121}$$

where

$\sigma_C$ = circumferential (hoop) stress (N m$^{-2}$)

$D_I$ = internal diameter (m)

$t_C$ = thickness of the cylinder wall resisting the circumferential stress (m)

$P$ = internal pressure (N m$^{-2}$)

Rearranging Eq. (19.121) gives:

$$\sigma_C = \frac{PD_I}{2t_C} \tag{19.122}$$

The accuracy of this equation depends on the thin-walled assumption with $D_I \gg t_C$. As with the spherical shell, there must

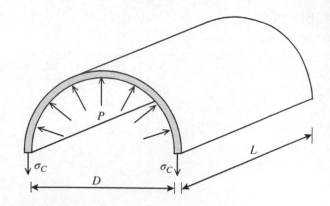

**Figure 19.36**

Circumferential (hoop) stress for a cylindrical vessel under internal pressure.

be a radial stress at the inside surface equal to the pressure $P$ and zero at the outside surface, as it is a free surface. However, the circumferential (hoop) stress in Eq. (19.122) is the product of $P$ times $D_I/2t_C$. When this ratio $D_I/2t_C$ is large, it can be assumed that the radial stress is small compared with the circumferential stress and can be neglected. Equation (19.122) can be rearranged to obtain the wall thickness for a maximum allowable stress $\sigma_{ALL}$:

$$t_C = \frac{PD_I}{2\sigma_{ALL}} \tag{19.123}$$

As with spherical shells, a thick-wall theory can be developed to account for the variation in stress across the wall (Ryder 1969). The thin-wall equation can be modified empirically to account for some variation in stress. For a moderately thick shell employing thin-shell theory, use the mean diameter instead of the internal diameter:

$$t_C = \frac{P(D_I + t_C)}{2\sigma_{ALL}} \tag{19.124}$$

Rearranging Eq. (19.124):

$$t_C = \frac{PD_I}{2\sigma_{ALL} - P} \tag{19.125}$$

This equation is modified empirically in the design codes to be (ASME 2019; BSI Standards Publications 2019):

$$t_C = \frac{PD_I}{2\sigma_{ALL} - 1.2P} \tag{19.126}$$

Including an efficiency for the joints used in the fabrication $\eta_J$ and a corrosion allowance $C$ gives (ASME 2019; BSI Standards Publications 2019):

$$t_C = \frac{PD_I}{2\sigma_{ALL}\eta_J - 1.2P} + C \tag{19.127}$$

This equation is applicable when the thickness does not exceed one quarter of the inside diameter $D_I$, or $P$ does not exceed $0.385\sigma_{ALL}\eta_J$ (ASME 2019; BSI Standards Publications 2019).

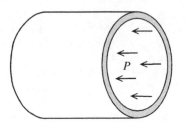
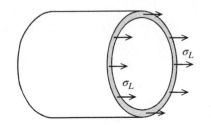

## Figure 19.37

Longitudinal stress for a cylindrical vessel under internal pressure.

Alternatively, rearranging Eq. (19.127) for $C = 0$ gives the stress for a given thickness:

$$\sigma_C = \frac{PD_I + 1.2Pt}{2\eta_J t} \tag{19.128}$$

Equation (19.128) refers to the uncorroded state. After corrosion $D_I$ will increase by $2C$ and $t$ will decrease by $C$. When $\eta_J = 1$, Eqs (19.122) and (19.128) give very similar results for thin-walled pressure vessels.

Now consider a cylindrical section cut in a transverse plane, as shown in Figure 19.37. The pressure $P$ acts on a projected area of $\pi D_I^2 / 4$, irrespective of the shape of the closing heads. This creates a longitudinal stress $\sigma_L$, which neglecting the curvature acts on an area $\pi(D_I + t)t$. If it is assumed that $D_I \gg t$, the area can be assumed to be $\pi D_I t$. Equating the forces:

$$P\frac{\pi D_I^2}{4} = \pi D_I t_L \sigma_L \tag{19.129}$$

where

$\sigma_L$ = longitudinal stress (N m$^{-2}$)

$t_L$ = thickness of the cylinder wall resisting the longitudinal stress (m)

Rearranging gives:

$$\sigma_L = \frac{PD_I}{4t_L} \tag{19.130}$$

Equation (19.130) can be rearranged to obtain the wall thickness for a maximum allowable stress $\sigma_{ALL}$:

$$t_L = \frac{PD_I}{4\sigma_{ALL}} \tag{19.131}$$

This equation can again be modified for moderately thick walls using the mean diameter with a joint efficiency and corrosion allowance included. The final equation, after being modified empirically, is given in the design codes as (ASME 2019; BSI Standards Publications 2019):

$$t_L = \frac{PD_I}{4\sigma_{ALL}\eta_J + 0.8P} + C \tag{19.132}$$

The minimum thickness of the cylindrical shell must not be less than calculated by either Eq. (19.127) or Eq. (19.132) (ASME 2019; BSI Standards Publications 2019). The circumferential stress will normally be the greater than the longitudinal stress, and the circumferential stress will dominate. Alternatively, rearranging Eq. (19.132) for $C = 0$ to give the stress for a given thickness:

$$\sigma_L = \frac{PD_I - 0.8Pt}{4\eta_J t} \tag{19.133}$$

Again, Eq. (19.133) refers to the uncorroded state. After corrosion, $D_I$ will increase by $2C$ and $t$ will decrease by $C$. When $\eta_J = 1$, Eqs (19.130) and (19.133) give very similar results for thin-walled pressure vessels.

**Example 19.9**   For the horizontal vessel in Example 19.7 with an internal diameter of 2 m and length 6 m, the process pressure has been set to 10 bara. If it is assumed that the design pressure should be 10% above the normal working gauge pressure, this gives a design pressure of 9.9 barg. Take the basis for the design pressure to be 10 barg. Assume the joint efficiency to be 0.85 and the corrosion allowance to be 2 mm.

a) Calculate the minimum shell thickness from the circumferential stress caused by the internal pressure to ensure a reasonable shell thickness has been chosen.

b) Calculate the circumferential and longitudinal stress caused by the internal pressure.

c) Combine the stresses from internal pressure and bending from Example 19.7 and calculate the difference between the principal stresses half way along the vessel on its upper and lower surfaces and above the supports midway between the top and bottom, corresponding with the neutral axis.

d) Repeat the calculation assuming the thickness is decreased by the corrosion allowance.

(Note: this exercise will neglect the strength of the ends of the vessel, which will be discussed later.)

**Solution**

a) First calculate the shell thickness to withstand in internal pressure. The design codes specify that the shell thickness must be greater than that calculated for both the circumferential and longitudinal stresses. The circumferential stress will normally be the greater. The allowable stress for carbon steel

SA 516-70 at 100 °C from Table 19.4 is $138 \times 10^6$ N m$^{-2}$. From Eq. (19.127):

$$t_C = \frac{PD_I}{2\sigma_{ALL}\eta_J - 1.2P} + C$$

$$= \frac{10 \times 10^5 \times 2.0}{2 \times 138 \times 10^6 \times 0.85 - 1.2 \times 10 \times 10^5} + 0.002$$

$$= 0.0106 \text{ m}$$

$$= 10.6 \text{ mm}$$

A thickness of 12 mm seams a reasonable starting point. Note that the ratio of shell thickness to internal diameter is well below 1/20 as required for a thin-walled pressure vessel.

**b)** The circumferential stress is given by Eq. (19.128):

$$\sigma_C = \frac{PD_I + 1.2Pt}{2\eta_J t}$$

$$= \frac{10 \times 10^5 \times 2.0 + 1.2 \times 10 \times 10^5 \times 0.012}{2 \times 0.85 \times 0.012}$$

$$= 98.75 \times 10^6 \text{ N m}^{-2}$$

The longitudinal stress is given by Eq. (19.133):

$$\sigma_L = \frac{PD_I - 0.8Pt}{4\eta_J t}$$

$$= \frac{10 \times 10^5 \times 2.0 - 0.8 \times 10 \times 10^5 \times 0.012}{4 \times 0.85 \times 0.012}$$

$$= 48.78 \times 10^6 \text{ N m}^{-2}$$

**c)** Figure 19.38 shows the stresses at the three locations on the vessel. Yielding will occur when the difference between the principal stresses exceeds the maximum allowable stress for the material at the working temperature.

For Location $A$ at the top mid-point of the vessel, the shear stresses are zero and the stresses are the principal stresses. The difference between the principal stresses is given by:

$$(\sigma_1 - \sigma_2) = (98.75 \times 10^6 - 45.63 \times 10^6) = 53.12 \times 10^6 \text{ N m}^{-2}$$

For Location $B$ at the bottom mid-point of the vessel, again the shear stresses are zero and the stresses are the principal stresses. The difference between the principal stresses is given by:

$$(\sigma_1 - \sigma_2) = (98.75 \times 10^6 - 51.93 \times 10^6) = 46.82 \times 10^6 \text{ N m}^{-2}$$

For Location $C$ above the supports midway between the top and bottom, corresponding with the neutral axis, the principal stresses are given by Eq. (19.103):

$$\sigma_1 = \frac{\sigma_x + \sigma_y}{2} + \sqrt{\left(\frac{\sigma_x - \sigma_y}{2}\right)^2 + \tau_{xy}^2}$$

$$\sigma_2 = \frac{\sigma_x + \sigma_y}{2} - \sqrt{\left(\frac{\sigma_x - \sigma_y}{2}\right)^2 + \tau_{xy}^2}$$

It should be noted that the shear stress, as shown in Figure 19.38, is negative in accordance with the sign convention in Figure 19.32:

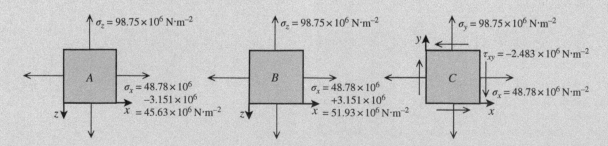

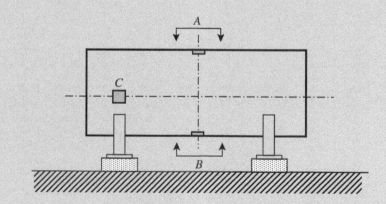

**Figure 19.38**

Stresses on the vessel for Example 19.9.

$$\sigma_1 = \frac{48.78 \times 10^6 + 98.75 \times 10^6}{2}$$

$$+ \sqrt{\left(\frac{48.78 \times 10^6 - 98.75 \times 10^6}{2}\right)^2 + (-2.483 \times 10^6)^2}$$

$$= 98.87 \times 10^6 \, \text{N m}^{-2}$$

$$\sigma_2 = \frac{48.78 \times 10^6 + 98.75 \times 10^6}{2}$$

$$- \sqrt{\left(\frac{48.78 \times 10^6 - 98.75 \times 10^6}{2}\right)^2 + (-2.483 \times 10^6)^2}$$

$$= 48.66 \times 10^6 \, \text{N m}^{-2}$$

The difference between the principal stresses is given by:

$$(\sigma_1 - \sigma_2) = (98.87 \times 10^6 - 48.66 \times 10^6) = 50.21 \times 10^6 \, \text{N m}^{-2}$$

**d)** For the corroded state, decrease the thickness by 0.002 m and increase the internal diameter by 0.004 m.

Location A: $(\sigma_1 - \sigma_2) = 63.55 \times 10^6 \, \text{N m}^{-2}$
Location B: $(\sigma_1 - \sigma_2) = 56.22 \times 10^6 \, \text{N m}^{-2}$
Location C: $(\sigma_1 - \sigma_2) = 60.16 \times 10^6 \, \text{N m}^{-2}$

Thus, the maximum difference between the principal stresses is $53.12 \times 10^6 \, \text{N m}^{-2}$ at the top midpoint of the vessel. This increases to $60.16 \times 10^6 \, \text{N m}^{-2}$ in the corroded state, but still well below the maximum permissible stress given in Table 19.4. It should also be noted that this assumes that the shell maintains its shape to act as a beam. Further consideration of the stresses, particularly around the saddle supports, is necessary for the final design (Zick 1951; Brownell and Young 1959; Bednar 1991; Escoe 1994). The analysis also does not consider the stresses in the flat heads that are assumed. The design of the vessel heads will be considered next.

# 19.5 Design of Heads for Cylindrical Vessels Under Internal Pressure

Different heads can be used to enclose the ends of cylindrical pressure vessels. Figure 19.39 shows four of the commonly used head shapes.

### 19.5.1 Hemispherical Heads

Figure 19.39a shows a *hemispherical head* in which the depth of the head is half the diameter. The volume of a single hemispherical head is given by:

$$V_{HS} = \frac{\pi D_I^3}{12} \tag{19.134}$$

where

$V_{HS}$ = volume of hemispherical head (m³)
$D_I$ = internal diameter (m)

The volume of a cylindrical vessel with hemispherical heads is given by:

$$V = \frac{\pi D_I^2}{4}L + \frac{\pi D_I^3}{6} \tag{19.135}$$

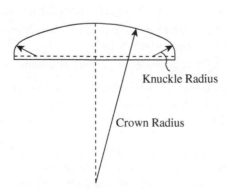

(a) Hemispherical head.

(b) Semi-ellipsoidal head.

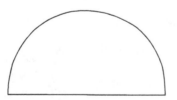

Knuckle Radius

Crown Radius

(c) Torispherical head.

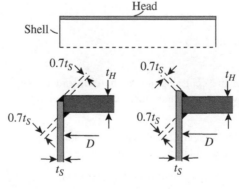

(d) Flat head.

**Figure 19.39**

Heads for cylindrical pressure vessels.

where

$V$ = volume of vessel with hemispherical heads ($m^3$)

$D_I$ = internal diameter (m)

$L$ = tangent to tangent length (m)

(The tangent point is the point along the length where the curvature of the head starts.)

To calculate the weight of a single hemispherical head requires the surface area at the mean diameter to be multiplied by the thickness and density, assuming a thin wall, and is given by:

$$W_{HS} = A_{HS}t_{HS}\rho_{MOC}g$$
$$= \frac{\pi(D_I + t_{HS})^2}{2}t_{HS}\rho_{MOC}9.81 \qquad (19.136)$$
$$= 4.905\pi\rho_{MOC}t_{HS}(D_I + t_{HS})^2$$

where

$W_{HS}$ = weight of a single hemispherical head (N m$^{-2}$)

$A_{HS}$ = surface area of a single hemispherical head based on the mean diameter (m$^2$)

$t_{HS}$ = thickness of the hemispherical head (m)

$\rho_{MOC}$ = density of the material of construction (kg m$^{-3}$)

The thickness required for a hemispherical head can be determined for hemispherical shells from (ASME 2019; BSI Standards Publications 2019):

$$t_{HS} = \frac{PD_I}{4\sigma_{ALL}\eta_J - 0.4P} + C \qquad (19.137)$$

The joint efficiency is necessary because hemispherical heads cannot usually be formed from a flat sheet. Instead, they are formed from welded sections.

### 19.5.2 Semi-ellipsoidal Heads

Figure 19.39b shows a *semi-ellipsoidal head*. The most common ratio for the semi-ellipsoid is based on a 2:1 ellipsoid; i.e. the depth of the semi-ellipsoidal head is one quarter that of the diameter. Other ratios are used but are less common. The volume of an ellipsoid is given by:

$$V_{ELL} = \frac{4}{3}\pi R_I^3 h_{ELL} \qquad (19.138)$$

where

$V_{ELL}$ = volume of ellipsoid (m$^3$)

$R_I$ = radius (m)

(distance from the central point of the ellipsoid to the surface following a plane that creates a circular cross-section)

$h_{ELL}$ = distance from the center of the ellipsoid to the surface perpendicular to the circular cross-sections (m)

For a 2:1 ellipsoid, $h_{ELL} = R_I/2$; thus, from Eq. (19.138):

$$V_{ELL} = \frac{4}{3}\pi R_I^2 \frac{R_I}{2} = \frac{2}{3}\pi R_I^3 = \frac{\pi D_I^3}{12} \qquad (19.139)$$

where

$D_I$ = internal diameter of the vessel (m)

The volume of a single a 2:1 semi-ellipsoidal head is given by:

$$V_{SE} = \frac{\pi D_I^3}{24} \qquad (19.140)$$

where

$V_{SE}$ = volume of a single semi-ellipsoidal head (m$^3$)

The volume of a cylindrical vessel with 2:1 semi-ellipsoidal heads is given by:

$$V = \frac{\pi D_I^2}{4}L + \frac{\pi D_I^3}{12} \qquad (19.141)$$

where

$L$ = tangent-to-tangent length of the vessel (m)

To calculate the weight of a semi-ellipsoidal head, or the amount of insulation necessary, requires its surface area to be known. The surface area of a 2:1 semi-ellipsoidal can be approximated with an error of around 1% to be 1.382 times the cross-section area of the circular opening:

$$A_{SE} = 1.382\frac{\pi D_I^2}{4} \qquad (19.142)$$

where

$A_{SE}$ = inside surface area of a 2:1 semi-ellipsoidal head (m$^2$)

The weight of a semi-ellipsoidal head based on the mean diameter is given by:

$$W_{SE} = A_{SE}t_{SE}\rho_{MOC}g$$
$$= 1.382\frac{\pi(D_I + t_{SE})^2}{4}t_{SE}\rho_{MOC}9.81 \qquad (19.143)$$
$$= 3.389\pi\rho_{MOC}t_{HS}(D_I + t_{SE})^2$$

where

$W_{SE}$ = weight of a single 2:1 hemispherical head (N m$^{-2}$)

$A_{SE}$ = surface area of a single 2:1 hemispherical head based on the mean diameter (m$^2$)

$t_{SE}$ = thickness of a 2:1 semi-ellipsoidal head (m)

Using the inside diameter will give the inside surface area of the head. The outside diameter will give the outside surface area. To estimate the weight of the head, the mean of the inside and outside diameter is used.

For a semi-ellipsoidal head with a depth of one quarter of that of the diameter, the required thickness can be determined from (ASME 2019; BSI Standards Publications 2019):

$$t_{SE} = \frac{PD_I}{2\sigma_{ALL}\eta_J - 0.2P} + C \qquad (19.144)$$

### 19.5.3   Torispherical Heads

Figure 19.39c shows a *torispherical head*. The torispherical head approximates a semi-ellipsoidal head using a combination of radii, as shown in Figure 19.39c. The largest radius is in the *crown* and the smallest (*knuckle*) radius at the outside diameter. In this way, it can be more straightforward to fabricate than a semi-ellipsoidal head. The standard geometry for a torispherical head is such that the inside crown radius equals the outside diameter and the knuckle radius is 6% of the inside crown radius. For a preliminary mechanical design, the volume and surface area of a standard torispherical head can be approximated to be that of a 2:1 ellipsoidal head, but substituting using the thickness of the torispherical head for the semi-ellipsoidal thickness. The required thickness of the head can be determined from (ASME 2019):

$$t_{TOR} = \frac{0.885PR_C}{\sigma_{ALL}\eta_J - 0.1P} + C \qquad (19.145)$$

where

$t_{TOR}$ = thickness of torispherical head (m)

$R_C$ = crown radius (m)

### 19.5.4   Flat Heads

The required thickness of *flat circular heads* can be calculated from (ASME 2019; BSI Standards Publications 2019):

$$t_{FH} = D_I\sqrt{\frac{KP}{\sigma_{ALL}\eta_J}} + C \qquad (19.146)$$

where

$t_{FH}$ = thickness of flat head (m)

$K$ = constant depending on the method of attachment of the head and shell dimensions

A number of different methods can be used to attach a head to a shell (ASME 2019). The constant $K$ in Eq. (19.146) depends on the method of attachment of the head to the shell and the shell dimensions, and varies between 0.1 and 0.33. Figure 19.39d shows two of the many methods of attachment in which the head is attached to the shell. For these two designs, $K$ can vary between 0.2 and 0.33 depending on the attachment of the head to the shell and the shell dimensions (ASME 2019). A value of 0.33 will be conservative.

Although the process design will often specify the internal diameter of the vessel, sometimes specifying the outside diameter is preferred. Whilst most heads greater than 0.6 m in diameter are custom-made, smaller diameter heads can be off-the-shelf and specified by the outside diameter pipe size. On the other hand, as heads get thicker, hot forming is necessary and the forming is based on the internal diameter.

**Example 19.10**  For the horizontal vessel in Examples 19.7 and 19.9, assuming a joint efficiency $\eta_J$ of unity and corrosion allowance of 2 mm, calculate the head thickness required for the following head designs:

**a)** Hemispherical
**b)** Semi-ellipsoid
**c)** Torispherical
**d)** Flat

**Solution**

**a)** Thickness of a hemispherical head is given by Eq. (19.137):

$$t_{HS} = \frac{PD_I}{4\sigma_{ALL}\eta_J - 0.4P} + C$$

$$= \frac{10 \times 10^5 \times 2.0}{4 \times 138 \times 10^6 \times 1.0 - 0.4 \times 10 \times 10^5} + 0.002$$

$$= 0.00563 \text{ m}$$

$$= 5.6 \text{ mm}$$

Say a 6 mm or 8 mm plate.

**b)** Thickness of a semi-ellipsoidal head is given by Eq. (19.144):

$$t_{SE} = \frac{PD_I}{2\sigma_{ALL}\eta_J - 0.2P} + C$$

$$= \frac{10 \times 10^5 \times 2.0}{2 \times 138 \times 10^6 \times 1.0 - 0.2 \times 10 \times 10^5} + 0.002$$

$$= 0.00925 \text{ m}$$

$$= 9.3 \text{ mm}$$

Say a 10 mm or 12 mm plate.

**c)** Thickness of a torispherical head is given by Eq. (19.145):

$$t_{TOR} = \frac{0.885PR_C}{\sigma_{ALL}\eta_J - 0.1P} + C$$

The standard geometry for a torispherical head is such that the inside crown radius equals the outside diameter.

$$t_{TOR} = \frac{0.885 \times 10 \times 10^5 \times 2.0}{138 \times 10^6 \times 1.0 - 0.1 \times 10 \times 10^5} + 0.002$$

$$= 0.0148 \text{ m}$$

$$= 14.8 \text{ mm}$$

Say a 15 mm, 16 mm, or 18 mm plate.

**d)** Thickness of a flat head is given by Eq. (19.146):

$$t_{FH} = D_I\sqrt{\frac{KP}{\sigma_{ALL}\eta_J}} + C$$

$K$ constant depends on the method of attachment of the head and the shell dimensions. Assume $K = 0.33$:

$$t_{FH} = 2.0\sqrt{\frac{0.33 \times 10 \times 10^5}{138 \times 10^6 \times 1.0}} + 0.002$$

$$= 0.0998 \text{ m}$$

$$= 99.9 \text{ mm}$$

Say a 100 mm, 110 mm, or 115 mm plate.

The thickness of the head increases from hemispherical to semi-ellipsoidal to torispherical to flat head. It is clear that the hemispherical head is the most efficient at withstanding the internal pressure with the flat head by far the least efficient. The hemispherical head is in pure tension and flat head in pure bending. Semi-ellipsoidal and torispherical heads are subject to a combination of tension and bending components. In this example, the hemispherical and semi-ellipsoidal heads are thinner than the thickness of the cylinder. If the head is thinner than the cylinder then the wall is tapered at the transition between the two. However, it is important not to taper the cylinder, as this needs the full wall thickness. The taper should be part of the head.

The choice of head to use is an economic decision. Torispherical heads are cheaper to fabricate than hemispherical and semi-elliptical heads, but thicker for a given pressure, involving greater material costs. The semi-ellipsoidal head is more expensive to fabricate than the torispherical, but is thinner for the same pressure. The hemispherical head is the most costly to fabricate, as it usually cannot be fabricated from a single sheet and is fabricated from pieces welded together, but the hemispherical head is the thinnest for a given pressure. Thus, the material cost decreases from torispherical to semi-ellipsoidal to hemispherical because the head gets thinner, but the fabrication costs increase. Hemispherical heads are commonly used for large-diameter or high-pressure applications where material savings compensate for the fabrication costs. Flat heads by contrast require much more material than the others.

Torispherical heads are typically used for cylindrical vessels up to 20 barg, semi-ellipsoidal heads used in the range 20–100 barg, and hemispherical heads used above 100 barg (Hall 2020).

# 19.6  Design of Vertical Cylindrical Pressure Vessels Under Internal Pressure

Vertical cylindrical vessels occur commonly as distillation and absorption columns. Although these tend to be the tallest vessels, there are many other instances of smaller vertical cylindrical pressure vessels. For vessels with a diameter less than 600 mm it is possible to fabricate the vessel from seamless pipe. Greater than 600 mm, the vessel is fabricated from standard flat plate, which is rolled and welded together in sections or *courses*. Fabricating from seamless pipe avoids the radiographic examination of welds. The process design will most often specify the internal diameter of the vessel. This simplifies the calculation of the flow conditions in the operation. It also simplifies the fabrication and installation of internal hardware (e.g. trays, support rings for the trays, distributors, etc.). Figure 19.40 illustrates the two general ways in which distillation and absorption columns might be installed.

Smaller columns are often supported, or partially supported, by a structure. Figure 19.40a illustrates a distillation column supported by a skirt and a plinth at the base, but with the structure

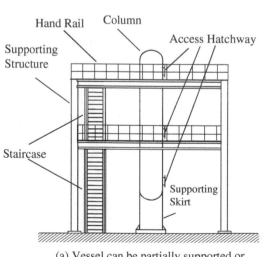

(a) Vessel can be partially supported or fully supported by a structure.

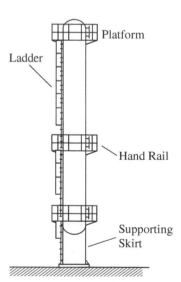

(b) Self-supporting stand-alone vessel.

**Figure 19.40**

Vertical cylindrical vessels can be supported by a structure or be self-supporting.

supporting lateral forces, e.g. wind. Ancillary equipment required for the operation, such as the condenser, distillate receiver, pumps, and so on will be laid out and supported within the structure.

Taller columns, as illustrated in Figure 19.40b are completely self-supporting. Access is in this instance available through ladders and platforms, rather than the structure. The ancillary equipment, such as condensers, will most often be either at ground level or supported by a small structure adjacent to the column close to ground level.

A self-supporting column must be capable of mechanically withstanding:

a) The internal pressure, or external pressure if operated under vacuum.

b) The stresses created by *dead loads* due to the self-weight of the vessel, its contents, and any ancillary equipment supported by the column, often referred to as *deadweight*.

c) Steady bending stresses created by the wind load.

d) Oscillations from shedding vortices created by the wind. Designing for these factors is outside the scope of the current text (see Brownell and Young 1959; Bednar 1991; Escoe 1994).

e) Bending stresses created by any ancillary equipment supported by the column that have a center of gravity not in line with the center of gravity of the column itself.

f) Seismic forces from earthquakes that create sudden and erratic motion of the ground on which the vessel is supported. The effect is transient and dynamic in nature. The horizontal forces create a horizontal shear force at the base of the vessel, creating shear and bending moments at different elevations, as well as an overturning moment at the base. Designing for these factors is outside the scope of the current text (see Brownell and Young 1959; Bednar 1991; Escoe 1994).

The dead load for distillation and absorption columns will be greatest for a hydrostatic pressure test with the vessel in the vertical position, for which the vessel would be filled with water and the pressure raised to typically 1.3 times the maximum allowable working pressure (MAWP) to test the integrity of the vessel. In designing the shell, the compressive stress resulting from the weight of the liquid in the hydrostatic test is carried by the bottom head of the shell and then transferred directly to the skirt (Brownell and Young 1959). This will not affect the design of the shell, but will affect the design of the skirt and the bottom head. Although the water in a hydrostatic test will not affect the compressive stress on the shell, it will affect the pressure at the base of the shell by virtue of the hydraulic head. Under normal operation the same is true for the deadweight of the liquid in the column sump. The components of the deadweight that need to be accounted for in the design of the shell include the following weights:

a) The column shell itself above the joint with the skirt.

b) Insulation applied to the column shell and ancillary components above the joint with the skirt.

c) Column internals, including trays, packing, liquid distributors, and so on, together with the weight of any processing materials held up within the column. The weight of trays, together with the liquid loading, can be approximated to be $1200\,N\,m^{-2}$ (Brownell and Young 1959).

d) Ladders. The weight of the caged ladders normally used can be approximated to be $370\,N\,m^{-1}$ (Brownell and Young 1959).

e) Platforms. The weight of platforms can be approximated to be $1700\,N\,m^{-2}$ (Brownell and Young 1959).

f) Pipework feeding or drawing process fluids to or from the column. These will include the vapor line from the top of the column, feed and reflux lines, and possibly others.

g) Insulation on the pipework supported by the column.

h) Process materials inside the pipework supported by the column.

i) Access hatchways for maintenance and inspection.

The weight of insulation will normally be small compared with the other weights and can often be neglected. Most external attachments are eccentric to the longitudinal axis, which in principle creates bending moments, but these moments are often negligible (Brownell and Young 1959). However, large heavy attachments, such as condensers, will require the bending moment to be considered. Once the deadweight has been calculated, this becomes a compressive stress to the shell that must be supported by the thickness of the shell.

Shells with a diameter greater than 600 mm are constructed in a series of cylindrical *courses* (sections) rolled from plate and welded together to create the long cylindrical shell. Because the bending and shear forces increase from the top of the column to the bottom for a self-supporting column, the thickness of the courses is typically increased from top to the bottom.

The choice of design pressure for the mechanical design will depend on the maximum allowable pressure during operation. Process design specifies the operating pressure for a specified temperature under normal operating conditions. The maximum operating pressure might extend beyond normal operating conditions to include start-up, shut-down, and other off-design conditions. The maximum allowable working pressure (MAWP) is the maximum gauge pressure that the weakest component of the system can safely withstand for a designated temperature. As shown in Figure 15.16, the maximum allowable working pressure would typically be 10% greater than the maximum operating pressure, which would correspond with the maximum allowable set pressure for single pressure relief devices. Thus, a design pressure 10% greater than the maximum allowable operating pressure is a reasonable basis for the mechanical design.

**Example 19.11**  Example 19.6 considered a self-supporting distillation column sized to have a height of 36.8 m and an inside diameter of 2.5 m. The column is fitted with 54 trays each with an assumed weight of 1200 N m$^{-2}$ spaced 0.6 m apart. The total height of the column includes an allowance of 1.5 m above the top tray for liquid disengagement and 3.5 m below the bottom tray for liquid sump and liquid disengagement. The shell is to be fitted with 2:1 semi-ellipsoidal heads. The preliminary design will assume the column is mounted on a cylindrical skirt support with a height of 3 m and the same outside diameter as the column with a thickness of 20 mm welded to the column shell. It can be assumed that the joint between the shell and the skirt will be at the base of the vertical cylindrical shell. The column has an operating pressure of 14 bara at the top of the column. Assuming the pressure drop across each tray is 0.007 bar, this gives a bottom pressure of 14.4 bara (13.4 barg). Assuming a design pressure 10% above the working pressure gives a pressure of 14.7 barg at the bottom. Assume a design pressure of 15 barg. The shell is constructed in a series of cylindrical *courses* (sections) 2.5 m long rolled from plate and welded together. The vessel is to be fabricated from carbon steel SA 516-70 with a density of 7850 kg m$^{-3}$ with a corrosion allowance of 2 mm. The working temperature is below 250 °C. For a preliminary design in this example, it will be assumed initially that the shell thickness of the top half of the column is 18 mm and that of the bottom half 20 mm. In practice the shell would most likely be constructed using more courses than two for such a tall column. Assume that welds for the column undergo 100% radiographical examination and $\eta_J = 1.0$. The distillation column is insulated for its entire area (not the skirt) with 75 mm of mineral wool covered in aluminum cladding, increasing the outside diameter exposed to the wind to 2.69 m. The density of the column insulation is 100 kg m$^{-3}$, corresponding with a weight of 981 N m$^{-3}$. The reboiler, condenser, and distillate receiver for the column are to be supported on a structure adjacent to the column. The details of the pipework supported by the column are given in Table 19.5.

The column is fitted with three platforms, one at the bottom of the column above the skirt, one midway, and the third just below the top of the column. Each of the three platforms have an area of 9 m$^2$. Safety considerations for access might dictate more platforms through the height. Caged ladders with a total length of 37.8 m allow access to the three platforms. Meteorological records for the location indicate that a maximum wind speed of 50 m s$^{-1}$ can be assumed. This wind speed can be assumed to apply at all heights. In practice the wind speed increases with height from zero at the ground. Air density can be assumed to be 1.3 kg m$^{-3}$.

a) Calculate the minimum shell thickness from the circumferential stress caused by the internal pressure to ensure a reasonable shell thickness has been chosen.
b) Calculate the thickness of the semi-ellipsoidal heads at the top and bottom of the shell.
c) Calculate the circumferential and longitudinal stress caused by the internal pressure for the top and bottom sections of the column before corrosion.
d) Calculate the bending moment and maximum bending stress at the base of the top and the bottom sections of the column before corrosion, assuming a wind load of 4409 N m$^{-1}$ calculated in Example 19.6.
e) Calculate the deadweight of the vessel before corrosion for the top section and the bottom section of the column above the joint to the cylindrical skirt.
f) Combine the stresses from internal pressure, bending and deadweight and calculate the difference between the principal stresses at the base of the top and the bottom sections on the upwind and downwind sides before corrosion.
g) Repeat the calculation for the after-corrosion condition, assuming the shell thickness is decreased by the corrosion allowance.

**Solution**

a) First calculate the shell thickness to withstand the internal pressure. The design codes specify that the shell thickness must be greater than that calculated for both the circumferential and longitudinal stresses. The circumferential stress will normally be the greater. The allowable stress for carbon steel SA 516-70 below 250 °C from Table 19.4 is $138 \times 10^6$ N m$^{-2}$. From Eq. (19.127):

$$t_C = \frac{PD_I}{2\sigma_{ALL}\eta_J - 1.2P} + C$$

$$= \frac{15 \times 10^5 \times 2.5}{2 \times 138 \times 10^6 \times 1.0 - 1.2 \times 15 \times 10^5} + 0.002$$

$$= 0.0157 \text{ m}$$

$$= 15.7 \text{ mm}$$

## Table 19.5

Pipe details for the distillation column in Example 19.11.

| Pipe | Pipe size[a] | Pipe weight (N·m$^{-1}$) | Weight of pipe insulation (N·m$^{-1}$) | Weight of pipe contents (N·m$^{-1}$) | Total weight of pipe (N·m$^{-1}$) | Approximate pipe length (m) |
|---|---|---|---|---|---|---|
| Overhead Vapor | DN 250 Schedule 40 | 591.6 | 106.5 | 17.4 | 715.5 | 33 |
| Reflux | DN 150 Schedule 40 | 277.2 | 70.8 | 87.1 | 435.1 | 37 |
| Feed | DN 150 Schedule 40 | 277.2 | 70.8 | 87.7 | 435.7 | 20 |

[a]Pipe sizes are specified by a nominal diameter (DN), and a schedule. The larger the schedule number, the thicker the pipe wall. Appendix H gives dimensions and weights for a number of commonly used carbon steel pipe sizes.

In addition to the stresses from the internal pressure there will also be stresses from bending due to the wind load and compression stresses due to the deadweight of the vessel and the ancillary equipment supported by the vessel. The stress from the internal pressure will be constant throughout the vessel if the internal pressure drop is assumed to be zero. The bending and compression stresses will increase from the top of the shell to the bottom and be maximum at the base of the shell. Thus, the shell will need to be thicker in the bottom half than in the top half. Assume initially a plate thickness of 18 mm in the top of the shell and 20 mm in the bottom half. In practice for a tall column, there would be more courses from top to bottom with different shell thicknesses for a tall column.

**b)** Thickness of the semi-ellipsoidal heads can be calculated from Eq. (19.144):

$$t = \frac{PD_I}{2\sigma_{ALL}\eta_J - 0.2P} + C$$

$$= \frac{15 \times 10^5 \times 2.5}{2 \times 138 \times 10^6 \times 1.0 - 0.2 \times 15 \times 10^5} + 0.002$$

$$= 0.0156 \text{ m}$$

$$= 15.6 \text{ mm}$$

Take the head thickness to be 18 mm.

**c)** For the top section of the column the circumferential stress can be calculated from Eq. (19.128) assuming a plate thickness of 18 mm:

$$\sigma_C = \frac{PD_I + 1.2Pt}{2\eta_J t}$$

$$= \frac{15 \times 10^5 \times 2.5 + 1.2 \times 15 \times 10^5 \times 0.018}{2 \times 1.0 \times 0.018}$$

$$= 105.1 \times 10^6 \text{ N m}^{-2}$$

The longitudinal stress can be calculated from Eq. (19.133) assuming a plate thickness of 18 mm:

$$\sigma_L = \frac{PD_I - 0.8Pt}{4\eta_J t}$$

$$= \frac{15 \times 10^5 \times 2.5 - 0.8 \times 15 \times 10^5 \times 0.018}{4 \times 1.0 \times 0.018}$$

$$= 51.78 \times 10^6 \text{ N m}^{-2}$$

For the bottom section of the column the circumferential stress can be calculated from Eq. (19.128), assuming a plate thickness of 20 mm:

$$\sigma_C = \frac{PD_I + 1.2Pt}{2\eta_J t}$$

$$= \frac{15 \times 10^5 \times 2.5 + 1.2 \times 15 \times 10^5 \times 0.02}{2 \times 1.0 \times 0.02}$$

$$= 94.65 \times 10^6 \text{ N m}^{-2}$$

The longitudinal stress can be calculated from Eq. (19.133) assuming a plate thickness of 20 mm:

$$\sigma_L = \frac{PD_I - 0.8Pt}{4\eta_J t}$$

$$= \frac{15 \times 10^5 \times 2.5 - 0.8 \times 15 \times 10^5 \times 0.02}{4 \times 1.0 \times 0.02}$$

$$= 46.58 \times 10^6 \text{ N m}^{-2}$$

**d)** The wind load was calculated in Example 19.6 to be $w = 4409$ N m$^{-1}$. This neglects the additional 0.625 m for the semi-ellipsoidal head at the top of the column. However, the shape of the head does not give as much resistance as an additional 0.625 m of cylindrical shell, and thus will be neglected. The resulting bending moment at the base of the top section is given by Eq. (19.77):

$$M_x = \frac{wx^2}{2}$$

$$= \frac{4409 \times (36.8/2)^2}{2}$$

$$= 7.464 \times 10^5 \text{ N m}$$

The bending stress at the base of the top section is given by Eq. (19.54):

$$\sigma_{B\,max} = \pm \frac{32M\,D_O}{\pi(D_O^4 - D_I^4)}$$

$$= \pm \frac{32 \times 7.464 \times 10^5 \times 2.536}{\pi(2.536^4 - 2.5^4)}$$

$$= \pm 8.386 \times 10^6 \text{ N m}^{-2}$$

The bending stress will be positive (in tension) on the upwind side and negative (in compression) on the downwind side. The bending moment at the base of the bottom section is given by Eq. (19.77):

$$M_x = \frac{wx^2}{2}$$

$$= \frac{4409 \times (36.8)^2}{2}$$

$$= 2.985 \times 10^6 \text{ N m}$$

The bending stress at the base of the bottom section is given by Eq. (19.54):

$$\sigma_{B\,max} = \pm \frac{32M\,D_O}{\pi(D_O^4 - D_I^4)}$$

$$= \pm \frac{32 \times 2.985 \times 10^6 \times 2.54}{\pi(2.54^4 - 2.5^4)}$$

$$= \pm 30.160 \times 10^6 \text{ N m}^{-2}$$

Again, the bending stress will be positive (in tension) on the upwind side and negative (in compression) on the downwind side.

**e)** *Deadweight for the top section.*
Weight of the cylindrical shell is given by:

$$= \frac{\pi}{4}(2.536^2 - 2.5^2) \times \frac{36.8}{2} \times 7850 \times 9.81$$

$$= 2.018 \times 10^5 \text{ N}$$

Weight of the top semi-ellipsoidal head is calculated from the area given by Eq. (19.143) using a mean diameter and a plate thickness of 18 mm:

$$= 1.382 \frac{\pi(D_I + t)^2}{4} \times t \times \rho_{MOC} \times 9.81$$

$$= 1.382 \frac{\pi(2.5 + 0.018)^2}{4} \times 0.018 \times 7850 \times 9.81$$

$$= 9539 \text{ N}$$

Weight of cylindrical shell insulation:

$$= \frac{\pi}{4} \left[ (2.536 + 2 \times 0.075)^2 - 2.536^2 \right] \times \frac{36.8}{2} \times 981$$

$$= 11\,105 \text{ N}$$

Weight of head insulation, taking the area for insulation for the head from Eq. (19.143) using the mean diameter of the insulation to be $(2.536 + 0.075)$ m:

$$= 1.382 \left( \frac{\pi \times (2.536 + 0.075)^2}{4} \right) \times 0.075 \times 981$$

$$= 544 \text{ N}$$

For the weight of the trays, the distribution of the trays between the top half and the bottom half of the column will be slightly different. If it is assumed that the disengagement height at the top of the column is 1.5 m and the sump has a height of 3.5 m, then the distribution of the trays can be assumed to be 29 in the top half and 25 in the bottom. The weight of the trays in the top section (including process liquids) is given to be $1200 \text{ N m}^{-2}$ (Brownell and Young 1959):

$$= 29 \times \frac{\pi}{4} \times 2.5^2 \times 1200$$

$$= 1.708 \times 10^5 \text{ N}$$

Weight of two platforms assuming $1700 \text{ N m}^{-2}$ (Brownell and Young 1959):

$$= 2 \times 9 \times 1700$$
$$= 30\,600 \text{ N}$$

Weight of ladders assuming 16 m and $370 \text{ N m}^{-1}$ (Brownell and Young 1959):

$$= 16 \times 370$$
$$= 5920 \text{ N}$$

Weight of pipes, assuming 21 m for the vapor line, 17 m for reflux, and 6 m for feed:

$$= 715.5 \times 21 + 435.1 \times 17 + 435.7 \times 6$$
$$= 25\,036 \text{ N}$$

Total weight of top section:

$$= 2.018 \times 10^5 + 9539 + 11\,105 + 544 + 1.708 \times 10^5$$
$$\quad + 30\,600 + 5920 + 25\,036$$
$$= 4.553 \times 10^5 \text{ N}$$

Compressive stress at the base of the top section:

$$= \frac{4.553 \times 10^5}{\frac{\pi}{4} \left( 2.536^2 - 2.5^2 \right)}$$

$$= 3.198 \times 10^6 \text{ N m}^{-2}$$

*Deadweight for the top and bottom sections together.*
Weight of the cylindrical shell is given by:

$$= 2.018 \times 10^5 + \frac{\pi}{4} \left( 2.54^2 - 2.5^2 \right) \times \frac{36.8}{2} \times 7850 \times 9.81$$

$$= 4.261 \times 10^5 \text{ N}$$

Weight of top semi-ellipsoidal head from above:

$$= 9539 \text{ N}$$

Weight of cylindrical shell insulation:

$$= 11\,105 + \frac{\pi}{4} \left[ (2.54 + 2 \times 0.075)^2 - 2.54^2 \right] \times \frac{36.8}{2} \times 981$$

$$= 22\,226 \text{ N}$$

Weight of top head insulation from above:

$$= 544 \text{ N}$$

Weight of trays, including process liquids (Brownell and Young 1959):

$$= 54 \times \frac{\pi}{4} \times 2.5^2 \times 1200$$

$$= 3.181 \times 10^5 \text{ N}$$

Weight of three platforms (Brownell and Young 1959):

$$= 3 \times 9 \times 1700$$

$$= 45\,900 \text{ N}$$

Weight of ladders, assuming 37.8 m (Brownell and Young 1959):

$$= 37.8 \times 370$$

$$= 13\,986 \text{ N}$$

Weight of pipes assuming 33 m for the vapor line, 37 m for reflux, and 20 m for feed:

$$= 715.5 \times 33 + 435.1 \times 37 + 435.7 \times 20$$

$$= 48\,427 \text{ N}$$

Total weight of top and bottom sections above the joint for the skirt:

$$= 4.261 \times 10^5 + 9539 + 22\,226 + 544 + 3.181 \times 10^5$$

$$\quad + 45\,900 + 13\,986 + 48\,427$$

$$= 8.848 \times 10^5 \text{ N}$$

Compressive stress at the base of the bottom section:

$$= \frac{8.848 \times 10^5}{\frac{\pi}{4} \left( 2.54^2 - 2.5^2 \right)}$$

$$= 5.588 \times 10^6 \text{ N m}^{-2}$$

As might be expected, the weight of the insulation makes only a minor contribution to the total deadweight. It should be noted that the weight of the process liquid in the sump has not been included as this is carried by the bottom head of the shell and then transferred directly to the skirt (Brownell and Young 1959). The weight of the bottom head and its insulation and load from bottom pipework below the joint with the skirt has also not been included.

f) Yielding will occur when the difference between the principal stresses exceeds the maximum allowable stress for the material at the working temperature. If the shear stresses are zero, the circumferential and longitudinal stresses are the principal stresses. For the base of the top section of the vessel, the difference between the principal stresses is given by:

Upwind

$$(\sigma_1 - \sigma_2) = 105.1 \times 10^6 - (51.78 \times 10^6 + 8.386 \times 10^6 - 3.198 \times 10^6)$$

$$= 48.13 \times 10^6 \, \text{N m}^{-2}$$

Downwind

$$(\sigma_1 - \sigma_2) = 105.1 \times 10^6 - (51.78 \times 10^6 - 8.386 \times 10^6 - 3.198 \times 10^6)$$

$$= 64.90 \times 10^6 \, \text{N m}^{-2}$$

For the base of the bottom section of the vessel, the difference between the principal stresses is given by:

Upwind

$$(\sigma_1 - \sigma_2) = 94.65 \times 10^6 - (46.58 \times 10^6 + 30.16 \times 10^6 - 5.588 \times 10^6)$$

$$= 23.50 \times 10^6 \, \text{N m}^{-2}$$

Downwind

$$(\sigma_1 - \sigma_2) = 94.65 \times 10^6 - (46.58 \times 10^6 - 30.16 \times 10^6 - 5.588 \times 10^6)$$

$$= 83.82 \times 10^6 \, \text{N m}^{-2}$$

These stresses are all below the maximum allowable stress of $138 \, \text{N m}^{-2}$.

g) After corrosion the internal diameter becomes $D_I = 2.5 + 2 \times 0.002 = 2.504$ m and $t = 18 - 2 = 16$ mm for the top section and $20 - 2 = 18$ mm for the bottom section. Repeating the calculation after corrosion.

For the base of the top section of the vessel, the difference between the principal stresses is given by:

Upwind

$$(\sigma_1 - \sigma_2) = 53.90 \times 10^6 \, \text{N m}^{-2}$$

Downwind

$$(\sigma_1 - \sigma_2) = 72.72 \times 10^6 \, \text{N m}^{-2}$$

For the base of the bottom section of the vessel, the difference between the principal stresses is given by:

Upwind

$$(\sigma_1 - \sigma_2) = 25.82 \times 10^6 \, \text{N m}^{-2}$$

Downwind

$$(\sigma_1 - \sigma_2) = 92.70 \times 10^6 \, \text{N m}^{-2}$$

Even after the corrosion allowance has been used up the increased stresses are all still below the maximum allowable stress of $138 \, \text{N m}^{-2}$.

The calculations in this example evaluated the possibility of failure of the material from tension and compression exceeding the strength of the material. However, the structure of the cylindrical shell can fail without the material failing through collapse of the shape in buckling. Buckling happens when compression forces prevent the vessel from supporting its shape and it suddenly collapses to take up a smaller volume. This will be explored later in this chapter.

# 19.7 Design of Horizontal Cylindrical Pressure Vessels Under Internal Pressure

Horizontal cylindrical vessels such as that illustrated in Figure 19.26, are very common in process plant for processing and storing liquids and separating gas–liquid and liquid–liquid mixtures. For the vessel dimensions, the length to diameter ratio $L/D$ is usually in the range 3–5 in the process industries. The optimum $L/D$ depends on the pressure, where the higher the pressure, the smaller the optimum diameter and the greater the optimum $L/D$. Such horizontal cylindrical vessels are typically constructed with hemispherical, semi-ellipsoidal, or torispherical heads.

To calculate the bending stress on the vessel requires the weight of the vessel and contents to be determined. The required relationships are detailed in Section 19.5. Figure 19.41 shows a horizontal cylindrical vessel with two supports represented as a beam. The vessel is subject to shear force and bending moments created by the weight of the vessel and its contents. Consider the analysis of the shear force and bending moments in a horizontal cylindrical vessel assumed to be full of liquid, as illustrated in Figure 19.41 (Zick 1951; Brownell and Young 1959; Bednar 1991; Escoe 1994; Chattopadhyay 2005). It is assumed here that the vessel acts effectively as a beam (to be discussed later in this chapter). Figure 19.41 shows the shear force diagram. Figure 19.42 shows the corresponding bending moment diagram.

Between the supports in Figure 19.41 there is a bending moment created by uniform load of the weight of the vessel and its contents. At each end of the vessel there is a cantilever that has three loads imposed on each end (Zick 1951; British Standards Institution 1982):

1) A bending moment is created by the uniform load of the cylindrical shell and its contents from the saddle to the tangent point, with a length of $a$.

2) A bending moment is created by the weight of the head and its contents, which acts through the center of gravity of the head. For a head filled with liquid, this can be regarded as a "solid" shape. For a hemispherical head the horizontal distance to the center of gravity would be $3D/16$ from the tangent point, where $D$ is the diameter of the vessel. For a semi-ellipsoidal head, the center of gravity is located at a horizontal distance 3/8 of $h_{HEAD}$, where $h_{HEAD}$ is the distance from the tangent point to the top of the head along the center line of the vessel. For a 2:1 semi-ellipsoidal head $h_{CG} = 3h_{HEAD}/8 = 3D/32$.

3) A bending moment is created by the hydraulic head of liquid in the vessel (Zick 1951; British Standards Institution 1982). The pressure at the bottom of the vessel is higher than that at the top, due to the hydraulic head of the fluid. This creates a bending moment $M_{HYD}$ in Figures 19.41 and 19.42. The magnitude of this bending moment is given by (see Appendix I):

$$M_{HYD} = -\frac{\pi \rho_L g D_I^4}{64} \tag{19.147}$$

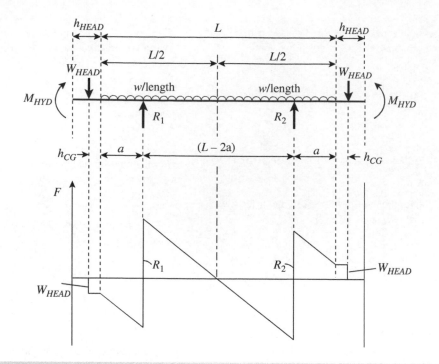

## Figure 19.41

Shear force diagram for horizontal cylindrical vessels with saddle supports.

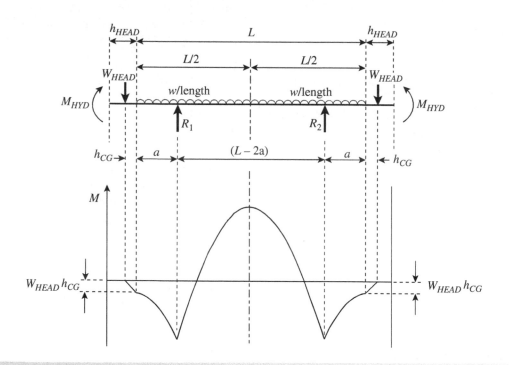

## Figure 19.42

Bending moments on horizontal cylindrical vessels.

where

$M_{HYD}$ = bending moment created by the hydraulic head (N m)

$\rho_L$ = density of the liquid in the vessel (kg m$^{-3}$)

$g$ = gravitational acceleration (m s$^{-2}$)

$D_I$ = internal diameter (m)

In the shear force diagram in Figure 19.41, the reaction on the supports $R_1$ and $R_2$ for a symmetrical design is given by:

$$R_1 = R_2 = \frac{1}{2}(wL + 2W_{HEAD}) \qquad (19.148)$$

For the vessel to act effectively as a beam, the saddle supports will need to be close to the heads, which means that the maximum bending stress will be at the center of the vessel. Taking moments around the center of the beam in Figure 19.42:

$$M = -\left[\frac{wL}{2} \times \frac{L}{4} + W_{HEAD} \times \left(\frac{L}{2} + h_{CG}\right) - \frac{\pi \rho_L g D_I^4}{64}\right] + R_1\left(\frac{L}{2} - a\right)$$

$$(19.149)$$

where

$M$ = bending moment at the center of the vessel (N m)

$L$ = tangent-to-tangent length of the vessel (m)

$W_{HEAD}$ = weight of the head and its contents (N)

$h_{CG}$ = distance from the tangent point to the center of gravity of the head along the centerline of the vessel (m)

$a$ = distance from the support to the tangent point (m)

Substituting $R_1$ from Eq. (19.148):

$$M = -\left[\frac{wL^2}{8} + W_{HEAD}\left(\frac{L}{2} + h_{CG}\right) - \frac{\pi \rho_L g D_I^4}{64}\right]$$

$$+ \frac{1}{2}(wL + 2W_{HEAD})\left(\frac{L}{2} - a\right)$$

$$(19.150)$$

$$= \frac{wL^2}{8} - \frac{wLa}{2} - W_{HEAD}h_{CG} - W_{HEAD}a + \frac{\pi \rho_L g D_I^4}{64}$$

Equation (19.150) requires knowledge of $h_{CG}$. Assuming a head filled with liquid can be regarded as a "solid" shape, for a hemispherical head $h_{CG} = 3D/16$ and for a 2:1 semi-ellipsoidal head $h_{CG} = 3D/32$.

In the region above each saddle, circumferential bending moments are introduced allowing the shell to deform and making it ineffective as a beam, as in Figure 19.43 (Zick 1951). The maximum stress occurs at the horn of the saddle. This reduces the effective cross-section from acting as a beam just as though the shell was split along a horizontal line at a level above the saddle, as in Figure 19.43 (Zick 1951). There are four ways to overcome this problem (Brownell and Young 1959; Bednar 1991; Escoe 1994).

- A *wear plate* with a thickness typically the same as the vessel can be welded to the outside of the shell above the saddle, as illustrated in Figure 19.44.

- The contact angle between the vessel and the saddle, $\theta$ in Figure 19.44, can be increased. The contact angle should not be less than 120° and is usually in the range 120–150°.

- The saddles can be moved closer to the heads to allow the heads to stiffen the vessel. The distance between the saddle and the *tangent point* along the length (where the curvature of the head starts) should not normally be greater than 20% of the tangent-to-tangent length of the vessel, otherwise the

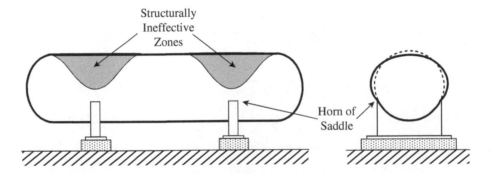

**Figure 19.43**

Circumferential bending moments above the saddles cause the upper section of the shell to deform, making it ineffective as a beam.

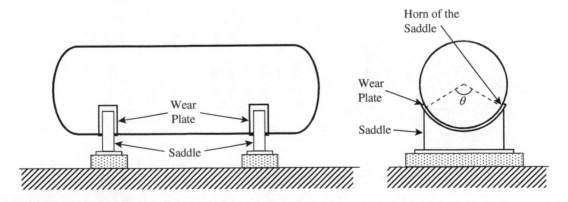

**Figure 19.44**

Stresses on the vessel at the saddle can be decreased by welding a wear plate to the vessel and increasing the angle of contact with the saddle.

cantilever action will be excessive. The distance between the saddle and the tangent point is often taken to be 0.4 times the radius (Brownell and Young 1959; Escoe 1994).

● Stiffening rings can be welded to the vessel internally or externally. Figure 19.45 illustrates two possible ways in which *internal stiffening rings* can be welded to the inside surface of the vessel. Such rings would normally be located close to the positions of the saddles. Typically for each saddle two internal rings might be welded to the inside of the vessel, each adjacent to the line of the saddle. Alternatively, *external stiffening rings* can be welded to the exterior of the vessel. A common design is to weld a stiffening ring to the exterior of the vessel in the line of each of the saddles. Figure 19.46 illustrates two possible alternative designs for external stiffening rings. Internal stiffening rings are normally preferred to

external (Brownell and Young 1959; Bednar 1991; Escoe 1994). However, the process function (e.g., settling device) might demand the use of external stiffening rings.

Given that the saddle supports would normally be located close to the ends of the vessel, this means that the maximum bending stress for symmetrical saddle supports will normally be at the center of the vessel between the supports.

As with the design of vertical pressure vessels discussed in the previous section, a design pressure 10% greater than the maximum allowable operating pressure is a reasonable basis for the mechanical design. The analysis here assumes that the horizontal vessel is full of liquid. If the vessel contains a gas, rather than a liquid, then the bending moment created by the hydraulic head can be neglected and left out of Eq. (19.150).

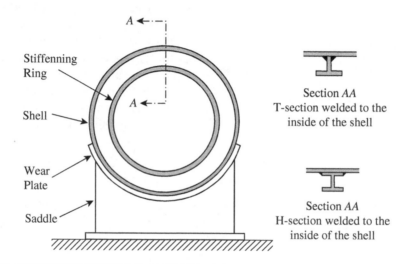

**Figure 19.45**

Two examples of how stiffening rings can be fitted internally to prevent deformation of the shell.

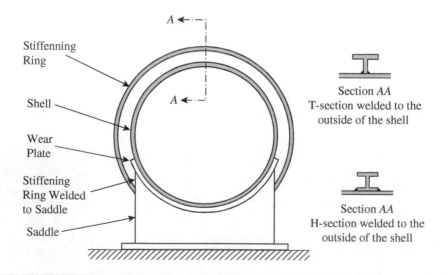

**Figure 19.46**

Example of how external stiffening rings can be attached to the saddles to prevent deformation of the shell.

**Example 19.12** A horizontal cylindrical pressure vessel with 2:1 semi-ellipsoidal heads is required to hold a volume of 20 m³ of water at a temperature of 100 °C, with a density of 957.9 kg m⁻³. The vessel is insulated with 75 mm of mineral wool with a density of 100 kg m⁻³ and weight of 981 N m⁻³, covered in aluminum cladding. The process pressure has been set to a maximum of 10 bara. As with Example 19.9, taking the design pressure to be 10% above the normal working gauge pressure gives, say, 10 barg as a reasonable basis for the design. The vessel is to be fabricated from carbon steel SA 516-70 with a density of 7850 kg m⁻³ and a corrosion allowance of 2 mm. Assume the vessel has a length to diameter ratio of 3. Take a welded joint efficiency of 0.85 for the cylindrical shell, but 1.0 for the heads. Assume there are two saddle supports located 0.4R from the tangent point, i.e., $a = 0.4R$ (as recommended by Brownell and Young 1959). It is assumed that the vessel acts effectively as a beam, as a result of the location of the saddles close to the heads, suitable design of saddle supports, and possibly the inclusion of stiffening rings. Carry out a preliminary mechanical design:

a) Calculate suitable dimensions for the vessel.
b) Calculate the minimum thickness for the shell and choose a reasonable thickness for the cylindrical shell.
c) Calculate the minimum thickness for the heads and choose a reasonable thickness of plate.
d) Calculate the circumferential and longitudinal stress caused by the internal pressure before corrosion.
e) Calculate the deadweight of the cylindrical shell with contents and lagging before corrosion.
f) Calculate the deadweight of the vessel heads with contents and lagging before corrosion.
g) Calculate the bending moment and maximum bending normal stress at the top and bottom of the vessel at the longitudinal mid-point before corrosion.
h) Combine the stresses from internal pressure and bending and calculate the difference between the principal stresses at the top and bottom of the vessel at the longitudinal mid-point before corrosion.
i) Calculate bending shear stress and the difference between the principal stresses above the supports midway between the top and bottom, corresponding with the neutral axis before corrosion.

**Solution**

a) The volume of the vessel is given by Eq. (19.141):

$$V = \frac{\pi D_I^2}{4}L + \frac{\pi D_I^3}{12}$$

where

$L$ = tangent-to-tangent length of the vessel (m)

For a length-to-diameter ratio of 3:

$$V = \frac{3\pi D_I^3}{4} + \frac{\pi D_I^3}{12}$$

$$= \frac{5\pi D_I^3}{6}$$

For a volume of 20 m³ solving for the diameter, $D_I = 1.97$ m. Take $D_I = 2$ m. To create a volume of 20 m³ for $D_I = 2$ m requires $L = 5.7$ m from Eq. (19.141). For the saddle supports, assume $a = 0.4R = 0.4$ m.

b) Calculate the shell thickness to withstand the internal pressure. The design codes specify that the shell thickness must be greater than that calculated for both the circumferential and longitudinal stresses. The circumferential stress will normally be the greater. The allowable stress for carbon steel SA 516-70 for 100 °C from Table 19.4 is $138 \times 10^6$ N m⁻². From Eq. (19.127):

$$t_C = \frac{PD_I}{2\sigma_{ALL}\eta_J - 1.2P} + C$$

$$= \frac{10 \times 10^5 \times 2.0}{2 \times 138 \times 10^6 \times 0.85 - 1.2 \times 10 \times 10^5} + 0.002$$

$$= 0.0106 \text{ m}$$

$$= 10.6 \text{ mm}$$

A thickness of 12 mm seams a reasonable starting point.

c) Thickness of the semi-ellipsoidal heads can be calculated from Eq. (19.144):

$$t = \frac{PD_I}{2\sigma_{ALL}\eta_J - 0.2P} + C$$

$$= \frac{10 \times 10^5 \times 2.0}{2 \times 138 \times 10^6 \times 1.0 - 0.2 \times 10 \times 10^5} + 0.002$$

$$= 0.00925 \text{ m}$$

$$= 9.25 \text{ mm}$$

Take the head thickness to be 10 mm.

d) The circumferential stress for the cylindrical section can be calculated from Eq. (19.128) assuming a plate thickness of 12 mm:

$$\sigma_C = \frac{PD_I + 1.2Pt}{2\eta_J t}$$

$$= \frac{10 \times 10^5 \times 2.0 + 1.2 \times 10 \times 10^5 \times 0.012}{2 \times 0.85 \times 0.012}$$

$$= 98.75 \times 10^6 \text{ N m}^{-2}$$

The longitudinal stress can be calculated from Eq. (19.133) assuming a plate thickness of 12 mm:

$$\sigma_L = \frac{PD_I - 0.8Pt}{4\eta_J t}$$

$$= \frac{10 \times 10^5 \times 2.0 - 0.8 \times 10 \times 10^5 \times 0.012}{4 \times 0.85 \times 0.012}$$

$$= 48.78 \times 10^6 \text{ N m}^{-2}$$

e) Weight of the cylindrical shell

$$= \frac{\pi}{4}\left(2.024^2 - 2.0^2\right) \times 5.7 \times 7850 \times 9.81$$

$$= 33\,295 \text{ N}$$

Weight of the contents of the cylindrical shell

$$= \left[\frac{\pi}{4} \times D_I^2 \times L\right]\rho_L \times g$$

$$= \left[\frac{\pi}{4} \times 2^2 \times 5.7\right]957.9 \times 9.81$$

$$= 168\,273 \text{ N}$$

Weight of shell insulation for the cylindrical shell

$$= \frac{\pi}{4}\left[(2.024 + 2 \times 0.075)^2 - 2.024^2\right] \times 5.7 \times 981$$

$$= 2765 \text{ N}$$

Total weight of the cylindrical shell:

$$= 33\,295 + 168\,273 + 2765$$

$$= 204\,333 \text{ N}$$

**f)** Weight of head from Eq. (19.143):

$$= 1.382 \frac{\pi(D_I + t)^2}{4} \times t \times \rho_{MOC} \times 9.81$$

$$= 1.382 \times \frac{\pi \times (2.0 + 0.01)^2}{4} \times 0.01 \times 7850 \times 9.81$$

$$= 3377 \text{ N}$$

Weight of head contents from Eq. (19.140):

$$= \frac{\pi D_I^3}{24} \times \rho_L \times g$$

$$= \frac{\pi \times 2.0^3}{24} \times 957.9 \times 9.81$$

$$= 9841 \text{ N}$$

Weight of head insulation, taking the area for insulation for the head from Eq. (19.143) using the mean diameter of the insulation:

$$= 1.382 \times \left(\frac{\pi \times (2.0 + 2 \times 0.01 + 0.075)^2}{4}\right) \times 0.075 \times 981$$

$$= 351 \text{ N}$$

Total weight of head:

$$= 3377 + 9841 + 351$$

$$= 13\,569 \text{ N}$$

**g)** Load per unit length from the weight of the cylindrical section and contents:

$$w = \frac{204\,333}{5.7}$$

$$= 35\,848 \text{ N}$$

Bending moment at the center of the vessel is given by Eq. (19.150); see Figure 19.42:

$$M = \frac{wL^2}{8} - \frac{wLa}{2} - W_{HEAD}h_{CG} - W_{HEAD}a + \frac{\pi \rho_L g D_I^4}{64}$$

$$= \frac{wL^2}{8} - \frac{wLa}{2} - W_{HEAD}\frac{3D_O}{32} - W_{HEAD}a + \frac{\pi \rho_L g D_I^4}{64}$$

$$= \frac{35\,848 \times 5.7^2}{8} - \frac{35\,848 \times 5.7 \times 0.4}{2} - 13\,569$$

$$\times \frac{3 \times (2.0 + 2 \times 0.01)}{32} - 13\,569 \times 0.4$$

$$+ \frac{\pi \times 957.9 \times 9.81 \times 2.0^4}{64}$$

$$= 104\,104 \text{ N m}$$

The maximum normal bending stress is given by Eqs (19.37) and (19.54):

$$\sigma_{B\,\max} = \pm \frac{M\,y_{\max}}{I} = \pm \frac{32M\,D_O}{\pi(D_O^4 - D_I^4)}$$

$$= \pm \frac{32 \times 104\,104 \times 2.024}{\pi(2.024^4 - 2^4)}$$

$$= \pm 2.745 \times 10^6 \text{ N m}^{-2}$$

Thus, the bending normal stress at the bottom of the vessel is $+2.745 \times 10^6 \text{ N m}^{-2}$ tensile and at the top of the vessel $-2.745 \times 10^6 \text{ N m}^{-2}$ compressive.

**h)** At the top mid-point of the vessel, the shear stresses are zero and the circumferential and longitudinal stresses are the principal stresses. The difference between the principal stresses is given by:

$$(\sigma_1 - \sigma_2) = 98.75 \times 10^6 - \left(48.78 \times 10^6 - 2.745 \times 10^6\right)$$

$$= 52.71 \times 10^6 \text{ N m}^{-2}$$

At the bottom mid-point of the vessel, again the shear stresses are zero and the stresses are the principal stresses. The difference between the principal stresses is given by:

$$(\sigma_1 - \sigma_2) = 98.75 \times 10^6 - \left(48.78 \times 10^6 + 2.745 \times 10^6\right)$$

$$= 47.22 \times 10^6 \text{ N m}^{-2}$$

**i)** The bending shear stress on the sides of the vessel above the supports midway between the top and bottom, corresponding with the neutral axis, is given by Eq. (19.76). The shear force is given by (Figure 19.41):

$$F = R_1 - wa - W_{HEAD}$$

$$= \frac{1}{2}(Lw + 2W_{HEAD}) - wa - W_{HEAD}$$

$$= \frac{Lw}{2} - wa$$

$$= \frac{5.7 \times 35\,848}{2} - 35\,848 \times 0.4$$

$$= 87\,827 \text{ N}$$

The area of the cross-section is given by:

$$A = \frac{\pi}{4}\left(D_O^2 - D_I^2\right)$$

$$= \frac{\pi}{4}\left(2.024^2 - 2^2\right)$$

$$= 0.07585 \text{ m}^2$$

From Eq. (19.76):

$$\tau_{B\,\max} = \frac{4F}{3A}\left(\frac{D_O^2 + D_O D_I + D_I^2}{D_O^2 + D_I^2}\right)$$

$$= \frac{4 \times 87\,827}{3 \times 0.07585}\left(\frac{2.024^2 + 2.024 \times 2.0 + 2.0^2}{2.024^2 + 2.0^2}\right)$$

$$= 2.316 \times 10^6 \text{ N m}^{-2}$$

For a location above the supports midway between the top and bottom, corresponding with the neutral axis, the principal stresses are given by Eq. (19.103):

$$\sigma_1 = \frac{\sigma_x + \sigma_y}{2} + \sqrt{\left(\frac{\sigma_x - \sigma_y}{2}\right)^2 + \tau_{xy}^2}$$

$$\sigma_2 = \frac{\sigma_x + \sigma_y}{2} - \sqrt{\left(\frac{\sigma_x - \sigma_y}{2}\right)^2 + \tau_{xy}^2}$$

It should be noted that the shear stress, as noted in Figure 19.38, is negative in accordance with the sign convention in Figure 19.32:

$$\sigma_1 = \frac{48.78 \times 10^6 + 98.75 \times 10^6}{2}$$
$$+ \sqrt{\left(\frac{48.78 \times 10^6 - 98.75 \times 10^6}{2}\right)^2 + \left(-2.316 \times 10^6\right)^2}$$
$$= 98.85 \times 10^6 \, \text{N m}^{-2}$$

$$\sigma_2 = \frac{48.78 \times 10^6 + 98.75 \times 10^6}{2}$$
$$- \sqrt{\left(\frac{48.78 \times 10^6 - 98.75 \times 10^6}{2}\right)^2 + \left(-2.316 \times 10^6\right)^2}$$
$$= 48.68 \times 10^6 \, \text{N m}^{-2}$$

The difference between the principal stresses is given by:

$$(\sigma_1 - \sigma_2) = \left(98.85 \times 10^6 - 48.68 \times 10^6\right)$$
$$= 50.18 \times 10^6 \, \text{N m}^{-2}$$

These stresses are all below the maximum allowable stress of 138 N m$^{-2}$ for the material for a working temperature below 250 °C.

# 19.8 Buckling of Cylindrical Vessels Due to External Pressure and Axial Compression

Rather than the cylindrical vessel operating under internal pressure, it might be required to operate under external pressure. This could be, for example, a distillation column being operated under vacuum conditions or a tube in a heat exchanger shell that has elevated pressure on both the inside and the outside of the tube, but the outside pressure is greater than the internal pressure. The failure mechanism of external pressure is different from the failure due to internal pressure. Internal pressure failure can be understood as a vessel failing after stresses exceed the strength of the material. By contrast, with external pressure failure, the vessel can no longer support its shape and suddenly, irreversibly, *collapses*, taking a new lower volume. Figure 19.47 shows a cylindrical tank that has suffered catastrophic collapse due to excessive external pressure. If the initial shape has imperfections, this collapse can occur at less extreme pressures.

Thin cylinders longer than a *critical length* will buckle at stresses below the yield point of the material. Beyond the critical length, the critical pressure at which buckling occurs is a function

**Figure 19.47**

Catastrophic collapse of a vessel from external pressure. Source: Reproduced with permission of American Institute of Chemical Engineers, February 2007.

only of the thickness to diameter ratio $t/D$ and the properties of the material. At lengths below the critical length, the critical pressure at which collapse occurs is a function of the length to diameter ratio $L/D$ as well as the thickness to diameter ratio $t/D$ and the properties of the material. The critical length is given by (Brownell and Young 1959):

$$L_C = \frac{4\pi\sqrt{6}}{27}\left(\sqrt[4]{1 - \nu^2}\right)\left(D\sqrt{\frac{D}{t}}\right) \tag{19.151}$$

where

$L_C$ = critical length of the vessel (m)
$\nu$ = Poisson's ratio (–)
$D$ = diameter of the vessel (m)
$t$ = thickness of the vessel wall (m)

Assuming $\nu = 0.3$, which is reasonable for carbon steel, Eq. (19.151) becomes:

$$L_C = 1.11D\sqrt{\frac{D}{t}} \tag{19.152}$$

The design codes dealing with design for external pressure tend to involve complex and iterative procedures (ASME 2019;

BSI Standards Publications 2019). An approximate thickness for a cylindrical shell under external radial pressure for long cylinders greater than the critical length is given by (Timoshenko and Gere 1961; Chattopadhyay 2005):

$$P_{CR} = \frac{2E}{(1-\nu^2)} \left(\frac{t}{D_O}\right)^3 \qquad (19.153)$$

where

$P_{CR}$ = critical buckling pressure (N m$^{-2}$)

$E$ = Modulus of elasticity for the material of construction (N m$^{-2}$)

$\nu$ = Poisson's ratio for the material of construction (–)

$t$ = thickness of the cylindrical shell (m)

$D_O$ = outside diameter of the cylindrical shell (m)

Equation (19.153) can be rearranged to give the required minimum thickness to avoid buckling collapse for a given pressure:

$$t = D_O \left[\frac{P(1-\nu^2)}{2E}\right]^{\frac{1}{3}} \qquad (19.154)$$

where

$P$ = external pressure difference across the cylinder (N m$^{-2}$)

$t$ = minimum thickness to withstand buckling collapse (m)

It must be emphasized that Eqs (19.153) and (19.154) do not account for a number of important factors and the ultimate design must be performed with the appropriate design codes. When applying Eq. (19.154), even for a preliminary design, it would be advisable to assume a generous safety factor by assuming the external pressure difference to be higher than the process conditions by a factor of at least 3 or 4.

The preliminary design of the vessel heads under external pressure can be approximated to the collapse of a sphere under external pressure. The critical buckling of spheres subject to external pressure is given by (Timoshenko and Gere 1961):

$$P_{CR} = \frac{2Et_{SPH}^2}{R_O^2 \sqrt{3(1-\nu^2)}} \qquad (19.155)$$

where

$R_O$ = external radius of the sphere (m)

$t_{SPH}$ = wall thickness of the sphere (m)

Although Eq. (19.155) is soundly based on theory (Timoshenko and Gere 1961), in practice the effects of shape imperfections lead to a lower value of the buckling load. After making some simplifying assumptions and assuming $\nu = 0.3$, von Kármán and Tsien (1939) suggested for a spherical shell:

$$P_{CR} = \frac{0.3652Et_{SPH}^2}{R_O^2} = \frac{1.461Et_{SPH}^2}{D_O^2} \qquad (19.156)$$

where

$R_O$ = external radius of the sphere (m)

$D_O$ = external diameter of the sphere (m)

Equation (19.156) can be assumed to apply to a hemispherical head and rearranged to give the required minimum thickness of a hemispherical head to avoid buckling collapse:

$$t_{HS} = 1.655R_O \sqrt{\frac{P}{E}} = 0.8275D_O \sqrt{\frac{P}{E}} \qquad (19.157)$$

The use of Eqs (19.156) and (19.157) can be extrapolated to semi-ellipsoid and torispherical heads by taking the largest radius for the head, which gives the most conservative result. For semi-ellipsoidal heads, the maximum radius of curvature from the geometry of an ellipse is $D_O^2/4h_{HEAD}$, where $h_{HEAD}$ is the height of the head. For a 2:1 semi-ellipsoidal head $h_{HEAD} = D_O/4$. Thus, combining with Eq. (19.157), the thickness of a 2:1 semi-ellipsoidal head is given by:

$$t_{SE} = 1.655D_O \sqrt{\frac{P}{E}} \qquad (19.158)$$

For a torispherical head, the standard geometry is such that the inside crown radius equals the outside diameter. Thus, combining with Eq. (19.157), the thickness of a standard torispherical head is given by:

$$t_{TOR} = 1.655D_O \sqrt{\frac{P}{E}} \qquad (19.159)$$

Although Eq. (19.156) is in broad agreement with experimental data for steel spheres, it is advisable to assume a generous safety factor by assuming the external pressure to be higher than the process pressure by a factor of at least 3 or 4 and up to 6 when applying Eqs (19.157) to (19.159).

It must be emphasized that the equations presented here are for preliminary design only. The design codes dealing with design for external pressure need to be consulted for the final design (ASME 2019; BSI Standards Publications 2019).

The thickness of shell required to withstand external pressure depends on the diameter of the shell, effective length of the shell and the material properties. Instead of making the shell thick enough to handle the external pressure, thinner shells can be used with the addition of stiffening rings. The stiffening rings can be added externally or internally around the circumference of the cylinder to prevent collapse under external pressure with a thinner shell. Figure 19.48 shows the different cross-sections that can be used. The stiffening ring acts to reduce the effective length of the cylinder. Placing a properly designed stiffening ring halfway along a cylindrical shell halves the effective length. In some cases, the internal features of the vessels required for the process design can be used as stiffeners. For example, a distillation column requires support rings for the distillation trays. These can be designed to also act as stiffening rings. Equations (19.153) and (19.154) assume that the distance between the stiffeners is greater than the critical length. The detailed design with stiffening rings requires reference to the design codes (ASME 2019; BSI Standards Publications 2019).

Rather than external pressure, cylindrical vessels can suffer from buckling due to axial compressive loads. This might be from

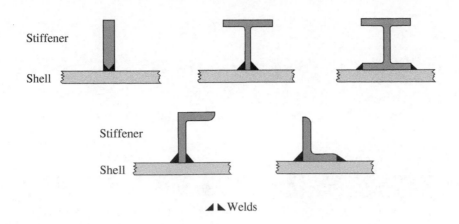

Welds

## Figure 19.48

Different profiles of stiffening rings can be used to strengthen the shell subject to external pressure.

the weight of the vessel, or bending from the wind, or seismic activity or a combination. The buckling stress from axially compressed cylinders is given by (von Kármán and Tsien 1941; Timoshenko and Gere 1961):

$$\sigma_{CR} = \frac{E}{\sqrt{3(1-\nu^2)}} \left(\frac{t}{R_O}\right) \qquad (19.160)$$

where

$\sigma_{CR}$ = critical buckling stress (N m$^{-2}$)

$E$ = modulus of elasticity for the material of construction (N m$^{-2}$)

$\nu$ = Poisson's ratio for the material of construction (–)

$t$ = thickness of a cylindrical shell (m)

$R_O$ = outside radius of a cylindrical shell (m)

Whilst Eq. (19.160) is soundly based on theory, it cannot be used directly for design because cylindrical shells are extremely sensitive to imperfections under axial compression. The experimental collapse is a function of shell geometry, loading conditions, initial imperfections, and other factors. Experimental measurements indicate that the collapse can occur between 0.7 and 0.1 of the stress predicted by of Eq. (19.160) (Harris et al. 1957). This suggests that if Eq. (19.160) is to be used, a safety factor of 10 should be applied. For carbon steel the modulus of elasticity varies with the grade of steel but it is reasonable to assume and $E = 200$ GPa and $\nu = 0.3$. Together with a safety factor of 10, Eq. (19.160) becomes:

$$\sigma_{CR} = 1.21 \times 10^{10} \left(\frac{t}{R_O}\right) = 2.42 \times 10^{10} \left(\frac{t}{D_O}\right) \qquad (19.161)$$

**Example 19.13**   For the distillation column in Example 19.11, the bending moment at the base of the top section was calculated to be $7.464 \times 10^5$ N m and at the base of the bottom section to be $2.985 \times 10^6$ N m. The weight at the base of the top section of the column was calculated in Example 19.11 to be $4.553 \times 10^5$ N and the weight of the top and bottom sections combined to be $8.848 \times 10^5$ N. The top section has an inside diameter of 2.5 m before corrosion and an outside diameter of 2.536 m. The bottom section has an inside diameter of 2.5 m before corrosion and an outside diameter of 2.54 m. The corrosion allowance can be assumed to be 2 mm. Assuming that the maximum compressive load applies across the whole diameter of the vessel and the maximum occurs at the base of each column section, check that the compressive load is below the critical stress for buckling for the top and bottom sections of the column in the uncorroded and corroded states.

### Solution

The compressive stress $\sigma_W$ from the weight $W$ is given by:

$$\sigma_W = -\frac{W}{\pi(D_O^2 - D_I^2)/4}$$

The bending stress $\sigma_{B\max}$ at the base of each section on the downwind side created by the bending moment $M$ is given by Eq. (19.54):

$$\sigma_{B\max} = -\frac{32M\, D_O}{\pi(D_O^4 - D_I^4)}$$

The combined compressive stress from weight and bending $(\sigma_W + \sigma_{B\max})$ must be less than the critical stress for buckling. The critical buckling stress $\sigma_{CR}$ for a shell thickness $t$ is given by Eq. (19.161):

$$\sigma_{CR} = 2.42 \times 10^{10} \left(\frac{t}{D_O}\right)$$

For the uncorroded state:

| | $D_O$ (m) | $D_I$ (m) | $\sigma_W$ (N m$^{-2}$) | $\sigma_{Bmax}$ (N m$^{-2}$) | $(\sigma_W + \sigma_{Bmax})$ (N m$^{-2}$) | $\sigma_{CR}$ (N m$^{-2}$) |
|---|---|---|---|---|---|---|
| Top | 2.536 | 2.5 | $-3.198 \times 10^6$ | $-8.386 \times 10^6$ | $-11.584 \times 10^6$ | $171.77 \times 10^6$ |
| Bottom | 2.54 | 2.5 | $-5.588 \times 10^6$ | $-30.160 \times 10^6$ | $-35.748 \times 10^6$ | $190.55 \times 10^6$ |

For the corroded state:

| | $D_O$ (m) | $D_I$ (m) | $\sigma_W$ (N m$^{-2}$) | $\sigma_{Bmax}$ (N m$^{-2}$) | $(\sigma_W + \sigma_{Bmax})$ (N m$^{-2}$) | $\sigma_{CR}$ (N m$^{-2}$) |
|---|---|---|---|---|---|---|
| Top | 2.536 | 2.504 | $-3.594 \times 10^6$ | $-9.4122 \times 10^6$ | $-13.006 \times 10^6$ | $152.68 \times 10^6$ |
| Bottom | 2.54 | 2.504 | $-6.204 \times 10^6$ | $-33.432 \times 10^6$ | $-39.636 \times 10^6$ | $171.50 \times 10^6$ |

This shows that the bottom section is limiting, as expected. However, in all cases the compressive stress is significantly lower than the critical stress for buckling. Even in the corroded condition, when the difference between the total compressive stress and the critical stress narrows, the critical stress for buckling is still significantly higher than the compressive stress.

# 19.9  Welded and Bolted Joints

Welded and bolted joints are required for the construction of a chemical process plant. Welding joins metals (or thermoplastics) using high-temperature heat to melt the parts together and allowing them to cool causing fusion. In addition to melting the base metal, a filler material is typically added to the joint to form a pool of molten material that cools to form a joint. Welding is distinct from lower-temperature metal-joining techniques, such as brazing and soldering, which do not melt the base metal. Welding generally results in a higher integrity system, providing the welding is carried out by a skilled welder to the appropriate standards and radiographically inspected, but it is permanent. Bolted joints are used when equipment needs to be dismantled for maintenance, cleaning, or inspection (ASME 2017b).

Figure 19.49 shows a number of different types of welded joint. The welds can be classified as *butt welds* or *fillet welds*. Butt welds are welds where the two pieces of metal to be joined are in the same plane. Fillet welds join two pieces of metal, but not in the same plane (Figure 19.49). There are many types of butt welds, with some examples shown in Figure 19.50. The joint can be prepared before welding by removing material from the edges, typically forming a "V" or "U", but other edge preparations are also used. The most appropriate edge preparation depends on the type of material and its thickness, the depth of weld penetration required, and the welding process. Butt welds can be single welded or double welded. A single welded butt joint, as its name implies, is only welded from one side. A double welded butt joint is created when the joint is welded from both sides. The final weld can be inspected using radiography. This uses penetrating radiation from an X-ray or gamma-ray source to view the internal structure of the weld and identify any defects. Should any defects be identified, then remedial action will normally require removal

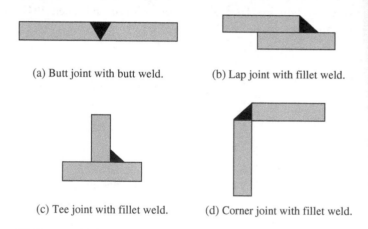

(a) Butt joint with butt weld.    (b) Lap joint with fillet weld.

(c) Tee joint with fillet weld.    (d) Corner joint with fillet weld.

**Figure 19.49**

Types of weld.

of the affected area of the weld by gouging or grinding to sound metal, followed by re-welding.

Cylindrical vessels are fabricated from plates of the required thickness that are rolled to the required curvature and then welded together to create a vessel with the required length. As shown in Figure 19.51, the plates are welded together such that the welds do not join up at the corners of the plates, which could create points that might be more vulnerable to failure.

By contrast, when equipment needs to be dismantled for maintenance, cleaning, or inspection, bolted joints are used. Figure 19.52 shows three of the more common types of bolted joints. Bolts provide a compressive force between the *flanges* of the joint. The seal is created by inserting a *gasket* or *seal ring* in the joint made from compressible material that is chemically resistant to the process materials. If a metal or metal-reinforced gasket is used it should also not create any galvanic couples to promote corrosion. It should be noted that gaskets and seal rings are

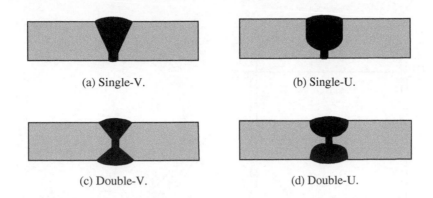

(a) Single-V.                    (b) Single-U.

(c) Double-V.                    (d) Double-U.

**Figure 19.50**

Types of butt weld.

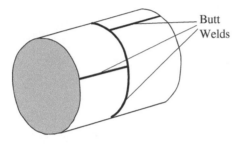

Butt
Welds

**Figure 19.51**

Cylindrical vessels are fabricated from plates rolled and welded together in such a way that the welds do not join up at the corner of the plates.

single-use items. When the pressure on the gasket or seal ring is relieved, the material recovers to some extent, but not to the original shape.

Figure 19.53 shows an example of a process vessel with a bolted joint for the lid. This allows access to the vessel. The different heads for cylindrical vessels discussed earlier can all be bolted to the cylinder, rather than welded. Another example of a bolted joint is shown in Figure 19.54. This shows a bolted hatchway for a large cylindrical vessel. Removal of the hatchway cover allows access for maintenance or inspection of the vessel.

The most common bolted joints in process plant are in pipelines. Figure 19.55 shows an example of a bolted flanged pipe joint. These are used for connecting pipes together and connecting pipes to valves, pumps, compressors, vessels, and heat exchangers. Flanges are standard designs (ASME 2017b). The number of bolts required depends on the force required to compress the gasket to create a seal against internal (or external) pressure, stresses from the weight of surrounding pipework, the diameter of pipe and flange, thickness of the flange, the diameter of the bolts at the narrowest point, operating temperature of the pipe, and the yield strength of the bolt material.

Figure 19.56 shows four examples of different designs of pipe flanges.

a) Figure 19.56a shows a *welding neck flange*. Welding neck flanges have a long tapered body that provides reinforcement and high strength for applications involving high pressure, high temperatures, and cryogenic temperatures. The design has advantages under conditions of variable force caused by, for example, thermal expansion or vibration. The flange is joined to the pipe with a single butt weld. Welding neck flanges are often used as nozzles on pressure vessels.

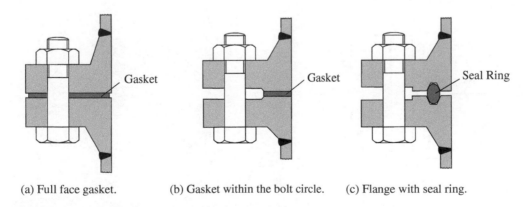

Gasket                    Gasket                    Seal Ring

(a) Full face gasket.        (b) Gasket within the bolt circle.        (c) Flange with seal ring.

**Figure 19.52**

Examples of bolted flange arrangements.

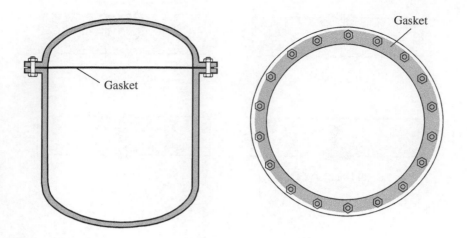

**Figure 19.53**

Cylindrical vessel with a bolted lid.

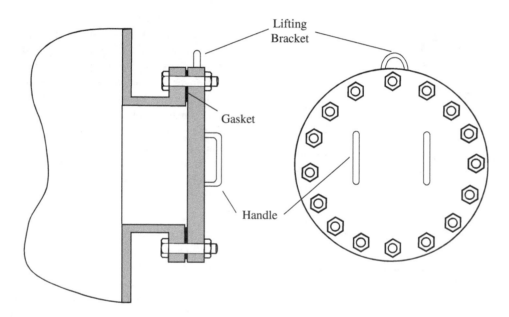

**Figure 19.54**

Bolted inspection and maintenance hatchway.

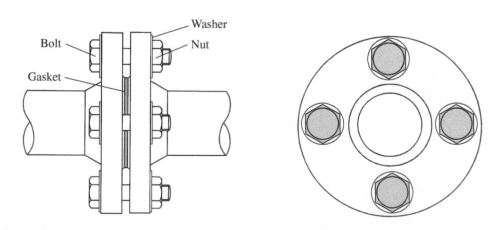

**Figure 19.55**

Example of a bolted flange pipe joint.

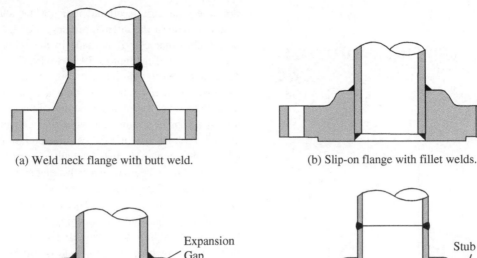

(a) Weld neck flange with butt weld.

(b) Slip-on flange with fillet welds.

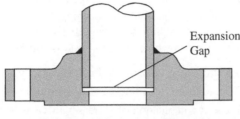

Expansion Gap

(c) Socket weld flange with fillet weld.

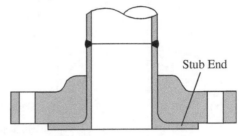

Stub End

(d) Lap joint flange with butt weld.

**Figure 19.56**

Examples of different pipe flanges.

**b)** Figure 19.56b shows a *slip-on flange*. Slip-on flanges are less expensive than welding neck flanges but have a strength of the order of two-thirds that of welding neck flanges. They are also more susceptible to fatigue than welding neck flanges. The connection between the flange and the pipe is achieved with fillet welds outside and inside the flange. Slip-on flanges are widely used in pipe systems.

**c)** Figure 19.56c shows a *socket weld flange*. Socket weld flanges were developed for use on small-size high-pressure piping. They have strength similar to slip-on flanges, but their fatigue strength is greater than slip-on flanges. The connection with the pipe is achieved using one fillet weld on the outside of the flange. For corrosive process fluids the crack between the pipe and the flange can lead to corrosion problems.

**d)** Figure 19.56d shows a *lap joint flange*. Lap joint flanges are used in conjunction with a lap joint stub end, which are not welded together. Pressure from the bolts creates a seal with the gasket against the face of the stub end. Lap joint flanges are similar to slip-on flanges but have a radius at the intersection of the flange face and the stub end. The ability to contain a given pressure is similar to a slip-on flange. However, the fatigue life is significantly lower than welding neck flanges. A major advantage of lap joint flanges is that the flange can be made from cheaper materials than the pipe when the process requires expensive materials of construction.

A number of points need to be noted regarding the strength of the bolts. It was pointed out in Section 19.1 earlier that bolts have a *major* cross-sectional area corresponding with the major diameter across the unthreaded part of the bolt and a *minor* cross-sectional area corresponding with the minor diameter across the threaded part between the roots of the threads. The minor area is the weakest part in the bolt. In tensile tests on bolts it has been found that the area resisting tension is lower than the major area, but not as small as the minor area. The actual area resisting the tensile force is termed the *tensile stress area*. Using the minor area in design will be conservative, being smaller than the actual tensile stress area. Table 19.6 gives the dimensions of some commonly used bolt sizes.

The strength of the material from which the bolt is fabricated will have a *yield strength*, as discussed in Chapter 18. At the yield point, the force becomes strong enough such that the material will stretch and not return to its original shape. If the force is low enough, the steel will elastically return to its original shape when the force is removed. *Proof strength* measures the force that a bolt must be able to withstand without permanently deforming. Proof strength tests are carried out on bolts, rather than material specimens. The length of the bolt is measured before and after the proof load test to ensure compliance. Table 19.7 shows the strength of various classes of steel bolts.

# 19.10  Opening Reinforcements

All pressure vessels need nozzles and connections for material to enter and leave the vessel and for instrumentation of the vessel contents. The most common practice is to connect nozzles and pipelines with standard bolted flanges. In order to attach a nozzle to a pressure vessel it is necessary to make a hole in the cylindrical shell or the head, which needs material to be removed. This

## Table 19.6

Dimensions of bolts.

| Nominal (major) diameter (mm) | Major area (mm²) | Tensile stress area (mm²) | Minor area (mm²) |
|---|---|---|---|
| 5 | 19.63 | 14.18 | 12.68 |
| 6 | 28.27 | 20.12 | 17.89 |
| 8 | 50.27 | 36.61 | 32.84 |
| 10 | 78.54 | 57.99 | 52.29 |
| 12 | 113.1 | 84.27 | 76.25 |
| 14 | 153.9 | 115.4 | 104.7 |
| 16 | 201.1 | 156.7 | 144.1 |
| 20 | 314.2 | 244.8 | 225.2 |
| 24 | 452.4 | 352.5 | 324.3 |
| 30 | 706.9 | 560.6 | 519.0 |
| 36 | 1018 | 816.7 | 759.3 |
| 42 | 1385 | 1121 | 1045 |
| 48 | 1810 | 1473 | 1377 |
| 56 | 2463 | 2030 | 1905 |
| 64 | 3217 | 2676 | 2520 |
| 72 | 4072 | 3460 | 3282 |
| 80 | 5027 | 4344 | 4144 |
| 90 | 6362 | 5591 | 5364 |
| 100 | 7854 | 6995 | 6740 |

decreases the cross-sectional area of the shell locally that is required to resist the internal pressure. For example, if a circular hole with a diameter $D$ is made in the spherical or cylindrical shells illustrated in Figures 19.34 to 19.37 with shell thickness $t$, the hole creates a decrease in area of the wall locally required to resist the internal pressure, with a maximum cross-sectional area of $D \times t$ (neglecting the curvature of the vessel wall). This creates stress paths around the hole that concentrate the stresses around the hole. The high stress concentrations that exist at the edge of an opening decrease radially outward from the opening, becoming small beyond twice the diameter from the center of the opening. *Reinforcement* is required to avoid failure of the area around the opening. The reinforcing material should be placed immediately adjacent to the opening, but with a profile and contour so as not to introduce a significant stress concentration itself. Some ways in which this can be accomplished are:

a) Increase the vessel wall thickness locally.
b) Increase the wall thickness of the nozzle.
c) Use a combination of increases in vessel wall thickness and nozzle thickness.

These are subject to limits on nozzle thickness, spacing between nozzles, and external pipe loads (ASME 2019; BSI Standards Publications 2019). Figure 19.57a shows a typical nozzle configuration with a *reinforcing ring* or *reinforcing pad*. The material removed for the opening is effectively relocated to an area within a boundary around the opening. For example, if the hole in the shell has a diameter $D$ and the shell has a thickness $t$, the lost area from the shell *to resist the internal pressure* is $Dt$. Welding an annular disc of thickness $t$ with an inner diameter of $D$ and outer diameter $2D$ around the hole replaces the lost area by providing an extra area $Dt/2$ on either side of the lost area $Dt$, thus providing *area compensation*. In some applications, e.g., high pressure, high temperature, or high thermal gradients, it is preferred to avoid a non-integral nozzle configuration and reinforcing pads. These are replaced with forged nozzles. Alternative reinforcement designs using forged nozzles are shown in Figure 19.57b and c. Figure 19.57d shows a nozzle design with

## Table 19.7

Property classes of steel bolts.

| Property class | Proof strength (MPa) | Minimum yield strength (MPa) | Minimum tensile strength (MPa) | Material |
|---|---|---|---|---|
| 4.6 | 225 | 240 | 400 | Low or medium carbon steel |
| 4.8 | 310 | 340 | 420 | Low or medium carbon steel; fully or partially annealed |
| 5.8 | 380 | 420 | 520 | Low or medium carbon steel; cold worked |
| 8.8 | 660 | 640 | 800 | Medium carbon steel; quenched and tempered |
| 8.8 | 580 | 660 | 830 | Medium carbon steel; quenched and tempered |
| 9.8 | 650 | 720 | 900 | Medium carbon steel; quenched and tempered |
| 10.9 | 830 | 940 | 1040 | Alloy steel; quenched and tempered |
| 12.9 | 970 | 1100 | 1220 | Alloy steel; quenched and tempered |

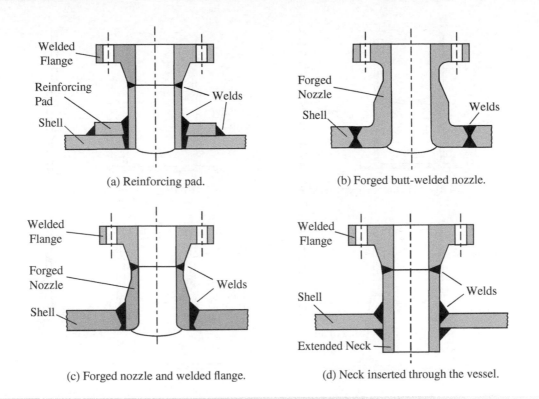

(a) Reinforcing pad.

(b) Forged butt-welded nozzle.

(c) Forged nozzle and welded flange.

(d) Neck inserted through the vessel.

**Figure 19.57**

Openings in pressure vessels require compensation for the weakness created by the opening.

the neck inserted into the vessel. Not all the nozzle walls in Figure 19.57d can be devoted to compensation since the internally pressurized branch is a cylinder under pressure in its own right (ASME 2019; BSI Standards Publications 2019).

# 19.11    Vessel Supports

Cylindrical vessels with shaped heads require support, whether mounted vertically or horizontally. Consider vertical cylindrical vessels first. The type of support required depends particularly on how tall the vessel is. It has already been illustrated in Figure 19.40 that tall cylindrical vessels, such as distillation and absorption columns, can be supported on a cylindrical *skirt*. Figure 19.58 shows four possible skirt designs for a tall cylindrical vessel. Figure 19.58a shows a right circular cylindrical skirt where the outside diameter of the skirt is the same outside diameter as the vessel. The vessel and skirt are welded together at the base of the vessel. An alternative design is illustrated in Figure 19.58b, which has a vertical skirt, but this time the skirt overlaps the bottom of the vessel and is joined to the vessel with a lap weld. Figure 19.58c shows a tapered skirt in the shape of a conical frustum. The design in Figure 19.58c shows a design in which the outside diameter at the top of the skirt and the outside diameter at the bottom of the vessel are the same and welded together. The angle of the skirt is limited to be less than 15° and is often taken to be 10°. It is also possible to have a lapped flared skirt, as illustrated in Figure 19.58d. The skirt is joined to the vessel with a lap weld. The circumferential weld attaching

the vessel to the skirt is a critical line. The weld transmits the weight of the vessel and contents to the skirt and resists bending moments. It must also resist thermal stresses created by the temperature differences between the vessel and the skirt. Details of the design of the joints between the vessel and skirt can be found elsewhere (Bednar 1991).

It is more common to use skirts with an outside diameter equal to the outside diameter of the vessel, rather than lapped designs. Flared skirts are more expensive than right circular skirts and less common. Flared skirts tend to be used when the holding-down bolts used to fix the skirt to the foundations cannot be kept apart to the required spacing between bolts, necessitating a larger bolt circle diameter. This will be discussed later in this section.

Figure 19.59 shows typical features of a right circular cylindrical skirt. An opening is required to allow pipes to connect to the bottom of the vessel. There also needs to be at least one opening in the skirt to permit inspection of the bottom of the vessel and possibly maintenance. Vent holes are required in many applications to ensure there is no build-up of hazardous vapors at the top of the enclosure under the vessel. These openings in the skirt need to be compensated (reinforced), as discussed in Section 19.10.

The bottom of the skirt is attached to a *bearing plate* or *base plate*. This plate is fixed to the concrete foundations by means of *anchor bolts* to prevent the overturning of the vessel due to wind and seismic loads. Figure 19.60 shows three examples of different designs of bearing plates. Figure 19.60a shows a double ring bearing plate strengthened with *gussets*. These tend to be used for the larger sizes of vessels. Smaller vessels tend to use a single ring

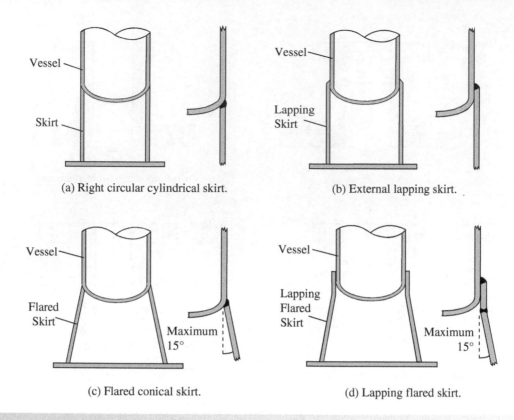

(a) Right circular cylindrical skirt.

(b) External lapping skirt.

(c) Flared conical skirt.

(d) Lapping flared skirt.

**Figure 19.58**

Typical skirt designs.

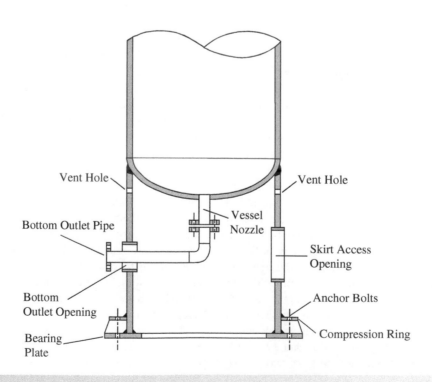

**Figure 19.59**

Typical features of a support skirt for a vertical cylindrical vessel.

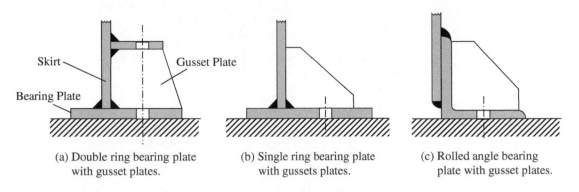

(a) Double ring bearing plate with gusset plates.

(b) Single ring bearing plate with gussets plates.

(c) Rolled angle bearing plate with gusset plates.

**Figure 19.60**

Examples of different designs of bearing plates.

bearing plate with gussets, as illustrated in Figure 19.60b or a rolled angle bearing plate with gussets illustrated in Figure 19.60c.

Figure 19.61 shows three examples of a typical arrangement for anchor bolts using *cast-in anchors*, where the concrete is poured around the anchors. The bolts can be *pre-loaded* or *pre-tensioned*. When a load is placed on a bolt, there is a limit to load that can be restrained by the bolt before failing. However, when a bolt is tightened against a plate, it allows the bolt to distribute some of the force through the plate, such that the bolt itself only holds part of the load. This means that a bolt can restrain a higher load when the correct amount of tension is applied. That tension is known as the *preload*. Pre-loaded bolts are put under tension such that they will never go above the allowable tensile stress of the bolt, but the bearing plate always exerts pressure on the concrete. This avoids stress reversal and

fatigue as conditions change. Anchor bolts may or may not be pre-loaded.

Figure 19.61a shows a square plate bolted to the end of the anchor bolt, which is cast in the concrete. Figure 19.61b shows an L-shape bolt and Figure 19.61c a J-shape bolt cast in the concrete. The minimum bolt diameter is normally 25 mm. The bolts need to be clean and free from oil so that the cement in the concrete can bond with the embedded surface of the bolt. When either a tensile load (from the bending moment) or compressive load (from the bending moment or deadweight) is applied to the anchor bolts, the load is transferred from the bolts to the concrete through the bond with the concrete. Surface irregularities, welded plates, or bends in the bolts transfer loads from the bolts to the concrete. The concrete foundation must have adequate reinforcing steel to carry the tensile loads in the foundation.

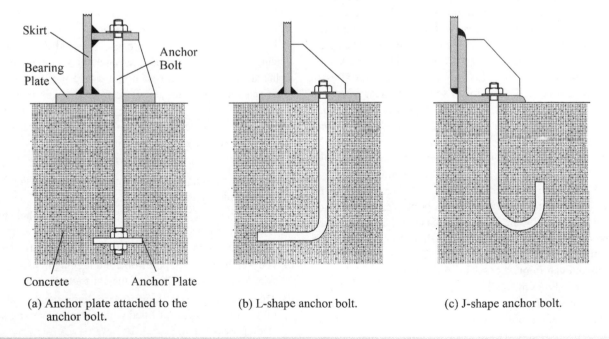

(a) Anchor plate attached to the anchor bolt.

(b) L-shape anchor bolt.

(c) J-shape anchor bolt.

**Figure 19.61**

Some typical arrangements for anchor bolts for self-supporting vertical cylindrical vessels.

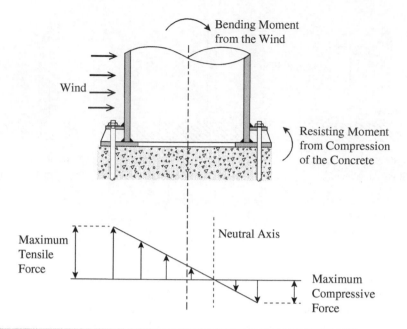

**Figure 19.62**
Anchor bolt loading distribution.

Figure 19.62 shows the influence of the transverse force from wind on the stresses in the bearing plate and concrete foundation. The wind creates a bending moment on the bearing plate and concrete foundation. Assume in the first instance that both the bearing plate and the concrete foundation are rigid (inelastic) and do not undergo any significant deformation from the bending moment. If the bearing plate is assumed to be symmetrical and any deadweight is assumed to be evenly distributed, then it would be expected that there would be a tensile force on the upwind side and an equal compressive force on the downwind side distributed by a neutral axis along the center of the bearing plate. In practice, both the bearing plate and concrete have some elasticity. However, the modulus of elasticity (ratio of stress/strain) of the concrete is very much lower than that of steel, resulting in a given stress creating a much greater strain in the concrete than the steel. The compression of the concrete on the downwind side creates an additional resisting moment, increasing the tensile force, which shifts the neutral axis away from the center to the downwind side on the bearing plate. Procedures to determine the location of the neutral axis have been given by Brownell and Young (1959). In principle, there is also a shear force created by the wind at the base of the column that should be accounted for. However, this shear force is counteracted by the friction between the base plate and the concrete foundation and can in most cases be neglected.

Figure 19.63 shows the different loadings for the anchor bolts around a bolt ring. Figure 19.63 shows an example of a bearing plate with 12 anchor bolts. The number of anchor bolts is increased or decreased in multiples of 4, and should be a minimum of 4. The bending moment of the wind means that some of the bolts are in tension and some in compression. The load on the individual bolts is linearly dependent on the distance from the neutral axis if the bearing plate is assumed to be rigid. In Figure 19.63, Bolt 5 to Bolt 9 make no contribution to resisting the tensile load on the upwind side of the bearing plate. Bolt 1 carries the greatest tensile load and Bolt 7 the greatest compressive load.

For small vessels the anchor bolts can be designed on the assumption that the neutral axis lies along a diameter of the bearing plate. However, this assumption leads to overdesign in the case of tall vessels with large overturning moments because the effect of the elasticity of the foundation is neglected.

For cast-in anchors that are not pre-loaded, the minimum spacing (center-to-center) between anchors is 4 times the diameter of the anchor bolt and for anchors that are pre-loaded the minimum spacing (center-to-center) between anchors is 6 times the diameter of the anchor bolt. If the required spacing between the bolts cannot be achieved by using a right cylindrical skirt, then it might be necessary to use a flared skirt.

The skirt must be capable of withstanding:

a) The load from applied moments and forces created by wind and seismic loads.

b) The weight of vessel, internals (including liquid held up in the internals), pipework, and lagging transmitted in compression to the skirt by the shell above the level of the skirt attachment.

c) The weight of vessel, pipework, and lagging transmitted to the skirt from below the level of the skirt attachment.

d) The weight of process liquid in the column sump, or the weight of water in the vessel if filled with water undergoing a hydrostatic pressure test.

e) The reactive force from the bearing plate and foundations illustrated in Figure 19.62.

f) Buckling of the cylinder.

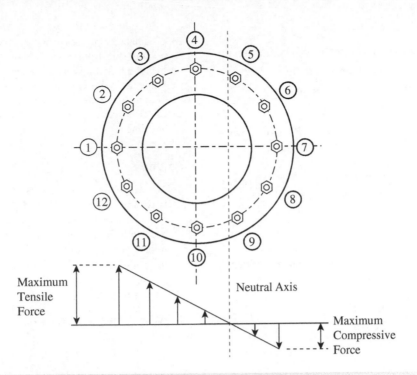

**Figure 19.63**

Distribution of the load on the anchor bolts.

The stress in the skirt from the weight of the vessel and contents and the bending stresses (but neglecting reactive forces from the bearing plate and foundations) are given by (Brownell and Young 1959; Chattopadhyay 2005; Escoe 1994):

$$\sigma_{SKIRT} = -\frac{W}{A} \pm \frac{M}{Z} \qquad (19.162)$$

where

$\sigma_{SKIRT}$ = longitudinal stress in the skirt (N m$^{-2}$)

$W$ = deadweight of the vessel and contents (N)

$A$ = cross-sectional area of the skirt shell (m$^2$)

$M$ = bending moment created by wind (or seismic force) (N m)

$Z$ = section modulus of the skirt (m$^3$)

For a right circular cylindrical skirt, substituting the cross-sectional area of the skirt and the section modulus given by Eq. (19.53) into Eq. (19.162) gives:

$$\sigma_{SKIRT} = -\frac{4W}{\pi\left(D_O^2 - D_I^2\right)} \pm \frac{32D_O M}{\pi\left(D_O^4 - D_I^4\right)} \qquad (19.163)$$

where

$D_O$ = outside diameter of the skirt (m)

$D_I$ = inside diameter of the skirt (m)

If a flared skirt is to be used in the form of a conical frustum, a preliminary design can be developed using the *equivalent cylinder* approximation. This is based on calculating the diameter of an

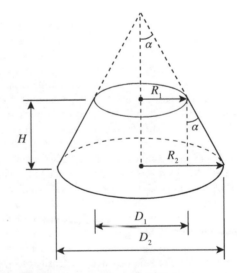

**Figure 19.64**

Flared column skirt in the form of a conical frustum.

equivalent cylinder with the same surface area to the conical frustum. The equivalent cylinder is used to calculate the section modulus and moment of inertia (Escoe 1994). Figure 19.64 shows a conical frustum. The radius of an equivalent cylinder is given by (Appendix J):

$$R_{EQ} = \frac{(R_2 + R_1)}{2\cos\alpha} \qquad (19.164)$$

where

$R_{EQ}$ = radius of the equivalent cylinder to the conical frustum (m)

$R_2$ = radius of the larger circle of the frustum (m)

$R_1$ = radius of the smaller circle of the frustum (m)

$\alpha$ = angle of the flare to the vertical (degrees)

The radius $R_2$ can be written in terms of $R_1$ and $\alpha$, where $R_2 = R_1 + H \tan \alpha$. Substituting in Eq. (19.164) gives:

$$R_{EQ} = \frac{2R_1 + H \tan \alpha}{2 \cos \alpha} \quad (19.165)$$

where

$H$ = height of to the skirt (m)

Alternatively, Eqs (19.164) and (19.165) can be written in terms of the equivalent cylinder diameter:

$$D_{EQ} = \frac{(D_2 + D_1)}{2 \cos \alpha} \quad (19.166)$$

or

$$D_{EQ} = \frac{D_1 + H \tan \alpha}{\cos \alpha} \quad (19.167)$$

The design codes dictate that the skirt should be not less than 6 mm thick (ASME 2019; BSI Standards Publications 2019).

**Example 19.14** For Examples 19.6 and 19.11, the skirt is to be a right circular cylindrical skirt butt-welded to the base of the distillation column, as in Figure 19.58a. From the previous examples, the outside and inside diameters of the column and skirt are 2.54 m and 2.5 m, corresponding with a skirt shell thickness of 20 mm. The tangent-to-tangent length of the column is 36.8 m and height of the skirt is 3 m. From Example 19.5 the wind creates a force of 4409 N m$^{-1}$ along the shell and skirt, creating a bending moment at the base of the skirt of $3.492 \times 10^6$ N m$^{-1}$. The weight of the column above the skirt was calculated in Example 19.11 to be $8.848 \times 10^5$ N. However, the skirt also needs to support the weight of the bottom head, weight of the head insulation, weight of liquid in the sump, weight of the skirt itself and possibly the weight of some of the bottom pipework (depending on the details of the pipe supports at the bottom of the column). The column sump can be assumed to have a height of liquid above the bottom of the cylindrical shell of 2 m, with a liquid density of 483 kg m$^{-3}$. In this case, no allowance will be made for weight of the bottom pipework. The allowable stress in the skirt is $135 \times 10^6$ N m$^{-2}$. Assume that the joint efficiency for the butt-welded skirt is 0.55 (Bednar 1991). The upwind side of the column suffers from a tensile force created by the bending moment. This tensile stress is countered by a compressive stress from the weight of the vessel and its contents. The downwind side of the column suffers from a compressive stress created by the bending moment from the wind. This compressive stress is enhanced by the weight of the vessel and contents. In this calculation the reactive force from the skirt bearing plate and foundations is to be neglected:

a) Calculate the stresses on the upwind and downwind sides of the column skirt from the weight of the column and contents and bending from the wind when the column is operating.

b) Calculate the stresses on the upwind and downwind sides of the column skirt from the weight of the column and contents and bending from the wind when the column is full of water undergoing a hydrostatic pressure test in the upright position. Assume the density of water is 998 kg m$^{-3}$.

c) Check the compressive stress is lower than the critical stress for buckling.

**Solution**

a) The weight of the column above the skirt joint was calculated in Example 19.11 to be $8.848 \times 10^5$ N. The weight supported at the base of the skirt also includes the weight of the bottom head with insulation, weight of the liquid in the sump, and the weight of the skirt itself.

Weight of the bottom head from Example 19.11:

$$= 9539 \text{ N}$$

Weight of the head insulation from Example 19.11:

$$= 544 \text{ N}$$

Weight of liquid in the sump given by Eq. (19.141) with one head volume:

$$= \left( \frac{\pi D_I^2}{4} L + \frac{\pi D_I^3}{24} \right) \rho_L g$$

$$= \left( \frac{\pi \times 2.5^2}{4} \times 2.0 + \frac{\pi \times 2.5^3}{24} \right) \times 483 \times 9.81$$

$$= 56\,209 \text{ N}$$

Weight of skirt:

$$= \frac{\pi \left( D_O^2 - D_I^2 \right)}{4} \rho_{MOC} g$$

$$= \frac{\pi \left( 2.54^2 - 2.5^2 \right)}{4} \times 7850 \times 9.81$$

$$= 12\,193 \text{ N}$$

Total weight of column and skirt:

$$= 8.848 \times 10^5 + 9539 + 544 + 56\,209 + 12\,193$$

$$= 9.633 \times 10^5 \text{ N}$$

The stress in the skirt from the weight of the vessel and contents and bending moment is given by Eq. (19.163):

$$\sigma_{SKIRT} = -\frac{4W}{\pi \left( D_O^2 - D_I^2 \right)} \pm \frac{32 D_O M}{\pi \left( D_O^4 - D_I^4 \right)}$$

$$= -\frac{4 \times 9.633 \times 10^5}{\pi \left( 2.54^2 - 2.5^2 \right)} \pm \frac{32 \times 2.54 \times 3.492 \times 10^6}{\pi \left( 2.54^4 - 2.5^4 \right)}$$

$$= -6.084 \times 10^6 \pm 35.282 \times 10^6$$

For the upwind side of the column:

$$\sigma_{SKIRT} = -6.084 \times 10^6 + 35.282 \times 10^6$$

$$= 29.198 \times 10^6 \text{ N m}^{-2}$$

For the downwind side of the column:

$$\sigma_{SKIRT} = -6.084 \times 10^6 - 35.282 \times 10^6$$

$$= -41.366 \times 10^6 \, \mathrm{N\,m^{-2}}$$

The allowable stress for a joint efficiency of 0.55 is given by:

$$\sigma_{ALL} = 0.55 \times 138 \times 10^6$$

$$= 75.9 \times 10^6 \, \mathrm{N\,m^{-2}}$$

**b)** For the case in which the vessel is full of water for a hydrostatic pressure test, the volume of water in the vessel can be calculated from Eq. (19.141) neglecting the volume occupied by the distillation internals:

$$V = \frac{\pi D_I^2}{4}L + \frac{\pi D_I^3}{12}$$

$$= \frac{\pi \times 2.5^2}{4}36.8 + \frac{\pi \times 2.5^3}{12}$$

$$= 184.7 \, \mathrm{m^3}$$

The weight of the vessel full of water (neglecting the volume occupied by the distillation internals and any process hold-up material on the trays and in the pipes, but subtracting the weight of process liquid in the sump of 56 209 N) is given by:

$$W = 9.633 \times 10^5 + 184.7 \times 998 \times 9.81 - 56\,209$$

$$= 2.715 \times 10^6 \, \mathrm{N}$$

From Eq. (19.163):

$$\sigma_{SKIRT} = -\frac{4W}{\pi(D_O^2 - D_I^2)} \pm \frac{32 D_O M}{\pi(D_O^4 - D_I^4)}$$

$$= -\frac{4 \times 2.715 \times 10^6}{\pi(2.54^2 - 2.5^2)} \pm \frac{32 \times 2.54 \times 3.492 \times 10^6}{\pi(2.54^4 - 2.5^4)}$$

$$= -17.147 \times 10^6 \pm 35.282 \times 10^6$$

For the upwind side of the column:

$$\sigma_{SKIRT} = -17.147 \times 10^6 + 35.282 \times 10^6$$
$$= 18.135 \times 10^6 \, \mathrm{N\,m^{-2}}$$

For the downwind side of the column:

$$\sigma_{SKIRT} = -17.147 \times 10^6 - 35.282 \times 10^6$$

$$= -52.429 \times 10^6 \, \mathrm{N\,m^{-2}}$$

Both the upwind and downwind stresses are below the allowable stress of $74.25 \times 10^6 \, \mathrm{N\,m^{-2}}$.

**c)** The combined compressive stress from weight and bending $(\sigma_W + \sigma_{Bmax})$ must be less than the critical stress for buckling. The critical buckling stress $\sigma_{CR}$ for a shell thickness $t$ is given by Eq. (19.161):

$$\sigma_{CR} = 2.42 \times 10^{10}\left(\frac{t}{D_O}\right)$$

$$= 2.42 \times 10^{10}\left(\frac{0.02}{2.54}\right)$$

$$= 190.6 \times 10^6 \, \mathrm{N}$$

Thus, the maximum compressive stress is significantly below the critical stress for buckling of the skirt.

The elastic strength in tension and in compression for ductile steels is usually considered to be the same. This assumption is sufficiently accurate for many purposes. However, this is not true for brittle materials. It should be noted that the limiting case for the skirt design *on the upwind side* is when the column is empty. *On the downwind side* the limiting case is when the column is full of water during a hydrostatic pressure test. The limiting case overall is on the downwind side. Both the upwind and downwind stresses are below the allowable stress for both normal operation and during a hydrostatic test. However, this cannot be considered the final design for the skirt. It needs to be checked for the reactive force from the bearing plate and foundations when the design of the bearing plate and foundations has been completed (Brownell and Young 1959).

**Example 19.15**  For the same column as Example 19.14, a flared skirt is to be used with an angle of 10° instead of a straight cylindrical skirt. Assume the outside diameter of the upper face of the skirt to be 2.54 m, the same as the outside diameter of the column at the base. As in Example 19.14, assume the plate thickness to be 20 mm with a height of 3 m and the joint efficiency to be 0.55. Assume also that the skirt flare does not change the bending moment at the base of the column and that the weights are the same as in Example 19.14.

**a)** Calculate the dimensions of the equivalent cylinder to the flared skirt.
**b)** Calculate the stresses in the skirt for the upwind side of the column when the column is in normal operation.
**c)** Calculate the stresses in the skirt for the downwind side of the column when the column is full of water undergoing a hydrostatic test in the upright position.

**Solution**

**a)** The diameter of an equivalent cylinder to the flared skirt is given by Eq. (19.167).

$$D_{EQ} = \frac{D_1 + H\tan\alpha}{\cos\alpha}$$

$$= \frac{2.54 + 3\tan 10°}{\cos 10°}$$

$$= 3.12 \, \mathrm{m}$$

Thus, for the equivalent cylinder:

$$D_O = 3.12 \, \mathrm{m}$$

$$D_I = 3.12 - 2 \times 0.02 = 3.08 \, \mathrm{m}$$

**b)** When the column is in normal operation, from Eq. (19.163):

$$\sigma_{SKIRT} = -\frac{4W}{\pi(D_O^2 - D_I^2)} \pm \frac{32 D_O M}{\pi(D_O^4 - D_I^4)}$$

$$= -\frac{4 \times 9.633 \times 10^5}{\pi(3.12^2 - 3.08^2)} \pm \frac{32 \times 3.12 \times 3.492 \times 10^6}{\pi(3.12^4 - 3.08^4)}$$

$$= -4.946 \times 10^6 \pm 23.281 \times 10^6$$

For the upwind side:

$$\sigma_{SKIRT} = -4.946 \times 10^6 + 23.281 \times 10^6$$
$$= 18.335 \times 10^6 \, \mathrm{N\,m^{-2}}$$

For the downwind side:

$$\sigma_{SKIRT} = -4.946 \times 10^6 - 23.281 \times 10^6$$
$$= -28.227 \times 10^6 \, \mathrm{N\,m^{-2}}$$

**c)** When the column is full of water undergoing a hydrostatic test in the upright position:

$$\sigma_{SKIRT} = -\frac{4W}{\pi\left(D_O^2 - D_I^2\right)} \pm \frac{32 D_O M}{\pi\left(D_O^4 - D_I^4\right)}$$

$$= -\frac{4 \times 2.715 \times 10^6}{\pi\left(3.12^2 - 3.08^2\right)} \pm \frac{32 \times 3.12 \times 3.492 \times 10^6}{\pi\left(3.12^4 - 3.08^4\right)}$$

$$= -13.939 \times 10^6 \pm 23.281 \times 10^6$$

For the upwind side:

$$\sigma_{SKIRT} = -13.939 \times 10^6 + 23.281 \times 10^6$$
$$= 9.342 \times 10^6 \, \mathrm{N\,m^{-2}}$$

For the downwind side:

$$\sigma_{SKIRT} = -13.939 \times 10^6 - 23.281 \times 10^6$$
$$= -37.220 \times 10^6 \, \mathrm{N\,m^{-2}}$$

Thus, both the upwind and downwind sides of the column skirt are below the allowable stress, and below the stresses for a straight cylindrical skirt with the same thickness.

Self-supporting columns must be fixed to the concrete foundation with the adequate number and size of anchor bolts. The minimum size of an anchor bolt should be 25 mm. Various methods are available to design the anchor bolts (Brownell and Young 1959; Bednar 1991; Escoe 1994; Jawad and Farr 2019). A simple approximate method is based on the application of Eqs (19.162) and (19.38) to the bolt circle:

$$\sigma_{BOLTS} = -\frac{W}{A} \pm \frac{My}{I} \qquad (19.168)$$

where

$\sigma_{BOLTS}$ = stress in the bolts (N m$^{-2}$)

$W$ = deadweight of the vessel and contents (N)

$A$ = cross-sectional area of the bolt circle (m$^2$)

$M$ = bending moment created by wind (or seismic force) (N m)

$y$ = distance from the neutral axis (m)

$I$ = moment of inertia of the bolt circle (m$^4$)

If the difference between the outside and inside diameter of the bolt circle is small:

$$A_{BC} = \pi D_M t \qquad (19.169)$$

where

$A_{BC}$ = area of the bolt circle (N m$^{-2}$)

$D_M$ = mean diameter (pitch circle diameter) of the bolt circle (m)

$t$ = width of the bolt circle (m)

The moment of inertia for the bolt circle can be approximated from Eq. (19.55):

$$I = \frac{\pi D_M^3 t}{8} \qquad (19.170)$$

Combining with Eq. (19.169):

$$I = \frac{A_{BC} D_M^2}{8} \qquad (19.171)$$

The bolt circle area can be expressed in terms of the number of bolts and their individual areas:

$$A_{BC} = N_B A_B \qquad (19.172)$$

where

$A_{BC}$ = cross-sectional area of the bolt circle (m$^2$)

$N_B$ = number of bolts in the bolt circle (−)

$A_B$ = cross-sectional area of the bolts at the bolt narrowest point (m$^2$)

Substituting Eq. (19.172) into Eq. (19.171) gives:

$$I = \frac{N_B A_B D_M^2}{8} \qquad (19.173)$$

Substituting for $A_{BC}$ and $I$ in Eq. (19.168) gives:

$$\sigma_{BOLTS} = -\frac{W}{N_B A_B} \pm \frac{8My}{N_B A_B D_M^2} \qquad (19.174)$$

The stress becomes a maximum for $y = D_M/2$:

$$\sigma_{BOLTS} = -\frac{W}{N_B A_B} \pm \frac{4M}{N_B A_B D_M} \qquad (19.175)$$

If it is assumed that the base plate transfers the compressive load on the downwind side of the column to the concrete foundation, then the number and diameter of the bolts is determined by the tensile force on the upwind side. The maximum tensile load on an individual bolt $F_B$ assuming the neutral axis is across the diameter of the bolt circle is given by:

$$F_B = \frac{4M}{N_B D_M} - \frac{W}{N_B} \qquad (19.176)$$

where

$F_B$ = maximum tensile load on the bolts (N)

For a given bolt proof stress (Table 19.7), the required area of the anchor bolts is given by:

$$A_B = \frac{F_B}{\sigma_{PROOF}} = \frac{1}{\sigma_{PROOF} N_B}\left[\frac{4M}{D_M} - W\right] \qquad (19.177)$$

where

$\sigma_{PROOF}$ = proof stress of the bolts (N m$^{-2}$)

It should be noted that this simplified analysis does not account for the fact that the stress on the bolts will vary around the bolt circle and will depend on the distance from the neutral axis. The simplified analysis also assumes the neutral axis to be perpendicular to the direction of the wind and across the diameter of the bolt circle. In practice, the neutral axis is offset from the center, as discussed previously (Brownell and Young 1959). It should also be noted that it neglects any pre-loading of the bolts. Finally, it should be noted that the anchor bolt considerations here are restricted to tensile failure of the metal bolt. Failure can also occur from failure of the concrete, e.g. pull-out of the anchor bolt (see Brownell and Young 1959; Bednar 1991; Escoe 1994; Jawad and Farr 2019).

**Example 19.16** For the vertical column from Example 19.14, determine the number of anchor bolts required for a right circular cylindrical skirt at the base of the distillation column. The tensile load on the bolts is a maximum when the column is empty. From the previous examples, the outside diameter of the column and the skirt are 2.54 m. The bolt circle diameter can be assumed to be 2.7 m. From Example 19.6, the wind creates a force of 4409 N m$^{-1}$ along the column shell and skirt, creating a bending moment at the base of the skirt of $3.492 \times 10^6$ N m. The weight of the distillation column and skirt when in operation was calculated in Example 19.14 to be $9.633 \times 10^5$ N. Bolts with a property class of 4.6 (Table 19.7) are to be used with a load on the anchor bolts less than 0.75 of the proof strength to allow for variations in material quality, poor installation procedures, etc. Calculate the number of anchor bolts required.

**Solution**

As a first iteration try 12 bolts. From Eq. (19.177):

$$A_B = \frac{1}{\sigma_{PROOF} N_B} \left[ \frac{4M}{D_M} - W \right]$$

$$= \frac{1}{0.75 \times 225 \times 10^6 \times 12} \left[ \frac{4 \times 3.492 \times 10^6}{2.7} - 9.633 \times 10^5 \right]$$

$$= 2.079 \times 10^{-3} \text{ m}^2$$

$$= 2079 \text{ mm}^2$$

From Table 19.6, for 12 bolts the diameter would need to be 64 mm. The number of bolts can be increased or decreased in multiples of 4. Try 20 bolts for comparison:

$$A_B = \frac{1}{0.75 \times 225 \times 10^6 \times 20} \left[ \frac{4 \times 3.492 \times 10^6}{2.7} - 9.633 \times 10^5 \right]$$

$$= 1.247 \times 10^{-3} \text{ m}^2$$

$$= 1247 \text{ mm}^2$$

From Table 19.6, for 20 bolts the diameter would need to be 48 mm. The minimum bolt spacing needs to be maintained. Check the bolt spacing for 20 bolts:

$$\text{Bolt spacing} = \frac{\pi D_{BC}}{20} = \frac{\pi \times 2.7}{20} = 0.42 \text{ m}$$

For 48 mm bolts, this would be a spacing of 8.75 bolt diameters. The minimum spacing for cast-in anchors that are pre-loaded is 6 times the diameter of the anchor. Note that the calculation neglects any pre-loading of the bolts.

Whilst tall vertical cylindrical vessel walls are normally supported on a skirt, smaller vertical cylindrical vessels are typically supported on *brackets* or *lugs*. These brackets can be supported on legs. Different cross-sections of leg can be used. As an example, Figure 19.65a illustrates a vessel supported by four brackets (lugs), which are mounted on vertical H-section stanchions. Rather than four legs, different numbers of legs can be used, but are equally spaced around the circumference. Depending on the vertical stresses, bending stresses, and the height, number, and cross-section of the legs, the legs might need to be braced to resist diagonal stresses that would cause the structure to twist.

Alternatively, rather than use legs, the brackets can be supported by a structure, as illustrated in Figure 19.65b. It should be noted that the stresses due to the weight of the vessel and its contents are different when supported on brackets rather than a skirt. When the vessel is mounted on a skirt the weight of the vessel and contents creates only compressive forces in the vessel. In Figure 19.65 the weight of the vessel and contents creates

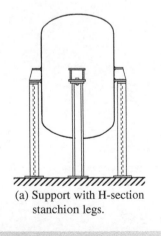

(a) Support with H-section stanchion legs.

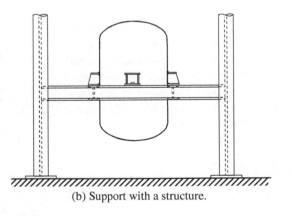

(b) Support with a structure.

**Figure 19.65**

Cylindrical vessel supports using legs and structures.

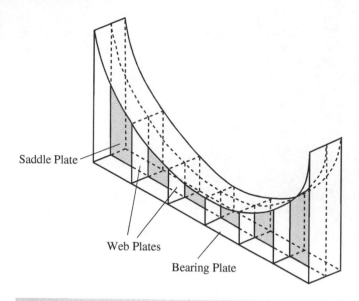

**Figure 19.66**

Saddle support for a horizontal cylindrical vessel.

compressive forces above the brackets, but tensile forces below the brackets.

Horizontal cylindrical vessels are normally supported on saddles, as illustrated in Figure 19.43. The use of more than two saddles is normally avoided. If more than two saddles are used it is difficult to ensure that the load is uniformly distributed across the saddles when the vessel is installed. Figure 19.66 illustrates a saddle support for a horizontal cylindrical vessel. The contact angle between the saddle and the vessel, as illustrated in Figure 19.44, should be a minimum of 120° and a maximum of 150°. Expansion and contraction of the vessel can create stresses in the saddles if both saddles are welded to the vessel and the saddles are anchored. If this is a problem, then one of the saddles is attached to the vessel and the vessel is allowed to slide across the other saddle. The sliding end may have a slide plate between the saddle and the vessel to reduce friction caused by thermal expansion and contraction. The slide plate might be fabricated from a soft metal, e.g. bronze, or from PTFE. The saddles are normally fabricated from steel and bolted to the foundation using anchor bolts. The longitudinal location of the saddles has a large effect on the magnitude of the stresses, as discussed previously. The distance between the saddle and the tangent point for the head should not normally be greater than 20% of the tangent-to-tangent length of the vessel. The distance between the saddle and the tangent point is often taken to be 0.4 times the radius of the vessel (Brownell and Young 1959). Design methods for saddle supports have been given by Brownell and Young (1959).

# 19.12 Design of Flat-bottomed Cylindrical Vessels

Flat-bottomed cylindrical tanks used for storing liquids and liquefied gases under refrigerated conditions were discussed in Chapter 16. Figure 16.28a and b illustrate fixed roof designs featuring permanently welded roofs that can be conical or dome-shaped. Conical roofs can be self-supporting or supported by internal frames. Dome roof tanks normally have a self-supporting roof. Such fixed roof flat-bottomed cylindrical storage tanks normally operate at atmospheric pressure or slightly above atmospheric pressure. Atmospheric storage tanks are defined to operate at pressures upto 17.2 kPa gauge (API Standard 650 2020). Fixed roof tanks might need to be designed with a slight positive pressure if the tank cannot be vented directly to atmosphere, or the vapor space needs to be purged with nitrogen, or the tank is storing a liquefied gas. Dome roof tanks are normally preferred for the storage of liquefied gases. Chapter 16 also discussed floating roof tanks. Figure 16.28c shows an *external floating roof* design. Such designs must feature drains to allow rainwater to run away from the surface of the roof. Tanks with *internal floating roofs* were also discussed in Chapter 16. Such internal designs would be preferred in locations where the accumulation of snow can cause problems, or leakage from the seals at the edges of the roof would be hazardous.

The height (measured by the vertical side of the vessel) to diameter ratio for flat-bottomed cylindrical vessels is typically in the range 1:1 and 1.5:1 (Hall 2020). Like the other cylindrical tanks discussed earlier, the cylindrical sides of flat-bottomed tanks are fabricated by rolling metal sheets to the required radius and then welding the sheets together, as shown in Figure 19.67. The structure is welded together to be both liquid and vapor tight. The vertical cylindrical wall is constructed in a series of *courses*, as with the vertical cylindrical vessels discussed earlier. The courses will feature decreasing thickness from the base of the tank, where the pressure is highest due to the head of the liquid, to the top, where the pressure is lowest. Figure 19.67 illustrates two

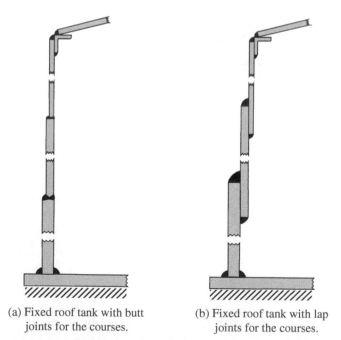

(a) Fixed roof tank with butt joints for the courses.

(b) Fixed roof tank with lap joints for the courses.

**Figure 19.67**

Different constructions of the sides for flat-bottomed cylindrical tanks (Brownell and Young 1959).

different methods of fabrication involving butt-welded plates in Figure 19.67a and lap-welded plates in Figure 19.67b (Brownell and Young 1959). If the plates are lap-welded, each course of the tank must be inside the course below it for the lap-welded joints.

A number of different stresses occur in the cylindrical shell:

**a)** Circumferential stress resulting from pressure within the vessel.

**b)** Longitudinal stress resulting from pressure within the vessel.

**c)** Stresses from wind load.

**d)** Stresses resulting from the weight of auxiliary equipment (e.g. vent and safety valves), snow and ice on the roof.

**e)** Thermal stresses resulting from differences in temperature in different parts of the shell due to differences in temperature between the stored liquid and the exterior.

**f)** Residual stresses frozen into the structure created by localized heating from the welding during fabrication.

The circumferential stress on the wall is given by Eq. (19.122):

$$\sigma_C = \frac{PD_I}{2t_C} \qquad (19.178)$$

where

$\sigma_C$ = circumferential (hoop) stress (N m$^{-2}$)

$P$ = internal pressure (N m$^{-2}$)

$D_I$ = internal diameter (m)

$t_C$ = thickness of the cylinder wall resisting the circumferential stress (m)

The shell thickness for circumferential stress is given by Eq. (19.123) with addition of allowances for joint efficiency and corrosion:

$$t_C = \frac{PD_I}{2\sigma_{ALL}\eta_J} + C \qquad (19.179)$$

where

$\sigma_{ALL}$ = allowable stress (N m$^{-2}$)

$\eta_J$ = joint efficiency (–)

$C$ = corrosion allowance (m)

The longitudinal stress on the wall is given by Eq. (19.130):

$$\sigma_L = \frac{PD_I}{4t_L} \qquad (19.180)$$

where

$\sigma_L$ = longitudinal stress (N m$^{-2}$)

$t_L$ = thickness of the cylinder wall resisting the longitudinal stress (m)

The shell thickness for longitudinal stress is given by Eq. (19.131) with addition of allowances for joint efficiency and corrosion:

$$t_L = \frac{PD_I}{4\sigma_{ALL}\eta_J} + C \qquad (19.181)$$

The pressure at any level in the tank depends on the head of liquid above that point plus the pressure in the vapor space above the liquid $P_{VS}$:

$$P_H = (H_{max} - H)\rho_L g + P_{VS} \qquad (19.182)$$

where

$P_H$ = gauge pressure at height $H$ above the base of the tank (N m$^{-2}$)

$H_{max}$ = maximum height of liquid (up to the tank overflow) (m)

$H$ = height above the base (m)

$\rho_L$ = liquid density (kg m$^{-3}$)

$g$ = acceleration due to gravity (m s$^{-2}$)

$P_{VS}$ = maximum gauge pressure in the vapor space above the liquid (N m$^{-2}$)

Whilst the pressure at the base of the tank will be the highest, and normally limiting, because thinner courses are used at higher levels, these need to be checked to ensure that the thickness of the course is adequate at the bottom of each course where the pressure is maximum.

# 19.13 Shell-and-Tube Heat Exchangers

By far the most commonly used type of heat exchanger is the shell-and-tube design. An example of the design with a *fixed tube-sheet*, a single shell pass, and two tube passes is illustrated in Figure 19.68. The fixed tubesheet arrangement in Figure 19.68 might require an expansion joint as a result of the differential expansion between the shell and the tubes. Many other heat exchanger arrangements are possible (Smith 2016). The use of U-tubes, floating head, and pull-through floating head arrangements avoids the necessity for an expansion joint (Smith 2016). The thermal design of shell and tube heat exchangers is dealt with elsewhere (Thulukkanam 2013; Smith 2016). The mechanical design requires consideration of design of:

- shell thickness;
- shell expansion joint if thermal stresses caused by the differential expansion of the shell and tubes are excessive;
- shell flanges;
- heads for the ends;
- pipe nozzles;
- tubesheet thickness;
- tube thickness;
- baffles;
- design of external supports for either horizontal or vertical orientation.

The design codes for pressure vessels also cover heat exchangers (ASME 2019; BSI Standards Publications 2019). In addition, Escoe (1986) provides guidance, and The Tubular Exchanger Manufacturers Association provides rules for mechanical design (TEMA 2019).

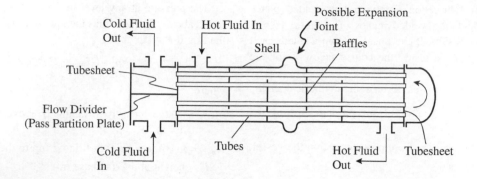

**Figure 19.68**

Example of a shell-and-tube heat exchanger arrangement. Source: (Smith 2016), *Chemical Process Design and Integration*, 2nd Edition, John Wiley & Sons Ltd.

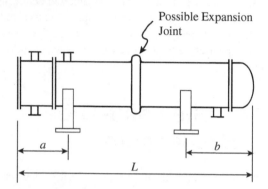

**Figure 19.69**

A horizontally mounted shell-and-tube heat exchanger supported by saddles.

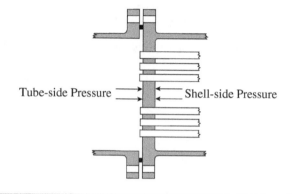

**Figure 19.70**

Shell-and-tube heat exchanger tubesheet.

The shell can be designed based on the principles of thin-walled pressure vessels discussed earlier. A horizontally mounted heat exchanger, as illustrated in Figure 19.69, will be subject to the same bending stresses as the pressure vessels discussed previously. The saddle supports for a horizontally mounted heat exchanger are likely to be asymmetric with the bending moment given by Eq. (19.88). The shell will normally be under internal pressure. However, there is a possibility that the shell could be under external pressure if, for example, the shell is performing vacuum condensation. The shell heads can be designed by the same methods as those described earlier for pressure vessels.

The tubes can be considered to be thin-walled pressure vessels. The tubes might be under internal or external pressure depending on the relative pressures inside the tubes and in the shell. The tubes are also subject to bending stresses, but the baffles provide intermediate support. Tubes at the edge of the bundle in the baffle windows will have the largest unsupported span. Flow-induced vibration of the tubes can lead to tube failure. Various mechanisms contribute to the tube vibration, but principally by vortex shedding resulting from cross-flow across the tubes on the shell-side (TEMA 2019).

A typical tube sheet arrangement is illustrated in Figure 19.70. The difference in pressure between the tube side and shell side creates a bending stress with a direction depending on the relative pressures of the tube side ($P_T$) and the shell side ($P_S$). For a mechanical design, this is a circular plate with perforations subject to a bending stress. Methods for the mechanical design can be found in the design codes (ASME 2019; BSI Standards Publications 2019; TEMA 2019).

## 19.14   Mechanical Design – Summary

Mechanical forces that can induce deformation of a solid body include:

- tensile;
- compressive;
- shearing;
- torsional;
- bending;
- buckling.

Also, differences in temperature can induce forces to create deformation and possibly failure. Normal stresses occur when opposing forces act in line through the body. Shear forces act when opposing forces are in parallel, but not in the same line.

Shear force and bending moment diagrams can be used to analyze the bending of beams. The bending force on a beam creates bending stresses that are a maximum at the maximum distance from the neutral axis. The forces change from tensile on one side of the beam to compressive on the other. The neutral axis is the longitudinal plane where the stresses change from tensile to compressive and vice versa. The bending force also creates shear stresses, which are a maximum at the neutral axis and zero at a maximum distance from the neutral axis.

Tall vertical self-supporting cylindrical vessels are subject to bending forces created by wind load. The bending stress is a maximum at the base. Such vessels are normally supported on a skirt, which is subject to the bending stresses and stresses created by the weight of the vessel and its contents. The design also needs to be checked for the effect of any seismic forces and fatigue. Horizontal cylindrical vessels are normally supported on two saddles. The weight of the vessel and contents creates bending and shear stresses that depend on the location of the saddles.

For ductile materials, failure normally occurs when the shear stress reaches a critical value. In two dimensions, this is given by the difference between the principal stresses.

Spheres and cylindrical pressure vessels can be analyzed using thin-walled pressure vessel theory if the ratio thickness of the vessel wall to the internal diameter is less than 1/20. Cylindrical vessels are subject to circumferential and longitudinal stresses from internal pressure, with the circumferential stresses being the greater. Enclosing heads for cylindrical vessels have a thickness for the same internal pressure in the order hemispherical < semi-ellipsoidal < torispherical < flat heads. Cylindrical vessels subject to external pressure can fail from buckling, by which the vessel wall does not fail due to the strength of the material being exceeded, but rather the vessel can no longer support the shape and collapses to take up a smaller volume.

The skirts for vertical pressure vessels are fixed to a base plate that is in turn fixed to concrete foundations using anchor bolts. The saddles for horizontal pressure vessels if located at grade will similarly be fixed to concrete foundations using anchor bolts.

Flat-bottomed cylindrical tanks are widely used for storing liquids and liquefied gases in the process industries. Stresses in the walls are created by the hydraulic head of liquid being stored, any pressure above the surface of the liquid, wind load, weight of the vessel and ancillary equipment, weight of snow and ice, and thermal stresses.

# Exercises

1. A simply supported beam with a length of 6 m supports a 10 kN load 2 m from the left support.

   a) Sketch the shear force and bending moment diagrams.
   b) Calculate the location and magnitude of the maximum shear force and bending moment.

2. A simply supported rectangular beam with a length of 5 m has a depth of 20 cm and a width of 10 cm. The beam carries a uniform load of 500 kg m$^{-1}$. Calculate the bending and shear stresses at the midpoint of the span at the top, bottom, and neutral axis of the beam.

3. A hollow rectangular beam with a cross-section illustrated in Figure 19.71 is to be used to support a pipe. The beam has a width of 4 cm, a depth of 8 cm, and a thickness of 0.5 cm. The beam can be considered to be simply supported with a span of 1 m with a point load at the middle of the span. The yield strength of the steel can be assumed to be 250 N mm$^{-2}$ with a safety factor of 1.67. Calculate the maximum load that can be supported by the maximum bending stress.

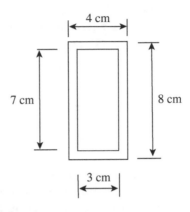

**Figure 19.71**

A hollow rectangular cross-section beam.

4. A solid cylindrical stainless steel shaft with a diameter of 8 cm is driving a process machine, requiring a power of 2 MW. The yield strength of the material is $215 \times 10^6$ Pa. Assume a safety factor of 2. What should be the maximum speed of the shaft?

5. Figure 19.72 shows three steel plates each with a thickness of 2 cm joined by 3 steel bolts. The three bolts are in double shear and need to resist a shearing force of 600 kN. The bolts are

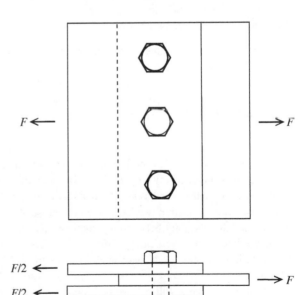

**Figure 19.72**

A joint between flat plates using three bolts.

assumed to have threads in the shearing plane. Assuming an allowable stress of 160 N mm$^{-2}$, determine the diameter of bolts required assuming the bolts to be in shearing only.

**6.** A flat-bottomed cylindrical tank with an open top and diameter of 5 m is used to store water with a density of 1000 kg m$^{-3}$. If the level in the tank is assumed to be 5 m, calculate at the base of the vertical wall:

   **a)** the axial stress;
   **b)** the hoop stress;
   **c)** the thickness of plate required for the stress from pressure (neglect wind load and deadweight).

Assume the material of construction to be carbon steel SA 516-70 below 40 °C.

**7.** A spherical storage tank with a diameter of 10 m contains a gas at 25 barg. Calculate the thickness required for the wall. Assume the material of construction to be carbon steel SA 516-70 below 40 °C.

**8.** A horizontal cylindrical tank with hemispherical heads contains gas with a pressure of 25 barg. If the diameter of the tank is 1 m, calculate the minimum thickness of the cylindrical section and the hemispherical heads to withstand the internal pressure. Assume the material of construction to be carbon steel SA 516-70 below 40 °C.

**9.** A process uses the design of the vessel illustrated in Figure 19.53 with semi-ellipsoidal heads. The internal diameter is 0.7 m and the vessel operates with a pressure of 60 barg. The material of construction has an allowable stress of $280 \times 10^6$ N m$^{-2}$. The lid of the vessel is attached using 36 bolts on a bolt circle diameter of 0.8 m. The allowable tensile stress of the bolts is $245 \times 10^6$ N m$^{-2}$.

   **a)** Calculate the minimum allowable wall thickness of the cylindrical shell and semi-ellipsoidal heads.
   **b)** Calculate a suitable bolt diameter.

**10.** A self-supporting distillation column has been sized to have a height of 26 m and an inside diameter of 1.5 m. The column is fitted with 36 trays each with an assumed weight of 1200 N m$^{-2}$ spaced 0.6 m apart. An additional 5 m height is added to the cylindrical shell height, 1.5 m above the top tray for liquid disengagement and 3.5 m below the bottom tray for liquid sump and liquid disengagement. The shell is to be fitted with 2:1 semi-ellipsoidal heads. The preliminary design will assume the column is mounted on a cylindrical skirt support with a height of 2.5 m and the same outside diameter as the column with a thickness of 10 mm welded to the column shell. It can be assumed that the joint between the shell and the skirt will be at the base of the vertical cylindrical shell. The column has a design pressure of 10 barg at the top of the column. The shell is constructed in a series of cylindrical courses. For a preliminary design it will be assumed initially that the shell has a uniform thickness of 10 mm across the total height. The vessel is to be fabricated from carbon steel SA 516-70 with a density of 7850 kg m$^{-3}$ with a corrosion allowance of 2 mm. The working temperature is below 250 °C. Assume that welds for the column undergo 100% radiographic examination and

$\eta_J = 1.0$. The distillation column is insulated for its entire area (not the skirt) with 75 mm of mineral wool covered in aluminum cladding. The density of the column insulation is 100 kg m$^{-3}$. The reboiler, condenser, and distillate receiver for the column are to be supported on a structure adjacent to the column. Assume that the pipework is effectively all supported by the adjacent structure. The column is fitted with three platforms, one at the bottom of the column above the skirt, one midway, and the third just below the top of the column. Each of the three platforms has an area of 8 m$^2$. Caged ladders with a total length of 26.5 m allow access to the three platforms. Meteorological records for the location indicate that a maximum wind speed of 50 m s$^{-1}$ can be assumed. This wind speed can be assumed to apply at all heights. In practice, the wind speed increases with height from zero at the ground. Air density can be assumed to be 1.3 kg m$^{-3}$.

   **a)** Calculate the bending moment at the base of the skirt, assuming an allowance for the influence platforms, ladders, and external pipes on the outside diameter.
   **b)** Calculate the minimum shell thickness from the circumferential stress caused by the internal pressure to ensure a reasonable shell thickness has been chosen.
   **c)** Calculate the thickness of the semi-ellipsoidal heads at the top and bottom of the shell.
   **d)** Calculate the circumferential and longitudinal stress caused by the internal pressure before corrosion.
   **e)** Calculate the bending moment and maximum bending stress at the base of the column before corrosion.
   **f)** Calculate the deadweight of the vessel before corrosion of the column above the joint to the cylindrical skirt.
   **g)** Combine the stresses from internal pressure, bending, and deadweight and calculate the difference between the principal stresses at the base of the column on the upwind and downwind sides before corrosion.
   **h)** Repeat the calculation after corrosion, assuming that the shell thickness is decreased by the corrosion allowance.

**11.** A horizontal cylindrical pressure vessel with 2:1 semi-ellipsoidal heads is required to hold a volume of 15 m$^3$ of liquid propane at a temperature of 25 °C, with a density of 495 kg m$^{-3}$. The process pressure has been set to 9.36 bara. Taking the design pressure to be 10% above the normal working pressure gives 9.2 barg as a reasonable basis for the design. The vessel is to be fabricated from carbon steel SA 516-70 with a density of 7850 kg m$^{-3}$. Assume the vessel has a length to diameter ratio of 3. Assume a welded joint efficiency of 0.85 for the cylindrical shell, but 1.0 for the heads. Assume there are two saddle supports located $0.4R$ from the tangent point (i.e., $a = 0.4R$ in Figure 19.41). It is assumed that the vessel acts effectively as a beam, as a result of the location of the saddles close to the heads, suitable design of saddle supports, and inclusion of stiffening rings. Carry out a preliminary mechanical design:

   **a)** Calculate suitable dimensions for the vessel.
   **b)** Calculate the minimum thickness for the shell and choose a reasonable thickness of plate for the cylindrical shell.
   **c)** Calculate the minimum thickness for the heads and choose a reasonable thickness of plate.

d) Calculate the circumferential and longitudinal stress caused by the internal pressure.

e) Calculate the deadweight of the vessel with contents.

f) Calculate the deadweight of the vessel heads with contents.

g) Calculate the bending moment and maximum bending normal stress at the top and bottom of the vessel at the longitudinal mid-point before corrosion.

h) Combine the stresses from internal pressure and bending and calculate the difference between the principal stresses at the top and bottom of the vessel at the longitudinal mid-point.

i) Calculate bending shear stress and the difference between the principal stresses above the supports midway between the top and bottom, corresponding with the neutral axis.

# References

API Standard 650 (2020). *Welded Tanks for Oil Storage*, 13e.

ASME (2017a). Boiler and Pressure Vessel Code Section II. Part D *Properties* (*Metric*).

ASME (2017b). Pipe Flanges and Flanged Fittings: NPS 1/2 through NPS 24 Metric/Inch Standard B16.5.

ASME Boiler and Pressure Vessel Code Section VIII Division 1 (2019). *Rules of Construction of Pressure Vessels*.

Bednar, H.H. (1991). *Pressure Vessel Design Handbook*, 2e. Malabar, Florida: Krieger Publishing Company.

British Standards Institution (1982). Stresses in horizontal cylindrical pressure vessels supported on twin saddles: a derivation of the basic equations and constants used in G.3.3 of BS 5500:1982. PD 6497.

Brownell, L.E. and Young, E.H. (1959). *Process Equipment Design*. Wiley.

BSI Standards Publications (2019). *Specification for Unfired Fusion Welded Pressure Vessels*, PD 5500:2018+A2:2019

Chattopadhyay, S. (2005). *Pressure Vessels Design and Practice*. CRC Press.

EN 1991-1-4: 2005+A1 (2010). Eurocode 1: Actions on Structures – Part 1–4: General Actions - Wind Actions.

Escoe, A.K. (1986). *Mechanical Design of Process Systems, Shell-and-Tube Heat Exchangers, Rotating Equipment,Bins, Silos, Stacks*, 2e, vol. 2. Gulf Publishing Company.

Escoe, A.K. (1994). *Mechanical Design of Process Systems, Piping and Pressure Vessels*, 2e, vol. 1. Gulf Publishing Company.

Gere, J.M. (2004). *Mechanics of Materials*, 6e. Thomson Learning Inc.

Hall, S. (2020). Rules of thumb 1. Tanks and vessels. *The Chemical Engineer* 952: 38.

Harris, L., Suer, H.S., Skene, W.T., and Benjamin, R.J. (1957). The stability of thin-walled unstiffened circular cylinders under axial compression including the effects of internal pressure. *Journal of Aeronautical Sciences* 24: 587–596.

Jawad, M.N. and Farr, J.R. (2019). *Structural Analysis and Design of Process Equipment*, 3e. Wiley.

von Kármán, T. and Tsien, H.S. (1939). The buckling of spherical shells by external pressure. *Journal of Aeronautical Sciences* 7: 43–50.

von Kármán, T. and Tsien, H.S. (1941). The buckling of thin cylindrical shells under axial compression. *Journal of Aeronautical Sciences* 8: 303–312.

Knudsen, J.G. and Katz, D.L. (1958). *Fluid Dynamics and Heat Transfer*. McGraw-Hill.

Ryder, G.H. (1969). *Strength of Materials*. Macmillan.

Smith, R. (2016). *Chemical Process Design and Integration*. Wiley.

TEMA (2019). *Standards of the Tubular Exchanger Manufacturers Association*, 10e.

Thulukkanam, K. (2013). *Heat Exchanger Design Handbook*, 2e. CRC Press, Taylor & Francis Group.

Timoshenko, S. (1956). *Strength of Materials, Part I, Elementary Theory and Problems*, 3e. D Van Nostrand Company.

Timoshenko, S.P. and Gere, J.M. (1961). *Theory of Elastic Stability*, 2e. New York: McGraw Hill.

Willems, N., Easley, J.T., and Rolfe, S.T. (1981). *Strength of Materials*. McGraw-Hill Book Company.

Zick, L.P. (1951). Stresses in large horizontal cylindrical pressure vessels on two saddle supports. *Welding Journal Research Supplement* 30: 556.

# Chapter 20

# Process Plant Layout – Site Layout

## 20.1 Site, Process, and Equipment Layout

Process plant layout can be considered at three levels:

1) *Site layout*. Site layout determines:

   a) Location of:
   - process plant areas;
   - raw materials delivery and product dispatch;
   - raw materials, product, and intermediate storage;
   - utilities plant (steam, cogeneration, cooling water, refrigeration, etc.);
   - effluent treatment plant;
   - workshops, stores, laboratories, offices, medical center, fire station.

   b) Flow of materials and utilities around the site:
   - materials from storage to processes and utilities, from processes to storage, and flow of materials between processes;
   - steam and other utilities (cooling water, electricity, nitrogen, compressed air, etc.).

   c) Access to the site for:
   - delivery of raw materials and dispatch of products;
   - operation, maintenance, and construction;
   - mobile cranes and maintenance vehicles;
   - emergency vehicles.

   d) Separation of:
   - surrounding community from hazards;
   - site personnel from hazards;
   - potential hazards from other hazards so that hazards can be isolated and not spread to a chain of events.

     Any hazard should be able to be contained at source. Particularly hazardous processes (e.g. containing toxic materials) should be isolated. Undesirable interactions between processes (e.g. sources of ignition close to plants containing flammables) must be avoided.

2) *Process layout*. Process layout considers:

   a) Flow of process materials within process areas:
   - from and to storage;
   - between operations.

   b) Flow of heat within process areas:
   - from and to the steam system;
   - from fired heaters;
   - to cooling systems (e.g., cooling water, air coolers);
   - heat exchange between operations directly or indirectly;

   c) Flow of utilities within process areas:
   - steam, cooling water, electricity, nitrogen, compressed air, etc.

   d) Access for:
   - safe evacuation of all staff in an emergency;
   - operations, maintenance, and construction staff;
   - maintenance operations (including heavy lifting equipment) and dismantling of equipment for maintenance;
   - removal of equipment from the process area for maintenance in workshops or off-site.

e) Elevation of equipment:

- use of structures to save plot space where the plot area is restricted;
- to allow gravity flow of liquids and solids where appropriate;
- to provide head for thermosiphon reboilers;
- to provide net positive suction head for pumps;
- to provide barometric head for liquids in vacuum operations;
- to provide pump priming.

3) *Equipment layout*. Equipment layout considers:

a) Mechanical construction and support:

- mechanical support of equipment, pipes, and valves associated with the equipment;
- foundations and anchoring of equipment.

b) Access for:

- safe operation of the equipment and ease of operation;
- maintenance operations (including heavy lifting equipment) and dismantling of equipment for maintenance;
- removal of equipment and components for maintenance in workshops or off-site.

c) Drainage, venting, and leaks:

- drainage and venting of process materials for operation and maintenance;
- safe disposed of drained or vented material;
- safe capture of liquid leaks via a solid floor to drain or drip tray to drain.

This chapter will concentrate on site layout and the following chapter on process layout. For equipment layout, see Mecklenburgh (1985) and Moran (2017).

# 20.2 Separation Distances

Safe mutual separation distances between process plant areas, storage, raw materials delivery, product dispatch, utilities plant (steam, cogeneration, cooling water, refrigeration, etc.), effluent treatment plant, workshops, stores, laboratories, offices, medical center, and, on larger sites, the fire station need to be established. Such safe separation distances are required for hazards as much as possible to be contained at source and to minimize danger to site personnel and the surrounding community. Separation distances are typically determined through either *spacing tables* based on

experience, or modeling of fire, explosion, and toxic release consequences. Spacing tables give preliminary guidance to the minimum distances between different site areas and are intended to limit fire exposure damage. Tables 20.1 and 20.2 present preliminary spacing distances (based on Global Asset Protection Services 2015; Moran 2017; Center for Chemical Process Safety 2018). For spacing tables, processes and operations can be qualitatively classified into three broad classes for preliminary analysis according to their fire and explosion hazards (Center for Chemical Process Safety 2018):

- *Moderate hazard.* This category includes processes, operations, or materials having a limited explosion hazard and a moderate fire hazard. This generally involves endothermic reactions and non-reactive operations, such as distillation, absorption, mixing, and blending of flammable liquids. Exothermic reactions involving no flammable liquids or gases also fit into this hazard group.
- *Intermediate hazard.* This category includes processes, operations, or materials having an appreciable explosion hazard and a moderate fire hazard. This class generally involves mildly exothermic reactions.
- *High hazard.* This category includes processes, operations, or materials having a high explosion hazard and moderate to heavy fire hazard. This class involves highly exothermic or potential runaway reactions and high hazard products handling.

Distances from spacing tables might need to be increased, or possibly decreased, based on risk analysis and implementation of layers of protection (from layer of protection analysis, LOPA). Explosion and toxic considerations might require distances to be increased significantly from those suggested in spacing tables. Highly reactive chemicals, such as metal alkyls or hydrazine, might also require greater spacing. If the required separation cannot be adhered to, then additional measures, such as fire proofing and blast proofing, may be necessary. If this cannot deliver the required hazard mitigation, then the basic process design might need to be changed.

It should be noted that the spacing guidelines are mostly applicable to open structures. Open-air designs favor vapor dispersion, provide ventilation, reduce the size of electrically classified areas (see later) and increase accessibility for firefighting. Operations within buildings will require more detailed analysis, accounting for building ventilation.

The spacing to a property boundary that is adjacent to a populated area is of greater importance than spacing to a property boundary adjacent to a minimally populated area. Toxic release hazards will also require further analysis. Distances may be reduced or increased based on risk analysis of specific conditions, when additional fire protection, safety measures, or layers of protection are considered.

## Table 20.1

Typical horizontal spacing for fire consequences (Global Asset Protection Services, 2015; Moran, 2017; Center for Chemical Process Safety, 2018).

| | Site boundary | Administrative and service buildings | Unloading and loading racks | Atmospheric storage tanks | Pressure storage tanks | Refrigerated storage tanks (dome roof) | Electrical substations | Utilities areas | Cooling towers | Control rooms | Fire station | Compressors | Pumps handling flammables | Fire water pumps | Process unit moderate hazard | Process unit intermediate hazard | Process unit high hazard | Fired heaters, thermal oxidisers | Flares |
|---|---|---|---|---|---|---|---|---|---|---|---|---|---|---|---|---|---|---|---|
| Site boundary | NM | 8 | 30 | 30 | 75 | 75 | 15 | 30 | 30 | 8 | 8 | 60 | 60 | 8 | 60 | 60 | 120 | 60 | 90 |
| Administrative and service buildings | | NM | 60 | 75 | 105 | 105 | 15 | 50 | 50 | NM | 15 | 30 | 30 | 15 | 30 | 60 | 120 | 75 | 90 |
| Unloading and loading racks | | | 15 | 75 | 105 | 105 | 60 | 60 | 60 | 60 | 60 | 60 | 60 | 60 | 60 | 60 | 90 | 75 | 90 |
| Atmospheric storage tanks | | | | * | * | * | 75 | 75 | 75 | 75 | 105 | 75 | 75 | 105 | 75 | 90 | 105 | 105 | 90 |
| Pressure storage tanks | | | | | * | * | 105 | 105 | 105 | 105 | 105 | 105 | 105 | 105 | 105 | 105 | 105 | 105 | 120 |
| Refrigerated storage tanks (dome roof) | | | | | | * | 105 | 105 | 105 | 105 | 105 | 105 | 105 | 105 | 105 | 105 | 105 | 105 | 120 |
| Electrical substations | | | | | | | NM | 15 | 15 | NM | 15 | 30 | 30 | 15 | 30 | 30 | 60 | 15 | 90 |
| Utilities areas | | | | | | | | NM | 30 | 30 | 15 | 30 | 30 | 15 | 30 | 30 | 60 | 30 | 90 |
| Cooling towers | | | | | | | | | 15 | 30 | 15 | 30 | 30 | 15 | 30 | 30 | 60 | 30 | 90 |
| Control rooms | | | | | | | | | | NM | 15 | 30 | 30 | 15 | 30 | 60 | 90 | 30 | 90 |
| Fire station | | | | | | | | | | | NM | 60 | 60 | NM | 60 | 90 | 90 | 30 | 90 |
| Compressors | | | | | | | | | | | | 9 | 9 | 60 | 9 | 15 | 30 | 15 | 90 |
| Pumps handling flammables | | | | | | | | | | | | | 9 | 60 | 9 | 15 | 30 | 15 | 90 |
| Fire water pumps | | | | | | | | | | | | | | NM | 60 | 90 | 90 | 60 | 90 |
| Process unit moderate hazard | | | | | | | | | | | | | | | 15 | 30 | 60 | 15 | 90 |
| Process unit intermediate hazard | | | | | | | | | | | | | | | | 30 | 60 | 15 | 90 |
| Process unit high hazard | | | | | | | | | | | | | | | | | 60 | 15 | 90 |
| Fired heaters, thermal oxidisers | | | | | | | | | | | | | | | | | | 7.5 | 90 |
| Flares | | | | | | | | | | | | | | | | | | | NM |

MN, No minimum distance specified.
* Spacing given in Storage Spacing Table.

## Table 20.2

Typical horizontal spacing for liquid storage tanks (Global Asset Protection Services, 2015, Moran, 2017, Center for Chemical Process Safety, 2018).

| Floating and cone roof tanks less than 480 m³ | Floating and cone roof tanks 480–1600 m³ | Floating roof tanks 1600–48 000 m³ | Cone roof tanks, Inerted Class I product 1600–48 000 m³ | Cone roof tanks, Class II and III product 1600–48 000 m³ | Floating and cone roof tanks greater than 48 000 m³ | Low pressure storage (up to 1 bar) less than 40 m³ | Low pressure storage (up to 1 bar) Greater than 40 m³ | High pressure storage Bullets and spheres | Refrigerated dome roof Storage tanks |
|---|---|---|---|---|---|---|---|---|---|
| 0.5 × D | | | | | | | | | |
| 0.5 × D | 0.5 × D | | | | | | | | |
| 1 × D | 1 × D | 1 × D | | | | | | | |
| 1 × D | 1 × D | 1 × D | 1 × D | | | | | | |
| 0.5 D | 0.5 D | 1 × D | 1 × D | 0.5 D | | | | | |
| 1 × D | 1 × D | 1 × D | 1 × D | 1 × D | 1 × D | | | | |
| 1 × D 15 m min | 1 × D 15 m min | 1 × D 15 m min | 1 × D 15 m min | 1 × D 15 m min | 1 × D 15 m min | 1 × D 15 m min | | | |
| 1.5 × D 30 m min | 1.5 × D 30 m min | 1.5 × D 30 m min | 1.5 × D 30 m min | 1.5 × D 30 m min | 2 × D | 1 × D 15 m min | 1 × D 15 m min | | |
| 2 × D 30 m min | 2 × D 30 m min | 2 × D 30 m min | 2 × D 30 m min | 2 × D 30 m min | 2 × D | 2 × D 30 m min | 2 × D 30 m min | 1 × D 30 m min | |
| 2 × D 50 m min | 2 × D 50 m min | 2 × D 50 m min | 2 × D 50 m min | 2 × D 50 m min | 2 × D | 2 × D 30 m min | 2 × D 30 m min | 1 × D 30 m min | 1 × D 30 m min |

D – Larger diameter of the two tanks (m)
Class I – Liquids with flash point below 38 °C
Class II – Liquids with flash point 38–60 °C
Class III – Liquids with flash point above 60 °C

**Example 20.1**  Using separation tables makes a preliminary assessment for the minimum distance for fire protection:

a) between two intermediate hazard processes;
b) between an intermediate hazard process and the site boundary;
c) between the control room and atmospheric storage tanks.
d) between two cone roof cylindrical storage tanks with a diameter of 10 m, each storing 750 m³ of a liquid with a flash point of 50 °C.

**Solution**

a) From Table 20.1 minimum spacing between two intermediate hazard processes is 30 m.
b) From Table 20.1, minimum spacing between an intermediate hazard process and the site boundary is 60 m.
c) From Table 20.1, minimum spacing between the control room and atmospheric storage tanks is 75 m.
d) From Table 20.2 minimum spacing between two cone roof cylindrical storage tanks, each storing 750 m³ of a liquid with a flash point of 50 °C is $0.5 \times D$. For a diameter of 10 m, spacing between the tanks is 5 m.

# 20.3  Separation for Vapor Cloud Explosions

For the layout it is also necessary to consider the possible consequences of explosions. An explosion occurs when a pressure (blast) wave referred to as an *overpressure* is generated in air by a rapid release of energy. This energy may have been stored in the form of chemical or pressure energy. The pressure wave generated can cause injury or death to humans, or damage to the process plant and buildings or damage to the process plant leading to further energy release and explosions via the *domino effect* or damage to the surrounding community. Generally, the magnitude of the pressure wave decreases as the distance from the explosion source increases.

Explosions that occur in the open air are *unconfined explosions*. An *unconfined vapor cloud explosion* is one of the most serious hazards in the process industries. Although a large toxic release may have a greater disaster potential, unconfined vapor explosions tend to occur more frequently (Mannan 2012). The problem of the explosion of an unconfined vapor cloud is not only that it is potentially very destructive, but also that it may occur some distance from the point of vapor release and may thus threaten a considerable area. An unconfined vapor cloud explosion results from the ignition of a flammable mixture of vapor or gas. Droplets of liquid carried from a release increase the mass of flammable material after vaporization in the air. For the mixture to be combustible, the flammable material must be at a concentration in air between the *lower flammability limit* and the *upper flammability limit*. Below the lower flammability limit, the mixture is too "lean" to burn. Above the upper flammability limit, it is too "rich" to burn. Concentrations between these limits constitute the *flammable range*. Combustion of a flammable mixture occurs if the composition of the mixture lies in the flammable range *and* if there is a source of ignition. Alternatively, the combustion of the mixture occurs without a source of ignition if the mixture is heated to its *autoignition temperature*, allowing spontaneous ignition without an external source of ignition. A release of the large cloud of vapor or gas required for an unconfined vapor cloud explosion would normally require a source of ignition to ignite.

An important distinction for the release of chemical energy from combustible material is whether it is characterized as a *deflagration* or a *detonation*. In a deflagration a flame front moves subsonically through the mixture and the overpressure caused by the combustion is small. By contrast, in a detonation the flame front accelerates to supersonic velocities, producing a shock wave with much higher overpressures. A deflagration can transition to a detonation by the combustion process, generating a hot expanding gas that forces the flame front to accelerate. The accelerated flame front generates gaseous combustion products more quickly, forcing the flame front to accelerate further until it becomes a shock wave capable of causing autoignition of the mixture ahead of the deflagration. In general, six conditions have a major influence on whether a vapor cloud explosion with damaging overpressure through deflagration can occur:

1) The released material must be flammable and at suitable conditions to form a vapor cloud. Some portion of the resulting cloud must mix with air such that concentrations are within the flammable range for the material.

2) The size of the vapor cloud must be above a minimum threshold. Mannan (2012) indicates that a release of typically 10 tons is required. However, detonations have been known to occur for smaller releases of perhaps 5 tons. The main point is that, the larger the release, the greater is the possibility of a detonation occurring.

3) Ignition of the flammable vapor cloud must be delayed until a cloud of sufficient size has formed and with at least part of the cloud within the flammable range.

4) An ignition source is needed to initiate the explosion in that part of the cloud that is in the flammable range. Alternatively, the flammable mixture may come into contact with a surface above its autoignition temperature. The location of the ignition source within the cloud (at the center or at the periphery) has an important influence on whether a detonation is likely to occur.

5) Turbulence is required for the flame front to accelerate to the speeds required for a detonation. This turbulence is typically formed by the interaction between the flame front and obstacles such as process structures or equipment. In the absence of turbulence, under laminar or near-laminar conditions, flame speeds are too low to produce a significant blast overpressure.

6) Confinement of the cloud by obstacles can result in rapid increases in pressure during combustion. Conversely, absence of confining obstacles allows unlimited outward expansion of the cloud during combustion, limiting the pressure increases. Confined clouds are more likely to lead to significant overpressures occurring. The degree of confinement in process plants, with their congested equipment layout and built-up structures, is generally high.

In most cases, the ignition of a vapor cloud release will lead to flame propagation by deflagration. However, a conservative approach to the analysis of the effects of a sizeable vapor cloud release should be to assume that a mixture that is capable of ignition is also capable of detonation.

To quantify the magnitude of overpressures, a simple approach is to model the explosion characteristics relating vapor cloud explosions to the explosion of an equivalent mass of trinitrotoluene (TNT). It should be noted that there are significant differences between the characteristics of TNT explosions and those of unconfined vapor clouds. However, this approach is used because of the amount of information available on the explosion of explosives. The peak overpressure of a vapor cloud explosion is related to the equivalent peak overpressure of TNT. The *TNT equivalent ratio* given in Eq. (20.1) refers to the mass ratio of the explosive in question that will produce an equal peak overpressure to the equivalent mass of TNT:

$$m_{TNT} = \eta_{VCE} m_{COMB} \frac{\Delta H_{COMB}}{\Delta H_{TNT}} \qquad (20.1)$$

where

$m_{TNT}$ = equivalent mass of TNT (kg)

$\eta_{VCE}$ = yield of the explosion (–)

$m_{COMB}$ = mass of flammable vapor (kg)

$\Delta H_{COMB}$ = heat of combustion of the flammable vapor (kJ kg$^{-1}$)

$\Delta H_{TNT}$ = energy released per kg for a TNT explosion (kJ kg$^{-1}$)

= 4650 kJ kg$^{-1}$ (Mannan 2012)

The yield of energy from a combustible cloud varies typically between a release of 0.01 (1%) and 0.1 (10%) of the total energy. Yields higher than 0.1 have been recorded, but most are below 0.1 (Gugan 1979; Mannan 2012). A reasonable worst-case assumption for the prediction of overpressure is that the yield is 0.1, although lower values are often assumed (Mannan 2012).

The overpressure caused by the explosion of $m$ kg of explosive at a distance $R$ meters from the center of the explosion can be assumed to obey the relationship (Mannan 2012):

$$\Delta P = \Delta P(\Gamma) \qquad (20.2)$$

where

$\Delta P$ = explosion overpressure (N m$^2$, kPa, bar)

$\Gamma$ = scaled range (m kg$^{-1/3}$)

The scaled range is defined as:

$$\Gamma = \frac{R}{m^{1/3}} \qquad (20.3)$$

where

$R$ = distance from the center of the explosion (m)

$m$ = mass of explosive (kg)

The relationship between the explosion overpressure and the scaled range has been determined by many observations and is shown in Figure 20.1. The effects of overpressure are summarized in Table 20.3.

As noted previously, the larger the release, the more likely it will give rise to a detonation. In addition to the quantity of the material, the condition of the material released is also important. Three broad categories for the condition of the released material can be identified:

1) *Gas.* Release of gas from the process can lead directly to a flammable cloud. However, if the release is through, for example, a hole in a vessel or pipe, then the rate of release is limited by the maximum velocity of the gas through the hole. Liquids might flow through the same size of hole at a lower velocity but be able to achieve a greater mass flowrate.

2) *Liquid below its atmospheric boiling point.* Liquid released below its atmospheric boiling point will form a pool of liquid from which evaporation to the atmosphere can form a flammable cloud. The evaporation will depend both on atmospheric conditions and the physical properties of the liquid. However, this is a relatively slow process and would be unlikely to create a large flammable cloud.

3) *Liquid above its atmospheric boiling point.* Escape of a liquid above its atmospheric boiling point from the process is in many cases the most hazardous situation. This is because liquid can escape through a hole in a vessel or pipe at a higher mass flowrate than a gas and then, if above its atmospheric boiling point, turn into vapor as soon as it is released, thereby creating the potential for the formation of a large vapor cloud. The liquid escaping above its atmospheric boiling point might have a temperature above or below atmospheric temperature. In both

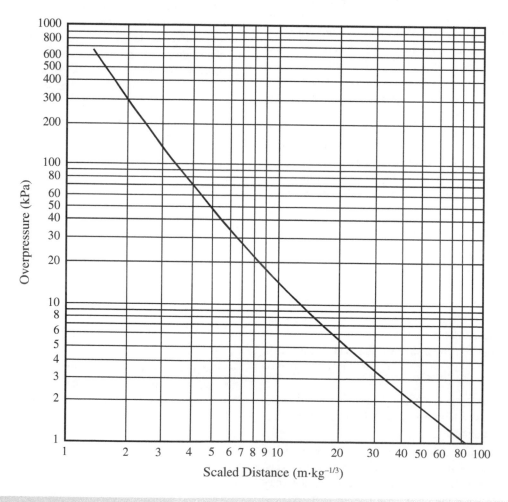

**Figure 20.1**

Overpressure from vapor cloud explosions.

## Table 20.3

The effects of explosion overpressure (Gugan 1979; Khan and Abbasi 2001; Mannan 2012).

| Damage effect | Approximate overpressure (kPa) |
|---|---|
| Human LD$_{50}$[a] | 1000 |
| Lung damage | 400 |
| Floating roof steel tank 99% damage | 138 |
| Spherical steel tank 99% damage | 110 |
| Ear damage | 100 |
| Vertical cylindrical steel pressure vessel 99% damage | 97 |
| Vertical cylindrical steel pressure vessel 20% damage | 83 |
| Severe damage to chemical plant | 70 |
| Total building destruction | 69 |
| Spherical steel tank 20% damage | 55 |
| Loaded rail wagons overturned | 48 |
| Oil storage tanks rupture | 27 |
| Floating roof steel tank 20% damage | 24 |

[a]Lethal dose to 50% of human beings

cases the liquid will flash as it is released, creating vapor. However, the behavior of the unvaporized liquid will be different in both cases. If the liquid is above ambient temperature, it will form a liquid pool from which there will be evaporation that depends both on the physical properties of the liquid and atmospheric conditions. If the liquid is below ambient temperature, then there will be rapid boiling on contact of the liquid with its relatively hot surroundings. As the temperature of the pool rises this will be followed by evaporation from the pool at a lower rate, again depending both on the physical properties of the liquid and atmospheric conditions.

The fraction of liquid vaporized on release is calculated from a heat balance (Crowl and Louvar 1990). The sensible heat above saturated conditions at atmospheric pressure provides the heat of vaporization. The excess heat in the superheated liquid is given by:

$$m\, C_P\, (T_{SUP} - T_{BPT}) \qquad (20.4)$$

where

$m$ = mass of liquid (kg)
$C_P$ = liquid heat capacity (kJ kg$^{-1}$ K$^{-1}$)
$T_{SUP}$ = temperature of the superheated liquid (°C, K)
$T_{BPT}$ = normal boiling point (°C, K)

If the mass of liquid vaporized is $m_V$, then:

$$m_V = \frac{mC_P(T_{SUP} - T_{BPT})}{\Delta H_{VAP}} \qquad (20.5)$$

where

$m_V$ = mass of liquid vaporized (kg)
$\Delta H_{VAP}$ = latent heat of vaporization (kJ kg−1)

Thus, the vapor fraction (VF) is given by:

$$VF = \frac{m_V}{m} = \frac{C_p(T_{SUP} - T_{BPT})}{\Delta H_{VAP}} \qquad (20.6)$$

Equation (20.6) allows the fraction of liquid escaping that flashes to vapor to be calculated. However, in addition to the flashing of the liquid to vapor, an aerosol of fine droplets will also be formed that will subsequently vaporize. This is very difficult to quantify. It is often assumed that the mass of aerosol formed is equal to the mass of vapor formed.

**Example 20.2**  A catastrophic failure of a vessel containing liquid propane at a pressure of 7.4 barg and temperature of 20 °C results in the release of 100 tons of propane to the atmosphere. The normal boiling point of propane is −42 °C, latent heat of vaporization is 427.8 kJ kg$^{-1}$, liquid specific heat capacity is 2.688 kJ kg$^{-1}$ K$^{-1}$ and heat of combustion 46 357 kJ kg$^{-1}$. The energy released per kg for a TNT explosion can be assumed to be 4650 kJ kg$^{-1}$.

a) Calculate the mass of propane in the resulting vapor cloud, assuming the mass of aerosol formed is equal to the mass of the flash vapor.
b) Assuming the vapor cloud explodes and detonates with a yield of the explosion of 0.1, calculate the distance from the explosion at which the following damage effects occur.

| Damage effect | Overpressure (kPa) |
|---|---|
| Vertical cylindrical steel pressure vessel 20% damage | 83 |
| Total building destruction | 69 |
| Spherical steel tank 20% damage | 55 |

## Solution

a) If the release occurs at a temperature above the normal boiling point, then flashing will occur. From Eq. (20.5):

$$VF = \frac{C_P(T_{SUP} - T_{BPT})}{\Delta H_{VAP}}$$

$$= \frac{2.688(20 - (-42))}{427.8}$$

$$= 0.390$$

Assuming an equal mass of aerosol is created as the flash vapor, mass of combustible material in a vapor cloud:

$$m_{COMB} = 0.390 \times 100 \times 2$$

$$= 78.0\,\text{t}$$

**b)** From Eq. (20.1):

$$m_{TNT} = \eta_{VCE} m_{COMB} \frac{\Delta H_{COMB}}{\Delta H_{TNT}}$$

$$= 0.1 \times 78.0 \times 10^3 \times \frac{46\,357}{4650}$$

$$= 77\,760\,\text{kg}$$

For an overpressure of 83 kPa, from Figure 20.1:

$$\Gamma = 3.7 = \frac{R}{m^{1/3}} = \frac{R}{(77,760)^{1/3}}$$

Thus

$$R = 158\,\text{m}$$

Repeating for the other damage cases:

| Damage effect | Overpressure (kPa) | $\Gamma$ | R (m) |
|---|---|---|---|
| Vertical cylindrical steel pressure vessel 20% damage | 83 | 3.7 | 158 |
| Total building destruction | 69 | 4 | 171 |
| Spherical steel tank 20% damage | 55 | 4.7 | 201 |

It should be noted that the distances calculated are sensitive to the assumptions that have been made. For example, the explosion yield has been assumed to be 0.1. Although on a few occasions this has been exceeded, a lower figure of 0.03 or 0.02 is often assumed. It has been assumed that the quantity of aerosol formed is the same as the quantity of flash vapor, which may be pessimistic. Also, it is always going to be difficult to estimate the actual quantity of combustible material released. Finally, it should be noted that the center of an explosion might be a significant distance from the point of release.

So far, the peak overpressure from an unconfined vapor explosion has been considered from a single source of leakage. When planning the consequences of unconfined vapor cloud explosions on layout, if there are multiple sources of potential release, each source can be analyzed to calculate the distance to a specified overpressure and the overpressure circles superimposed. The resulting *contour* or *petal* diagram shows areas affected by possible vapor cloud explosions for a specified overpressure. This is a series of intersecting arcs taken from each leak site, as illustrated in Figure 20.2. However, it should be noted that the center of any explosion will not necessarily be the point of release. The location of the vapor cloud depends on the mechanism of release and whether the release forms a high velocity jet. A vapor cloud can also drift some distance from the point of release before ignition. The ignition can in principle be anywhere in the vapor cloud where a flammable mixture comes into contact with a source of ignition, or a source of autoignition. For example, the center of the explosion of the unconfined vapor cloud explosion at Flixborough in 1974 has been estimated to be some 30 m from the point of release and not in the direction of

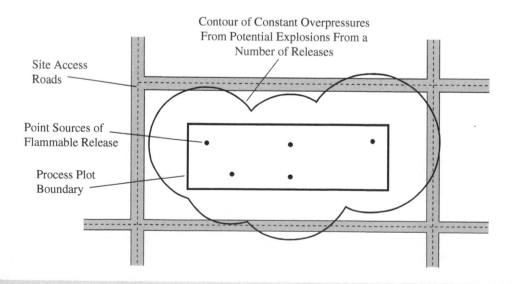

**Figure 20.2**

Contour of constant overpressures from a number of releases.

the wind because of the high-velocity release of boiling flammable liquid.

For a preliminary assessment of overpressure contours from vapor cloud explosions (Wells 1980):

- Equipment and buildings within a 0.2 bar overpressure contour should be designed to withstand the peak pressure at its location.
- All equipment and buildings in another area should be outside the 0.2 bar overpressure contour to prevent the domino effect.
- All equipment and buildings outside of a 0.2 bar overpressure contour but within a 0.07 bar overpressure contour should be designed to withstand a 0.14 bar overpressure.
- No large building or public roads should be allowed within the 0.07 bar overpressure contour.

The method used here to estimate the size of the overpressure circles from an assumed TNT equivalence explosion is simplistic and neglects a considerable amount of detail that could be included. Vapor cloud explosions have a shorter duration than TNT explosions. More elaborate procedures that take account of much more detail are available (Mannan 2012). Many factors will influence whether a release of combustible material will explode and detonate:

- quantity of material released needs to be large;
- mechanism of release (e.g., leak from a pipe flange, hole in a vessel of a certain size, catastrophic failure of the vessel);
- pressure of the source of release;
- mixing of the release with the atmosphere;
- how much of the vapor cloud is in the flammable range;
- atmospheric conditions (e.g., wind speed);
- release of the combustible cloud finding a source of ignition or source of autoignition temperature;
- location of the source of ignition or autoignition within the vapor cloud (e.g., at the center or at the periphery of the cloud);
- drift of the vapor cloud to a source of ignition;
- confinement of the cloud by obstacles and the local terrain.

Thus, many factors will influence whether a release will explode and detonate. Choosing a worst-case scenario is the safest option, but might lead to a very expensive design to mitigate the effect of the resulting overpressures. It should be emphasized again that the separation distances in Tables 20.2 and 20.3 do not necessarily represent safe distances, but only make a preliminary allowance for the effect of fire. Explosions and release of toxic material might require safe distances to be increased significantly. This will cause particular problems on sites that are restricted for space or are close to a surrounding community. If distances cannot be increased to provide adequate safety, then measures introduced from layer of protection analysis (LOPA) might be required. These could, for example, include additional safety instrumented systems or blast walls. Alternatively, some features of the process might need to be redesigned to reduce the inventory of hazardous material in the process or storage or to change the conditions under which it is processed or stored.

In the worst case, the process might need to be redesigned completely to eliminate or mitigate the hazard.

## 20.4  Separation for Toxic Emissions

As stated above, toxic emissions to atmosphere are generally regarded to have the greatest disaster potential (Mannan 2012), as well as creating potentially serious environmental consequences. Evaluation of the consequences of a toxic release to atmosphere requires dispersion modeling of toxic gas release, estimation of the toxic gas concentration spatially and temporally, and quantification of the effects of the toxic gas concentration on people. The consequences of such a release might go well beyond the boundary of the site. Different modeling techniques can be applied to predict the dispersion and concentration. In particular, computational fluid dynamics (CFD) is capable of predicting dispersion and concentration in a full three-dimensional analysis accounting for geometry, both outdoors and indoors. To be realistic, models should not be restricted to people being static, but also include the escape patterns in an emergency. The extent to which such modeling studies are justified depends on how hazardous are the materials being processed. The results of such studies can have a major influence on the layout of the site and might prompt changes in the process design to mitigate the hazard at source.

## 20.5  Site Access

Figure 20.3 shows an example of a site onto which four process plots need to be located. The site in this example is bounded on three of its sides by a river, a railway line, and a public road. The boundary of the site must be secure and surrounded by a fence, but with access managed for safety. A preliminary assessment in this example suggests that the raw materials can be delivered to the site by a combination of road and rail transport and products can be dispatched by barge to the river. Other major considerations are pipeline connections from outside the site. For example, these might be pipelines for import of raw materials from another site, or chemical products dispatched by pipeline, or natural gas supply. Figure 20.4 illustrates possible access for delivery of raw materials, dispatch of products and access within the site. A number of points need to be noted for access (Mecklenburgh 1985; Moran 2017):

- The road system within the site must not allow any area to be cut off in an emergency. The site should have a peripheral road with access to the public road system *at least at two points*. Emergency and rescue vehicles should be able to approach within a reasonable distance of any risk *from at least two directions*.
- Raw materials delivery facilities should be separate and close to the site entry point.

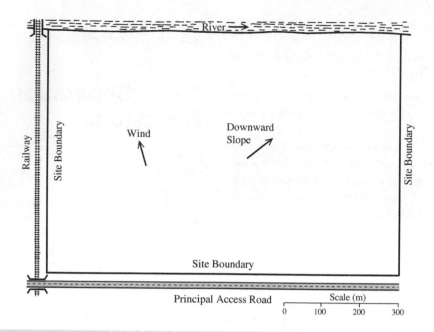

**Figure 20.3**

Pipe racks are required to connect the various operations on the site.

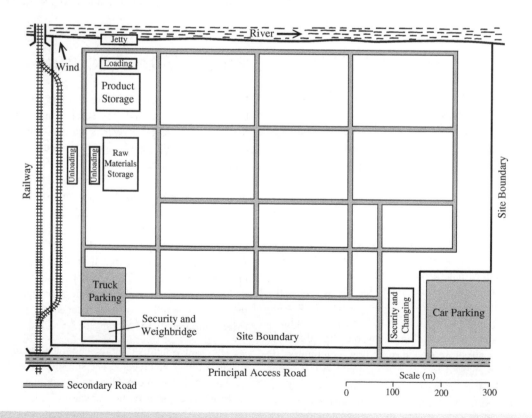

**Figure 20.4**

Site access, raw material delivery, and product dispatch.

- Product dispatch facilities should be separate and close to the site exit point.
- Storage areas are often close to raw materials delivery and product dispatch points.
- A good site layout minimizes the distances materials need to flow to and from storage and during processing.
- An access way should be at least 6 m wide. Access to different areas within the site should be available at least every 60 m.

Figure 20.4 shows a railway siding for the deliveries by rail and access from the public roads intended for delivery of materials by road. The deliveries must be guarded by security staff and for road deliveries require a weighbridge to measure the size of the deliveries. The other road access also has security facilities and includes a changing facility for the process operators and maintenance staff. Parking space has been allocated for both road trucks and cars.

Bulk liquid storage should normally be located outdoor in purpose-built tank farms and, where possible, at close to atmospheric pressure. Storage tanks are typically arranged in groups containing materials of similar flammability characteristics. Figure 20.5 illustrates a typical storage area. A retaining wall or *bund* is normally constructed around liquid storage tanks where flammable, corrosive, or toxic liquids are held to contain any liquid lost from containment. Alternatively, an *earth dike* might be used to surround the storage to contain any liquid lost from containment in the tanks. It is normal to limit the number of tanks in a single bund to 60 000 m$^3$ total capacity. Incompatible materials should have a separate bund. Bunds should be sized to hold 110% of the maximum capacity of the largest tank. Bund wall height is often 1–1.5 m to facilitate access for firefighting. Bunds are generally constructed from concrete, but bricks and mortar can be used, and corrosive-resistant cladding can be applied for corrosive materials.

Table 20.4 shows recommended distances for on-site access (Mecklenburgh 1985; Moran 2017).

**Table 20.4**

Recommended distances for on-site access (Mecklenburgh 1985; Moran 2017).

| On-site Access | | |
|---|---|---|
| **Item** | **Description** | **Clearance (m)** |
| Roads | 1) Headroom for primary access roads or major maintenance vehicles | 6.0 |
| | 2) Width of primary access roads | 6.0 |
| | 3) Headroom for secondary roads and pump access roads | 3.0–4.5 |
| | 4) Width of secondary roads and pump access roads | 3.0–4.5 |
| Railways | 5) Headroom over through railways from top of rail | 6.7 |

# 20.6 Site Topology, Groundwater, and Drainage

Figure 20.3 shows the general slope of the terrain on the site. Significant changes in elevation on a site can create challenges for layout. Gravity will cause liquids and heavy vapors to flow downhill to lower elevations on the site, possibly spreading the flow of flammable material to ignition sources, or toxic material to personnel. Extra pumping and more complex interconnecting pipework arrangements are other problems. Whenever practical, fired heaters and flares should be located at higher elevations such that any release of flammable liquid or heavy vapor from elsewhere on the site flows away from open flames. On the positive side, elevation differences can be used to facilitate gravity flow of materials and reduce energy costs.

Liquid effluent treatment should preferably be at a lower elevation on the site to allow gravity flow of liquid waste from the point of generation to treatment, as illustrated in Figure 20.6. It is also preferable to locate the effluent treatment close to the discharge point. The location is next to the river in the example in Figure 20.6. The effluent will need to be treated to the regulatory specifications before discharge is permitted. Drainage system requirements and water treatment systems will depend on rainwater volume and patterns, natural streams within the site boundary, wastewater generation from the process operations, and firewater usage. Protective dikes and spillways may be necessary depending on groundwater levels and potential flooding. Rainfall on the site will typically require water treatment for discharge, unless it can be safely segregated from contamination on the site for discharge. Firewater run-off will normally need to be collected and routed to treatment before discharge. In general, it is better to segregate effluent streams for treatment and discharge (Smith 2016). For process effluent, not only should the effluent be segregated, but hazardous effluent needs to be transported to material recovery or treatment in closed drains. Less hazardous effluent can be segregated and flow in drains open

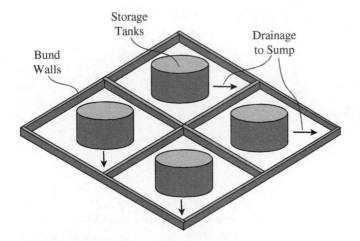

**Figure 20.5**

Liquid storage tanks are normally located in tank farms out-of-doors and protected by bunds.

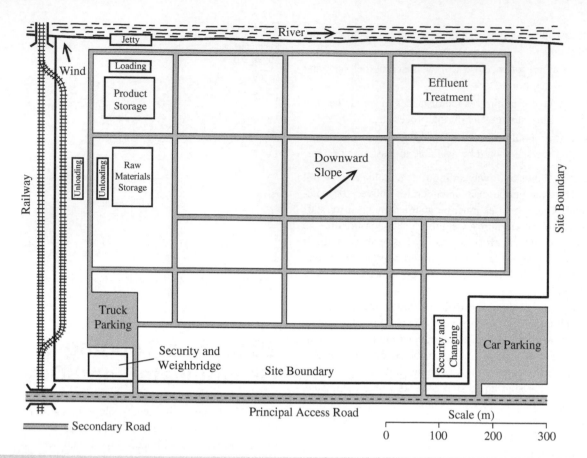

**Figure 20.6**

Location of effluent treatment and drainage is directed by the site topology.

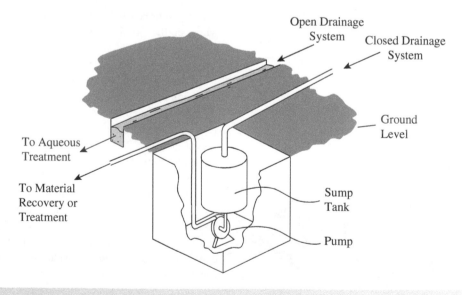

**Figure 20.7**

Liquid effluent management with open drains and closed drains to sump tanks. Source: Smith R (2016), *Chemical Process Design and Integration*, John Wiley & Sons Ltd.

to the atmosphere. Open drains in process and storage areas will normally be covered by an open metal grid for safety reasons. An example of a segregated process effluent system is illustrated in Figure 20.7. The less hazardous liquid flows under gravity in an open drain. In this example, the hazardous effluent is collected from a closed drainage system in a sump tank and then pumped away. Another example of a closed drainage system is shown in Figure 20.8 using only gravity flow.

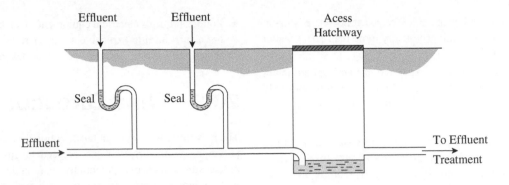

**Figure 20.8**

Example of a closed liquid drain flowing under gravity.

# 20.7 Geotechnical Engineering

Geotechnical engineering considers soil mechanics and rock mechanics to identify necessary earthworks and structure foundations. Foundations are needed to support structures by transferring the loads from the structure to the soil or rock. The layer at which the foundation transfers the load needs to have an adequate load-bearing capacity. Shallow foundations can be used when the load-bearing capacity of the surface soil is adequate to carry the loads imposed by the structure. Deep foundations are used when the bearing capacity of the surface soil is not sufficient to carry the loads imposed by the structure. If this is the case, the loads need to be transferred to a deeper level where the soil layer has a higher load-bearing capacity. Transfer of the load to a deeper level is achieved using *piles*. Piles are long, slender columns of steel or reinforced concrete and can be classified according to the mechanism of load transfer from the pile to the soil. End bearing piles have the bottom end of the pile resting on a strong layer of soil or rock. In this way the pile transfers the load to a strong layer. Friction piles transfer the load from the structure to the soil by the frictional force between the surface of the pile and the soil surrounding the pile. Friction might be developed for the entire length of the pile or only part of the length of the pile. Good ground load-bearing characteristics are required for storage areas and large structures, particularly multilevel structures.

# 20.8 Atmospheric Discharges

Gaseous discharges from process plant vents, whether under normal operation or emergency release, must be disposed of so as not to present a hazard on the site or to the surrounding community and to not create a nuisance to the surrounding community. Combustible waste from a vent can be combusted in a *thermal oxidizer*, *steam boiler*, or *flare*. As discussed in

Chapter 6, flares should only be used for abnormal operation or emergency upsets when there is a short-term requirement to deal with an abnormal discharge of combustible gases. Location of elevated flares is dictated by the flare height, load, and radiant heat permitted for neighboring equipment, storage, buildings, and personnel. Flares should not be used for discharges containing sulfur and halogenated compounds. For normal operation, other methods should be used, such as combustion in thermal oxidizers or steam boilers. Combustion of steady discharges of combustible material should allow heat recovery where possible to derive a benefit from the discharge (both economic and environmental). This could be, for example, a steam boiler or a thermal oxidizer equipped with heat recovery facilities. If the discharge cannot be combusted safely, then the discharge might be *absorbed* in a suitable solvent, or *adsorbed* on to a suitable adsorbent. It might be possible to vent directly to atmosphere. Whether the release is direct to atmosphere or after treatment, the release must be at a height that under *all* atmospheric conditions does not violate discharge regulations or become a public hazard or nuisance.

Small releases of flammable material can occur directly to atmosphere, even during normal operation that can possibly lead to the formation of an explosive gas mixture. Note that these releases are not in this case from the catastrophic failure of equipment (e.g., rupture of a pressure vessel), but from small releases that can happen during normal operation (e.g., releases from process equipment seals). The resulting hazardous areas can be classified with the objective of decreasing to an acceptable minimum probability of a flammable atmosphere occurring where there is a source of ignition. Sources of ignition might be electrical equipment, fired equipment, or hot surfaces above the autoignition temperature of the flammable mixture. This affects the selection and location of equipment (e.g., relocation of an electrical substation to avoid location in a hazardous area).

Hazardous areas are classified into zones based on an assessment of the frequency of the occurrence and duration of an explosive gas atmosphere:

- Zone 0: an area in which an explosive gas atmosphere is present continuously or for long periods (typically more than 1000 h y$^{-1}$).

- Zone 1: an area in which an explosive gas atmosphere is likely to occur in normal operation (typically more than $10 \, h \, y^{-1}$, but less than $1000 \, h \, y^{-1}$).
- Zone 2: an area in which an explosive gas atmosphere is not likely to occur in normal operation and, if it occurs, will only exist for a short time (typically less than $10 \, h \, y^{-1}$).

Rather than explosions from flammable gases, dusty atmospheres can be susceptible to *dust explosions*. In this case, zoning for a gas atmosphere is replaced by zoning for dust/air mixtures. The zone numbers used for dust/air mixtures are 20, 21, and 22, corresponding with 0, 1, and 2 used for gases.

The classification of hazardous areas requires the selection of electrical equipment that is appropriate for use in such hazardous areas:

- Zone 0: requires intrinsic safety – safe limiting of current, voltage, and the resulting power such that no sparks can be created capable of being a source of ignition.
- Zone 1: requires electrical equipment to be flameproof, pressurized, powder filled, oil immersed, or other design.

- Zone 2: requires electrical equipment to not be capable of igniting a surrounding explosive atmosphere through hermetically sealed, non-sparking, sealed, or other designs.

## 20.9 Wind Direction

The prevailing wind is from the direction that is predominant. The prevailing wind direction can be used to guide the location of process units such that potentially flammable and toxic releases will be carried away from ignition sources, personnel, or populated areas close to the site. However, it must be emphasized that the predominant wind direction is not the *only* wind direction. Thus, considerations based on the predominant wind direction will not always apply.

Fired heaters and flares as much is practicable should be located upwind of the prevailing wind from potential vapor releases to minimize the potential for fires and explosions. Storage tanks should be as much as practicable located downwind of potential ignition sources to minimize the risk of ignition of flammable vapor releases from storage. Figure 20.9 shows an example with the direction of the prevailing wind for the example. Storage tanks are downwind of ignition sources.

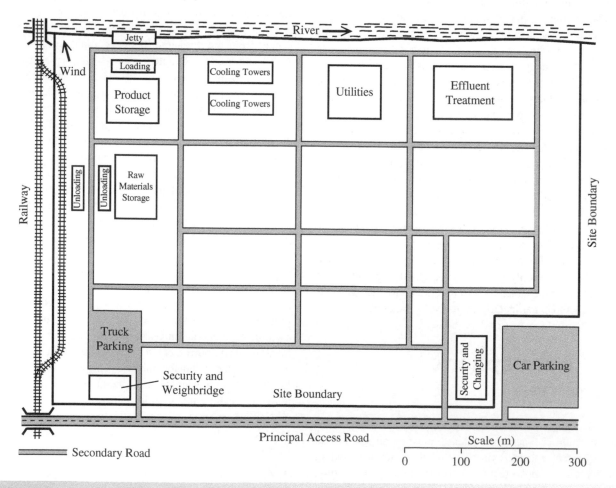

**Figure 20.9**

Adding process utilities.

# 20.10    Utilities

The site will normally require a steam system, which needs a boiler house (power station) for steam generation and possibly power generation from steam turbines or gas turbines. Boiler feed-water treatment facilities are also required for steam generation (see Chapter 4).

Cooling towers will most often be required (see Chapter 5). Care is needed for the location of cooling towers. Moisture-laden air formed from the evaporation of cooling water causes fog and mist to form around cooling towers. Thus, cooling towers need to be located such that mist and fog formed do not create a hazard on the site or in the surrounding areas (especially roads and residential areas). Induced draft cooling towers should if possible be perpendicular to the prevailing wind in order to maximize the intake of fresh air in hot weather. Although cooling towers are inherently wet, the wooden structures often used can burn. Wooden cooling towers can be easily damaged in an explosion, with the subsequent loss of site cooling capacity. Rather than using cooling water cooling, part, or all, of the site cooling might be provided by air coolers (see Chapter 5). These can in principle be distributed around the site more local to where the cooling is required.

In addition to the steam, power generation, and cooling towers, there will be requirements for inert gas (nitrogen), instrument air compression, electricity substations, etc.

Figure 20.9 shows a preliminary location for the utilities and cooling towers for the example.

# 20.11    Process Units

An early decision that needs to be made regarding a process unit is whether the unit will be housed in a building or be outdoors. Most process units are outdoors. However, if it is necessary to guard against the possibility of product contamination or to provide sterile conditions, for example food, pharmaceutical, and biochemical process plants, it might be necessary to house the unit in a building. Labor-intensive batch processes tend also to be housed in buildings. However, housing the unit in a building will be more expensive than outdoors and create more constraints on layout than if located outdoors. Ventilation of buildings also becomes a problem.

The size of plot required can be estimated from a similar existing plant. If no such data are available, then a preliminary process layout needs to be done to understand the area it requires and where best to locate it on the site. Process units are usually grouped together away from low hazard areas in order to minimize fire and explosion exposure to the site. On an existing mature site the location and area available for a new plant might be very constrained. Process equipment is normally located at grade level unless:

- plot space is limited, or
- height differences between process operations cannot be avoided, or
- equipment proximity is necessary (e.g., heat recovery).

Building a structure and installing equipment above grade reduces the area of the plot. Where equipment is located at an elevated level, drip trays and drainage must be provided to contain and drain releases so as not to pool under equipment or spill down to lower levels.

As stated previously, fired heaters and flares should as much as possible be located upwind of any potential release of flammable material. Locating fired heaters outside the process boundary can allow good access and the necessary separation distances.

Figure 20.10 shows a preliminary location for the four process plots for the example.

# 20.12    Control Room

For large process plants, control rooms are normally in separate buildings situated away from the process plant which they operate. For small plants, where the process materials are not at all hazardous, control rooms may be within the same building as the process plant. Control rooms should be designed as much as possible to ensure that risks to the occupants are within acceptable limits. The control room should also be able to maintain control of the process plant in the event of a serious safety incident within the plant, should the emergency response plan require it. The control room buildings should normally be at least 60 m from the process plant.

The construction of the control room should allow safe occupation and plant control to be maintained in the event of a toxic release. If exposure to a toxic release is possible, the building would need to have an air seal to prevent the ingress of toxic gases and vapors. The control room ventilation should ensure that the air intake is situated away from affected areas, or air intake is closed using an automatic valve operated from a gas analyzer during an incident, or by drawing the air intake through a suitable filtration system to remove toxic material down to a safe level of exposure.

The construction of the control building should feature measures for protection from fires. The building should be able to withstand the thermal radiation effects from a fire. Materials of construction should be fire resistant for the duration of any possible fire incident. Also, in the event of a fire, smoke ingress should be controlled in a similar way to the ingress of toxic gas and vapor.

The structural design of the control room building should be capable of withstanding, without collapse or serious damage, any plausible predicted overpressure from an explosion.

Figure 20.11 shows a possible location of the control room for the example plant layout. This maintains a safe distance from the process plant, storage areas, and utilities. In Figure 20.11 the location is central to the site, allowing easy access for the process operators to the process plant, delivery, and dispatch facilities, storage and utilities. There is a compromise between maintaining safe distance for the control room and easy access from the control room for process operators to carry out regular inspections of the process plant and possible rapid access necessary in the event of an emergency. There is also a compromise between the structural design of the control room and its location relative to hazardous areas. The closer to hazardous areas, the more robust the construction must be and the greater the necessity of controlling the ingress of toxic materials and smoke from a fire to the control room.

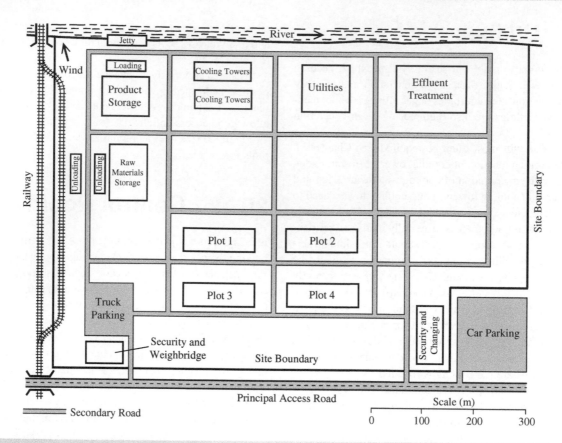

**Figure 20.10**

Adding process plots.

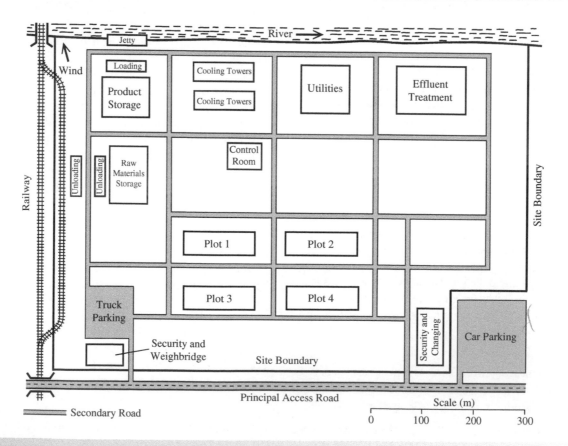

**Figure 20.11**

Location of the control room requires a safe distance from the process plant and storage.

## 20.13   Ancillary Buildings

In addition to the process plant, the site will include ancillary buildings that need to be located:

- maintenance workshops;
- stores for maintenance and operating supplies;
- laboratories for process control;
- offices for general administration;
- fire station on large sites;
- canteens, changing facilities, medical centers, etc.

Offices for general administration might be occupied by large numbers of personnel. Offices should be in a safe area upwind of potential releases. Location of administration offices, canteens, changing facilities, and medical centers should be close to the site entrance. Figure 20.12 shows a preliminary location for the ancillary buildings for the example site.

## 20.14   Pipe Racks

Pipes connecting process units to storage, connecting one unit to another and utility pipes are generally cheaper to run at grade on *sleepers*. Sleepers are ground-level concrete or steel supports for pipes. However, it is often not possible to run pipes on sleepers around the site as this can create severe restrictions for safe access. Sleepers are mainly restricted for use outside of battery limits in areas where the resulting access restrictions are acceptable. Instead, pipes are usually elevated in structures referred to as *pipe racks* to support elevated pipes. Figure 20.13 illustrates a pipe rack support. The pipes are supported on the beams in Figure 20.13. In addition to supporting pipes, pipe racks are also used to run electrical cables and instrument lines. Multiple tiers of 1 to 4 levels are used, as illustrated in Figure 20.14. Pipe racks are typically 6 m, 8 m, or 10 m wide, with spacing typically 6 to 8 m between portals. Minimum clearance provided below the rack for main roads is typically 7 m, with 6 m minimum clearance provided for secondary roads. As shown in Figure 20.14, instrumentation and electrical cable trays are normally located on the top tier. Process lines are located on the bottom tier as much as possible. Utility lines and hot process lines are normally located above the bottom tier. Larger diameter pipes are kept to the outside of the rack near the rack columns to minimize the bending load on the beams. Lines requiring thermal expansion loops are located near the edge of the rack. Small-bore line sizes are typically increased to DN50 on the rack for mechanical strength in order to avoid intermediate pipe support between the racks.

Figure 20.15 shows a preliminary layout for the pipe racks for the example site.

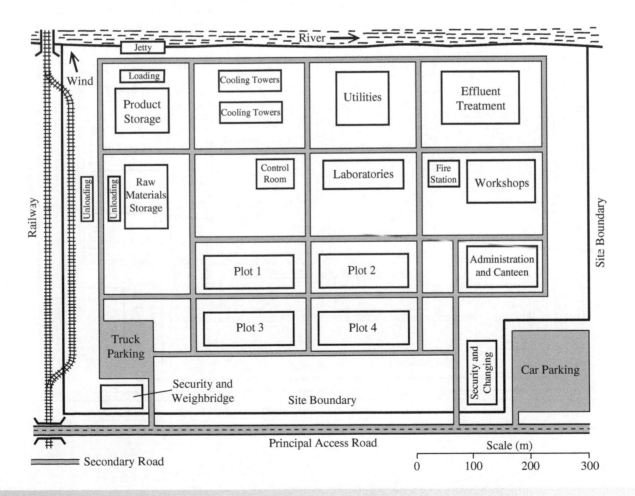

**Figure 20.12**

Adding the ancillary buildings.

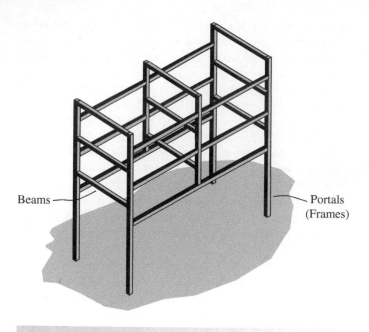

Beams

Portals
(Frames)

**Figure 20.13**

Pipe racks are required to route process and utility pipes, cables, and control signals around the site.

Intstrumentation and
Electrical CableTrays

Utility Pipes

Process Pipes

**Figure 20.14**

Different levels of the pipe racks carry different services.

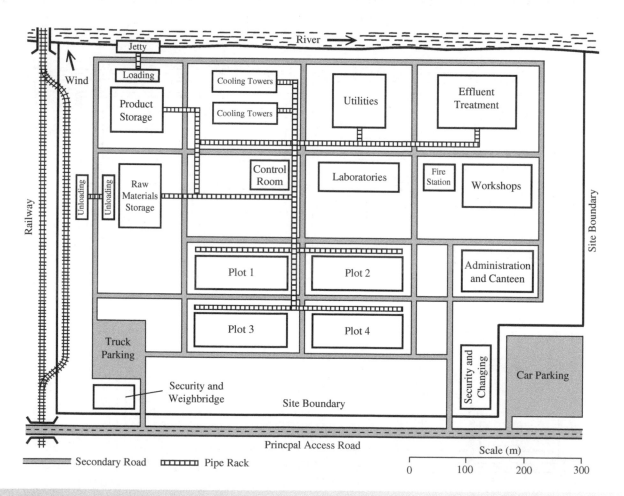

River

Jetty

Wind

Loading

Cooling Towers

Cooling Towers

Utilities

Effluent
Treatment

Product
Storage

Control
Room

Laboratories

Fire
Station

Workshops

Unloading

Unloading

Raw
Materials
Storage

Railway

Plot 1

Plot 2

Administration
and Canteen

Plot 3

Plot 4

Truck
Parking

Security and
Weighbridge

Security and Changing

Car Parking

Site Boundary

Site Boundary

Princpal Access Road

Secondary Road    Pipe Rack

Scale (m)

0    100    200    300

**Figure 20.15**

Pipe racks are required to connect the various operations on the site.

## 20.15 Constraints on Site Layout

Layout options are often constrained by existing process plant, infrastructure, or buildings. The site boundary and services, e.g., public roads, rail entry, sewers, power supplies, water supplies, effluent discharge, and sewers, can create constraints on layout. On an existing mature site, the location and area available for a new plant might be very constrained. A constrained plot area might force the use of a multilevel structure that might otherwise not be necessary. The existence of utility infrastructure at the proposed site will also constrain layout options, but will reduce project cost. If pipelines are to be used for raw material delivery or product dispatch, routing will constrain the layout of the site equipment through both the physical location and the risks associated with pipeline leaks.

Firewater supply should be laid as a ring main in an independent system with water available from two sources. Firewater pipes should be buried with fire hydrants spaced no more than 60 m apart.

Administrative buildings, medical centers, laboratories, workshops, fire stations, and emergency response equipment should be located outside of potential fire or vapor cloud explosion damage areas. Administrative buildings should be located as close to the main entrance as possible.

Environmental conditions relating to adjacent properties, e.g., residential property and neighboring plant hazardous operations, can constrain layout options. Finally, legal requirements, e.g., planning and building laws, effluent and pollution regulations, fire and other safety legislation, create the ultimate constraints.

## 20.16 The Final Site Layout

In most cases a variety of different site layouts can be developed for the same problem. There are many issues that need to be considered simultaneously. Guidelines are applied to help with the decision making, but the guidelines for different aspects of the layout might be in conflict. That means priority needs to be given to certain aspects of the layout. Production companies will have different preferences for different aspects of the layout. When priorities are changed slightly, significantly different layouts can result. If there are options in each stage of the development of the layout, then it would be good practice to develop alternative layouts. Once developed, these alternative layouts can be evaluated in detail for the hazards associated with fire, explosion, and toxic release and the response to emergencies.

## 20.17 Site Layout – Summary

Process plant layout can be considered at the site, process, and equipment levels. Good site layout should provide:

- safe and economical flow of process materials, utilities, and people;
- minimizing the potential for fire, explosion, and toxic release and their impact.

The site should have a peripheral road with access to the public road system at least at two points. Emergency and rescue vehicles should be able to approach within a reasonable distance of any risk from at least two directions. Any hazard should be able to be contained at source. Layout needs to consider the site topography, groundwater and drainage, prevailing wind direction, and geotechnical engineering. Particularly hazardous processes (e.g., containing toxic materials) should be isolated. Undesirable interactions between processes (e.g., sources of ignition close to plants containing flammables) must be avoided. Minimum separation distances should be used to mitigate potential undesirable interactions between storage, process operations, utilities, and people. Minimum separation distances should be used to the surrounding community.

## References

Center for Chemical Process Safety (2018). *Guidelines for Facility Siting and Layout*, 2e, American Institute of Chemical Engineers. Wiley.

Crowl, D.A. and Louvar, J.F. (1990). *Chemical Process Safety-Fundamentals with Applications*. Prentice Hall.

Global Asset Protection Services, GAP 2.5.2 (2015) *Oil and Chemical Plant Layout and Spacing*.

Gugan, K. (1979). *Unconfined Vapor Cloud Explosions*. The Institution of Chemical Engineers and George Godwin Ltd.

Khan, F.I. and Abbasi, S.A. (2001). An assessment of the likelihood of occurrence, and the damage potential of domino effect (chain of accidents) in a typical cluster of industries. *Journal of Loss Prevention in the Process Industries* 14: 283.

Mannan, S. (2012). *Lees Loss Prevention in the Process Industries*, 4e. Butterworth-Heinemann.

Mecklenburgh, J.C. (1985). *Process Plant Layout*. George Godwin.

Moran, S. (2017). *Process Plant Layout*, 2ee. Butterworth-Heinemann.

Smith, R. (2016). *Chemical Process Design and Integration*. Wiley.

Wells, G.L. (1980). *Safety in Process in Process Plant Design*. IChemE, Wiley.

# Chapter 21

# Process Plant Layout – Process Layout

## 21.1 Process Access

The most economic process plant layout is that in which the spacing of the main items of equipment minimizes interconnecting pipework and structural steelwork. However, there must be adequate clearances for access, both for process operation and maintenance, and to prevent and isolate hazardous incidents. Importantly the layout must consider safe evacuation of all staff in an emergency. Pathways should have at least two access and exit routes in the event of an emergency (Mecklenburgh 1985; Moran 2017). For normal operation there should be unrestricted access for operations, maintenance, and construction. Maintenance not only requires unrestricted access for maintenance staff, but adequate space for dismantling of equipment and, if appropriate, use of heavy lifting equipment. In some cases large items of equipment will need to be removed from the process area for maintenance to workshops or off-site. There should be a clear route to remove such items from the plant. This might be overhead using a mobile crane or through the process on a trolley.

To prevent contamination or provide sterile conditions, food, pharmaceutical, and biochemical process plants may be required to be in a building. Labor-intensive batch processes also tend to be housed in buildings. The layout restrictions can become more complex in such situations. For buildings, there are ventilation issues, as well as access for operations, maintenance, and construction.

Table 21.1 shows typical in-process clearances (adapted from Mecklenburgh 1985; Moran 2017).

## 21.2 Process Structures

As a general rule all equipment should aim to be located at ground level, but a structure often needs to be provided:

- because height differences between process operations cannot be avoided or

*Process Plant Design*, First Edition. Robin Smith.
© 2024 John Wiley & Sons Ltd. Published 2024 by John Wiley & Sons Ltd.
Companion website: www.wiley.com/go/processplantdesign

- the plot area is restricted because of layout constraints, perhaps on an existing site.

Structures are most often steel structures, but might also be reinforced concrete. Figure 21.1 shows an example of a self-supporting distillation column where a structure is required for the distillation ancillary equipment. Pipes to and from the distillation are supported by pipe racks. The reboiler is a natural circulation thermosiphon reboiler. This needs a certain head for the circulation relative to the height of liquid in the distillation column sump. The pumps for the overhead product, bottom product, and reflux require a certain *net positive suction head* (Smith 2016). There is also a requirement for minimum headroom for access and maintenance procedures. These dictate the height of the first level in the structure. The level above supports the condenser. The height of this level is dictated by both gravity flow requirements from the condenser to the distillate receiver and the minimum headroom for access and maintenance procedures. The location of the higher level of the structure in Figure 21.1 is shown to be located away from the column for clarity. In practice, the higher level of the structure would be most likely to be closer to the distillation column, as pipes between the structure and distillation column require support. Ladders and platforms would be required on the exterior of the distillation column for access, but are not shown in Figure 21.1. Also not shown are the stairways for access to the structure and hand rails, but would be required.

Another example is shown in Figure 21.2, which shows a vacuum condenser arrangement. The vacuum condenser is at a lower pressure than the condensate receiver, which is often vented to atmosphere or a low-pressure vent system. This creates a situation where the condensate from the condenser is under vacuum and trying to flow to a condensate receiver that is under a positive pressure. The pressure difference is against the direction of flow required. To overcome this pressure difference, the condenser must be located higher than the condensate receiver to allow enough static head pressure for the condensate to exceed the pressure difference. The piping between the condenser and the receiver tank is referred to as the *barometric leg*. If the condensate

## Table 21.1

Typical in-process access, walkways, and maintenance clearances.

| Access feature | Clearance (m) |
|---|---|
| 1) Headroom for primary access roads and major maintenance vehicles | 6.0 |
| 2) Width of primary access roads | 6.0 |
| 3) Headroom for secondary access roads | 3.0–4.5 |
| 4) Width of secondary access roads | 3.0–4.5 |
| 5) Headroom over platforms, walkways, access ways, maintenance areas | 2.5 |
| 6) Width of stairways | 0.75 |
| 7) Width of landings in direction of stairway | 0.9 |
| 8) Width of walkways at grade, or elevated walkways | 0.75 |
| 9) Vertical rise of stairways – one flight | 4.0 |
| 10) Vertical rise of ladders – single run | 6.0 |

*Source:* adapted from Mecklenburgh 1985; Moran 2017.

is water with a density of 1000 kg m$^{-3}$, then a height of 10.33 m is required to give a pressure difference of 1 atm. Also, the condensate pump needs to be at a sufficient height below the condensate receiver to allow for the net positive suction head. Again, as with

Figure 21.1, no access stairways or hand rails have been shown in Figure 21.2 but would be required.

Figure 21.3 shows two alternative design arrangements for distillation columns. In Figure 21.3a the columns are supported, at least to some extent, against lateral bending forces by a structure, whereas in Figure 21.3b the columns are completely self-supporting. The spacing between the columns in the structure should be a minimum of 1 m between the plinths of the columns (Moran 2017). For the self-supporting columns there should be a spacing of at least 3 m (Moran 2017). Access to process structures is either via stairways, ladders, or maintenance elevators. Table 21.1 shows that if long stairways are required (as shown in Figure 21.3a), the stairway should include intermediate platforms such that a single flight of stairs should have a rise of no more than 4.0 m. If ladders are to be used for access (as shown in Figure 21.3b), these should feature a hooped cage as a safety measure with hoops commencing at a height of 2.5 m above the base of the ladder. Ladder runs of more than 6.0 m should normally have a landing at every 6.0 m point, with successive ladder runs out of alignment to reduce the distance personnel might fall.

Shell-and-tube heat exchangers should where possible be located at ground level, mounted on legs and supported by concrete plinths, as shown in Figure 21.4. Horizontal shell-and-tube heat exchangers are normally at the edge of the plant with their axis oriented perpendicular to the plant boundary, as shown in Figure 21.5. This allows the tube bundles to be withdrawn for cleaning and maintenance. Figure 21.6 shows examples of how horizontal heat exchangers may be located in structures. As pointed out previously, they are preferably located at ground level.

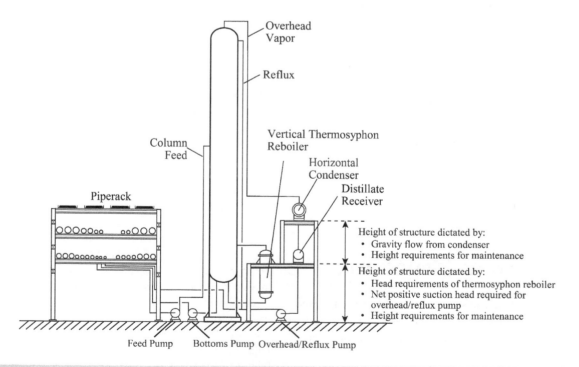

## Figure 21.1

Stand-alone distillation column with a support structure for ancillary equipment.

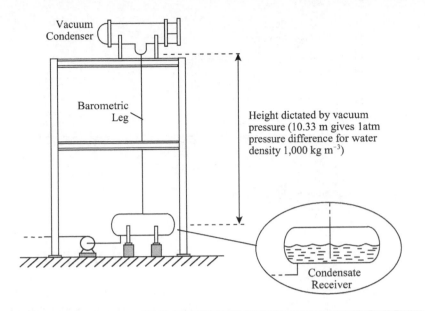

**Figure 21.2**

A vacuum condenser with a barometric leg requires a structure.

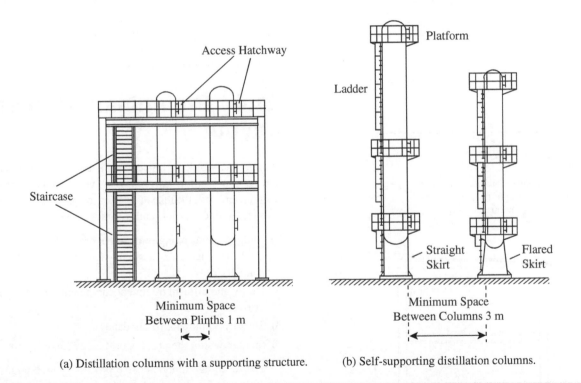

(a) Distillation columns with a supporting structure.    (b) Self-supporting distillation columns.

**Figure 21.3**

Distillation columns can be supported by a structure or be self-supporting.

Figure 21.6 shows an example of two horizontal heat exchangers stacked together to a maximum height of 3.7 m (Moran 2017). The spacing between the concrete plinths at ground level should be a minimum of 0.9 m (Moran 2017). Clearance between the bottom of the heat exchanger and ground level should be at least 0.3 m (Moran 2017). Figure 21.6 also shows a heat exchanger located at an upper level, which has been anchored directly to the steel structure.

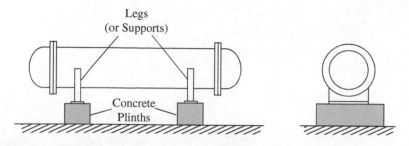

**Figure 21.4**

Heat exchangers are normally mounted on legs supported by concrete plinths.

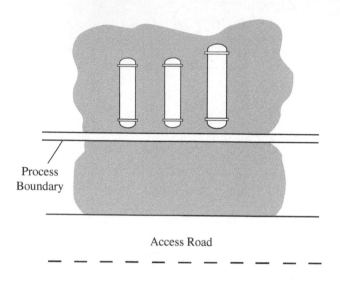

**Figure 21.5**

Horizontal heat exchangers are normally oriented perpendicular to the plant boundary at the edge of the plant to allow the tube bundles to be withdrawn for cleaning and maintenance.

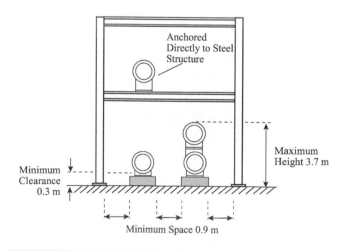

**Figure 21.6**

Heat exchangers mounted in a structure.

## 21.3   Hazards

The plant layout should be arranged to reduce the risk of disaster spreading, provide adequate means of escape, access for fire fighting, rescue, etc. As discussed in Chapter 20, hazardous areas are classified into zones based on an assessment of the frequency of the occurrence and duration of an explosive gas atmosphere. Hazardous areas (such as flame-proof areas) must be clearly defined. It is normally economic to put only hazardous operations in a hazardous area to avoid unnecessary protection of equipment. Solid floors should, where possible, be used in structures for plant liable to leak liquids. Otherwise equipment liable to leak should have a drip tray below the equipment with a drainage system for liquid spills to drain to a safe location.

## 21.4   Preliminary Process Layout

A preliminary process layout can be developed from the *comprehensive process flow diagram*. This takes the basic process flow diagram and adds the missing steps and equipment from raw materials in, to products out, and includes:

1) Multiple items of equipment that are shown in the basic process flow diagram as a single step (e.g., multiple reactors in parallel).
2) Utility and effluent treatment equipment directly linked to the process.
3) Standby equipment (redundancy).
4) Storage within the process or directly linked to the process.

The resulting comprehensive process flow diagram provides the basis for the development of a layout in which the process can be located on one or more plots. For smaller processes a single process plant from raw inlets to product outlets might be located on a single plot. However, for larger process plants, increasing the area of plots with a large amount of equipment introduces access problems and hazards. Safe access and emergency exit for personnel becomes more of a problem the larger and more complex is the plot. Large and complex plots also have a greater potential for the spread of fire and explosion hazards, and create difficulties for firefighting in the event of an incident. Thus, larger process plants tend to be divided into sections and located on multiple process plots adjacent to each other, but separated for access and to minimize the potential of hazardous incidents spreading. Operations should be grouped to be within a plot to:

- Follow operations in the sequence of the comprehensive process flow diagram, unless there are good reasons to do otherwise, for example to group hazardous operations (*tending to minimize the pipework, ease of control, and simplify process operation*).
- Operations and equipment in a plot should constrain the amount of equipment in a plot to limit the hazard in an area of the process (*tending to give some separation of the hazards, making the design inherently safer*).
- Maintain a degree of operational independence of the plots, especially if there can be intermediate storage between the plots (*tending to simplify start-up, process control, shut-down, and process availability if there is intermediate storage*).
- Allow heat integration between operations (*tending to minimize pipework and increase energy efficiency*).
- Group operations with common utility requirements (*tending to minimize utility pipework in the process and pipe racks*).
- Limit hazardous inventories in each plot (*tending to make the design inherently safer*).
- Maintain a consistent hazardous area classification, requiring the same electrical equipment specification, and avoid unnecessary overprotection of electrical equipment (*tending to reduce costs of fire and explosion protection*).
- Satisfy a requirement to be in a building because the process is small scale and labor intensive, or to prevent product contamination, or to maintain sterile conditions (*tending to minimize capital and operating costs*).

The area of the process plots depends on:

- Number and size of items of equipment in the process (or section of the process).
- Spacing between equipment for access, maintenance, and safety.
- Whether a structure is required because height differences are required between process operations. If a structure is necessary, then other equipment than those requiring height difference can be housed in the structure to reduce the plot area required.
- Site constraints that force the use of a structure because of plot area restrictions.
- Whether a building is required for the process.

Having grouped the operations to be within a plot, the detailed layout arrangement for each plot needs to be created. For the layout within a plot, as a general rule equipment should aim to be at ground level, but a structure needs to be provided for:

- Pump net positive suction head requirements.
- Head requirements for natural circulation reboilers and vaporizers.
- Vacuum operations requiring a barometric leg.

- Gravity flow of liquids and solids required between operations.
- Gravity flow of liquid required for pump priming.
- Limited plot space constrains location of equipment at ground level (e.g. brownfield constraints).

Operations and equipment in a plot should be located close to the plot boundary and an access road when:

- Mobile cranes are needed for maintenance.
- Equipment needs replacement of internals.
- Reactors need catalyst changes.
- Equipment needs regular cleaning.
- Operations need frequent operator intervention (e.g., filters). These should be located to obtain the shortest and most direct routes from the control room.
- Heat exchangers that will require cleaning and maintenance.

Operations and equipment in a plot should be located close to a pipe rack when:

- Operations have pipes entering or leaving the plot.
- Operations within the plot are separated, but need to be connected by pipework.

Operations and equipment in a plot should be located mid-plot when:

- Vessels require little maintenance or operator access.
- Equipment needs little operator access and maintenance does not need lifting equipment or significant space.

Consider now an example.

# 21.5  Example – Preliminary Process Layout

Figure 21.7 shows a basic process flow diagram for the manufacture of phthalic anhydride by the controlled oxidation of *o*-xylene via the reaction:

$$C_8H_{10} \ + \ 3O_2 \ \longrightarrow \ C_8H_4O_3 \ + \ 3H_2O$$
*o*-xylene        phthalic anhydride    water

A number of side reactions occur including:

$$C_8H_{10} \ + \ 15\!/\!2O_2 \ \longrightarrow \ C_4H_2O_3 \ + \ 4CO_2 + 4H_2O$$
*o*-xylene        maleic anhydride

$$C_8H_{10} \ + \ 5/2O_2 \ \longrightarrow \ 8CO_2 \ + \ 5H_2O$$
*o*-xylene        carbon dioxide    water

$$C_4H_2O_3 \ + \ 3O_2 \ \longrightarrow \ 4CO_2 \ + \ H_2O$$
maleic anhydride        carbon dioxide    water

Other byproducts formed in smaller quantities include benzoic and toluic acids. The reaction uses a fixed-bed vanadium

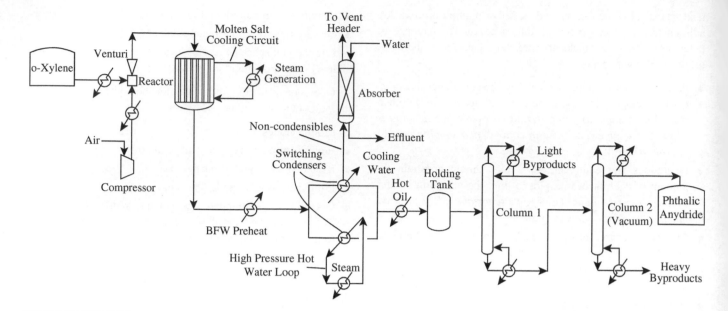

## Figure 21.7

Process flow diagram for a phthalic anhydride process.

pentoxide–titanium dioxide catalyst that gives good selectivity for phthalic anhydride, providing temperature is controlled within relatively narrow limits. The reaction is carried out in the vapor phase with reactor temperatures typically in the range 380–400 °C. The reaction is exothermic, and a multi-tubular reactor is employed with direct cooling of the reactor, using in this case a molten heat transfer salt, which is a eutectic mixture of potassium nitrate, sodium nitrate, and sodium nitrite.

Figure 21.7 shows air and *o*-xylene being heated and mixed in a venturi, where the *o*-xylene vaporizes. The reaction mixture enters the tubular catalytic reactor where the heat of reaction is removed by recirculation of molten salt. The molten salt is cooled by the generation of steam and returned to the reactor.

The gaseous reactor product is cooled first by heating boiler feedwater before entering a cooling water condenser. The phthalic anhydride forms as a solid on the tube walls in the cooling water condenser and is cooled to 70 °C. Periodically, the on-line condenser is taken off-line and the phthalic anhydride is melted off the surfaces by recirculation of high-pressure hot water. Two condensers are used in parallel, one on-line performing the condensation duty and one off-line recovering the phthalic anhydride. The non-condensable gases contain small quantities of byproducts and traces of phthalic anhydride and are scrubbed before being vented to the atmosphere. Depending on the components, their concentrations in the vent, and the legislative restrictions, the vent might need to be subjected to thermal oxidation before release to the atmosphere.

The crude phthalic anhydride is heated and held at 260 °C to allow some byproduct reactions to go to completion. Purification is by continuous distillation in two columns. In the first column, maleic anhydride and benzoic and toluic acids are removed overhead to form a crude maleic anhydride byproduct. In the second column, pure phthalic anhydride is removed overhead.

High-boiling residues are removed from the bottom of the second column and sent for use as fuel.

Figure 21.8 shows the comprehensive process flow diagram corresponding to the basic process flow diagram in Figure 21.7:

1) Standby equipment has been added.

2) The reactor cooling from molten salt shown in Figure 21.7 as generating steam requires extra features in the cooling circuit for start-up and shut-down. If the reactor is to be shut down, water is added in a molten salt holding tank to cool the salt and prevent it from solidifying by creating a solution/slurry of the salt in water. When the reactor is started up, the water needs to be evaporated and the salt melted before the reactor can be started, which requires a fired heater for start-up purposes. Once the reactor is operational the furnace is shut down. Then to shut down the reactor requires the feed to be stopped and the furnace started to replace the heat of reaction. Water is then gradually fed to the salt holding tank and the fired heating is gradually turned down to cool down the reactor and create a stable salt solution/slurry ready for start-up.

3) Additional features have been added to the switch condensers.

4) The distillation reboilers require heating above steam temperature. This is being supplied by a hot oil circuit (see Chapter 4). The hot oil circuit also supplies heat to the holding tank.

5) Storage tanks have been added.

Figure 21.9 shows the comprehensive process flow diagram divided into plots:

1) Reaction section.

2) Separation section.

3) Storage of *o*-xylene feed, naphthalene product, and maleic anhydride byproduct.

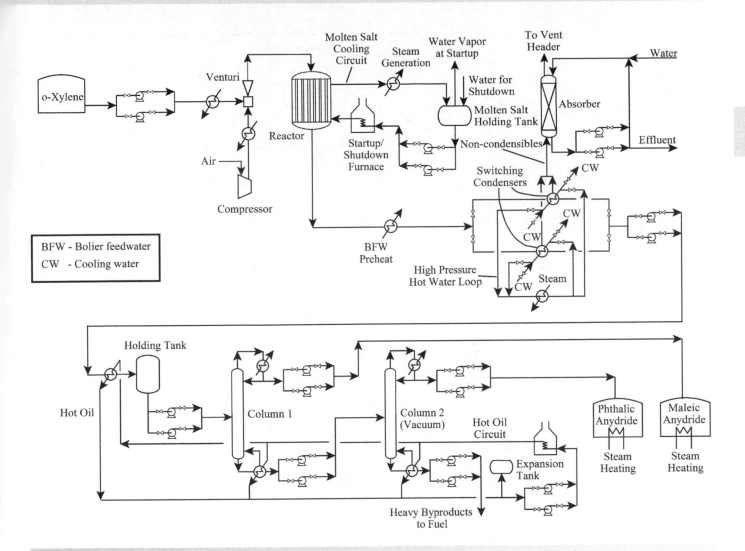

## Figure 21.8

Comprehensive process flow diagram for a phthalic anhydride process.

The reaction and separation sections can be operated with some degree of independence by exploiting the residence time in the holding tank. Table 21.2 shows separation distances from Table 20.1 for fire consequences assuming both the reaction and separation sections are classified as intermediate hazards (Center for Chemical Process Safety 2018).

Figure 21.10 shows one possible preliminary plot layout corresponding with the separation distances in Table 21.2. The fired heaters have been located upwind of the process plots and storage areas. Many other process layouts are possible. In practice, existing site features and infrastructure will tend to direct acceptable options in the layout.

If the reaction section is classified as high hazard and the separation section as intermediate hazard (Center for Chemical Process Safety 2018), then the corresponding separation distances from Table 20.1 are given in Table 21.3.

Figure 21.11 shows one possible preliminary plot layout corresponding with the separation distances in Table 21.3. The layout has been spaced out significantly compared with Figure 21.10.

The preliminary layouts in Figures 21.10 and 21.11 need to be evaluated further in terms of the consequences of explosion or toxic release. This might require the separation distances to be increased. On an existing site the location and area available for a new plant might be very constrained. A constrained plot area might force the use of a multilevel structure that might otherwise not be necessary. The existing infrastructure on the site, such as location of utility plant and pipe racks, will also constrain layout options. If distances cannot be increased to provide adequate safety, then measures introduced from a layer of protection analysis might be required. These could, for example, include additional safety instrumented systems in the process or blast walls. It might even be necessary for some features of the process to be redesigned to reduce the inventory of hazardous material in the process or storage, or change the process conditions.

The detailed plot layout then needs to be developed using the guidelines in the previous section for location of operations and equipment relative to the plot boundary, pipe racks, and mid-plot (Moran 2017).

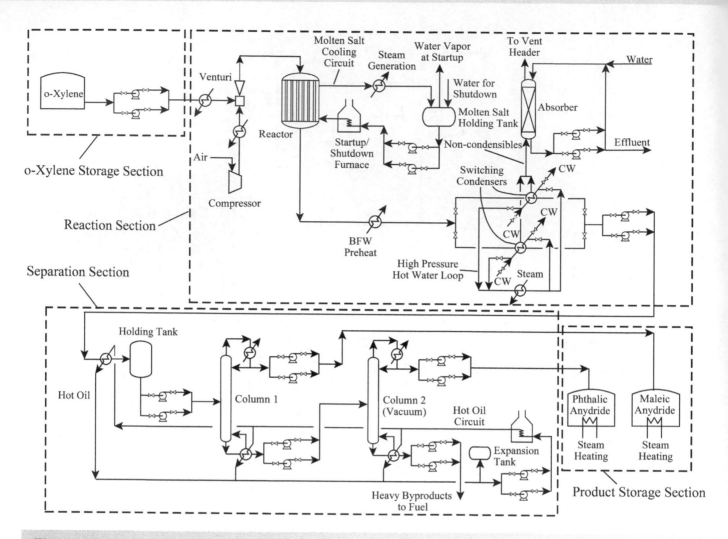

## Figure 21.9

Comprehensive process flow diagram for a phthalic anhydride process.

## Table 21.2

Separations used for the plot layout assuming both the reaction and separation sections are classified as intermediate hazard.

| From | To | Distances (m) |
|------|-----|--------------|
| Process | Process | 30 |
| Process | Fired heater | 15 |
| Process | Control room | 60 |
| Process | Storage | 90 |
| Fired heater | Fired heater | 7.5 |
| Fired heater | Storage | 105 |
| Control room | Storage | 75 |
| Pipe rack | Main road | 7 |
| Pipe rack | Secondary road | 5 |

## Table 21.3

Separations used for the plot layout assuming the reaction section is classified as high hazard and separation section intermediate hazard.

| From | To | Distances for reaction section (m) | Distances for separation section (m) |
|------|-----|-----------------------------------|--------------------------------------|
| Process | Process | 60 | 30 |
| Process | Fired heater | 15 | 15 |
| Process | Control room | 90 | 60 |
| Process | Storage | 105 | 90 |

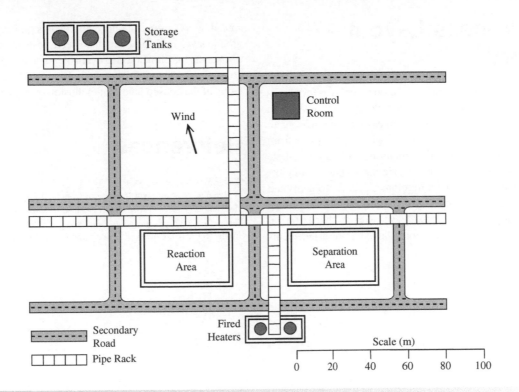

## Figure 21.10

A possible layout of plots for the phthalic anhydride process with the reaction area classified as medium hazard.

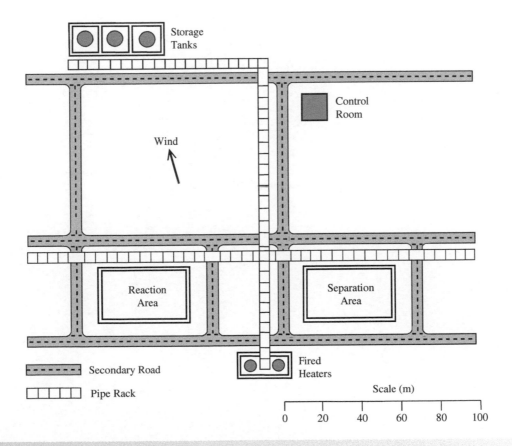

## Figure 21.11

A possible layout of plots for the phthalic anhydride process with the reaction area classified as high hazard.

# 21.6 Process Layout – Summary

The comprehensive flow diagram can be used as the basis for a process layout. The process layout should follow operations in the sequence of the process flow diagram, unless there are reasons to do otherwise. For smaller processes a single process plant from raw material inlets to product outlets might be located on a single plot. However, for larger process plants, increasing the area of plots with a large amount of equipment introduces access problems and hazards. Thus, larger process plants tend to be divided into sections and located on multiple process plots adjacent to each other, but separated for access and to minimize the potential of hazardous incidents spreading. As a general rule, equipment is preferred to be located at ground level, but a structure often needs to be provided. On an existing site the location and area available for a new plant might be very constrained. The preliminary layouts need to be evaluated further in terms of the consequences of explosion or toxic release.

# References

Center for Chemical Process Safety (2018). *Guidelines for Facility Siting and Layout*, 2e, American Institute of Chemical Engineers. Wiley.

Mecklenburgh, J.C. (1985). *Process Plant Layout*. George Godwin.

Moran, S. (2017). *Process Plant Layout*. Butterworth-Heinemann.

Smith, R. (2016). *Chemical Process Design and Integration*. Wiley.

# Appendix A

# Weibull Reliability Function

The Weibull failure rate $\lambda$ is assumed to be a function of time $t$:

$$\lambda(t) = \frac{\beta}{\theta}\left(\frac{t}{\theta}\right)^{\beta-1} \quad \sigma > 0, \beta > 0, \; t \geq 0 \qquad (A.1)$$

where

$\beta, \theta$ = fitting parameters

$\quad \beta$ = shape parameter

$\quad \beta = 1$ reduces to exponential distribution

The Weibull reliability is then defined as:

$$
\begin{aligned}
R(t) &= \exp\left[-\int_0^{t'} \lambda(t)\,\mathrm{d}t\right] \\
&= \exp\left[-\int_0^{t'} \frac{\beta}{\theta}\left(\frac{t}{\theta}\right)^{\beta-1}\,\mathrm{d}t\right] \\
&= \exp\left[-\frac{\beta}{\theta^\beta}\int_0^{t'} t^{\beta-1}\,\mathrm{d}t\right] \qquad (A.2) \\
&= \exp\left[-\frac{\beta}{\theta^\beta}\frac{t^\beta}{\beta}\right]_0^{t'} \\
&= \exp\left[-\left(\frac{t'}{\theta}\right)^\beta\right]
\end{aligned}
$$

*Process Plant Design*, First Edition. Robin Smith.
© 2024 John Wiley & Sons Ltd. Published 2024 by John Wiley & Sons Ltd.
Companion website: www.wiley.com/go/processplantdesign

# Appendix B

# *MTTF* for the Weibull Distribution

By definition:

$$MTTF = \int_0^\infty R(t)\ dt \tag{B.1}$$

The Weibull Distribution for reliability is given by:

$$R(t) = \exp\left[-\left(\frac{t}{\theta}\right)^\beta\right] \tag{B.2}$$

Substitute Eq. (B.2) into Eq. (B.1):

$$MTTF = \int_0^\infty \exp\left[-\left(\frac{t}{\theta}\right)^\beta\right]\ dt \tag{B.3}$$

Let $y = \left(\frac{t}{\theta}\right)^\beta$ \hfill (B.4)

Thus: $\quad t = \theta y^{\frac{1}{\beta}}$ and $\dfrac{dt}{dy} = \dfrac{\theta}{\beta} y^{\frac{1}{\beta}-1}$ \hfill (B.5)

$$MTTF = \int_0^\infty \frac{\theta}{\beta} y^{\frac{1}{\beta}-1} e^{-y} dy$$

$$= \frac{\theta}{\beta} \int_0^\infty y^{\frac{1}{\beta}-1} e^{-y} dy \tag{B.6}$$

$$= \frac{\theta}{\beta} \Gamma\left(\frac{1}{\beta}\right)$$

where $\Gamma$ is the gamma function, defined by:

$$\Gamma\left(\frac{1}{\beta}\right) = \int_0^\infty y^{\frac{1}{\beta}-1} e^{-y} dy \tag{B.7}$$

An alternative form of the expression for *MTTF* can be obtained by transforming the gamma function. By definition of the gamma function:

$$\Gamma\left(1 + \frac{1}{\beta}\right) = \int_0^\infty y^{\frac{1}{\beta}} e^{-y} dy \tag{B.8}$$

Integrating by parts:

$$\int u\left(\frac{dv}{dy}\right) dy = uv - \int v\left(\frac{du}{dy}\right) dy \tag{B.9}$$

Let $u = y^{\frac{1}{\beta}}$ and $\dfrac{dv}{dy} = e^{-y}$ \hfill (B.10)

Thus: $\quad v = -e^{-y}$ and $\dfrac{du}{dy} = \dfrac{1}{\beta} y^{\frac{1}{\beta}-1}$ \hfill (B.11)

$$\Gamma\left(1 + \frac{1}{\beta}\right) = [uv]_0^\infty - \int_0^\infty v\left(\frac{du}{dy}\right) dy$$

$$= \left[-y^{\frac{1}{\beta}} e^{-y}\right]_0^\infty - \int_0^\infty -e^{-y} \frac{1}{\beta} y^{\frac{1}{\beta}-1} dy$$

$$= \left[-y^{\frac{1}{\beta}} e^{-y}\right]_0^\infty + \frac{1}{\beta} \int_0^\infty y^{\frac{1}{\beta}-1} e^{-y} dy \tag{B.12}$$

$$= \frac{1}{\beta} \int_0^\infty y^{\frac{1}{\beta}-1} e^{-y} dy$$

$$= \frac{1}{\beta} \Gamma\left(\frac{1}{\beta}\right)$$

This assumes in the $uv$ term that $e^{-y} \to 0$ faster than $-y^{\frac{1}{\beta}} \to \infty$ as $y \to \infty$. Thus:

$$MTTF = \frac{\theta}{\beta} \Gamma\left(\frac{1}{\beta}\right)$$

$$= \theta\, \Gamma\left(1 + \frac{1}{\beta}\right) \tag{B.13}$$

where

$$\Gamma\left(\frac{1}{\beta}\right) = \int_0^\infty y^{\frac{1}{\beta}-1} e^{-y} dy$$

$$\Gamma\left(1 + \frac{1}{\beta}\right) = \int_0^\infty y^{\frac{1}{\beta}} e^{-y} dy$$

$$y = \left(\frac{t}{\theta}\right)^\beta$$

*Process Plant Design*, First Edition. Robin Smith.
© 2024 John Wiley & Sons Ltd. Published 2024 by John Wiley & Sons Ltd.
Companion website: www.wiley.com/go/processplantdesign

# Appendix C

# Reliability of Cold Standby Systems

Consider an arrangement of identical units in which one unit is on-line and carries the full load. Given failure of the on-line unit it is assumed that there is a perfect switch to bring one of the standby units on-line to take the full load. Assume that the on-line unit is subject to a constant failure rate and the off-line units are not subject to failure. In a small time interval $\Delta t$ the probability of one failure is $\lambda \Delta t$. It is assumed that the probability of more than one failure in $\Delta t$ is negligible.

Let $Pr(k, t)$ = probability of exactly $k$ failures in a time interval of length $t$. Then the probability of zero failures in time length $\Delta t$ is $Pr(0, \Delta t) = 1 - \lambda \Delta t$. If the number of failures in different time intervals are independent:

$$Pr(0, t + \Delta t) = Pr(0, t)Pr(0, \Delta t)$$
$$= Pr(0, t)(1 - \lambda t) \quad \text{(C.1)}$$

Rearranging:

$$\frac{Pr(0, t + \Delta t) - Pr(0, t)}{\Delta t} = -\lambda Pr(0, t) \quad \text{(C.2)}$$

As $\Delta t \to 0$:

$$\frac{dPr(0, t)}{dt} = -\lambda Pr(0, t) \quad \text{(C.3)}$$

This is a first-order differential equation that can be solved for the initial condition $Pr(0, 0) = 1$ to give:

$$Pr(0, t) = \exp(-\lambda t) \quad \text{(C.4)}$$

For one failure $Pr(1, t)$:

$$Pr(1, t + \Delta t) = Pr(1, t)Pr(0, \Delta t) + Pr(0, t)Pr(1, \Delta t)$$
$$= Pr(1, t)(1 - \lambda \Delta t) + \exp(-\lambda t)\lambda \Delta t \quad \text{(C.5)}$$

Rearranging:

$$\frac{Pr(1, t + \Delta t) - Pr(1, t)}{\Delta t} = -\lambda Pr(1, t) + \lambda \exp(-\lambda t) \quad \text{(C.6)}$$

As $\Delta t \to 0$:

$$\frac{dPr(1, t)}{dt} = -\lambda Pr(1, t) + \lambda \exp(-\lambda t) \quad \text{(C.7)}$$

This can be solved for the initial condition $Pr(1, 0) = 0$ to give:

$$Pr(1, t) = \lambda t \exp(-\lambda t) \quad \text{(C.8)}$$

In a similar way, $Pr(2, t)$ can be solved for the initial condition $Pr(2, 0) = 0$ to give:

$$Pr(2, t) = \frac{(\lambda t)^2 \exp(-\lambda t)}{2!} \quad \text{(C.9)}$$

Then, generally for $Pr(k, t)$:

$$Pr(k, t) = \frac{(\lambda t)^k \exp(-\lambda t)}{k!} \quad \text{(C.10)}$$

This represents the probability of $k$ failures in time $t$.

If there are $n$ units, then the system can withstand $(n - 1)$ failures. The reliability of a system of $n$ units with $(n - 1)$ on cold standby with no failures in the standby mode and perfect switching on failure of the on-line component is given by:

$$R(t) = \exp(-\lambda t) \sum_{k=0}^{n-1} \frac{(\lambda t)^k}{k!} \quad \text{(C.11)}$$

It is common to operate with two identical units, one on cold standby. Noting $0! = 1$, the reliability is given by:

$$R(t) = (1 + \lambda t) \exp(-\lambda t) \quad \text{(C.12)}$$

*Process Plant Design*, First Edition. Robin Smith.
© 2024 John Wiley & Sons Ltd. Published 2024 by John Wiley & Sons Ltd.
Companion website: www.wiley.com/go/processplantdesign

The reliability of three identical units, with two on cold standby is given by:

$$R(t) = \left(1 + \lambda t + \frac{(\lambda t)^2}{2!}\right) \exp(-\lambda t) \quad \text{(C.13)}$$

The mean time to failure for two units with one on cold standby is given by:

$$MTTF = \int_0^\infty (1 + \lambda t) \exp(-\lambda t)\, dt \quad \text{(C.14)}$$

Given the definite integral (Dwight 1969):

$$\int_0^\infty t^a \exp(-\lambda t)\, dt = \frac{a!}{\lambda^{a+1}} \quad \text{(C.15)}$$

Thus, Eq. (C.14) becomes:

$$
\begin{aligned}
MTTF &= \int_0^\infty (1 + \lambda t) \exp(-\lambda t)\, dt \\
&= \int_0^\infty \exp(-\lambda t)\, dt + \int_0^\infty \lambda t \exp(-\lambda t)\, dt \\
&= \frac{1}{\lambda} + \frac{1}{\lambda} \\
&= \frac{2}{\lambda}
\end{aligned}
$$

$$\text{(C.16)}$$

The mean time to failure for three units with two on cold standby is given by:

$$
\begin{aligned}
MTTF &= \int_0^\infty \left(1 + \lambda t + \frac{(\lambda t)^2}{2!}\right) \exp(-\lambda t)\, dt \\
&= \int_0^\infty \exp(-\lambda t)\, dt + \int_0^\infty \lambda t \exp(-\lambda t)\, dt + \int_0^\infty \frac{\lambda^2 t^2}{2} \exp(-\lambda t)\, dt \\
&= \frac{1}{\lambda} + \frac{1}{\lambda} + \frac{1}{\lambda} \\
&= \frac{3}{\lambda}
\end{aligned}
$$

$$\text{(C.17)}$$

In general, for $n$ units ($n - 1$ on standby):

$$
\begin{aligned}
MTTF &= \int_0^\infty \exp(-\lambda t) \sum_{k=0}^{n-1} \frac{(\lambda t)^k}{k!}\, dt \\
&= \frac{n}{\lambda}
\end{aligned}
$$

$$\text{(C.18)}$$

# Reference

Dwight, H.B. (1969). *Tables of Integrals and Other Mathematical Data*, 4e. The Macmillan Company.

# Appendix D

# Corrosion Resistance Table

| | Metal | | | | | | | Polymer | | |
|---|---|---|---|---|---|---|---|---|---|---|
| Fluid | Carbon Steel | 304 Stainless Steel | 316 Stainless Steel | Monel | Hastelloy B | Hastelloy C | Titanium | Poly Vinyl Chloride | Polyethylene | Polypropylene |
| Acetaldehyde | 2 | 1 | 1 | 1 | NA | 1 | 1 | 3 | 2 | |
| Acetic acid | 2 | 2 | 1 | 2 | 2 | 1 | 1 | 1 | 2 | 1 |
| Acetic acid 80% | NA | 2 | 1 | NA | 2 | 1 | 1 | 3 | NA | 2 |
| Acetic acid 20% | NA | 2 | 1 | NA | NA | 1 | 1 | 2 | NA | 1 |
| Acetic anhydride | 3 | 1 | 1 | NA | 2 | 1 | 1 | 3 | 1 | 1 |
| Acetone | 1 | 1 | 1 | 1 | 1 | 1 | 1 | 3 | 2 | 2 |
| Acetylene | 1 | 1 | 1 | 1 | 1 | 1 | 2 | 2 | NA | 3 |
| Acrylonitrile | NA | 1 | 2 | NA | NA | 2 | 2 | 3 | 3 | 2 |
| Alcohols | 2 | 1 | 1 | 1 | 1 | 1 | 1 | 1 | 2 | 2 |
| Aluminum sulfate | 3 | 2 | 2 | 2 | 1 | 1 | 1 | 1 | 2 | 1 |
| Ammonia, anhydrous | 2 | 2 | 1 | NA | 2 | 1 | 2 | 1 | 2 | 1 |
| Ammonia 10% | NA | 1 | 1 | NA | 2 | 1 | 1 | 1 | NA | 1 |
| Ammonium chloride | 3 | 2 | 2 | 2 | 1 | 1 | 3 | 1 | 2 | 1 |
| Ammonium hydroxide | 2 | 1 | 1 | 3 | 3 | 1 | 2 | 1 | 2 | 1 |
| Ammonium nitrate | 3 | 1 | 1 | 3 | 1 | 1 | 3 | 1 | 2 | 1 |
| Ammonium phosphate | 3 | 1 | 1 | 2 | 1 | 1 | 1 | 1 | 2 | 1 |
| Ammonium sulfate | 2 | 3 | 2 | 1 | 1 | 1 | 1 | 1 | 2 | 1 |

Corrosion Resistance (1) Potentially Usable (2) Use with Caution (3) Not Usable (NA) Data Not Available

*(continued)*

*Process Plant Design*, First Edition. Robin Smith.
© 2024 John Wiley & Sons Ltd. Published 2024 by John Wiley & Sons Ltd.
Companion website: www.wiley.com/go/processplantdesign

(*Continued*)

| | Metal | | | | | | Polymer | | |
|---|---|---|---|---|---|---|---|---|---|---|
| **Corrosion Resistance (1) Potentially Usable (2) Use with Caution (3) Not Usable (NA) Data Not Available** | | | | | | | | | | |
| **Fluid** | **Carbon Steel** | **304 Stainless Steel** | **316 Stainless Steel** | **Monel** | **Hastelloy B** | **Hastelloy C** | **Titanium** | **Poly Vinyl Chloride** | **Polyethylene** | **Polypropylene** |
| Aniline | 3 | 1 | 1 | 2 | 1 | 2 | 1 | 3 | 2 | 2 |
| Asphalt | 1 | 2 | 1 | 1 | 1 | 1 | NA | 1 | NA | 1 |
| Benzene | 3 | 1 | 1 | 1 | 1 | 2 | 1 | 3 | 3 | 3 |
| Benzoic acid | 3 | 1 | 1 | 1 | 1 | 1 | 1 | 1 | 2 | 3 |
| Butadiene | 2 | 1 | 1 | NA | 2 | 2 | NA | 1 | NA | 1 |
| Butane | 2 | 1 | 1 | 1 | 1 | 1 | 1 | 1 | 2 | 3 |
| Butanol | 2 | 1 | 1 | 1 | NA | 1 | 2 | 1 | 2 | 2 |
| Calcium carbonate | NA | 1 | 1 | NA | 1 | 1 | 1 | 1 | 2 | 1 |
| Calcium chloride | 2 | 3 | 3 | 1 | 1 | 1 | 1 | 1 | 2 | 1 |
| Calcium hypochlorite | 3 | 3 | 2 | 2 | 3 | 2 | 1 | 3 | 2 | 1 |
| Carbon dioxide, dry | 1 | 1 | 1 | 1 | 1 | 1 | 1 | NA | NA | NA |
| Carbon dioxide, wet | 3 | 1 | 1 | 1 | 1 | 1 | 1 | NA | NA | NA |
| Carbon disulfide | 2 | 2 | 1 | 2 | 1 | 1 | 1 | 3 | 3 | 3 |
| Carbon tetrachloride | 3 | 2 | 2 | 1 | 2 | 1 | 1 | 2 | 3 | 3 |
| Carbonic acid | 3 | 1 | 2 | 1 | 1 | 1 | NA | 1 | 2 | 1 |
| Chlorine gas | 1 | 2 | 2 | 1 | 1 | 1 | 3 | NA | NA | NA |
| Chlorine gas, wet | 3 | 3 | 3 | 3 | 3 | 2 | 1 | 1 | NA | 3 |
| Chlorine liquid | 3 | 3 | 3 | 3 | 3 | 1 | 3 | 3 | 3 | 3 |
| Chlorobenzene | 2 | 1 | 1 | NA | 1 | 1 | NA | 3 | 3 | 3 |
| Chromic acid 50% | 3 | 2 | 2 | 1 | 3 | 1 | 1 | 2 | 2 | 2 |
| Chromic acid 10% | NA | 2 | NA | NA | NA | 1 | 1 | 1 | NA | 1 |
| Citric acid | 3 | 2 | 1 | 2 | 1 | 1 | 1 | 1 | 2 | 2 |
| Copper sulfate | 3 | 2 | 2 | 3 | 2 | 1 | 1 | 1 | NA | 1 |
| Cyclohexane | 1 | 1 | 2 | 1 | NA | 2 | 1 | NA | NA | 3 |
| Dichloroethane | NA | 1 | 1 | 1 | 2 | 1 | NA | 3 | 3 | NA |
| Diesel | 1 | 1 | 1 | 1 | NA | 1 | 1 | 1 | NA | 3 |
| Ethane | 1 | 1 | 1 | 1 | 1 | 1 | 1 | NA | NA | NA |
| Ethanolamine | 2 | 1 | 1 | 1 | NA | 2 | 1 | 3 | NA | 3 |
| Ethyl chloride | 3 | 1 | 1 | 1 | 1 | 2 | 1 | 3 | 3 | 3 |
| Ethylene | 1 | 1 | 1 | 1 | 1 | 1 | 1 | NA | NA | NA |
| Ethylene glycol | 2 | 1 | 1 | 1 | NA | 1 | 1 | 1 | 2 | 1 |
| Ethylene oxide | NA | 2 | 1 | 1 | NA | 1 | NA | 3 | NA | 3 |

(*Continued*)

| | Metal | | | | | | | Polymer | | |
|---|---|---|---|---|---|---|---|---|---|---|
| **Fluid** | Carbon Steel | 304 Stainless Steel | 316 Stainless Steel | Monel | Hastelloy B | Hastelloy C | Titanium | Poly Vinyl Chloride | Polyethylene | Polypropylene |
| Ferric chloride | 3 | 3 | 3 | 3 | 3 | 2 | 1 | NA | NA | NA |
| Fluorine | 3 | 3 | 3 | 1 | | 1 | 3 | 2 | 2 | NA |
| Formaldehyde | 1 | 1 | 1 | 1 | 1 | 1 | 1 | 1 | 2 | 1 |
| Formic acid | 3 | 2 | 2 | 1 | 1 | 1 | 3 | 3 | 2 | 1 |
| Gasoline | 1 | 1 | 1 | 1 | 1 | 1 | 3 | 2 | 3 | 2 |
| Hydrochloric acid, dry gas | 3 | 2 | 1 | NA | NA | 1 | NA | 1 | NA | NA |
| Hydrochloric acid 100% | NA | 3 | 3 | 3 | 2 | 2 | 3 | 1 | 1 | NA |
| Hydrochloric acid 20% | NA | 3 | 3 | 2 | 2 | 3 | 2 | 1 | 1 | 1 |
| Hydrofluoric acid, aerated | 2 | 3 | 2 | 3 | 1 | 1 | 3 | NA | NA | NA |
| Hydrofluoric acid, air free | 1 | 3 | 2 | 1 | 1 | 1 | 3 | NA | NA | NA |
| Hydrogen | 2 | 1 | 1 | 1 | 1 | 1 | 1 | 1 | NA | 1 |
| Hydrogen peroxide 100% | 3 | 2 | 2 | 3 | 2 | 2 | 2 | 1 | 2 | 1 |
| Hydrogen peroxide 30% | NA | 2 | 2 | NA | NA | 2 | 2 | 1 | NA | 1 |
| Hydrogen sulfide, dry gas | 2 | 2 | 1 | 2 | NA | 1 | NA | 1 | NA | 1 |
| Magnesium hydroxide | 2 | 1 | 1 | 1 | 1 | 1 | 1 | 1 | 2 | 1 |
| Maleic acid | 2 | 1 | 1 | 2 | 2 | 1 | 1 | 1 | NA | 2 |
| Maleic anhydride | NA | 1 | NA | NA | 1 | 1 | NA | NA | NA | NA |
| Mercury | 1 | 1 | 1 | 2 | 1 | 1 | 2 | 1 | 2 | 1 |
| Methanol | 1 | 1 | 1 | 1 | 1 | 1 | 1 | 2 | 2 | 1 |
| Methyl acetate | 2 | 1 | 1 | 2 | 2 | 1 | NA | 3 | NA | 2 |
| Methyl acrylate | NA | 1 | 1 | 2 | NA | NA | NA | NA | NA | NA |
| Methyl ethyl ketone | 1 | 1 | 1 | 1 | 1 | 1 | 1 | 3 | 3 | 1 |
| Natural gas | 1 | 1 | 1 | 1 | 1 | 1 | 1 | 1 | NA | 1 |
| Nitric acid, concentrated | NA | 3 | 2 | 3 | NA | 2 | 1 | 3 | 3 | 3 |
| Nitric acid 50% | NA | 1 | 1 | 3 | NA | 1 | 1 | 1 | 2 | 3 |
| Nitric acid 20% | NA | 1 | 1 | 3 | NA | 1 | 1 | 1 | 2 | 1 |
| Nitric acid 10% | 3 | 1 | 1 | 3 | 2 | 1 | 1 | 1 | 2 | 1 |
| Nitrobenzene | 2 | 1 | 2 | 2 | 2 | 1 | 1 | 3 | 3 | 2 |
| Oleum | 2 | 1 | 1 | NA | 2 | 1 | NA | 3 | NA | 3 |
| Oxalic acid | 3 | 2 | 2 | 2 | 1 | 2 | 2 | 1 | 1 | 1 |
| Oxygen | 1 | 1 | 1 | 1 | 1 | 1 | 1 | 1 | NA | 1 |
| Pentane | 2 | 2 | 2 | 1 | 2 | 2 | NA | NA | NA | NA |

The table title spanning all columns reads: Corrosion Resistance (1) Potentially Usable (2) Use with Caution (3) Not Usable (NA) Data Not Available

D

(*Continued*)

| | Metal | | | | | | | Polymer | | |
|---|---|---|---|---|---|---|---|---|---|---|
| **Fluid** | **Carbon Steel** | **304 Stainless Steel** | **316 Stainless Steel** | **Monel** | **Hastelloy B** | **Hastelloy C** | **Titanium** | **Poly Vinyl Chloride** | **Polyethylene** | **Polypropylene** |
| Phenol | 3 | 1 | 1 | 1 | 1 | 1 | 2 | 1 | 3 | 2 |
| Phosphoric acid 40–100% | NA | 2 | 2 | 1 | 2 | 1 | 2 | 2 | 2 | 1 |
| Phosphoric acid, below 40% | NA | 2 | 2 | 1 | 2 | 1 | 1 | 1 | 2 | 1 |
| Phthalic anhydride | 2 | 1 | 2 | 3 | 1 | 1 | NA | 3 | NA | 3 |
| Potassium carbonate | 2 | 1 | NA | 1 | 2 | 1 | 1 | 1 | 2 | 1 |
| Potassium chloride | 2 | 1 | 1 | 2 | 1 | 1 | 1 | 1 | 2 | 1 |
| Potassium hydroxide | 2 | 2 | 2 | 1 | 1 | 1 | 2 | 1 | 2 | 1 |
| Propane | 2 | 1 | 1 | 1 | 1 | 1 | 1 | 3 | NA | 3 |
| Propylene glycol | 2 | 2 | 1 | 2 | NA | 2 | 2 | NA | 2 | NA |
| Sodium acetate | 2 | 2 | 1 | 1 | 1 | 1 | 1 | 1 | 2 | 1 |
| Sodium carbonate | 2 | 1 | 2 | 1 | 1 | 1 | 1 | 1 | 2 | 1 |
| Sodium chloride | 3 | 2 | 2 | 1 | 1 | 1 | 1 | 1 | 2 | 1 |
| Sodium chromate | 2 | 1 | 1 | 1 | 1 | 2 | 1 | 1 | 2 | 1 |
| Sodium hydroxide 80% | NA | 1 | 3 | NA | NA | 2 | 1 | 1 | 2 | 1 |
| Sodium hydroxide 50% | NA | 1 | 2 | NA | NA | 1 | 1 | 1 | 2 | 1 |
| Sodium hydroxide 20% | NA | 1 | 1 | NA | NA | 1 | 1 | 1 | 2 | 1 |
| Sodium hypochlorite | 3 | 3 | 3 | 3 | 3 | 1 | 1 | 1 | NA | 1 |
| Sodium hypochlorite, up to 20% | NA | 2 | 2 | 3 | NA | 1 | 1 | 1 | 2 | 3 |
| Sulfur | 1 | 1 | 1 | 1 | 1 | 1 | 1 | 1 | NA | 3 |
| Sulfur dioxide, dry | 2 | 1 | 1 | 1 | 2 | 1 | 1 | 3 | 3 | 1 |
| Sulfur trioxide, dry | 2 | 1 | 2 | 1 | 2 | 1 | 1 | 1 | NA | 3 |
| Sulfuric acid 75–100% | 3 | 3 | 3 | 3 | 1 | 2 | 3 | 2 | NA | 2 |
| Sulfuric acid 10–75% | 3 | 3 | 3 | 2 | 1 | 2 | 2 | 1 | 2 | 1 |
| Sulfuric acid up to 10% | NA | 3 | 2 | 2 | 3 | 1 | 1 | 1 | 2 | 1 |
| Sulfurous acid | 3 | 2 | 2 | 3 | 1 | 1 | 1 | 1 | 2 | 1 |
| Trichloroethylene | 2 | 2 | 1 | 1 | 1 | 1 | 1 | 3 | 3 | 3 |
| Water, boiler feedwater | 2 | 1 | 1 | 1 | 1 | 1 | 1 | NA | NA | NA |
| Water, distilled | 1 | 1 | 1 | 1 | 1 | 1 | 1 | 1 | NA | 1 |
| Water, sea | 2 | 2 | 2 | 1 | 1 | 1 | 1 | 1 | NA | 1 |
| Xylenes | 2 | 1 | 1 | 1 | 2 | 1 | NA | 3 | 3 | 3 |
| Zinc chloride | 3 | 3 | 3 | 3 | 1 | 2 | 1 | 1 | 2 | 1 |
| Zinc sulfate | 3 | 1 | 1 | 1 | 1 | 2 | 1 | 2 | 2 | 1 |

Corrosion Resistance (1) Potentially Usable (2) Use with Caution (3) Not Usable (NA) Data Not Available

# Appendix E

# Moment of Inertia and Bending Stress for Common Beam Cross-Sections

## E.1 Solid Rectangular Cross-Section

Consider the rectangular cross-section shown in Figure E.1. Writing the moment of inertia (second moment of area) about the $XX$ axis through the centroid, the neutral axis:

$$I_{XX} = \int_{-d/2}^{d/2} y^2 \, dA = \int_{-d/2}^{d/2} y^2 b \, dy = b\left[\frac{y^3}{3}\right]_{-d/2}^{d/2} = \frac{bd^3}{12} \quad (E.1)$$

where

$I_{XX}$ = second moment of area around the $x$-axis (m$^4$)

$b$ = breadth of the beam (m)

$d$ = depth of the beam (m)

The section modulus is now given by the moment of inertia divided by the distance to the most extreme fiber from the neutral axis:

$$Z = \frac{I}{y_{\max}} = \frac{bd^3}{12} \times \frac{2}{d} = \frac{bd^2}{6} \quad (E.2)$$

where

$Z$ = section modulus (m$^3$)

$y_{\max}$ = distance of the most extreme fiber from the neutral axis (m)

Thus, for a rectangular beam the maximum normal bending stress is given by Eq. (19.37) or Eq. (19.39):

$$\sigma_{B\max} = \frac{M\,y_{\max}}{I} = \frac{Md/2}{bd^3/12} = \frac{6M}{bd^2} \quad (E.3)$$

where

$\sigma_{B\max}$ = longitudinal normal stress (N m$^{-2}$)

$M$ = bending moment (N m)

$y_{\max}$ = distance to the most extreme fiber from the neutral axis (m)

## E.2 Hollow Rectangular Cross-Section

A hollow rectangular section (rectangular tube) is illustrated in Figure E.2. The simplest way to determine the moment of inertia is by using Eq. (19.42). The section can be decomposed into an outer "positive" rectangle and a concentric "inner" negative rectangle. If both rectangles are concentric and have the same centroid, then the two moments of inertia can be subtracted:

$$I_{XX} = \frac{b_O d_O^3}{12} - \frac{b_I d_I^3}{12} \quad (E.4)$$

where

$b_O$ = outer breadth of the beam (m)

$d_O$ = outer depth of the beam (m)

$b_I$ = inner breadth of the beam (m)

$d_I$ = inner depth of the beam (m)

The section modulus is now given by the moment of inertia divided by the distance to the most extreme fiber from the neutral axis:

$$Z = \frac{I}{y_{\max}} = \frac{b_O d_O^3 - b_I d_I^3}{12} \times \frac{2}{d_O} \quad (E.5)$$

$$= \frac{b_O d_O^3 - b_I d_I^3}{6d_O}$$

*Process Plant Design*, First Edition. Robin Smith.
© 2024 John Wiley & Sons Ltd. Published 2024 by John Wiley & Sons Ltd.
Companion website: www.wiley.com/go/processplantdesign

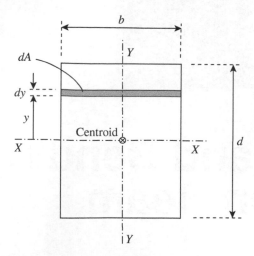

**Figure E.1**

Second moment of area (moment of inertia) for a rectangular cross-section.

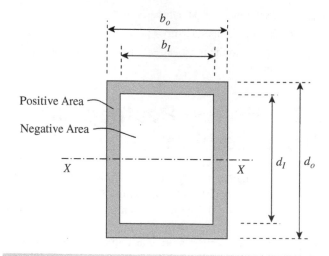

**Figure E.2**

Cross-section of a hollow rectangular beam separated into positive and negative areas.

The maximum normal bending stress is given by Eq. (19.37) or Eq. (19.39):

$$\sigma_{B\,max} = \frac{M\,y_{max}}{I}$$

$$= M \times \frac{d_O}{2} \times \frac{12}{b_O d_O^3 - b_I d_I^3} \qquad (E.6)$$

$$= \frac{6 d_O M}{b_O d_O^3 - b_I d_I^3}$$

# E.3   Solid Circular Cylinder

Figure E.3 shows a solid circular cross-section. The moment of inertia about an axis through $O$ is called the *polar moment of inertia*:

$$J = \int_A r^2 \, dA$$

$$= \int_0^R r^2 2\pi r \, dr$$

$$= 2\pi \int_0^R r^3 \, dr \qquad (E.7)$$

$$= 2\pi \left[ \frac{r^4}{4} \right]_0^R$$

$$= \frac{\pi R^4}{4}$$

$$= \frac{\pi D^4}{32}$$

where

$J$ = polar moment of inertia (m$^4$)

$A$ = area (m$^2$)

$R$ = radius of the beam (m)

$D$ = diameter of the beam (m)

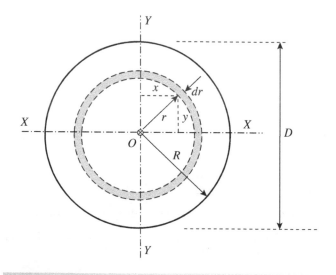

**Figure E.3**

Second moment of area (moment of inertia) for a circular cross-section.

This result can be related to $I_{XX}$ and $I_{YY}$ with reference to Figure E.3:

$$
\begin{aligned}
J &= \int_A r^2 \mathrm{d}A \\
&= \int_A \left(x^2 + y^2\right)\mathrm{d}A \\
&= \int_A x^2 \mathrm{d}A + \int_A y^2 \mathrm{d}A \\
&= I_{XX} + I_{YY}
\end{aligned}
\tag{E.8}
$$

For a solid circular beam, combining Eqs (E.7) and (E.8) gives:

$$
I_{XX} = I_{YY} = \frac{\pi D^4}{64}
\tag{E.9}
$$

The section modulus is now given by:

$$
Z = \frac{I}{y_{\max}} = \frac{\pi D^3}{32}
\tag{E.10}
$$

Thus, for a solid circular beam the maximum normal bending stress is given by Eq. (19.37) or Eq. (19.39):

$$
\sigma_{B\,\max} = \frac{M\, y_{\max}}{I} = \frac{MD/2}{\pi D^4/64} = \frac{32M}{\pi D^3}
\tag{E.11}
$$

# E.4  Hollow Circular Cross-Section

A hollow circular cross-section is shown in Figure E.4. The section can be decomposed into an outer "positive" circle and a concentric "inner" negative circle. If both circles are concentric and have the same centroid, then the two moments of inertia can be subtracted:

$$
J = \frac{\pi}{32}\left(D_O^4 - D_I^4\right)
\tag{E.12}
$$

where

$D_O$ = outside diameter of the beam (m)
$D_I$ = inside diameter of the beam (m)

From Eqs (E.8) and (E.12):

$$
I_{XX} = I_{YY} = \frac{\pi}{64}\left(D_O^4 - D_I^4\right)
\tag{E.13}
$$

Thus, the section modulus is given by

$$
Z = \frac{I}{y_{\max}} = \frac{\pi}{32 D_O}\left(D_O^4 - D_I^4\right)
\tag{E.14}
$$

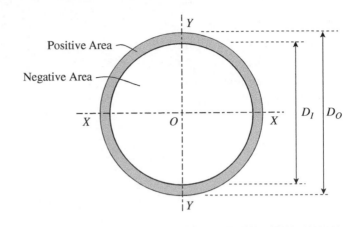

**Figure E.4**

Cross-section of a hollow circular beam separated into positive and negative areas.

Thus, the maximum normal bending stress is given by Eq. (19.37) or Eq. (19.39):

$$
\sigma_{B\,\max} = \frac{M\, y_{\max}}{I} = \frac{M D_O/2}{\frac{\pi}{64}\left(D_O^4 - D_I^4\right)} = \frac{32M\, D_O}{\pi\left(D_O^4 - D_I^4\right)}
\tag{E.15}
$$

# E.5  Approximate Expressions for Thin-Walled Cylinders

For the special case when the hollow circular cross-section is a thin-walled cylinder, approximate expressions can be developed for Eqs (E.12) to (E.15). Starting from Eq. (E.12):

$$
\begin{aligned}
J &= \frac{\pi}{32}\left(D_O^4 - D_I^4\right) \\
&= \frac{\pi}{32}\left[D_O^4 - (D_O - 2t)^4\right]
\end{aligned}
\tag{E.16}
$$

where $t$ = thickness of the cylinder (m).
From Eq. (E.16):

$$
\begin{aligned}
J &= \frac{\pi}{32}\left[D_O^4 - \left(D_O^4 - 8D_O^3 t + 24D_O^2 t^2 - 32D_O t^3 + 16t^4\right)\right] \\
&= \frac{\pi}{32}\left[8D_O^3 t - 24D_O^2 t^2 + 32D_O t^3 - 16t^4\right] \\
&= \frac{\pi D_O^3}{32}\left[8t - \frac{24t^2}{D_O} + \frac{32t^3}{D_O^2} - \frac{16t^4}{D_O^3}\right]
\end{aligned}
\tag{E.17}
$$

For thin-walled cylinders where $D_O \gg t$, only the first term in the bracket is significant and the diameter can be expressed as a mean diameter to give:

$$
J \approx \frac{\pi D_M^3 t}{4} = 2\pi R_M^3 t
\tag{E.18}
$$

E

where

$J$ = polar moment of inertia (m$^4$)

$D_M$ = mean diameter of the cylinder (m)

$R_M$ = mean radius of the cylinder (m)

$t$ = thickness of the cylinder (m)

Following this approximation:

$$I_{XX} = I_{YY} = \frac{\pi D_M^3 t}{8} = \pi R_M^3 t \tag{E.19}$$

$$Z = \frac{\pi D_M^2 t}{4} = \pi R_M^2 t \tag{E.20}$$

$$\sigma_{B\,\text{max}} = \frac{4M}{\pi D_M^2 t} = \frac{M}{\pi R_M^2 t} \tag{E.21}$$

# Appendix F

# First Moment of Area and Shear Stress for Common Beam Cross-Sections

## F.1 Solid Rectangular Cross-Section

For a rectangular cross-section as shown in Figure F.1, from Eq. (19.65):

$$
\begin{aligned}
Q &= \int_A y\, b\, dy \\
&= b \int_{y_0}^{d/2} y\, dy \\
&= b \left[ \frac{y^2}{2} \right]_{y_0}^{d/2} \\
&= \frac{b}{2} \left[ \frac{d^2}{4} - y_0^2 \right]
\end{aligned}
\tag{F.1}
$$

where

$Q$ = first moment of area (m$^3$)

$b$ = width of the rectangular section (m)

$d$ = depth of the rectangular section (m)

$y$ = distance of incremental area from the neutral axis (m)

$dy$ = depth of incremental area (m)

$y_0$ = distance of area from the neutral axis (m)

Substituting Eq. (F.1) into Eq. (19.66):

$$
\tau_B = \frac{FQ}{zI} = \frac{6F}{bd^3} \left( \frac{d^2}{4} - y_0^2 \right)
\tag{F.2}
$$

where

$\tau_B$ = transverse bending shear stress acting over the longitudinal plane (N m$^{-2}$)

$F$ = force acting on the beam (N)

$I$ = second moment of area (m$^4$)

$z$ = length of the cutting plane (m)

Equation (F.2) shows $\tau_B$ to exhibit a parabolic variation with $y_0$. The maximum shear stress occurs when $y_0 = 0$. The maximum value of $Q$ occurs when $y_0 = 0$:

$$
Q_{max} = \frac{bd^2}{8}
\tag{F.3}
$$

where $Q_{max}$ = maximum first moment of area (m$^3$).

The maximum bending shear stress can be obtained from Eq. (19.67):

$$
\begin{aligned}
\tau_{B\,max} &= \frac{FQ_{max}}{zI} \\
&= \frac{F \dfrac{bd^2}{8}}{b \dfrac{bd^3}{12}} \\
&= \frac{3F}{2A}
\end{aligned}
\tag{F.4}
$$

where

$\tau_{Bmax}$ = maximum transverse shear stress acting over the longitudinal plane (N m$^{-2}$)

$A$ = area of the rectangular cross-section (m)

= $bd$

## F.2 Hollow Rectangular Cross-Section

A hollow rectangular section (rectangular tube) is illustrated in Figure F.2. The simplest way to determine the first moment is using Eq. (19.68). The section can be decomposed into an outer

*Process Plant Design*, First Edition. Robin Smith.
© 2024 John Wiley & Sons Ltd. Published 2024 by John Wiley & Sons Ltd.
Companion website: www.wiley.com/go/processplantdesign

F

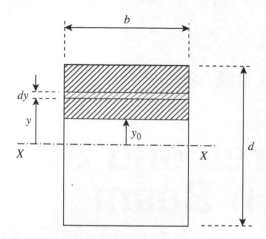

## Figure F.1

Cross-section of a rectangular beam subject to a transverse load.

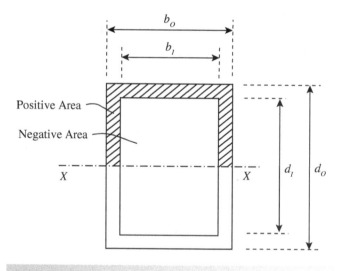

## Figure F.2

Cross-section of a hollow rectangular beam subject to a transverse load.

"positive" rectangle and a concentric "inner" negative rectangle. If both rectangles are concentric and have the same centroid, the two moments of area can be subtracted:

$$Q_{max} = \frac{b_O d_O^2}{8} - \frac{b_I d_I^2}{8} \qquad (F.5)$$

where

$b_O$ = outer breadth of the beam (m)
$d_O$ = outer depth of the beam (m)
$b_I$ = inner breadth of the beam (m)
$d_I$ = inner depth of the beam (m)

The maximum bending shear stress can be obtained from Eq. (19.67):

$$
\begin{aligned}
\tau_{B\,max} &= \frac{F Q_{max}}{zI} \\
&= \frac{F\left(b_O d_O^2 - b_I d_I^2\right)}{8} \times \frac{12}{(b_O - b_I)\left(b_O d_O^3 - b_I d_I^3\right)} \\
&= \frac{2F\left(b_O d_O^2 - b_I d_I^2\right)}{3(b_O - b_I)\left(b_O d_O^3 - b_I d_I^3\right)} \qquad (F.6)
\end{aligned}
$$

# F.3   Solid Circular Cross-Section

A solid circular cross-section is shown in Figure F.3. The area of an increment distance $y$ from the neutral axis has an area $2xdy$. Thus the first moment of area is defined by:

$$Q_{XX} = \int_A (2xdy)y \qquad (F.7)$$

Integration across the shaded area in Figure F.3, with the length $x$ defined by the right-angle triangle:

$$Q_{XX} = \int_{R\sin\theta}^{R} \left(R^2 - y^2\right)^{1/2} ydy \qquad (F.8)$$

where

$R$ = radius of the cylinder (m)
$y$ = distance from the neutral axis (m)
$\theta$ = angle to the neutral axis (radians)

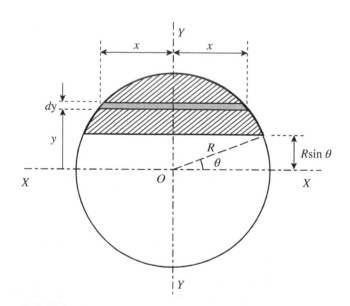

## Figure F.3

Cross-section of a solid circular beam subject to a transverse load.

From tables of integrals (Dwight 1969):

$$Q_{XX} = -2\left[\frac{\left(R^2-y^2\right)^{3/2}}{3}\right]_{R\sin\theta}^{R}$$

$$= \frac{2}{3}\left[\left(R^2-y^2\right)^{3/2}\right]_{R}^{R\sin\theta}$$

$$= \frac{2}{3}\left[\left(R^2\right)^{3/2}\left(1-\sin^2\theta\right)^{3/2}\right] \tag{F.9}$$

$$= \frac{2R^3}{3}\left(1-\sin^2\theta\right)^{3/2}$$

$$= \frac{2R^3}{3}\left(\cos^2\theta\right)^{3/2}$$

$$= \frac{2R^3\cos^3\theta}{3}$$

The maximum value of $Q_{XX}$ occurs when $\theta = 0$:

$$Q_{\max} = \frac{2R^3}{3} = \frac{D^3}{12} \tag{F.10}$$

Substituting $Q_{\max}$ in Eq. (19.67):

$$\tau_{B\max} = \frac{FQ_{\max}}{zI}$$

$$= \frac{F\dfrac{2R^3}{3}}{2R \times \dfrac{\pi R^4}{4}} \tag{F.11}$$

$$= \frac{4F}{3A}$$

where

$\tau_{B\max}$ = maximum transverse bending shear stress acting over the longitudinal plane (N m$^{-2}$)

$F$ = force acting on the beam (N)

$I$ = second moment of area (m$^4$)

$z$ = length of the cutting plane (m$^2$)

$A$ = area of the cylinder cross-section (m$^2$)

$$= \pi R^2 = \frac{\pi D^2}{4}$$

# F.4 Hollow Circular Cross-Sections

A hollow cylinder is illustrated in Figure F.4. The simplest way to determine the first moment is using Eq. (19.68). The section can be decomposed into an outer "positive" circle and a concentric inner "negative circle". If both circles are concentric and have the same centroid, then the two moments of area can be subtracted:

$$Q_{\max} = \frac{2}{3}\left(R_O^3-R_I^3\right) = \frac{1}{12}\left(D_O^3-D_I^3\right) \tag{F.12}$$

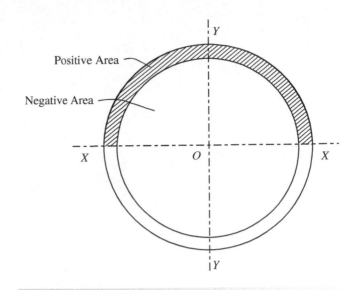

**Figure F.4**

Cross-section of a hollow circular beam subject to a transverse load.

where

$R_O$ = outside radius of the cylinder (m)

$R_I$ = inside radius of the cylinder (m)

$D_O$ = outside diameter of the cylinder (m)

$D_I$ = inside diameter of the cylinder (m)

Substituting $Q_{\max}$ in Eq. (19.67):

$$\tau_{B\max} = \frac{FQ_{\max}}{zI}$$

$$= \frac{\dfrac{2F}{3}\left(R_O^3-R_I^3\right)}{2(R_O-R_I) \times \dfrac{\pi}{4}\left(R_O^4-R_I^4\right)}$$

$$= \frac{4F}{3\pi}\frac{\left(R_O^3-R_I^3\right)}{\left(R_O^4-R_I^4\right)(R_O-R_I)}$$

$$= \frac{4F}{3\pi}\frac{(R_O-R_I)\left(R_O^2+R_OR_I+R_I^2\right)}{\left(R_O^2+R_I^2\right)\left(R_O^2-R_I^2\right)(R_O-R_I)}$$

$$= \frac{4F}{3A}\left(\frac{R_O^2+R_OR_I+R_I^2}{R_O^2+R_I^2}\right) = \frac{4F}{3A}\left(\frac{D_O^2+D_OD_I+D_I^2}{D_O^2+D_I^2}\right) \tag{F.13}$$

where

$A$ = area of the hollow cylinder cross-section (m$^2$)

$$= \pi\left(R_O^2-R_I^2\right) = \frac{\pi}{4}\left(D_O^2-D_I^2\right)$$

# Reference

Dwight, H.B. (1969). *Tables of Integrals and Other Mathematical Data*, 4e. The Macmillan Company.

# Appendix G

# Principal Stresses

To understand situations in which both normal (direct) stress and shear stress act together, consider the body under tensile force $F$ in Figure G.1. In Figure G.1a the body is shown cut by an imaginary inclined plane $AC$, which defines the axes $x'$ and $y'$ for the inclined plane. In Figure G.1b a normal stress $\sigma_{x'}$ and shear stress $\tau_{x'y'}$ are necessary to maintain the equilibrium relative to the inclined plane.

Figure G.1b shows the shear stress $\tau_{x'y'}$ on the inclined plane $AC$ resolved into $xy$ components. For each component, there must be an equal complementary shear stress component to balance the turning couple (see Section 19.1). The normal stress $\sigma_{x'}$ in Figure G.1b is also shown resolved into $\sigma_x$ and $\sigma_y$ components.

Carrying out a force balance in the $x'$ axis:

$$0 = \Delta A \sigma_{x'} - \Delta A \left(\tau_{xy} \sin \theta\right) \cos \theta - \Delta A \left(\tau_{xy} \cos \theta\right) \sin \theta \\ - \Delta A (\sigma_x \cos \theta) \cos \theta - \Delta A (\sigma_y \sin \theta) \sin \theta \tag{G.1}$$

where $\Delta A$ = area of the inclined plane
Rearranging Eq. (G.1):

$$\sigma_{x'} = \sigma_x \cos^2 \theta + \sigma_y \sin^2 \theta + \tau_{xy}(2\sin \theta \cos \theta) \tag{G.2}$$

Given the following trigonometric relations:

$$\cos^2 \theta = (1 + \cos 2\theta)/2$$

$$\sin^2 \theta = (1 - \cos 2\theta)/2 \tag{G.3}$$

$$2 \sin \theta \cos \theta = \sin 2\theta$$

Equation (G.2) becomes:

$$\sigma_{x'} = \frac{\sigma_x + \sigma_y}{2} + \frac{\sigma_x - \sigma_y}{2} \cos 2\sigma + \tau_{xy} \sin 2\theta \tag{G.4}$$

For the stresses in the $y'$ axis, substitute $\theta + 90°$ for $\theta$. Given the following trigonometric relations:

$$\cos(2\theta + 180°) = \cos 2\theta \cos 180° - \sin 2\theta \sin 180° \\ = \cos 2\theta(-1) - \sin 2\theta(0) \tag{G.5} \\ = \cos 2\theta$$

$$\sin(2\theta + 180°) = \sin 2\theta \cos 180° + \cos 2\theta \sin 180° \\ = \sin 2\theta(-1) + \cos 2\theta(0) \tag{G.6} \\ = - \sin 2\theta$$

Equation (G.4) for $\theta + 90°$ becomes:

$$\sigma_{y'} = \frac{\sigma_x + \sigma_y}{2} - \frac{\sigma_x - \sigma_y}{2} \cos 2\theta - \tau_{xy} \sin 2\theta \tag{G.7}$$

Carrying out a force balance in the $y'$ direction:

$$0 = \Delta A \tau_{x'y'} + \Delta A \left(\tau_{xy} \sin \theta\right) \sin \theta - \Delta A \left(\tau_{xy} \cos \theta\right) \\ \cos \theta - \Delta A (\sigma_x \sin \theta) \cos \theta + \Delta A (\sigma_x \cos \theta) \sin \theta \tag{G.8}$$

Rearranging Eq. (G.8):

$$\tau_{x'y'} = - \left(\sigma_x - \sigma_y\right) \sin \theta \cos \theta + \tau_{xy} \left(\cos^2 \theta - \sin^2 \theta\right) \tag{G.9}$$

Combining with the trigonometric relations in Eq. (G.3) gives:

$$\tau_{x'y'} = - \left(\frac{\sigma_x - \sigma_y}{2}\right) \sin 2\theta + \tau_{xy} \cos 2\theta \tag{G.10}$$

To determine the maximum and minimum normal stresses (the principal stresses) requires Eq. (G.4) to be differentiated and the derivative equated to zero:

$$\frac{d\sigma_{x'}}{d\theta} = \left(\frac{\sigma_x - \sigma_y}{2}\right)(-2 \sin \theta) + \tau_{xy}(2\cos \theta) = 0 \tag{G.11}$$

*Process Plant Design*, First Edition. Robin Smith.
© 2024 John Wiley & Sons Ltd. Published 2024 by John Wiley & Sons Ltd.
Companion website: www.wiley.com/go/processplantdesign

G

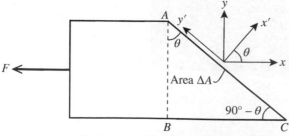

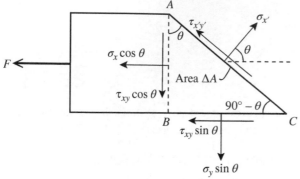

(a) A body subjected to a normal (tensile) force can be cut by an inclined plane.

(b) Resolving the stresses around the inclined plane

## Figure G.1

Stresses on an inclined plane.

Rearranging:

$$\tan 2\theta_P = \frac{2\tau_{xy}}{\sigma_x - \sigma_y} \qquad (G.12)$$

where $\theta_P$ is the angle $\theta$ corresponding with that for the principal stresses. Given the trigonometric relation:

$$\tan(\theta + 180°) = \frac{\tan\theta + \tan 180°}{1 - \tan\theta\tan 180°}$$

$$= \frac{\tan\theta + 0}{1 - \tan\theta(0)} \qquad (G.13)$$

$$= \tan\theta$$

Thus:

$$\tan 2(\theta + 90°) = \tan 2\theta \qquad (G.14)$$

This means that Eq. (G.12) can be solved for two values of $2\theta_p$ differing by 180°, or two values of $\theta_p$ differing by 90°. This in turn means that the planes of the two principal stresses are mutually perpendicular. To determine the principal stresses requires the value of $\theta_P$ from Eq. (G.12) to be substituted in Eqs (G.4) and (G.7). Figure G.2 shows a triangle based on Eq. (G.12) to determine the values of $\cos 2\theta_P$ and $\sin\theta_P$ from Eq. (G.12). The value of $\tan 2\theta_P$ from Eq. (G.12) fixes the sides of the triangle. Thus:

$$\cos 2\theta_P = \frac{\left(\frac{\sigma_x - \sigma_y}{2}\right)}{\sqrt{\left(\frac{\sigma_x - \sigma_y}{2}\right)^2 + \tau_{xy}^2}} \qquad (G.15)$$

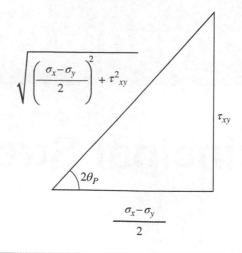

## Figure G.2

Evaluation of $\cos 2\theta_P$ and $\sin 2\theta_P$ from $\tan 2\theta_P$.

$$\sin 2\theta_p = \frac{\tau_{xy}}{\sqrt{\left(\frac{\sigma_x - \sigma_y}{2}\right)^2 + \tau_{xy}^2}} \qquad (G.16)$$

Substituting Eqs (G.15) and (G.16) into Eq. (G.4) gives:

$$\sigma_1 = \frac{\sigma_x + \sigma_y}{2} + \sqrt{\left(\frac{\sigma_x - \sigma_y}{2}\right)^2 + \tau_{xy}^2} \qquad (G.17)$$

where

$\sigma_x$ = first (maximum) principal stress

$\sigma_x$ = normal stress acting on the $x$-plane (N m$^{-2}$)

$\sigma_y$ = normal stress acting on the $y$-plane (N m$^{-2}$)

$\tau_{xy}$ = shear stress acting on the $x$-plane in the direction of the $y$-axis (N m$^{-2}$)

To obtain $\sigma_{y'}$, first add Eqs (G.4) and (G.7) to give:

$$\sigma_{x'} + \sigma_{y'} = \sigma_x + \sigma_y \qquad (G.18)$$

Thus, combining Eqs. (G.17) and (G.18) gives:

$$\sigma_2 = \frac{\sigma_x + \sigma_y}{2} - \sqrt{\left(\frac{\sigma_x - \sigma_y}{2}\right)^2 + \tau_{xy}^2} \qquad (G.19)$$

where $\sigma_2$ = second (minimum) principal stress

To determine the maximum and minimum shear stresses requires Eq. (G.10) to be differentiated and the derivative equated to zero:

$$\frac{d\tau_{x'y'}}{d\theta} = -2\left(\frac{\sigma_x - \sigma_y}{2}\right)\cos 2\theta - 2\tau_{xy}\sin 2\theta = 0 \qquad (G.20)$$

Rearranging Eq. (G.20):

$$\tan 2\theta_S = -\left(\frac{\sigma_x - \sigma_y}{2\tau_{xy}}\right) \qquad (G.21)$$

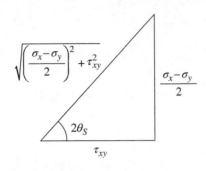

**Figure G.3**

Evaluation of $\cos 2\theta_S$ and $\sin 2\theta_S$ from $\tan 2\theta_S$.

where $\theta_S$ is now the angle $\theta$ corresponding with that for the maximum shear stress. As before, there are two values of $2\theta$ in the range 0–180°, with two values of $\theta$ differing by 90°. Thus, the planes of maximum shear stress are mutually perpendicular. The shear stresses on perpendicular planes have equal magnitudes. The maximum and minimum shear stresses differ only in sign. As before, the values of $\cos 2\theta_S$ and $\sin 2\theta_S$ in Eq. (G.10) can be determined from the properties of a triangle based on Eq. (G.21), as shown in Figure G.3. Thus:

$$\sin 2\theta_S = -\frac{\sigma_x - \sigma_y}{2\sqrt{\left(\frac{\sigma_x - \sigma_y}{2}\right)^2 + \tau_{xy}^2}} \qquad \text{(G.22)}$$

$$\cos 2\theta_S = \frac{\tau_{xy}}{\sqrt{\left(\frac{\sigma_x - \sigma_y}{2}\right)^2 + \tau_{xy}^2}} \qquad \text{(G.23)}$$

Substituting Eqs (G.22) and (G.23) in Eq. (G.10) gives:

$$\tau_{max} = \sqrt{\left(\frac{\sigma_x - \sigma_y}{2}\right)^2 + \tau_{xy}^2} \qquad \text{(G.24)}$$

$$\tau_{min} = -\tau_{max} = -\sqrt{\left(\frac{\sigma_x - \sigma_y}{2}\right)^2 + \tau_{xy}^2} \qquad \text{(G.25)}$$

where $\tau_{max}$, $\tau_{min}$ = maximum and minimum shear stresses

The relationship between the principal stresses and the maximum shear stress can be determined by subtracting Eq. (G.19) from Eq. (G.17) to give:

$$\sigma_1 - \sigma_2 = 2\sqrt{\left(\frac{\sigma_x - \sigma_y}{2}\right)^2 + \tau_{xy}^2} \qquad \text{(G.26)}$$

Substituting Eq. (G.24) into Eq. (G.26) and rearranging gives:

$$\tau_{max} = \frac{\sigma_1 - \sigma_2}{2} \qquad \text{(G.27)}$$

The angle between the principal stresses and the maximum shear stress can be determined by combining Eqs (G.12) and (G.21):

$$\tan 2\theta_P = -\frac{1}{\tan 2\theta_S} \qquad \text{(G.28)}$$

Rearranging:

$$\frac{\sin 2\theta_P}{\cos 2\theta_P} + \frac{\cos 2\theta_S}{\sin 2\theta_S} = 0 \qquad \text{(G.29)}$$

Thus:

$$\sin 2\theta_P \sin 2\theta_S + \cos 2\theta_P \cos 2\theta_S = 0 \qquad \text{(G.30)}$$

From trigonometric relations:

$$\cos(2\theta_P - 2\theta_S) = 0$$
$$2\theta_P - 2\theta_S = \pm 90° \qquad \text{(G.31)}$$
$$\theta_P - \theta_S = \pm 45°$$

Thus, the planes of maximum and minimum shear stress occur at 45° to the principal planes. Also, the value of $\tau_{x'y'}$ acting on the principal plane is zero. This can be shown by equating Eq. (G.10) to zero:

$$\tau_{x'y'} = 0 = -\left(\frac{\sigma_x - \sigma_y}{2}\right)\sin 2\theta_S + \tau_{xy}\cos 2\theta_S \quad \text{(G.32)}$$

Rearranging:

$$\tan 2\theta_S = \frac{2\tau_{xy}}{(\sigma_x - \sigma_y)}$$
$$= \tan 2\theta_P \text{ from Eq.}(G.12)$$

In summary, there are two perpendicular planes for which the shear stress is zero. These are the principal planes where the normal stress is a maximum. The normal stresses acting on the planes of zero shear stress are either maximum or minimum. The maximum principal stress is the most tensile (least compressive) and the minimum tensile stress the least tensile (most compressive).

G

# Appendix H

# Dimensions and Weights of Carbon Steel Pipes

| Nominal size | OD | Wall thickness in mm of welded and seamless carbon steel pipe to ASME | | | | | | | | | | | | |
| | | Schedule number | | | | | | | | | | | | |
| DN | mm | 10 | 20 | 30 | 40 | Std | 60 | 80 | XS | 100 | 120 | 140 | 160 | XXS |
|---|---|---|---|---|---|---|---|---|---|---|---|---|---|---|
| 6 | 10.3 | 1.24 | | 1.45 | 1.73 | 1.73 | | 2.41 | 2.41 | | | | | |
| 8 | 13.7 | 1.65 | | 1.85 | 2.24 | 2.24 | | 3.02 | 3.02 | | | | | |
| 10 | 17.1 | 1.65 | | 1.85 | 2.31 | 2.31 | | 3.20 | 3.20 | | | | | |
| 15 | 21.3 | 2.11 | | 2.41 | 2.77 | 2.77 | | 3.73 | 3.73 | | | | 4.78 | 7.47 |
| 20 | 26.7 | 2.11 | | 2.41 | 2.87 | 2.87 | | | 3.91 | | | | 5.56 | 7.82 |
| 25 | 33.4 | 2.77 | | 2.90 | 3.38 | 3.38 | | 4.55 | 4.55 | | | | 6.35 | 9.09 |
| 32 | 42.2 | 2.77 | | 2.97 | 3.56 | 3.56 | | 4.85 | 4.85 | | | | 6.35 | 9.70 |
| 40 | 48.3 | 2.77 | | 3.18 | 3.68 | 3.68 | | 5.08 | 5.08 | | | | 7.14 | 10.15 |
| 50 | 60.3 | 2.77 | | 3.18 | 3.91 | 3.91 | | 5.54 | 5.54 | | | | 8.74 | 11.07 |
| 65 | 73.0 | 3.05 | | 4.78 | 5.16 | 5.16 | | 7.01 | 7.01 | | | | 9.53 | 14.02 |
| 80 | 88.9 | 3.05 | | 4.78 | 5.49 | 5.49 | | 7.62 | 7.62 | | | | 11.13 | 15.24 |
| 90 | 101.6 | 3.05 | | 4.78 | 5.74 | 5.74 | | 8.08 | 8.08 | | | | | |
| 100 | 114.3 | 3.05 | | 4.78 | 6.02 | 6.02 | | 8.56 | 8.56 | | 11.13 | | 13.49 | 17.12 |
| 125 | 141.3 | 3.40 | | | 6.55 | 6.55 | | 9.53 | 9.53 | | 12.70 | | 15.88 | 19.05 |
| 150 | 168.3 | 3.40 | | | 7.11 | 7.11 | | 10.97 | 10.97 | | 14.27 | | 18.26 | 21.95 |
| 200 | 219.1 | 3.76 | 6.35 | 7.04 | 8.18 | 8.18 | 10.31 | 12.70 | 12.70 | 15.09 | 18.26 | 20.62 | 23.01 | 22.23 |
| 250 | 273.0 | 4.19 | 6.35 | 7.80 | 9.27 | 9.27 | 12.70 | 15.09 | 12.70 | 18.26 | 21.44 | 25.40 | 28.58 | 25.40 |
| 300 | 323.8 | 4.57 | 6.35 | 8.38 | 10.31 | 9.53 | 14.27 | 17.48 | 12.70 | 21.44 | 25.40 | 28.58 | 33.32 | 25.40 |
| 350 | 355.6 | 6.35 | 7.92 | 9.53 | 11.13 | 9.53 | 15.09 | 19.05 | 12.70 | 23.83 | 27.79 | 31.75 | 35.71 | |

*(continued)*

*Process Plant Design*, First Edition. Robin Smith.
© 2024 John Wiley & Sons Ltd. Published 2024 by John Wiley & Sons Ltd.
Companion website: www.wiley.com/go/processplantdesign

(*Continued*)

| Nominal size | OD | Wall thickness in mm of welded and seamless carbon steel pipe to ASME | | | | | | | | | | | | |
|---|---|---|---|---|---|---|---|---|---|---|---|---|---|---|
| | | Schedule number | | | | | | | | | | | | |
| DN | mm | 10 | 20 | 30 | 40 | Std | 60 | 80 | XS | 100 | 120 | 140 | 160 | XXS |
| 400 | 406.4 | 6.35 | 7.92 | 9.53 | 12.70 | 9.53 | 16.66 | 21.44 | 12.70 | 26.19 | 30.96 | 36.53 | 40.49 | |
| 450 | 457 | 6.35 | 7.92 | 11.13 | 14.27 | 9.53 | 19.05 | 23.83 | 12.70 | 29.36 | 34.93 | 39.67 | 45.24 | |
| 500 | 508 | 6.35 | 9.53 | 12.70 | 15.09 | 9.53 | 20.62 | 26.19 | 12.70 | 32.54 | 38.10 | 44.45 | 50.01 | |
| 550 | 559 | 6.35 | 9.53 | 12.70 | | 9.53 | 22.23 | 28.58 | 12.70 | 34.93 | 41.28 | 47.63 | 53.98 | |
| 600 | 610 | 6.35 | 9.53 | 14.27 | 17.48 | 9.53 | 24.61 | 30.96 | 12.70 | 38.89 | 46.02 | 52.37 | 59.54 | |
| 650 | 660 | 7.92 | 12.70 | | | 9.53 | | | 12.70 | | | | | |
| 700 | 711 | 7.92 | 12.70 | 15.88 | | 9.53 | | | 12.70 | | | | | |
| 750 | 762 | 7.92 | 12.70 | 15.88 | | 9.53 | | | 12.70 | | | | | |
| 800 | 813 | 7.92 | 12.70 | 15.88 | 17.48 | 9.53 | | | 12.70 | | | | | |
| 850 | 864 | 7.92 | 12.70 | 15.88 | 17.48 | 9.53 | | | 12.70 | | | | | |
| 900 | 914 | 7.92 | 12.70 | 15.88 | 19.05 | 9.53 | | | 12.70 | | | | | |
| 950 | 965 | | | | | 9.53 | | | 12.70 | | | | | |
| 1000 | 1016 | | | | | 9.53 | | | 12.70 | | | | | |
| 1050 | 1067 | | | | | 9.53 | | | 12.70 | | | | | |
| 1100 | 1118 | | | | | 9.53 | | | 12.70 | | | | | |
| 1150 | 1168 | | | | | 9.53 | | | 12.70 | | | | | |
| 1200 | 1219 | | | | | 9.53 | | | 12.70 | | | | | |

| Nominal size | OD | Weights in kg m$^{-1}$ of welded and seamless carbon steel pipe to ASME | | | | | | | | | | | | |
|---|---|---|---|---|---|---|---|---|---|---|---|---|---|---|
| | | Schedule | | | | | | | | | | | | |
| DN | mm | 10 | 20 | 30 | 40 | Std | 60 | 80 | XS | 100 | 120 | 140 | 160 | XXS |
| 6 | 10.3 | 0.28 | | 0.32 | 0.37 | 0.37 | | 0.47 | 0.47 | | | | | |
| 8 | 13.7 | 0.49 | | 0.54 | 0.63 | 0.63 | | 0.80 | 0.80 | | | | | |
| 10 | 17.1 | 0.63 | | 0.70 | 0.84 | 0.84 | | 1.10 | 1.10 | | | | | |
| 15 | 21.3 | 1.00 | | 1.12 | 1.27 | 1.27 | | 1.62 | 1.62 | | | | 1.95 | 2.55 |
| 20 | 26.7 | 1.28 | | 1.44 | 1.69 | 1.69 | | 2.20 | 2.20 | | | | 2.90 | 3.64 |
| 25 | 33.4 | 2.09 | | 2.18 | 2.50 | 2.50 | | 3.24 | 3.24 | | | | 4.24 | 5.45 |
| 32 | 42.2 | 2.70 | | 2.87 | 3.39 | 3.39 | | 4.47 | 4.47 | | | | 561 | 7.77 |
| 40 | 48.3 | 3.11 | | 3.53 | 4.05 | 4.05 | | 5.41 | 5.41 | | | | 7.25 | 9.56 |
| 50 | 60.3 | 3.93 | | 4.48 | 5.44 | 5.44 | | 7.48 | 7.48 | | | | 11.11 | 13.44 |
| 65 | 73.0 | 5.26 | | 8.04 | 8.63 | 8.63 | | 11.41 | 11.41 | | | | 14.92 | 20.39 |
| 80 | 88.9 | 6.45 | | 9.92 | 11.29 | 11.29 | | 15.27 | 15.27 | | | | 21.35 | 27.67 |
| 90 | 101.6 | 7.40 | | 11.41 | 13.57 | 13.57 | | 18.63 | 18.63 | | | | | |
| 100 | 114.3 | 8.36 | | 12.91 | 16.07 | 16.07 | | 22.32 | 22.32 | | 28.32 | | 33.54 | 41.03 |

(*Continued*)

| Nominal size | OD | Weights in kg m⁻¹ of welded and seamless carbon steel pipe to ASME | | | | | | | | | | | | |
|---|---|---|---|---|---|---|---|---|---|---|---|---|---|---|
| | | Schedule | | | | | | | | | | | | |
| DN | mm | 10 | 20 | 30 | 40 | Std | 60 | 80 | XS | 100 | 120 | 140 | 160 | XXS |
| 125 | 141.3 | 11.57 | | | 21.77 | 21.77 | | 30.97 | 30.97 | | 40.28 | | 49.11 | 57.43 |
| 150 | 168.3 | 13.84 | | | 28.26 | 28.26 | | 42.56 | 42.56 | | 54.20 | | 67.56 | 79.22 |
| 200 | 219.1 | 19.96 | 33.31 | 36.81 | 42.55 | 42.55 | 53.08 | 64.64 | 64.64 | 75.92 | 90.44 | 100.92 | 111.27 | 107.92 |
| 250 | 273.0 | 27.78 | 41.77 | 51.03 | 60.31 | 60.31 | 81.55 | 96.01 | 81.55 | 114.75 | 133.06 | 155.15 | 172.33 | 155.15 |
| 300 | 323.8 | 36.00 | 49.73 | 65.20 | 79.73 | 73.88 | 108.96 | 132.08 | 186.97 | 159.91 | 186.97 | 208.14 | 238.76 | 186.97 |
| 350 | 355.6 | 54.69 | 67.90 | 81.33 | 94.55 | 81.33 | 126.71 | 158.10 | 107.39 | 194.96 | 224.65 | 253.56 | 281.70 | |
| 400 | 406.4 | 62.64 | 77.83 | 93.27 | 123.30 | 93.27 | 160.12 | 203.53 | 123.30 | 245.56 | 286.64 | 333.19 | 365.35 | |
| 450 | 457 | 70.57 | 87.71 | 122.38 | 155.80 | 105.16 | 205.74 | 254.55 | 139.15 | 309.62 | 363.56 | 408.26 | 459.37 | |
| 500 | 508 | 78.55 | 117.15 | 155.12 | 183.42 | 117.15 | 247.83 | 311.17 | 155.12 | 381.53 | 441.49 | 508.11 | 564.81 | |
| 550 | 559 | 86.54 | 129.13 | 171.09 | | 129.13 | 294.25 | 373.83 | 171.09 | 451.42 | 527.02 | 600.63 | 672.26 | |
| 600 | 610 | 94.53 | 141.12 | 209.64 | 255.41 | 141.12 | 355.26 | 442.08 | 187.06 | 547.71 | 640.03 | 720.15 | 808.22 | |
| 650 | 660 | 127.36 | 202.72 | | | 152.87 | | | 202.72 | | | | | |
| 700 | 711 | 137.32 | 218.69 | 271.21 | | 164.85 | | | 218.69 | | | | | |
| 750 | 762 | 147.28 | 234.67 | 292.18 | | 176.84 | | | 234.67 | | | | | |
| 800 | 813 | 157.24 | 250.64 | 312.15 | 342.91 | 188.82 | | | 250.64 | | | | | |
| 850 | 864 | 167.20 | 266.61 | 332.12 | 364.90 | 200.31 | | | 266.61 | | | | | |
| 900 | 914 | 176.96 | 282.27 | 351.70 | 420.42 | 212.56 | | | 282.27 | | | | | |
| 950 | 965 | | | | | 224.54 | | | 298.24 | | | | | |
| 1000 | 1016 | | | | | 236.53 | | | 314.22 | | | | | |
| 1050 | 1067 | | | | | 248.52 | | | 330.19 | | | | | |
| 1100 | 1118 | | | | | 260.50 | | | 346.16 | | | | | |
| 1150 | 1168 | | | | | 272.25 | | | 351.82 | | | | | |
| 1200 | 1219 | | | | | 284.24 | | | 377.79 | | | | | |

H

# Appendix I

# Bending Moment on Horizontal Cylindrical Vessels Resulting from a Liquid Hydraulic Head

The pressure difference between the top and bottom of a horizontal cylindrical vessel due to the hydraulic head of a liquid content creates a bending moment (Zick 1951; British Standards Institution 1982). The higher pressure at the bottom of the vessel relative to the top creates a bending moment $M_{HYD}$, shown in Figure I.1. To calculate the size and location of this bending moment, consider Figure I.2. This shows a cross-section through the cylindrical vessel and an incremental layer in the vessel at a location with an angle $\theta$ radians from the center of the cross-section at Point $O$. Figure I.2 shows the variation of the hydraulic head from the top of the vessel to the bottom. The pressure at any point depends on the height $R(1 - \sin\theta)$, where $R$ is the radius of the vessel, giving a pressure at that point of $\rho_L g R(1 - \sin\theta)$, where $\rho_L$ is the density of the liquid in the vessel and $g$ the gravitational acceleration. The pressure at any height acts on an area at the end of the vessel, which varies from the top of the vessel to the bottom. To define the incremental area at any height requires the depth of the incremental layer in Figure I.2 to be determined. Consider Figure I.3, which shows the geometry of the incremental layer created by the differential angle $d\theta$. The length of the arc $AB$ is $R\,d\theta$, with $\theta$ in radians. As $d\theta$ tends to zero, the arc $AB$ tends to a straight line and the angle $OBA$ in Figure I.3 tends to a right angle. This allows the angles in the triangle $ABC$ in Figure I.3 to be defined, giving the angle $BAC$ to be $\theta$. This allows the depth of the incremental layer to be defined as $R\,d\theta \cos\theta$. It is now possible to define the horizontal pressure force acting on the end of the vessel as:

$$\int_{-\pi/2}^{\pi/2} 2R\cos\theta\,\rho_L g R(1 - \sin\theta)\,R\cos\theta\,d\theta$$

$$= 2\rho_L g R^3 \int_{-\pi/2}^{\pi/2} \cos\theta(1 - \sin\theta)\cos\theta\,d\theta \qquad (I.1)$$

$$= 2\rho_L g R^3 \int_{-\pi/2}^{\pi/2} \left(\cos^2\theta - \sin\theta\cos^2\theta\right)d\theta$$

The integral can be found in tables of integrals (Dwight 1969):

$$= 2\rho_L g R^3 \left[\left(\frac{\theta}{2} + \frac{\sin 2\theta}{4}\right) - \left(-\frac{\cos^3\theta}{3}\right)\right]_{-\pi/2}^{\pi/2} \qquad (I.2)$$

$$= \pi\rho_L g R^3$$

As the vertical distance from Point $O$ is given by $R\sin\theta$, the moment about the horizontal axis through Point $O$ is given by:

$$\int_{-\pi/2}^{\pi/2} 2R\cos\theta\,\rho_L g R(1 - \sin\theta)\,R\cos\theta\,R\sin\theta\,d\theta$$

$$= 2\rho_L g R^4 \int_{-\pi/2}^{\pi/2} \cos\theta(1 - \sin\theta)\cos\theta\,\sin\theta\,d\theta \qquad (I.3)$$

$$= 2\rho_L g R^4 \int_{-\pi/2}^{\pi/2} \left(\sin\theta\cos^2\theta - \sin^2\theta\cos^2\theta\right)d\theta$$

The integral can be found in tables of integrals (Dwight 1969):

$$= 2\rho_L g R^4 \left[\left(-\frac{\cos^3\theta}{3}\right) - \left(\frac{\theta}{8} - \frac{\sin 4\theta}{32}\right)\right]_{-\pi/2}^{\pi/2} \qquad (I.4)$$

$$= -\frac{\pi\rho_L g R^4}{4} = -\frac{\pi\rho_L g D^4}{64}$$

where $D$ = diameter of vessel

Comparing Eqs. (I.2) and (I.4), the bending moment created by the hydraulic head of a liquid acts at a distance $R/4$ below the horizontal axis through Point $O$.

*Process Plant Design*, First Edition. Robin Smith.
© 2024 John Wiley & Sons Ltd. Published 2024 by John Wiley & Sons Ltd.
Companion website: www.wiley.com/go/processplantdesign

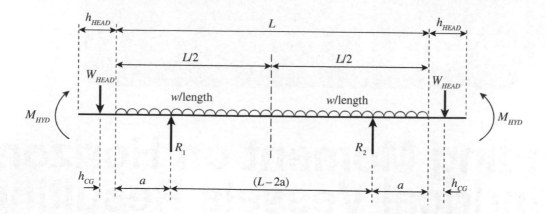

## Figure I.1

Bending moments on horizontal cylindrical vessels.

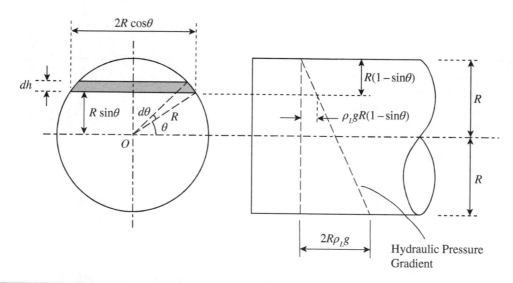

## Figure I.2

Variation of hydraulic pressure in a cylindrical vessel.

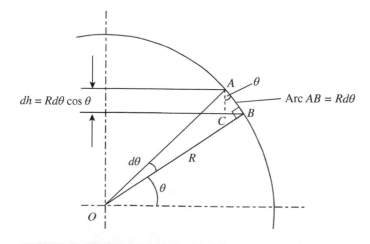

## Figure I.3

Thickness of an incremental layer of fluid.

# References

British Standards Institution (1982). Stresses in horizontal cylindrical pressure vessels supported on twin saddles: a derivation of the basic equations and constants used in G.3.3 of BS 5500:1982. PD 6497.

Dwight, H.B. (1969). *Tables of Integrals and Other Mathematical Data*, 4e. The Macmillan Company.

Zick, L.P. (1951). Stresses in large horizontal cylindrical pressure vessels on two saddle supports. *Welding Journal Research Supplement* 30: 556.

# Appendix J

# Equivalent Cylinder Approximation

The equivalent cylinder approximation can be used for the preliminary design of conical sections. The method replaces the conical section with a cylinder having the same surface area. The stresses are then calculated using the equivalent cylinder.

Figure J.1a shows a conical frustum. Figure J.1b shows the surface area unwrapped as a two-dimensional area. The areas at the base and top of the frustum in Figure J.1a are not included. After unwrapping the surface, the angle of the resulting section is defined to be $\theta$. From basic geometry:

$$\theta = \frac{\text{Arc length}}{\text{Radius}}$$
$$= \frac{2\pi R_2}{L} \tag{J.1}$$

Also:

$$\theta = \frac{2\pi R_1}{(L-l)} \tag{J.2}$$

where

$\theta$ = angle of circular sections (radians)
$R_2$ = radius of the larger circle of the frustum (m)
$R_1$ = radius of the smaller circle of the frustum (m)
$L$ = length of the side of the cone (m)
$l$ = length of the side of the conical frustum (m)

The surface area of the frustum is the difference between the major and minor circular sections:

$$A = \frac{\theta}{2\pi}\pi L^2 - \frac{\theta}{2\pi}\pi (L-l)^2$$
$$= \frac{\theta}{2}\left[L^2 - (L-l)^2\right] \tag{J.3}$$
$$= \frac{\theta l}{2}[2L - l]$$

where $A$ = area of the conical frustum (m)

Eliminating $\theta$ from Eqs (J.1) and (J.2):

$$L = \frac{lR_2}{(R_2 - R_1)} \tag{J.4}$$

Substituting $L$ into Eq. (J.1):

$$\theta = \frac{2\pi R_2}{L} = 2\pi R_2 \frac{(R_2 - R_1)}{lR_2} = \frac{2\pi(R_2 - R_1)}{l} \tag{J.5}$$

Substituting Eqs (J.4) and (J.5) into Eq. (J.3):

$$A = \frac{\theta l}{2}[2L - l]$$
$$= \frac{2\pi(R_2 - R_1)}{l}\frac{l}{2}\left[\frac{2lR_2}{(R_2 - R_1)} - l\right] \tag{J.6}$$
$$= \pi l(R_2 + R_1)$$

Equating this area to that of an equivalent cylinder:

$$A = \pi l(R_2 + R_1) = 2\pi R_{EQ}H \tag{J.7}$$

where

$R_{EQ}$ = radius of the equivalent cylinder to the conical frustum (m)
$H$ = height of the conical frustum and the equivalent cylinder (m)

Rearranging Eq. (J.7):

$$R_{EQ} = \frac{l(R_2 + R_1)}{2H} \tag{J.8}$$

Therefore, from Figure J.1a, $l = H/\cos \alpha$:

$$R_{EQ} = \frac{(R_2 + R_1)}{2\cos \alpha} \tag{J.9}$$

*Process Plant Design*, First Edition. Robin Smith.
© 2024 John Wiley & Sons Ltd. Published 2024 by John Wiley & Sons Ltd.
Companion website: www.wiley.com/go/processplantdesign

J

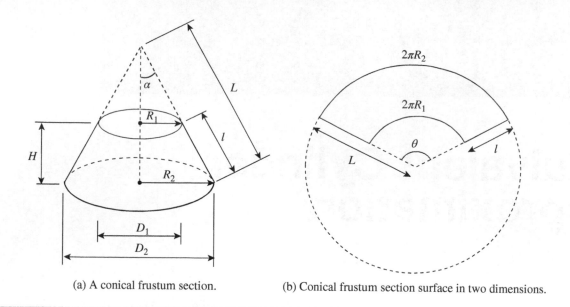

(a) A conical frustum section.          (b) Conical frustum section surface in two dimensions.

## Figure J.1
Surface area of a conical frustum section.

where $\alpha$ = angle of side of the conical frustum to the vertical (degrees)

Alternatively:

$$D_{EQ} = \frac{(D_2 + D_1)}{2 \cos \alpha} \qquad (J.10)$$

where

$D_{EQ}$ = diameter of the equivalent cylinder to the conical frustum (m)

$D_2$ = diameter of the larger circle of the frustum (m)

$D_1$ = diameter of the smaller circle of the frustum (m)

Equation (J.10) can be used to predict the diameter of a cylinder with an equivalent area to a conical frustum.

*Process Plant Design*, First Edition. Robin Smith.
© 2024 John Wiley & Sons Ltd. Published 2024 by John Wiley & Sons Ltd.
Companion website: www.wiley.com/go/processplantdesign